HANDBOOK OF
CHEMICAL
ENGINEERING
CALCULATIONS

ABOUT THE EDITORS

TYLER G. HICKS, P.E., is a consulting engineer and a successful engineering book author. He has worked in plant design and operation in a variety of industries, taught at several engineering schools, and lectured both in the United States and abroad. Mr. Hicks is the author of more than 20 books, including *Civil Engineering Formulas*, Second Edition, and *Handbook of Mechanical Engineering Calculations*, Second Edition.

NICHOLAS P. CHOPEY was Editor-in-Chief at *Chemical Engineering* magazine, with more than 45 years' experience in engineering and publishing.

HANDBOOK OF CHEMICAL ENGINEERING CALCULATIONS

Tyler G. Hicks, P.E.   Editor

International Engineering Associates
Member: American Society of Mechanical Engineers
United States Naval Institute

Nicholas P. Chopey   Editor

Editor-in-Chief of Chemical Engineering Magazine
Member: American Institute of Chemical Engineers,
American Society for Engineering Education

Fourth Edition

New York   Chicago   San Francisco   Lisbon   London   Madrid
Mexico City   Milan   New Delhi   San Juan   Seoul
Singapore   Sydney   Toronto

The McGraw·Hill Companies

Cataloging-in-Publication Data is on file with the Library of Congress.

McGraw-Hill books are available at special quantity discounts to use as premiums and sales promotions, or for use in corporate training programs. To contact a representative please e-mail us at bulksales@mcgraw-hill.com.

Handbook of Chemical Engineering Calculations, Fourth Edition

1 2 3 4 5 6 7 8 9 0   DOC/DOC   1 9 8 7 6 5 4 3 2

ISBN 978-0-07-176804-7
MHID 0-07-176804-1

This book is printed on acid-free paper.

Sponsoring Editor
Larry S. Hager

Acquisitions Coordinator
Bridget Thoreson

Editorial Supervisor
David E. Fogarty

Project Manager
Sheena Uprety, Cenveo Publisher Services

Copy Editor
Cenveo Publisher Services

Proofreader
Cenveo Publisher Services

Production Supervisor
Richard C. Ruzycka

Composition
Cenveo Publisher Services

Art Director, Cover
Jeff Weeks

CONTENTS

CONTRIBUTORS

Robert M. Baldwin, Ph.D. *Professor, Chemical Engineering Department, Colorado School of Mines, Golden, CO* (REACTION KINETICS AND REACTOR DESIGN)

James R. Beckman, Ph.D. *Associate Professor and Associate Chair, Department of Chemical and Materials Engineering, Arizona State University, Tempe, AZ* (CRYSTALLIZATION)

Gerald D. Button, *Marketing Manager, Pharmaceuticals, Rohm & Haas Co., Philadelphia, PA*

David S. Dickey, Ph.D. *Senior Consultant, Mix-Tech, Inc., Dayton, OH* (LIQUID AGITATION)

Robert L. Dream, Ph.D. *Director, Pharmaceutical and Biotechnology, Europe, Lockwood Greene, Somerset, NJ* (BIOTECHNOLOGY)

Otto Frank, *Consultant (retired), Air Products and Chemicals Inc., Allentown, PA* (DISTILLATION)

James H. Gary, Ph.D. *Professor Emeritus, Chemical and Petroleum-Refining Engineering Department, Colorado School of Mines, Golden, CO* (STOICHIOMETRY)

Michael S. Graboski, Ph.D. *Research Professor, Chemical Engineering Department, Colorado School of Mines, Golden, CO* (REACTION KINETICS AND REACTOR DESIGN)

Avinash Gupta, Ph.D. *Senior Principal Chemical Engineer, Chevron Lummus Global, Bloomfield, NJ* (PHYSICAL AND CHEMICAL PROPERTIES)

Wenfang Leu, Ph.D. *Research Scientist, Chemical Engineering Department, University of Houston, Houston, TX* (FILTRATION)

Kenneth J. McNulty, Sc.D. *Technical Director of R&D, Koch Membrane Systems, Wilmington, MA* (ABSORPTION AND STRIPPING)

Paul E. Minton, *Principal Engineer (retired), Union Carbide., South Charleston, WV* (HEAT TRANSFER)

Edward S.S. Morrison, *Senior Staff Engineer (retired), Union Carbide, Houston, TX* (HEAT TRANSFER)

A. K. S. Murthy, Eng.Sc.D. *Technology Fellow, BOC, Murray Hill, NJ* (PHASE EQUILIBRIUM)

E. Dendy Sloan, Jr., Ph.D. *Professor, Chemical Engineering Department, Colorado School of Mines, Golden, CO* (CHEMICAL EQUILIBRIUM)

Ross W. Smith, Ph.D. *Professor, Department of Chemical and Metallurgical Engineering, Mackay School of Mines, University of Nevada at Reno, Reno, NV* (SIZE REDUCTION)

Louis Theodore, Eng.Sc.D. *Professor, Chemical Engineering Department, Manhattan College, Riverdale, NY* (AIR POLLUTION CONTROL)

Frank M. Tiller, Ph.D. *Professor Emeritus, Chemical Engineering Department, University of Houston, Houston, TX* (FILTRATION)

William Vatavuk, P.E., *President, Vatavuk Engineering, Durham, NC* (COST ENGINEERING)

Frank H. Verhoff, Ph.D. *Vice President, USTech, Cincinnati, OH* (EXTRACTION AND LEACHING)

PREFACE

This handbook presents some 1000 calculation procedures for the field of chemical engineering. Today, in technical literature, the chemical engineering field is often referred to as the CPI— Chemical Process Industries. These industries comprise the practical application of chemical engineering to the production of millions of different products important in the lives of consumers and industries worldwide.

This fourth edition of the handbook includes numerous new applied calculation procedures important in the CPI. Each section of the handbook contains appropriate new applied hands-on type calculation procedures.

Since the CPI is facing new challenges from environmentalists, special attention is paid to the "green" aspects of the CPI. Numerous calculations have been added covering energy conservation, air and water pollution reduction, clean energy design procedures, alternative sources of energy, and design calculations for reducing energy consumption in the CPI.

New calculation procedures presented in this edition include: Section 1: Analysis of a saturated solution; ternary liquid system analysis; determining the heat of mixing of chemicals; chemical equation material balance; batch physical process balance. Section 2: Steady-state continuous physical balance with recycle and bypass; steady-state physical process balance. Section 3: Liquid-liquid separation analysis; determining the characteristics of an immiscible solution. Section 4: Equilibrium constant and conversion achieved in chemical reactions; determining the heat of solution for chemical compounds; determining the heat of reaction of chemical compounds. Section 5: Thermodynamic analysis of a Linde system; sizing reactor desuperheater condensers economically; design of a complete-mix activated sludge reactor. Section 6: Pump selection for chemical plants; determining the friction factor for flow of Bingham plastics. Section 7: Heat-exchanger choice for specific chemical-plant applications; sizing shell-and-tube heat exchangers for chemical plants; temperature determination in heat-exchanger operation; selecting and sizing heat exchangers based on fouling factors; chemical-plant electric process heater selection and application. Section 8: Sizing rupture disks for gases and liquids. Section 9: Determining the number of stages for a countercurrent extractor. Section 10: Heat removal required in crystallization. Section 11: Preliminary process design of an absorber; absorption tower flow and absorption rate. Section 12: Temperature analysis of heated agitated vessel. Section 13: Crusher power input determination; cooling-water flow rate for chemical-plant mixers; making a preliminary choice of size-reduction machinery. Section: 14: Sizing of a traveling-bridge filter; sizing a polymer dilution/feed system; design of a plastic media trickling filter. Section 15: Estimating size and cost of venturi scrubbers; sizing vertical liquid-vapor separators; sizing a horizontal liquid-vapor separator; effective stack height for disposing plant gases and vapors. Section 16: Design of an aerobic digester; design of a rapid-mix basin and flocculation basin; design of an anaerobic digester; design of a chlorination system for wastewater disinfection. Section 17: Fractionating column and condenser analysis for biotechnology. Section 18: Cost estimation of chemical-plant heat exchangers and storage tanks via correlations; estimating chemical-plant centrifugal-pump and electric-motor cost by using correlations. Section 19: Flash-steam heat recovery for cogeneration in chemical processing plants; energy conservation and cost reduction design for flash-steam usages in chemical processing plants; heat recovery energy and fuel savings in chemical processing plants; capital cost of cogeneration heat-recovery boilers in chemical processing plants; explosive-vent sizing for chemical processing plants; ventilation design for chemical processing plant environmental safety; environmental and safety-regulation ranking of equipment criticality in chemical processing plants; energy process-control system selection for chemical processing; process-energy temperature control system selection; energy process control valve selection;

steam-control and pressure-reducing valve sizing for maximum energy savings in chemical processing plants; atmospheric control system investment analysis for chemical plants; environmental pollution project selection for chemical plants; saving energy loss from storage tanks and vessels; energy savings from vapor recompression; savings possible from using low-grade waste heat for refrigeration; energy design analysis for shell-and-tube heat exchangers for chemical plants.

The additional calculation procedures, along with the modernization of many of the existing procedures, thoroughly updates this handbook to today's standards in the CPI. With an increasing focus on environmental and energy savings, the engineering user of this handbook will be well prepared to meet today's calculation challenges.

And with the CPI getting into almost every aspect of consumer and industrial life, good calculation ability is more important than ever before. Thus, we see the CPI working on underground conversion of brown coal to refinery feed stock. (One ton of coal produces one barrel of oil feedstock.)

We also see the CPI producing an extractor that reduces waste-treatment sludge volume by up to 50 percent, saving space and cost. Other CPI cutting-edge technologies that are being developed, as reported in *Chemical Engineering* magazine, include:

- *Process to produce* liquid fuels and chemicals from municipal waste.
- *Recovery of sulfur* from sulfur dioxide emissions from CPI plants.
- *Development of high-purity piping* for processes requiring hygienic piping.
- *Production of high-purity lithium* used in electric-vehicle batteries.
- *Manufacture of thin-film zeolite membranes* at much reduced energy and investment cost.
- *Conversion of biomass to sugar* for use in biofuels and bio-based chemicals.
- *Production of high-performance* engineering plastics for a variety of operations.
- Plus thousands of more important and unique developments.

As technical advances rush ahead, "green chemical engineering" has come on the scene. In the area of green chemical engineering there are many opportunities for chemical engineers to produce excellent results. The areas for these significant results in the CPI include:

- Green research and development (R&D)
- Green chemical-plant design
- Green plant operations

Today there is great demand for green (clean) chemical engineering from a host of areas starting with powerful regulating agencies and moving on to

- Consumer customers
- Corporate stockholders
- General public
- Educators at all levels
- Health providers giving public and private care
- Safety regulators and experts
- Energy suppliers and consultants
- Waste-disposal regulators and inspectors
- Environmentalists of many kinds and interests
- Emission compliance regulators
- Plus many others

One unexpected benefit of green design and operation is that CPI plant profit and efficiency often rise as green activities are introduced and implemented. Relations with the local community often improve as well. And the green aspects of individual chemical companies often result in an improved image for the firm and sales for the brand to both consumer and industrial customers.

So the benefits of green chemical engineering are many and widespread. Since the CPI is responsible for emissions of many kinds–waste formation, greenhouse-gas release, large energy consumption with attendant major fuel usage, plus other "nongreen" or "nonclean" operations—the move toward environmentally acceptable manufacturing and procedures is rapidly accelerating throughout the CPI. Today's CPI research, design, and operations must proceed with one eye on green considerations and the other on chemical engineering excellence.

The current editor of this edition of this unique handbook worked with its original editor, Nicholas Chopey, on each edition, starting with the first. Nick was a superb chemical engineer with the highest standards for the profession. It is the hope of the current editor that this edition hews to Nick's high standards and that he would be proud of the many new calculations and the updating of the earlier procedures.

While every effort has been made to produce a fully accurate handbook, errors can occur. If a reader finds an error, he or she can contact the editor in care of the publisher and an immediate correction will be made. Further, if any reader believes that one or more important calculations have been omitted from the handbook, the editor would appreciate having this called to his attention. The calculation will be added to the next edition of the handbook if it deserves inclusion.

The editor would like to thank his wife, Mary Shanley Hicks, a publishing professional, for her help in preparing the large manuscript needed to complete this revision.

Further, the editor thanks his sponsoring editor, Larry Hager, Senior Editor, Technical Group, for his guidance and help in revising this important handbook. Without Larry's instructions, the job would have taken much longer and would not have been as thorough or complete.

TYLER G. HICKS, P.E.

IMPORTANT NOTE: Where calculations have been added by the handbook editor to an existing section in this handbook, each such calculation procedure is identified by an asterisk (*) in its title. Thus, calculation *1.35 ANALYSIS OF A SATURATED SOLUTION* was added to the section by the handbook editor.

SECTION 1
PHYSICAL AND CHEMICAL PROPERTIES

Avinash Gupta, Ph.D.
Senior Principal Chemical Engineer
Chevron Lummus Global
Bloomfield, NJ

1.1   *MOLAR GAS CONSTANT*

Calculate the molar gas constant R in the following units:

a. $(atm)(cm^3)/(g \cdot mol)(K)$
b. $(psia)(ft^3)/(lb \cdot mol)(°R)$
c. $(atm)(ft^3)/(lb \cdot mol)(K)$
d. $kWh/(lb \cdot mol)(°R)$
e. $hp \cdot h/(lb \cdot mol)(°R)$
f. $(kPa)(m^3)/(kg \cdot mol)(K)$
g. $cal/(g \cdot mol)(K)$

Calculation Procedure

1. *Assume a basis.*
Assume gas is at standard conditions, that is, 1 g · mol gas at 1 atm (101.3 kPa) pressure and 0°C (273 K, or 492°R), occupying a volume of 22.4 L.

2. *Compute the gas constant.*
Apply suitable conversion factors and obtain the gas constant in various units. Use $PV = RT$; that is, $R = PV/T$. Thus,

a. $R = (1 \text{ atm})[22.4 \text{ L}/(g \cdot mol)](1000 \text{ cm}^3/\text{L})/273 \text{ K} = 82.05 \text{ (atm)}(cm^3)/(g \cdot mol)(K)$
b. $R = (14.7 \text{ psia})[359 \text{ ft}^3/(lb \cdot mol)]/492°R = 10.73 \text{ (psia)}(ft^3)/(lb \cdot mol)(°R)$
c. $R = (1 \text{ atm})[359 \text{ ft}^3/(lb \cdot mol)]/273 \text{ K} = 1.315 \text{ (atm)}(ft^3)/(lb \cdot mol)(K)$
d. $R = [10.73 \text{ (psia)}(ft^3)/(lb \cdot mol)(°R)](144 \text{ in}^2/ft^2)[3.77 \times 10^{-7} \text{ kWh}/(ft \cdot lbf)] = 5.83 \times 10^{-4} \text{ kWh}/(lb \cdot mol)(°R)$
e. $R = [5.83 \times 10^{-4} \text{ kWh}/(lb \cdot mol)(°R)](1/0.746 \text{ hp} \cdot h/\text{kWh}) = 7.82 \times 10^{-4} \text{ hp} \cdot h/(lb \cdot mol)(°R)$
f. $R = (101.325 \text{ kPa/atm})[22.4 \text{ L}/(g \cdot mol)][1000 \text{ g} \cdot mol/(kg \cdot mol)]/(273 \text{ K})(1000 \text{ L/m}^3) = 8.31 \text{ (kPa)}(m^3)/(kg \cdot mol)(K)$
g. $R = [7.82 \times 10^{-4} \text{ hp} \cdot h/(lb \cdot mol)(°R)][6.4162 \times 10^5 \text{ cal}/(hp \cdot h)][1/453.6 \text{ lb} \cdot mol/(g \cdot mol)](1.8°R/K) = 1.99 \text{ cal}/(g \cdot mol)(K)$

1.2   *ESTIMATION OF CRITICAL TEMPERATURE FROM EMPIRICAL CORRELATION*

Predict the critical temperature of (a) *n*-eicosane, (b) 1-butene, and (c) benzene, using the empirical correlation of Nokay. The Nokay relation is

$$\log T_c = A + B \log SG + C \log T_b$$

where T_c is critical temperature in kelvins, T_b is normal boiling point in kelvins, and SG is specific gravity of liquid hydrocarbons at 60°F relative to water at the same temperature. As for A, B, and C, they are correlation constants given in Table 1.1.

Calculation Procedure

1. *Obtain normal boiling point and specific gravity.*
Obtain T_b and SG for these three compounds from, for instance, Reid, Prausnitz, and Sherwood [1]. These are (a) for *n*-eicosane ($C_{20}H_{42}$), $T_b = 617$ K and $SG = 0.775$; (b) for 1-butene (C_4H_8), $T_b = 266.9$ K and $SG = 0.595$; and (c) for benzene (C_6H_6), $T_b = 353.3$ K and $SG = 0.885$.

TABLE 1.1 Correlation Constants for Nokay's Equation

Family of compounds	A	B	C
Alkanes (paraffins)	1.359397	0.436843	0.562244
Cycloalkanes (naphthenes)	0.658122	−0.071646	0.811961
Alkenes (olefins)	1.095340	0.277495	0.655628
Alkynes (acetylenes)	0.746733	0.303809	0.799872
Alkadienes (diolefins)	0.147578	−0.396178	0.994809
Aromatics	1.057019	0.227320	0.669286

2. Compute critical temperature using appropriate constants from Table 1.1.
Thus (a) for n-eicosane:

$$\log T_c = 1.359397 + 0.436843 \log 0.775 + 0.562244 \log 617 = 2.87986$$

so T_c = 758.3 K (905°F). (b) For l-butene:

$$\log T_c = 1.095340 + 0.277495 \log 0.595 + 0.655628 \log 266.9 = 2.62355$$

so T_c = 420.3 K (297°F). (c) For benzene:

$$\log T_c = 1.057019 + 0.22732 \log 0.885 + 0.669286 \log 353.3 = 2.75039$$

so T_c = 562.8 K (553°F)

Related Calculations.   This procedure may be used to estimate the critical temperature of hydrocarbons containing a single family of compounds, as shown in Table 1.1. Tests of the equation on
paraffins in the range C_1–C_{20} and various other hydrocarbon families in the range C_3–C_{14} have shown
average and maximum deviations of about 6.5 and 35°F (3.6 and 19 K), respectively.

1.3 CRITICAL PROPERTIES FROM GROUP-CONTRIBUTION METHOD

Estimate the critical properties of p-xylene and n-methyl-2-pyrrolidone using Lydersen's method of
group contributions.

Calculation Procedure

1. Obtain molecular structure, normal boiling point T_b, and molecular weight MW.
From handbooks, for p-xylene (C_8H_{10}), MW = 106.16, T_b = 412.3 K, and the structure is

For *n*-methyl-2-pyrrolidone (C_5H_9NO), MW = 99.1, T_b = 475.0 K, and the structure is

2. Sum up structural contributions of the individual property increments from Table 1.2, pp. 1.6 and 1.7.

The calculations can be set out in the following arrays, in which N stands for the number of groups. For *p*-xylene:

Group type	N	ΔT	ΔP	ΔV	$(N)(\Delta T)$	$(N)(\Delta P)$	$(N)(\Delta V)$
—CH_3 (nonring)	2	0.020	0.227	55	0.04	0.454	110
—C≡ (ring)	2	0.011	0.154	36	0.022	0.308	72
HC≡ (ring)	4	0.011	0.154	37	0.044	0.616	148
Total					0.106	1.378	330

For *n*-methyl-2-pyrrolidone:

Group type	N	ΔT	ΔP	ΔV	$(N)(\Delta T)$	$(N)(\Delta P)$	$(N)(\Delta V)$
—CH_3 (nonring)	1	0.020	0.227	55	0.020	0.227	55
—CH_2— (ring)	3	0.013	0.184	44.5	0.039	0.552	133.5
C≡O (ring)	1	0.033	0.2	50	0.033	0.20	50
—N— (ring)	1	0.007	0.13	32	0.007	0.13	32
Total					0.099	1.109	270.5

3. Compute the critical properties.

The formulas are

$$T_c = T_b\{[(0.567) + \Sigma(N)(\Delta T) - [\Sigma(N)(\Delta T)]^2\}^{-1}$$
$$P_c = MW[0.34 + (N)(\Delta P)]^{-2}$$
$$V_c = [40 + (N)(\Delta V)]$$
$$Z_c = P_cV_c/RT_c$$

where T_c, P_c, V_c, and Z_c are critical temperature, critical pressure, critical volume, and critical compressibility factor, respectively. Thus, for *p*-xylene,

$$T_c = 412.3[0.567 + 0.106 - (0.106)^2]^{-1}$$
$$= 623.0 \text{ K}(661.8°F) \text{ (literature value is 616.2 K)}$$
$$P_c = 106.16(0.34 + 1.378)^{-2} = 35.97 \text{ atm (3644 kPa) (literature value is 34.7 atm)}$$

$$V_c = 40 + 330$$
$$= 370 \text{ cm}^3/(\text{g} \cdot \text{mol}) [5.93 \text{ ft}^3/(\text{lb} \cdot \text{mol})] \text{ [literature value} = 379 \text{ cm}^3/(\text{g} \cdot \text{mol})]$$

Since $R = 82.06$ (cm^3)(atm)/(g · mol)(K),

$$Z_c = (35.97)(370)/(82.06)(623) = 0.26$$

For *n*-methyl-2-pyrrolidone,

$$T_c = 475[0.567 + 0.099 - (0.099)^2]^{-1} = 723.9 \text{ K (843°F)}$$
$$P_c = 99.1(0.34 + 1.109)^{-2} = 47.2 \text{ atm (4780 kPa)}$$
$$V_c = 40 + 270.5 = 310.5 \text{ cm}^3/(\text{g} \cdot \text{mol})[4.98 \text{ ft}^3/(\text{lb} \cdot \text{mol})]$$
$$Z_c = (47.2)(310.5)/(82.06)(723.9) = 0.247$$

Related Calculations.   Extensive comparisons between experimental critical properties and those estimated by several other methods have shown that the Lydersen group-contribution method is the most accurate. This method is relatively easy to use for both hydrocarbons and organic compounds in general, provided that the structure is known. Unlike Nokay's correlation (see Procedure 1.2), it can be readily applied to hydrocarbons containing characteristics of more than a single family, such as an aromatic with olefinic side chains. A drawback of the Lydersen method, however, is that it cannot distinguish between isomers of similar structure, such as 2,3-dimethylpentane and 2,4-dimethylpentane.

Based on tests with paraffins in the C_1–C_{20} range and other hydrocarbons in the C_3–C_{14} range, the average deviation from experimental data for critical pressure is 18 lb/in^2 (124 kPa), and the maximum error is around 70 lb/in^2 (483 kPa). In general, the accuracy of the correlation is lower for unsaturated compounds than for saturated ones. As for critical temperature, the typical error is less than 2 percent; it can range up to 5 percent for nonpolar materials of relatively high molecular weight (e.g., 7100). Accuracy of the method when used with multifunctional polar groups is uncertain.

1.4   *REDLICH-KWONG EQUATION OF STATE*

Estimate the molar volume of isopropyl alcohol vapor at 10 atm (1013 kPa) and 473 K (392°F) using the Redlich-Kwong equation of state. For isopropyl alcohol, use 508.2 K as the critical temperature T_c and 50 atm as the critical pressure P_c. The Redlich-Kwong equation is

$$P = RT/(V - b) - a/T^{0.5}V(V - b)$$

where P is pressure, T is absolute temperature, V is molar volume, R is the gas constant, and a and b are equation-of-state constants given by

$$a = 0.4278 \, R^2 T_c^{2.5}/P_c \quad \text{and} \quad b = 0.0867 \, RT_c/P_c$$

when the critical temperature is in kelvins, the critical pressure is in atmospheres, and R is taken as 82.05 (atm)(cm^3)/(g · mol)(K).

In an alternate form, the Redlich-Kwong equation is written as

$$Z = 1/(1 - h) - (A/B)[h/(1 + h)]$$

where $h = b/V = BP/Z$, $B = b/RT$, $A/B = a/bRT^{1.5}$, and Z, the compressibility factor, is equal to PV/RT.

TABLE 1.2   Critical-Property Increments—Lydersen's Structural Contributions

Symbols	ΔT	ΔP	ΔV
Nonring increments			
—CH₃	0.020	0.227	55
—CH₂	0.020	0.227	55
—CH	0.012	0.210	51
—C—	0.00	0.210	41
=CH₂	0.018	0.198	45
=CH	0.018	0.198	45
=C—	0.0	0.198	36
=C=	0.0	0.198	36
≡CH	0.005	0.153	(36)
≡C—	0.005	0.153	(36)
Ring increments			
—CH₂—	0.013	0.184	44.5
—CH	0.012	0.192	46
—C—	(−0.007)	(0.154)	(31)
=CH	0.011	0.154	37
=C—	0.011	0.154	36
=C=	0.011	0.154	36
Halogen increments			
—F	0.018	0.221	18
—Cl	0.017	0.320	49
—Br	0.010	(0.50)	(70)
—I	0.012	(0.83)	(95)
Oxygen increments			
—OH (alcohols)	0.082	0.06	(18)
—OH (phenols)	0.031	(−0.02)	(3)
—O— (nonring)	0.021	0.16	20
—O— (ring)	(0.014)	(0.12)	(8)
—C=O (nonring)	0.040	0.29	60
—C=O (ring)	(0.033)	(0.2)	(50)
HC=O (aldehyde)	0.048	0.33	73
—COOH (acid)	0.085	(0.4)	80
—COO— (ester)	0.047	0.47	80
=O (except for combinations above)	(0.02)	(0.12)	(11)

TABLE 1.2   Critical-Property Increments—Lydersen's Structural Contributions (*Continued*)

Symbols	ΔT	ΔP	ΔV
Nitrogen increments			
—NH$_2$	0.031	0.095	28
—NH (nonring)	0.031	0.135	(37)
—NH (ring)	(0.024)	(0.09)	(27)
—N— (nonring)	0.014	0.17	(42)
—N— (ring)	(0.007)	(0.13)	(32)
—CN	(0.060)	(0.36)	(80)
—NO$_2$	(0.055)	(0.42)	(78)
Sulfur increments			
—SH	0.015	0.27	55
—S— (nonring)	0.015	0.27	55
—S— (ring)	(0.008)	(0.24)	(45)
=S	(0.003)	(0.24)	(47)
Miscellaneous			
—Si—	0.03	(0.54)	
—B—	(0.03)		

Note: There are no increments for hydrogen. All bonds shown as free are connected with atoms other than hydrogen. Values in parentheses are based on too few experimental data to be reliable.
   Source: A. L. Lydersen, U of Wisconsin Eng. Exp, Station, 1955.

Calculation Procedure

1. *Calculate the compressibility factor Z.*

Since the equation is not explicit in Z, solve for it by an iterative procedure. For Trial 1, assume that $Z = 0.9$; therefore,

$$h = 0.0867(P/P_c)/Z(T/T_c) = \frac{0.087(10/50)}{(0.9)(473/508.2)} = 0.0208$$

Substituting for the generalized expression for A/B in the Redlich-Kwong equation,

$$Z = \frac{1}{1-h} - \left[\frac{\left(0.4278\,R^2 T_c^{2.5}/P_c \right)}{\left(0.0867\,RT_c/P_c \right)\left(RT^{1.5} \right)} \right] \left(\frac{h}{1+h} \right)$$

$$= \frac{1}{1-h} - (4.9343)(T_c/T)^{1.5}\left(\frac{h}{1+h} \right)$$

$$= \frac{1}{1-0.0208} - \left[(4.9343)\left(\frac{508.2}{473} \right)^{1.5} \right]\left[\frac{0.0208}{1+0.0208} \right]$$

$$= 0.910$$

For Trial 2, then, assume that $Z = 0.91$; therefore,

$$h = \frac{0.0867(10/50)}{0.91(473/508.2)} = 0.0205$$

and

$$Z = \frac{1}{1 - 0.0205} - (4.9343)(508.2/473)^{1.5}\frac{0.0205}{1 + 0.0205} = 0.911$$

which is close enough.

2. Calculate molar volume.
By the definition of Z,

$$V = ZRT/P$$
$$= (0.911)(82.05)(473)/(10)$$
$$= 3535.6\ \text{cm}^3/(\text{g} \cdot \text{mol})\ [3.536\ \text{m}^3/(\text{kg} \cdot \text{mol})\ \text{or}\ 56.7\ \text{ft}^3/(\text{lb} \cdot \text{mol})]$$

Related Calculations.   This two-constant equation of Redlich-Kwong is extensively used for engineering calculations and enjoys wide popularity. Many modifications of the Redlich-Kwong equations of state, such as those by Wilson, Barnes-King, Soave, and Peng-Robinson, have been made and are discussed in Reid et al. [1]. The constants for the equation of state may be obtained by least-squares fit of the equation to experimental P-V-T data. However, such data are often not available. When this is the case, estimate the constants on the basis of the critical properties, as shown in the example.

1.5   P-V-T PROPERTIES OF A GAS MIXTURE

A gaseous mixture at 25°C (298 K) and 120 atm (12,162 kPa) contains 3% helium, 40% argon, and 57% ethylene on a mole basis. Compute the volume of the mixture per mole using the following: (a) ideal-gas law, (b) compressibility factor based on pseudoreduced conditions (Kay's method), (c) mean compressibility factor and Dalton's law, (d) van der Waal's equation and Dalton's law, and (e) van der Waal's equation based on averaged constants.

Calculation Procedure

1. Solve the ideal-gas law for volume.
By definition, $V = RT/P$, where V is volume per mole, T is absolute temperature, R is the gas constant, and P is pressure. Then,

$$V = [82.05\ (\text{cm}^3)(\text{atm})/(\text{g} \cdot \text{mol})(\text{K})]\ 298\ \text{K}/120\ \text{atm} = 203.8\ \text{cm}^3/(\text{g} \cdot \text{mol})$$

2. Calculate the volume using Kay's method.
In this method, V is found from the equation $V = ZRT/P$, where Z, the compressibility factor, is calculated on the basis of pseudocritical constants that are computed as mole-fraction-weighted averages of the critical constants of the pure compounds. Thus, $T_c' = \Sigma Y_i T_{c,i}$ and similarly for P_c' and Z_c', where the subscript c denotes critical, the prime denotes pseudo, the subscript i pertains to the ith component, and Y is mole fraction. Pure-component critical properties can be obtained from handbooks. The calculations can then be set out as a matrix:

Component, i	Y_i	$T_{c,i}$ (K)	$Y_i T_{c,i}$ (K)	$P_{c,i}$ (atm)	$Y_i P_{c,i}$ (atm)	$Z_{c,i}$	$Y_i Z_{c,i}$
He	0.03	5.2	0.16	2.24	0.07	0.301	0.009
A	0.40	150.7	60.28	48.00	19.20	0.291	0.116
C_2H_4	0.57	283.0	161.31	50.50	28.79	0.276	0.157
$\Sigma =$	1.00		221.75		48.06		0.282

Then the reduced temperature $T_r = T/T_c' = 298/221.75 = 1.34$, and the reduced pressure $P_r = P/P_c' = 120/48.06 = 2.50$. Now $Z_c' = 0.282$. Refer to the generalized compressibility plots in Figs. 1.2 and 1.3, which pertain respectively to Z_c' values of 0.27 and 0.29. Figure 1.2 gives a Z of 0.71, and

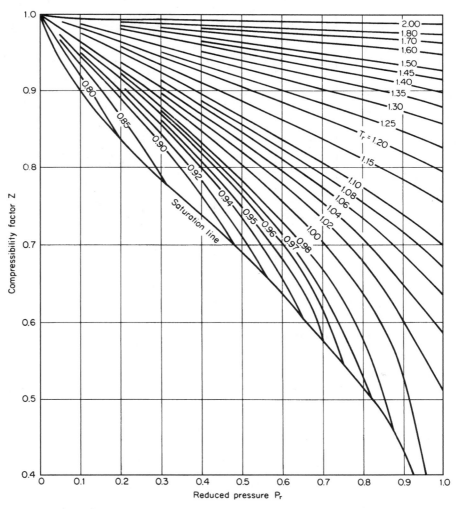

FIGURE 1.1   Generalized compressibility factor; $Z_c = 0.27$; low-pressure range. (*Lydersen et al., University of Wisconsin Engineering Experiment Station, 1955.*)

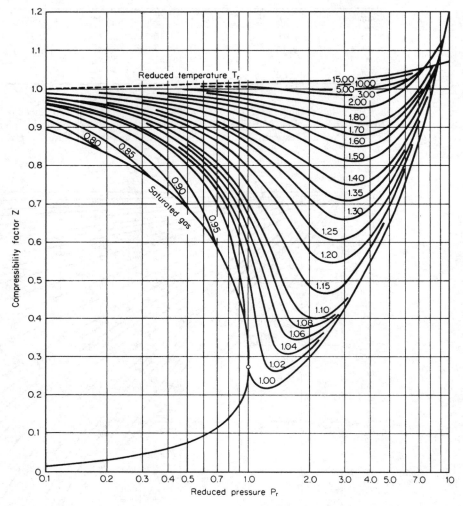

FIGURE 1.2   Generalized compressibility factor; $Z_c = 0.27$; high-pressure range. (*Lydersen et al., University of Wisconsin Engineering Experiment Station, 1955.*)

Fig. 1.3 gives a Z of 0.69. By linear interpolation, then, Z for the present case is 0.70. Therefore, the mixture volume is given by

$$V = ZRT/P = (0.70)(82.05)(298)/120 = 138.8 \text{ cm}^3/(\text{g} \cdot \text{mol})$$

3. *Calculate the volume using the mean compressibility factor and Dalton's law.*

Dalton's law states that the total pressure exerted by a gaseous mixture is equal to the sum of the partial pressures. In using this method, assume that the partial pressure of a component of a mixture is equal to the product of its mole fraction and the total pressure. Thus the method consists of calculating the partial pressure for each component, calculating the reduced pressure and reduced temperature, finding the corresponding compressibility factor for each component (from a conventional

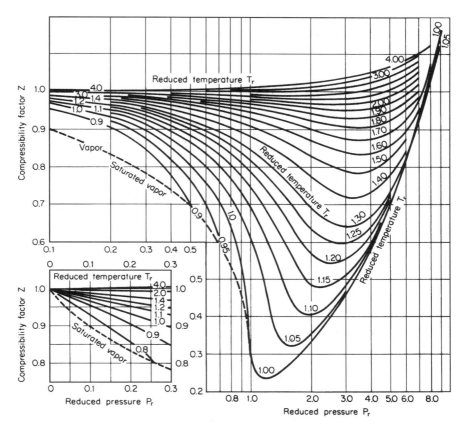

FIGURE 1.3 Generalized compressibility factor; $Z_c = 0.29$. (*Lydersen et al., University of Wisconsin Engineering Experiment Station, 1955.*)

compressibility-factor chart in a handbook), and then taking the mole-fraction-weighted average of those compressibility factors and using that average value to find V. The calculations can be set out in matrix form, employing the critical properties from the matrix in step 2:

Component (i)	Y_i	Partial pressure ($p_i = PY_i$)	Reduced pressure ($p_i/P_{c,i}$)	Reduced temperature ($T/T_{c,i}$)	Compressibility factor (Z_i)	Z_iY_i
Helium	0.03	3.6	1.61	57.3	1.000	0.030
Argon	0.40	48.0	1.00	1.98	0.998	0.399
Ethylene	0.57	68.4	1.35	1.05	0.368	0.210
Total	1.00	120.0				0.639

Therefore

$$V = ZRT/P = (0.639)(82.05)(298)/120 = 130.2 \text{ cm}^3/(\text{g} \cdot \text{mol})$$

4. Calculate the volume using van der Waal's equation and Dalton's law.
Van der Waal's equation is

$$P = RT/(V - b) - a/V^2$$

where a and b are van der Waal constants, available from handbooks, that pertain to a given substance. The values for helium, argon, and ethylene are as follows (for calculations with pressure in atmospheres, volume in cubic centimeters, and quantity in gram-moles):

Component	van der Waal constant	
	a	b
Helium	0.0341×10^6	23.7
Argon	1.350×10^6	32.3
Ethylene	4.480×10^6	57.2

For a mixture obeying Dalton's law, the equation can be rewritten as

$$P = RT \left[\frac{Y_{He}}{(V - Y_{He}b_{He})} + \frac{Y_A}{(V - Y_A b_A)} + \frac{Y_{Eth}}{(V - Y_{Eth}b_{Eth})} \right] - \left(\frac{1}{V^2} \right) \left(Y_{He}^2 a_{He} + Y_A^2 a_A + Y_{Eth}^2 a_{Eth} \right)$$

Upon substitution,

$$120 = (82.05)(298) \left[\frac{0.03}{V - (0.03)(23.7)} + \frac{0.40}{V - (0.4)(32.3)} + \frac{0.57}{V - (0.57)(57.2)} \right]$$

$$- \frac{1}{V^2} [(0.0341)(10^6)(0.03^2) + (1.35)(10^6)(0.4^2) + (4.48)(10^6)(0.57^2)]$$

Solving for volume by trial and error,

$$V = 150.9 \, \text{cm}^3/(\text{g} \cdot \text{mol})[2.42 \, \text{ft}^3/(\text{lb} \cdot \text{mol})]$$

5. Calculate the volume using van der Waal's equation with averaged constants.
In this method it is convenient to rearrange the van der Waal equation into the form

$$V^3 - (b_{avg} + RT/P)V^2 + a_{avg}V/P - a_{avg}b_{avg}/P = 0$$

For a_{avg}, take the expression $[\Sigma Y_i(a_i)^{0.5}]^2$; for b_{avg}, use the straightforward mole-fraction-weighted linear average $\Sigma Y_i b_i$. Thus, taking the values of a_i and b_i from the martix in step 4,

$$a_{avg} = [(0.03)(0.0341 \times 10^6)^{0.5} + (0.40)(1.350 \times 10^6)^{0.5} + (0.57)(4.48 \times 10^6)^{0.5}]^2$$

$$= 2.81 \times 10^6$$

$$b_{avg} = (0.03)(23.7) + (0.4)(32.3) + (0.57)(57.2)$$

$$= 46.23$$

Upon substitution,

$$V^3 - [46.23 + (82.05)(298)/120]V^2 + (2.81 \times 10^6)V/120 - (2.81 \times 10^6)(46.23)/120 = 0$$

Trial-and-error solution gives

$$V = 137 \text{cm}^3/(\text{g} \cdot \text{mol})[2.20 \text{ ft}^3/(\text{lb} \cdot \text{mol})]$$

Related Calculations. This illustration outlines various simple techniques for estimating *P-V-T* properties of gaseous mixtures. Obtain the compressibility factor from the generalized corresponding-state correlation, as shown in step 2.

The ideal-gas law is a simplistic model that is applicable to simple molecules at low pressure and high temperature. As for Kay's method, which in general is superior to the others, it is basically suitable for nonpolar/nonpolar mixtures and some polar/polar mixtures, but not for nonpolar/polar ones. Its average error ranges from about 1 percent at low pressures to 5 percent at high pressures and to as much as 10 percent when near the critical pressure.

For a quick estimate one may compute the pseudocritical parameters for the mixture using Kay's mole-fraction-averaging mixing rule and obtain the compressibility factor from the generalized corresponding-state correlation as shown in step 2.

1.6 *DENSITY OF A GAS MIXTURE*

Calculate the density of a natural gas mixture containing 32.1% methane, 41.2% ethane, 17.5% propane, and 9.2% nitrogen (mole basis) at 500 psig (3,550 kPa) and 250°F (394 K).

Calculation Procedure

1. *Obtain the compressibility factor for the mixture.*
Employ Kay's method, as described in step 2 of Procedure 1.5. Thus *Z* is found to be 0.933.

2. *Calculate the mole-fraction-weighted average molecular weight for the mixture.*
The molecular weights of methane, ethane, propane, and nitrogen are 16, 30, 44, and 28, respectively. Therefore, average molecular weight $M' = (0.321)(16) + (0.412)(30) + (0.175)(44) + (0.092)(28) = 27.8$ lb/mol.

3. *Compute the density of the mixture.*
Use the formula

$$\rho = M'P/ZRT$$

where ρ is density, P is pressure, R is the gas constant, and T is absolute temperature. Thus,

$$\rho = (27.8)(500 + 14.7)/(0.933)(10.73)(250 + 460)$$
$$= 2.013 \text{ lb/ft}^3 \ (32.2 \text{ kg/m}^3)$$

Related Calculations. Use of the corresponding-states three-parameter graphic correlation developed by Lydersen, Greenkorn, and Hougen (Figs. 1.1, 1.2, and 1.3) gives fairly good results for predicting the gas-phase density of nonpolar pure components and their mixtures. Errors are within 4 to 5 percent. Consequently, this generalized correlation can be used to perform related calculations, except in the regions near the critical point. For improved accuracy in estimating P-V-T properties of pure components and their mixtures, use the Soave-modified Redlich-Kwong equation or the Lee-Kesler form of the Bendict-Webb-Rubin (B-W-R) generalized equation. For hydrocarbons, either of the two are accurate to within 2 to 3 percent, except near the critical point; for nonhydrocarbons, the Lee-Kesler modification of the B-W-R equation is recommended, the error probably being within a few percent except for polar molecules near the critical point. However, these equations are fairly complex and therefore not suitable for hand calculation. For a general discussion of various corresponding-state and analytical equations of state, see Reid et al. [1].

1.7 ESTIMATION OF LIQUID DENSITY

Estimate the density of saturated liquid ammonia at 37°C (310 K or 99°F) using (a) the Gunn-Yamada generalized correlation and (b) the Rackett equation. The Gunn-Yamada correlation [16] is

$$V/V_{Sc} = V_r^{(0)}(1 - \omega\Gamma)$$

where V is the liquid molar specific volume in cubic centimeters per gram-mole; ω is the acentric factor; Γ is as defined below; V_{Sc} is a scaling parameter equal to $(RT_c/P_c)(0.2920 - 0.0967\,\omega)$, where R is the gas constant, P is pressure, and the subscript c denotes a critical property; and $V_r^{(0)}$ is a function whose value depends on the reduced temperature T/T_c:

$$V_r^{(0)} = 0.33593 - 0.33953(T/T_c) + 1.51941(T/T_c)^2 - 2.02512(T/T_c)^3$$
$$+ 1.11422(T/T_c)^4 \quad \text{for } 0.2 \le T/T_c \le 0.8$$

or

$$V_r^{(0)} = 1.0 + 1.3(1 - T/T_c)^{0.5} \log(1 - T/T_c) - 0.50879(1 - T/T_c)$$
$$- 0.91534(1 - T_r)^2 \quad \text{for } 0.8 \le T/T_c \le 1.0$$

and

$$\Gamma = 0.29607 - 0.09045(T/T_c) - 0.04842(T/T_c)^2 \quad \text{for } 0.2 \le T/T_c \le 1.0$$

The Rackett equation [17] is

$$V_{\text{sat liq}} = V_c Z_c^{(1 - T/T_c)^{0.2857}}$$

where $V_{\text{sat liq}}$ is the molar specific volume for saturated liquid, V_c is the critical molar volume, and Z_c is the critical compressibility factor. Use these values for ammonia: $T_c = 405.6$ K. $P_c = 111.3$ atm, $Z_c = 0.242$, $V_c = 72.5$ cm^3(g · mol), and $\omega = 0.250$.

Calculation Procedure

1. Compute saturated-liquid density using the Gunn-Yamada equation

$$V_{Sc} = (82.05)(405.6)[0.2920 - (0.0967)(0.250)]/111.3$$
$$= 80.08 \text{ cm}^3(\text{g} \cdot \text{mol})$$

and the reduced temperature is given by

$$T/T_c = (37 + 273)/405.6$$
$$= 0.764$$

Therefore,

$$V_r^{(0)} = 0.33593 - (0.33953)(0.764) + (1.51941)(0.764)^2 - (2.02512)(0.764)^3 + (1.11422)(0.764)^4$$
$$= 0.4399$$
$$\Gamma = 0.29607 - 0.09045(0.764) - 0.04842(0.764)^2 = 0.1987$$

and the saturated liquid volume is given by

$$V = (0.4399)(80.08)[1 - (0.250)(0.1987)]$$
$$= 33.48 \, cm^3/(g \cdot mol)$$

Finally, letting M equal the molecular weight, the density of liquid ammonia is found to be

$$\rho = M/V = 17/33.48 = 0.508 \, g/cm^3 \, (31.69 \, lb/ft^3)$$

(The experimental value is 0.5834 g/cm³; so the error is 12.9 percent.)

2. Compute saturated-liquid density using the Rackett equation

$$V_{sat} = (72.5)(0.242)^{(1-0.764)^{0.2857}} = 28.34 \, cm^3/(g \cdot mol)$$

So

$$\rho = 17/28.34 = 0.5999 \, g/cm^3 \, (37.45 \, lb/ft^3) \qquad (error = 2.8 \, percent)$$

Related Calculations. Both the Gunn-Yamada and Rackett equations are limited to saturated liquids. At or below a T_r of 0.99, the Gunn-Yamada equation appears to be quite accurate for nonpolar as well as slightly polar compounds. With either equation, the errors for nonpolar compounds are generally within 1 percent. The correlation of Yen and Woods [18] is more general, being applicable to compressed as well as saturated liquids.

1.8 ESTIMATION OF IDEAL-GAS HEAT CAPACITY

Estimate the ideal-gas heat capacity C_p^o of 2-methyl-1,3-butadiene and n-methyl-2-pyrrolidone at 527°C (800 K, or 980°F) using the group-contribution method of Rihani and Doraiswamy. The Rihani-Doraiswamy method is based on the equation

$$C_p^o = \sum_i N_i a_i + \sum_i N_i b_i T + \sum_i N_i c_i T^2 + \sum_i N_i d_i T^3$$

where N_i is the number of groups of type i, T is the temperature in kelvins, and a_i, b_i, c_i, and d_i are the additive group parameters given in Table 1.3.

TABLE 1.3 Group Contributions to Ideal-Gas Heat Capacity

Symbol	Coefficients			
	a	$b \times 10^2$	$c \times 10^4$	$d \times 10^6$
Aliphatic hydrocarbon groups				
—CH$_3$	0.6087	2.1433	−0.0852	0.01135
H\C=CH$_2$ /	0.2773	3.4580	−0.1918	0.004130
\C=CH$_2$ /	−0.4173	3.8857	−0.2783	0.007364
H\C=C/ H (cis)	−3.1210	3.8060	−0.2359	0.005504
H\C=C/ /H	0.9377	2.9904	−0.1749	0.003918
H\C=C/	−1.4714	3.3842	−0.2371	0.006063
\C=C/	0.4736	3.5183	−0.3150	0.009205
H\C=C=CH$_2$ /	2.2400	4.2896	−0.2566	0.005908
\C=C=CH$_2$ /	2.6308	4.1658	−0.2845	0.007277
H\C=C=C/ H	−3.1249	6.6843	−0.5766	0.017430
≡CH	2.8443	1.0172	−0.0690	0.001866
—C≡	−4.2315	7.8689	−0.2973	0.00993
Aromatic hydrocarbon groups				
HC	−1.4572	1.9147	−0.1233	0.002985
—C	−1.3883	1.5159	−0.1069	0.002659
↔C	0.1219	1.2170	−0.0855	0.002122

TABLE 1.3 Group Contributions to Ideal-Gas Heat Capacity (*Continued*)

Symbol	Coefficients			
	a	$b \times 10^2$	$c \times 10^4$	$d \times 10^6$
Oxygen-containing groups				
—OH	6.5128	−0.1347	0.0414	−0.001623
—O—	2.8461	−0.0100	0.0454	−0.002728
$\overset{H}{\underset{}{\mid}}$ —C=O	3.5184	0.9437	0.0614	−0.006978
$\overset{\backslash}{\underset{/}{C}}$ =O	1.0016	2.0763	−0.1636	0.004494
$\overset{O}{\underset{}{\parallel}}$ —C—O—H	1.4055	3.4632	−0.2557	0.006886
—C $\overset{\nearrow O}{\underset{\searrow O—}{}}$	2.7350	1.0751	0.0667	−0.009230
O $\overset{\nwarrow}{\underset{\swarrow}{}}$	−3.7344	1.3727	−0.1265	0.003789
Nitrogen-containing groups				
—C≡N	4.5104	0.5461	0.0269	−0.003790
—N≡C	5.0860	0.3492	0.0259	−0.002436
—NH₂	4.1783	0.7378	0.0679	−0.007310
$\overset{\backslash}{\underset{/}{N}}$ H	−1.2530	2.1932	−0.1604	0.004237
$\overset{\backslash}{\underset{/}{N}}$ —	−3.4677	2.9433	−0.2673	0.007828
N $\overset{\nwarrow}{\underset{\swarrow}{}}$	2.4458	0.3436	0.0171	−0.002719
—NO₂	1.0898	2.6401	−0.1871	0.004750

Source: Reprinted with permission from D. N. Rihani and L. K. Doraiswamy, *Ind. Eng. Chem. Fund.* 4:17, 1965. Copyright 1965. American Chemical Society.

Calculation Procedure

1. *Obtain the molecular structure from a handbook, and list the number and type of groups.* For 2-methyl-1,3-butadiene, the structure is

$$H_2C{=}CH{-}\underset{\underset{CH_3}{|}}{C}{=}CH_2$$

and the groups are

$$-CH_3 \qquad \overset{H}{\underset{/}{\diagdown}}C=CH_2 \quad \text{and} \quad \overset{\diagdown}{\underset{/}{}}C=CH_2$$

For *n*-methyl-2-pyrrolidone, the structure is

$$
\begin{array}{ccc}
H_2C & \!\!\!\!—\!\!\!\! & CH_2 \\
| & & | \\
H_2C & & C=O \\
\diagdown & & / \\
 & N & \\
 & | & \\
 & CH_3 &
\end{array}
$$

and the groups are

$$-CH_3 \qquad -\overset{|}{C}H_2 \qquad -\overset{|}{C}=O \qquad \overset{\diagdown}{\underset{/}{}}N-$$

and a 5-membered (pentene) ring.

2. Sum up the group contributions for each compound.
Obtain the values of a, b, c, and d from Table 1.3, and set out the calculations in a martix:

	N	a	$b \times 10^2$	$c \times 10^4$	$d \times 10^6$
2-Methyl-1,3-butadiene:					
$-CH_3$	1	0.6087	2.1433	−0.0852	0.01135
$\overset{\diagdown}{HC}=CH_2$	1	0.2773	3.4580	−0.1918	0.004130
$\overset{\diagdown}{\underset{/}{C}}=CH_2$	1	−0.4173	3.8857	−0.2783	0.007364
$\Sigma^{(N)}$ (group parameter)		0.4687	9.4870	−0.5553	0.02284
n-Methyl-2-pyrrolidone:					
5-membered (pentene) ring	1	−6.8813	0.7818	−0.0345	0.000591
$-CH_3$	1	0.6087	2.1433	−0.0852	0.01135
$-CH_2$	3	0.3945	2.1363	−0.1197	0.002596
$\overset{\diagdown}{\underset{}{}}C=O$	1	1.0016	2.0763	−0.1636	0.004494
$\overset{\diagdown}{\underset{/}{}}N-$	1	−3.4677	2.9433	−0.2673	0.007828
$\Sigma^{(N)}$ (group parameter)		−7.5552	14.3536	−0.9097	0.026859

3. Compute the ideal-gas heat capacity for each compound.
Refer to the equation in the statement of the problem. Now, $T = 527 + 273 = 800$ K. Then, for 2-methyl-l,3-butadiene,

$$C_p^\circ = 0.4687 + (9.4870 \times 10^{-2})(800) + (-0.5553 \times 10^{-4})(800)^2 + (0.02284 \times 10^{-6})(800)^3$$
$$= 52.52 \text{ cal/(g} \cdot \text{mol)(K) } [52.52 \text{ Btu/(lb} \cdot \text{mol)(°F)]}$$

And for *n*-methyl-2-pyrrolidone,

$$C_p^\circ = -7.5552 + (14.3536 \times 10^{-2})(800) + (-0.9097 \times 10^{-4})(800)^2 + (0.02686 \times 10^{-6})(800)^3$$
$$= 62.81 \text{ cal/(g} \cdot \text{mol)(K) [62.81 Btu/(lb} \cdot \text{mol)(°F)]}$$

Related Calculations.   The Rihani-Doraiswamy method is applicable to a large variety of compounds, including heterocyclics; however, it is not applicable to acetylenics. It predicts to within 2 to 3 percent accuracy. Accuracy levels are somewhat less when predicting at temperatures below about 300 K (80°F). Good accuracy is obtainable using the methods of Benson [25] and of Thinh [26].

1.9   HEAT CAPACITY OF REAL GASES

Calculate the heat capacity C_p of ethane vapor at 400 K (260°F) and 50 atm (5065 kPa). Also estimate the heat-capacity ratio C_p/C_v at these conditions. The ideal-gas heat capacity for ethane is given by

$$C_p^\circ = 2.247 + (38.201 \times 10^{-3})T - (11.049 \times 10^{-6})T^2$$

where C_p° is in cal/(g · mol)(K), and T is in kelvins. For ethane, critical temperature $T_c = 305.4$ K and critical pressure $P_c = 48.2$ atm.

Calculation Procedure

1. Compute reduced temperature T_r and reduced pressure P_r.
Thus $T_r = T/T_c = 400/305.4 = 1.310$, and $P_r = P/P_c = 50/48.2 = 1.04$.

2. Obtain ΔC_p from Fig. 1.4.
Thus $\Delta C_p = C_p - C_p^\circ = 3$ cal/(g · mol)(K) at $T_r = 1.31$ and $P_r = 1.04$.

3. Calculate ideal-gas heat capacity.

$$C_p^\circ = 2.247 + (38.201 \times 10^{-3})(400) - (11.049 \times 10^{-6})(400^2)$$
$$= 15.76 \text{ cal/(g} \cdot \text{mol)(K)}$$

4. Compute real-gas heat capacity.

$$C_p = \Delta C_p + C_p^\circ = 3 + 15.76 = 18.76 \text{ cal/(g} \cdot \text{mol)(K)[18.76 Btu/(lb} \cdot \text{mol)(°F)]}$$

5. Estimate heat-capacity ratio.
From Fig. 1.5, $C_p - C_v = 4$ at $T_r = 1.31$ and $P_r = 1.04$. So the real-gas heat-capacity ratio is

$$\frac{C_p}{C_v} = \frac{C_p}{C_p - (C_p - C_v)} = \frac{18.76}{18.76 - 4} = 1.27$$

Note that the ideal-gas heat-capacity ratio is

$$\frac{C_p^\circ}{C_v^\circ} = \frac{C_p^\circ}{(C_p^\circ - R)} = 15.76/(15.76 - 1.987) = 1.144$$

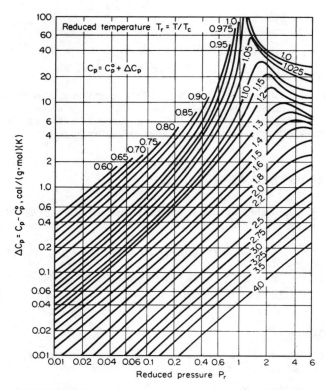

FIGURE 1.4 Isothermal pressure correction to the molar heat capacity of gases. (*Perry and Chilton—Chemical Engineers' Handbook, McGraw-Hill, 1973.*)

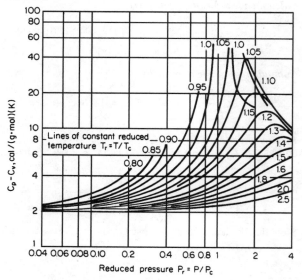

FIGURE 1.5 Generalized heat-capacity differences, $C_p - C_v$. (*Perry and Chilton—Chemical Engineers' Handbook, McGraw-Hill, 1973.*)

Related Calculations. This graphic correlation may be used to estimate the heat-capacity ratio of any nonpolar or slightly polar gas. The accuracy of the correlation is poor for highly polar gases and (as is true for correlations in general) near the critical region. For polar gases, the Lee-Kesler method [27] is suggested.

1.10 LIQUID HEAT CAPACITY—GENERALIZED CORRELATION

Estimate the saturated-liquid heat capacity of (a) *n*-octane and (b) ethyl mercaptan at 27°C (80.6°F) using the Yuan-Stiel corresponding-states correlation [19], given as

$$C_{\sigma,L} - C_p^o = (\Delta C_{\sigma,L})^{(0)} + \omega(\Delta C_{\sigma,L})^{(1)}$$

for nonpolar liquids, or

$$C_{\sigma,L} - C_p^o = (\Delta C_{\sigma,L})^{(0p)} + \omega(\Delta C_{\sigma,L})^{(1p)} + X(\Delta C_{\sigma,L})^{(2p)} + X^2(\Delta C_{\sigma,L})^{(3p)}$$
$$+ \omega^2(\Delta C_{\sigma,L})^{(4p)} + X\omega(\Delta C_{\sigma,L})^{(5p)}$$

for polar liquids, where $C_{\sigma L}$ is saturated-liquid heat capacity and C_p^o is ideal-gas heat capacity, both in calories per gram-mole kelvin; ω is the Pitzer acentric factor; the $\Delta C_{\sigma L}$ terms are deviation functions for saturated-liquid heat capacity (given in Table 1.4); and X is the Stiel polarity factor (from Table 1.5).
For *n*-octane, $T_c = 568.8$ K, $\omega = 0.394$, $X = 0$ (nonpolar liquid), and

$$C_p^o = -1.456 + (1.842 \times 10^{-1})T - (1.002 \times 10^{-4})T^2 + (2.115 \times 10^{-8})T^3$$

where T is in kelvins.
For ethyl mercaptan, $T_c = 499$ K, $\omega = 0.190$, and $X = 0.004$ (slightly polar), and

$$C_p^o = 3.564 + (5.615 \times 10^{-2})T - (3.239 \times 10^{-5})T^2 + (7.552 \times 10^{-9})T^3$$

where T is in kelvins.

TABLE 1.4 Yuan and Stiel Deviation Functions for Saturated-Liquid Heat Capacity

Reduced temperature	$(\Delta C_\sigma)^{(0)}$	$(\Delta C_\sigma)^{(1)}$	$(\Delta C_\sigma)^{(0p)}$	$(\Delta C_\sigma)^{(1p)}$	$(\Delta C_\sigma)^{(2p)}$	$(\Delta C_\sigma)^{(3p)} \times 10^{-2}$	$(\Delta C_\sigma)^{(4p)}$	$(\Delta C_\sigma)^{(5p)}$
0.70	6.01	21.7	6.01	24.5	−110	1.31	−10.9	−29.1
0.68	5.91	21.6	5.94	25.7	−113	2.36	−15.2	−22.8
0.66	5.83	21.8	5.79	27.2	−118	3.06	−20.0	−7.94
0.64	5.74	22.2	5.57	29.3	−124	3.24	−25.1	14.8
0.62	5.64	22.8	5.33	31.8	−132	2.87	−30.5	43.0
0.60	5.54	23.5	5.12	34.5	−141	1.94	−36.3	73.1
0.58	5.42	24.5	4.92	37.6	−151	0.505	−42.5	102
0.56	5.30	25.6	4.69	41.1	−161	−1.37	−49.2	128
0.54	5.17	26.9	4.33	45.5	−172	−3.58	−56.3	149

Source: R. C. Reid, J. M. Prausnitz, and T. K. Sherwood, *Properties of Gases and Liquids*, McGraw-Hill, New York, 1977.

TABLE 1.5   Stiel Polarity Factors of Some Polar Materials

Material	Polarity factor	Material	Polarity factor
Methanol	0.037	Water	0.023
Ethanol	0.0	Hydrogen chloride	0.008
n-Propanol	−0.057	Acetone	0.013
Isopropanol	−0.053	Methyl fluoride	0.012
n-Butanol	−0.07	Ethylene oxide	0.012
Dimethylether	0.002	Methyl acetate	0.005
Methyl chloride	0.007	Ethyl mercaptan	0.004
Ethyl chloride	0.005	Diethyl ether	−0.003
Ammonia	0.013		

Source: R. C. Reid, J. M. Prausnitz, and T. K. Sherwood, *Properties of Gases and Liquids*, McGraw-Hill, New York, 1977.

Calculation Procedure

1. *Estimate the deviation functions.*
For n-octane, $T_r = (273 + 27)/568.8 = 0.527$. From Table 1.4, using linear interpolation and the nonpolar terms,

$$(\Delta C_{\sigma,L})^{(0)} = 5.08 \quad \text{and} \quad (\Delta C_{\sigma,L})^{(1)} = 27.9$$

For ethyl mercaptan, $T_r = (273 + 27)/499 = 0.60$. From Table 1.4, for polar liquids,

$$(\Delta C_{\sigma,L})^{(0p)} = 5.12 \quad (\Delta C_{\sigma,L})^{(1p)} = 34.5 \quad (\Delta C_{\sigma,L})^{(2p)} = -141$$
$$(\Delta C_{\sigma,L})^{(3p)} = 0.0194 \quad (\Delta C_{\sigma,L})^{(4p)} = -36.3 \quad \text{and} \quad (\Delta C_{\sigma,L})^{(5p)} = 73.1$$

2. *Compute ideal-gas heat capacity.*
For n-octane,

$$C_p^o = -1.456 + (1.842 \times 10^{-1})(300) - (1.002 \times 10^{-4})(300^2) + (2.115 \times 10^{-8})(300^3)$$
$$= 45.36 \text{ cal/(g} \cdot \text{mol)(K)}$$

For ethyl mercaptan,

$$C_p^o = 3.564 + (5.615 \times 10^{-2})(300) - (3.239 \times 10^{-5})(300^2) + (7.552 \times 10^{-9})(300^3)$$
$$= 17.7 \text{ cal/(g} \cdot \text{mol)(K)}$$

3. *Compute saturated-liquid heat capacity.*
For n-octane,

$$C_{\sigma,L} = 5.08 + (0.394)(27.9) + 45.36 = 61.43 \text{ cal/(g} \cdot \text{mol)(K)}$$

The experimental value is 60 cal/(g · mol)(K), so the error is 2.4 percent.

For ethyl mercaptan,

$$C_{\sigma,L} = 5.12 + (0.19)(34.5) + (0.004)(-141) + (0.004^2)(1.94)(10^{-2})$$
$$+ (0.190^2)(-36.3) + (0.004)(0.19)(73.1) + 17.7$$
$$= 27.6 \text{ cal/(g} \cdot \text{mol)(K) } [27.6 \text{ Btu/(lb} \cdot \text{mol)(°F)}]$$

The experimental value is 28.2 cal/(g · mol)(K), so the error is 2.1 percent.

1.11 ENTHALPY DIFFERENCE FOR IDEAL GAS

Compute the ideal-gas enthalpy change for *p*-xylene between 289 and 811 K (61 and 1000°F), assuming that the ideal-gas heat-capacity equation is (with *T* in kelvins)

$$C_p^\circ = -7.388 + (14.9722 \times 10^{-2})T - (0.8774 \times 10^{-4})T^2 + (0.019528 \times 10^{-6})T^3$$

Calculation Procedure

1. Compute the ideal-gas enthalpy difference.
The ideal-gas enthalpy difference $(H_2^\circ - H_1^\circ)$ obtained by integrating the C_p° equation between two temperature intervals:

$$
\begin{aligned}
(H_2^\circ - H_1^\circ) &= \int_{T_1}^{T_2} C_p^\circ \, dt \\
&= \int_{T_1}^{T_2} [-7.388 + (14.9772)(10^{-2})T - (0.8774)(10^{-4})T^2 + (0.019528)(10^{-6})T^3] dT \\
&= (-7.388)(811 - 289) + (14.9772 \times 10^{-2})(811^2 - 289^2)/2 \\
&\quad - (0.8774 \times 10^{-4})(811^3 - 289^3)/3 + (0.019528 \times 10^{-6})(811^4 - 289^4)/4 \\
&= 26{,}327 \text{ cal/(g} \cdot \text{mol)}[47{,}400 \text{ Btu/(lb} \cdot \text{mol)}]
\end{aligned}
$$

The literature value is 26,284 cal/(g · mol).

Related Calculations.   Apply this procedure to compute enthalpy difference for any ideal gas. In absence of the ideal-gas heat-capacity equation, estimate C_p° using the Rihani-Doraiswamy group-contribution method, Procedure 1.8.

1.12 ESTIMATION OF HEAT OF VAPORIZATION

Estimate the enthalpy of vaporization of acetone at the normal boiling point using the following relations, and compare your results with the experimental value of 7230 cal/(g · mol).

1. Clapeyron equation and compressibility factor [20]:

$$\Delta H_{v,b} = (RT_c \Delta Z_v T_{b,r} \ln P_c)/(1 - T_{b,r})$$

2. Chen method [21]:

$$\Delta H_{v,b} = RT_c\, T_{b,r}\left(\frac{3.978\,T_{b,r} - 3.938 + 1.555\ln P_c}{1.07 - T_{b,r}}\right)$$

3. Riedel method [22]:

$$\Delta H_{v,b} = 1.093\, RT_c\left[T_{b,r}\frac{(\ln P_c - 1)}{0.930 - T_{b,r}}\right]$$

4. Pitzer correlation:

$$\Delta H_{v,b} = RT_c[7.08(1 - T_{b,r})^{0.354} + 10.95\omega(1 - T_{b,r})^{0.456}]$$

where $\Delta H_{v,b}$ = enthalpy of vaporization at the normal boiling point in cal/(g · mol)
T_c = critical temperature in kelvins
P_c = critical pressure in atmospheres
ω = Pitzer acentric factor
R = gas constant = 1.987 cal/(g · mol)(K)
$T_{b,r} = T_b/T_c$, reduced temperature at the normal boiling point T_b
$\Delta Z_v = Z_v - Z_L$, the difference in the compressibility factor between the saturated vapor and saturated liquid at the normal boiling point, given in Table 1.6.

Also estimate the heat of vaporization of water at 300°C (572°F) by applying the Watson correlation:

$$\frac{\Delta H_2}{\Delta H_1} = \left(\frac{1 - T_{r,2}}{1 - T_{r,1}}\right)^{0.38}$$

where ΔH_1 and ΔH_2 are the heats of vaporization at reduced temperatures of $T_{r,1}$ and $T_{r,2}$, respectively. Data for acetone are $T_b = 329.7$ K, $T_c = 508.7$ K, $P_c = 46.6$ atm, and $\omega = 0.309$. Data for water are $T_b = 373$ K, $T_c = 647.3$ K, and $\Delta H_{v,b} = 9708.3$ cal/(g · mol).

TABLE 1.6   Values of ΔZ_v as a Function of Reduced Pressure

P_r	$Z_v - Z_L$	P_r	$Z_v - Z_L$	P_r	$Z_v - Z_L$
0	1.0	0.25	0.769	0.80	0.382
0.01	0.983	0.30	0.738	0.85	0.335
0.02	0.968	0.35	0.708	0.90	0.280
0.03	0.954	0.40	0.677	0.92	0.256
0.04	0.942	0.45	0.646	0.94	0.226
0.05	0.930	0.50	0.612	0.95	0.210
0.06	0.919	0.55	0.578	0.96	0.192
0.08	0.899	0.60	0.542	0.97	0.170
0.10	0.880	0.65	0.506	0.98	0.142
0.15	0.838	0.70	0.467	0.99	0.106
0.20	0.802	0.75	0.426	1.00	0.000

Calculation Procedure

1. *Calculate reduced temperature $T_{b,r}$ and reduced pressure P_r for the acetone at normal-boiling-point conditions (1 atm) and obtain ΔZ_v.*
Thus,

$$T_{b,r} = \frac{329.7}{508.7} = 0.648 \quad \text{and} \quad P_r = \frac{1}{46.6} = 0.0215$$

From Table 1.6, by extrapolation, $\Delta Z_v = 0.966$.

2. *Compute the heat of vaporization of acetone using the Clapeyron equation.*
From the preceding equation,

$$\Delta H_{v,b} = (1.987)(508.7)(0.966)(0.648)(\ln 46.6)/(1 - 0.648)$$
$$= 6905 \text{ cal/(g} \cdot \text{mol)[12,430 Btu/(lb} \cdot \text{mol)]}$$

Percent error is $100(7230 - 6905)/7230$, or 4.5 percent.

3. *Compute the heat of vaporization using the Chen method.*
Thus,

$$\Delta H_{v,b} = (1.987)(508.7)(0.648)\left[\frac{(3.978)(0.648) - (3.938) + (1.555)\ln 46.6}{1.07 - 0.648}\right]$$
$$= 7160 \text{ cal/(g} \cdot \text{mol)[12,890 Btu/lb} \cdot \text{mol)]}$$

Error is 1.0 percent.

4. *Compute the heat of vaporization using the Riedel method.*
Thus,

$$\Delta H_{v,b} = (1.093)(1.987)(508.7)\left\{0.648\left[\frac{\ln (46.6) - 1}{0.930 - 0.648}\right]\right\}$$
$$= 7214 \text{ cal/(g} \cdot \text{mol)[12,985 Btu/(lb} \cdot \text{mol)]}$$

Error is 0.2 percent.

5. *Compute the heat of vaporization using the Pitzer correlation.*
Thus,

$$\Delta H_{v,b} = (1.987)(508.7)[7.08(1 - 0.648)^{0.354} + (10.95)(0.309)(1 - 0.648)^{0.456}]$$
$$= 7069 \text{ cal/(g} \cdot \text{mol)[12,720 Btu/(lb} \cdot \text{mol)]}$$

Error is 2.2 percent.

6. *Compute the heat of vaporization of the water.*

Now, $T_{r,1} = (100 + 273)/647.3 = 0.576$ and $T_{r,2} = (300 + 273)/647.3 = 0.885$, where the subscript 1 refers to water at its normal boiling point and the subscript 2 refers to water at 300°C. In addition, $\Delta H_{v,b} \, (= \Delta H_1)$ is given above as 9708.3 cal/(g · mol). Then, from the Watson correlation,

$$\Delta H_v (\text{at } 300°\text{C}) = 9708.3 \left(\frac{1 - 0.885}{1 - 0.576} \right)^{0.38}$$

$$= 5913 \text{ cal/(g · mol)} \, [10{,}640 \text{ Btu/(lb · mol)}]$$

The value given in the steam tables is 5949 cal/(g · mol), so the error is 0.6 percent.

Related Calculations. This illustration shows several techniques for estimating enthalpies of vaporization for pure liquids. The Clapeyron equation is inherently accurate, especially if ΔZ_v is obtained from reliable *P-V-T* correlations. The other three techniques yield approximately the same error when averaged over many types of fluids and over large temperature ranges. They are quite satisfactory for engineering calculations. A comparison of calculated and experimental results for 89 compounds has shown average errors of 1.8 and 1.7 percent for the Riedel and Chen methods, respectively. For estimating ΔH_v at any other temperature from a single value at a given temperature, use the Watson correlation. Such a value is normally available at some reference temperature.

1.13 PREDICTION OF VAPOR PRESSURE

Estimate the vapor pressure of 1-butene at 100°C (212°F) using the vapor-pressure correlation of Lee and Kesler [23]. Also compute the vapor pressure of ethanol at 50°C (122°F) from the Thek-Stiel generalized correlation [24]. The Lee-Kesler equation is

$$(\ln P_r^*) = (\ln P_r^*)^{(0)} + \omega (\ln P_r^*)^{(1)}$$

at constant T_r, and the Thek-Stiel correlation for polar and hydrogen-bonded molecules is

$$\ln P_r^* = \frac{\Delta H_{vb}}{RT_c (1 - T_{b,r})^{0.375}} \times \left(1.14893 - 0.11719 T_r - 0.03174 T_r^2 - \frac{1}{T_r} - 0.375 \ln T_r \right)$$

$$+ \left[1.042 \alpha_c - \frac{0.46284 H_{vb}}{RT_c (1 - T_{b,r})^{0.375}} \right] \left[\frac{(T_r)^A - 1}{A} + 0.040 \left(\frac{1}{T_r} - 1 \right) \right]$$

where $P_r^* = P^*/P_c$, reduced vapor pressure, $P^* =$ vapor pressure at T_r, $P_c =$ critical pressure, $\omega =$ acentric factor, $T_r = T/T_c$, reduced temperature, $(\ln P_r^*)^{(0)}$ and $(\ln P_r^*)^{(1)}$ are correlation functions given in Table 1.7, $\Delta H_{vb} =$ heat of vaporization at normal boiling point T_b, α_c, = a constant obtained from the Thek-Stiel equation from conditions $P^* = 1$ atm at $T = T_b$,

$$A = \left[5.2691 + \frac{2.0753 \Delta H_{vb}}{RT_c (1 - T_{b,r})^{0.375}} - \frac{3.1738 T_{b,r} \ln P_c}{1 - T_{b,r}} \right]$$

and $T_{b,r}$ is the reduced normal boiling point.

TABLE 1.7   Correlation Terms for the Lee-Kesler Vapor-Pressure Equation

T_r	$-\ln(P_r^*)^{(0)}$	$-\ln(P_r^*)^{(1)}$	T_r	$-\ln(P_r^*)^{(0)}$	$-\ln(P_r^*)^{(1)}$
1.00	0.000	0.000	0.60	3.568	3.992
0.98	0.118	0.098	0.58	3.876	4.440
0.96	0.238	0.198	0.56	4.207	4.937
0.94	0.362	0.303	0.54	4.564	5.487
0.92	0.489	0.412	0.52	4.951	6.098
0.90	0.621	0.528	0.50	5.370	6.778
0.88	0.757	0.650	0.48	5.826	7.537
0.86	0.899	0.781	0.46	6.324	8.386
0.84	1.046	0.922	0.44	6.869	9.338
0.82	1.200	1.073	0.42	7.470	10.410
0.80	1.362	1.237	0.40	8.133	11.621
0.78	1.531	1.415	0.38	8.869	12.995
0.76	1.708	1.608	0.36	9.691	14.560
0.74	1.896	1.819	0.34	10.613	16.354
0.72	2.093	2.050	0.32	11.656	18.421
0.70	2.303	2.303	0.30	12.843	20.820
0.68	2.525	2.579			
0.66	2.761	2.883			
0.64	3.012	3.218			
0.62	3.280	3.586			

Source: R. C. Reid, J. M. Prausnitz, and T. K. Sherwood, *Properties of Gases and Liquids*, McGraw-Hill, New York, 1977.

Calculation Procedure

1. Obtain critical properties and other necessary basic constants from Reid, Prausnitz, and Sherwood [1].
For l-butene, $T_c = 419.6$ K, $P_c = 39.7$ atm, and $\omega = 0.187$. For ethanol, $T_b = 351.5$ K, $T_c = 516.2$ K, $P_c = 63$ atm, and $\Delta H_{vb} = 9260$ cal/(g · mol).

2. Obtain correlation terms in the Lee-Kesler equation.
From Table 1.7, interpolating linearly at $T_r = (273 + 100)/419.6 = 0.889$, $(\ln P_r^*)^{(0)} = -0.698$ and $(\ln P_r^*)^{(1)} = -0.595$.

3. Compute the vapor pressure of l-butene.
Using the Lee-Kesler equation, $(\ln P_r^*) = -0.698 + (0.187)(-0.595) = -0.8093$, so

$$P_r^* = 0.4452 \quad \text{and} \quad P^* = (0.4452)(39.7) = 17.67 \text{ atm (1790 kPa)}$$

The experimental value is 17.7 atm, so the error is only 0.2 percent.

4. Compute the constant α_c for ethanol.
Now, when $T_{b,r} = 351.5/516.2 = 0.681$, $P^* = 1$ atm. So, in the Thek-Stiel equation, the A term is

$$5.2691 + \left[\frac{(2.0753)(9260)}{(1.987)(516.2)(1 - 0.681)^{0.375}} - \frac{(3.1738)(0.681)(\ln 63)}{(1 - 0.681)} \right] = 5.956$$

at those conditions. Substituting into the full Thek-Stiel equation,

$$\ln\frac{1}{63} = \frac{9260}{(1.987)(516.2)(1 - 0.681)^{0.375}}$$
$$\times [1.14893 - (0.11719)(0.681) - (0.03174)(0.681)^2 - 1/0.681 - (0.375)\ln 0.681]$$
$$+ \left[1.042\alpha_c - \frac{(0.46284)(9260)}{(1.987)(516.2)(1 - 0.681)^{0.375}} \right] \times \left[\frac{(0.681)^{5.956} - 1}{5.956} + (0.040)(1/0.681 - 1) \right]$$

Solving for α_c, we find it to be 9.078.

5. Compute the vapor pressure of ethanol.
Now, $T_r = (273 + 50)/516.2 = 0.626$. Substituting into the Thek-Stiel correlation,

$$\ln P_r^* = \frac{9260}{(1.987)(516.2)(1 - 0.681)^{0.375}} \left[1.14893 - 0.11719(0.626) - 0.03174(0.626)^2 \right.$$
$$\left. - \frac{1}{0.626} - 0.375\ln 0.626 \right] + \left[1.042(9.078) - \frac{(0.46284)(9260)}{(1.987)(516.2)(1 - 0.681)^{0.375}} \right]$$
$$\times \left[\frac{(0.626)^{5.956} - 1}{5.956} + 0.040\left(\frac{1}{0.626} - 1 \right) \right]$$
$$= -5.37717$$

Therefore, P_r^* 0.00462, so $P^* = (0.00462)(63) = 0.2911$ atm (29.5 kPa). The experimental value is 0.291 atm, so the error in this case is negligible.

Related Calculations. For nonpolar liquids, use the Lee-Kesler generalized correlation. For polar liquids and those having a tendency to form hydrogen bonds, the Lee-Kesler equation does not give satisfactory results. For predicting vapor pressure of those types of compounds, use the Thek-Stiel correlation. This method, however, requires heat of vaporization at the normal boiling point, besides critical constants. If heat of vaporization at the normal boiling point is not available, estimate using the Pitzer correlation discussed in Procedure 1.12. If a heat-of-vaporization value at any other temperature is available, use the Watson correlation (Procedure 1.12) to obtain the value at the normal boiling point.

The Lee-Kesler and Thek-Stiel equations each can yield vapor pressures whose accuracy is within ±1 percent. The Antoine equations, based on a correlation with three constants, is less accurate; in some cases, the accuracy is within ±4 or 5 percent. For an example using the Antoine equation, see Procedure 3.1.

1.14 ENTHALPY ESTIMATION—GENERALIZED METHOD

Calculate (a) enthalpy H_v of ethane vapor at 1000 psia (6900 kPa) and 190°F (360 K), (b) enthalpy H_L of liquid ethane at 50°F (283 K) and 450 psia (3100 kPa). Use generalized enthalpy departure charts (Figs. 1.6 through 1.9) to estimate enthalpy values, and base the calculations relative to H = 0 for saturated liquid ethane at −200°F. The basic constants for ethane are molecular weight MW = 30.07, critical temperature T_c = 550°R, critical pressure P_c = 709.8 psia, and critical compressibility factor Z_c = 0.284. Ideal-gas enthalpy $H°$ (relative to saturated liquid ethane at −200°F) at 190°F = 383 Btu/lb, and at 50°F = 318 Btu/lb.

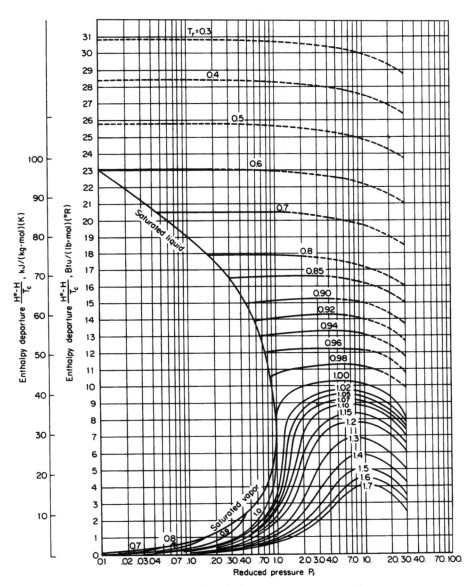

FIGURE 1.6   Enthalpy departure from ideal-gas state; $Z_c = 0.23$. (*Yen and Alexander—AICHE Journal 11:334, 1965.*)

Calculation Procedure

1. *Compute reduced temperature T_r and reduced pressure P_r.*

a. For the vapor, $T_r = (190 + 459.7)/550 = 1.18$, and $P_r = 1000/709.8 = 1.41$.
b. For the liquid, $T_r = (50 + 459.7)/550 = 0.927$, and $P_r = 450/709.8 = 0.634$.

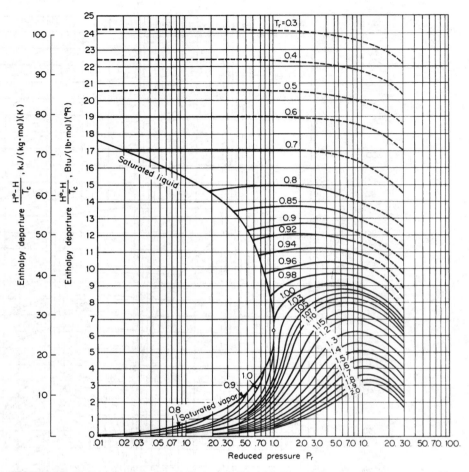

FIGURE 1.7 Enthalpy departure from ideal-gas state; $Z_c = 0.25$. (*Yen and Alexander—AICHE Journal 11:334, 1965.*)

2. Obtain the enthalpy departure function.

a. From Fig. 1.9, for $Z_c = 0.29$, $T_r = 1.18$, $P_r = 1.41$, $(H° - H)/T_c = 2.73$ Btu/(lb · mol)(°R); and from Fig. 1.8, for $Z_c = 0.27$, $T_r = 1.18$, $P_r = 1.41$, $(H° - H)/T_c = 2.80$ Btu/(lb · mol)(°R). Interpolating linearly for $Z_c = 0.284$, $(H° - H)/T_c = 2.75$.

b. From Fig. 1.9, for $Z_c = 0.29$, $P_r = 0.634$, and $T_r = 0.927$, in the liquid region, $(H° - H)/T_c = 7.83$ Btu/ (lb · mol)(°R); and from Fig. 1.8, for $Z_c = 0.27$, $P_r = 0.634$, and $T_r = 0.927$, in the liquid region, $(H° - H)/ T_c = 9.4$ Btu/(lb · mol)(°R). Interpolating for $Z_c = 0.284$, $(H° - H)/T_c = 8.3$.

3. Compute enthalpy of vapor and liquid.

a. Enthalpy of ethane vapor at 190°F and 1000 psia:

$$H_v = H°_{190°F} - \left(\frac{H° - H}{T_c}\right)\left(\frac{T_c}{MW}\right) = 383 - (2.75)(550)/(30.07)$$

$$= 332.7 \text{ Btu/lb } (773,800 \text{ J/kg})$$

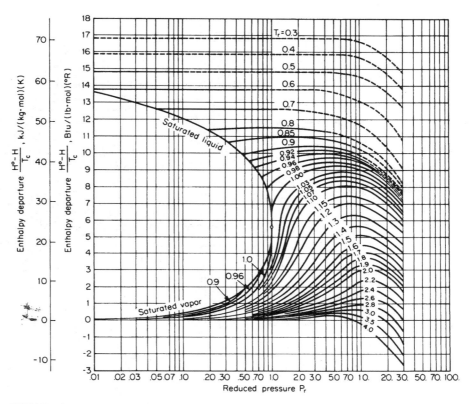

FIGURE 1.8 Enthalpy departure from ideal-gas state; $Z_c = 0.27$. (*Yen and Alexander—AICHE Journal 11:334, 1965.*)

b. Enthalpy of ethane liquid at 50°F and 450 psia:

$$H_L = H^o_{50°F} - \left(\frac{H^o - H}{T_c} \right) \left(\frac{T_c}{MW} \right) = 318 - (8.3)(550)/(30.07)$$

$$= 166 \text{ Btu/lb } (386,100 \text{ J/kg})$$

Related Calculations. This procedure may be used to estimate the enthalpy of any liquid or vapor for nonpolar or slightly polar compounds. Interpolation is required if Z_c values lie between 0.23, 0.25, 0.27, and 0.29. However, extrapolation to Z_c values less than 0.23 or higher than 0.29 should not be made, because serious errors may result. When estimating enthalpy departures for mixtures, estimate mixture pseudocritical properties by taking mole-fraction-weighted averages. Do not use this correlation for gases having a low critical temperature, such as hydrogen, helium, or neon.

1.15 *ENTROPY INVOLVING A PHASE CHANGE*

Calculate the molar entropies of fusion and vaporization for benzene. Having a molecular weight of 78.1, benzene melts at 5.5°C with a heat of fusion of 2350 cal/(g · mol). Its normal boiling point is 80.1°C, and its heat of vaporization at that temperature is 94.1 cal/g.

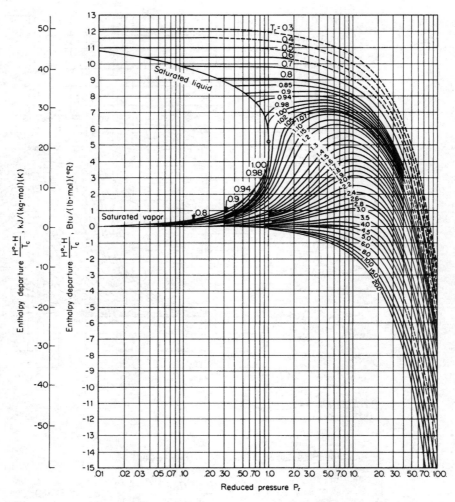

FIGURE 1.9 Enthalpy departure from ideal-gas state; $Z_c = 0.29$. (*Garcia-Rangel and Yen—Paper presented at 159th National Meeting of American Chemical Society, Houston, Tex., 1970.*)

Calculation Procedure

1. Calculate the entropy of fusion ΔS_{fusion}.
By definition, $\Delta S_{fusion} = \Delta H_{fusion}/T_{fusion}$, where the numerator is the heat of fusion and the denominator is the melting point in absolute temperature.
Thus

$$\Delta S_{fusion} = \frac{2350}{(5.5 + 273)} = 8.44 \text{ cal/(g · mol)(K) } [8.44 \text{ Btu/(lb · mol)(°F)}]$$

2. Calculate the entropy of vaporization ΔS_{vap}.
By definition, $\Delta S_{vap} = \Delta H_{vap}/T_{vap}$, where the numerator is the heat of vaporization and the denominator is the absolute temperature at which the vaporization takes place. Since the heat of vaporization

is given on a weight basis, it must be multiplied by the molecular weight to obtain the final result on a molar basis. Thus

$$\Delta S_{vap} = \frac{(94.1)(78.1)}{(80.1 + 273)} = 20.81 \text{ cal/(g} \cdot \text{mol)(K) [20.81 Btu/(lb} \cdot \text{mol)(°F)]}$$

Related Calculations.   This procedure can be used to obtain the entropy of phase change for any compound. If heat-of-vaporization data are not available, the molar entropy of vaporization for non-polar liquids can be estimated via an empirical equation of Kistyakowsky:

$$\Delta S_{vap} = 8.75 + 4.571 \log T_b$$

where T_b is the normal boiling point in kelvins and the answer is in calories per gram-mole per kelvin. For benzene, the calculated value is 20.4, which is in close agreement with the value found in step 2.

1.16   *ABSOLUTE ENTROPY FROM HEAT CAPACITIES*

Calculate the absolute entropy of liquid n-hexanol at 20°C (68°F) and 1 atm (101.3 kPa) from these heat-capacity data:

Temperature, K	Phase	Heat capacity, cal/(g · mol)(K)
18.3	Crystal	1.695
27.1	Crystal	3.819
49.9	Crystal	8.670
76.5	Crystal	15.80
136.8	Crystal	24.71
180.9	Crystal	29.77
229.6	Liquid	46.75
260.7	Liquid	50.00
290.0	Liquid	55.56

The melting point of n-hexanol is −47.2°C (225.8 K), and its enthalpy of fusion is 3676 cal/(g · mol). The heat capacity of crystalline n-hexanol $(C_p)_{crystal}$ at temperatures below 18.3 K may be estimated using the Debye-Einstein equation:

$$(C_p)_{crystal} = aT^3$$

where a is an empirical constant and T is the temperature in kelvins. The absolute entropy may be obtained from

$$S^o_{liq, 20°C} = \underbrace{\int_0^{18.3} (C_p/T)dT}_{(A)} + \underbrace{\int_{18.3}^{225.8} (C_p/T)dT}_{(B)} + \underbrace{\Delta H_{fusion}/225.8}_{(C)} + \underbrace{\int_{225.8}^{(273+20)} (C_p/T)dT}_{(D)} \quad (1.1)$$

where $S^o_{liq, 20°C}$ = absolute entropy of liquid at 20°C (293 K)

(A) = absolute entropy of crystalline n-hexanol at 18.3 K, from the Debye-Einstein equation

(**B**) = entropy change between 18.3 K and fusion temperature, 225.8 K

(**C**) = entropy change due to phase transformation (melting)

(**D**) = entropy change of liquid *n*-hexanol from melting point to the desired temperature (293 K)

Calculation Procedure

1. Estimate absolute entropy of crystalline n-hexanol.

Since no experimental data are available below 18.3 K, estimate the entropy change below this temperature using the Debye-Einstein equation. Use the crystal entropy value of 1.695 cal/(g · mol) (°K) at 18.3 K to evaluate the coefficient *a*. Hence $a = 1.695/18.3^3 = 0.2766 \times 10^{-3}$. The "**A**" term in Eq. 1.1 therefore is

$$\int_0^{18.3} [(0.2766 \times 10^{-3})T^3/T]dT = (0.2766 \times 10^{-3})18.3^3/3 = 0.565 \text{ cal/(g · mol)(K)}$$

2. Compute entropy change between 18.3 K and fusion temperature.

Plot the given experimental data on the crystal heat capacity versus temperature in kelvins and evaluate the integral (the "**B**" term in Eq. 1.1) graphically. See Fig. 1.10. Thus,

$$\int_{18.3}^{225.8} (C_p/T)\,dT = 38.0 \text{ cal/(g · mol)(K)}$$

3. Compute the entropy change due to phase transformation.

In this case,

$$\Delta H_{\text{fusion}}/225.8 = 3676/225.8 = 16.28 \text{ cal/(g · mol)(K)}$$

This is the "**C**" term in Eq. 1.1.

4. Compute the entropy change between 225.8 K and 293 K.

Plot the given experimental data on the liquid heat capacity versus temperature in kelvins and evaluate the integral (the "**D**" term in Eq. 1.1) graphically. See Fig. 1.10. Thus,

$$\int_{225.8}^{293} (C_p/T)\,dT = 13.7 \text{ cal/(g · mol)(K)}$$

5. Calculate the absolute entropy of liquid n-hexanol.

The entropy value is obtained from the summation of the four terms; that is, $S_{\text{liq}}^{\circ}, 20°C = 0.565 + 38.0 + 16.28 + 13.7 = 68.5$ cal/(g · mol)(K) [68.5 Btu/(lb · mol)(°F)].

Related Calculations. This general procedure may be used to calculate entropy values from heat-capacity data. However, in many situations involving practical computations, the entropy changes rather than absolute values are required. In such situations, the "**A**" term in Eq. 1.1 may not be needed. Entropy changes associated with phase changes, such as melting and vaporization, can be evaluated from the $\Delta H/T$ term (see Procedure 1.15).

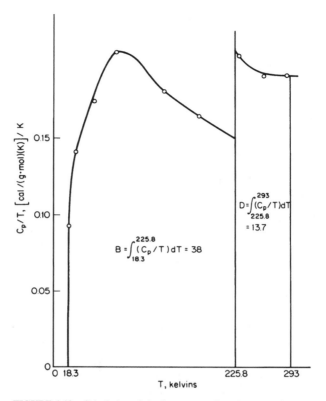

FIGURE 1.10   Calculation of absolute entropy from heat-capacity data (Procedure 1.16).

1.17   EXPANSION UNDER ISENTROPIC CONDITIONS

Calculate the work of isentropic expansion when 1000 lb · mol/h of ethylene gas at 1500 psig (10,450 kPa) and 104°F (313 K) is expanded in a turbine to a discharge pressure of 150 psig (1135 kPa). The ideal-gas heat capacity of ethylene is

$$C_p^o = 0.944 + (3.735 \times 10^{-2})T - (1.993 \times 10^{-5})T^2$$

where C_p^o is in British thermal units per pound-mole per degree Rankine, and T is temperature in degrees Rankine. Critical temperature T_c is 282.4 K (508.3°R), critical pressure P_c is 49.7 atm, and critical compressibility factor Z_c is 0.276.

Calculation Procedure

1. *Estimate degree of liquefaction, if any.*
Since the expansion will result in cooling, the possibility of liquefaction must be considered. If it does occur, the final temperature will be the saturation temperature T_{sat} corresponding to 150 psig.

On the assumption that liquefaction has occurred from the expansion, the reduced saturation temperature T_{rs} at the reduced pressure P_r of $(150 + 14.7)/(49.7)(14.7)$, or 0.23, is 0.81 from Fig. 1.1. Therefore, $T_{sat} = (0.81)(508.3)$ or 412°R. Use the following equation to estimate the mole fraction of ethylene liquefied:

$$\Delta S = 0 = (S_1^\circ - S_1) + \int_{T_1}^{T_{sat}} (C_p^\circ/T)\,dT - R\ln(P_2/P_1) - (S_2^\circ - S_2)_{SG} - x\Delta S_{vap}$$

where $(S_1^\circ - S_1)$ and $(S_2^\circ - S_2)_{SG}$ are the entropy departure functions for gas at inlet conditions and saturation conditions, respectively; ΔS_{vap} is the entropy of vaporization at the saturation temperature; x equals moles of ethylene liquefied per mole of ethylene entering the turbine; and P_1 and P_2 are the inlet and exhaust pressures, respectively. Obtain entropy departure functions from Fig. 1.11 at inlet and exhaust conditions. Thus, at inlet conditions, $T_{r1} = (104 + 460)/508.3 = 1.11$, and $P_{r1} = (1500 + 14.7)/(14.7)(49.7) = 2.07$, so $(S_1^\circ - S_1) = 5.0$ Btu/(lb · mol)(°R). At outlet conditions, $T_{r2} = 0.81$, and $P_{r2} = 0.23$, so $(S_2^\circ - S_2)_{SG} = 1.0$ Btu/(lb · mol)(°R) and $(S_2^\circ - S_2)_{SL} = 13.7$ Btu/(lb · mol)(°R). The difference between these last two values is the entropy of vaporization at the saturation temperature S_{vap}.

Substituting in the entropy equation, $\Delta S = 0 = 5.0 + 0.944\ln(412/564) + (3.735 \times 10^{-2})(412 - 564) - (1/2)(1.993 \times 10^{-5})(412^2 - 564^2) - 1.987\ln(164.7/1514.7) - 1.0 - x(13.7 - 1.0)$. Upon solving, $x = 0.31$. Since the value of x is between 0 and 1, the assumption that liquefaction occurs is valid.

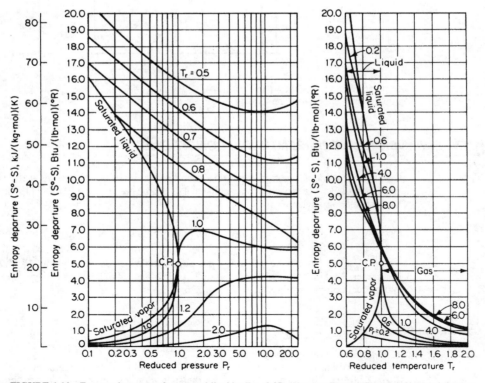

FIGURE 1.11   Entropy departure of gases and liquids; $Z_c = 0.27$. (*Hougen, Watson, Ragatz—Chemical Process Principles, Part II, Wiley, 1959.*)

2. Compute the work of isentropic expansion.
The work of isentropic expansion is obtained from enthalpy balance equation:

$$\text{Work} = -\Delta H$$

$$= -\left\{ T_c[(H_1^\circ - H_1)/T_c] + \int_{T_1}^{T_{\text{sat}}} C_p^\circ \, dT - T_c[(H_2^\circ - H_2)_{SG}/T_c] - x\Delta H_{\text{vap}} \right\}$$

where $(H_1^\circ - H_1)/T_c$ and $(H_2^\circ - H_2)_{SG}/T_c$ are enthalpy departure functions for gas at inlet conditions and saturation conditions, respectively; and ΔH_{vap} is enthalpy of vaporization at the saturation temperature $(= T\Delta S_{\text{vap}})$.

The enthalpy departure functions, obtained as in Procedure 1.14, are

$$(H_1^\circ - H_1)/T_c = 6.1 \text{ Btu/(lb} \cdot \text{mol)(}^\circ\text{R)}$$

$$(H_2^\circ - H_2)_{SG}/T_c = 0.9 \text{ Btu/(lb} \cdot \text{mol)(}^\circ\text{R)}$$

Substituting these in the enthalpy balance equation,

$$\text{Work} = -\Delta H$$

$$= -[(508.3)(6.1) + (412 - 564)(0.944) + (1/2)(3.735)(10^{-2})(412^2 - 564^2)$$

$$- (1/3)(1.993)(10^{-5})(412^3 - 564^3) - (508.3)(0.9) - (0.31)(412)(13.7 - 1.0)]$$

$$= 1165.4 \text{ Btu/(lb} \cdot \text{mol)}$$

The power from the turbine equals [1165.4 Btu/(lb · mol)](1000 lb · mol/h) (0.000393 hp · h/Btu) = 458 hp (342 kW).

Related Calculations. When specific thermodynamic charts, namely, enthalpy-temperature, entropy-temperature, and enthalpy-entropy, are not available for a particular system, use the generalized enthalpy and entropy charts to perform expander-compressor calculations, as shown in this example.

1.18 CALCULATION OF FUGACITIES

Calculate fugacity of (a) methane gas at 50°C (122°F) and 60 atm (6080 kPa), (b) benzene vapor at 400°C (752°F) and 75 atm (7600 kPa), (c) liquid benzene at 428°F (493 K) and 2000 psia (13,800 kPa), and (d) each component in a mixture of 20% methane, 40% ethane, and 40% propane at 100°F (310 K) and 300 psia (2070 kPa), assuming ideal-mixture behavior. The experimental pressure-volume data for benzene vapor at 400°C (752°F) from very low pressures up to about 75 atm are represented by

$$Z = \frac{PV}{RT} = 1 - 0.0046 P$$

where Z is the compressibility factor and P is the pressure in atmospheres.

Calculation Procedure

1. Obtain the critical-property data.
From any standard reference, the critical-property data are

Compound	T_c, K	P_c, atm	P_c, kPa	Z_c
Benzene	562.6	48.6	4924	0.274
Methane	190.7	45.8	4641	0.290
Ethane	305.4	48.2	4884	0.285
Propane	369.9	42.0	4256	0.277

2. Calculate the fugacity of the methane gas.

Now, reduced temperature $T_r = T/T_c = (50 + 273)/190.7 = 1.69$, and reduced pressure $P_r = P/P_c = 60/45.8 = 1.31$. From the generalized fugacity-coefficient chart (Fig. 1.12), the fugacity coefficient f/P at the reduced parameters is 0.94. (Ignore the fact that Z_c differs slightly from the standard value of 0.27). Therefore, the fugacity $f = (0.94)(60) = 56.4$ atm (5713 kPa).

3. Calculate the fugacity of the benzene vapor using the P-V-T relationship given in the statement of the problem.

Start with the equation

$$d \ln (f/P)_T = [(Z - 1)d \ln P]_T$$

Upon substituting the given P-V-T relationship,

$$d \ln (f/P) = (1 - 0.0046P)(d \ln P) - d \ln P = -0.0046P(d \ln P)$$

Since $d P/P = d \ln P$, $d \ln (f/P) = -0.0046\, d P$. Upon integration,

$$\ln (f_2/P_2) - \ln (f_1/P_1) = -0.0046(P_2 - P_1)$$

From the definition of fugacity, f/P approaches 1.0 as P approaches 0; therefore, $\ln (f_1/P_1)$ approaches $\ln 1$ or 0 as P_1 approaches 0. Hence $\ln (f_2/P_2) = -0.0046\, P_2$, and $f_2 = P_2 \exp(-0.0046\, P_2)$. So, when $P_2 = 75$ atm, fugacity $f = 75 \exp [(-0.0046)(75)] = 53.1$ atm (5380 kPa).

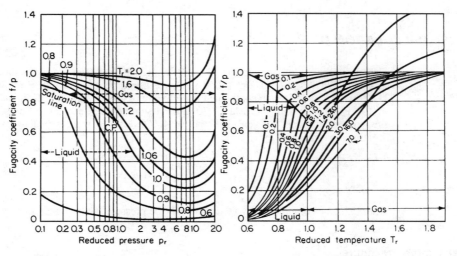

FIGURE 1.12   Fugacity coefficients of gases and liquids; $Z_c = 0.27$. (*Hougen, Watson, Ragatz—Chemical Process Principles, Part II, Wiley, 1959.*)

4. Calculate the fugacity of the benzene vapor from the generalized correlation.
Now $T_r = (400 + 273)/562.6 = 1.20$, and $P_r = 75/48.6 = 1.54$. Ignoring the slight difference in Z_c from Fig. 1.12, $f/P = 0.78$. So fugacity $f = (0.78)(75) = 58.5$ atm (5926 kPa).

5. Calculate the fugacity of the liquid benzene from the generalized correlation.
Now $T_r = (428 + 460)/[(562.6)(1.8)] = 0.88$, and $P_r = 2000/[(14.69)(48.6)] = 2.80$. From Fig. 1.12, $f/P = 0.2$. So fugacity $f = (0.2)(2000) = 400$ psia (2760 kPa).

6. Calculate the fugacity of each component in the mixture.
The calculations for this step can be set out as follows:

	Methane	Ethane	Propane
Mole fraction, Y_i	0.2	0.4	0.4
Reduced temperature, $T_r = T/T_c$	1.63	1.02	0.84
Reduced pressure, $P_r = P/P_c$	0.45	0.42	0.49
Fugacity coefficient f_i°/P (from Fig. 1.12)	0.98	0.89	0.56
Fugacity of pure component, $f_i^\circ = (f_i^\circ/P)(300)$, psia (kPa)	294 (2027)	267 (1841)	168 (1158)
Fugacity of component in the mixture, $f_i = Y_i f_i^\circ$, (ideal mixture), psia (kPa)	58.8 (405)	106.8 (736)	67.2 (463)

Related Calculations. If experimental P-V-T data are available, either as an analytical expression or as tabular values, the fugacity coefficient may be calculated by integrating the data (numerically or otherwise) as shown in step 3 above. However, if such data are not available, use the generalized fugacity coefficient chart to estimate fugacity values. Refer to Hougen, Watson, and Ragatz [4] for deviation-correction terms for values of Z_c above and below the standard value of 0.27.

The generalized-correlation method used in this example is fast and adequate for calculations requiring typical engineering accuracy. Fugacities can also be calculated by thermodynamically rigorous methods based on equations of state. Although these are cumbersome for hand calculation, they are commonly used for estimating vapor-phase nonidealities and making phase-equilibrium calculations. (Examples are given in Section 3, on phase equilibrium; in particular, see Procedures 3.2 and 3.3.)

In the present example involving a mixture (step 4), ideal behavior was assumed. For handling nonideal gaseous mixtures, volumetric data are required, preferably in the form of an equation of state at the temperature under consideration and as a function of composition and density, from zero density (lower integration limit) to the density of interest. These computations often require trial-and-error solutions and consequently are tedious for hand calculation.

1.19 ACTIVITY COEFFICIENTS FROM THE SCATCHARD-HILDEBRAND EQUATION

Experimental vapor-liquid-equilibrium data for benzene(1)/n-heptane(2) system at 80°C (176°F) are given in Table 1.8. Calculate the vapor compositions in equilibrium with the corresponding liquid compositions, using the Scatchard-Hildebrand regular-solution model for the liquid-phase activity coefficient, and compare the calculated results with the experimentally determined composition. Ignore the nonideality in the vapor phase. Also calculate the solubility parameters for benzene and n-heptane using heat-of-vaporization data.

TABLE 1.8   Vapor-Liquid Equilibrium Data for Benzene(1)/n-Heptane(2) System at 80°C

Mole fraction of benzene in liquid phase, x_1	0.000	0.0861	0.2004	0.3842	0.5824	0.7842	0.8972	1.000
Mole fraction of benzene in vapor phase, y_1	0.000	0.1729	0.3473	0.5464	0.7009	0.8384	0.9149	1.000
Total pressure, P								
mmHg:	427.8	476.25	534.38	613.53	679.74	729.77	748.46	757.60
kPa:	57.0	63.49	71.24	81.80	90.62	97.29	99.79	101.01

The following data are available on the two components:

Compound	Vapor pressure at 80°C, mmHg	Normal boiling point, °C	Heat of vaporization at normal boiling point, cal/(g · mol)	Solubility parameter, $(\text{cal/cm}^3)^{1/2}$	Liquid molar volume at 25°C, [cm³/(g · mol)]	Critical temperature T_c, K
Benzene	757.6	80.3	7352	9.16	89.4	562.1
n-heptane	427.8	98.6	7576	7.43	147.5	540.2

The Scatchard-Hildebrand regular-solution model expresses the liquid activity coefficients γ_i in a binary mixture as

$$\ln \gamma_1^L = \frac{V_1^L \phi_2^2}{RT}(\delta_1 - \delta_2)^2$$

and

$$\ln \gamma_2^L = \frac{V_2^L \phi_1^2}{RT}(\delta_1 - \delta_2)^2$$

and the activity coefficient of liquid component i in a multicomponent mixture as

$$\ln \gamma_i^L = \frac{V_i^L}{RT}(\delta_i - \bar{\delta})^2$$

where V_i^L = liquid molar volume of component i at 25°C
 R = gas constant
 T = system temperature
 ϕ_i = molar volume fraction of component i at 25°C = $(x_i V_i^L)/(\Sigma x_i V_i^L)$
 δ_i = solubility parameter of component i
 $\bar{\delta}$ = a molar volume fraction average of δ_i = $\Sigma \phi_i \delta_i$

The solubility parameter is defined as

$$\delta_i = \left(\frac{\Delta H_i^V - 298.15R}{V_i^L} \right)^{1/2}$$

where ΔH_i^V = heat of vaporization of component i from saturated liquid to the ideal-gas state at 25°C, cal/(g · mol)
 R = gas constant, 1.987 cal/(g · mol)(K)
 V_i^L = liquid molar volume of component i at 25°C, cm³/(g · mol)

TABLE 1.9 Experimental and Calculated Vapor-Liquid Equilibrium Data for the Benzene(1)/*n*-Heptane(2) System at 80°C

x_1	y_1^{exp}	γ_1	γ_2	y_1^{calc}	Percent deviation $(y_1^{calc} - y_1^{exp})(100)/y_1^{exp}$
0.0000	0.0000	1.4644	1.0000	0.0000	—
0.0861	0.1729	1.4070	1.002	0.1927	11.5
0.2004	0.3473	1.3330	1.0110	0.3787	9.0
0.3842	0.5464	1.2224	1.0485	0.5799	6.1
0.5824	0.7009	1.1185	1.1412	0.7260	3.6
0.7842	0.8384	1.0379	1.3467	0.8450	0.8
0.8972	0.9194	1.0097	1.5607	0.9170	−0.3
1.0000	1.0000	1.0000	1.8764	1.0000	0.00
					(Overall = 5.1%)

Note: Superscript ^{exp} means experimental; superscript ^{calc} means calculated.

Calculation Procedure

1. Compute the liquid-phase activity coefficients.
Using the given values of liquid molar volumes and solubility parameters, the activity coefficients are calculated for each of the eight liquid compositions in Table 1.8. For instance, when $x_1 = 0.0861$ and $x_2 = 1 - x_1 = 0.914$, then $\phi_1 = (0.0861)(89.4)/[(0.0861)(89.4) + (0.914)(147.5)] = 0.0540$ and $\phi_2 = 1 - \phi_1 = 0.9460$. Therefore,

$$\ln \gamma_1 = \frac{(89.4)(0.9460)^2}{(1.987)(273 + 80)}(9.16 - 7.43)^2 = 0.341$$

so $\gamma_1 = 1.407$, and

$$\ln \gamma_2 = \frac{(147.5)(0.054)^2}{(1.987)(273 + 80)}(9.16 - 7.43)^2 = 0.002$$

so $\gamma_2 = 1.002$.
The activity coefficients for other liquid compositions are calculated in a similar fashion and are given in Table 1.9.

2. Compute the vapor-phase mole fractions.
Assuming ideal vapor-phase behavior, $y_1 = x_1 P_1^° \gamma_1 / P$ and $y_2 = x_2 P_2^° \gamma_2 / P = 1 - y_1$, where $P_i^°$ is the vapor pressure of component *i*. From Table 1.8, when $x_1 = 0.0861$, $P = 476.25$ mmHg. Therefore, the *calculated* vapor-phase mole fraction of benzene is $y_1 = (0.0861)(757.6)(1.407)/476.25 = 0.1927$. The mole fraction of *n*-heptane is $(1 - 0.1927)$ or 0.8073. The vapor compositions in equilibrium with other liquid compositions are calculated in a similar fashion and are tabulated in Table 1.9. The last column in the table shows the deviation of the calculated composition from that determined experimentally.

3. Estimate heats of vaporization at 25°C.
Use the Watson equation:

$$\Delta H_{vap, 25°C} = \Delta H_{vap, TNBP}\left[\frac{1 - (273 + 25)/T_c}{1 - T_{NBP}/T_c}\right]^{0.38}$$

where ΔH_{vap} is heat of vaporization, and T_{NBP} is normal boiling-point temperature. Thus, for benzene,

$$\Delta H_{vap, 25°C} = 7353\left[\frac{1 - 298/562.1}{1 - (273 + 80.3)/562.1}\right]^{0.38} = 8039 \text{ cal/(g} \cdot \text{mol)}$$

And for *n*-heptane,

$$\Delta H_{vap,25°C} = 7576 \left[\frac{1 - 298/540.2}{1 - (273 + 98.6)/540.2} \right]^{0.38} = 8694 \text{ cal/(g · mol)}$$

TABLE 1.10   Selected Values of Solubility Parameters at 25°C

Formula	Substance	Liquid molar volume, cm³	Molar heat of vaporization at 25°C, kcal	Solubility parameter (cal/cm³)^{1/2}
	Aliphatic hydrocarbons			
C_5H_{12}	*n*-Pentane	116	6.40	7.1
	2-Methyl butane (isopentane)	117	6.03	6.8
	2,2-Dimethyl propane (neopentane)	122	5.35	6.2
C_6H_{14}	*n*-Hexane	132	7.57	7.3
C_7H_{16}	*n*-Heptane	148	8.75	7.4
C_8H_{18}	*n*-Octane	164	9.92	7.5
	2,2,4-Trimethylpentane ("isooctane")	166	8.40	6.9
	Aromatic hydrocarbons			
C_6H_6	Benzene	89	8.10	9.2
C_7H_8	Toluene	107	9.08	8.9
C_8H_{10}	Ethylbenzene	123	10.10	8.8
	o-Xylene	121	10.38	9.0
	m-Xylene	123	10.20	8.8
	p-Xylene	124	10.13	8.8
C_9H_{12}	*n*-Propyl benzene	140	11.05	8.6
	Mesitylene	140	11.35	8.8
	Fluorocarbons			
C_6F_{14}	Perfluoro-*n*-hexane	205	7.75	5.9
C_7F_{16}	Perfluoro-*n*-heptane (pure)	226	8.69	6.0
	Perfluoroheptane (mixture)			5.85
C_6F_{12}	Perfluorocyclohexane	170	6.9	6.1
C_7F_{14}	Perfluoro (methylcyclohexane)	196	7.9	6.1
	Other fluorochemicals			
$(C_4F_9)_3N$	Perfluoro tributylamine	360	13.0	5.9
$C_4Cl_2F_6$	Dichlorohexafluorocyclobutane	142		7.1
$C_4Cl_3F_7$	2,2,3-Trichloroheptafluorobutane	165	8.51	6.9
$C_2Cl_3F_3$	1,1,2-Trichloro, 1,2,2-trifluoroethane	120	6.57	7.1
$C_7F_{15}H$	Pentadecafluoroheptane	215	9.01	6.3
	Other aliphatic halogen compounds			
CH_2Cl_2	Methylene chloride	64	6.84	9.8
$CHCl_3$	Chloroform	81	7.41	9.2
CCl_4	Carbon tetrachloride	97	7.83	8.6
$CHBr_3$	Bromoform	88	10.3	10.5
CH_3I	Methyl iodide	63	6.7	9.9
CH_2I_2	Methylene iodide	81		11.8
C_2H_5Cl	Ethyl chloride	74	5.7	8.3
C_2H_5Br	Ethyl bromide	75	6.5	8.9
C_2H_5I	Ethyl iodide	81	7.7	9.4

Source: J. Hildebrand, J. Prausnitz, and R. Scott, *Regular and Related Solutions,* copyright 1970 by Litton Educational Publishing Inc. Reprinted with permission of Van Nostrand Co.

4. Compute the solubility parameters.
By definition,

$$\delta_{\text{benzene}} = \left[\frac{8039 - (298.15)(1.987)}{89.4} \right]^{1/2} = 9.13 \, (\text{cal/cm}^3)^{1/2}$$

and

$$\delta_{n\text{-heptane}} = \left[\frac{8694 - (298.15)(1.987)}{147.5} \right]^{1/2} = 7.41 \, (\text{cal/cm}^3)^{1/2}$$

These calculated values are reasonably close to the true values given in the statement of the problem.

Related Calculations. The regular-solution model of Scatchard and Hildebrand gives a fair representation of activity coefficients for many solutions containing nonpolar components. This procedure is suggested for estimating vapor-liquid equilibria if experimental data are not available. The solubility parameters and liquid molar volumes used as characteristic constants may be obtained from Table 1.10. For substances not listed there, the solubility parameters may be calculated from heat of vaporization and liquid molar volume data as shown in step 4.

For moderately nonideal liquid mixtures involving similar types of compounds, the method gives activity coefficients within ±10 percent. However, extension of the correlation to hydrogen-bonding compounds and highly nonideal mixtures can lead to larger errors.

When experimental equilibrium data on nonideal mixtures are not available, methods such as those based on Derr and Deal's analytical solution of groups (ASOG) [28] or the UNIFAC correlation (discussed in Procedure 3.4) may be used. Activity-coefficient estimation methods are also available in various thermodynamic-data packages, such as Chemshare. Further discussion may be found in Prausnitz [3] and in Reid, Prausnitz, and Sherwood [1].

The following two examples show how to use experimental equilibrium data to obtain the *equation* coefficients (as opposed to the *activity* coefficients themselves) for activity-coefficient correlation equations. Use of these correlations to calculate the activity coefficients and make phase-equilibrium calculations is discussed in Section 3.

1.20 ACTIVITY-COEFFICIENT-CORRELATION EQUATIONS AND LIQUID-LIQUID EQUILIBRIUM DATA

Calculate the coefficients of Van Laar equations and the three-suffix Redlich-Kister equations from experimental solubility data at 70°C (l58°F, or 343 K) for the water(1)/trichloroethylene(2) system. The Van Laar equations are

$$\log \gamma_1 = \frac{A_{1,2}}{\left(1 + \dfrac{A_{1,2}X_1}{A_{2,1}X_2} \right)^2} \qquad \log \gamma_2 = \frac{A_{2,1}}{\left(1 + \dfrac{A_{2,1}X_2}{A_{1,2}X_1} \right)^2}$$

The Redlich-Kister equation is

$$g^E/RT = X_1 X_2 [B_{1,2} + C_{1,2}(2X_1 - 1)]$$

From the relation $\ln \gamma_i = [\delta(g^E/RT)/\delta X_i]$

$$\ln \gamma_i = X_1 X_2 [B_{1,2} + C_{1,2}(X_1 - X_2)] + X_2 [B_{1,2}(X_2 - X_1) + C_{1,2}(6X_1 X_2 - 1)]$$

In these equations, γ_1 and γ_2 are activity coefficients of components 1 and 2, respectively; X_1 and X_2 are mole fractions of components 1 and 2, respectively; $A_{1,2}$ and $A_{2,1}$ are Van Laar coefficients; $B_{1,2}$ and $C_{1,2}$ are Redlich-Kister coefficients; R is the gas constant; and T is the absolute system temperature.

The solubility data at 70°C in terms of mole fraction are: mole fraction of water in water-rich phase $X_1^W = 0.9998$; mole fraction of trichloroethylene in water-rich phase $X_2^W = 0.0001848$; mole fraction of water in trichloroethylene-rich phase $X_1^O = 0.007463$; and mole fraction of trichloroethylene in trichloroethylene-rich phase $X_2^O = 0.9925$.

Calculation Procedure

1. Estimate the Van Laar constants.
In the water-rich phase,

$$\log \gamma_1^W = \frac{A_{1,2}}{\left[1 + \dfrac{(A_{1,2})(0.9998)}{(A_{2,1})(0.0001848)} \right]^2} = \frac{A_{1,2}}{[1 + (5410.2)(A_{1,2})/A_{2,1}]^2}$$

$$\log \gamma_2^W = \frac{A_{2,1}}{\left[1 + \dfrac{(A_{2,1})(0.0001848)}{(A_{1,2})(0.9998)} \right]^2} = \frac{A_{2,1}}{[1 + (0.0001848)(A_{2,1})/A_{1,2}]^2}$$

In the trichloroethylene-rich phase,

$$\log \gamma_1^O = \frac{A_{1,2}}{\left[1 + \dfrac{(A_{1,2})(0.007463)}{(A_{2,1})(0.9925)} \right]^2} = \frac{A_{1,2}}{[1 + (0.007520)(A_{1,2})/A_{2,1}]^2}$$

$$\log \gamma_2^O = \frac{A_{2,1}}{\left[1 + \dfrac{(A_{2,1})(0.9925)}{(A_{1,2})(0.007463)} \right]^2} = \frac{A_{2,1}}{[1 + (132.99)(A_{2,1})/A_{1,2}]^2}$$

Now, at liquid-liquid equilibria, the partial pressure of component i is the same both in water-rich and organic-rich phases, and partial pressure $\bar{P}_i = P_i^* \gamma_i X_i$, where P_i^* is the vapor pressure of component i. Hence $\gamma_1^O/\gamma_1^W = K_1 = X_1^W/X_1^O = 0.9998/0.007463 = 134$, and $\gamma_2^O/\gamma_2^W = K_2 = X_2^W/X_2^O = 0.0001848/0.9925 = 0.0001862$.

From the Van Laar equation,

$$\log \gamma_1^O - \log \gamma_1^W = \log K_1$$

$$= \frac{A_{1,2}}{[1 + (0.007520)(A_{1,2})/A_{2,1}]^2} - \frac{A_{1,2}}{[1 + (5410.2)(A_{1,2})/A_{2,1}]^2}$$

Substituting for K_1 and rearranging,

$$A_{1,2}^2 + 0.0246 A_{2,1}^2 + 132.9788 A_{1,2} A_{2,1} - 62.5187 A_{1,2}^2 A_{2,1} = 0 \qquad (1.2)$$

Similarly,

$$\log \gamma_2^O - \log \gamma_2^W = \log K_2$$

$$= \frac{A_{2,1}}{[1 + (132.99)(A_{2,1})/A_{1,2}]^2} - \frac{A_{2,1}}{[1 + (0.0001848)(A_{2,1})/A_{1,2}]^2}$$

Substituting for K_2 and rearranging,

$$A_{1,2}^2 + 0.02456 A_{2,1}^2 + 132.8999 A_{1,2} A_{2,1} + 35.6549 A_{1,2} A_{2,1}^2 = 0 \qquad (1.3)$$

Solving Eqs. 1.2 and 1.3 simultaneously for the Van Laar constants, $A_{1,2} = 2.116$ and $A_{2,1} = -3.710$.

2. Estimate the Redlich-Kister coefficients

$$\log \left(\gamma_1^O / \gamma_1^W \right) = \log K_1$$

$$= B_{1,2}\left(X_2^{2,O} - X_2^{2,W} \right) + C_{1,2}\left[\left(X_2^{2,O} \right)\left(4X_1^O - 1 \right) - \left(X_2^{2,W} \right)\left(4X_1^W - 1 \right) \right] \qquad (1.4)$$

$$\log \left(\gamma_2^O / \gamma_2^W \right) = \log K_2$$

$$= B_{1,2}\left(X_1^{2,O} - X_1^{2,W} \right) + C_{1,2}\left[\left(X_1^{2,O} \right)\left(1 - 4X_2^O \right) - \left(X_1^{2,W} \right)\left(1 - 4X_2^W \right) \right] \qquad (1.5)$$

Substituting numerical values and simultaneously solving Eqs. 1.4 and 1.5 for the Redlich-Kister constants, $B_{1,2} = 2.931$ and $C_{1,2} = 0.799$.

Related Calculations. This illustration outlines the procedure for obtaining coefficients of a liquid-phase activity-coefficient model from mutual solubility data of partially miscible systems. Use of such models to calculate activity coefficients and to make phase-equilibrium calculations is discussed in Section 3. This leads to estimates of phase compositions in liquid-liquid systems from limited experimental data. At ordinary temperature and pressure, it is simple to obtain experimentally the composition of two coexisting phases, and the technical literature is rich in experimental results for a large variety of binary and ternary systems near 25°C (77°F) and atmospheric pressure. Procedure 1.21 shows how to apply the same procedure with vapor-liquid equilibrium data.

1.21 ACTIVITY-COEFFICIENT-CORRELATION EQUATIONS AND VAPOR-LIQUID EQUILIBRIUM DATA

From the isothermal vapor-liquid equilibrium data for the ethanol(l)/toluene(2) system given in Table 1.11, calculate (a) vapor composition, assuming that the liquid phase and the vapor phase obey Raoult's and Dalton's laws, respectively, (b) the values of the infinite-dilution activity coefficients, γ_1^∞ and γ_2^∞, copyright Van Laar parameters using data at the azeotropic point as well as from the infinite-dilution activity coefficients, and (d) Wilson parameters using data at the azeotropic point as well as from the infinite-dilution activity coefficients.

For a binary system, the Van Laar equations are

$$\log \gamma_1 = \frac{A_{1,2}}{\left(1 + \dfrac{A_{1,2}X_1}{A_{2,1}X_2}\right)^2} \quad \text{and} \quad \log \gamma_2 = \frac{A_{2,1}}{\left(1 + \dfrac{A_{2,1}X_2}{A_{1,2}X_1}\right)^2}$$

or

$$A_{1,2} = \log \gamma_1 \left(1 + \frac{X_2 \log \gamma_2}{X_1 \log \gamma_1}\right)^2 \quad \text{and} \quad A_{2,1} = \log \gamma_2 \left(1 + \frac{X_1 \log \gamma_1}{X_2 \log \gamma_2}\right)^2$$

The Wilson equations are

$$G^E/RT = -X_1 \ln(X_1 + X_2 G_{1,2}) - X_2 \ln(X_2 + X_1 G_{1,2})$$

$$\ln \gamma_1 = -\ln(X_1 + X_2 G_{1,2}) + X_2 \left[\frac{G_{1,2}}{(X_1 + X_2 G_{1,2})} - \frac{G_{2,1}}{(X_2 + X_1 G_{2,1})}\right]$$

$$\ln \gamma_2 = -\ln(X_2 + X_1 G_{2,1}) - X_1 \left[\frac{G_{1,2}}{(X_1 + X_2 G_{1,2})} - \frac{G_{2,1}}{(X_2 + X_1 G_{2,1})}\right]$$

$$G^E/X_1 X_2 RT = \ln \gamma_1/X_2 + \ln \gamma_2/X_1$$

TABLE 1.11   Vapor-Liquid Equilibria for Ethanol(1)/Toluene(2) System at 55°C

P, mmHg (1)	X_1^{exp} (2)	Y_1^{exp} (3)	$Y_1 P$, mmHg (4)	$Y_2 P$, mmHg (5)	Y_1, ideal (6)	$\ln \gamma_1^{exp}$ (7)	$\ln \gamma_2^{exp}$ (8)	$\left(\dfrac{G^E}{X_1 X_2 RT}\right)^{exp}$ (9)
114.7	0.0000	0.0000	0.0000	114.7	0.0000	(2.441)[†]	0.0000	(2.441)[†]
144.2	0.0120	0.2127	30.7	113.5	0.0233	2.213	0.002	2.407
194.6	0.0400	0.4280	83.3	111.3	0.0575	2.010	0.011	2.369
243.0	0.1000	0.5567	135.3	107.7	0.1151	1.577	0.043	2.182
294.5	0.4000	0.6699	197.3	97.2	0.3798	0.568	0.345	1.809
308.2 (Azeotrope)	0.7490	0.7490	230.8	77.4	0.6795	0.097	0.988	1.706
305.7	0.8400	0.7994	244.4	61.3	0.7683	0.040	1.206	1.686
295.2	0.9400	0.8976	265.0	30.2	0.8903	0.008	1.480	1.708
279.6	1.000	1.0000	279.6	0.0000	1.0000	0.000	(1.711)[†]	(1.711)[†]

Note: Superscript[exp] means experimental.

[†]Values determined graphically by extrapolation.

In these equations, γ_1 and γ_2 are activity coefficients of components 1 and 2, respectively; G^E is Gibbs molar excess free energy; $A_{1,2}$ and $A_{2,1}$ are Van Laar parameters; $G_{1,2}$ and $G_{2,1}$ are Wilson parameters, that is,

$$G_{i,j} = \frac{V_j}{V_i} e^{-a_{i,j}/RT} \qquad i \neq j$$

$a_{i,j}$ are Wilson constants ($a_{i,j} \neq a_{j,i}$ and $G_{i,j} \neq G_{j,i}$); X_1 and X_2 are liquid-phase mole fractions of components 1 and 2, respectively; Y_1 and Y_2 are vapor-phase mole fractions of components 1 and 2, respectively; R is the gas constant; and T is the absolute system temperature.

Calculation Procedure

1. Compute vapor compositions, ignoring liquid- and vapor-phase nonidealities.
By Raoult's and Dalton's laws, $Y_1 = X_1 P_1^O/P$, $Y_2 = X_2 P_2^O/P$ or $Y_2 = 1 - Y_1$, where P^O is vapor pressure and P is total system pressure. Now, $P_1^O = 279.6$ mmHg (the value of pressure corresponding to $X_1 = Y_1 = 1$), and $P_2^O = 114.7$ mmHg (the value of pressure corresponding to $X_2 = Y_2 = 1$).

The vapor compositions calculated using the preceding equations are shown in col. 6 of Table 1.11.

2. Compute logarithms of the activity coefficients from experimental X-Y data.
Assuming an ideal vapor phase, the activity coefficients are given as $\gamma_1 = PY_1/X_1 P_1^O$ and $\gamma_2 = PY_2/X_2 P_2^O$. The natural logarithms of the activity coefficients calculated using the preceding equations are shown in cols. 7 and 8 of Table 1.11.

3. Compute infinite-dilution activity coefficients.
First, calculate $G^E/X_1 X_2 RT = \ln \gamma_1/X_2 + \ln \gamma_2/X_1$. The function $G^E/X_1 X_2 RT$ calculated using the preceding equation is tabulated in col. 9 of Table 1.11.

Next, obtain the infinite-dilution activity coefficients graphically by extrapolating $G^E/X_1 X_2 RT$ values to $X_1 = 0$ and $X_2 = 0$ (Fig. 1.13). Thus, $\ln \gamma_1^\infty = 2.441$ and $\ln \gamma_2^\infty = 1.711$, so $\gamma_1^\infty = 11.48$ and $\gamma_2^\infty = 5.53$.

4. Calculate Van Laar parameters from azeotropic data.
At the azeotropic point, $X_1 = 0.7490$, $X_2 = 1 - X_1 = 0.251$, and $\ln \gamma_1 = 0.097$ and $\ln \gamma_2 = 0.988$ (see Table 1.11). Therefore,

$$A_{1,2} = \frac{0.097}{2.303}\left[1 + \frac{(0.251)(0.988)(2.303)}{(0.749)(0.097)(2.303)}\right]^2$$
$$= 0.820 \qquad (\ln \gamma_i = 2.303 \log \gamma_i)$$

and

$$A_{2,1} = \frac{0.988}{2.303}\left[1 + \frac{(0.749)(0.097)(2.303)}{(0.251)(0.988)(2.303)}\right]^2$$
$$= 0.717$$

5. Calculate Van Laar parameters from the infinite-dilution activity coefficients.
By definition, $A_{1,2} = \log \gamma_1^\infty$ and $A_{2,1} = \log \gamma_2^\infty$. Therefore, $A_{1,2} = 2.441/2.303 = 1.059$, and $A_{2,1} = 1.711/2.303 = 0.743$.

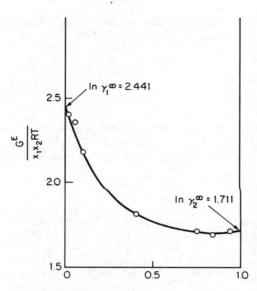

FIGURE 1.13   Graphic solution for infinite-dilution activity coefficients (Procedure 1.21).

6. Calculate Wilson parameters from azeotropic data.
From the azeotropic data,

$$0.097 = -\ln(0.749 + 0.251G_{1,2}) + 0.251\left[\frac{G_{1,2}}{(0.749 + 0.251G_{1,2})} - \frac{G_{2,1}}{(0.251 + 0.749G_{2,1})}\right]$$

and

$$0.988 = -\ln(0.251 + 0.749G_{2,1}) - 0.749\left[\frac{G_{1,2}}{(0.749 + 0.251G_{1,2})} - \frac{G_{2,1}}{(0.251 + 0.749G_{2,1})}\right]$$

Solving for $G_{1,2}$ and $G_{2,1}$ by trial and error gives $G_{1,2} = 0.1260$ and $G_{2,1} = 0.4429$.

7. Calculate Wilson parameters from the infinite-dilution activity coefficients.
The equations are $\ln \gamma_1^\infty = 2.441 = -\ln G_{1,2} + 1 - G_{2,1}$, and $\ln \gamma_2^\infty = 1.711 = -\ln G_{2,1} + 1 - G_{1,2}$. Solving by trial and error, $G_{1,2} = 0.1555$ and $G_{2,1} = 0.4209$.

Related Calculations.   These calculations show how to use vapor-liquid equilibrium data to obtain parameters for activity-coefficient correlations such as those of Van Laar and Wilson. (Use of liquid-liquid equilibrium data for the same purpose is shown in Procedure 1.20.) If the system forms an azeotrope, the parameters can be obtained from a single measurement of the azeotropic pressure and the composition of the constant boiling mixture. If the activity coefficients at infinite dilution are available, the two parameters for the Van Laar equation are given directly, and the two in the case of the Wilson equation can be solved for as shown in the example.

In principle, the parameters can be evaluated from minimal experimental data. If vapor-liquid equilibrium data at a series of compositions are available, the parameters in a given excess-free-energy model can be found by numerical regression techniques. The goodness of fit in each case depends on the suitability of the form of the equation. If a plot of G^E/X_1X_2RT versus X_1 is nearly linear, use the Margules equation (see Section 3). If a plot of X_1X_2RT/G^E is linear, then use the Van Laar equation. If neither plot approaches linearity, apply the Wilson equation or some other model with more than two parameters.

The use of activity-coefficient-correlation equations to calculate activity coefficients and make phase-equilibrium calculations is discussed in Section 3. For a detailed discussion, see Prausnitz [3].

1.22 CONVERGENCE-PRESSURE VAPOR-LIQUID EQUILIBRIUM K VALUES

In a natural-gas processing plant (Fig. 1.14), a 1000 lb · mol/h (453.6 kg · mol/h) stream containing 5 mol % nitrogen, 65% methane, and 30% ethane is compressed from 80 psia (552 kPa) at 70°F (294 K) to 310 psia (2137 kPa) at 260°F (400 K) and subsequently cooled in a heat-exchanger chilling train (a cooling-water heat exchanger and two refrigeration heat exchangers) to partially liquefy the feed stream. The liquid is disengaged from the vapor phase in a separating drum at 300 psia (2070 kPa) and −100°F (200 K) and pumped to another part of the process. The vapor from the drum is directed to a 15-tray absorption column.

In the column, the vapor is countercurrently contacted with 250 mol/h of liquid propane, which absorbs from the vapor feed 85% of its ethane content, 9% of its methane, and a negligible amount of nitrogen. The pressure of the bottom stream of liquid propane with its absorbed constituents is raised to 500 psia (3450 kPa), and the stream is heated to 50°F (283 K) in a heat exchanger before being directed to another section of the process. The vapor, leaving the overhead tower at −60°F (222 K), consists of unabsorbed constituents plus some vaporized propane from the liquid-propane absorbing medium.

a. In which heat exchanger of the chilling train does condensation start? Assume that the convergence pressure is 800 psia (5520 kPa).
b. How much propane is in the vapor leaving the overhead of the column? Assume again that the convergence pressure is 800 psia (5520 kPa).
c. Does any vaporization take place in the heat exchanger that heats the bottoms stream from the column?

Use convergence-pressure vapor-liquid equilibrium *K*-value charts (Figs. 1.15 through 1.17).

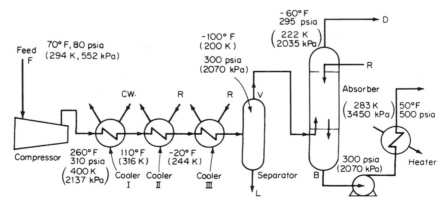

FIGURE 1.14   Flow diagram for portion of natural-gas-processing plant (Procedure 1.22).

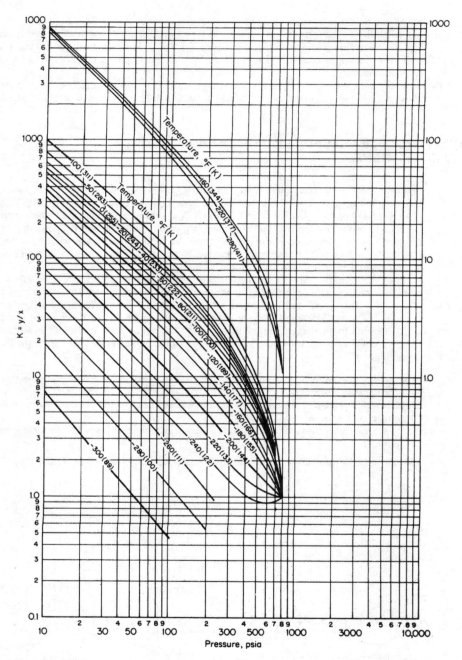

FIGURE 1.15 *K* values for nitrogen; convergence pressure = 800 psia. (Note: 1 psi = 6.895 kPa.) (*Courtesy of NGPA.*)

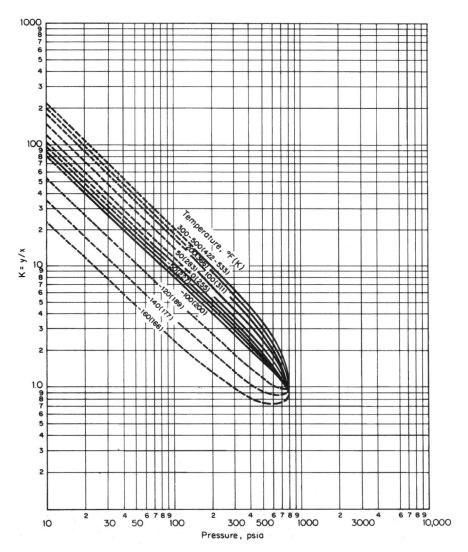

FIGURE 1.16 *K* values for methane; convergence pressure = 800 psia. (Note: 1 psi = 6.895 kPa.) *(Courtesy of NGPA.)*

Calculation Procedure

1. *Calculate dew-point equilibrium for the feed.*

A vapor is at its dew-point temperature when the first drop of liquid forms upon cooling the vapor at constant pressure and the composition of the vapor remaining is the same as that of the initial vapor mixture. At dew-point conditions, $Y_i = N_i = K_i X_i$, or $X_i = N_i / K_i$, and $\Sigma N_i / K_i = 1.0$, where Y_i is the mole fraction of component i in the vapor phase, X_i is the mole fraction of component i in the liquid phase, N_i, is the mole fraction of component i in the original mixture, and K_i is the vapor-liquid equilibrium K value.

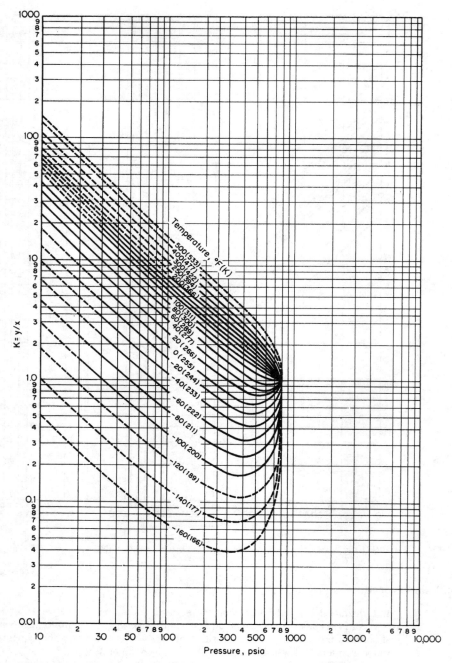

FIGURE 1.17   *K* values for ethane; convergence pressure = 800 psia. (Note: 1 psi = 6.895 kPa.) *(Courtesy of NGPA.)*

TABLE 1.12   Calculation of Dew Point for Feed Gas in Procedure 1.22 (convergence pressure = 800 psia)

Component i	Mole fraction N_i	0°F		−50°F		−100°F		−60°F	
		K_i	N_i/K_i	K_i	N_i/K_i	K_i	N_i/K_i	K_i	N_i/K_i
Methane	0.65	3.7	0.176	3.1	0.210	2.6	0.250	3.0	0.217
Nitrogen	0.05	18.0	0.003	16.0	0.003	12.0	0.004	15.0	0.003
Ethane	0.30	0.8	0.375	0.45	0.667	0.17	1.765	0.35	0.857
Total	1.00		0.554		0.880		2.019		1.077

Calculate the dew-point temperature of the mixture at a pressure of 310 psia and an assumed convergence pressure of 800 psia. At various assumed temperatures and at a pressure of 310 psia, the K values for methane, nitrogen, and ethane, as obtained from Figs. 1.15, 1.16, and 1.17, are listed in Table 1.12, as are the corresponding values of N_i/K_i. At the dew-point temperature, the latter will add up to 1.0. It can be seen from the table that the dew point lies between −60 and −50°F (222 and 227 K). Therefore, the condensation will take place in the last heat exchanger in the train, because that one lowers the stream temperature from −20°F (244 K) to −100°F (200 K).

2. Estimate the compositions of the liquid and vapor phases leaving the separator.
At the separator conditions, −100°F and 310 psia, estimate the mole ratio of vapor to liquid V/L from the relationship

$$L_i = F_i/[1 + (V/L)K_i]$$

where L_i is moles of component i in the liquid phase, F_i is moles of component i in the feed (given in the statement of the problem), and K_i is the vapor-liquid equilibrium K value for that component, obtainable from Figs. 1.15, 1.16, and 1.17. Use trial and error for finding V/L: Using the known F_i and K_i, assume a value of V/L and calculate the corresponding L_i; add up these L_i to obtain the total moles per hour of liquid, and subtract the total from 1000 mol/h (given in the statement of the problem) to obtain the moles per hour of vapor; take the ratio of the two, and compare it with the assumed V/L ratio. This trial-and-error procedure, leading to a V/L of 2.2, is summarized in Table 1.13. Thus the composition of the liquid phase leaving the separator (i.e., the L_i corresponding to a V/L of 2.2) appears as the next-to-last column in that table. The vapor composition (expressed as moles per hour of each component), given in the last column, is found by subtracting the moles per hour in the liquid phase from the moles per hour of feed.

3. Estimate the amount of propane in the vapor leaving the column overhead.
The vapor stream leaving the column overhead may be assumed to be at its dew point; in other words, the dew point of the overhead mixture is −60°F and 295 psia. At the dew point, $\Sigma(Y_i/K_i) = 1.0$.

TABLE 1.13   Equilibrium Flash Calculations for Procedure 1.22 (convergence pressure = 800 psia; feed rate = 1000 mol/h)

Component i	F_i, lb · mol/h	K_i at −100°F, 300 psia	L_i, lb · mol/h, at assumed V/L ratio of					V_i, lb · mol/h, at $V/L = 2.2$
			1	2	2.5	2.3	2.2	
Methane	650	2.6	180.6	104.8	86.7	93.1	96.7	553.3
Nitrogen	50	12.0	3.8	2.0	1.6	1.7	1.8	48.2
Ethane	300	0.18	254.2	220.6	206.9	212.2	214.9	85.1
Total	1000		438.6	327.4	295.2	307	313.4	686.6
$V = 1000 − L$			561.4	672.6	704.8	693	686.6	
Calculated V/L			1.280	2.054	2.388	2.257	2.191 (close enough)	

TABLE 1.14   Overhead from Absorption Column in Procedure 1.22 (convergence pressure = 800 psia)

Component i	Amount in feed to absorption column F_i, lb · mol/h	Fraction absorbed	Quantity absorbed L_i, lb · mol/h	Quantity in overhead D_i, lb · mol/h	K_i at −60°F, 295 psia	D_i / K_i
Nitrogen	48.2	0	0	48.2	15	3.21
Methane	553.3	0.09	49.8	503.5	3	167.8
Ethane	85.1	0.85	72.3	12.8	0.36	35.6
Propane	0	—	—	D_{pr}	0.084	$11.9D_{pr}$
Total	686.6			$(564.5 + D_{pr})$		$(206.6 + 11.9D_{pr})$

Designate Y_i, the mole fraction of component i in the column overhead, as D_i/D, where D_i is the flow rate of component i in the overhead stream and D is the flow rate of the total overhead stream. Then the dew-point equation can be rearranged into $\Sigma(D_i/K_i) = D$. Then the material balance and distribution of components in the vapor leaving the overhead can be summarized as in Table 1.14. (From a graph not shown, K_i for propane at the dew point and 800 psia convergence pressure is 0.084.) Since $\Sigma D_i = D$, the following equation can be written and solved for D_{pr}, the flow rate of propane:

$$206.6 + 11.9D_{pr} = 564.5 + D_{pr}$$

Thus D_{pr} is found to be 32.8 lb · mol/h (14.9 kg · mol/h). (The amount of propane leaving the column bottom, needed for the next step in the calculation procedure, is 250 − 32.8, or 217.2 lb · mol/h.)

4. Compute the convergence pressure at the tower bottom.
The convergence pressure is the pressure at which the vapor-liquid K values of all components in the mixture converge to a value of $K = 1.0$. The concept of convergence pressure is used empirically to account for the effect of composition. Convergence pressure can be determined by the critical locus of the system: For a binary mixture, the convergence pressure is the pressure corresponding to the system temperature read from the binary critical locus. Critical loci of many hydrocarbon binaries are given in Fig 1.18. This figure forms the basis for determining the convergence pressure for use with the K-value charts (Figs. 1.15 through 1.17).

Strictly speaking, the convergence pressure of a binary mixture equals the critical pressure of the mixture only if the system temperature coincides with the mixture critical temperature. For multicomponent mixtures, furthermore, the convergence pressure depends on both the temperature and the liquid composition of mixture. For convenience, a multicomponent mixture is treated as a pseudobinary mixture in this K-value approach. The pseudobinary mixture consists of a light component, which is the lightest component present in not less than 0.001 mol fraction in the liquid, and a pseudoheavy component that represents the remaining heavy components. The critical temperature and the critical pressure of the pseudoheavy component are defined as $T_{c,\text{heavy}} = \Sigma W_i T_{c,i}$ and $P_{c,\text{heavy}} = \Sigma W_i P_{c,i}$, where W_i is the weight fraction of component i in the liquid phase on a lightest-component-free basis, and $T_{c,i}$ and $P_{c,i}$ are critical temperature and critical pressure of component i, respectively. The pseudocritical constants computed by following this procedure (outlined in Table 1.15) are $T_{c,\text{heavy}} = 186.7°F$ and $P_{c,\text{heavy}} = 633.1$ psia. Locate points $(T_{c,\text{light}}, P_{c,\text{light}})$ and $(T_{c,\text{heavy}}, P_{c,\text{heavy}})$ of the pseudobinary mixture on Fig. 1.18, and construct the critical locus by interpolating between the adjacent loci as shown by the dotted line. Read off convergence-pressure values corresponding to the system temperature from this critical locus. Thus, at 50 and 75°F (283 and 297 K), the convergence pressures are about 1400 and 1300 psia (9653 and 8964 kPa), respectively.

5. Obtain convergence-pressure K values.
Table 1.16 lists K values of each component from the appropriate charts corresponding to given temperature and pressure at two convergence pressures, namely, 1000 and 1500 psia. Values are interpolated linearly for convergence pressures of 1400 psia (corresponding to 50°F) and 1300 psia (corresponding to 75°F).

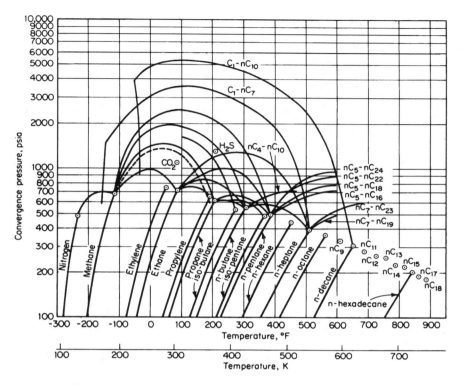

FIGURE 1.18   Critical loci of binary mixtures. (Note: 1 psi = 6.895 kPa.) *(Courtesy of NGPA.)*

TABLE 1.15   Pseudocritical Constants for Tower Bottoms in Procedure 1.22

Component	Flow rate B_i, mol/h	Molecular weight, lb/ (lb · mol)	Flow rate, lb/h	Critical temperature $T_{c,i}$, °F	Critical pressure $P_{c,i}$, psia	Weight fraction W_i	$W_i T_{c,i}$	$W_i P_{c,i}$
Methane	49.8	—	—	—	—	—	—	—
Ethane	72.3	30.07	2174	90.3	709.8	0.169	15.3	120.0
Propane	217.2	49.09	10,662	206.3	617.4	0.831	171.4	513.1
Total	339.3		12,836			1.000	186.7	633.1

TABLE 1.16   Convergence-Pressure K Values for Procedure 1.22

	K values					
	50°F, 500 psia			75°F, 500 psia		
Component	Conv. press. = 1000 psia	Conv. press. = 1500 psia	Conv. press.[†] = 1400 psia	Conv. press. = 1000 psia	Conv. press. = 1500 psia	Conv. press.[†] = 1300 psia
CH_4	2.9	4.1	3.86	3.05	4.4	3.86
C_2H_6	0.95	0.93	0.93	1.18	1.2	1.19
C_3H_8	0.36	0.32	0.33	0.46	0.38	0.41

[†]Values interpolated linearly.

6. *Estimate bubble-point temperature of tower bottom stream after it leaves the pump and heat exchanger.*
At the bubble point, $\Sigma K_i X_i = 1.0$, where X_i is the mole fraction of component i in the liquid mixture. Designate X_i as B_i/B, where B_i is the flow rate of component i and B is the flow rate of the total bottoms stream. Then the bubble-point equation can be rearranged into $\Sigma K_i B_i = B$. Using the K_i values that correspond to temperatures of 50 and 75°F (from the previous step), we find $\Sigma K_i B_i$ at 50°F to be 331.1 and $\Sigma K_i B_i$ at 75°F to be 367.3. However, at the bubble point, $\Sigma K_i B_i$ should be 339.3 mol/h, i.e., the amount of the total bottom stream. By linear interpolation, therefore, the bubble-point temperature is 56°F (286 K). Since the temperature of the stream leaving the heat exchanger is 50°F (286 K), no vaporization will take place in that exchanger.

Related Calculations.   The convergence-pressure K-value charts provide a useful and rapid graphical approach for phase-equilibrium calculations. The Natural Gas Processors Suppliers Association has published a very extensive set of charts showing the vapor-liquid equilibrium K values of each of the components methane to n-decane as functions of pressure, temperature, and convergence pressure. These charts are widely used in the petroleum industry. The procedure shown in this illustration can be used to perform similar calculations. See Procedures 3.10 and 3.11 for straightforward calculation of dew points and bubble points, respectively.

1.23   *HEAT OF FORMATION FROM ELEMENTS*

Calculate the values of standard heat of formation ΔH_f° and standard free energy of formation ΔG_f° of 2-methyl propene (isobutene) from the elements at 400 K (260°F):

$$4C(s) + 4H_2(g) \rightarrow C_4H_8(g)$$

The standard heat of formation of a compound relative to its elements, all in their standard state of unit activities, is expressed as

$$\Delta H_{f,T}^\circ = [(H_T^\circ - H_0^\circ) + \Delta H_{f,0}^\circ]_{\text{compound}} - [\Sigma(H_T^\circ - H_0^\circ)]_{\text{elements}}$$

where $\Delta H_{f,T}^\circ$ = standard heat of formation at temperature T
$\quad\;\; H_T^\circ$ = enthalpy of compound or element at temperature T
$\quad\;\; H_0^\circ$ = enthalpy of compound or element at 0 K
$\quad\; \Delta H_{f,0}^\circ$ = standard heat of formation at 0 K

The standard free energy of formation of a compound at a temperature T from the elements at the same temperature is expressed as

$$\left(\frac{\Delta G_f^\circ}{T}\right)_T = \left[\left(\frac{G_T^\circ - H_0^\circ}{T}\right) + \frac{\Delta H_{f,0}^\circ}{T}\right]_{\text{compound}} - \left[\sum \frac{G_T^\circ - H_0^\circ}{T}\right]_{\text{elements}}$$

where ΔG_f° = standard free energy of formation at temperature T
$\quad\;\; G_T^\circ$ = standard free energy of a compound or element at temperature T

Free-energy functions and enthalpy functions are given in Tables 1.17 and 1.18.

TABLE 1.17 Enthalpy above 0 K

	State	$(H_T^\circ - H_0^\circ)$, kg · cal/(g · mol)						
		298.16 K	400 K	500 K	600 K	800 K	1000 K	1500 K
Methane	g	2.397	3.323	4.365	5.549	8.321	11.560	21.130
Ethane	g	2.856	4.296	6.010	8.016	12.760	18.280	34.500
Ethene (ethylene)	g	2.525	3.711	5.117	6.732	10.480	14.760	27.100
Ethyne (acetylene)	g	2.3915	3.5412	4.7910	6.127	8.999	12.090	20.541
Propane	g	3.512	5.556	8.040	10.930	17.760	25.670	48.650
Propene (propylene)	g	3.237	4.990	7.076	9.492	15.150	21.690	40.570
n-Butane	g	4.645	7.340	10.595	14.376	23.264	33.540	63.270
2-Methylpropane (isobutane)	g	4.276	6.964	10.250	14.070	23.010	33.310	63.050
1-Butane	g	4.112	6.484	9.350	12.650	20.370	29.250	54.840
cis-2-Butene	g	3.981	6.144	8.839	12.010	19.510	28.230	53.620
trans-2-Butene	g	4.190	6.582	9.422	12.690	20.350	29.190	54.710
2-Methylpropene (isobutene)	g	4.082	6.522	9.414	12.750	20.490	29.370	55.000
n-Pentane	g	5.629	8.952	12.970	17.628	28.568	41.190	77.625
2-Methylbutane (isopentane)	g	5.295	8.596	12.620	17.300	28.300	41.010	77.740
2,2-Dimethylpropane (neopentane)	g	5.030	8.428	12.570	17.390	28.640	41.510	78.420
n-Hexane	g	6.622	10.580	15.360	20.892	33.880	48.850	92.010
2-Methylpentane	g	6.097	10.080	14.950	20.520	33.600	48.700	
3-Methylpentane	g	6.622	10.580	15.360	20.880	33.840	48.800	
2,2-Dimethylbutane	g	5.912	9.880	14.750	20.340	33.520	48.600	
2,3-Dimethylbutane	g	5.916	9.833	14.610	20.170	33.230	48.240	
Graphite	s	0.25156	0.5028	0.8210	1.1982	2.0816	3.0750	5.814
Hydrogen, H_2	g	2.0238	2.7310	3.4295	4.1295	5.5374	6.9658	10.6942
Water, H_2O	g	2.3677	3.1940	4.0255	4.8822	6.6896	8.6080	13.848
CO	g	2.0726	2.7836	3.4900	4.2096	5.7000	7.2570	11.3580
CO_2	g	2.2381	3.1948	4.2230	5.3226	7.6896	10.2220	17.004
O_2	g	2.0698	2.7924	3.5240	4.2792	5.8560	7.4970	11.7765
N_2	g	2.07227	2.7824	3.4850	4.1980	5.6686	7.2025	11.2536
NO	g	2.1942	2.9208	3.6440	4.3812	5.9096	7.5060	11.6940

Note: H_T° = molal enthalpy of the substance in its standard state, at temperature T.

H_0° = molal enthalpy of the substance in its standard state, at 0 K.

Source: A. Hougen, K.M. Watson, and R.A. Ragatz, *Chemical Process Principles*, part III. Wiley, New York, 1959.

Calculation Procedure

1. *Tabulate free-energy and enthalpy functions.*
From Tables 1.17 and 1.18:

Energy function	$C_4H_8(g)$	C(s)	$H_2(g)$
$(G_T^\circ - H_0^\circ)/T$, cal/(g · mol)(K)	−60.90	−0.824	−26.422
$\Delta H_{f,0}^\circ$, cal/(g · mol)	980	0	0
$(H_T^\circ - H_0^\circ)_{400K}$, cal/(g · mol)	6522	502.8	2731

TABLE 1.18 Free-Energy Function and Standard Heat of Formation at 0 K

	State	\multicolumn{7}{c}{$-(G_T^\circ - H_0^\circ)/T$}	$(\Delta H_f^\circ)_0$
		298.16 K	400 K	500 K	600 K	800 K	1000 K	1500 K	
Methane	g	36.46	38.86	40.75	42.39	45.21	47.65	52.84	−15.987
Ethane	g	45.27	48.24	50.77	53.08	57.29	61.11	69.46	−16.517
Ethene (ethylene)	g	43.98	46.61	48.74	50.70	54.19	57.29	63.94	+14.522
Ethyne (acetylene)	g	39.976	42.451	44.508	46.313	49.400	52.005	57.231	+54.329
Propane	g	52.73	56.48	59.81	62.93	68.74	74.10	85.86	−19.482
Propene (propylene)	g	52.95	56.39	59.32	62.05	67.04	71.57	81.43	+8.468
n-Butane	g	58.54	63.51	67.91	72.01	70.63	86.60	101.95	−23.67
2-Methylpropane (isobutane)	g	56.08	60.72	64.95	68.95	76.45	83.38	98.64	−25.30
1-Butene	g	59.25	63.64	67.52	71.14	77.82	83.93	97.27	+4.96
cis-2-Butene	g	58.67	62.89	66.51	69.94	76.30	82.17	95.12	+3.48
trans-2-Butene	g	56.80	61.31	65.19	68.84	75.53	81.62	94.91	+2.24
2-Methylpropene (isobutene)	g	56.47	60.90	64.77	68.42	75.15	81.29	94.66	+0.98
n-Pentane	g	64.52	70.57	75.94	80.96	90.31	98.87	117.72	−27.23
2-Methylbutane (isopentane)	g	64.36	70.67	75.28	80.21	89.44	97.96	116.78	−28.81
2,2-Dimethylpropane (neopentane)	g	56.36	61.93	67.04	71.96	81.27	89.90	108.91	−31.30
n-Hexane	g	70.62	77.75	84.11	90.06	101.14	111.31	133.64	−30.91
2-Methylpentane	g	70.50	77.2	83.3	89.1	100.1	110.3	132.5	−32.08
3-Methylpentane	g	68.56	75.69	82.05	88.0	99.08	109.3	131.6	−31.97
2,2-Dimethylbutane	g	65.79	72.3	78.3	84.1	95.0	105.1	127.4	−34.65
2,3-Dimethylbutane	g	67.58	74.06	80.05	85.77	96.54	106.57	128.70	−32.73
Graphite	s	0.5172	0.824	1.146	1.477	2.138	2.771	4.181	0
Hydrogen, H_2	g	24.423	26.422	27.950	29.203	31.186	32.738	35.590	0
H_2O	g	37.165	39.505	41.293	42.766	45.128	47.010	50.598	−57.107
CO	g	40.350	42.393	43.947	45.222	47.254	48.860	51.864	−27.2019
CO_2	g	43.555	45.828	47.667	49.238	51.895	54.109	58.481	−93.9686
O_2	g	42.061	44.112	45.675	46.968	49.044	50.697	53.808	0
N_2	g	38.817	40.861	42.415	43.688	45.711	47.306	50.284	0
NO	g	42.980	45.134	46.760	48.090	50.202	51.864	54.964	+21.477

Note: G_T° = molal free energy of the substance in its standard state, at temperature T, g · cal/(g · mol).

H_0° = molal enthalpy of the substance in its standard state, at 0 K, g · cal/(g · mol).

$(\Delta H_f^\circ)_0$ = standard modal heat of formation at 0 K, kg · cal/(g · mol).

Source: O. A. Hougen, K. M. Watson, and R. A. Ragatz, *Chemical Process Principles*, part III, Wiley, New York, 1959.

2. Calculate standard heat of formation.

Thus, $\Delta H_{f,400K}^\circ = (6522 + 980) - [(4)(502.8) + (4)(2731)] = -5433.2$ cal/(g · mol)[−9779.8 Btu/(lb · mol)].

3. Calculate standard free energy of formation.

Thus, $\Delta G_f^\circ/T = (-60.90 + 980/400) - [(4)(-0.824) + (4)(-26.422)] = 50.534$ cal/(g·mol)(K). So, $\Delta G_f^\circ = (50.534)(400) = 20{,}213.6$ cal/(g · mol) [36,384 Btu/(lb · mol)].

Related Calculations. Use this procedure to calculate standard heats and free energies of formation of any compound relative to its elements. The functions $(H_T^\circ - H_0^\circ)$, $(G_T^\circ - H_0^\circ)/T$, and $\Delta H_{f,0}^\circ/T$ not listed in Tables 1.17 and 1.18 may be found in other sources, such as Stull and Prophet [15].

1.24 STANDARD HEAT OF REACTION, STANDARD FREE-ENERGY CHANGE, AND EQUILIBRIUM CONSTANT

Calculate the standard heat of reaction ΔH_T°, standard free-energy ΔG_T°, and the reaction equilibrium constant K_T for the water-gas shift reaction at 1000 K (1340°F):

$$CO(g) + H_2O(g) = CO_2(g) + H_2(g)$$

The standard heat of reaction is expressed as

$$\Delta H_T^\circ = \Sigma[(H_T^\circ - H_0^\circ) + \Delta H_{f,0}^\circ]_{\text{products}} - \Sigma[(H_T^\circ - H_0^\circ) + \Delta H_{f,0}^\circ]_{\text{reactants}}$$

The standard free-energy change is expressed as

$$\Delta G^\circ /T = \Sigma[(G_T^\circ - H_0^\circ)/T + \Delta H_{f,0}^\circ/T]_{\text{products}} - \Sigma[(G_T^\circ - H_0^\circ)/T + \Delta H_{f,0}^\circ/T]_{\text{reactants}} \cdot$$

The equilibrium constant is expressed as $K = e^{-\Delta G/RT}$, where the terms are as defined in Procedure 1.23.

Calculation Procedure

1. Tabulate free-energy and enthalpy functions.
From Tables 1.17 and 1.18:

Energy function	$CO_2(g)$	$H_2(g)$	$CO(g)$	$H_2O(g)$
$(G_T^\circ - H_0^\circ)/T$, cal/(g · mol)(K)	−54.109	−32.738	−48.860	−47.010
$\Delta H_{f,0}^\circ$, cal/(g · mol)	−93,968.6	0	−27,201.9	−57,107
$(H_T^\circ - H_0^\circ)$, cal/(g · mol)	10,222	6965.8	7257	8608

2. Calculate standard heat of reaction.
Thus, $\Delta H_{1000\,K}^\circ = [(10,222.0) + (−93,968.6) + (6,965.8) + (0)] − [(7,257.0) + (−27,201.9) + (8,608) + (−57,107)] = −8336.9$ cal/(g · mol) [−15,006.4 Btu/(lb · mol)].

3. Calculate standard free-energy change.
Thus, $\Delta G^\circ/1000 = [−54.109 + (−93,968.6/1000) + (−32.738) + 0/1000] − [−48.860 + (−27201.9/1000) + (−47.010) + (−57,107.0/1000)] = −0.638$ cal/(g · mol)(K). Therefore, $\Delta G_{1000\,K}^\circ = −638.0$ cal/(g · mol) [−1148.4 Btu/(lb · mol)].

4. Calculate reaction equilibrium constant.
Thus, $K_{1000\,K} = e^{638/(1.987)(1000)} = 1.379$.

Related Calculations. Use this procedure to calculate heats of reaction and standard free-energy changes for reactions that involve components listed in Tables 1.17 and 1.18.

Heat of reaction, free-energy changes, and reaction equilibrium constants are discussed in more detail in Section 4 in the context of chemical-reaction equilibrium.

1.25 STANDARD HEAT OF REACTION FROM HEAT OF FORMATION—AQUEOUS SOLUTIONS

Calculate the standard heat of reaction $\Delta H°$ for the following acid-base-neutralization reaction at standard conditions [25°C, 1 atm (101.3 kPa)]:

$$2NaOH(aq) + H_2SO_4(l) = Na_2SO_4(aq) + 2H_2O(l)$$

Calculation Procedure

1. Calculate the heat of reaction.
The symbol *(aq)* implies that the sodium hydroxide and sodium sulfate are in infinitely dilute solution. Therefore, the heat of solution must be included in the calculations. Data on both heat of formation and heat of solution at the standard conditions (25°C and 1 atm) are available in Table 1.19. Since the answer sought is also to be at standard conditions, there is no need to adjust for differences in temperature (or pressure), and the equation to be used is simply

$$\Delta H° = \Sigma(\Delta H_F°)_{products} + \Sigma(\Delta H_s°)_{dissolved\ products} - \Sigma(\Delta H_F°)_{reactants} - \Sigma(\Delta H_s°)_{dissolved\ reactants}$$

TABLE 1.19   Standard Heats of Formation and Standard Integral Heats of Solution at Infinite Dilution (25°C, 1 atm)

Compound	Formula	State	$\Delta H_f°$, cal/(g · mol)	$\Delta H_s°$, cal/(g · mol)
Ammonia	NH_3	g	−11,040	−8,280
		l	−16.060	−3,260
Ammonium nitrate	NH_4NO_3	s	−87,270	6,160
Ammonium sulfate	$(NH_4)_2SO_4$	s	−281,860	1,480
Calcium carbide	CaC_2	s	−15,000	
Calcium carbonate	$CaCO_3$	s	−288,450	
Calcium chloride	$CaCl_2$	s	−190,000	−19,820
Calcium hydroxide	$Ca(OH)_2$	s	−235,800	−3,880
Hydrochloric acid	HCl	g	−22,063	−17,960
Hydrogen sulfide	H_2S	g	−4,815	−4,580
Iron oxide	Fe_3O_4	s	−267,000	
Iron sulfate	$FeSO_4$	s	−220.500	−15,500
Potassium sulfate	K_2SO_4	s	−342,660	5,680
Sodium carbonate	Na_2CO_3	s	−270,300	−5,600
	$Na_2CO_3 \cdot 10H_2O$	s	−975,600	16,500
Sodium chloride	NaCl	s	−98,232	930
Sodium hydroxide	NaOH	s	−101,990	−10,246
Sodium nitrate	$NaNO_3$	s	−101,540	−5,111
Sodium sulfate	Na_2SO_4	s	−330,900	−560
	$Na_2SO_4 \cdot 10H_2O$	s	−1,033,480	18,850
Sulfur dioxide	SO_2	g	−70,960	9,900
Sulfur trioxide	SO_3	g	−94,450	−54,130
Sulfuric acid	H_2SO_4	l	−193,910	−22,990
Water	H_2O	g	−57,798	
		l	−68,317	
Zinc sulfate	$ZnSO_4$	s	−233,880	−19,450

Source: F. D. Rossini et al., *Selected Values of Chemical Thermodynamic Properties*, National Bureau of Standards, Circular 500, 1952.

where ΔH_F° is standard heat of formation, and ΔH_s° standard integral heat of solution at infinite dilution.
Thus, from Table 1.19, and taking into account that there are 2 mol each of water and sodium
hydroxide,

$$\Delta H^\circ = [-330,900 + 2(-68,317)] + (-560) - [2(-101,990) + (-193,910)] - 2(-10,246)$$
$$= -49,712 \text{ cal/(g} \cdot \text{mol)[}-89,482 \text{ Btu/(lb} \cdot \text{mol)]}$$

The reaction is thus exothermic. To maintain the products at 25°C, it will be necessary to remove
49,712 cal of heat per gram-mole of sodium sulfate produced.

Related Calculations. Use this general procedure to calculate heats of reaction for other aqueous-
phase reactions. Calculation of heat of reaction from standard heat of reaction is covered in Section 4
in the context of chemical-reaction equilibrium; see in particular Procedure 4.1.

1.26 STANDARD HEAT OF REACTION FROM HEAT OF COMBUSTION

Calculate the standard heat of reaction ΔH° of the following reaction using heat-of-combustion data:

$$CH_3OH(l) + CH_3COOH(l) = CH_3OOCCH_3(l) + H_2O(l)$$

(Methanol)   (Acetic acid)   (Methyl acetate)   (Water)

Calculation Procedure

1. Obtain heats of combustion.
Data on heats of combustion for both organic and inorganic compounds are given in numerous reference works. Thus:

Compound	State	$\Delta H_{combustion}$, kcal/(g $\cdot$ mol)
CH_3OH	Liquid	-173.65
CH_3COOH	Liquid	-208.34
CH_3OOCCH_3	Liquid	-538.76
H_2O	Liquid	0

2. Calculate ΔH°.
The heat of reaction is the difference between the heats of combustion of the reactants and of the
products:

$$\sum \Delta H_{combustion, reactants}^\circ - \sum \Delta H_{combustion, products}^\circ$$

So, $\Delta H^\circ = (-173.65) + (-208.34) - (-538.76) = 156.77$ kcal/(g $\cdot$ mol).

Related Calculations. For a reaction between organic compounds, the basic thermochemical data
are generally available in the form of standard heats of combustion. Use the preceding procedure for
calculating the standard heats of reaction when organic compounds are involved, using the standard
heats of combustion directly instead of standard heats of formation. Heat-of-formation data must be
used when organic and inorganic compounds both appear in the reaction.
Calculation of heat of reaction from standard heat of reaction is covered in Section 4 in the context of chemical-reaction equilibrium; see in particular Procedure 4.1.

TABLE 1.20 Heat of Formation from Heat of Combustion

Final products of the combustion (1)	Equation for heat of formation H_f^o, cal/(g· mol) (2)
$CO_2(g)$, $H_2O(l)$, $Br_2(g)$, $HCl(aq)$, $I(s)$, $HNO_3(aq)$, $H_2SO_4(aq)$	$H_f^o = -\Delta H_c - 94{,}051.8a - 34{,}158.7b$ $+ 3670c - 5864.3d - 44{,}501e - 15{,}213.3g$ $-148{,}582.6i$ (A)
$CO_2(g)$, $H_2O(l)$, $Br_2(g)$, $HF(aq)$, $I(s)$, $N_2(g)$, $SO_2(g)$	$H_f^o = -\Delta H_c - 94{,}051.8a - 34{,}158.7b + 3670c$ $-44{,}501e - 70{,}960i$ (B)

Note: In the above equations, ΔH_c = heat of combustion corresponding to the final products in col. 1, a = atoms of carbon in the compound, b = atoms of hydrogen, c = atoms of bromine, d = atoms of chlorine, e = atoms of fluorine, g = atoms of nitrogen, and i = atoms of sulfur.

1.27 STANDARD HEAT OF FORMATION FROM HEAT OF COMBUSTION

Calculate the standard heats of formation of benzene(*l*), methanol(*l*), aniline(*l*), methyl chloride(*g*), and ethyl mercaptan(*l*) using heat-of-combustion data, knowledge of the combustion products, and the equations in Table 1.20.

Calculation Procedure

1. Obtain data on the heats of combustion, note the corresponding final combustion products, and select the appropriate equation in Table 1.20.

From standard reference sources, the standard heats of combustion and the final products are as follows. The appropriate equation, A or B, is selected based on what the final combustion products are, i.e., where they are within col. 1 of Table 1.20.

Compound (and number of constituent atoms)	Combustion products	Heat of combustion, kcal/(g · mol)	Equation to be used
Benzene, $C_6H_6(l)$	$CO_2(g)$, $H_2O(l)$	780.98	A or B
Methanol, $CH_4O(l)$	$CO_2(g)$, $H_2O(l)$	173.65	A or B
Aniline, $C_6H_7N(l)$	$CO_2(g)$, $H_2O(l)$, $N_2(g)$	812.0	B
Methyl chloride, $CH_3Cl(g)$	$CO_2(g)$, $H_2O(l)$, $HCl(aq)$	182.81	A
Ethyl mercaptan, $C_2H_6S(l)$	$CO_2(g)$, $H_2O(l)$, $SO_2(g)$	448.0	B

2. Calculate the heats of formation and compare with the literature values.

This step can be set out in matrix form, as follows:

Compound	Equation	Heat of formation, cal/(g · mol) Calculated value	Literature value
Benzene(*l*)	A	$= -(-780.98)(10^3) - (94{,}051.8)(6)$ $-(34{,}158.7)(6) = 11{,}717$	11,630
Methanol(*l*)	A	$= -(-173.65)(10^3) - (94{,}051.8)(1)$ $-(34{,}158.7)(4) = -57{,}036$	−57,040
Aniline(*l*)	B	$= -(-812.0)(10^3) - (94{,}051.8)(6)$ $-(34{,}158.7)(7) = 8{,}578$	8,440
Methyl chloride(*l*)	A	$= -(-182.81)(10^3) - (94{,}051.8)(1)$ $-(34{,}158.7)(3) - (5{,}864.3)(1) = -19{,}582$	−19,580
Ethyl mercaptan(*l*)	B	$= -(-448.0)(10^3) - (94{,}051.8)(2)$ $-(34{,}158.7)(6) - (70{,}960)(1) = -16{,}015.8$	−16,000

Related Calculations.   This approach may be used to find the heat of formation of a compound all of whose constituent atoms are among the following: carbon, hydrogen, bromine, chlorine, fluorine, iodine, nitrogen, oxygen, and sulfur.

1.28 HEAT OF ABSORPTION FROM SOLUBILITY DATA

Estimate the heat of absorption of carbon dioxide in water at 15°C (59°F, or 288 K) from these solubility data:

Temperature,°C	0	5	10	15	20
Henry constant, atm/mol-fraction	728	876	1040	1220	1420

The formula is

$$\Delta H_{abs} = (\bar{h}_i - H_i) = R[d \ln H / d(1/T)]$$

where ΔH_{abs} is heat of absorption, $\bar{h}_i$ is partial molar enthalpy of component i at infinite dilution in the liquid at a given temperature and pressure. H_i is molar enthalpy of pure gas i at the given temperature and pressure, H is the Henry constant, the partial pressure of the gas divided by its solubility, R is the gas constant, and T is absolute temperature.

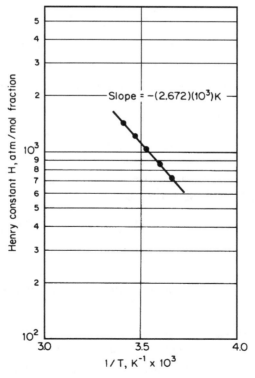

FIGURE 1.19 Henry constant for carbon-dioxide/water system at 15°C (Procedure 1.28). Note: 1 atm = 101.3 kPa.

Calculation Procedure

1. Determine d ln H/d(1/T).
This is the slope of the plot of ln H against $1/T$. A logarithmic plot based on the data given in the statement of the problem is shown in Fig. 1.19. The required slope is found to be -2.672×10^3 K.

2. Calculate the heat of absorption.
Substitute directly into the formula. Thus,

$$\Delta H_{abs} = R[d \ln H/d(1/T)] = 1.987(-2.672 \times 10^3)$$
$$= -5.31 \times 10^3 \text{cal/(g} \cdot \text{mol)}[9.56 \text{ Btu/(lb} \cdot \text{mol})]$$

Related Calculations. This procedure may be used for calculating the heats of solution from low-pressure solubility data where Henry's law in its simple form is obeyed.

1.29 ESTIMATION OF LIQUID VISCOSITY AT HIGH TEMPERATURES

Use the Letsou-Stiel high-temperature generalized correlation to estimate the viscosity of liquid benzene at 227°C (500 K, 440°F).
The correlation is expressed as

$$\mu_L\psi = (\mu_L\psi)^{(0)} + \omega(\mu_L\psi)^{(1)}$$

where $(\mu_L\psi)^{(0)} = 0.015174 - 0.02135T_R + 0.0075T_R^2$
$(\mu_L\psi)^{(1)} = 0.042552 - 0.07674T_R + 0.0340T_R^2$

μ_L = liquid viscosity at the reduced temperature, in centipoise (cP)

$$\psi = \frac{T_c^{1/6}}{(M^{1/2}P_c^{2/3})}$$

ω = Pitzer accentric factor
T_c = critical temperature, K
P_c = critical pressure, atm
M = molecular weight

Use these values for benzene: $T_c = 562.6$ K, $P_c = 48.6$ atm, $M = 78.1$, and $\omega = 0.212$.

Calculation Procedure

1. Calculate the correlating parameters for the Letsou-Stiel method.
By use of the equations outlined above,

$$\psi = T_c^{1/6}/(M^{1/2}P_c^{2/3}) = (562.6)^{1/6}/(78.1)^{1/2}(48.6)^{2/3} = 0.0244$$
$$T_R = T/T_c = (227 + 273)/562.6 = 0.889$$
$$(\mu_L\psi)^{(0)} = 0.015174 - (0.02135)(0.889) + (0.0075)(0.889)^2 = 0.0021$$
$$(\mu_L\psi)^{(1)} = 0.042552 - (0.07674)(0.889) + (0.0340)(0.889)^2 = 0.0012$$

2. Calculate the viscosity of liquid benzene.
Thus, upon dividing through by ψ, we have the equation

$$\mu_L = (1/\psi)\left[(\mu_L\psi)^{(0)} + \omega(\mu_L\psi)^{(1)}\right]$$
$$= (1/0.024)[(0.0021) + (0.212)(0.0012)] = 0.0981 \text{ cP}$$

Related Calculations. The Letsou-Stiel correlation is a fairly accurate method for estimating viscosities of liquids at relatively high temperatures, $T_R = 0.75$ or higher. It has been tested on a large number of compounds and is reported to fit most materials up to a reduced temperature of about 0.9, with an average error of ±3 percent.

1.30 VISCOSITY OF NONPOLAR AND POLAR GASES AT HIGH PRESSURE

Use the Jossi-Stiel-Thodos generalized correlation to estimate the vapor viscosity of (1) methane (a nonpolar gas) at 500 psig (35 atm abs) and 250°F (394 K), and (2) ammonia (a polar gas) at 340°F (444.4 K) and 1980 psig (135.8 atm abs). Experimentally determined viscosities for those two gases at low pressure and the same temperatures are 140 μP for methane and 158 μP for ammonia.

The Jossi-Stiel-Thodos correlation is summarized in these equations:
For nonpolar gases,

$$[(\mu - \mu^0)\psi + 1]^{0.25} = 1.0230 + 0.2336\rho_R + 0.58533\rho_R^2 - 0.040758\rho_R^3 + 0.093324\rho_R^4$$

For polar gases,

$$(\mu - \mu^0)\psi = 1.656\rho_R^{1.111} \quad \text{if } \rho_R \leq 0.1$$
$$(\mu - \mu^0)\psi = 0.0607(9.045\rho_R + 0.63)^{1.739} \quad \text{if } 0.1 \leq \rho_R \leq 0.9$$
$$\log\{4 - \log[(\mu - \mu^0)\psi]\} = 0.6439 - 0.1005\rho_R - D \quad \text{if } 0.9 \leq \rho_R < 2.6$$

where $D = 0$ if $0.9 \leq \rho_R \leq 2.2$, or
$$D = (0.000475)(\rho_R^3 - 10.65)^2 \text{ if } 2.2 < \rho_R \leq 2.6$$
μ = viscosity of high-pressure (i.e., dense) gas, μP
μ^0 = viscosity of gas at low pressure, μP
ρ_R = reduced gas density, ρ/ρ_c (which equals V_c/V)
ρ = density
ρ_c = critical density
V = specific volume
V_c = critical specific volume
$\psi = T_c^{1/6}/(M^{1/2}P_c^{2/3})$
T_c = critical temperature, K
P_c = critical pressure, atm
M = molecular weight

Use these values for methane: $T_c = 190.6$ K; $P_c = 45.4$ atm; $V_c = 99.0$ cm³/(g·mol); $Z_c = 0.288$; $M = 16.04$. And use these values for ammonia: $T_c = 405.6$ K; $P_c = 111.3$ atm; $V_c = 72.5$ cm³/(g·mol); $Z_c = 0.242$; $M = 17.03$.

Calculation Procedure

1. Calculate the nonideal compressibility factor for the methane and the ammonia.
Use the generalized correlations shown in Figs. 1.1 through 1.3 in this section. For methane, $T_R = 394/190.6 = 2.07$, and $P_R = 35/45.4 = 0.771$. From the statement of the problem, $Z_c = 0.288$. Thus, by interpolation from Fig. 1.2 (for $Z_c = 0.27$) and Fig. 1.3 (for $Z_c = 0.29$), Z is found to be 0.98.
Similarly for ammonia, $T_R = 444.4/405.6 = 1.10$, $P_R = 135.8/111.3 = 1.22$, and $Z_c = 0.242$. By extrapolation from Figs. 1.2 and 1.3, $Z = 0.65$.

2. Calculate the reduced density for the high-pressure gas.
For methane, $\rho_R = \rho/\rho_c = \rho V_c = PV_c/ZRT = (35)(99.0)/(0.98)(82.07)(394) = 0.109$.
Similarly for ammonia, $\rho_R = PV_c/ZRT = (135.8)(72.5)/(0.65)(82.07)(444.4) = 0.415$.

3. Calculate the parameter ψ.
For methane, $\psi = T_c^{1/6}/(M^{1/2}P_c^{2/3}) = (190.6)^{1/6}/(16.04)^{1/2} \times (45.4)^{2/3} = 0.0471$.
And for ammonia, $\psi = (405.6)^{1/6}/(17.03)^{1/2}(111.3)^{2/3} = 0.0285$.

4. Calculate the viscosity for the high-pressure methane, using the nonpolar equation.
Thus, $[(\mu - \mu^0)\psi + 1]^{0.25} = 1.023 + (0.2336)(0.109) + (0.58533)(0.109)^2 - (0.040758)(0.109)^3 + (0.093324)(0.109)^4 = 1.0550$. Accordingly,

$$\mu - \mu^0 + \left[(1.0550)^{1/0.25} - 1\right]/\psi = 140 + \left[(1.0550)^{1/0.25} - 1\right]/0.0471 = 145.07 \ \mu P$$

This correlation thus indicates that the pressure effect raises the viscosity of the methane by about 4 percent.

5. Calculate the viscosity for the high-pressure ammonia, using the appropriate polar equation.
Because the reduced density (0.415) lies between 0.1 and 0.9, the appropriate equation is

$$(\mu - \mu^0)\psi = 0.0607(9.045\rho_R + 0.63)^{1.739}$$

Thus, $(\mu - \mu^0)\psi = 0.0607(9.045)(0.415) + 0.63]^{1.739} = 0.7930$, so $\mu = \mu_0 + 0.7930/\psi = 158 + 0.7930/0.0258 = 185.8 \ \mu P$. This correlation thus predicts that the pressure effect raises the viscosity of the ammonia by about 18 percent.

1.31 THERMAL CONDUCTIVITY OF GASES

Use the modified Eucken correlation to estimate the thermal conductivity of nitric oxide (NO) vapor at 300°C (573 K, 572°F) and 1 atm (101.3 kPa). At that temperature, the viscosity of nitric oxide is 32.7×10^{-5} P.
The modified Eucken correlation is

$$\lambda M/\mu = 3.52 + 1.32C_v^o = 3.52 + 1.32C_p^o/\gamma$$

where M = molecular weight
μ = viscosity, P
C_p^o, C_v^o = ideal-gas heat capacity at constant pressure and constant volume, respectively, cal/(g · mol)K
γ = heat capacity ratio, equal to C_p^o/C_v^o
λ = thermal conductivity, cal/(cm)(s)(K)

Use these values and relationships for nitric oxide: $M = 30.01$, $C_p^o = 6.461 + 2.358 \times 10^{-3}\ T$ (where T is in kelvins), and $C_v^o = C_p^o - R = C_p^o - 1.987$.

Calculation Procedure

1. Calculate C_v^o.
Thus, $C_v^o = C_p^o - R = 6.461 + (2.358 \times 10^{-3})(300 + 273) - 1.987 = 5.826$ cal/(g · mol)(K).

2. Calculate the thermal conductivity.
Thus, upon rearranging the Eucken equation,

$$\lambda = (\mu/M)(3.52 + 1.32 C_v^o)$$
$$= [(32.7 \times 10^{-5})/30.01][3.52 + (1.32)(5.826)]$$
$$= 1.22 \times 10^{-4}\, \text{cal/(cm)(s)(K)}$$

The experimentally determined value is reported as 1.07×10^{-4} cal/(cm)(s)(K). The error is thus 14 percent.

1.32 THERMAL CONDUCTIVITY OF LIQUIDS

Estimate the thermal conductivity of carbon tetrachloride at 10°C (283 K, 50°F) using the Robbins and Kingrea correlation:

$$\lambda_L = [(88 - 4.94\text{H})(0.001)/\Delta S][(0.55/T_R)^N]\left[C_p \rho^{4/3} \right]$$

where λ_L = liquid thermal conductivity, cal/(cm)(s)(K)

$\quad\quad T_R$ = reduced temperature, equal to T/T_c where T_c is critical temperature

$\quad\quad C_p$ = molal heat capacity of liquid, cal/(g · mol)(K)

$\quad\quad \rho$ = molal liquid density, (g · mol)/cm^3

$\quad\quad \Delta S$ = $(\Delta H_{vb}/T_b) + R \ln (273/T_b)$, cal/(g · mol)(K)

$\quad\quad \Delta H_{vb}$ = molal heat of vaporization at normal boiling point, cal/(g · mol)

$\quad\quad T_b$ = normal boiling point, K

$\quad\quad H$ = empirical parameter whose value depends on molecular structure and is obained from Table 1.21

$\quad\quad N$ = empirical parameter whose value depends on liquid density at 20°C (It equals 0 if the density is greater than 1.0 g/cm^3, and 1 otherwise.)

Use these values for carbon tetrachloride: $T_c = 556.4$ K, molecular weight = 153.8, liquid density = 1.58 g/cm^3, molal heat capacity = 31.37 cal/(g · mol)(K), T_b, = 349.7 K, $\Delta H_{vb} = 7170$ cal/(g · mol).

Calculation Procedure

1. Obtain the structural constant H from Table 1.21.
Since carbon tetrachloride is an unbranched hydrocarbon with three chlorine substitutions, H has a value of 3.

TABLE 1.21   *H*-factors for Robbins-Kingrea Correlation

Functional group	Number of groups	*H*
Unbranched hydrocarbons:		
Paraffins		0
Olefins		0
Rings		0
CH_3 branches	1	1
	2	2
	3	3
C_2H_5 branches	1	2
i-C_3H_7 branches	1	2
C_4H_9 branches	1	2
F substitutions	1	1
	2	2
Cl substitutions	1	1
	2	2
	3 or 4	3
Br substitutions	1	4
	2	6
I substitutions	1	5
OH substitutions	1 (iso)	1
	1 (normal)	−1
	2	0
	1 (tertiary)	5
Oxygen substitutions:		
$-\overset{\mid}{C}=O$ (ketones, aldehydes)		0
$-\overset{\overset{O}{\parallel}}{C}-O-$ (acids, esters)		0
$-O-$ (ethers)		2
NH_2 substitutions	1	1

Source: Reprinted by permission from "Estimate Thermal Conductivity," *Hydrocarbon Processing*, May 1962, copyright 1962 by Gulf Publishing Co., all rights reserved.

2. Calculate ΔS.
Thus, $\Delta S = (\Delta H_{vb}/T_b) + R[\ln(273/T_b)] = 7170/349.7 + 1.987[\ln(273/349.7)] = 20.0$ cal/(g · mol)(K).

3. Calculate the thermal conductivity.
Thus,

$$\lambda_L = [(88 - 4.94H)(0.001)/\Delta S][(0.55/T_R)^N]\left[C_p\rho^{4/3}\right]$$
$$= \{[88 - (4.94)(3)][0.001]/[20.0]\}\{[0.55/(283/556.4)]^0\}\left\{[31.37][1.58/153.8]^{4/3}\right\}$$
$$= 2.564 \times 10^{-6} \text{cal}/(\text{cm})(\text{s})(\text{K})$$

The experimentally determined value is reported as 2.510×10^{-6} cal/(cm)(s)(K). The error is thus 2.2 percent.

TABLE 1.22 Atomic Diffusion Volumes

Molecule	Diffusion volume	Molecule	Diffusion volume
H_2	7.07	CO	18.9
He	2.88	CO_2	26.9
N_2	17.9	NH_3	14.9
O_2	16.6	Cl_2	37.7
Air	20.1	Br_2	67.2
Ar	16.1	SO_2	41.1
H_2O	12.7	CCl_2F_2	114.8

Atom	Volume increment, V	Atom	Volume increment, V
C	16.5	Cl	19.5
H	1.98	S	17.0
O	5.48	Aromatic ring	−20.2
N	5.69	Heterocyclic ring	−20.2

1.33 DIFFUSION COEFFICIENTS FOR BINARY GAS SYSTEMS AT LOW PRESSURES

Estimate the diffusion coefficient of benzene vapor diffusing into air at 100°F and at atmospheric pressure using the following empirical correlation[†] for binary air–hydrocarbon systems at low pressures (less than 30 atm):

$$D_{AB} = \frac{0.0204\, T^{1.75} \left(\dfrac{1}{M_A} + \dfrac{1}{M_B} \right)^{1/2}}{[(\Sigma_A V_i)^{1/3} + (\Sigma_B V_i)^{1/3}]^2}$$

where D_{AB} = diffusion component of component A (benzene) into component B (air), square feet per hour

T = temperature, degrees Rankine

M_A, M_B = molecular weights of components A and B

P = system pressure, pounds per square inch absolute

ΣV_i = sum of atomic diffusion volumes shown in Table 1.22

Calculation Procedure

1. Determine the diffusion volumes for benzene and air.
Obtain the diffusion volumes for carbon, hydrogen, and an aromatic ring from Table 1.22. Weight the carbon and hydrogen in accordance with the number of atoms in the benzene molecule (six each), then add the diffusion increment for the aromatic ring (note that it is a negative number). The sum is the diffusion volume for benzene:

$$\text{benzene}\, \Sigma V_i = \underset{\text{(carbon)}}{(6)(16.5)} + \underset{\text{(hydrogen)}}{(6)(1.98)} + \underset{\text{(aromatic ring)}}{(1)(-20.2)} = 90.7$$

As for air, its diffusion volume can be read directly from the table: 20.1.

[†] This correlation was developed by Fuller, Schettler, and Giddings [31].

2. *Obtain the molecular weights*

$$\text{Component A (benzene):} \quad M_A = 78.1$$
$$\text{Component B (air):} \quad M_B = 28.9$$

3. *Calculate the diffusion coefficient.*

Substitute the values from steps 1 and 2 into the equation presented in the problem statement. Thus,

$$D_{AB} = \frac{(0.0204)(100 + 459.7)^{1.75}(1/78.1 + 1/28.9)^{1/2}}{(14.69)[(90.7)^{1/3} + (20.1)^{1/3}]^2}$$

$$= 0.374 \text{ ft}^2/\text{h}$$

The experimentally determined value is reported as 0.372 ft²/h. The calculated value represents a deviation of 0.5 percent.

1.34 *ESTIMATION OF SURFACE TENSION OF A PURE LIQUID*

Estimate the surface tension, σ, of *n*-butane at 20°C using the generalized corresponding state correlation of Brock and Bird [32] and the Miller relationship [33]. The correlation and the relationship are as follows:

$$\sigma = \left[\left(P_c^{2/3}\right)\right]\left[T_c^{1/3}\right]\left[(1 - T/T_c)^{11/9}\right][K]$$

where K is defined as follows:

$$K = 0.1196\{1 + [(T_b/T_c)\ln(P_c/1.01325)]/[1 - T_b/T_c]\} - 0.279$$

where σ = surface tension in dynes/cm
P_c = critical pressure, bar
T_c = critical temperature, K
T_b = normal boiling point, K

Use the following physical data: $T_c = 425.5$ K, $P_c = 37.5$ bar, and $T_b = 272.7$ K.

Calculation Procedure

1. *Calculate K*

$$K = 0.1196\left[1 + \frac{(272.7/425.7)\ln(37.5/1.01325)}{(1 - 272.7/425.2)}\right] - 0.279$$

$$= 0.612$$

2. *Calculate the surface tension*

$$\sigma = (37.5)^{2/3} \times (425.2)^{1/3}(1 - 293/425.2)^{11/9} \times 0.612$$

$$= 12.36 \text{ dyne/cm}$$

The experimentally determined value is reported as 12.46. The calculated value represents an error of +0.9 percent.

*1.35 ANALYSIS OF A SATURATED SOLUTION

If 1000 gal (3785.4 L) of water is saturated with potassium chlorate ($KClO_3$) at 80°C (176°F), determine (a) the weight, in pounds, of $KClO_3$ that will precipitate if the solution is cooled to 30°C (86°F) and (b) the weight of $KClO_3$ that will precipitate if one-half the 1000 gal (3785.4 L) of water is evaporated at 100°C (212°F).

Calculation Procedure

1. Compute the precipitate when the solution is cooled.
When a solid is dissolved in water (or any other solvent liquid), the resulting solution is termed *saturated* when at a given temperature the solvent cannot dissolve any more of the solid. Most solvents dissolve (hold) more solids at higher temperatures than at lower temperatures. Thus, when the solution temperature is lowered or a portion of the solvent is evaporated, the solution becomes *supersaturated* and solid material may precipitate. This is the basis of *crystallization*, a chemical engineering operation frequently used to produce a purer or more crystalline product.

Referring to Fig. 1.20, obtain these solubilities: at 80°C (176°F), $KClO_3$ solubility = 38.5 g per 100 g H_2O; at 30°C (86°F), $KClO_3$ solubility = 10.5 g per 100 g of H_2O.

The weight of the water at 80°C (176°F) = (1000 gal H_2O)(0.97183 g H_2O per cm^3 H_2O) × (1 lb/454 g) = 8103 lb (3683.2 kg). Now, the weight of $KClO_3$ that any solvent can dissolve at a given temperature = weight of solvent at the given temperature, lb (solubility of $KClO_3$ at the given temperature, g per 100 g of the solvent), Or, at 80°C (176°F), weight of $KClO_3$ dissolved by the water = (8103 lb of water)(38.5 g $KClO_3$ per 100 g of H_2O) = 3119 lb (1417.7 kg) of $KClO_3$. And at 30°C (86°F) with the same quantity of water but the reduced solubility, the weight of $KClO_3$ that can be dissolved = (8103)(10.5 g per 100 g) = 851 lb (386.8 kg) of $KClO_3$.

When the temperature of the water (solvent) is reduced from 80 to 30°C (176 to 86°F), the weight of $KClO_3$ precipitated = weight of $KClO_3$ dissolved at 80°C (176°F) − weight of $KClO_3$ dissolved at 30°C (86°F), or 3119 − 851 = 2271 lb (1032.3 kg) of $KClO_3$ precipitated.

Note that the same procedure can be followed for any similar solution, i.e., any similar solvent and solid. Neither the solvent nor the solid need be the ones considered here.

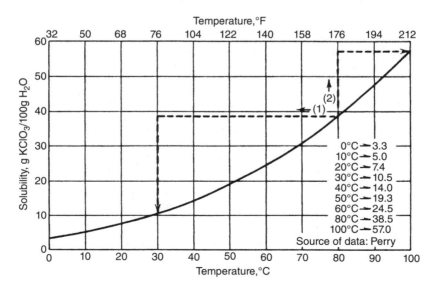

FIGURE 1.20   Solubility of $KClO_3$.

2. Compute the precipitate when a portion of the solvent is evaporated.
Since half the solvent (water in this case) is evaporated, the weight of the water remaining = 8103/2 = 4051.5 lb (1841.6 kg). Using the solubility of $KClO_3$ as before, except that the solvent temperature is 100°C (212°F), we see the weight of $KClO_3$ dissolved = 4051.5(57 g $KClO_3$ per 100 g H_2O) = 2309 lb (1047.3 kg) of $KClO_3$. Then the weight of $KClO_3$ precipitated by the evaporation = weight of $KClO_3$ dissolved in 1000 gal (3785.0 L) of water at 80°C (176°F) − weight of $KClO_3$ dissolved in 500 gal (1892.5 L) of water at 100°C (212°F) = 3119 − 2309 = 810 lb (367.4 kg) of $KClO_3$ precipitated.

*1.36 TERNARY LIQUID SYSTEM ANALYSIS

For a liquid mixture of 20 wt % water, 30 wt % acetic acid, and 50 wt % isopropyl ether, determine the composition of the two phases (e.g., the ether layer and the water layer) and the amount of acetic acid that must be added to the system to form a one-phase (single-layer) solution.

Calculation Procedure

1. Compute the composition of the two layers.
When two pure liquids are mixed, they will dissolve in each other to some degree. If they are completely soluble in each other, such as water and acetic acid, they are *miscible*.

If their mutual solubilities are zero, they are called *immiscible*. Between these extremes, liquids are partially miscible.

Addition of a third liquid component often affects the mutual solubilities of the two original liquids. The third liquid may be more soluble in one liquid than in another. This difference in solubilities is the basis of the chemical engineering operation termed *liquid-liquid extraction*.

The third liquid may cause immiscible liquids to become completely miscible, or the third liquid may produce miscibility only in certain concentration ranges. Such interrelationships can be shown graphically, as with the two parts in Fig. 1.21

The *phase envelope*, Fig. 1.21, separates the two-phase region from the one-phase region. Note, Fig. 1.21, that the acetic acid and water are completely miscible, as indicated by the phase envelope not touching the horizontal axis at any point. Likewise, the isopropyl ether and acetic acid are completely miscible. But water and isopropyl ether are virtually immiscible, as indicated by little of the vertical axis being free of the two-phase region of the phase envelope, Fig. 1.21.

The composition of the two phases, for the mixture in the two-phase region, is found on the phase envelope line itself, Fig. 1.21. Toward the lower part of the phase envelope, Fig. 1.21, is the water-rich layer, and toward the top of the phase envelope line is the ether-rich layer.

Plot on the upper portion of Fig. 1.21 the given values of acetic acid and isopropyl ether, that is, 30 wt %, 50 wt %. Through this point, draw a tie line to intersect the phase envelope at two points, line 1, Fig. 1.21. Read the values: *lower point*—acetic acid in water layer = 20 wt %; *upper point*—acetic acid in isopropyl ether layer = 31.5 wt %.

Transferring the lower intersection point to the bottom diagram for tie line 1 shows that equilibrium exists between a layer that is 20 wt % acetic acid in water and a layer that is 9 wt % (not 31.5 wt %) acetic acid in isopropyl ether.

Draw a second tie line, 2, Fig. 1.21, as shown. Line 2 gives a check between the upper and lower diagrams, giving *water layer* − $x_a = 0.415$; $x_c = 0.065$; $x_w = 0.520$; *ether layer* − $y_a = 0.270$; $y_c = 0.650$; $x_w = 0.080$.

2. Compute the amount of acetic acid that must be added to form a one-phase system.
The water/ether ratio remains unchanged at water/ether = 0.20/0.50 = 0.40. Then the total system is: water + ether + acid = 1.000; so, ether (weight percent) = [1.000 − acid (weight percent)]/1.40.

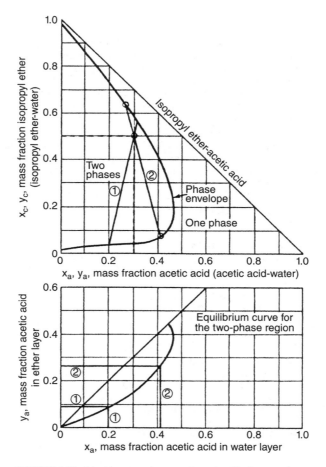

FIGURE 1.21 Liquid-system phase-envelope plot. (*Anderson and Wenzel—Introduction to Chemical Engineering, McGraw-Hill.*)

Assume that the acid = 0.350. Then ether = $(1.000 - 0.350)/1.40 = 0.464$. Checking against the upper diagram in Fig. 1.21, this point ($x_a = 0.350$, $x_c = 0.464$) falls inside the two-phase region. Hence, the assumption was incorrect.

As a second trial, assume acid = 0.380. Then ether = $(1.000 - 0.380)/1.40 = 0.443$. Checking $x_a = 0.380$, $x_c = 0.443$ in the upper diagram of Fig. 1.21 shows that the point falls exactly on the phase envelope line. Hence, it is at the minimum one-phase region.

*1.37 DETERMINING THE HEAT OF MIXING OF CHEMICALS

How many Btu's of heat are released (generated) when 1000 lb (453.6 kg) of water at 80°C (176°F) is mixed with (1) 500 lb (226.8 kg) of aluminum bromide, $AlBr_3$; (2) 750 lb (340.3 kg) of barium nitrate, $Ba(NO_3)_2$; and (3) 1000 lb (453.6 kg) of dextrin, $C_{12}H_{20}O_{10}$?

Calculation Procedure

1. *Compute the heat released when AlBr₃ is dissolved in water.*

When two or more substances are mixed, heat is usually generated or absorbed. The heat released or absorbed may be small when two similar organic liquids are mixed or very large when strong acids are mixed in water. The heat evolved (or absorbed) during the mixing of liquids is often called the *heat of dilution,* whereas the heat from the mixing of solids is often termed the *heat of solution.* Data for heats of solution for both organic and inorganic liquids and solids are given in Perry, Lange, and similar reference works.

Thus, at 80°C (176°F) the solubility of AlBr₃ in water is 126 g per 100 g of water. The weight of AlBr₃ that will dissolve in 1000 lb (454.5 kg) of water = (126/100)1000 = 1260 lb (572.7 kg). Standard references show that the heat of solution for AlBr₃ is 85.3 kg · cal per g · mol AlBr₃. Since the total Btu/(lb · mol) = 1.8 g · cal/g, the AlBr₃ in this solution can release [85.3 kg · cal/(g · mol)] [1000 cal/(kg · cal)]1.8 = 153,540 Btu/(lb · mol) (357.1 kJ/mol).

The weight of 1 lb·mol of AlBr₃ = (27 + 79.9 × 3) = 266.7 lb (121.2 kg). Hence, the heat evolved when 500 lb (227.3 kg) of AlBr₃ is dissolved in water is {500 lb AlBr₃/[266.7 lb/(lb·mol)]} [153,540 Btu/(lb · mol)] = 287,800 Btu (303.6 kJ).

2. *Compute the heat released in dissolving Ba(NO₃)₂ in water.*

At 80°C (176°F), the solubility of Ba(NO₃)₂ in water is 27.0 lb/lb (12.3 kg/kg) of water. The weight of Ba(NO₃)₂ that will dissolve in 1000 lb (454.5 kg) of water = (27/100)1000 = 270 lb (122.7 kg). Since 750 lb (340.9 kg) of Ba(NO₃)₂ is available for dissolving, the weight that will not dissolve is 750 − 270 = 480 lb (218.2 kg).

The heat of solution of Ba(NO₃)₂ is −10.2 kg·cal/(g·mol) of Ba(NO₃)₂. As in step 1, the weight of 1 lb · mol of Ba(NO₃)₂ = 137.34 + 2(14.0 + 3 × 16) = 261.4. Then, as in step 1, the heat released = [480 lb Ba(NO₃)₂/261.4 lb/(lb · mol)][−10.2 kg · cal/(g · mol) × 1000 g · cal/(kg · cal) × 1.8 [Btu/ (lb · mol)/[g · cal/(g · mol)] = −33,300 Btu (−35,131.5 J).

The negative heat release means that 33,300 Btu (35,131.5 J) of heat must be added to the system to maintain the solution temperature at 80°C (176°F) because a fall in temperature would reduce the solubility of the Ba(NO₃)₂ in water and thus change the resulting solution.

3. *Compute the heat released in dissolving C₁₂H₂₀O₁₀ in water.*

Perry indicates that there is no solubility limit for C₁₂H₂₀O₁₀ in water. By following the same procedure as in step 1, the heat released = [1000 lb C₁₂H₂₀O₁₀/324.2 lb/(lb · mol)] × {268 g · cal/(g · mol) × 1.8[Btu/(lb · mol)]/[g · cal/(g · mol)]} = 1488 Btu (1569.8 J) released.

Related Calculations.   Use the general procedure to determine the heat of mixing of any material dissolved in another.

*1.38   *CHEMICAL EQUATION MATERIAL BALANCE*

Ethylene oxide is produced by the catalytic reaction of ethylene and oxygen: $C_2H_4 + \frac{1}{2}O_2 \rightarrow$ $(CH_2)_2O$. For each 100 lb (45.5 kg) of ethylene, (1) how much ethylene oxide is produced, (2) how much oxygen is required, and (3) what are the quantities of ethylene oxide and ethylene in the product if there is a 20 percent deficiency of oxygen?

Calculation Procedure

1. *Compute the quantity of ethylene oxide produced.*

The two most frequently met calculations in day-to-day chemical engineering are the *material balance* (discussed here) and the *energy balance,* discussed later. In a chemical process, a balance is the same as any other type of balance, i.e., an equating of input, output, and accumulation or loss: Input − output = ± accumulation.

Such a balance may be written around a single item of chemical process equipment, a portion of a process, or an entire chemical plant. A balance may be used to check experimental data or to determine an unknown quantity of some process stream.

For purposes of balance calculations, chemical processes are classified as *steady state,* i.e., input = output, no accumulation; *unsteady state,* i.e., input ≠ output, a ± accumulation; *batch process,* i.e., system is loaded, no further ± accumulation; or continuous process, i.e., continuous input and output. Chemical processes may be further classed as physical, in which there is no chemical reaction, or chemical, in which a chemical change occurs. To analyze chemical reactions, the principles of chemical equation balances must be understood.

A *stoichiometrically balanced reaction* is one in which the reactants are exactly proportioned to give a product free of excess reactants, as in $C_2H_4 + \frac{1}{2}O_2 \rightarrow (CH_2)_2O$. An *excess reactant* is one present in excess of the stoichiometric quantity, such as if there were more than 0.5 mol of oxygen in the above equation.

The *degree of completion* is the percentage of the limiting reactant that reacts. The *limiting reactant* is the one present in less than stoichiometric proportion, so that the other reactant is in excess.

To determine how much ethylene oxide is produced, find the molecular weight of $C_2H_4 = 2(12) + 4(1) = 28$. The moles of $C_2H_4 = 100/28 = 3.571$.

Referring to the reaction equation shows that for each mole of C_2H_4, 1 mol of $(CH_2)_2O$ is produced, having a molecular weight of $2(12) + 4(1) + 16 = 44$. Then the weight of $(CH_2)_2O = 44(3.571) = 157.14$ lb (71.4 kg).

2. Compute the amount of oxygen required.

The molecular weight of $O_2 = 16(2) = 32$. Referring to the reaction equation shows that ½ mol of oxygen is needed for each mole of ethylene, C_2H_4. Hence, the weight of oxygen needed = ½(32) (3.571) = 57.14 lb (25.9 kg).

3. Compute the product mix for a reactant deficiency.

Referring to step 2, we see that a 20 percent oxygen deficiency means that there was $0.80(\frac{1}{2}) = 0.40$ mol of oxygen available. Rewriting the equation gives $0.2C_2H_4 + 0.8C_2H_4 + 0.4O_2 \rightarrow 0.8(CH_2)O + 0.2C_2H_4$. Hence, the ethylene oxide $(CH_2)_2O$ in the product = 0.8(157.14) = 125.71 lb (57.14 kg). And the ethylene, C_2H_4, in the product = 0.2(100) = 20 lb (9.1 kg).

Related Calculations. Use this general procedure for any chemical equation balance similar to that analyzed here.

*1.39 BATCH PHYSICAL PROCESS BALANCE

A load of clay containing 35 percent moisture on a wet basis weighs 2000 lb (909.1 kg). If the clay is dried to a 15 percent moisture content (on a wet basis), how much water is evaporated in the drying process?

Calculation Procedure

1. Compute the initial moisture content.

The 2000 lb (909.1 kg) of wet clay contains 35 percent moisture, or 2000 (0.35) = 700 lb (318.2 kg) of water. Thus, the dry clay weighs 2000 − 700 = 1300 lb (590.9 kg).

2. Compute the weight after drying.

Set up the relation y lb of wet clay + x lb of water = 1300 lb (590.9 kg) of dry clay. But the final batch contains 15 percent moisture. Hence, the water = $0.15y$. Therefore, the dry clay = $(1.00 - 0.15)$ $y = 0.85y = 1300$. Solving, we find $y = 1529$ lb (694.9 kg) of wet (15 percent moisture) clay. Since $y + x = 2000$ lb (909.1 kg), $x = 2000 - y = 2000 - 1529 = 471$ lb (214.1 kg) of water evaporated.

Related Calculations.   Use this general procedure for any batch physical process balance involving evaporation or drying of a solid.

Where the rate of feed is given, a steady-state physical process balance can be analyzed. Thus, if the 2000 lb (909.1 kg) of clay in the above process were fed to the dryer in 1 h, the rate of evaporation would be 471 lb/h (214.1 kg/h) of water.

REFERENCES

1. Reid, Prausnitz, and Sherwood—*The Properties of Gases and Liquids*, McGraw-Hill.

2. Lewis, Randall, and Pitzer—*Thermodynamics*, McGraw-Hill.

3. Prausnitz—*Molecular Thermodynamics of Fluid-Phase Equilibria*, Prentice-Hall.

4. Hougen, Watson, and Ragatz—*Chemical Process Principles*, parts I and II, Wiley.

5. Bretsznajder—*Prediction of Transport and Other Physical Properties of Fluids*, Pergamon.

6. Smith and Van Ness—*Chemical Engineering Thermodynamics,* McGraw-Hill.

7. Perry and Chilton—*Chemical Engineers' Handbook*, McGraw-Hill.

8. The Chemical Rubber Company—*Handbook of Chemistry and Physics.*

9. Bland and Davidson—*Petroleum Processing Handbook*, McGraw-Hill.

10. American Petroleum Institute—*Technical Data Book—Petroleum Refining.*

11. Natural Gas Processors Suppliers Association—*Engineering Data Book.*

12. Dreisbach—*Physical Properties of Chemical Compounds,* vols. 1–3, American Chemical Society.

13. Timmermans—*Physico-Chemical Constants of Binary Systems and Pure Organic Compounds*, Elsevier.

14. Rossini—*Selected Values of Properties of Chemical Compounds*, Thermodynamics Research Center, Texas A&M University.

15. Stull and Prophet—*JANAF Thermochemical Tables*, NSRDS, NBS-37.

16. Gunn and Yamada—*AIChE Journal 17*:1341, 1971.

17. Rackett—*J. Chem. Eng. Data 15*:514, 1970.

18. Yen and Woods—*AIChE Journal 12*:95, 1966.

19. Yuan and Stiel—*Ind. Eng. Chem. Fund. 9*:393, 1970.

20. *J. Chem. Eng. Data 10*:207, 1965.

21. *Chem. Eng. Tech. 26*:679, 1954.

22. *J. Am. Chem. Soc. 77*:3433, 1955.

23. *AIChE Journal 21*:510, 1975.

24. *AIChE Journal 13*:626, 1967.

25. *J. Chem. Phys. 29*:546, 1958.

26. *Ind. Eng. Chem. Process Des. Dev. 10*:576, 1971.

27. *AIChE Journal 15*:615, 1969.

28. Palmer—*Chemical Engineering 82*:80, 1975.

29. Letsou and Stiel—*AIChE Journal 19*:409, 1973.

30. Jossi et al.—*AIChE Journal 8*:59, 1962.

31. Fuller, Schettler, and Giddings—*Industrial and Engineering Chemistry 58*:5, 1966.

32. Brock and Bird—*AIChE Journal 1*:174, 1955.

33. Miller—*Industrial and Engineering Chemistry Fundamentals 2*:78. 1963.

SECTION 2
STOICHIOMETRY†

James H. Gary, Ph.D.
Professor Emeritus
Chemical and Petroleum-Refining
Engineering Department
Colorado School of Mines
Golden, CO

The first law of thermodynamics is the basis for material- or energy-balance calculations. Usually there is no significant transformation of mass to energy, and for a material balance, the first law can be reduced to the form

$$\text{Mass in} = \text{mass out} + \text{accumulation}$$

A similar equation can be used to express the energy balance

$$\text{Energy in (above datum)} + \text{energy generated} = \text{energy out (above datum)}$$

Energy balances differ from mass balances in that the total mass is known but the total energy of a component is difficult to express. Consequently, the heat energy of a material is usually expressed relative to its standard state at a given temperature. For example, the heat content, or enthalpy, of steam is expressed relative to liquid water at 273 K (32°F) at a pressure equal to its own vapor pressure.

---

†Procedure 2.8 is adapted from Smith, *Chemical Process Design*, McGraw-Hill; Procedures 2.9, 2.10, and 2.11 are taken from T.G. Hicks, *Standard Handbook of Engineering Calculations*, McGraw-Hill.

Regardless of how complicated a material-balance system may appear, the use of a systematic approach can resolve it into a number of independent equations equal to the number of unknowns. One suitable stepwise approach is (1) state the problem, (2) list available data, (3) draw a sketch of the system, (4) define the system boundaries, (5) establish the bases for the system parameters, (6) write component material balances, (7) write an overall material balance, (8) solve the equations, and (9) make another mass balance as a check.

2.1 MATERIAL BALANCE—NO CHEMICAL REACTIONS INVOLVED

A slurry containing 25 percent by weight of solids is fed into a filter. The filter cake contains 90 percent solids and the filtrate contains 1 percent solids. Make a material balance around the filter for a slurry feed rate of 2000 kg/h (4400 lb/h). For that feed rate, what are the corresponding flow rates for the cake and the filtrate?

Calculation Procedure

1. Sketch the system, showing the available data, indicating the unknowns, defining the system boundary, and establishing the basis for the calculations.
When no chemical reactions are involved, the balances are based on the masses of individual chemical compounds appearing in more than one incoming or outgoing stream. Components appearing in only one incoming and one outgoing stream can be lumped together as though they are one component to simplify calculations and increase precision. A convenient unit of mass is selected, usually the kilogram or pound, and all components are expressed in that unit.

As the basis for a continuous process, always choose a unit of time or a consistent set of flow rates per unit of time. For batch processes, the appropriate basis is 1 batch. In the present (continuous) process, let the basis be 1 h. Let C be the mass flow rate of filter cake and F the mass flow rate of filtrate, in kilograms per hour. Figure 2.1 is a sketch of the system.

2. Set up and solve the material-balance equations.
This is a steady-state operation, so accumulation equals zero and the amount of mass in equals the amount of mass out (per unit of time). Since there are two unknowns, C and F, two independent equations must be written. One will be an overall balance; the other can be either a liquid balance (the option chosen in this example) or a solids balance.

Overall balance: Filtrate out + cake out = slurry in, or $F + C = 2000$ kg/h (4400 lb/h).

Liquid balance: Liquid in filtrate + liquid in cake = liquid in slurry, or (wt fraction liquid in filtrate)(mass of filtrate) + (wt fraction liquid in cake)(mass of cake) = (wt fraction liquid in slurry)(mass of slurry), or $(1.0 - 0.01)F + (1.0 - 0.90)C = (1.0 - 0.25)(2000)$.

Simultaneous solving of the two equations, $F + C = 2000$ and $0.99F + 0.1C = 1500$, gives F to be 1460.7 kg/h (3214 lb/h) of filtrate and C to be 539.3 kg/h (1186 lb/h) of cake.

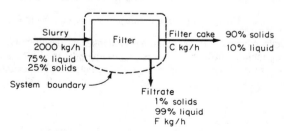

FIGURE 2.1 Material balance for filter (Procedure 2.1).

3. *Check the results.*
It is convenient to check the answers by substituting them into the equation not used above, namely, the solids balance. Thus, solid in filtrate + solid in cake = solid in slurry, or 0.01 (1460.7) + 0.9(539.3) = 0.25 (2000). The answers, therefore, are correct.

2.2 *MATERIAL BALANCE—CHEMICAL REACTIONS INVOLVED*

Natural gas consisting of 95% methane and 5% nitrogen by volume is burned in a furnace with 15% excess air. How much air at 289 K (61°F) and 101.3 kPa (14.7 psia) is required if the fuel consumption is 10 m³/s (353 ft³/s) measured at 289 K and 101.3 kPa? Make an overall material balance and calculate the quantity and composition of the flue gas.

Calculation Procedure

1. *Sketch the system, setting out the available data, indicating the unknowns, defining the system boundary, and establishing the basis for the calculations.*
For a process involving chemical reactions, the usual procedure is to express the compositions of the streams entering and leaving the process in molar concentrations. The balances are made in terms of the largest components remaining unchanged in the reactions. These can be expressed as atoms (S), ions (SO_4^{2-}), molecules (O_2), or other suitable units.

Whenever the reactants involved are not present in the proper stoichiometric ratios, the limiting reactant should be determined and the excess quantities of the other reactants calculated. Unconsumed reactants and inert materials exit with the products in their original form.

By convention, the amount of excess reactant in a reaction is always defined on the basis of the reaction going to 100 percent completion for the limiting reactant. The degree of completion is not a factor in determining or specifying the excess of reactants. For example, if methane is burned with 10 percent excess air, the volume of air needed to burn the methane is calculated as though there is total combustion of methane to carbon dioxide and water.

In the present problem, let the basis be 1 s. Let A and F be the volumetric flow rates for air and flue gas, in cubic meters per second. Figure 2.2 is a sketch of the system. The data are as follows:

Natural gas at 289 K and 101.3 kPa = 10 m³/s
    95% CH₄   MW = 16
    5% N₂   MW = 28

Air at 289 K and 101.3 kPa = A m³/s
    21% O₂   MW = 32
    79% N₂   MW = 28

$R = 8.314$ kJ/(kg · mol)(K)   (ideal-gas constant)

2. *Convert the natural gas flow rate to kilogram-moles per second.*
At the conditions of this problem, the ideal-gas law can be used. Thus, $n = PV/RT$, where n is number of moles, P is pressure, V is volume, R is the gas constant, and T is absolute temperature.

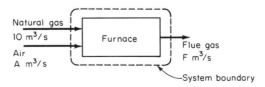

FIGURE 2.2   Material balance for furnace (Procedure 2.2).

For CH_4, n = (101.3 kPa)[(10 m^3/s)(0.95)]/[8.314 kJ/(kg · mol)(K)](289 K) = 0.40 kg · mol/s. For N_2, since the volumetric composition is 95% CH_4 and 5% N_2, n = (0.05/0.95)(0.40) = 0.02 kg · mol/s.

3. Determine the amount of oxygen required and the airflow rate.

The combustion reaction for CH_4 is $CH_4 + 2O_2 \rightarrow CO_2 + 2H_2O$. Thus 0.40 mol/s of CH_4 requires 2(0.40), or 0.80, mol/s of O_2 for stoichiometric combustion. Since 15 percent excess air is specified, the number of moles of oxygen in the air is (1.15)(0.80), or 0.92, kg · mol/s. The amount of nitrogen in with the air is [(0.79 mol N_2/mol air)/(0.21 mol O_2/mol air)](0.92 kg · mol/s of O_2) = 3.46 kg. mol/s. Total moles in the incoming air are 0.92 + 3.46 = 4.38 kg · mol/s. Finally, using the ideal-gas law to convert to volumetric flow rate, $V = nRT/P$ = (4.38)(8.314)(289)/101.3 = 103.9 m^3/s (3671 ft^3/s) of air.

4. Set up the material balance and calculate the composition and quantity of the flue gas.

Convert to a mass basis because it is always true (unless there is a conversion between mass and energy) that from a mass standpoint the input equals the output plus the accumulation. In the present problem, there is no accumulation. The output (the flue gas) includes nitrogen from the air and from the natural gas, plus the 15 percent excess oxygen, plus the reaction products, namely, 0.40 mol/s CO_2 and 2(0.40) = 0.80 mol/s water.

Select 1 s as the basis. Then the inputs and outputs are as follows:

Component	Kilogram-moles	×	Kilograms per mole = Kilograms
		Inputs	
Natural gas:			
CH_4	0.40	16	6.40
N_2	0.02	28	0.56
Air:			
N_2	3.46	28	96.88
O_2	0.92	32	29.44
Total			133.28
		Outputs	
Flue gas:			
N_2	(3.46 + 0.02)	28	97.44
O_2	(0.92 − 0.80)	32	3.84
CO_2	0.40	44	17.60
H_2O	0.80	18	14.40
Total			133.28

The accumulation is zero. The overall material balance is 133.28 = 133.28 + 0. The total quantity of flue gas, therefore, is 133.28 kg/s (293 lb/s).

Related Calculations.   The composition of the flue gas as given above is by weight. If desired, the composition by volume (which, indeed, is the more usual basis for expressing gas composition) can readily be obtained by calculating the moles per second of nitrogen (molar and volumetric compositions are equal to each other).

For more complex chemical reactions, it may be necessary to make mass balances for each molecular or atomic species rather than for the compounds.

2.3 MATERIAL BALANCE—INCOMPLETE DATA ON COMPOSITION OR FLOW RATE

Vinegar with a strength of 4.63% (by weight) acetic acid is pumped into a vat to which 1000 kg (2200 lb) of 36.0% acetic acid is added. The resulting mixture contains 8.50% acid. How much of this 8.50% acid solution is in the vat?

Calculation Procedure

1. *List the available data, establish a basis for the calculations, and assign letters for the unknown quantities.*

In many situations, such as in this example, some streams entering or leaving a process may have incomplete data to express their compositions or flow rates. The usual procedure is to write material balances as in the preceding examples but to assign letters to represent the unknown quantities. There must be one independent material balance written for each unknown in order to have a unique solution.

The present problem is a batch situation, so let the basis be 1 batch. There are two inputs: an unknown quantity of vinegar having a known composition (4.63% acetic acid) and a known amount of added acetic acid [1000 kg (2200 lb)] of known composition (36.0% acid). There is one output: a final batch of unknown quantity but known composition (8.50% acid). Let T represent the kilograms of input vinegar and V the size of the final batch in kilograms.

2. *Set up and solve the material-balance equations.*

Two independent material balances can be set up, one for acetic acid or for water and the other for the overall system.

Acetic acid: Input = output, or $0.0463T + 0.360(1000 \text{ kg}) = 0.0850V$
Water: Input = output, or $(1 - 0.0463)T + (1 - 0.360)(1000 \text{ kg}) = (1 - 0.0850)V$
Overall: Input = output, or $T + 1000 = V$

Use the overall balance and one of the others, say, the one for acetic acid. By substitution, then, $0.0463T + 0.360(1000) = 0.0850(T + 1000)$; so T is found to be 7106 kg vinegar, and $V = T + 1000 = 8106$ kg (17,833 lb) solution in the vat.

3. *Check the results.*

It is convenient to make the check by substituting into the equation not used above, namely, the water balance:

$$(1 - 0.0463)(7106) + (1 - 0.360)(1000) = (1 - 0.0850)(8106)$$

Thus, the results are checked.

2.4   USE OF A TIE ELEMENT IN MATERIAL-BALANCE CALCULATIONS

The spent catalyst from a catalytic-cracking reactor is taken to the regenerator for reactivation. Coke deposited on the catalyst in the reactor is removed by burning with air, and the flue gas is vented. The coke is a mixture of carbon and high-molecular-weight tars considered to be hydrocarbons. For the following conditions, calculate the weight percent of hydrogen in the coke. Assume that the coke on the regenerated catalyst has the same composition as the coke on the spent catalyst:

Carbon on spent catalyst   1.50 wt %

Carbon on regenerated catalyst   0.80 wt %

Air from blower   150,000 kg/h (330,000 lb/h)

Hydrocarbon feed to reactor   300,000 kg/h (660,000 lb/h)

Flue gas analysis (dry basis):

CO_2	12.0 vol %
CO	6.0 vol %
O_2	0.7 vol %
N_2	81.3 vol %
	100.0

Assume that all oxygen not reported in flue gas analysis reacted with hydrogen in the coke to form water. All oxygen is reported as O_2 equivalent. Assume that air is 79.02% nitrogen and 20.98% oxygen.

Calculation Procedure

1. Select a basis and a tie component, and write out the relevant equations involved.
Select 100 kg · mol dry flue gas as the basis. Since nitrogen passes through the system unreacted, select it as the tie component. That is, the other components of the system can be referred to nitrogen as a basis, thus simplifying the calculations.

Since dry flue gas is the basis, containing CO_2 and CO, the relevant reactions and the quantities per 100 mol flue gas are

$$C \quad + \quad O_2 \quad \rightarrow \quad C_2$$
$$12 \text{ kg} \cdot \text{mol} \quad 12 \text{ kg} \cdot \text{mol} \quad 12 \text{ kg} \cdot \text{mol}$$

$$C \quad + \quad \tfrac{1}{2}O_2 \quad \rightarrow \quad CO$$
$$6 \text{ kg} \cdot \text{mol} \quad 3 \text{ kg} \cdot \text{mol} \quad 6 \text{ kg} \cdot \text{mol}$$

2. Calculate the amount of oxygen in the entering air.
The total moles of nitrogen in the entering air must equal the total moles in the flue gas, namely, 81.3 kg · mol. Since air is 79.02% nitrogen and 20.98% oxygen, the oxygen amounts to (20.98/79.02)(81.3), or 21.59 kg · mol.

3. Calculate the amount of oxygen that leaves the system as water.
The number of kilogram-moles of oxygen in the regenerator exit gases should be the same as the number in the entering air, that is, 21.59. Therefore, the oxygen not accounted for in the dry analysis of the flue gas is the oxygen converted to water. The dry analysis accounts for 12 mol oxygen as CO_2, 3 mol as CO, and 0.7 mol as unreacted oxygen. Thus the water leaving the system in the (wet) flue gas accounts for (21.59 − 12 − 3 − 0.7) = 5.89 mol oxygen.

4. Calculate the weight percent of hydrogen in the coke.
Since 2 mol water is produced per mole of oxygen reacted, the amount of water in the wet flue gas is 2(5.89) = 11.78 mol. This amount of water contains 11.78 mol hydrogen or (11.78)[2.016 kg/(kg · mol)] = 23.75 kg hydrogen.

Now, the amount of carbon associated with this is the 12 mol that reacted to form CO_2 plus the 6 mol that reacted to form CO or (12 + 6)[12.011 kg/(kg · mol)] = 216.20 kg carbon. Therefore, the weight percent of hydrogen in the coke is (100)(23.75)/(23.75 + 216.0) = 9.91 percent.

2.5   MATERIAL BALANCE—CHEMICAL REACTION AND A RECYCLE STREAM INVOLVED

In the feed-preparation section of an ammonia plant, hydrogen is produced from methane by a combination steam-reforming/partial-oxidation process. Enough air is used in partial oxidation to give a 3:1 hydrogen-nitrogen molar ratio in the feed to the ammonia unit. The hydrogen-nitrogen mixture is heated to reaction temperature and fed into a fixed-bed reactor where 20 percent conversion of reactants to ammonia is obtained per pass. After leaving the reactor, the mixture is cooled and the ammonia is removed by condensation. The unreacted hydrogen-nitrogen mixture is recycled and

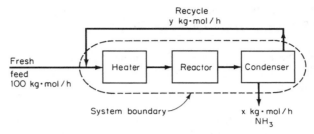

FIGURE 2.3   Material balance for ammonia plant with no purge stream
(Procedure 2.5).

mixed with fresh feed. On the basis of 100 kg · mol/h (220 lb · mol/h) of fresh feed, make a material balance and determine the ammonia-production and recycle rates.

Calculation Procedure

1. Sketch the system, showing the available data, indicating the unknowns, and establishing the system boundary.
Since one of the answers sought is the recycle rate, the system boundary must be selected in such a way as to be crossed by the recycle stream.

Since the feed is a 3:1 ratio, the 100 mol/h will consist of 25 mol/h of nitrogen and 75 mol/h of hydrogen. Let x equal the moles per hour of ammonia produced and y the moles per hour of recycle. The sketch and the system boundary are shown in Fig. 2.3.

2. Determine the amount of ammonia produced.
Set out the ammonia-production reaction, namely, $N_2 + 3H_2 \rightarrow 2NH_3$. Thus 4 mol hydrogen-nitrogen mixture in a 3:1 ratio (as is the case for the feed in this example) will yield 2 mol ammonia. Since the system boundary is drawn in such a way that the exiting and reentering recycle streams offset each other algebraically, the net output from the system, consisting of liquid ammonia, must equal the net input, consisting of fresh feed. The amount x of ammonia produced per hour thus can be determined by straightforward stoichiometry: $x = [100 \text{ kg} \cdot \text{mol/h } (H_2 + N_2)] [(2 \text{ mol } NH_3)/4 \text{ mol } (H_2 + N_2)]$, or $x = 50$ kg · mol/h (110 lb · mol/h) NH_3.

3. Determine the recycle rate.
The total feed to the heater and reactor consists of $(100 + y)$ kg · mol/h. Twenty percent of this feed is converted to ammonia, and during that conversion, 2 mol ammonia is produced per 4 mol feed (inasmuch as it consists of 3:1 hydrogen-nitrogen mixture). Therefore, the amount of ammonia produced equals $[0.20(100 + y)][(2 \text{ mol } NH_3)/4 \text{ mol } (H_2 + N_2)]$. Ammonia production is 50 kg · mol/h, so solving this equation for y gives a recycle rate of 400 kg · mol/h (880 lb · mol/h).

4. Check the results.
A convenient way to check is to set up an overall mass balance. Since there is no accumulation, the input must equal the output. Input = (25 kg · mol/h N_2)(28 kg/mol) + (75 kg · mol/h H_2)(2 kg/mol) = 850 kg/h, and output = (50 kg · mol/h NH_3)(17 kg/mol) = 850 kg/h. The results are thus checked.

Related Calculations.   It is also possible to calculate the recycle rate in the preceding example by making a material balance around the reactor-condenser system.

The ratio of the quantity of material recycled to the quantity of fresh feed is called the "recycle ratio." In the preceding problem, the recycle ratio is 400/100, or 4.

2.6 MATERIAL BALANCE—CHEMICAL REACTION, RECYCLE STREAM, AND PURGE STREAM INVOLVED

In the Procedure 2.5 for producing ammonia, the amount of air fed is set by the stoichiometric ratio of hydrogen to nitrogen for the ammonia feed stream. In addition to nitrogen and oxygen, the air contains inert gases, principally argon, that gradually build up in the recycle stream until the process is affected adversely. It has been determined that the concentration of argon in the reactor must be no greater than 4 mol argon per 100 mol hydrogen-nitrogen mixture. Using the capacities given in the preceding example, calculate the amount of the recycle stream that must be vented to meet the concentration requirement. The fresh feed contains 0.31 mol argon per 100 mol hydrogen-nitrogen mixture. Also calculate the amount of ammonia produced.

Calculation Procedure

1. Select the basis for calculation and sketch the system, showing the available data and indicating the unknowns.
For ease of comparison with the preceding example, let the basis be 100.31 kg · mol/h total fresh feed, consisting of 100 mol $(H_2 + N_2)$ and 0.31 mol argon (A). Let x equal the moles of NH_3 produced per hour, y the moles of $H_2 + N_2$ recycled per hour, w the moles of A recycled per hour, and z the moles of $H_2 + N_2$ purged per hour. The sketch is Fig. 2.4.

2. Calculate the amount of recycle stream that must be vented.
As noted in the preceding example, the conversion per pass through the reactor is 20 percent. Therefore, for every 100 mol $(H_2 + N_2)$ entering the heater-reactor-condenser train, 20 mol will react to form ammonia and 80 mol will leave the condenser to be recycled or purged. All the argon will leave with this recycle-and-purge stream. Since the maximum allowable argon level in the reactor input is 4 mol argon per 100 mol $(H_2 + N_2)$, there will be 4 mol argon per 80 mol $(H_2 + N_2)$ in the recycle-and-purge stream.

Under steady-state operating conditions, the argon purged must equal the argon entering in the fresh feed. The moles of argon in the purge $(4/80)z = 0.05z$. Therefore, $0.05z = 0.31$, so $z = 6.2$ mol/h $(H_2 + N_2)$ purged. The total purge stream consists of 6.2 mol/h $(H_2 + N_2)$ plus 0.31 mol/h argon.

3. Calculate the amount of $H_2 + N_2$ recycled.
The moles of $H_2 + N_2$ in the feed to the reactor is 100 mol fresh feed plus y mol recycle. Of this, 80 percent is to be either purged or recycled; that is, $0.80(100 + y) = y + z = y + 6.2$. Therefore $y = 369$ mol/h $(H_2 + N_2)$ recycled.

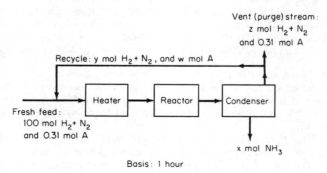

FIGURE 2.4   Material balance for ammonia plant with purge stream (Procedure 2.6).

Although not needed for the solution of this problem, the amount of argon in the recycle can be calculated as a matter of interest. Total argon entering the reactor is $0.31 + w$ mol; then, according to the argon limitation stipulated, $0.31 + w = 0.04(100 + 369)$, so $w = 18.45$ mol/h.

4. Calculate the amount of ammonia produced.
Of the $100 + y$ mol ($H_2 + N_2$) (in the stoichiometric 3:1 ratio) entering the reactor, 20 percent is converted to ammonia. The reaction is $N_2 + 3H_2 \rightarrow 2NH_3$, so 4 mol reactants yields 2 mol ammonia. Therefore, total ammonia production is $0.20(100 + y)(2/4) = 0.20(100 + 369)(2/4) = 46.9$ mol/h.

5. Check the results.
It is convenient to check by making an overall mass balance. The input per hour consists of 100 mol ($H_2 + N_2$) in a 3:1 ratio plus 0.31 mol A; that is, (75 mol H_2)(2 kg/mol) + (25 mol N_2)(28 kg/mol) + (0.31 mol A)(40 kg/mol) = 862.4 kg. The output per hour consists of 46.9 mol ammonia plus a vent-stream mixture of 6.2 mol ($H_2 + N_2$) (in a 3:1 ratio) and 0.31 mol A; that is, (46.9 mol NH_3) (17 kg/mol) + (3/4)[6.2 mol ($H_2 + N_2$)](2 kg H_2/mol) + (1/4)(6.2)(28 kg N_2/mol) + (0.31 mol A) (40 kg/mol) = 862.4 kg.

There is no accumulation in the system, so input should equal output. Since 862.4 kg = 862.4 kg, the results are thus checked.

2.7   USE OF ENERGY BALANCE WITH MATERIAL BALANCE

A particular crude oil is heated to 510 K (458°F) and charged at 10 L/h (0.01 m³/h, or 2.6 gal/h) to the flash zone of a laboratory distillation tower. The flash zone is at an absolute pressure of 110 kPa (16 psi). Determine the percent vaporized and the amounts of the overhead and bottoms streams. Assume that the vapor and liquid are in equilibrium.

Calculation Procedure

1. Select the approach to be employed.
In this problem there is not enough information available to employ a purely material-balance approach. Instead, use an energy balance as well. Such an approach is especially appropriate in cases such as this one in which some of the components undergo a phase change.

From the American Petroleum Institute's (API) *Technical Data Book—Petroleum Refining,* specific heats, specific gravities, latent heats of vaporization, and percent vaporization can be obtained, for a given oil, as a function of flash-zone temperature (percent vaporization and flash-zone temperature are functionally related because the flash vaporization takes place adiabatically). This suggests a trial-and-error procedure: Assume a flash-zone temperature and the associated percent vaporization; then make an energy balance to check the assumptions. Finally, complete the material balance.

2. Assume a flash-zone temperature and percent vaporization, and obtain the data for the system at those conditions.
Assume, for a first guess, that 30 percent (by volume) of the feed is vaporized. The *API Data Book* indicates that for this oil, the corresponding flash-zone temperature is 483 K (410°F); the fraction vaporized has a latent heat of vaporization of 291 kJ/kg (125 Btu/lb) and a density of 0.750 kg/L (750 kg/m³, or 47.0 lb/ft³) and a specific heat of 2.89 kJ/(kg)(K) [0.69 Btu/(lb)(°F)]. The unvaporized portion has a density of 0.892 kg/L (892 kg/m³, or 55.8 lb/ft³) and a specific heat of 2.68 kJ/(kg)(K) [0.64 Btu/(lb)(°F)]. In addition, the feed has a density of 0.850 kg/L (850 kg/m³, or 53.1 lb/ft³) and a specific heat of 2.85 kJ/(kg)(K) [0.68 Btu/(lb)(°F)].

3. Make an energy balance.
For convenience, use the flash temperature, 483 K, as the datum temperature. The energy brought into the system by the feed, consisting of sensible-heat energy with reference to the datum

temperature, must equal the energy in the vapor stream (its latent heat plus its sensible heat) plus the energy in the bottoms stream (its sensible heat). However, since the flash temperature is the datum, and since both the vapor and the bottoms streams are at the datum temperature, neither of those product streams has a sensible-heat term associated with it. Thus the energy balance on the basis of 1 h (10 L) is as follows:

$$(10 \text{ L})(0.850 \text{ kg/L})[2.85 \text{ kJ/(kg)(K)}](510 \text{ K} - 483 \text{ K}) = (3 \text{ L})(0.750 \text{ kg/L})\{(291 \text{ kJ/kg})$$
$$+ [2.89 \text{ kJ/(kg)(K)}] (483 \text{ K} - 483 \text{ K})\} + (7 \text{ L})(0.892 \text{ kg/L})[2.68 \text{ kJ/(kg)(K)}](483 \text{ K} - 483 \text{ K})$$

Or, 654 = 655 + 0. Since this is within the limits of accuracy, the assumption of 30 percent vaporized is correct.

4. *Make the material balance to determine the amount in the overhead and bottoms streams.*
On the basis of 1 h, the mass in is (10 L)(0.850 kg/L) = 8.5 kg. The mass out consists of the mass that becomes vaporized (the overhead) plus the mass that remains unvaporized (the bottoms). The overhead is (3 L)(0.750 kg/L) = 2.25 kg (4.96 lb). The bottoms stream is (7 L)(0.892 kg/L) = 6.24 kg (13.76 lb). Thus, 8.5 kg = (2.25 + 6.24) kg. The material balance is consistent, within the limits of accuracy.

Related Calculations.   The problem can be worked in similar fashion using values from enthalpy tables. In this case, the datum temperature is below the flash-zone temperature: therefore, sensible heat in the two exiting streams must be taken into account.

2.8 MATERIAL BALANCE—LINKING RECYCLE AND PURGE WITH REACTION SELECTIVITY AND CONVERSION

Benzene is to be produced from toluene according to the reaction

$$\underset{\text{toluene}}{C_6H_5CH_3} + \underset{\text{hydrogen}}{H_2} \rightarrow \underset{\text{benzene}}{C_6H_6} + \underset{\text{methane}}{CH_4}$$

The reaction is carried out in the gas phase and normally operates at around 700°C and 40 bar. Some of the benzene formed undergoes a secondary reversible reaction to an unwanted byproduct, diphenyl, according to the reaction

$$\underset{\text{benzene}}{2C_6H_6} \rightleftarrows \underset{\text{diphenyl}}{C_{12}H_{10}} + \underset{\text{hydrogen}}{H_2}$$

Laboratory studies indicate that a hydrogen-toluene ratio of 5 at the reactor inlet is required to prevent excessive coke formation in the reactor. Even with a large excess of hydrogen, the toluene cannot be forced to complete conversion. The laboratory studies indicate that the selectivity (i.e., fraction of toluene reacted that is converted to benzene) is related to the conversion (i.e., fraction of toluene fed that is reacted) according to

$$S = 1 - \frac{0.0036}{(1 - X)^{1.544}}$$

where S = selectivity
   X = conversion

The reactor effluent is thus likely to contain hydrogen, methane, benzene, toluene, and diphenyl. Because of the large differences in volatility of these components, it seems likely that partial condensation will allow the effluent to be split into a vapor stream containing predominantly hydrogen and methane and a liquid stream containing predominantly benzene, toluene, and diphenyl.

The hydrogen in the vapor stream is a reactant and hence should be recycled to the reactor inlet (see flow diagram). The methane enters the process as a feed impurity, is also a byproduct from the primary reaction, and must be removed from the process. The hydrogen-methane separation is likely to be expensive, but the methane can be removed from the process by means of a purge (see flow diagram).

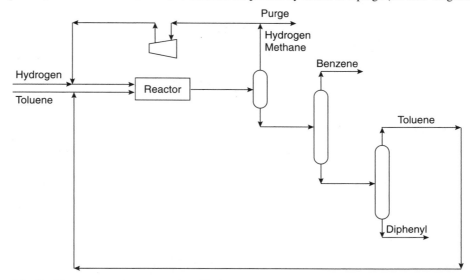

The liquid stream can be separated readily into pure components by distillation, with the benzene taken off as product, the diphenyl as an unwanted byproduct, and the toluene recycled. It is possible to recycle the diphenyl to improve selectivity, but we will assume that is not done here.

The hydrogen feed contains methane as an impurity at a mole fraction of 0.05. The production rate of benzene required is 265 kmol h^{-1}.

Assume initially that a phase split can separate the reactor effluent into a vapor stream containing only hydrogen and methane and a liquid stream containing only benzene, toluene, and diphenyl, and that the liquid separation system can produce essentially pure products.

For a conversion in the reactor of 0.75:

1. Determine the relation between the fraction of vapor from the phase split sent to purge (α) and the fraction of methane in the recycle and purge (y).

2. Given the assumptions, estimate the composition of the reactor effluent for fraction of methane in the recycle and purge of 0.4.

Calculation Procedure

1. *Calculate the benzene selectivity from toluene.*
Let P_B be the production rate of benzene. Then,

$$C_6H_5CH_3 + H_2 \rightarrow C_6H_6 + C_6H_6 \qquad + CH_4$$

$$\frac{P_B}{S} \qquad \frac{P_B}{S} \qquad P_B \qquad P_B\left(\frac{1}{S}-1\right) \qquad \frac{P_B}{S}$$

$$2C_6H_6 \quad \rightleftarrows \quad C_{12}H_{10} \qquad + H_2$$

$$P_B\left(\frac{1}{S}-1\right) \qquad \frac{P_B}{2}\left(\frac{1}{S}-1\right) \quad + \frac{P_B}{2}\left(\frac{1}{S}-1\right)$$

For $X = 0.75$,

$$S = 1 - \frac{0.0036}{(1 - 0.75)^{1.544}}$$

$$= 0.9694$$

For more on calculation of selectivities, see Section 4 in this book.

2. *Calculate the amount of toluene to be fed to the reactor.*

$$\text{Fresh toluene feed} = \frac{P_B}{S}$$

and

$$\text{toluene recycle} = R_T$$

Then

$$\text{toluene entering the reactor} = \frac{P_B}{S} + R_T$$

and

$$\text{toluene in reactor effluent} = \left(\frac{P_B}{S} + R_T \right)(1 - X) = R_T$$

For $P_B = 265$ kmol h^{-1}, $X = 0.75$, and $S = 0.9694$,

$$R_T = 91.12 \text{ kmol h}^{-1}$$

Therefore,

$$\text{toluene entering the reactor} = \frac{265}{0.9694} + 91.12$$

$$= 364.5 \text{ kmol h}^{-1}$$

3. *Find the amount of hydrogen to be fed, as a function of the phase split that is sent to purge, α.*

$$\text{Hydrogen entering the reactor} = 5 \times 364.5$$

$$= 1823 \text{ kmol h}^{-1}$$

$$\text{Net hydrogen consumed in reaction} = \frac{P_B}{S} - \frac{P_B}{2}\left(\frac{1}{S} - 1 \right)$$

$$= \frac{P_B}{S}\left(1 - \frac{1 - S}{2} \right)$$

$$= 269.2 \text{ kmol h}^{-1}$$

Therefore,

$$\text{hydrogen in reactor effluent} = 1823 - 269.2$$

$$= 1554 \text{ kmol h}^{-1}$$

$$\text{hydrogen lost in purge} = 1554\alpha$$

Thus,

hydrogen feed to the process $= 1554\alpha + 269.2$

4. *Find the fraction y of methane in the purge and recycle, as a function of α.*

Methane feed to process as an impurity $= (1554\alpha + 269.2)\dfrac{0.05}{0.95}$

and

$$\text{methane produced by reactor} = \frac{P_B}{S}$$

So,

$$\text{methane in purge} = \frac{P_B}{S} + (1554\alpha + 269.2)\frac{0.05}{0.95}$$
$$= 81.79\alpha + 287.5$$
$$\text{total flow rate of purge} = 1554\alpha + 81.79\alpha + 287.5$$
$$= 1636\alpha + 287.5$$

The fraction of methane in the purge (and recycle) as a function of α is thus

$$y = \frac{81.79\alpha + 287.5}{1636\alpha + 287.5}$$

The graph shows a plot of this equation. As the purge fraction α is increased, the flow rate of purge increases, but the concentration of methane in the purge and recycle decreases. This variation (along with reactor conversion) is an important degree of freedom in the optimization of reaction and separation systems.

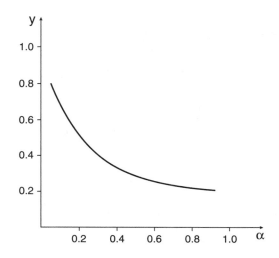

5. *For an α of 0.4, find the amount of methane in the reactor effluent.*

$$\text{Mole fraction of methane in vapor from phase separator} = 0.4$$

Therefore,

$$\text{methane in reactor effluent} = \frac{0.4}{0.6} \times 1554$$
$$= 1036 \text{ kmol h}^{-1}$$

6. *Find the amount of diphenyl in the reactor effluent.*

$$\text{Diphenyl in reactor effluent} = \frac{P_B}{2}\left(\frac{1}{S} - 1\right)$$
$$= 4 \text{ kmol h}^{-1}$$

7. *Summarize the composition of the reactor effluent.*
The estimated composition of the reactor effluent is given in the table. These results assume that all separations in the phase split are sharp.

Component	Flow rate (kmol h^{-1})
Hydrogen	1554
Methane	1036
Benzene	265
Toluene	91
Diphenyl	4

Source: Adapted from Smith, *Chemical Process Design*, McGraw-Hill.

2.9   COMBUSTION OF COAL FUEL IN A FURNACE

A coal has the following ultimate analysis: C = 0.8339, H_2 = 0.0456, O_2 = 0.0505, N_2 = 0.0103. S = 0.0064, ash = 0.0533, and total = 1.000. This coal is burned in a steam-boiler furnace. Determine the weight of air required for theoretically perfect combustion, the weight of gas formed per pound of coal burned, and the volume of flue gas at the boiler exit temperature of 600°F (589 K) per pound of coal burned; the air required with 20 percent excess air and the volume of gas formed with this excess; and the CO_2 percentage in the flue gas on a dry and wet basis.

Calculation Procedure

1. *Compute the weight of oxygen required per pound of coal.*
To find the weight of oxygen required for theoretically perfect combustion of coal, set up the following tabulation, based on the ultimate analysis of the coal:

Element	×	Molecular-weight ratio	=	Pounds O_2 required
C; 0.8339	×	32/12	=	2.2237
H_2; 0.0456	×	16/2	=	0.3648
O_2; 0.0505; decreases external O_2 required			=	−0.0505
N_2; 0.0103; is inert in combustion and is ignored				
S; 0.0064	×	32/32	=	0.0064
Ash; 0.0533; is inert in combustion and is ignored				
Total 1.0000				
Pounds external O_2 per lb fuel			=	2.5444

Note that of the total oxygen needed for combustion, 0.0505 lb is furnished by the fuel itself and is assumed to reduce the total external oxygen required by the amount of oxygen present in the fuel. The molecular-weight ratio is obtained from the equation for the chemical reaction of the element with oxygen in combustion. Thus, for carbon, $C + O_2 \rightarrow CO_2$, or $12 + 32 = 44$, where 12 and 32 are the molecular weights of C and O_2, respectively.

2. Compute the weight of air required for perfect combustion.
Air at sea level is a mechanical mixture of various gases, principally 23.2% oxygen and 76.8% nitrogen by weight. The nitrogen associated with 2.5444 lb of oxygen required per pound of coal burned in this furnace is the product of the ratio of the nitrogen and oxygen weights in the air and 2.5444, or (2.5444) (0.768/0.232) = 8.4219 lb. Then the weight of air required for perfect combustion of 1 lb coal = sum of nitrogen and oxygen required = 8.4219 + 2.5444 = 10.9663 lb of air per pound of coal burned.

3. Compute the weight of the products of combustion.
Find the products of combustion by addition:

Fuel constituents	+	Oxygen	→	Products of combustion
C; 0.8339	+	2.2237	→	CO_2 = 3.0576 lb
H; 0.0456	+	0.3648	→	H_2O = 0.4104
O_2; 0.0505; this is *not* a product of combustion				
N_2; 0.0103; inert but passes through furnace				= 0.0103
S; 0.0064	+	0.0064	→	SO_2 = 0.0128
Outside nitrogen from step 2				= N_2 = 8.4219
Pounds of flue gas per pound of coal burned				= 11.9130

4. Convert the flue-gas weight to volume.
Use Avogadro's law, which states that under the same conditions of pressure and temperature, 1 mol (the molecular weight of a gas expressed in pounds) of any gas will occupy the same volume.

At 14.7 psia and 32°F, 1 mol of any gas occupies 359 ft^3. The volume per pound of any gas at these conditions can be found by dividing 359 by the molecular weight of the gas and correcting for the gas temperature by multiplying the volume by the ratio of the absolute flue-gas temperature and the atmospheric temperature. To change the weight analysis (step 3) of the products of combustion to volumetric analysis, set up the calculation thus:

Products	Weight, lb	Molecular weight	Temperature correction	Volume at 600°F, ft^3
CO_2	3.0576	44	(359/44)(3.0576)(2.15)	= 53.8
H_2O	0.4104	18	(359/18)(0.4104)(2.15)	= 17.6
Total N_2	8.4322	28	(359/28)(8.4322)(2.15)	= 233.0
SO_2	0.0128	64	(359/64)(0.0128)(2.15)	= 0.17
Cubic feet of flue gas per pound of coal burned				= 304.57

In this calculation, the temperature correction factor $2.15 = $ (absolute flue-gas temperature)/(absolute atmospheric temperature), $R = (600 + 460)/(32 + 460)$. The total weight of N_2 in the flue gas is the sum of the N_2 in the combustion air and the fuel, or $8.4219 + 0.0103 = 8.4322$ lb. This value is used in computing the flue-gas volume.

5. Compute the CO_2 content of the flue gas.
The volume of CO_2 in the products of combustion at 600°F is 53.8 ft^3 as computed in step 4, and the total volume of the combustion products is 304.57 ft^3. Therefore, the percent CO_2 on a wet basis (i.e., including the moisture in the combustion products) $= ft^3$ CO_2/total $ft^3 = 53.8/304.57 = 0.1765$, or 17.65 percent.

The percent CO_2 on a dry, or Orsat, basis is found in the same manner except that the weight of H_2O in the products of combustion, 17.6 lb from step 4, is subtracted from the total gas weight, or, percent CO_2, dry, or Orsat, basis $= (53.8)/(304.57 - 17.6) = 0.1875$, or 18.75 percent.

6. Compute the air required with the stated excess flow.
With 20 percent excess air, the airflow required $= (0.20 + 1.00)$(airflow with no excess) $= 1.20(10.9663) = 13.1596$ lb of air per pound of coal burned. The airflow with no excess is obtained from step 2.

7. Compute the weight of the products of combustion.
The excess air passes through the furnace without taking part in the combustion and increases the weight of the products of combustion per pound of coal burned. Therefore, the weight of the products of combustion is the sum of the weight of the combustion products without the excess air and the product of (percent excess air)(air for perfect combustion, lb); or using the weights from steps 3 and 2, respectively, $= 11.9130 + (0.20)(10.9663) = 14.1063$ lb of gas per pound of coal burned with 20 percent excess air.

8. Compute the volume of the combustion products and the percent CO_2.
The volume of the excess air in the products of combustion is obtained by converting from the weight analysis to the volumetric analysis and correcting for temperature as in step 4, using the air weight from step 2 for perfect combustion and the excess-air percentage, or $(10.9663)(0.20)(359/28.95)$ $(2.15) = 58.5$ ft^3 (1.66 m^3). In this calculation, the value 28.95 is the molecular weight of air. The total volume of the products of combustion is the sum of the column for perfect combustion, step 4, and the excess-air volume, above, or $304.57 + 58.5 = 363.07$ ft^3 (10.27 m^3).

Using the procedure in step 5, the percent CO_2, wet basis, $= 53.8/363.07 = 14.8$ percent. The percent CO_2, dry basis, $= 53.8/(363.07 - 17.6) = 15.6$ percent.

Related Calculations.   Use the method given here when making combustion calculations for any type of coal—bituminous, semibituminous, lignite, anthracite, cannel, or coking—from any coal field in the world used in any type of furnace—boiler, heater, process, or waste-heat. When the air used for combustion contains moisture, as is usually true, this moisture is added to the combustion-formed moisture appearing in the products of combustion. Thus, for 80°F (300 K) air of 60 percent relative humidity, the moisture content is 0.013 lb per pound of dry air. This amount appears in the products of combustion for each pound of air used and is a commonly assumed standard in combustion calculations.

2.10 COMBUSTION OF FUEL OIL IN A FURNACE

A fuel oil has the following ultimate analysis: $C = 0.8543$, $H_2 = 0.1131$, $O_2 = 0.0270$, $N_2 = 0.0022$, $S = 0.0034$, and total $= 1.0000$. This fuel oil is burned in a steam-boiler furnace. Determine the weight of air required for theoretically perfect combustion, the weight of gas formed per pound of oil burned, and the volume of flue gas at the boiler exit temperature of 600°F (589 K) per pound of oil burned; the air required with 20 percent excess air and the volume of gas formed with this excess; and the CO_2 percentage in the flue gas on a dry and wet basis.

Calculation Procedure

1. Compute the weight of oxygen required per pound of oil.

The same general steps as given in the previous Calculation Procedure will be followed. Consult that procedure for a complete explanation of each step. Using the molecular weight of each element, the following table can be set up:

Element	$\times$	Molecular-weight ratio	=	Pounds O_2 required
C; 0.8543	$\times$	32/12	=	2.2781
H_2; 0.1131	$\times$	16/2	=	0.9048
O_2; 0.0270; decreases external O_2 required			=	−0.0270
N_2; 0.0022; is inert in combustion and is ignored				
S; 0.0034	$\times$	32/32	=	0.0034
Total 1.0000				
Pounds of external O_2 per pound fuel			=	3.1593

2. Compute the weight of air required for perfect combustion.

The weight of nitrogen associated with the required oxygen = (3.1593)(0.768/0.232) = 10.4583 lb. The weight of air required = 10.4583 + 3.1593 = 13.6176 lb per pound of oil burned.

3. Compute the weight of the products of combustion.

As before:

Fuel constituents + oxygen =			Products of combustion
C; 0.8543 + 2.2781	=	3.1324	CO_2
H_2; 0.1131 + 0.9148	=	1.0179	H_2O
O_2; 0.270; *not* a product of combustion			
N_2; 0.0022; inert but passes through furnace	=	0.0022	N_2
S; 0.0034 + 0.0034	=	0.0068	SO_2
Outside N_2 from step 2	=	10.458	N_2
Pounds of flue gas per pound of oil burned	=	14.6173	

4. Convert the flue-gas weight to volume.

As before:

Products	Weight, lb	Molecular weight	Temperature correction		Volume at 600°F, ft^3
CO_2	3.1324	44	(359/44)(3.1324)(2.15)	=	55.0
H_2O	1.0179	18	(359/18)(1.0179)(2.15)	=	43.5
N_2 (total)	10.460	28	(359/28)(10.460)(2.15)	=	288.5
SO_2	0.0068	64	(359/64)(0.0068)(2.15)	=	0.82
Cubic feet of flue gas per pound of oil burned				=	387.82

In this calculation, the temperature correction factor 2.15 = (absolute flue-gas temperature)/(absolute atmospheric temperature) = (600 + 460)/(32 + 460). The total weight of N_2 in the flue gas is the sum of the N_2 in the combustion air and the fuel, or 10.4583 + 0.0022 = 10.4605 lb.

5. Compute the CO_2 content of the flue gas.

The CO_2, wet basis, = 55.0/387.82 = 0.142, or 14.2 percent. The CO_2, dry basis, = 55.0/(387.2 − 43.5) = 0.160, or 16.0 percent.

6. Compute the air required with stated excess flow.

The pounds of air per pound of oil with 20 percent excess air = (1.20)(13.6176) = 16.3411 lb air per pound of oil burned.

7. *Compute the weight of the products of combustion.*
The weight of the products of combustion = product weight for perfect combustion, lb + (percent excess air)(air for perfect combustion, lb) = 14.6173 + (0.20)(13.6176) = 17.3408 lb flue gas per pound of oil burned with 20 percent excess air.

8. *Compute the volume of the combustion products and the percent CO_2.*
The volume of excess air in the products of combustion is found by converting from the weight to the volumetric analysis and correcting for temperature as in step 4, using the air weight from step 2 for perfect combustion and the excess-air percentage, or (13.6176)(0.20)(359/28.95)(2.15) = 72.7 ft³ (2.06 m³). Add this to the volume of the products of combustion found in step 4, or 387.82 + 72.70 = 460.52 ft³ (13.03 m³).

Using the procedure in step 5, the percent CO_2, wet basis, = 55.0/460.52 = 0.1192, 11.92 percent. The percent CO_2, dry basis, = 55.0/(460.52 − 43.5) = 0.1318, or 13.18 percent.

Related Calculations.   Use the method given here when making combustion calculations for any type of fuel oil—paraffin-base, asphalt-base, Bunker C, No. 2, 3, 4, or 5—from any source, domestic or foreign, in any type of furnace—boiler, heater, process, or waste-heat. When the air used for combustion contains moisture, as is usually true, this moisture is added to the combustion-formed moisture appearing in the products of combustion. Thus, for 80°F air of 60 percent relative humidity, the moisture content is 0.013 lb per pound of dry air. This amount appears in the products of combustion for each pound of air used and is a commonly assumed standard in combustion calculations.

2.11   COMBUSTION OF NATURAL GAS IN A FURNACE

A natural gas has the following volumetric analysis at 60°F: CO_2 = 0.004, CH_4 = 0.921, C_2H_6 = 0.041, N_2 = 0.034, and total = 1.000. This natural gas is burned in a steam-boiler furnace. Determine the weight of air required for theoretically perfect combustion, the weight of gas formed per pound of natural gas burned, and the volume of the flue gas at the boiler exit temperature of 650°F per pound of natural gas burned; the air required with 20 percent excess air and the volume of gas formed with this excess; and the CO_2 percentage in the flue gas on a dry and wet basis.

Calculation Procedure

1. *Compute the weight of oxygen required per pound of gas.*
The same general steps as given in the previous Calculation Procedures will be followed, except that they will be altered to make allowances for the differences between natural gas and coal.

The composition of the gas is given on a volumetric basis, which is the usual way of express-ing a fuel-gas analysis. To use the volumetric-analysis data in combustion calculations, they must be converted to a weight basis. This is done by dividing the weight of each component by the total weight of the gas. A volume of 1 ft³ of the gas is used for this computation. Find the weight of each component and the total weight of 1 ft³ as follows, using the properties of the combustion elements and compounds given in Table 2.1.

Component	Percent by volume	Density, lb/ft³	Component weight, lb = col. 2 × col. 3
CO_2	0.004	0.1161	0.0004644
CH_4	0.921	0.0423	0.0389583
C_2H_6	0.041	0.0792	0.0032472
N_2	0.034	0.0739	0.0025026
Total	1.000		0.0451725 lb/ft³

TABLE 2.1   Properties of Combustion Elements

Element or compound	Formula	Molecular weight	At 14.7 psia, 60°F		Nature		Heat value, Btu		
			Weight, lb/ft^3	Volume, ft^3/lb	Gas or solid	Combustible	Per pound	Per ft^3 at 14.7 psia, 60°F	Per mole
Carbon	C	12	—	—	S	Yes	14,540	—	174,500
Hydrogen	H$_2$	2.02*	0.0053	188	G	Yes	61,000	325	123,100
Sulfur	S	32	—	—	S	Yes	4,050	—	129,600
Carbon monoxide	CO	28	0.0739	13.54	G	Yes	4,380	323	122,400
Methane	CH$_4$	16	0.0423	23.69	G	Yes	24,000	1,012	384,000
Acetylene	C$_2$H$_2$	26	0.0686	14.58	G	Yes	21,500	1,483	562,000
Ethylene	C$_2$H$_4$	28	0.0739	13.54	G	Yes	22,200	1,641	622,400
Ethane	C$_2$H$_6$	30	0.0792	12.63	G	Yes	22,300	1,762	668,300
Oxygen	O$_2$	32	0.0844	11.84	G				
Nitrogen	N$_2$	28	0.0739	13.52	G				
Air†	—	29	0.0765	13.07	G				
Carbon dioxide	CO$_2$	44	0.1161	8.61	G				
Water	H$_2$O	18	0.0475	21.06	G				

*For most practical purposes, the value of 2 is sufficient.
†The molecular weight of 29 is merely the weighted average of the molecular weight of the constituents.
Source: P. W. Swain and L. N. Rowley, "Library of Practical Power Engineering" (collection of articles published in *Power*).

$$\text{Percent CO}_2 = 0.0004644/0.0451725 = 0.01026, \quad \text{or 1.03 percent}$$

$$\text{Percent CH}_4 \text{ by weight} = 0.0389583/0.0451725 = 0.8625, \quad \text{or 86.25 percent}$$

$$\text{Percent C}_2\text{H}_6 \text{ by weight} = 0.0032472/0.0451725 = 0.0718, \quad \text{or 7.18 percent}$$

$$\text{Percent N}_2 \text{ by weight} = 0.0025026/0.0451725 = 0.0554, \quad \text{or 5.54 percent}$$

The sum of the weight percentages = 1.03 + 86.25 + 7.18 + 5.54 = 100.00. This sum checks the accuracy of the weight calculation, because the sum of the weights of the component parts should equal 100 percent.

Next, find the oxygen required for combustion. Since both the CO$_2$ and N$_2$ are inert, they do not take part in the combustion; they pass through the furnace unchanged. Using the molecular weights of the remaining compounds in the gas and the weight percentages, we have:

Compound	×	Molecular-weight ratio	=	Pounds O$_2$ required
CH$_4$; 0.8625	×	64/16	=	3.4500
C$_2$H$_6$; 0.0718	×	112/30	=	0.2920
Pounds of external O$_2$ required per pound fuel			=	3.7420

In this calculation, the molecular-weight ratio is obtained from the equation for the combustion chemical reaction, or CH$_4$ + 2O$_2$ = CO$_2$ + 2H$_2$O, that is, 16 + 64 = 44 + 36, and C$_2$H$_6$ + 7½O$_2$ = 2CO$_2$ + 3H$_2$O, that is, 30 + 112 = 88 + 54.

2. *Compute the weight of air required for perfect combustion.*
The weight of nitrogen associated with the required oxygen = (3.742)(0.768/0.232) = 12.39 lb. The weight of air required = 12.39 + 3.742 = 16.132 lb per pound of gas burned.

3. *Compute the weight of the products of combustion.*
Use the following relation:

Fuel constituents	+	Oxygen	=	Products of combustion
CO_2; 0.0103; inert but passes through the furnace			=	0.010300
CH_4: 0.8625	+	3.45	=	4.312500
C_2H_6: 0.003247	+	0.2920	=	0.032447
N_2; 0.0554; inert but passes through the furnace			=	0.055400
Outside N_2 from step 2			=	12.390000
Pounds of flue gas per pound of natural gas burned			=	16.800347

4. *Convert the flue-gas weight to volume.*
The products of complete combustion of any fuel that do not contain sulfur are CO_2, H_2O, and N_2. Using the combustion equation in step 1, compute the products of combustion thus: $CH_4 + 2O_2 = CO_2 + H_2O$; $16 + 64 = 44 + 36$; or the CH_4 burns to CO_2 in the ratio of 1 part CH_4 to 44/16 parts CO_2. Since, from step 1, there is 0.03896 lb CH_4 per ft^3 natural gas, this forms $(0.03896)(44/16) = 0.1069$ lb CO_2. Likewise, for C_2H_6, $(0.003247)(88/30) = 0.00952$ lb. The total CO_2 in the combustion products $= 0.00464 + 0.1069 + 0.00952 = 0.11688$ lb, where the first quantity is the CO_2 in the fuel.

Using a similar procedure for the H_2O formed in the products of combustion by CH_4, $(0.03896)(36/16) = 0.0875$ lb. For C_2H_6, $(0.003247)(54/30) = 0.005816$ lb. The total H_2O in the combustion products $= 0.0875 + 0.005816 = 0.093316$ lb.

Step 2 shows that 12.39 lb N_2 is required per pound of fuel. Since 1 ft^3 of the fuel weighs 0.04517 lb, the volume of gas that weighs 1 lb is $1/0.04517 = 22.1$ ft^3. Therefore, the weight of N_2 per cubic foot of fuel burned $= 12.39/22.1 = 0.560$ lb. This, plus the weight of N_2 in the fuel, step 1, is $0.560 + 0.0025 = 0.5625$ lb N_2 in the products of combustion.

Next, find the total weight of the products of combustion by taking the sum of the CO_2, H_2O, and N_2 weights, or $0.11688 + 0.09332 + 0.5625 = 0.7727$ lb. Now convert each weight to cubic feet at 650°F, the temperature of the combustion products, or:

Products	Weight, lb	Molecular weight	Temperature correction		Volume at 650°F, ft^3
CO_2	0.11688	44	(379/44)(0.11688)(2.255)	=	2.265
H_2O	0.09332	18	(379/18)(0.09332)(2.255)	=	4.425
N_2 (total)	0.5625	28	(379/28)(0.5625)(2.255)	=	17.190
Cubic feet of flue gas per cubic foot of natural-gas fuel				=	23.880

In this calculation, the value of 379 is used in the molecular-weight ratio because at 60°F and 14.7 psia the volume of 1 lb of any gas = 379/gas molecular weight. The fuel gas used is initially at 60°F and 14.7 psia. The ratio $2.255 = (650 + 460)/(32 + 460)$.

5. *Compute the CO_2 content of the flue gas.*
The CO_2, wet basis, $= 2.265/23.88 = 0.0947$, or 9.47 percent. The CO_2, dry basis, $= 2.265/(23.88 - 4.425) = 0.1164$, or 11.64 percent.

6. *Compute the air required with the stated excess flow.*
The pounds of air per pound of natural gas with 20 percent excess air $= (1.20)(16.132) = 19.3584$ lb air per pound of natural gas, or $19.3584/22.1 = 0.875$ lb of air per cubic foot of natural gas (14.02 kg/m^3). See step 4 for an explanation of the value 22.1.

7. *Compute the weight of the products of combustion.*
Weight of the products of combustion = product weight for perfect combustion, lb + (percent excess air)(air for perfect combustion, lb) $= 16.80 + (0.20)(16.132) = 20.03$ lb.

8. *Compute the volume of the combustion products and the percent CO_2.*
The volume of excess air in the products of combustion is found by converting from the weight to the volumetric analysis and correcting for temperature as in step 4, using the air weight from step 2 for perfect combustion and the excess-air percentage, or $(16.132/22.1)(0.20)(379/28.95)(2.255) = 4.31$ ft³. Add this to the volume of the products of combustion found in step 4, or $23.88 + 4.31 = 28.19$ ft³ (0.80 m³).

Using the procedure in step 5, the percent CO_2, wet basis, $= 2.265/28.19 = 0.0804$, or 8.04 percent. The percent CO_2, dry basis, $= 2.265/(28.19 - 4.425) = 0.0953$, or 9.53 percent.

Related Calculations. Use the method given here when making combustion calculations for any type of gas used as a fuel—natural gas, blast-furnace gas, coke-oven gas, producer gas, water gas, sewer gas—from any source, domestic or foreign, in any type of furnace—boiler, heater, process, or waste-heat. When the air used for combustion contains moisture, as is usually true, this moisture is added to the combustion-formed moisture appearing in the products of combustion. Thus, for 80°F (300 K) air of 60 percent relative humidity, the moisture content is 0.013 lb per pound of dry air. This amount appears in the products of combustion for each pound of air used and is a commonly assumed standard in combustion calculations.

*2.12 STEADY-STATE CONTINUOUS PHYSICAL BALANCE WITH RECYCLE AND BYPASS

Feed to a distillation tower is 1000 lb · mol/h (0.126 kg · mol/s) of a solution of 35 mole percent ethylene dichloride (EDC) in xylene. There is not any accumulation in the tower. The overhead distillate stream contains 90 mole percent ethylene dichloride, and the bottoms stream contains 15 mole percent ethylene dichloride. Cooling water to the overhead condenser is adjusted to give a reflux ratio of 10:1 (10 mol reenters the column for each mole of overhead product). Heat to the reboiler, Fig. 2.5, is adjusted so that the recycle ratio is 5:1 (5 mol reenters the column for each mole of bottom product), with a 2:15 bypass (2 mol bypasses the reboiler for each 15 mol that passes through the reboiler). Determine the flow rate of the overhead product, bottoms product, overhead reflux reentering the column, bottoms recycle reentering the column, bottoms bypassing the reboiler, and the total bottoms.

Calculation Procedure

1. *Compute the bottoms product.*
Since this is a physical system with no change in chemical composition and no accumulation, any component may be followed through the system. Having been given the important values of the ethylene dichloride (X_F, X_D, X_W), use them as the basis of the calculation, with X representing the moles of ethylene dichloride in each stream.

Set up a total material balance thus: $F = D + W = 1000$ lb · mol/h (0.126 kg · mol/s), Eq. 2.1, where F = feed; D = product (i.e., distillate); W = product (i.e., bottoms), all expressed in lb · mol/h as shown in Fig. 2.5.

An ethylene dichloride balance is $FX_F = DX_D + WX_W$. Substituting given values, we have $1000(0.35) = 0.90D + 0.15W = 350$, Eq. 2.1. Solving Eqs. 2.1 and 2.2 simultaneously gives $D = 1000 - W$; $W = (350 - 0.90D)/0.15$; $D = 1000 - (350 - 0.90D)/0.15$; $D = 266.67$ lb · mol/h (0.034 kg·mol/s) distillate product; $W = 733.33$ lb · mol/h (0.092 kg · mol/s) bottoms product.

2. *Compute the reflux flow rate.*
Taking the tower overhead as a separate system, Fig. 2.5, we find $X_L/X_D = 10 = L/D$. Hence, $L = 10D = 10(266.67) = 2666.7$ lb · mol/h (0.34 kg · mol/s) reflux.

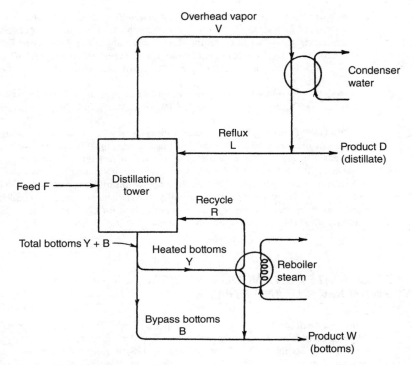

FIGURE 2.5   Distillation tower flow.

3. *Analyze the condenser.*
A total material balance around the condenser is input = output; or $V = D + L$, Fig. 2.5. Hence, $V = 266.67 + 2666.7 = 2933.37$ lb · mol/h (0.37 kg · mol/s) overhead vapor.

4. *Analyze the tower reboiler.*
Taking the tower reboiler as a separate system, Fig. 2.5, gives $X_R/X_W = 5 = R/W$; hence, $R = 5W = 5(733.33) = 3666.7$ lb · mol/h (0.46 kg · mol/s) bottoms recycle.

Also, $X_B/X_Y = 2/15 = B/Y$; $2Y = 15B$, Eq. 2.3. And a total material balance around the reboiler is input = output, or $Y = R + (W − B)$, Eq. 2.4. Solving Eqs. 2.3 and 2.4 simultaneously gives $B = 2/15(Y)$; $Y = R + [W − 2/15(Y)]$; $Y = [15(3666.7) + 15(733.3)]/17 = 3882.35$ lb · mol/h(0.49 kg · mol/s) reboiled bottoms. Then $Y + B = 4399.97$ lb · mol/h (0.55 kg · mol/s) total bottoms.

Related Calculations.   Use this general procedure to analyze distillation towers handling liquids similar to those considered here.

*2.13   STEADY-STATE CONTINUOUS PHYSICAL PROCESS BALANCE

The distillation tower of the previous calculation procedure has the temperature and thermal conditions shown in Fig. 2.6. The reboiler is heated by steam that condenses at 280°F (137.8°C). Cooling water enters the overhead condenser at 70°F (21.1°C) and leaves at 120°F (48.9°C). In the condenser, the overhead vapor condenses at 184°F (84.4°C) before being cooled at 175°F (79.4°C),

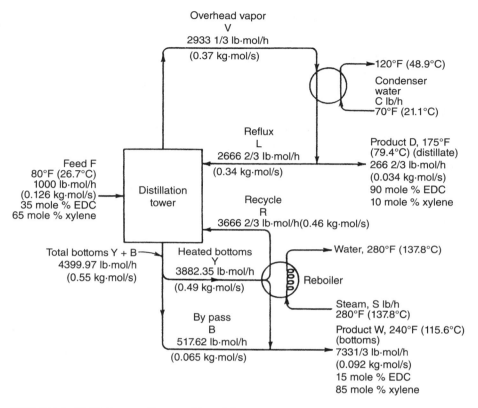

FIGURE 2.6 Distillation-tower flow quantities and flow rates.

the temperature of the liquid reflux and distillate product. The heat of condensation ΔH of the overhead vapor is 14,210 Btu/(lb·mol) [33.1 kJ/(kg·mol)], as given in a standard reference work, and 14,820 Btu/(lb·mol) [34.5 kJ/(kg·mol)] for the tower bottoms. The heat capacity of all liquid streams in this installation is 40 Btu/(lb·mol·°F) [167.4 kJ/(kg·mol·K)]. Determine the steam and cooling-water flow rates required.

Calculation Procedure

1. Set up a heat balance for the column.

Thus, heat in = heat in feed + heat in steam. Let the temperature basis for the calculation = 80°F (26.7°C) = t_b. Then, the heat in the feed = ΔH_F = (feed rate, lb/h) [heat capacity of the feed, Btu/(lb · mol · °F)](feed temperature, °F − temperature basis for the calculation, °F) = 1000(40)(80 − 80) = 0.

The enthalpy of vaporization of the steam is, from any standard properties, 924.74 Btu/lb (2151 kJ/kg) when there is complete condensation of the steam and the condensate leaves the reboiler at 280°F (137.8°C). Then the heat given up by the condensation of S lb (kg) of steam is ΔH_S = 924.74S Btu/h (271.0 W).

The heat out = heat in distillate + heat in bottoms + heat in water, all expressed in Btu/h (W). Using the same procedure as for the heat in the feed, we see the heat in the distillate = $\Delta H_D/DC\Delta t_d$, where C = distillate heat capacity, Btu/(lb·mol·°F) [kJ/(kg·°C)]; Δt_d = temperature change of the distillate, °F (°C). Or, $\Delta H_D = 266.7(40)(175 - 80) = 1,010,000$ Btu/h (295.9 kW). Likewise, for the bottoms, $\Delta H_W = 733.3(40)(240 - 80) = 4,690,000$ Btu/h (1374.2 kW).

For the water, ΔH_W = heat to condense overhead vapor + heat absorbed when cooling the condensed vapor from 184 to 175°F (84.4 to 79.4°C), all expressed in Btu/h (W). With a flow of 2933.3 lb/h (0.37 kg/s) of vapor, $\Delta H_W = 2933.3\ [14,210 + 40(184 - 175)] = 42,700,000$ Btu/h (12.5 MW).

2. *Analyze the heat balance.*

The heat balance is heat in = heat out, or $\Delta H_S = \Delta H_D + \Delta H_W + \Delta H_W$. Thus, $924.74S = 1,010,000 + 4,690,000 + 42,700,000$; $S = 52,300$ lb/h (6.59 kg/s) of steam.

For the water, $\Delta H_W = Cc\Delta t_c$, where C = water flow rate, lb/h; c = specific heat of water = 1.0 Btu/(lb·°F). Substituting the values gives $\Delta H_W = C(1)(120 - 70) = 42,700,000$; $C = 854,000$ lb/h (107.6 kg/s) of water.

SECTION 3
PHASE EQUILIBRIUM

A.K.S. Murthy, Eng. Sc.D.
Technology Fellow
BOC
Murray Hill, NJ

3.1 VAPOR-LIQUID EQUILIBRIUM RATIOS FOR IDEAL-SOLUTION BEHAVIOR

Assuming ideal-system behavior, calculate the K values and relative volatility for the benzene-toluene system at 373 K (212°F) and 101.3 kPa (1 atm).

Calculation Procedure

1. Determine the relevant vapor-pressure data.

Design calculations involving vapor-liquid equilibrium (VLE), such as distillation, absorption, or stripping, are usually based on vapor-liquid equilibrium ratios, or K values. For the ith species, K_i is defined as $K_i = y_i/x_i$, where y_i is the mole fraction of that species in the vapor phase and x_i is its mole fraction in the liquid phase. Sometimes the design calculations are based on relative volatility $\alpha_{i,j}$, which equals K_i/K_j, the subscripts i and j referring to two different species. In general, K values depend on temperature and pressure and the compositions of both phases.

When a system obeys Raoult's law and Dalton's law, it is known as an "ideal system" (see Related Calculations for guidelines). Then $K_i = P_i^\circ/P$, and $\alpha_{i,j} = P_i^\circ/P_j^\circ$, where P_i° is the vapor pressure of the (pure) ith component at the system temperature and P is the total pressure.

One way to obtain the necessary vapor-pressure data for benzene and toluene is to employ the Antoine equation:

$$\log_{10} P^\circ = A - B/(t + C)$$

(See also Section 1.) When P° is in millimeters of mercury and t is temperature in degrees Celsius, the following are the values for the constants A, B, and C:

	A	B	C
Benzene	6.90565	1211.033	220.790
Toluene	6.95464	1424.255	219.482

Then, at 373 K (i.e., 100°C),

$$\log_{10} P^\circ_{\text{benzene}} = 6.90565 - 1211.033/(100 + 220.790) = 3.1305$$

and P°_{benzene} therefore equals 1350.5 mmHg (180.05 kPa). Similarly,

$$\log_{10} P^\circ_{\text{toluene}} = 6.95464 - 1424.255/(100 + 219.482) = 2.4966$$

so P°_{toluene} equals 313.8 mmHg (41.84 kPa).

2. Divide vapor pressures by total pressure to obtain K values.
Total pressure P is 1 atm, or 760 mmHg. Therefore, $K_{\text{benzene}} = 1350.5/760 = 1.777$, and $K_{\text{toluene}} = 313.8/760 = 0.413$.

3. Calculate relative volatility of benzene with respect to toluene.
Divide the vapor pressure of benzene by that of toluene. Thus, $\alpha_{\text{benzene-toluene}} = 1350.5/313.8 = 4.304$.

Related Calculations.   Many systems deviate from the ideal solution behavior in either or both phases, so the K values given by $K_i = P_i^\circ/P$ are not adequate. The rigorous thermodynamic definition of K is

$$K_i = \gamma_i f_i^\circ / \phi_i P$$

where γ_i is the activity coefficient of the ith component in the liquid phase, f_i° is the fugacity of pure liquid i at system temperature T and pressure P, and ϕ_i is the fugacity coefficient of the ith species in the vapor phase.

In this definition, the activity coefficient takes account of nonideal liquid-phase behavior; for an ideal liquid solution, the coefficient for each species equals 1. Similarly, the fugacity coefficient represents deviation of the vapor phase from ideal gas behavior and is equal to 1 for each species when the gas obeys the ideal gas law. Finally, the fugacity takes the place of vapor pressure when the pure vapor fails to show ideal gas behavior, either because of high pressure or as a result of vapor-phase association or dissociation. Methods for calculating all three of these follow.

The vapor-phase fugacity coefficient can be neglected when the system pressure is low [e.g., less than 100 psi (689.5 kPa), generally] and the system temperature is not below a reduced temperature of 0.8. The pure-liquid fugacity is essentially equal to the vapor pressure at system temperatures up to a reduced temperature of 0.7. Unfortunately, however, many molecules (among them hydrogen fluoride and some organic acids) associate in the vapor phase and behave nonideally even under the preceding conditions. There is no widely recognized listing of all such compounds. As for nonideality in the liquid phase, perhaps the most important cause of it is hydrogen bonding. For general rules of thumb for predicting hydrogen bonding, see R. H. Ewell, J. M. Harrison, and L. Berg, *Ind. Eng. Chem.* 36(10):871, 1944.

3.2 FUGACITY OF PURE LIQUID

Calculate the fugacity of liquid hydrogen chloride at 40°F (277.4 K) and 200 psia (1379 kPa). (The role of fugacity in phase equilibrium is discussed under Related Calculations in Procedure 3.1).

Calculations Procedure

1. Calculate the compressibility factor.
For components whose critical temperature is greater than the system temperature,

$$f° = vP° \exp[V(P - P°)/RT]$$

where $f°$ is the pure-liquid fugacity, v is the fugacity coefficient for pure vapor at the system temperature, $P°$ is the vapor pressure at that temperature, V is the liquid molar volume, P is the system pressure, and T is the absolute temperature.

Thermodynamically, the fugacity coefficient is given by

$$\ln v = \int_0^{P°} \frac{Z - 1}{P} dP$$

where Z is the compressibility factor. This integral has been evaluated for several equations of state. For instance, for the Redlich-Kwong equation, which is very popular in engineering design and is employed here, the relationship is

$$\ln v = (Z - 1) - \ln(Z - BP°) - \frac{A^2}{B} \ln\left(1 + \frac{BP°}{Z}\right)$$

where $A^2 = 0.4278/T_r^{2.5} P_c$
$B = 0.0867/T_r P_c$
T_r = reduced temperature (T/T_c)
T_c = critical temperature
P_c = critical pressure

The compressibility factor Z is calculated by solving the following cubic equation (whose symbols are as defined above):

$$Z^3 - Z^2 + [A^2P° - BP°(1 + BP°)]Z + (A^2P°)(BP°) = 0$$

The critical temperature of HCl is 584°R, so the reduced temperature at 40°F is $(460 + 40)/584$ or 0.85616. The critical pressure of HCl is 1206.9 psia. Then, $A^2 = 0.4278/[0.85616^{2.5}(1206.9)] = 5.226 \times 10^{-4}$, and $B = 0.0867/(0.85616 \times 1206.9) = 8.391 \times 10^{-5}$. The vapor pressure $P°$ of HCl at 40°F is 423.3 psia, so $A^2P° = 0.2212$, $BP° = 0.03552$, $A^2P° - BP°(1 + BP°) = 0.1844$, and $(A^2P)(BP°) = 7.857 \times 10^{-3}$. The cubic equation thus becomes $Z^3 - Z^2 + 0.1844Z + 7.856 \times 10^{-3} = 0$.

This equation can be solved straightforwardly or by trial and error. The largest real root is the compressibility factor for the vapor; in this case, $Z = 0.73431$.

2. Calculate the fugacity coefficient.
Using the preceding relationship based on the Redlich-Kwong equation,

$$\ln v = (0.73431 - 1) - \ln(0.73431 - 0.03552) - \frac{5.226 \times 10^{-4}}{8.391 \times 10^{-5}} \ln\left(1 + \frac{0.03552}{0.73431}\right)$$

$$= -0.2657 + 0.3584 - 0.2942 = -0.2015$$

Therefore, $v = 0.8175$.

3. Calculate the fugacity.
The density of saturated vapor at 40°F is 55 lb/ft³, and the molecular weight of HCl is 36.46. Therefore, the liquid molar volume V is $36.46/55 = 0.663$ ft³/(lb · mol). The gas constant R is 10.73 (psia)(ft³)/(lb · mol)(°R). Therefore,

$$\exp[V(P - P°)/RT] = \exp\left[\frac{0.663(200 - 423.3)}{10.73(460 + 40)}\right] = 0.9728$$

Finally, using the equation for fugacity at the beginning of this problem, $f° = 0.8175 \times 423.3 \times 0.9728 = 336.6$ psi (2321 kPa).

The exponential term, 0.9728 in this equation, is known as the "Poynting correction." It is greater than unity if system pressure is greater than the vapor pressure. The fugacity coefficient v is always less than unity. Depending on the magnitudes of v and the Poynting correction, fugacity of pure liquid $f°$ can thus be greater or less than the vapor pressure.

Related Calculations.   This procedure is valid only for those components whose critical temperature is above the system temperature. When the system temperature is instead above the critical temperature, generalized fugacity-coefficient graphs can be used. However, such an approach introduces the concept of hypothetical liquids. When accurate results are needed, experimental measurements should be made. The Henry constant, which can be experimentally determined, is simply $\gamma^\infty f°$, where γ^∞ is the activity coefficient at infinite dilution (see Procedure 3.8).

Use of generalized fugacity coefficients (e.g., see Procedure 1.18) eliminates some computational steps. However, the equation-of-state method used here is easier to program on a programmable calculator or computer. It is completely analytical, and use of an equation of state permits the computation of all the thermodynamic properties in a consistent manner.

3.3   NONIDEAL GAS-PHASE MIXTURES

Calculate the fugacity coefficients of the components in a gas mixture containing 80% HCl and 20% dichloromethane (DCM) at 40°F (277.4 K) and 200 psia (1379 kPa).

Calculation Procedure

1. Calculate the compressibility factor for the mixture.
In a manner similar to that used in the previous problem, an expression for the fugacity coefficient in vapor mixtures can be derived from any equation of state applicable to such mixtures. If the Redlich-Kwong equation of state is used, the expression is

$$\ln\phi_i = (Z - 1)\frac{B_i}{B} - \ln(Z - BP) - \frac{A^2}{B}\left(\frac{2A_i}{A} - \frac{B_i}{B}\right)\ln\left(1 + \frac{BP}{Z}\right)$$

where ϕ_i = the fugacity coefficient of the ith component in the vapor
$A_i^2 = 0.4278/T_{r,i}^{2.5}P_{c,i}$
$B_i = 0.0867/T_{r,i}P_{c,i}$
$A = \sum y_i A_i$
$B = \sum y_i B_i$
$T_{r,i}$ = the reduced temperature of component i ($T/T_{c,i}$)
$T_{c,i}$ = the critical temperature of component i
$P_{c,i}$ = the critical pressure of component i
y_i = the mole fraction of component i in the vapor mixture.

The compressibility factor Z is calculated by solving the cubic equation:

$$Z^3 - Z^2 + [A^2P - BP(1 + BP)]Z + (A^2P)(BP) = 0$$

The relevant numerical inputs are:

	HCl	DCM
Critical temperature $T_{c,i}$, °R	584	933
Critical pressure $P_{c,i}$, psia	1206.9	893
Reduced temperature $T_{r,i}$ at 40°F	0.8562	0.5359
A_i, equal to $(0.4278/T_{r,i}^{2.5}P_{c,i})^{1/2}$	0.02286	0.04774
B_i, equal to $0.0867/T_{r,i}P_{c,i}$	8.390×10^{-5}	1.812×10^{-4}
Mole fraction y_i	0.8	0.2

Then, $A = y_1A_1 + y_2A_2 = 0.0278$, $B = y_1B_1 + y_2B_2 = 1.0336 \times 10^{-4}$, $A^2P = (0.0278)^2(200) = 0.1546$, $BP = (1.0336 \times 10^{-4})(200) = 0.02067$, $A^2P - BP(1 + BP) = 0.1335$, and $(A^2P) \times (BP) = 0.003195$. The cubic equation becomes $Z^3 - Z^2 + 0.1335Z + 0.003195 = 0$.

This equation can be solved straightforwardly or by trial and error. The largest real root is the compressibility factor; in this case, $Z = 0.8357$.

2. Calculate the fugacity coefficients.
Substituting into the preceding relationship based on the Redlich-Kwong equation,

$$\ln\phi_i = (0.8357 - 1)B_i/1.0336 \times 10^{-4} - \ln(0.8357 - 0.02067)$$
$$- (0.0278^2/1.0336 \times 10^{-4})[(2A_i/0.0278)$$
$$- (B_i/1.0336 \times 10^{-4})][\ln(1 + 0.02067/0.8357)]$$
$$= -1589.6B_i + 0.2045 - 0.1827(71.94A_i - 9674.9B_i)$$

Letting subscript 1 correspond to HCl and subscript 2 to DCM, the equation yields $\ln \phi_1 = -0.08095$ and $\ln \phi_2 = -0.3905$. Therefore, $\phi_1 = 0.9222$ and $\phi_2 = 0.6767$.

Related Calculations. Certain compounds, such as acetic acid and hydrogen fluoride, are known to form dimers, trimers, or other oligomers by association in the vapor phase. Simple equations of state are not adequate for representing the nonideality in systems containing such compounds. Unfortunately, there is no widely recognized listing of all such compounds.

3.4 NONIDEAL LIQUID MIXTURES

Calculate the activity coefficients of chloroform and acetone at 0°C in a solution containing 50 mol % of each component, using the Wilson-equation model. The Wilson constants for the system (with subscript 1 pertaining to chloroform and subscript 2 to acetone) are

$$\lambda_{1,2} - \lambda_{1,1} = -332.23 \, \text{cal/(g} \cdot \text{mol)}[-1390 \, \text{kJ/(kg} \cdot \text{mol)}]$$
$$\lambda_{1,2} - \lambda_{2,2} = -72.20 \, \text{cal/(g} \cdot \text{mol)}[-302.1 \, \text{kJ/(kg} \cdot \text{mol)}]$$

Calculation Procedure

1. Compute the $G_{i,j}$ parameters for the Wilson equation.
General engineering practice is to establish liquid-phase nonideality through experimental measurement of vapor-liquid equilibrium. Models with adjustable parameters exist for adequately representing most nonideal-solution behavior. Because of these models, the amount of experimental information needed is not excessive (see Procedure 3.9, which shows procedures for calculating such parameters from experimental data).

One such model is the Wilson-equation model, which is applicable to multicomponent systems while having the attraction of entailing only parameters that can be calculated from binary data alone. Another attraction is that the Wilson constants are approximately independent of temperature. This model is

$$\ln \gamma_i = 1 - \ln \sum_{j=1}^{N} x_j G_{j,i} - \sum_{j=1}^{N} \left(x_j G_{i,j} / \sum_k x_k G_{k,j} \right)$$

where γ_i is the activity coefficient of the ith component, x_i is the mole fraction of that component, and $G_{i,j} = V_i / V_j \exp\left[-(\lambda_{i,j} - \lambda_{j,j})/RT\right]$, with V_i being the liquid molar volume of the ith component and $(\lambda_{i,j} - \lambda_{j,j})$ being the Wilson constants. Note that $\lambda_{i,j} = \lambda_{j,i}$, but that $G_{i,j}$ is not equal to $G_{j,i}$.

For a binary system, $N = 2$ and $(x_i + x_j) = 1.0$, and the model becomes

$$\ln \gamma_1 = -\ln (x_1 + x_2 G_{2,1}) + x_2 \left[\frac{G_{2,1}}{x_1 + x_2 G_{2,1}} - \frac{G_{1,2}}{x_2 + x_1 G_{1,2}} \right]$$

and

$$\ln \gamma_2 = -\ln (x_2 + x_1 G_{1,2}) + x_1 \left[\frac{G_{1,2}}{x_2 + x_1 G_{1,2}} - \frac{G_{2,1}}{x_1 + x_2 G_{2,1}} \right]$$

where

$$G_{1,2} = \frac{V_1}{V_2} \exp\left(-\frac{\lambda_{1,2} - \lambda_{2,2}}{RT} \right) \quad \text{and} \quad G_{2,1} = \frac{V_2}{V_1} \exp\left(-\frac{\lambda_{1,2} - \lambda_{1,1}}{RT} \right)$$

At 0°C, $V_1 = 71.48$ cc/(g · mol) and $V_2 = 78.22$ cc/(g · mol). Therefore, for the chloroform/acetone system,

$$G_{1,2} = \frac{71.48}{78.22} \exp\left(-\frac{(-72.20)}{(1.9872)(273.15)} \right) = 1.0438$$

and

$$G_{2,1} = \frac{78.22}{71.48} \exp\left(-\frac{(-332.23)}{(1.9872)(273.15)} \right) = 2.0181$$

2. Calculate the activity coefficients.
The mole fractions x_1 and x_2 each equal to 0.5. Therefore, $x_1 + x_2 G_{2,1} = 1.509$, and $x_2 + x_1 G_{1,2} = 1.022$. Substituting into the Wilson model,

$$\ln \gamma_1 = -\ln 1.509 + 0.5\left(\frac{2.0181}{1.509} - \frac{1.0438}{1.022} \right) = -0.2534$$

Therefore, $\gamma_1 = 0.7761$.

$$\ln \gamma_2 = -\ln 1.022 + 0.5\left(\frac{1.0438}{1.022} - \frac{2.0181}{1.509} \right) = -0.1798$$

Therefore, $\gamma_2 = 0.8355$.
This system, where the activity coefficients are less than unity, is an example of negative deviation from ideal behavior.

Related Calculations. When the Wilson model is used for systems with more than two components, it is important to remember that the $G_{i,j}$ summations must be made over every possible pair of components (also, remember that $G_{i,j} \neq G_{j,i}$).

A limitation on the Wilson equation is that it is not applicable to systems having more than one liquid phase. The NRTL model, which is similar to the Wilson equation, may be used for systems forming two liquid phases.

An older model for predicting liquid-phase activity coefficients is that of Van Laar, in which (for two components) $\ln \gamma_1 = A_{1,2}/(1 + A_{1,2} x_1/A_{2,1} x_2)^2$, with the As being constants to be determined from experimental data. This model can handle systems having more than one liquid phase. Another older model is that of Margules, available in "two-suffix" and "three-suffix" versions. These are (for two components), respectively, $\ln \gamma_1 = A x_2^2$ and $\ln \gamma_1 = x_2^2 [A_{1,2} + 2(A_{2,1} - A_{1,2})x_1]$. Ternary versions of these older models are available, but extending the Van Laar and Margules correlations to multicomponent systems is in general rather awkward.

Fredenslund and coworkers have developed the UNIFAC correlation, which is satisfactory for those systems covered by the extensive amount of experimental data in their work. For details, see A. Fredenslund, J. Gmehling, and P. Rasumssen, *Vapor-Liquid Equilibria Using UNIFAC*, Elsevier Scientific Publishing Co., Amsterdam, 1977.

3.5   *K VALUE FOR IDEAL LIQUID PHASE, NONIDEAL VAPOR PHASE*

Assuming the liquid phase but not the vapor phase to be ideal, calculate the K values for HCl and dichloromethane (DCM) in a system at 200 psia (1379 kPa) and 40°F (277.4 K) and whose vapor composition is 80 mol % HCl. Also calculate the relative volatility. Compare the calculated values with those which would exist if the system showed ideal behavior.

Calculation Procedure

1. *Set out the relevant form of the thermodynamic definition of K value.*
Refer to the rigorous definition of K value given under Related Calculations in Procedure 3.1. When the liquid phase is ideal, then $\gamma_i = 1$. Thus the relevant form is

$$K_i = f_i^\circ/\phi_i P$$

where f_i° is the fugacity of pure liquid i at system temperature and pressure, ϕ_i is the fugacity coefficient of the ith species in the vapor phase, and P is the system pressure. Let subscript 1 pertain to HCl and subscript 2 pertain to DCM.

2. *Determine the pure-liquid fugacities and the vapor-phase fugacity coefficients.*
The fugacity of HCl at 200 psia and 40°F was calculated in Procedure 3.2 to be 336.6 psi. The same procedure can be employed to find f_2°, the pure-liquid fugacity of DCM. With reference to that example, $A^2 = 0.04774$ and $B_2 = 1.812 \times 10^{-4}$; the vapor pressure of DCM at 40°F is 3.28 psi. The resulting cubic equation for compressibility factor yields a Z of 0.993; the Redlich-Kwong relationship yields a v_2 of 0.993. The liquid molar volume of DCM at 40°F is 1.05 ft³/(lb · mol), so the Poynting correction is calculated to be 1.039, and the resulting value f_2° emerges as 3.38 psi (23.3 kPa). As for the fugacity coefficients, they were calculated in Procedure 3.3 to be $\phi_1 = 0.9222$ and $\phi_2 = 0.6767$.

3. *Calculate the K values and relative volatility.*
Using the equation in step 1, $K_1 = 336.6/(0.9222)(200) = 1.82$, and $K_2 = 3.38/(0.6767)(200) = 0.025$. From the definition in Procedure 3.1, relative volatility $\alpha_{1,2} = K_1/K_2 = 1.82/0.025 = 72.8$.

4. *Compare these results with those which would prevail if the system were ideal.*
As indicated in Procedure 3.1, the K value based on ideal behavior is simply the vapor pressure of the component divided by the system pressure. Vapor pressures of HCl and DCM at 40°F are 423.3 and 3.28 psia, respectively. Thus $K_1^{\text{ideal}} = 423.3/200 = 2.12$, $K_2^{\text{ideal}} = 3.28/200 = 0.016$, and $\alpha_{1,2}^{\text{ideal}} = 2.12/0.016 = 133$.

It may be seen that under the system conditions that prevail in this case, the actual relative volatility is considerably lower than it would be if the system were ideal.

3.6   K VALUE FOR IDEAL VAPOR PHASE, NONIDEAL LIQUID PHASE

Assuming the vapor phase but not the liquid phase to be ideal, calculate the K values for ethanol and water in an 80% ethanol solution at 500 mmHg (66.7 kPa) and 70°C (158°F, 343 K). Also calculate the relative volatility, and compare the calculated values with those which would exist if the system showed ideal behavior.

Calculation Procedure

1. *Set out the relevant form of the thermodynamic definition of K value.*
At this low system pressure, the vapor-phase nonideality is negligible. Since neither component has a very high vapor pressure at the system temperature, and since the differences between the vapor pressures and the system pressure are relatively small, the pure-liquid fugacities can be taken to be essentially the same as the vapor pressures.

Refer to the rigorous definition of K value given under Related Calculations in Procedure 3.1. Taking into account the assumptions in the preceding paragraph, this definition simplifies into $K_i = \gamma_i P_i / P$, where γ_i is the activity coefficient of the ith component in the liquid phase, P_i is the vapor pressure of that component, and P is the system pressure.

2. *Determine the activity coefficients and the vapor pressures.*
The activity coefficients can be calculated from the Wilson-equation model, as discussed in Procedure 3.4, or by one of the other methods discussed in the same example under Related Calculations. For instance, the experimentally determined Van Laar constants for the ethanol/water system are $A_{1,2} = 1.75$ and $A_{2,1} = 0.91$, where subscripts 1 and 2 refer to ethanol and water, respectively. So the Van Laar model discussed there becomes $\ln \gamma_1 = 1.75/[1 + (1.75)(0.08)/(0.91)(0.2)]^2$, and (by interchanging subscripts) $\ln \gamma_2 = 0.91/[1 + (0.91)(0.2)/(1.75)(0.8)]^2$. Accordingly, $\gamma_1 = 1.02$, and $\gamma_2 = 2.04$.

From tables, the vapor pressures of ethanol and water at 70°C are 542 and 233 mmHg, respectively.

3. *Calculate the K values and relative volatility.*
Using the equation in step 1, $K_1 = (1.02)(542)/500 = 1.11$, and $K_2 = (2.04)(233)/500 = 0.95$.

From the definition in Procedure 3.1, relative volatility $\alpha_{1,2} = K_1/K_2 = 1.11/0.95 = 1.17$.

4. *Compare these results with those which would prevail if the system were ideal.*
As indicated in that example, the K value based on ideal behavior is simply the vapor pressure of the component divided by the system pressure. Thus $K_1^{ideal} = 542/500 = 1.084$, $K_2^{ideal} = 233/500 = 0.466$, and $\alpha_{1,2}^{ideal} = 1.084/0.466 = 2.326$. Under the system conditions that prevail in this case, actual relative volatility is considerably lower than the ideal-system value.

Related Calculations.   As computer-based computation has become routine, a growing trend in the determination of K values has been the use of cubic equations of state, such as the Peng-Robinson, for calculating the fugacities of the components in each phase. Such calculations are mathematically complex and involve iteration.

3.7   THERMODYNAMIC CONSISTENCY OF EXPERIMENTAL VAPOR-LIQUID EQUILIBRIUM DATA

Vapor-liquid equilibrium data for the ethanol/water system (subscripts 1 and 2, respectively) at 70°C (158°F, 343 K) are given in the three left columns of Table 3.1. Check to see if the data are thermodynamically consistent.

TABLE 3.1 Ethanol/Water System at 70°C (Procedures 3.7 and 3.8)

	Experimental data		Calculated results				
Pressure P, mmHg	Mole fraction ethanol in liquid x_1	Mole fraction ethanol in vapor y_1	γ_1, $Py_1/P_1^\circ x_1$	γ_2, $Py_2/P_2^\circ x_2$	$\ln(\gamma_1/\gamma_2)$	$\ln\gamma_1/x_2^2$	$-\ln\gamma_2/x_1^2$
362.5	0.062	0.374	4.034	1.038	1.357	1.585	—
399.0	0.095	0.439	3.402	1.062	1.165	1.495	—
424.0	0.131	0.482	2.878	1.085	0.976	1.400	—
450.9	0.194	0.524	2.247	1.143	0.676	1.246	—
468.0	0.252	0.552	1.891	1.203	0.452	1.139	—
485.5	0.334	0.583	1.564	1.305	0.181	1.008	—
497.6	0.401	0.611	1.399	1.387	0.009	0.936	—
525.9	0.593	0.691	1.131	1.714	−0.416	—	−1.532
534.3	0.680	0.739	1.071	1.870	−0.557	—	−1.354
542.7	0.793	0.816	1.030	2.070	−0.698	—	−1.157
543.1	0.810	0.826	1.022	2.135	−0.737	—	−1.156
544.5	0.943	0.941	1.002	2.419	−0.881	—	−0.993
544.5	0.947	0.945	1.002	2.425	−0.883	—	−0.988

Calculation Procedure

1. Select the criterion to be used for thermodynamic consistency.
Deviations from thermodynamic consistency arise as a result of experimental errors. Impurities in the samples used for vapor-liquid equilibrium measurements are often the source of error. A complete set of vapor-liquid equilibrium data includes temperature T, pressure P, liquid composition x_i, and vapor composition y_i. Usual practice is to convert these data into activity coefficients by the following equation, which is a rearranged form of the equation that rigorously defines K values (i.e., defines the ratio y_i/x_i under Related Calculations in Procedure 3.1):

$$\gamma_i = \phi_i P y_i / f_i^\circ x_i$$

The fugacity coefficients ϕ_i are estimated using procedures described in Procedure 3.3. Fugacity f_i° of pure liquid is calculated using procedures described in Procedure 3.2. Assuming that ϕ_i and f_i° are correctly calculated, the γ_i obtained using the preceding equation must obey the Gibbs-Duhem equation. The term "thermodynamically consistent data" is used to refer to data that obey that equation.

At constant temperature, the Gibbs-Duhem equation can be rearranged to give the approximate equality

$$\int_0^1 \ln \frac{\gamma_1}{\gamma_2} dx_1 = 0$$

(This would be an exact equality if both temperature and pressure were constant, but that would be inconsistent with the concept of vapor-liquid equilibrium.) In other words, the net area under the curve in (γ_1/γ_2) versus x_1 should be zero. This means that the area above and below the x axis must be equal. Since real data entail changes in system pressure and are subject to experimental errors and errors in estimating f_i° and ϕ_i, the preceding requirement cannot be expected to be exactly satisfied. However, the deviation should not be more than a few percent of the total absolute area.

2. Determine the system activity coefficients, and plot the natural logarithm of their ratio against liquid-phase ethanol content.
As in the previous example, and for the reasons discussed there, the system pressures and vapor pressures in the present example are such that the vapor pressure may be used for f° and $\phi_i = 1$.

Activity coefficients calculated from the experimental data, as well as ln (γ_1/γ_2) values, are also given in Table 3.1. A plot of ln (γ_1/γ_2) versus x_1 is shown in Fig. 3.1.

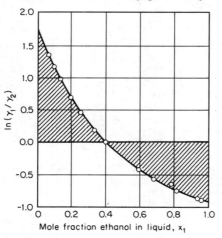

FIGURE 3.1   Area test for thermodynamic consistency of data on ethanol(1)/water(2) system (Procedure 3.7).

3. Determine and evaluate the net area under the plot.
By measurement, the area above the x axis is 0.295, and the area below it is 0.325. Net area is $0.295 - 0.325 = 0.03$. The total absolute area is $0.295 + 0.325 = 0.62$. Net area as a percent of total area is thus 5 percent. This figure is small enough that the data can be assumed to be thermodynamically consistent.

Related Calculations.   When vapor-liquid equilibrium data are taken under isobaric rather than isothermal conditions, as is often the case, the right-hand side of the preceding Gibbs-Duhem equation cannot as readily be taken to approximate zero. Instead, the equation should be taken as

$$\int_0^1 \ln\left(\frac{\gamma_1}{\gamma_2}\right)dx = \int_{x=0}^{x=1}\left(\frac{\Delta H}{RT^2}\right)dT$$

where ΔH is the heat of mixing.

For some systems, the integral on the right-hand side may indeed be neglected, i.e., set equal to zero. These include systems consisting of chemically similar components with low values for the heat of mixing and ones in which the boiling points of pure components are close together. In general, however, this will not be the case.

Since the enthalpy of mixing necessary for the evaluation of the integral is often not available, the integral can instead be estimated as $1.5(\Delta T_{max})/T_{min}$. Here, T_{min} is the lowest-boiling temperature in the isobaric system. Usually, this will be the boiling temperature of the lower-boiling component. In cases of low-boiling azeotropes, it is the boiling temperature of the azeotrope. Moreover, ΔT_{max} is the maximum difference of boiling points in the total composition range of the isobaric system. Usually, this will be the boiling-point difference of the pure components. For azeotropic systems, ΔT_{max} is the difference between the boiling temperature of the azeotrope and the component boiling most distant from it.

The preceding test for thermodynamic data treats the data set as a whole. It does not determine whether individual data points are consistent. A point-consistency test has been proposed, but it is too cumbersome for manual calculations.

3.8   ESTIMATING INFINITE-DILUTION ACTIVITY COEFFICIENTS

Estimate the infinite-dilution activity coefficients of ethanol and water at 70°C (158°F, 343 K) using the data given in the left three columns of Table 3.1.

Calculation Procedure

1. Determine the system activity coefficients, calculate the ratios ln γ_1/x_2^2, $-$ln γ_2/x_1^2, and ln (γ_1/γ_2), and plot them against x_1.
In a binary system, the infinite-dilution activity coefficients are defined as

$$\gamma_1^\infty = \lim_{x_1 \to 0} \gamma_1 \quad \text{and} \quad \gamma_2^\infty = \lim_{x_1 \to 1} \gamma_2$$

They are generally calculated for two reasons: First, activity coefficients in the very dilute range are experimentally difficult to measure but are commonly needed when designing separation systems, and estimation of infinite-dilution activity coefficients is needed for meaningful extrapolation of whatever experimental data are available. Second, these coefficients are useful when estimating the parameters that are required in several mathematical models used for determining activity coefficients.

One way to estimate infinite-dilution activity coefficients is related to the preceding example on thermodynamic consistency; in addition, it takes advantage of the fact that the plot of $\ln \gamma_1 / x_2^2$ versus x_1 is considerably more linear than is $\ln \gamma_i$ versus x_1.

The first step is to calculate $\ln \gamma_1 / x_2^2$, $-\ln \gamma_2 / x_1^2$, and $\ln (\gamma_1 / \gamma_2)$. The results are shown in Table 3.1 and plotted as a function of x_1 in Fig. 3.2.

2. Extrapolate the two curves relating to γ_1 so that they converge on a common point at $x_1 = 0$ and the two curves relating to γ_2 so that they converge at $x_1 = 1$. Determine these two points of convergence and find their antilogarithms.

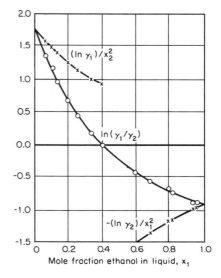

FIGURE 3.2   Determining infinite-dilution activity coefficients by extrapolation (Procedure 3.8).

The extrapolations are carried out in Fig. 3.2. The point of convergence along $x_1 = 0$ that corresponds to $\ln \gamma_1^{\infty}$ is 1.75, so γ_1^{∞} is 5.75. The point of convergence along $x_1 = 1$ that corresponds to $-\ln \gamma_2^{\infty}$ is -0.91, so γ_2^{∞} is 2.48.

3.9   ESTIMATING THE PARAMETERS FOR THE WILSON-EQUATION MODEL FOR ACTIVITY COEFFICIENTS

Calculate the Wilson constants for the ethanol/water system using the infinite-dilution activity coefficients calculated in the preceding example: $\gamma_1^{\infty} = 5.75$ and $\gamma_2^{\infty} = 2.48$, with subscript 1 pertaining to ethanol.

Calculation Procedure

1. Rearrange the Wilson-equation model so that its constants can readily be calculated from infinite-dilution activity coefficients.
In Procedure 3.4, given Wilson constants were employed in the Wilson-equation model to calculate the activity coefficients for the two components of a binary nonideal liquid mixture. The present example, in essence, reverses the procedure; it employs known activity coefficients (at infinite dilution) in order to calculate Wilson constants, so that these can be employed to determine activity coefficients in other situations concerning the same two components.

As shown in Procedure 3.4, $G_{1,2} = (V_1 / V_2) \exp [-(\lambda_{1,2} - \lambda_{2,2}) RT]$. This can be rearranged into $\lambda_{1,2} - \lambda_{2,2} = -RT \ln (G_{1,2} V_2 / V_1)$. Similarly, $\lambda_{1,2} - \lambda_{1,1} = -RT \ln(G_{2,1} V_1 / V_2)$. Moreover, at infinite dilution, when x_1 and x_2, respectively, equal zero, the Wilson-equation model becomes $G_{1,2} + \ln G_{2,1} = 1 - \ln \gamma_1^{\infty}$, and $G_{2,1} + \ln G_{1,2} = 1 - \ln \gamma_2^{\infty}$.

Thus the problem consists of first calculating $G_{1,2}$ and $G_{2,1}$ from the known values of γ_1^{∞} and γ_2^{∞} and then employing the calculated G values to determine the Wilson constants, $\lambda_{1,2} - \lambda_{1,1}$ and $\lambda_{1,2} - \lambda_{2,2}$.

2. *Calculate the G values.*
This calculation, involving a transcendental function, must be made by trial and error. Since $\ln \gamma_1^\infty = \ln 5.75 = 1.75$ and $\ln \gamma_2^\infty = \ln 2.48 = 0.91$, the preceding G equations become $G_{1,2} + \ln G_{2,1} = 1 - 1.75 = -0.75$, and $G_{2,1} + \ln G_{1,2} = 1 - 0.91 = 0.09$. The procedure consists of guessing a value for $G_{1,2}$, then using this value in the first equation to calculate a $G_{2,1}$, then using this $G_{2,1}$ in the second equation to calculate a $G_{1,2}$ (designated $G_{1,2}^*$), and repeating these steps until the difference between (guessed) $G_{1,2}$ and (calculated) $G_{1,2}^*$ becomes sufficiently small.

Start with $G_{1,2}$ guessed to be 1.0. Then $G_{2,1} = \exp(-0.75 - 1) = 0.1738$, $G_{1,2}^* = \exp(0.09 - 0.1738) = 0.9196$, and $G_{1,2}^* - G_{1,2} = 0.9196 - 1 = -0.0804$. Next, guess $G_{1,2}$ to be 0.9. Then $G_{2,1} = 0.192$, $G_{1,2}^* = 0.903$, and $G_{1,2}^* - G_{1,2} = 0.003$. Next, guess $G_{1,2}$ to be 0.9036 (by linear interpolation). Then $G_{2,1} = 0.1914$, $G_{1,2}^* = 0.9036$, and $G_{1,2}^* - G_{1,2} = 0$. Hence the solution is $G_{1,2} = 0.9036$ and $G_{2,1} = 0.1914$.

3. *Calculate the Wilson constants.*
At 70°C (343 K), the temperature at which the activity coefficients were determined, $V_1 = 62.3$ cc/(g · mol), and $V_2 = 18.5$ cc/(g · mol). Then $\lambda_{1,2} - \lambda_{2,2} = -RT \ln (G_{1,2} V_1/V_2) = -(1.987)343 \ln [0.9036(18.5)/62.3] = 897$ cal/(g · mol) [3724 kJ/(kg · mol)], and $\lambda_{1,2} - \lambda_{1,1} = -(1.987)343 \ln [0. 1914(62.3)/18.5] = 299$ cal/(g · mol) [1251 kJ/(kg · mol)].

Related Calculations. The constants for the binary Margules and Van Laar models for predicting activity coefficients (see Related Calculations under Procedure 3.4) are simply the natural logarithms of the infinite-dilution activity coefficients: $A_{1,2} = \ln \gamma_1^\infty$ and $A_{2,1} = \ln \gamma_2^\infty$

The Wilson, Margules, and Van Laar procedures described in this example are suitable for manual calculation. However, to take full advantage of all the information available from whatever vapor-liquid equilibrium data are at hand, statistical procedures for estimating the parameters should instead be employed, with the aid of digital computers. These procedures fall within the domain of nonlinear regression analysis.

In such an analysis, one selects a suitable objective function and then varies the parameters so as to maximize or minimize the function. Theoretically, the objective function should be derived using the statistical principles of maximum-likelihood estimation. In practice, however, it is satisfactory to use a weighted-least-squares analysis, as follows:

Usually, the liquid composition and the temperature are assumed to be error-free independent variables. Experimental values of vapor composition and total pressure are employed to calculate vapor-phase fugacity coefficients and pure-liquid fugacities. Then one uses trial and error in guessing parameters in the activity-coefficient model to be determined so as to arrive at parameter values that minimize the weighted sums of the squared differences between (1) experimental values of total pressure and of vapor mole fractions and (2) the values of pressure and mole fraction that are calculated on the basis of the model-generated activity coefficients. For several algorithms for minimizing the sum of squares, see D. M. Himmelblau, *Applied Nonlinear Programming*, McGraw-Hill, New York, 1972.

3.10 CALCULATING DEW POINT WHEN LIQUID PHASE IS IDEAL

Calculate the dew point of a vapor system containing 80 mol % benzene and 20 mol % toluene at 1000 mmHg (133.3 kPa).

Calculation Procedure

1. *Select a temperature, and test its suitability by trial and error.*
The dew point of a system at pressure P whose vapor composition is given by mole fractions y_i is that temperature at which there is onset of condensation. Mathematically, it is that temperature at which

$$\sum_{i=1}^{N} \frac{y_i}{K_i} = 1$$

where the K_i are vapor-liquid equilibrium ratios as defined in Procedure 3.1.

When the liquid phase is ideal, K_i depends only on the temperature, the pressure, and the vapor composition. The procedure for determining the dew point in such a case is to (1) guess a temperature; (2) calculate the K_i, which equal $f_i^o/\phi_i P$, where f_i^o is the fugacity of pure liquid i at the system temperature and pressure, ϕ_i is the fugacity coefficient of the ith species in the vapor phase, and P is the system pressure; and (3) check if the preceding dew-point equation is satisfied. If it is not, repeat the procedure with a different guess.

For this system, f_i^o may be assumed to be the same as the vapor pressure. The vapor phase can be assumed to be ideal, that is, $\phi_i = 1$. (For a discussion of the grounds for these assumptions, see Procedure 3.6).

The boiling points of benzene and toluene at 1000 mmHg are first calculated (for instance, by using the Antoine equation, as discussed in Procedure 3.1). They are 89°C and 141°C, respectively. As a first guess at the dew-point temperature, try a linear interpolation of these boiling points: $T = (0.8)(89) + (0.2)(141)$, which approximately equals 100. Let subscript 1 refer to benzene; subscript 2 to toluene.

At 100°C, $P_1^o = 1350$ mmHg and $P_2^o = 314$ mmHg (see Procedure 3.1). Therefore, $K_1 = 1350/1000 = 1.35$ and $K_2 = 314/1000 = 0.314$. Moreover, the dew-point equation becomes $(y_1/K_1) + (y_2/K_2) = (0.8)/(1.35) + (0.2)/(0.314) = 1.230$.

Since this sum is greater than 1, the K values are too low. So the vapor pressures and, accordingly, the assumed temperature are too low.

2. Repeat the trial-and-error procedure with the aid of graphic interpolation until the dew-point temperature is found.
As the next estimate of the dew point, try 110°C. At this temperature, $K_1 = 1.756$ and $K_2 = 0.428$, as found via the procedure outlined in the previous step. Then $(y_1/K_1) + (y_2/K_2) = 0.923$. Since the sum is less than 1, the assumed temperature is too high.

Next plot $(y_1/K_1) + (y_2/K_2)$ against temperature, as shown in Fig. 3.3. Linear interpolation of the first two trial values suggests 107.5°C as the next guess. At this temperature, $K_1 = 1.647$ and $K_2 = 0.397$, and $(y_1/K_1) + (y_2/K_2) = 0.989$.

The second and third guess points on Fig. 3.3 suggest that the next guess be 107.1°C. At this temperature, $K_1 = 1.630$ and $K_2 = 0.392$, and $(y_1/K_1) + (y_2/K_2) = 1.001$. Thus, the dew point can be taken as 107.1°C (380.3 K or 224.8°F).

As a matter of interest, the liquid that condenses at the dew point has the composition $x_i = y_i/K_i$, where the x_i are the liquid-phase mole fractions. In this case, $x_1 = 0.8/1.630 = 0.49$ and $x_2 = 0.2/0.392 = 0.51 (= 1 - 0.49)$.

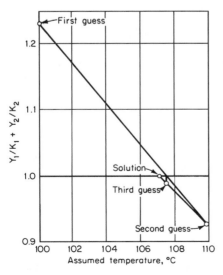

FIGURE 3.3   Trial-and-error solution for dew point (Procedure 3.10).

Related Calculations.   When the liquid phase is nonideal, the K values depend on the liquid composition, in this case the dew composition, which is not known. The procedure then is to guess both the liquid composition and the temperature and to check not only whether the dew-point equation is satisfied, but also whether the values of x_i calculated by the equations $x_i = y_i/K_i$ are the same as the guessed values. (When there is no other basis for the initial guess of the dew composition, an estimate of it may be obtained by dew-point calculations presuming an ideal liquid phase, as outlined above.) Such nonideal liquid-phase dew-point calculations are usually carried out using a digital computer. A logic diagram for dew-point determination is shown in Fig. 3.4.

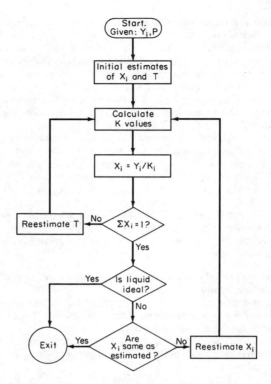

FIGURE 3.4   Logic diagram for determining dew point (Procedure 3.10).

3.11   CALCULATING BUBBLE POINT WHEN VAPOR PHASE IS IDEAL

Calculate the bubble point of a liquid system containing 80 mol % ethanol and 20 mol % water at 500 mmHg (66.7 kPa).

Calculation Procedure

1. Select a temperature, and test its suitability by trial and error.
The bubble point of a system at pressure P and whose liquid composition is given by mole fractions x_i is that temperature at which there is onset of vaporization. Mathematically, it is the temperature such that

$$\sum_{i=1}^{N} K_i x_i = 1$$

where the K_i are vapor-liquid equilibrium ratios.

When the vapor phase is ideal, the K_i are independent of the vapor composition. In such a case, the procedure for bubble-point determination is to (1) guess a temperature; (2) calculate the K_i, which equal $\gamma_i f_i^\circ / P$, where γ_i is the activity coefficient of the ith component in the liquid phase, f_i° is the fugacity of pure liquid i at system temperature and pressure, and P is the system pressure; and (3) check if the preceding bubble-point equation is satisfied. If it is not, repeat the procedure with a different guess.

For this system, f_i° may be assumed to be the same as the vapor pressure (for a discussion of the grounds for this assumption, see Procedure 3.6). Activity coefficients can be calculated using the Wilson, Margules, or Van Laar equations (see Procedure 3.4).

From vapor-pressure calculations, the boiling points of ethanol and water at 500 mmHg are 68°C and 89°C, respectively. As a first guess at the bubble-point temperature, try a linear interpolation of the boiling points: $T = (0.8)(68) + (0.2)(89)$, which is approximately equal to 72. Let subscript 1 refer to ethanol and subscript 2 to water. At 72°C, activity coefficients calculated via the Van Laar equation are 1.02 for ethanol and 2.04 for water. From tables or calculation, the vapor pressures are 589 and 254 mmHg. Then $K_1 = (1.02)(589)/500 = 1.202$ and $K_2 = (2.04)(254)/500 = 1.036$, and the bubble-point equation becomes $K_1 x_1 + K_2 x_2 = (1.202)(0.8) + (1.036)(0.2) = 1.169$.

Since this is greater than 1, the K values are too high. This means that the vapor pressures and, accordingly, the assumed temperature are too high.

2. Repeat the trial-and-error procedure until the bubble-point temperature is found.

As a second guess, try 67°C. Assume that because the temperature difference is small, the activity coefficients remain the same as in step 1. At 67°C, the vapor pressures are 478 and 205 mmHg. Then, $K_1 = 0.9746$ and $K_2 = 0.8344$, as found by the procedure outlined in the previous step. Then $K_1 x_1 + K_2 x_2 = 0.9466$. Since this sum is less than 1, the assumed temperature is too low.

As a next estimate, try 68°C. At this temperature, $K_1 = 1.039$ and $K_2 = 0.891$, and $K_1 x_1 + K_2 x_2 = 1.009$.

Finally, at 68.3°C, $K_1 = 1.030$ and $K_2 = 0.883$, and $K_1 x_1 + K_2 x_2 = 1.0006$. This is very close to 1.0, so 68.3°C (341.5 K or 154.9°F) can be taken as the bubble point.

As a matter of interest, the vapor that is generated at the bubble point has the composition $y_i = K_i x_i$, where the y_i are the vapor-phase mole fractions. In this case, $y_1 = (1.030)(0.8) = 0.82$ and $y_2 = (0.883)(0.2) = 0.18 \, (= 1.0 - 0.82)$.

Related Calculations. When the vapor phase is nonideal, the K values depend on the vapor composition, in this case the bubble composition, which is unknown. The procedure then is to guess both the vapor composition and the temperature and to check not only whether the bubble-point equation is satisfied, but also whether the values of y_i given by the equations $y_i = K_i x_i$ are the same as the guess values. (When there is no other basis for guessing the bubble composition, an estimate based on calculations that presume an ideal vapor phase may be used, as outlined above.) As in the case of nonideal dew points, nonideal bubble points are calculated using a digital computer. Figure 3.5 shows a logic diagram for computing bubble points.

3.12 BINARY PHASE DIAGRAMS FOR VAPOR-LIQUID EQUILIBRIUM

A two-phase binary mixture at 100°C (212°F or 373 K) and 133.3 kPa (1.32 atm or 1000 mmHg) has an overall composition of 68 mol % benzene and 32 mol % toluene. Determine the mole fraction benzene in the liquid phase and in the vapor phase.

Calculation Procedure

1. Obtain (or plot from data) a phase diagram for the benzene/toluene system.

Vapor-liquid equilibrium behavior of binary systems can be represented by a temperature-composition diagram at a given constant pressure (such as Fig. 3.6) or by a pressure-composition diagram at a given constant temperature (such as Fig. 3.7).

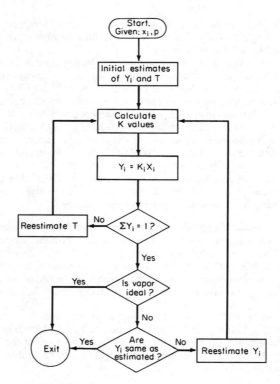

FIGURE 3.5 Logic diagram for determining bubble point (Procedure 3.11).

Curve ABC in each figure represents the states of saturated-liquid mixtures; it is called the "bubble-point curve" because it is the locus of bubble points in the temperature-composition diagram. Curve ADC represents the states of saturated vapor; it is called the "dew-point curve" because it is the locus of the dew points. The bubble- and dew-point curves converge at the two ends, which represent the saturation points of the two pure components. Thus in Fig. 3.6, point A corresponds to the boiling point of toluene at 133.3 kPa, and point C corresponds to the boiling point of benzene. Similarly, in Fig. 3.7, point A corresponds to the vapor pressure of toluene at 100°C, and point C corresponds to the vapor pressure of benzene.

The regions below ABC in Fig. 3.6 and above ABC in Fig. 3.7 represent subcooled liquid; no vapor is present. The regions above ADC in Fig. 3.6 and below ADC in Fig. 3.7 represent superheated vapor; no liquid is present. The area between the curves is the region where both liquid and vapor phases coexist.

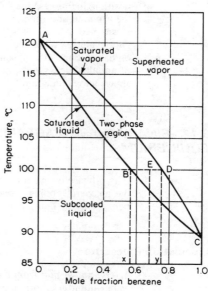

FIGURE 3.6 The benzene-toluene system at 1000 mmHg (133 kPa) (Procedure 3.12).

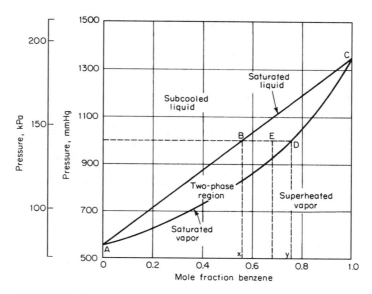

FIGURE 3.7 The benzene-toluene system at 100°C (Procedure 3.12).

2. Determine vapor and liquid compositions directly from the diagram.

If a system has an overall composition and temperature or pressure such that it is in the two-phase region, such as the conditions at point E in either diagram, it will split into a vapor phase whose composition is given by point D and a liquid phase whose composition is given by point B, where line BD is the horizontal (constant-temperature or constant-pressure) line passing through E.

Related Calculations. The relative amounts of vapor and liquid present at equilibrium are given by (moles of liquid)/(moles of vapor) = (length of line ED)/(length of line BE).

Figures 3.6 and 3.7 have shapes that are characteristic for ideal systems. Certain nonideal systems deviate so much from these as to form maxima or minima at an intermediate composition rather than at one end or the other of the diagram. Thus the dew-point and bubble-point curves meet at this intermediate composition as well as at the ends. Such a composition is called an "azeotropic composition."

The example discussed here pertains to binary systems. By contrast, multicomponent vapor-liquid equilibrium behavior cannot easily be represented on diagrams and instead is usually calculated at a given state by using the procedures described in the preceding two examples.

Phase equilibrium is important in design of distillation columns. Such design is commonly based on use of an xy diagram, a plot of equilibrium vapor composition y versus liquid composition x for a given binary system at a given pressure. An xy diagram can be prepared from a temperature-composition or a pressure-composition diagram, such as Figs. 3.6 and 3.7. Select a value of x; find the bubble-point condition (temperature on Fig. 3.6, for instance) for that value on curve ABC; then move horizontally to curve ADC in order to find the equilibrium value of y; and then plot that y as ordinate versus that x as abscissa. Figure 3.8 shows the xy diagram for the benzene/toluene system.

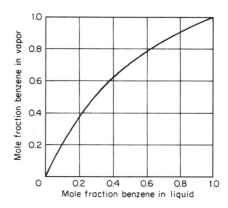

FIGURE 3.8 xy diagram for benzene-toluene system at 1 atm (Procedure 3.12).

3.13 MULTICOMPONENT LIQUID-LIQUID EQUILIBRIUM

Develop a design method for calculating multicomponent liquid-liquid equilibrium.

Calculation Procedure

1. Establish what equations must be satisfied.
Ideal solutions or solutions exhibiting negative deviation from ideal behavior cannot form two liquid phases. Instead, strong positive deviation is necessary for two or more liquid phases to exist together.

Calculations of multicomponent liquid-liquid equilibrium are needed in the design of liquid (solvent) extraction systems. Since these operations take place considerably below the bubble point, it is not necessary to consider the equilibrium-vapor phase. The equations to be solved are:
Material balances:

$$\phi x_i^{(1)} + (1 - \phi)x_i^{(2)} = Z_i$$

Equilibrium relationships:

$$\gamma_i^{(1)} x_i^{(1)} = \gamma_i^{(2)} x_i^{(2)}$$

Constraints:

$$\sum_{i=1}^{N} x_i^{(1)} = \sum_{i=1}^{N} x_i^{(2)} = 1$$

In these equations, ϕ is the mole fraction of phase 1 (i.e., the moles of phase 1 per total moles), Z_i is the mole fraction of the ith species in the total system, $x_i^{(1)}$ is the mole fraction of the ith species in phase 1, $x_i^{(2)}$ is the mole fraction of the ith species in phase 2, $\gamma_i^{(1)}$ is the activity coefficient of the ith species in phase 1, and $\gamma_i^{(2)}$ is its activity coefficient in phase 2.

2. Set out the calculation procedure.
The preceding equations are rearranged to facilitate the solution according to the following procedure: (1) guess the liquid compositions of both phases, (2) calculate the activity coefficients in both phases at the solution temperature, (3) solve the following nonlinear equation for ϕ:

$$F(\phi) = \sum_{i=1}^{N} \left[x_i^{(1)} - x_i^{(2)} \right]$$
$$= \sum_{i=1}^{N} \frac{z_i \left[1 - \gamma_i^{(1)}/\gamma_i^{(2)} \right]}{\phi + (1 - \phi)\gamma_i^{(1)}/\gamma_i^{(2)}}$$
$$= 0$$

(4) calculate the liquid compositions from

$$x_i^{(1)} = \frac{Z_i}{\phi + (1 - \phi)\gamma_i^{(1)}/\gamma_i^{(2)}}$$

and

$$x_i^{(2)} = x_i^{(1)} \gamma_i^{(1)}/\gamma_i^{(2)}$$

and (5) if the calculated values in step 4 are not the same as the guessed values from step 1, repeat the procedure with different guesses.

Any nonlinear equation-solving techniques can be used in the preceding procedure, which is usually carried out with a digital computer.

Related Calculations. Graphic representation of liquid-liquid equilibrium is convenient only for binary systems and isothermal ternary systems. Detailed discussion of such diagrams appears in A. W. Francis, *Liquid-Liquid Equilibrium*, Interscience, New York, 1963. Thermodynamic correlations of liquid-liquid systems using available models for liquid-phase nonideality are not always satisfactory, especially when one is trying to extrapolate outside the range of the data.

Of interest in crystallization calculations is solid-liquid equilibrium. When the solid phase is a pure component, the following thermodynamic relationship holds:

$$\ln \frac{1}{\gamma_i x_i} = \frac{\Delta H_i}{RT}\left(1 - \frac{T}{T_i^*}\right) - \frac{\Delta C_i}{R}\left[\left(\frac{T_i^*}{T} - 1\right) - \ln\left(\frac{T_i^*}{T}\right)\right]$$

where x_i is the saturation mole fraction of the *i*th component in equilibrium with pure solid, γ_i is the activity coefficient of the *i*th species with reference to pure solid at absolute temperature T, R is the universal gas constant, T_i^* is the triple-point temperature of the component, ΔH_i is its heat of fusion at T_i^*, and ΔC_i is the difference between the specific heats of pure liquid and solid. For many substances, the second term on the right-hand side of the equation is negligible and can be omitted. With such a simplification, the equation can be rewritten as

$$x_i = \frac{\exp\left[\dfrac{\Delta H_i}{RT_i^*}\left(1 - \dfrac{T_i^*}{T}\right)\right]}{\gamma_i}$$

Often the melting point and the heat of fusion at the melting point are used as estimates of T_i^* and ΔH_i. It should be noted that the latter equation is nonlinear, since γ_i on the right-hand side is a function of x_i. Hence the determination of x_i calls for an iterative numerical procedure, such as the Newton-Raphson or the secant methods.

3.14 *CALCULATIONS FOR ISOTHERMAL FLASHING*

A mixture containing 50 mol % benzene and 50 mol % toluene exists at 1 atm (101.3 kPa) and 100°C (212°F or 373 K). Calculate the compositions and relative amounts of the vapor and liquid phases.

Calculation Procedure

1. Calculate the mole fraction in the vapor phase.
When the components of the system are soluble in each other in the liquid state (as is the case for benzene and toluene), so that there is only one liquid phase, then the moles of vapor V per total moles can be calculated from the equation

$$\sum_{i=1}^{N}[Z_i(1 - K_i)/(1 - V + VK_i)] = 0$$

where Z_i is the mole fraction of the *i*th component in the total system and K_i is the K value of the *i*th component, as discussed in the first example in this section.

Since the pressure is not high, ideal-system K values can be used, and these are independent of composition. From the above-mentioned example, $K_1 = 1.777$ and $K_2 = 0.413$, where the subscript 1 refers to benzene and subscript 2 to toluene.

For a binary system, as in this case, the preceding summation equation becomes $Z_1(1 - K_1)/(1 - V + V K_1) + Z_2(1 - K_2)/(1 - V + V K_2) = 0$, or $0.5(1 - 1.777)/(1 - V + 1.777V) + 0.5(1 - 0.413)/(1 - V + 0.413V) = 0$. By algebra, $V = 0.208$, so 20.8 percent of the system will be in the vapor phase.

2. Calculate the liquid composition.
The liquid composition can be calculated from the equation $x_i = Z_i/(1 - V + V K_i)$. For benzene, $x_1 = 0.5/[1 - 0.208 + (0.208)(1.777)] = 0.430$, and for toluene, $x_2 = 0.5/[1 - 0.208 + (0.208)(0.413)] = 0.570$ (or, of course, for a binary mixture, $x_2 = 1 - x_1 = 0.570$).

3. Calculate the vapor composition.
The vapor composition can be calculated directly from the defining equation for K values in rearranged form: $y_i = K_i x_i$. For benzene, $y_1 = K_1 x_1 = 1.777(0.430) = 0.764$, and for toluene, $y_2 = K_2 x_2 = 0.413(0.570) = 0.236$ (or, of course, for a binary mixture, $y_2 = 1 - y_1 = 0.236$).

Related Calculations.   If the system is such that two liquid phases form, then the defining equation for V above must be replaced by a pair of equations in which the unknowns are V and ϕ, the latter being the moles of liquid phase 1 per total liquid moles:

$$\sum_{i=1}^{N} Z_i \left[1 - K_i^{(1)}\right] / \left[\phi(1 - V) + (1 - \phi)(1 - V)K_i^{(1)}/K_i^{(2)} + VK_i^{(1)}\right] = 0,$$

and

$$\sum_{i=1}^{N} Z_i \left[1 - K_i^{(1)}/K_i^{(2)}\right] / \left[\phi(1 - V) + (1 - \phi)(1 - V)K_i^{(1)}/K_i^{(2)} + VK_i^{(1)}\right] = 0$$

In addition, the equation for liquid composition becomes $x_i^{(1)} = Z_i / [\phi(1 - V) + (1 - \phi)(1 - V) K_i^{(1)}/K_i^{(2)} + VK_i^{(1)}]$. In these three equations, the superscripts 1 and 2 refer to the two liquid phases.

In systems where more than one liquid phase exists, the K values are not independent of composition. In such a case, it is necessary to use a trial-and-error procedure, guessing the composition of all the phases, calculating a K value for the equilibrium of each liquid phase with the vapor phase, solving the equations for V and ϕ, and then calculating the compositions of the phases by the equations for the x_i and y_i and seeing if these calculations agree with the guesses. This procedure should be carried out using a computer.

In the case of adiabatic rather than isothermal flashing, when the total enthalpy of the system rather than its temperature is specified, the equations associated with isothermal flash are solved jointly with an enthalpy-balance equation, treating the temperature as another variable. The general enthalpy balance is

$$H_F = V H_V [1 - V] \left[\phi H_L^{(1)} + (1 - \phi)H_L^{(2)}\right], \text{ where}$$

$$H_V = \sum_{i=1}^{N} h_{v,i} y_i + \Delta h_d \text{ and } H_L^{(k)} = \sum_{i=1}^{N} h_{L,i} x_i^{(k)} + \Delta h_m^{(k)}$$

where $h_{v,i}$ is ideal-gas enthalpy per mole of pure ith component, $h_{L,i}$ is the enthalpy per mole of pure liquid of ith component, Δh_d is enthalpy departure from ideal-gas state per mole of vapor of composition given by the y_i, and $\Delta h_m^{(k)}$ is the heat of mixing per mole of liquid phase having the composition given by the $x_i^{(k)}$, k taking the values 1 and 2 if there are two liquid phases or the value 1 if there is only one liquid phase. Very often the heat of mixing is negligible. A convenient procedure for adiabatic flash is to perform isothermal flashes at a series of temperatures to find the temperature at which the enthalpy balance is satisfied.

*3.15 LIQUID-LIQUID SEPARATION ANALYSIS

Size a liquid-liquid separator or decanter by using gravitational force for continuous separation of two liquids, the first of which has a density of 47 lb/ft³ (752.5 kg/m³) and the second liquid a density of 81 lb/ft³ (1296.8 kg/m³). Both liquids flow into the separator at a rate of 50 gal/min (189.3 L/min). The time required for settling is 35 min. What size separator is required to handle this flow? How far above the separator bottom should overflow of the heavier liquid be located?

Calculation Procedure

1. Compute the liquid holdup volume.
Since there are two liquids, a light one and a heavy one, entering the separator, the holdup volume = (number of liquids entering)(liquid flow rate into the separator, gal/min)(holdup time, min). Or for this separator, total holdup volume = 2(50)(35) = 3500 gal (13,247.5 L).

2. Determine the separator tank volume.
Usual design practice is to make the separator tank volume 10 to 25 percent greater than the required holdup volume. Using a volume 20 percent greater than the required holdup volume gives a required tank volume of 1.20(3500 gal) = 4200 gal (15,897.0 L).

3. Size the separator tank.
Most decanter-type separator tanks are sized so that the tank diameter and height are approximately equal. Selecting a 10-ft (3.05-m) diameter and 10-ft (3.05-m) high tank gives a total tank volume of (head area, ft²)(height, ft) = $(d^2\pi/4)h$, where d and h are the diameter and height of the tank, ft, respectively. Or, volume = $(10^2\pi/4)(10)$ = 785.4 ft³ (2.22 m³). Since 1 gal (3.8 L) of liquid occupies 0.13 ft³, the capacity of this tank = 785.4/0.13 = 5850 gal (22,142.2 L). This is sufficient to store the holdup liquid but somewhat oversize.

Try a 9-ft (2.74-m) diameter and high tank. By the same method, the tank capacity is 4250 gal (16,086.3 L). This is closer to the required holdup capacity. Hence, a 9-ft (2.74-m) tank will be used.

4. Compute the liquid depth in the tank.
Use the relation D_1 ft = 4(holdup volume, gal)/$7.48\pi d^2$, where D_1 = liquid depth, ft. So D_1 = 4(3500)/$7.48\pi 9^2$ = 7.34 ft (2.24 m).

5. Determine the height of the heavy-liquid overflow.
Assume that the two liquids interface midway between the vessel bottom and the liquid surface. Then the height of the heavy liquid = 7.34/2 = 3.67 ft (1.12 m).

To find the height of the heavy-liquid overflow, solve $H_h = H_1 + (D_1 - H_1)$(density of lighter liquid, lb/ft³)/(density of heavier liquid, lb/ft³), where H_h = height of heavy-liquid overflow above tank bottom, ft; H_1 = height of heavy liquid in tank, ft; other symbols as before. Solving gives H_h = 3.67 + (7.34 − 3.67)(47/81) = 5.80 ft (1.77 m). This is the distance measured to the inside lower surface of the overflow pipe from the tank bottom.

The *continuous decanter* is a popular type of static separator for immiscible liquids of many types. This type of separator is fed from the top and vented to the open air through both the light-and heavy-liquid overflow lines.

*3.16 DETERMINING THE CHARACTERISTICS OF AN IMMISCIBLE SOLUTION

For steam distillation of 2-bromoethylbenzene, the vapor temperature is 222.4°F (105.8°C). Analysis shows 0.16 lb (0.073 kg) of 2-bromoethylbenzene (BB) per lb (kg) of vapor. Saturated steam is used in the distillation process. Determine the pressure in the still and how far from ideal the actual conditions are.

Calculation Procedure

1. *Compute the pressure in the still.*
Each component of an immiscible mixture of liquids exerts a vapor pressure that is independent of its concentration and equal to the vapor pressure of the pure substance—but only if stratification is avoided by vigorous mixing or boiling. The major industrial application of immiscible systems is in steam distillation of high-molecular-weight heat-sensitive organic materials. The mixture of water (steam) and an organic substance will boil when the total solution pressure equals atmospheric pressure. Since the organic material must exert some vapor pressure, it vaporizes with the steam, at a greatly reduced temperature.

The relationship for immiscible components A and B is $w_A/w_B = y_A M_A/y_B M_B = P_{VA} M_A/P_{VB} M_B$, where $w_{A,B}$ = weight of component A, B in vapor; $M_{A,B}$ = molecular weight of component A, B; $y_{A,B}$ = vapor-phase mole fraction of component A, B; $P_{VA, VB}$ = vapor pressure of component A, B.

The vapor pressure of BB at 222.4°F (105.8°C) is, from Perry's—*Chemical Engineers' Handbook*, 20 mmHg, and the vapor pressure of water at 222.4°F (105.8°C) is 938 mmHg. Hence, the total pressure (ideal) in the still is 938 + 20 = 958 mmHg.

2. *Compare the ideal to the actual conditions.*
If conditions in the still were ideal (i.e., exactly according to theory), the weight of the BB in the vapor would be, according to step 1, $w_{BB} = (P_{V,BB}/P_{V,H_2O})(M_{BB} M_{H_2O}) = (20/938)(185/18) = 0.219$ lb (0.0995 kg), versus 0.16 lb (0.073 kg) actual, as given.

Or, by computing the ideal BB vapor pressure for 0.16 lb of BB per lb (0.07 kg/kg) of vapor, from the relation in step 1, $(P_{V,BB}/P_{H_2O})(185/18) = (P_{V,BB}/938)(185/18) = 0.16$ lb/lb (0.07 kg/kg). Solving, we find $P_{V,BB} = (0.16)(938)(18/185) = 14.6$ mmHg versus 20 mmHg actual.

The divergence between the actual and ideal most likely means that the time of contact between the steam and the BB is insufficient to reach equilibrium. Also, the total pressure should be 938 + 14.6 = 952.6 mmHg, not the 958 mmHg of the ideal case.

Related Calculations. This procedure is valid for immiscible solutions of all types resembling the one considered here.

Plant engineers and designers in the chemical processing industry must be extremely careful about making changes in chemical processes or waste disposal. Seemingly routine decisions changing a process or disposal method can run into trouble under the Toxic Substance Control Act (TSCA). This act gives the U.S. Environmental Protection Agency (EPA) information and control over commercial chemicals.

Fines as high as $23,000 a day can be levied when TSCA rules are not obeyed. EPA applies rigid formulas when enforcing the act. Violation can result in million-dollar assessments.

It is important that engineers submit a premanufacture notification (PMN) in accordance with Section 5 of TSCA *before* manufacturing or importing a new chemical substance. A new chemical substance, as defined by Matthew Kuryla in *Chemical Engineering* magazine, is one that does not appear on an EPA list known as the TSCA Inventory. This list is constantly changing. Further, a portion of the list is confidential. To make a comprehensive search of the list requires a written request to EPA certifying a bona fide intent to manufacture or import a chemical.

Unless an exemption applies, a manufacturer must file a PMN with EPA *before* commencing production or importation of a new chemical. (In certain instances, a PMN must also be filed for existing chemical production that falls under a regulation known as the "significant new use rule.") The PMN must include the identity of the chemical, information about its proposed use and quantity, its by-products, and all available data concerning potential worker exposure and environmental or public-health effects.

Ninety days after filing a PMN, a company may commence manufacture or import of the chemical if Notice of Commencement (NOC) is filed. When EPA receives an NOC, it places the chemical on the TSCA Inventory. The chemical is then no longer considered new, but an "existing" chemical subject to other TSCA rules.

A number of chemicals and processes are exempt from the PMN requirements of TSCA, including: (1) foods, drugs, and cosmetics (including their intermediates); (2) pesticides (but not their intermediates);

(3) chemicals used solely for research and development purposes, in small quantities; (4) chemicals manufactured solely for export; (5) impurities unintentionally present in another chemical; (6) by-products whose only commercial purpose is for burning as fuel, disposal as waste, or reclamation; (7) nonisolated intermediates (i.e., those mixed with other products and reactants) or incidental reaction by-products.

Besides these exemptions, TSCA provides a specialized PMN process for certain limited uses of a new chemical. These rules are known as the "test market," "low volume," and "polymer" exemptions. Such exemptions are subject to detailed rules of their own. They do not apply across the board, and frequently do not have specific, quantitative limits, notes Matthew Kuryla in *Chemical Engineering*.

REFERENCES

Vapor Pressure of Pure Compounds

1. Boublik, Fried, and Hala—*The Vapour Pressures of Pure Substances*, Elsevier.
2. Riddick and Bunger—*Organic Solvents*, 3d ed., vol. 2, Wiley-Interscience.
3. Wichterle and Linek—*Antoine Vapor Pressure Constants of Pure Compounds*, Academia.
4. Zwolinski and Wilhoit—*Vapor Pressures and Heats of Vaporization of Hydrocarbons and Related Compounds*, Thermodynamics Research Center, Texas A&M University.
5. Zwolinski et al.—*Selected Values of Properties of Hydrocarbons and Related Compounds*, API Research Project 44, Thermodynamics Research Center, Texas A&M University.
6. Ohe—*Computer Aided Data Book of Vapor Pressure*, Data Book Publishing Co.
7. Stull—*Ind. Eng. Chem. 39*:517, 1947.

Equations of State

8. Redlich and Kwong—*Chem. Rev. 44*:233, 1949.
9. Wohl—*Z. Phys. Chem. B2*:77, 1929.
10. Benedict, Webb, and Rubin—*Chem. Eng. Prog. 47*:419, 1951; *J. Chem. Physics 8*:334, 1940; *J. Chem. Physics 10*:747, 1942.
11. Soave—*Chem. Engr. Sci. 27*:1197, 1972.
12. Peng and Robinson—*Ind. Eng. Chem. Fund. 15*:59, 1976.

Fugacity of Pure Liquid

13. Pitzer and Curl—*J. Am. Chem. Soc. 79*:2369, 1957.
14. Chao and Seader—*AIChE Journal 7*:598, 1961.

Activity Coefficient

15. Wohl—*Trans. Am. Inst. Chem. Engrs. 42*:217, 1946.
16. Wilson—*J. Am. Chem. Soc. 86*:127, 1964.
17. Renon and Prausnitz—*AIChE Journal 14*:135, 1968.
18. Abrams and Prausnitz—*AIChE Journal 21*:116, 1975.
19. Fredenslund, Gmehling, and Rasmussen—*Vapor-Liquid Equilibria Using UNIFAC*, Elsevier.
20. Herington—*J. Inst. Petrol. 37*:457, 1951.
21. Van Ness, Byer, and Gibbs—*AIChE Journal 19*:238, 1973.

VLE Data

22. Chu, Wang, Levy, and Paul—*Vapor-Liquid Equilibrium Data*, Edwards.
23. Gmehling and Onken—*Vapor-Liquid Equilibrium Data Collection*. Chemistry Data Series, DECHEMA.
24. Hala, Wichterle, Polak, and Boublik—*Vapor-Liquid Equilibrium Data at Normal Pressures,* Pergamon Press.
25. Hirata, Ohe, and Nagahama—*Computer Aided Data Book of Vapor-Liquid Equilibria*, Kodansha/Elsevier.
26. Horsley—*Azeotropic Data*, American Chemical Society.
27. Wichterle, Linek, and Hala—*Vapor-Liquid Equilibrium Data Bibliography,* Elsevier; *Supplement I,* Elsevier. 1976.

Liquid-Liquid Equilibrium

28. Francis—*Liquid-Liquid Equilibrium*, Interscience.
29. Francis—*Handbook for Components in Solvent Extraction*, Gordon & Breach.
30. Seidel—*Solubilities of Organic Compounds*, 3d ed., Van Nostrand; *Supplement*, Van Nostrand.
31. Stephen and Stephen—*The Solubilities of Inorganic and Organic Compounds*, Pergamon.
32. Sorenson and Arlt—*Liquid-Liquid Equilibrium Data Collection*, Chemistry Data Series, DECHEMA.
33. Sorensen et al.—*Fluid Phase Equilibria* 2:297, 1979; *Fluid Phase Equilibria* 3:47, 1979.

General

34. Prausnitz—*Molecular Thermodynamics of Fluid-Phase Equilibria*, Prentice-Hall.
35. Hunter—*Ind. Eng. Chem. Fund.* 6:461, 1967.
36. Himmelblau—*Applied Nonlinear Programming*, McGraw-Hill.
37. Sandler—*Chemical and Engineering Thermodynamics*, Wiley.

SECTION 4
CHEMICAL REACTION EQUILIBRIUM[†]

E. Dendy Sloan, Ph.D.
Professor
Chemical Engineering Department
Colorado School of Mines
Golden, CO

Chemical engineers make reaction-equilibria calculations to find the potential yield of a given reaction, as a function of temperature, pressure, and initial composition. In addition, the heat of reaction is often obtained as an integral part of the calculations. Because the calculations are made from thermodynamic properties that are accessible for most common compounds, the feasibility of a large number of reactions can be determined without laboratory study. Normally, thermodynamic feasibility should be determined before obtaining kinetic data in the laboratory, and if a given set of reaction conditions does not yield a favorable final mixture of products and reactants, the reaction-equilibria principles indicate in what direction one or more of the conditions should be changed.

The basic sequence of equilibria calculations consists of (1) determining the standard-state Gibbs free-energy change $\Delta G°$ for the reaction under study at the temperature of interest (since this step commonly includes calculation of the standard enthalpy change of reaction $\Delta H°$, an example of that calculation is included below), (2) determining the equilibrium constant K from $\Delta G°$, (3) relating the equilibria compositions to K, and (4) evaluating how this equilibrium composition changes as a function of temperature and pressure. A frequent complication is the need to deal with simultaneous or heterogeneous reactions or both.

[†]Procedure 4.10, Calculating Reaction Conversions, Selectivities, and Yields, is from Smith, *Chemical Process Design*, McGraw-Hill. Procedures 4.11 through 4.13 are the work of Loyal Clarke, from *Manual for Process Engineering Calculations*, McGraw-Hill, 1951.

4.1 HEAT OF REACTION

Evaluate the heat of reaction of $CO(g) + 2H_2(g) \rightarrow CH_3OH(g)$ at 600 K (620°F) and 10.13 MPa (100 atm).

Calculation Procedure

1. Calculate the heat of reaction at 298 K, ΔH°_{298}.
Calculating heat of reaction is a multistep process. One starts with standard heats of formation at 298 K, calculates the standard heat of reaction, and then adjusts for actual system temperature and pressure.

The heat of reaction at 298 K is commonly referred to as the "standard heat of reaction." It can be calculated readily from the standard heats of formation (ΔH°_{f298}) of the reaction components; these standard heats of formation are widely tabulated. Thus, Perry and Chilton [7] show ΔH°_{298} for CH_3OH and CO to be −48.08 and −26.416 kcal/(g · mol), respectively (the heat of formation of hydrogen is zero, as is the case for all other elements). The standard heat of reaction is the sum of the heats of formation of the reaction products minus the heats of formation of the reactants. In this case,

$$\Delta H^\circ_{298} = -48.08 - [-26.416 + 2(0)]$$
$$= -21.664 \text{ kcal/(g · mol)}$$

2. Calculate the heat-capacity constants.
The heat capacity per mole of a given substance C_p can be expressed as a function of absolute temperature T by equations of the form $C_p = a + bT + cT^2 + dT^3$. Values of the constants are tabulated in the literature. Thus Sandler [10] shows the following values for the substances in this example:

	CH$_3$OH	CO	H$_2$
a	4.55	6.726	6.952
b (× 10^2)	2.186	0.04001	−0.04576
c (× 10^5)	−0.291	0.1283	0.09563
d (× 10^9)	−1.92	−0.5307	−0.2079

In this step, we are determining the constants for the equation $\Delta C_p = \Delta a + \Delta bT + \Delta cT^2 + \Delta dT^3$ where Δa (for instance) equals $\Sigma a_{\text{products}} - \Sigma a_{\text{reactants}}$. Thus, for the reaction in this example, $\Delta a = 4.55 - 6.726 - 2(6.952) = -16.08$; $\Delta b = [2.186 - 0.04001 + 2(0.04576)] \times 10^{-2} = 2.2375 \times 10^{-2}$; $\Delta c = [-0.291 - 0.1283 - 2(0.09563)] \times 10^{-5} = -0.61056 \times 10^{-5}$; and $\Delta d = [-1.92 + 0.5307 + 2(0.2079)] \times 10^{-9} = -0.9735 \times 10^{-9}$. The resulting ΔC_p is in cal/(g · mol)(K) (rather than in kcal).

3. Calculate the standard heat of reaction at 600 K, ΔH°_{600}.
This step combines the results of the previous two, via the equation

$$\Delta H^\circ_T = \Delta H^\circ_{298} + \int_{298}^{T} [\Delta a + \Delta bT + \Delta cT^2 + \Delta dT^3]\,dT$$

Integration gives the equation

$$\Delta H_T = \Delta H^\circ_{298} + \Delta a(T - 298) + \frac{\Delta b}{2}(T^2 - 298^2)$$
$$+ \frac{\Delta c}{3}(T^3 - 298^3) + \frac{\Delta d}{4}(T^4 - 298^4)$$

Therefore,

$$\Delta H^{\circ}_{600} = -21.664 - 16.08(600 - 298) + \left(\frac{2.2375 \times 10^{-2}}{2} \right)(600^2 - 298^2)$$

$$- \left(\frac{0.61056 \times 10^{-5}}{3} \right)(600^3 - 298^3) - \left(\frac{0.9735 \times 10^{-9}}{4} \right)(600^4 - 298^4)$$

$$= -2.39 \times 10^4 \text{ cal/(g} \cdot \text{mol)}[-1.00 \times 10^8 \text{ J/(kg} \cdot \text{mol)}$$

$$\text{or } -4.30 \times 10^4 \text{ Btu/(lb} \cdot \text{mol)}]$$

4. Calculate the true heat of reaction by correcting ΔH°_{600} for the effect of pressure.
The standard heats of formation, from which ΔH°_{600} was calculated, are based on a pressure of 1 atm. The actual pressure in this case is 100 atm. The correction equation is $\Delta H_{600,100 \text{ atm}} = \Delta H^{\circ}_{600} - \Delta H^1$, where the correction factor ΔH^1 is in turn defined as follows: $\Delta H^1 = \Delta H^{\circ} + \omega \Delta H'$. In the latter equation, ΔH° and $\Delta H'$ are parameters whose values depend on the reduced temperature and reduced pressure, while ω is the acentric factor. Generalized correlations that enable the calculation of ΔH° and $\Delta H'$ can be found, for instance, in Smith and Van Ness [11], and ω is tabulated in the literature. Analogously to the approach in step 2, the calculation is based on the sum of the values for the reaction products minus the sum of the values for the reactants.
For the present example, the values are as follows:

Component	ω	ΔH°	$\Delta H'$	ΔH^1, cal/(g · mol)
CH_3OH	0.556	1222	305.5	1392
CO	0.041	0	0	0
H_2	0	0	0	0

Therefore, ΔH^1 for this example is 1392. In turn,

$$\Delta H_{600,100 \text{ atm}} = -2.39 \times 10^4 - 1392$$

$$= -2.529 \times 10^4 \text{ cal/(g} \cdot \text{mol)}[-1.058 \times 10^8 \text{ J/(kg} \cdot \text{mol)}$$

$$\text{or } -4.55 \text{ Btu/(lb} \cdot \text{mol)}]$$

Related Calculations.   When reactants enter at a temperature different from that of the exiting products, the enthalpy changes that are due to this temperature difference should be computed independently and added to ΔH°_{298K}, along with the pressure deviations of enthalpy.
When heat capacities of enthalpies of formation are not tabulated, a group-contribution method such as that in the following example may be used.

4.2 HEAT OF FORMATION FOR UNCOMMON COMPOUNDS

Assuming that tabulated heat-of-formation data are not available for ethanol, *n*-propanol, and *n*-butanol, use the group-contribution approach of Benson to estimate these data at 298 K.

Calculation Procedure

1. *Draw structural formulas for the chemical compounds.*
Take the formulas from a chemical dictionary, chemistry textbook, or similar source. For this example, they are

2. *Identify all groups in each formula.*
A group is defined here as a polyvalent atom (ligancy ≥ 2) in a molecule together with all of its ligands. Thus the ethanol molecule has one each of three kinds of groups: (1) a carbon atom linked to one other carbon atom and to three hydrogen atoms, the notation for this group being $C-(C)(H)_3$; (2) a carbon atom linked to one other carbon atom and to an oxygen atom and to two hydrogen atoms, which can be expressed as $C-(C)(O)(H)_2$; and (3) an oxygen atom linked to a carbon atom and to a hydrogen atom, which can be expressed as $O-(C)(H)$.

The *n*-propanol molecule has one each of four kinds of groups: $C-(C)(H)_3$, $C-(C)_2(H)_2$, $C-(C)(O)(H)_2$, and $O-(C)(H)$. And the *n*-butanol molecule differs from the *n*-propanol molecule only in that it has two (rather than one) of the $C-(C)_2(H)_2$ groups.

3. *For each molecule, add up the partial heats of formation for its constituent groups.*
The partial heats of formation can be found in Benson [2]. Adding these up for a given molecule gives an estimate of the heat of formation for that molecule.

For ethanol, the result is as follows:

$C-(C)(H)_3$	-10.20 kcal/(g · mol)
$C-(C)(O)(H)_2$	-8.1
$O-(C)(H)$	-37.9
$\Delta H^\circ_{f\,298} =$	-56.2 kcal/(g · mol) [-2.35×10^8 J/(kg · mol) or -1.012×10^5 Btu/(lb · mol)]

For *n*-propanol:

$C-(C)(H)_3$	-10.20
$C-(C)_2(H)_2$	-4.93
$C-(C)(O)(H)_2$	-8.1
$O-(C)(H)$	-37.9
$\Delta H^\circ_{f\,298} =$	-61.13 kcal/(g · mol) [-2.556×10^8 J/(kg · mol) or -1.10×10^5 Btu/(lb · mol)]

And for *n*-butanol:

$C-(C)(H)_3$	-10.20
$2[C-(C)_2(H)_2] = 2(-4.93) =$	-9.86
$C-(C)(O)(H)_2$	-8.1
$O-(C)(H)$	-37.9
$\Delta H^\circ_{f\,298} =$	-66.06 kcal/(g · mol) [-2.762×10^8 J/(kg · mol) or -1.189×10^5 Btu/(lb · mol)]

Related Calculations. As an indicator of the degree of accuracy of this method, the observed values for ethanol, *n*-propanol, and *n*-butanol are -2.35×10^8, -2.559×10^8, and -2.83×10^8 J/(kg · mol), respectively. Deviations may be as high as $\pm 1.25 \times 10^7$ J/(kg · mol). Benson's method may also be used to estimate ideal-gas heat capacities.

4.3 STANDARD GIBBS FREE-ENERGY CHANGE OF REACTION

Calculate the standard Gibbs free-energy change $\Delta G°$ for the reaction $CH_3CH_2OH \rightarrow CH_2 = CH_2 + H_2O$ at 443 K (338°F).

Calculation Procedure

1. Calculate $\Delta H°_{298}$ for the reaction, and set out an equation for $\Delta H°_T$ as a function of temperature T.
This step directly employs the procedures developed in the preceding examples. Thus $\Delta H°_{298}$ is found to be 1.0728×10^4 cal/(g · mol), and the required equation is

$$\Delta H°_T = \Delta H°_{298} + \Delta a(T - 298) + (\Delta b/2)(T^2 - 298^2)$$
$$+ (\Delta c/3)(T^3 - 298^3) + (\Delta d/4)(T^4 - 298^4)$$

The calculated numerical values of the heat-capacity constants are $\Delta a = 3.894$, $\Delta b = -0.01225$, $\Delta c = 0.7381 \times 10^{-5}$, and $\Delta d = -1.4287 \times 10^{-9}$. Therefore, after arithmetic partial simplification, the equation becomes (with the answer emerging in calories per gram-mole)

$$\Delta H°_T = 1.0049 \times 10^4 + \Delta aT + (\Delta b/2)T^2 + (\Delta c/3)T^3 + (\Delta d/4)T^4$$

2. Calculate $\Delta G°_{298}$ for the reaction.
This can be determined in a straightforward manner from tabulated data on standard Gibbs free energies of formation at 298 K, $\Delta G°_{f298}$. The answer is the sum of the values for the reaction products minus the sum of the values for the reactants. Thus Perry and Chilton [7] show $\Delta G°_{f298}$ for ethanol, ethylene, and water to be, respectively, -40.23, 16.282, and -54.635 kcal/(g · mol). Accordingly, $\Delta G°_{298} = (16.282 - 54.635) - (-40.23) = 1.877$ kcal/(g · mol) [7.854×10^6 J/(kg · mol) or 3.3786×10^3 Btu/(lb · mol)].

3. Calculate $\Delta G°$ at the reaction temperature, 443 K.
The equation used is

$$\Delta G° = \Delta H_0 - \Delta aT \ln T - (\Delta b/2)T^2 - (\Delta c/6)T^3 - (\Delta d/12)T^4 - IRT$$

where ΔH_0 is the integration constant that is determined for the $\Delta H°_T$ equation in step 1 (that is, 1.0049×10^4), R is the universal gas constant, the heat-capacity constants (Δa, etc.) are as determined in step 1, and I is a constant yet to be determined.

To determine I, substitute into the equation the (known) values pertaining to $\Delta G°$ at 298 K. Thus 1.877×10^3 cal/(g · mol) $= 1.0049 \times 10^4 - (3.894)(298) \ln 298 - \frac{1}{2}(-0.01225)(298)^2 - (\frac{1}{6})(0.7381 \times 10^{-5})(298)^3 - (\frac{1}{12})(-1.4287 \times 10^{-9})(298)^4 - (1.987)(298)(I)$. Solving for I, it is found to be 3.503.

Finally, insert this value into the equation and solve for $\Delta G°$ at 443 K. Thus $\Delta G° = 1.0049 \times 10^4 - (3.894)(443) \ln 443 - \frac{1}{2}(-0.01225)(443)^2 - (\frac{1}{6})(0.7381 \times 10^{-5})(443)^3 - (\frac{1}{12})(-1.4287 \times 10^{-9})(443)^4 - (1.987)(443)(3.503) = -2.446 \times 10^3$ cal/(g · mol) [or -1.0237×10^7 J/(kg · mol) or -4.4019 Btu/(lb · mol)].

Related Calculations. ΔG°_{298} may also be calculated via the relationship $\Delta G^{\circ}_{298} = \Delta H^{\circ}_{298} - T\Delta S^{\circ}_{298}$, where S is entropy. Values for S can be found in the literature, or ΔS°_{298} can be estimated by Benson's group-contribution method (see example for estimating $\Delta H^{\circ}_{f\,298}$ by that method above). However, the deviations for entropy values calculated by Benson's method are high compared with those for $\Delta H^{\circ}_{f\,298}$, so when such group methods are used in the course of calculating ΔG°_{298}, only qualitative feasibility studies are possible, based on these rules-of-thumb: (1) if ΔG°_{T} is less than zero, the reaction is clearly feasible; (2) if the value is less than 4.184×10^{7} J/(kg · mol) [or 1.8×10^{4} Btu/ (lb · mol)], the reaction should be studied further; and (3) if the value is greater than 4.184×10^{7} J/ (kg · mol), the reaction is feasible only under exceptional circumstances. (Thus, in the present example, the reaction appears unfavorable at 298 K, but much more promising at 443 K.)

The Gibbs free-energy change of reaction may be related to the heat of reaction by $d(\Delta G^{\circ}/RT)/dT = -\Delta H^{\circ}/RT^{2}$. And if ΔH° is constant, or if changes in ΔG° are needed over a small temperature range, then a given ΔG° value at a known temperature T_1 may be used to obtain ΔG° at a different temperature T_v via the relationship. $\Delta G_v/RT_v = \Delta G^{\circ}/RT_1 - (\Delta H^{\circ}/R)[(1/T_v) - (1/T_1)]$.

4.4 ESTIMATION OF EQUILIBRIUM COMPOSITIONS

Estimate the equilibrium composition of the reaction *n*-pentane $\rightarrow$ neopentane, at 500 K (440°F) and 10.13 MPa (100 atm), if the system initially contains 1 g · mol *n*-pentane. Ignore other isomerization reactions.

Calculation Procedure

1. Calculate ΔG° at 500 K.
This step is carried out as outlined in the previous example. Accordingly ΔG°_{500} is found to be 330.54 cal/ (g · mol) [or 13.83×10^{6} J/(kg · mol) or 594.97 Btu/(lb · mol)].

2. Determine the equilibrium constant K.
This step consists of solving the equation $K = \exp(-\Delta G^{\circ}/RT)$. Thus $K = \exp[-330.54/(1.987)(500)] = 0.717$.

3. Express mole fractions in terms of the reaction-progress variable x.
The reaction-progress variable x is a measure of the extent to which a reaction has taken place. In the present example, x would be 0 if the "equilibrium" mixture consisted solely of *n*-pentane; it would be 1 if the mixture consisted solely of neopentane (i.e., if all the *n*-pentane had reacted). It is defined as $\int_{n_{o,i}}^{n_{f,i}} dn/v_i$, where n is number of moles of component i, subscripts o and f refer to initial and final states, respectively, and v_i is the stoichiometric number for component i, which is equal to its stoichiometric coefficient in that reaction but with the convention that reactants are given a minus sign. In the present example, $v_{n\text{-pentane}} = -1$ and $v_{\text{neopentane}} = 1$, and $n_{o,n\text{-pentane}} = 1$ and $n_{o,\text{neopentane}} = 0$. Since a given system (such as the system in this example) is characterized by a particular value of the reaction-progress variable, the integrals described above (an integral can be written for each component) must equal each other for that system.

In the present case,

$$\int_1^{nf,n\text{-pentane}} dn_{n\text{-pentane}}/-1 = \int_0^{nf,\text{neopentane}} dn_{\text{neopentane}}/1 = x$$

Carrying out the two integrations and solving for the two n_f values in terms of x, we find $n_{f,n\text{-pentane}} = (1 - x)$, and $n_{f,\text{neopentane}} = x$. Total moles at equilibrium equals $(1 - x) + x = 1$. Thus the mole fractions

y_i at equilibrium are $(1 - x)/1 = (1 - x)$ for *n*-pentane and $x/1 = x$ for neopentane. [This procedure is valid for any system, however complicated. In the present simple example, it is intuitively clear that at equilibrium, the two n_f values must be $(1 - x)$ and x.]

4. Estimate the final composition by using the Lewis-Randall rule.

By definition, $K = \prod \hat{a}_i^{\nu i}$, where the $\hat{a}_i$ are the activities of the components within the mixture, and the ν_i, are the stoichiometric numbers as defined above. Then, using 101.3 kPa (1 atm) as the standard-state fugacity, $K = (\hat{f}/f°)_{\text{neopentane}} / (\hat{f}/f°)_{\text{n-pentan}} = (y\hat{\phi}P)_{\text{neopentane}} / (y\,\hat{\phi}\,p)_{\text{n-pentane}}$, where f is fugacity, y is mole fraction, ϕ is fugacity coefficient, P is system pressure, the caret symbol ^ denotes the value of f or ϕ in solution, and the superscript ° denotes standard state, normally taken 1 atm.

The Lewis-Randall rule assumes that the fugacity of a component in solution is directly proportional to its mole fraction. Under this assumption, $\hat{\phi}_i = \phi_i$. Therefore (and noting that P cancels in numerator and denominator), $K = (y\phi)_{\text{neopentane}} / (y\phi)_{\text{n-pentane}}$. The ϕ can be calculated by methods outlined in Section 3 (or by use of Pitzer correlations); they are found to be 0.505 for neopentane and 0.378 for *n*-pentane. Therefore, substituting into the equation for K, $0.717 = 0.505\,y_{\text{neopentane}}/0.378\,y_{\text{n-pentane}} = 0.505x/0.378(1 - x)$ Solving for x, it is found to be 0.349. Therefore, the equilibrium composition consists of 0.349 mole fraction neopentane and $(1 - 0.349) = 0.651$ mole fraction *n*-pentane.

Related Calculations. If the standard heat of reaction $\Delta H°$ is known at a given T and K is known at that temperature, a reasonable estimate for K (designated K_1) at a not-too-distant temperature T_1 can be obtained via the equation $\ln (K/K_1) = (-\Delta H°/R)[(1/T) - (1/T_1)]$, where the heat of reaction is assumed to be constant over the temperature range of interest.

Occasionally, laboratory-measured values of K are available. These can be used to determine either or both constants of integration (ΔH_0 and I) in the calculations for ΔG.

As for the calculation of equilibrium composition, the Lewis-Randall-rule assumption in step 4 is a simplification. The rigorous calculation instead requires use of $\hat{\phi}$, the values of fugacity coefficient as they actually exist in the reaction mixture. This entails calculation from an equation of state. Initially, one would hope to use the virial equation, with *initial estimates* of mole fractions made as in step 4; however, in the present example, the pseudoreduced conditions are beyond the range of convergence of the virial equation. If the modified Soave-Redlich-Kwong equation is used instead, with a computer program, to calculate the $\hat{\phi}$, the resulting mole fractions are found to be 0.362 for neopentane and 0.638 for *n*-pentane. Thus the Lewis-Randall-rule simplification in this case leads to errors of 3.6 and 2.0 percent, respectively. The degree of accuracy required should dictate whether the simplified approach is sufficient.

4.5   ACTIVITIES BASED ON MIXED STANDARD STATES

For the reaction $2A \rightarrow B + C$ at 573 K (572°F), a calculation based on standard states of ideal gas at 101.32 kPa (1 atm) for A and B and pure liquid at its vapor pressure of 202.65 kPa (2 atm) for C produces a $\Delta G°/RT$ of -5, with G in calories per gram-mole. Calculate the equilibrium constant based on ratios of final mole fractions at 303.97 kPa (3 atm), assuming that all three components are ideal gases.

Calculation Procedure

1. Select the solution method to be employed.
This problem may be solved either by setting out the equilibrium-constant equation in terms of the mixed (two) standard states or by first converting the $\Delta G°$ to a common standard state. Both methods are shown here.

2. Solve the problem by using mixed standard states.

a. *Calculate the equilibrium constant K in terms of activities based on the given mixed standard states.* By definition, $K = \exp(-\Delta G^\circ/RT) = \exp-(-5) = 148.41$.

b. *Relate the K to compositions.* By definition (see previous problem $K = a_C a_B/a_A^2 = (f_C/f_C^\circ) \times (f_B/f_B^\circ)/(f_A/f_A^\circ)^2$. In the present case, the standard-state fugacities $f_A^\circ = f_B^\circ = 1$ atm, and $f_C^\circ = 2$ atm. Since the gases are assumed to be ideal, $f_i = y_i P$, where y is mole fraction and P is the system pressure. Therefore,

$$K = 148.41 = [y_C(3\text{ atm})/(2\text{ atm})][y_B(3\text{ atm})/(1\text{ atm})]/[y_A(3\text{ atm})/(1\text{ atm})]^2$$

c. *Define K in terms of mole fractions and solve for its numerical value.* In terms of mole fractions for this reaction, $K = y_C y_B/y_A^2$. Rearranging the final equation in the preceding step,

$$y_C y_B/y_A^2 = 148.41(2/3)(1/3)(3/1)^2 = 296.82 = K$$

3. Solve the problem again, this time using uniform standard states.

a. *Correct ΔG° to 1 atm for all the components.* For this reaction, $\Delta G^\circ = G_C^\circ + G_B^\circ - 2G_A^\circ$. Now, G_B° and G_A° are already based on 1 atm. To correct G_C° to 1 atm, use the relationship $\Delta G = RT \ln(f_2/f_1)$ where R is the gas constant, T is absolute temperature, and f is fugacity. Since we wish to convert from a basis of 2 atm to one of 1 atm, $f_1 = 2$ atm and $f_2 = 1$ atm; therefore,

$$\Delta G = (1.987)(573)[\ln(1/2)] = -789.2 \text{ cal/(g · mol)}$$

Thus $G_{C,1\text{ atm}}^\circ = G_{C,2\text{ atm}}^\circ - 789.2$. Therefore, to correct ΔG° to 1 atm for all components, 789.2 cal/(g · mol) must be subtracted from the ΔG° as given in the statement of the problem; that is, from $-5RT$. Thus $\Delta G^\circ = (-5)(1.987)(573) - 789.2 = -6481.96$, and $\Delta G^\circ/RT$ corrected to 1 atm becomes -5.693.

b. *Solve for K, and relate it to compositions and express it in terms of mole fractions.* By definition, $K = \exp(-\Delta G^\circ/RT) = \exp-(-5.693) = 296.82$. And as in step 2b, $K = (y_C P/f_C^\circ)(y_B P/f_B^\circ)/(y_A P/f_A^\circ)^2 = (3y_C/1)(3y_B/1)/(3y_A/1)^2$. Therefore, $y_C y_B/y_A^2 = K = 296.82$, the same result as via the method based on mixed standard states.

Related Calculations. As this example illustrates, one should know the standard state chosen for the Gibbs free energy of formation for each compound (usually, ideal gas at 1 atm) when considering the relation of mole fractions to K values.

When a solid phase occurs in a reaction, it is often pure; in such a case, its activity may be expressed by

$$a_i = f_i/f_i^\circ = \exp[(v_s/RT)(P - P^\circ)]$$

where v_s is the molar volume of the solid and P° is its vapor pressure. This activity is usually close to unity. If a reaction takes place in the liquid phase, the activity may be expressed by $a_i = \gamma_i x_i$, where γ_i is the activity coefficient for component i and x_i is its mole fraction.

4.6 EFFECTS OF TEMPERATURE AND PRESSURE ON EQUILIBRIUM COMPOSITION

The reaction $H_2O \rightarrow H_2 + \frac{1}{2}O_2$ appears unlikely at 298 K (77°F) and 101.3 kPa (1 atm) because its ΔG° is 2.286×10^8 J/(kg · mol) (see Related Calculations under Procedure 4.3). If one wishes to increase the equilibrium conversion to hydrogen and oxygen, (a) should the temperature be increased or decreased, and (b) should the pressure be increased or decreased?

Calculation Procedure

1. Calculate $\Delta H°$, the heat of reaction.
This is done as outlined in Procedure 4.1. (Actually, since 298 K is a standard temperature, $\Delta H°$ for the present example can be read directly from tables.) The value is found to be 2.42×10^8 J/(kg · mol) [1.04×10^5 Btu/(lb · mol)].

2. Calculate the algebraic sum of the stoichiometric numbers for the Σv_i.
As discussed in Procedure 4.4, stoichiometric numbers for the components in a given reaction are numerically equal to the stoichiometric coefficients, but with the convention that the reactants get minus signs. In the present case, the stoichiometric numbers for H_2O, H_2, and O_2 are, respectively, -1, $+1$, and $+\frac{1}{2}$. Their sum is $+\frac{1}{2}$.

3. Apply rules that pertain to partial derivatives of the reaction-progress variable.
As discussed in Procedure 4.4, the reaction-progress variable x is a measure of the extent to which a reaction has taken place. In the present case, the desired increase in equilibrium conversion to hydrogen and oxygen implies an increase in x.

It can be shown that $(\partial x/\partial T)_P$ = (a positive number)$(\Delta H°)$ for all reactions. In the present example, $\Delta H°$ is positive (i.e., the reaction is endothermic). Therefore, an increase in temperature will increase x and, thus, the conversion to hydrogen and oxygen.

It can also be shown that $(\partial x/\partial P)_T$ = (a negative number)(Σv_i) for all reactions. Since Σv_i is positive, an increase in pressure will decrease x. So, a *decrease* in pressure will favor the conversion to hydrogen and oxygen.

Related Calculations. While nothing is said above about kinetics, increasing the temperature very frequently changes the reaction rate favorably. Accordingly, in some exothermic-reaction situations, it may be worthwhile to sacrifice some degree of equilibrium conversion in favor of shorter reactor residence time by raising reaction temperature. Similarly, a pressure change may have an effect on kinetics that is contrary to its effect on equilibrium.

The reaction in this example illustrates another point. It happens that (at 1 atm) water will not appreciably dissociate into hydrogen and oxygen unless the temperature is raised about 1500 K (2240°F); but at such temperatures, molecules of H_2 and O_2 may dissociate into atomic H and O and enter into unexpected reactions. The engineer should keep such possibilities in mind when dealing with extreme conditions.

4.7 EQUILIBRIUM COMPOSITION FOR SIMULTANEOUS KNOWN CHEMICAL REACTIONS

Given an initial mixture of 1 g · mol each of CO and H_2, estimate the equilibrium composition that will result from the following set of simultaneous gas-phase reactions at 900 K (1160°F) and 101.3 kPa (1 atm).

$$CO + 3H_2 \rightarrow CH_4 + H_2O \tag{4.1}$$

$$CO + H_2O \rightarrow CO_2 + H_2 \tag{4.2}$$

$$CO_2 + 4H_2 \rightarrow CH_4 + 2H_2O \tag{4.3}$$

$$4CO + 2H_2O \rightarrow CH_4 + 3CO_2 \tag{4.4}$$

Calculation Procedure

1. Find the independent chemical reactions.
Write equations for the formation of each compound present:

$$C + \tfrac{1}{2}O_2 \rightarrow CO \tag{4.5}$$

$$C + 2H_2 \rightarrow CH_4 \tag{4.6}$$

$$C + O_2 \rightarrow CO_2 \tag{4.7}$$

$$H_2 + \tfrac{1}{2}O_2 \rightarrow H_2O \tag{4.8}$$

Next, algebraically combine these equations to eliminate all elements not present as elements in the system. For instance, eliminate O_2 by combining Eqs. 4.5 and 4.7

$$2CO \rightarrow CO_2 + C \tag{4.9}$$

and Eqs. 4.5 and 4.8

$$C + H_2O \rightarrow CO + H_2 \tag{4.10}$$

Then eliminate C by combining Eqs. 4.6 and 4.9

$$2CO + 2H_2 \rightarrow CO_2 + CH_4 \tag{4.11}$$

and Eqs. 4.6 and 4.10

$$CO + 3H_2 \rightarrow H_2O + CH_4 \tag{4.12}$$

The four initial equations are thus reduced into two independent equations, namely. Eqs. 4.11 and 4.12, for which we must find simultaneous equilibria. All components present in the original four equations are contained in Eqs. 4.11 and 4.12.

2.   Calculate $\Delta G°$ and K for each independent reaction.
This may be done as in the relevant examples earlier in this section, with determination of $\Delta G°$ as a function of temperature. An easier route, however, is to use the standard Gibbs free-energy change of formation $\Delta G_f°$ for each compound at the temperature of interest in the relationship

$$\Delta G° = \Sigma \Delta G_{f,\text{products}}° - \Sigma \Delta G_{f,\text{reactants}}°$$

Reid et al. [9] give the following values of $\Delta G_f°$ at 900 K: for CH_4, 2029 cal/(g · mol); for CO, −45,744 cal/(g · mol); for CO_2, −94,596 cal/(g · mol); and for H_2O, −47,352 cal/(g · mol) (since H_2 is an element, its $\Delta G_f°$ is zero).

For Eq. 4.11, $\Delta G° = 2029 + (−94,596) − 2(−45,744) = −1079$ cal/(g · mol). For Eq. 4.12, $\Delta G° = 2029 + (−47,352) − (−45,744) = 421$ cal/(g · mol).

Finally, since the equilibrium constant K is equal to $\exp(−\Delta G°/RT)$, the value of K for Eq. 4.11 is $\exp[−(−1079)/(1.987)(900)] = 1.8283$, and the value of K for Eq. 4.12 is $\exp[−421/(1.987)(900)] = 0.7902$.

3. Express the equilibrium mole fractions in terms of the reaction-progress variables.
See Procedure 4.4 for a discussion of the reaction-progress variable. Let x_K and x_L be the reaction-progress variables for Reactions 4.11 and 4.12, respectively. Since Reaction 4.11 has 1 mol CO_2 on the product side, the number of moles of CO_2 at equilibrium is x_K. Since Reactions 4.11 and 4.12 each have 1 mol CH_4 on the product side, the number of moles of CH_4 at equilibrium is $x_K + x_L$. Since Reaction 4.11 has 2 mol CO on the reactant side and Reaction 4.12 has 1 mol, the number of moles of CO at equilibrium is $1 − 2x − x$ (because 1 mol CO was originally present). This approach allows us to express the equilibrium mole fractions as follows:

Component	Number of moles		Equilibrium mole fraction y
	Initially	At equilibrium	
CO_2	0	x_K	$x_K/2(1 - x_K - x_L)$
CH_4	0	$x_K + x_L$	$(x_K + x_L)/2(1 - x_K - x_L)$
CO	1	$1 - 2x_K - x_L$	$(1 - 2x_K - x_L)/2(1 - x_K - x_L)$
H_2	1	$1 - 2x_K - 3x_L$	$(1 - 2x_K - 3x_L)/2(1 - x_K - x_L)$
H_2O	0	x_L	$x_L/2(1 - x_K - x_L)$
		$\overline{2 - 2x_K - 2x_L}$	

4. Relate equilibrium mole fractions to the equilibrium constants.
By definition, $K = \Pi \hat{a}_i^{v_i}$, where the $\hat{a}_i$ are the activities of the components with the mixture, and the v_i are the stoichiometric numbers for the reaction (see Procedure 4.4). The present example is at relatively large reduced temperatures and relatively low reduced pressures, so the activities can be represented by the equilibrium mole fractions y_i. For Reaction 4.11, $K_K = y_{CO_2} y_{CH_4} / (y_{CO})^2 (y_{H_2})^2$. Substituting the value for K from step 2 and the values for the y_i from the last column of the table in step 3 and algebraically simplifying,

$$1.8283 = \frac{x_K(x_K + x_L)[2(1 - x_K - x_L)]^2}{(1 - 2x_K - x_L)^2(1 - 2x_K - 3x_L)^2} \tag{4.13}$$

And for Reaction 4.12, $K_L = y_{H_2O} y_{CH_4} / y_{CO}(y_{H_2})^3$,   or

$$0.7902 = \frac{x_L(x_K + x_L)[2(1 - x_K - x_L)]^2}{(1 - 2x_K - x_L)(1 - 2x_K - 3x_L)^3} \tag{4.14}$$

5. Solve for the equilibrium conditions.
Equations 4.13 and 4.14 in step 4 must be solved simultaneously. These are nonlinear and have more than one set of solutions; however, this complication can be eased by imposing two restrictions to ensure that no more CO or H_2 is used than the amount of each that is available (1 g · mol). Thus,

$$2x_K + x_L \leq 1 \tag{4.15}$$

$$2x_K + 3x_L \leq 1 \tag{4.16}$$

Even with Restrictions 4.15 and 4.16, the solution to Eqs. 4.13 and 4.14 requires a sophisticated mathematical technique or multiple trial and error calculations, as done most easily on the computer. At the solution, $x_K = 0.189038$ and $x_L = 0.0632143$; therefore, $y_{CO_2} = 0.1264$, $y_{CH_4} = 0.1687$, $y_{CO} = 0.37360$, $y_{H_2} = 0.28906$, and $y_{H_2O} = 0.04227$.

Related Calculations.   If the gas is not ideal, the fugacity coefficients ϕ_i will not be unity, so the activities cannot be represented by the mole fractions. If the pressure is sufficient for a nonideal solution to exist in the gas phase, $\hat{\phi}_i$ will be a function of y_i, the solution to the problem. In this case, the y_i value obtained for the solution with $\hat{\phi}_i = 1$ should be used for the next iteration and so on until convergence. Alternatively, one could initially solve the problem using the Lewis-Randall rule for $\hat{\phi}_i(\hat{\phi}_i = \phi_i)$; the y_i obtained in that solution could be substituted back into $\hat{\phi}_1$ for the next estimate. Many times this is done most easily by computer.

4.8 EQUILIBRIUM COMPOSITION FOR SIMULTANEOUS UNSPECIFIED CHEMICAL REACTIONS

Given an initial mixture of 1 g · mol CO and 1 g · mol H_2, determine the equilibrium composition of a final system at 900 K (1160°F) and 101.3 kPa (1 atm) that contains CO, CO_2, CH_4, H_2, and H_2O.

Calculation Procedure

1. Determine the number of gram-atoms of each atom present in the system.
Here 1 g · mol CO contains 1 g · atom each of C and O, and 1 g · mol of H_2 contains 2 g · atoms of H. Let A_k equal the number of gram-atoms of element k present in the system. Then $A_C = 1$, $A_O = 1$, and $A_H = 2$.

2. Determine the number of gram-atoms of each element present per gram-mole of each substance.
Let $a_{i,k}$ equal the number of gram-atoms of element k per gram-mole of substance i. Then the following matrix can be set up:

i	$a_{i,C}$	$a_{i,O}$	$a_{i,H}$
CO	1	1	0
CO_2	1	2	0
CH_4	1	0	4
H_2	0	0	2
H_2O	0	1	2

3. Determine the Gibbs free energy of formation ΔG_f° for each compound at 900 K.
See step 2 of the previous problem. From Reid et al. [9], the values are: for CH_4, 2029 cal/(g · mol); for CO, −45,744 cal/(g · mol); for CO_2, −94,596 cal/(g · mol); for H_2O, −47,352 cal/(g · mol); and for H_2, zero.

4. Write equations for minimization of total Gibbs free energy.
This step employs the method of Lagrange undetermined multipliers for minimization under constraint; for a discussion of this method, refer to mathematics handbooks. As for its application to minimization of total Gibbs free energy, see Perry and Chilton [7] and Smith and Van Ness [11].
    Write the following equation for each substance i:

$$\Delta G_f^\circ + RT \ln \left(y_i \hat{\phi}_i P/f_f^\circ \right) + \sum_k (\lambda_k a_{i,k}) = 0$$

where R is the gas constant, T is temperature, y is mole fraction, $\hat{\phi}_i$ is the fugacity coefficient, P is pressure, f° is standard-state fugacity, and λ_k is the Lagrange undetermined multiplier for element k within substance i. Since T is high and P is low, gas ideality is assumed; $\hat{\phi}_i = 1.0$. Set f° at 1 atm.
    Then the five equations are
For CO: $-45,744 + RT \ln y_{CO} + \lambda_C + \lambda_O = 0$
For CO_2: $-94,596 + RT \ln y_{CO_2} + \lambda_C + 2\lambda_O = 0$
For CH_4: $2029 + RT \ln y_{CH_4} + \lambda_C + 4\lambda_H = 0$
For H_2: $RT \ln y_{H_2} + 2\lambda_H = 0$
For H_2O: $-47,352 + RT \ln y_{H_2O} + 2\lambda_H + \lambda_O = 0$
[In all cases, $RT = (1.987)(900)$.]

5. Write material-balance and mole-fraction equations.
A material-balance equation can be written for each element, based on the values found in steps 1 and 2.

$$\sum_i y_i a_{i,k} = A_k/n_T$$

where n_T is the total number of moles in the system. These three equations are

For O: $y_{CO} + 2y_{CO_2} + y_{H_2O} = 1/n_T$
For C: $y_{CO} + y_{CO_2} + y_{CH_4} = 1/n_T$
For H: $4y_{CH_4} + 2y_{H_2} + 2y_{H_2O} = 2/n_T$

In addition, the requirement that the mole fractions sum to unity yields $y_{CO} + y_{CO_2} + y_{H_4} + y_{H_2} + y_{H_2O} = 1$.

6. Solve the nine equations from steps 4 and 5 simultaneously to find $y_{CO}, y_{CO_2}, y_{CH_4}, y_{H_2}, y_{H_2O},$ $\lambda_C, \lambda_O, \lambda_H,$ **and** n_T.
This step should, of course, be done on a computer. The Lagrange multipliers have no physical significance and should be eliminated from the solution scheme. The equilibrium composition is thus found to be as follows: $y_{CO_2} = 0.122$; $y_{CH_4} = 0.166$; $y_{CO} = 0.378$; $y_{H_2} = 0.290$; and $y_{H_2O} = 0.044$.

Note that these results closely compare with those found in the previous example, which is based on the same set of reaction conditions.

4.9 HETEROGENEOUS CHEMICAL REACTIONS

Estimate the composition of the liquid and vapor phases when n-butane isomerizes at 311K (100°F). Assume that the reaction occurs in the vapor phase.

Calculation Procedure

1. Determine the number of degrees of freedom for the system.
Use the phase rule $F = C - P + 2 - r$, where F is degrees of freedom, C is number of components, P is number of phases, and r is the number of independent reactions. In this case, C is 2, P is 2, and r is 1 (namely, n-$C_4H_{10} \rightarrow$ iso-C_4H_{10}); therefore, $F = 1$. This means that we can choose either temperature or pressure alone to specify the system; when the temperature is given (311 K in this case), the system pressure is thereby established.

2. Calculate the equilibrium constant at the given temperature.
See Procedure 4.4. K is found to be 2.24.

3. Relate the equilibrium constant to compositions of the two phases.
Let subscripts 1 and 2 refer to n-C_4H_{10} and iso-C_4H_{10}, respectively. Then $K = a_2/a_1 = (\hat{\phi}_2 y_2 P/f_2^\circ)/(\hat{\phi}_1 y_1 P/f_1^\circ)$, where a is activity, ϕ is fugacity coefficient, y is mole fraction, P is system pressure, f is fugacity, the caret symbol $^\wedge$ denotes the value of ϕ in solution, and the superscript$^\circ$ denotes the standard state. If f_1° and f_2° are both selected to be 1 atm (101.3 kPa), the expression simplifies to

$$K = \hat{\phi}_2 y_2 P/\hat{\phi}_1 y_1 P \qquad (4.17)$$

4. *Relate the gas phase to the liquid phase.*
Using phase-equilibrium relationships (see Section 3), the following equation can be set out for each of the two components:

$$y_i \hat{\phi}_i P = x_i \gamma_i P_i^{sat} \phi_i^{sat}$$

where x_i is liquid-phase mole fraction, γ_i is activity coefficient, P_i^{sat} is vapor pressure, and ϕ_i^{sat} is fugacity of pure i in the vapor at the system temperature.

The two butane isomers form an ideal solution in the liquid phase at the system temperature, the molecules being quite similar. Therefore, both activity coefficients can be taken to be unity.

Via relationships discussed in Section 3, ϕ_i^{sat} is found to be 0.91 and ϕ_2^{sat} to be 0.89. And from vapor pressure data, P_1^{sat} is 3.53 atm and P_2^{sat} is 4.95 atm.

5. *Solve for chemical and phase equilibria simultaneously.*
Write the equation in step 4 for each component and substitute into Eq. 4.17 while taking into account that $x_1 + x_2 = 1$. This gives

$$K = x_2 P_2^{sat} \phi_2^{sat} / (1 - x_2) P_1^{sat} \phi_1^{sat}$$

Or, substituting the known numerical values,

$$2.24 = x_2 (4.95)(0.89)/(1 - x_2)(3.53)(0.91)$$

Thus the liquid composition is found to be $x_2 = 0.62$ and $x_i = 0.38$.

Finding the vapor composition y_1 and y_2 first requires trial-and-error solution for P and ϕ_i in the equations

$$y_1 \phi_1 P = x_1 P_1^{sat} \phi_1^{sat} \quad \text{and} \quad y_2 \phi_2 P = x_2 P_2^{sat} \phi_2^{sat}$$

Initially assuming that $\phi_1 = \phi_2 = 1$ and noting that $y_1 + y_2 = 1$, we can write a combined expression

$$\begin{aligned} P &= x_1 P_1^{sat} \phi_1^{sat} + x_2 P_2^{sat} \phi_2^{sat} \\ &= (0.38)(3.53)(0.91) + (0.62)(4.95)(0.89) \\ &= 3.95 \end{aligned}$$

Therefore, $P = 3.95$ atm.

Using this value of P to estimate the ϕ_i values (see Section 3), we obtain $\phi_1 = 0.903$ and $\phi_2 = 0.913$. Then we substitute these and solve for a second-round estimate of P in

$$\begin{aligned} (y_1 + y_2) = 1 &= (x_1 P_1^{sat} \phi_1^{sat} / \phi_1 P) + (x_2 P_2^{sat} \phi_2^{sat} / \phi_2 P) \\ &= (0.38)(3.53)(0.91)/0.903\, P + (0.62)(4.95)(0.89)/0.913\, P \end{aligned} \qquad (4.18)$$

Therefore, $P = 4.344$ atm in this second-round estimate.

Now, use $P = 4.344$ to correct the estimates of the ϕ_i values, to obtain $\phi_1 = 0.894$ and $\phi_2 = 0.904$.

Substituting in Eq. 4.18 yields $P = 4.38$ atm, which converges with the previous estimates. Finally, from the relationship $y_i = x_i P_i^{sat} \phi_i^{sat} / \phi_i P$, $y_1 = (0.38)(3.53)(0.91)/(0.894)(4.38) = 0.31$, and $y_2 = (0.62)(4.95)(0.89)/(0.904)(4.38) = 0.69$.

Thus the liquid composition is $x_1 = 0.38$ and $x_2 = 0.62$, and the vapor composition is $y_1 = 0.31$ and $y_2 = 0.69$. The system pressure at equilibrium is 4.38 atm (443.7 kPa).

Related Calculations.   Most frequently, liquid ideality cannot be assumed, and the engineer must use activity coefficients for the liquid phase. Activity coefficients are strong functions of liquid composition and temperature, so these calculations become trial-and-error, most easily done by computer.

In addition, there are frequently more than two nonideal-liquid components, which requires the use of a multicomponent activity-coefficient equation; see Prausnitz [8]. Often the parameters of the activity-coefficient equation have not been experimentally determined and must be estimated by a group-contribution technique; see Reid et al. [9].

4.10 CALCULATING REACTION CONVERSIONS, SELECTIVITIES, AND YIELDS

Benzene is to be produced from toluene according to the reaction

$$C_6H_5CH_3 + H_2 = C_6H_6 + CH_4$$

Some of the benzene formed undergoes a secondary reaction in series to an unwanted byproduct, diphenyl, according to the reaction

$$2C_6H_6 = C_{12}H_{10} + H_2$$

The compositions of the reactor feed and effluent streams appear in the table.

Component	Inlet flow rate, kmol h^{-1}	Outlet flow rate, kmol h^{-1}
H_2	1858	1583
CH_4	804	1083
C_6H_6	13	282
$C_6H_5CH_3$	372	93
$C_{12}H_{10}$	0	4

Calculate the conversion, selectivity, and reactor yield with respect to (a) the toluene feed and (b) the hydrogen feed.

Calculation Procedure

1. Determine the toluene conversion.
The toluene conversion equals (toluene consumed in the reactor)/(toluene fed to the reactor). Thus, toluene conversion = $(372 − 93)/372 = 0.75$.

2. Determine the benzene selectivity from toluene.
First, find the relevant stoichiometric factor, namely, the stoichiometeric moles of toluene required per mole of benzene produced. From the reaction to produce benzene from toluene, this stoichiometric factor is seen to equal 1.

Then apply the following relationship:

benzene selectivity from toluene = [(benzene produced in reactor)/(toluene consumed in reactor)] [stoichiometric factor] = $[(282 − 13)/(372 − 93)][1] = 0.96$

3. *Determine the reactor yield of benzene from toluene.*
The relationship is as follows:

reactor yield of benzene from toluene = [(benzene produced in reactor)/(toluene fed to reactor)]
[stoichiometric factor] = [(282 − 13)/372][1] = 0.72

4. *Determine the hydrogen conversion.*
Analogously to step 1,

hydrogen conversion = (hydrogen consumed in reactor)/(hydrogen fed to reactor)
= (1858 − 1583)/1858 = 0.15

5. *Determine the benzene selectivity from hydrogen.*
As in step 2, the stoichiometric factor with respect to hydrogen and benzene is, likewise, 1, from the
first reaction in this section.
Then,

benzene selectivity from hydrogen = [(benzene produced in reactor)/(hydrogen consumed in reactor)]
[stoichiometric factor] = [(282 − 13)/(1858 − 1583)][1] = 0.98

6. *Determine reactor yield of benzene from hydrogen.*
As in step 3,

reactor yield of benzene from hydrogen = [(benzene produced in reactor)/(hydrogen fed to reactor)]
[stoichiometric factor] = [(282 − 13)/1858][1] = 0.14

Related Calculations.   Because there are two feeds to this process, the reactor performance can be
calculated with respect to both feeds. However, the main concern is the performance with respect to
toluene, as it is presumably the more expensive feed.

*4.11 EQUILIBRIUM CONSTANT AND CONVERSION ACHIEVED IN CHEMICAL REACTIONS

Determine the equilibrium constant for the reaction $CH_4(g) + H_2O(g) = CO(g) + 3H_2(g)$ at 1700°F
(1200K), and find the amount of conversion achieved when 2 mol of steam per mole of methane are
reacted at 1 atm.

Calculation Procedure

1. *Find the free-energy formation of the products.*
Using free-energy data from engineering handbooks, the free energy of formation of products

$$CO: \Delta F°/T = -43.63 \text{ Btu/(lb mole)(°R)}$$
$$3H_2: \Delta F°/T = \underline{\quad 0.00 \quad}$$
$$\text{Total} = -43.63$$

2. Determine the free-energy formation of the reactants thus:

$$CH_4: \Delta F°/T = 8.14 \text{ Btu/(lb mole)(°R)}$$
$$H_2O: \Delta F°/T = \underline{-36.15}$$
$$\text{Total} = -28.01$$

3. Find the free energy of reaction.

$$\Delta F°/T = -43.63 - (-28.01) = -15.62$$

4. Compute the equilibrium constant.

$$\log K = \frac{1}{4.577} \times \frac{-\Delta F}{T} = \frac{15.62}{4.577} = 3.410$$

$$K = 2570 = \frac{(CO)(H_2)^3}{(CH_4)(H_2O)}$$

5. Determine the methane conversion.
Trial assumption, conversion is nearly complete. After reaction of 100 mol CH_4 + 200 mol H_2O, moles $CH_4 = y$ (to be computed)

$$H_2O = 200 - (100 - y) = 100 + y.... \quad 100$$
$$CO = 100 - y........................ \quad 100$$
$$\underline{H_2 = 300 - 3y...................... \quad 300}$$
$$\text{Total} = 500 - 3y...................... \quad 500$$

Partial pressure in atmospheres = mole fractions.

$$CH_4 = y/500 = 0.002y$$
$$H_2O = 100/500 = 0.2$$
$$CO = 100/500 = 0.2$$
$$H_2 = 300/500 = 0.6$$

$$2570 = \frac{(CO)(H_2)^3}{(CH_4)(H_2O)} = \frac{0.2 \times (0.6)^3}{0.002y \times 0.2} = \frac{113}{y}$$

or

$$y = \frac{113}{2570} = 0.4 \text{ mol/100 mol } CH_4 \text{ charged}$$

This is nearly complete conversion justifying the assumption made and the neglect of y in computing the number of moles of H_2O, CO, and H_2.

Related Calculations. This computation shows that *if* this reaction is rapid and *if* no side reactions are involved, complete conversion of methane may be effected at 1700°F (1200 K). This reaction is the basis of one commercial process for production of hydrogen. In practice a catalyst (such as nickel) is employed to speed the reaction. Some CO_2 is also formed in accordance with the reaction.

$$CO + H_2O = CO_2 + H_2$$

*4.12   DETERMINING THE HEAT OF SOLUTION FOR CHEMICAL COMPOUNDS

Find the heat of solution of carbon dioxide in sodium hydroxide solution (1 mol dissolved in 400 mol of water, that is 0.55% NaOH) to form sodium carbonate solution (1 mol dissolved in 401 mol of water).

Calculation Procedure

1. Write the reaction equation.

$$CO_2(g) + 2NaOH(aq, 400) = Na_2CO_3(aq, 401) + H_2O(l)$$

2. Compute the heat of formation of the products.
Using data from engineering handbooks, heat of formation of products

$Na_2CO_3(aq, 400)\Delta H°$...........................	−495,200 Btu/lb mol
$H_2O(l)\Delta H°$......................................	−123,070 Btu/lb mol
Total...	−618,270

3. Compute the heat of formation of the reactants.
Again, using data from engineering handbooks, heat of formation of reactants

$CO_2(g)\Delta H°$.............................. −169,300 Btu/lb mol
2NaOH $(aq, 400)\Delta H°$
$\qquad\qquad = -2 \times 201,700$ −403,400   Btu/2 lb mol
$\qquad$ Total................................−572,700

4. Find the change in heat content and heat of reaction.

$$\text{Change in heat content} = \Delta H = -618,270 + 572,700 = -45,570$$
$$\text{Heat of reaction} = -\Delta H = 45,570 \text{ Btu/lb mol of } CO_2$$

This is the heat of solution of 1 mol of carbon dioxide in 0.55% sodium hydroxide to form a solution of sodium carbonate.

Related Calculations.   Heats of reaction are conveniently summarized in tabular form by the heats of formation from the elements. The heat of formation of a compound is usually defined as the heat that would be absorbed by the spontaneous formation of 1 mol of that substance from the elements when the reaction is conducted at 18°C (64.4°F) and atmospheric pressure. Actually, of course, very few substances may be so formed, and most heats of formation are, therefore, determined indirectly. Heats of formation are frequently given at 25°C (77°F) instead of 18°C (64.4°F).

The heats of reaction computed by this method are precise, provided the reactants and products are at the standard conditions specified. If high pressures or concentrated solutions are involved, corrections may be necessary.

Very often a heat of reaction may be desired at temperatures other than 64.4°F (18°C). In this case correction is necessary but may be readily made from specific heat data by use of Hess' law of constant heat summation. This law states that the heat of reaction is independent of the manner in which it is conducted and depends only on the initial and final states.

If the heat of reaction is desired at some temperature, T, it may be computed in three steps: (1) compute the heat change, ΔH_1, required to heat or cool the reactants to 64.4°F (18°C); (2) compute the heat change accompanying reaction ΔH_2 at 64.4°F (18°C) from the heats of formation; and (3) compute the heat change, ΔH_3, required to restore the reaction products to T, the initial temperature. The heat of reaction, $-\Delta H$, will then be obtained by securing the algebraic sum

$$(\Delta H = \Delta H_1 + \Delta H_2 + \Delta H_3)$$

*4.13   DETERMINING THE HEAT OF REACTION OF CHEMICAL COMPOUNDS

Find the heat of the reaction:

$$C(s) + CO_2(g) = 2CO(g) \text{ at } 1500°F\,(816°C)$$

Calculation Procedure

1. Compute the cooling, using engineering handbook data for C.
Cooling C(s) from 1500 to 64.4°F (816°C to 18°C)

$$\Delta H = \frac{4.4}{40}8 - 506 = -505 \text{ Btu/lb}$$
$$= -12 \times 505 = -6060 \text{ Btu/lb mol}$$

2. Compute the cooling for CO.
Cooling $CO_2(g)$ from 1500 to 64.4°F (816°C to 18°C)

$$\Delta H = 0.1 - 45.7 = -45.6 \text{ Btu/SCF}$$
$$= -45.6 \times 378.7 = -17,300 \text{ Btu/lb mole}$$

3. Find the total change in heat content for cooling reactants.
Total change in heat content for cooling reactants

$$\Delta H_1 = -6060 - 17,300 = -23,360 \text{ Btu}$$

4. Calculate the reaction at 64.4°F (18°C).

Products, $2CO(g)$: $\Delta H° = +2 \times (-47,500)$........ $-95,000$ Btu
Reactants, $C(s) + CO_2(g)$: $\Delta H° = 0 - 169,300$ $\underline{-169,300}$ Btu
Subtracting, ΔH_2..   74,300 Btu

5. Find the heat input from 64.4°F (18°C) to 1500°F (816°C).
Heating CO product from 64.4 to 1500°F (18°C to 816°C)

$$\Delta H = 29.3 - 0.1 = 29.2 \text{ Btu/SCF}$$
$$\Delta H_3 = 2 \times 378.7 \times 29.2 = 22,100 \text{ Btu}$$

6. *Compute the reaction at 1500°F (816°C).*
Reaction at 1500°F (816°C)

$$\Delta H = \Delta H_1 + \Delta H_2 + \Delta H_3 = -23,360 + 22,100 + 74,300 = 73,040 \text{ Btu}$$

7. *Find the heat of reaction at 1500°F (816°C).*
Heat of reaction at 1500°F (816°C)

$$-\Delta H = -73,040 \text{ Btu}$$

REFERENCES

1. Balzhiser, Samuels, and Eliassen—*Chemical Engineering Thermodynamics,* Prentice-Hall.
2. Benson—*Thermochemical Kinetics*, Wiley.
3. Bett, Rowlinson, and Saville—*Thermodynamics for Chemical Engineers,* MIT Press.
4. Denbigh—*Principles of Chemical Equilibrium*, Cambridge Univ. Press.
5. Hougen, Watson, and Ragatz—*Chemical Process Principles*, Part II. Wiley.
6. Modell and Reid—*Thermodynamics and Its Applications*, Prentice-Hall.
7. Perry and Chilton—*Chemical Engineers' Handbook*, McGraw-Hill.
8. Prausnitz—*Molecular Thermodynamics of Fluid Phase Equilibria,* Prentice-Hall.
9. Reid, Prausnitz, and Sherwood—*The Properties of Gases and Liquids,* McGraw-Hill.
10. Sandler—*Chemical Engineering Thermodynamics,* Wiley.
11. Smith and Van Ness—*Introduction to Chemical Engineering Thermodynamics,* McGraw-Hill.
12. Soave—*Chem. Eng. Sci.* 27:1197, 1972.
13. Stull and Prophet—*JANAF Thermochemical Tables,* NSRDS-NBS 37, 1971.
14. Smith, *Chemical Process Design,* McGraw-Hill.

SECTION 5
REACTION KINETICS, REACTOR DESIGN, AND SYSTEM THERMODYNAMICS[†]

R. M. Baldwin, Ph.D.
Professor
Chemical Engineering Department
Colorado School of Mines
Golden, CO

M. S. Graboski, Ph.D.
Research Professor
Chemical Engineering Department
Colorado School of Mines
Golden, CO

[†]Procedure 5.10 is adapted from Smith, *Chemical Process Design*, McGraw-Hill.

5.1 DETERMINING A RATE EXPRESSION BY INTEGRAL ANALYSIS OF BATCH-REACTOR DATA

Saponification of ethyl acetate with sodium hydroxide, that is,

$$CH_3COOC_2H_5 + NaOH \rightarrow CH_3COONa + C_2H_5OH$$

has been investigated at 298 K (77°F) in a well-stirred isothermal batch reactor. The following data were collected:

Time, min	5	9	13	20	25	33	37
Concentration of NaOH, g · mol/L	0.00755	0.00633	0.00541	0.00434	0.00383	0.0032	0.00296

The run began with equimolar (0.1 g · mol/L) amounts of sodium hydroxide and ethyl acetate as the reactants. Calculate the overall order of the reaction and the value of the reaction rate constant at 298 K, and write the rate expression for the reaction.

Calculation Procedure

1. Assume a functional form for the rate expression.
The rate expression for this reaction may be given by the following equation, which relates the rate of disappearance r of sodium hydroxide to concentrations of reactants and products:

$$-r_{NaOH} = k_1[CH_3COOC_2H_5]^a[NaOH]^b - k_{-1}[CH_3COONa]^c[C_2H_5OH]^d$$

In this expression, a, b, c, and d are the unknown reaction orders, k_1 and k_{-1} are the forward and reverse rate constants, and the bracketed formulas denote the concentrations of the compounds. Integral analysis of the data requires an assumption as to the functional form of the reaction rate expression (e.g., zero-order, first-order, second-order with regard to a given reactant), which is then inserted into the appropriate reactor material balance. Literature values indicate that the equilibrium constant for this reaction is very large ($k_1/k_{-1} \rightarrow \infty$). As an initial guess, the reaction may be considered to be first-order in both reactants and irreversible. Thus $-r_{NaOH} = k[EtAc]^1[NaOH]^1$, where k is the reaction rate constant. Since the reactants are present in equimolar ratio initially, [EtAc] = [NaOH] throughout the run (if the initial guess is correct), and the rate expression can therefore be expressed as $-r_{NaOH} = k[NaOH]^2$.

2. Insert the rate expression into batch-reactor material balance.
A transient material balance for the NaOH on an isothermal batch reactor becomes (NaOH in) − (NaOH out) + (net NaOH generation) = NaOH accumulation. For this system,

$$+r_{NaOH}V = \frac{dN_{NaOH}}{dt}$$

where r = rate of generation
V = reactor volume
N = number of moles
t = time

Rearranging, referring to step 1, and noting that concentration may be given by N/V leads to the expression

$$\frac{-d[\text{NaOH}]}{dt} = k[\text{NaOH}]^2$$

3. Solve for concentration-versus-time profile.
The preceding expression may be separated and integrated to give a concentration-versus-time profile that may be tested against the experimental data. Integrating,

$$-\int_{[\text{NaOH}]_0}^{[\text{NaOH}]_t} \frac{d[\text{NaOH}]}{[\text{NaOH}]^2} = k \int_0^t dt$$

Carrying out the indicated integrations and simplifying leads to the expression

$$\frac{1}{[\text{NaOH}]_0} - \frac{1}{[\text{NaOH}]} = -kt$$

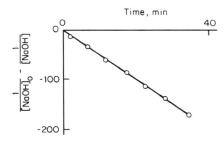

FIGURE 5.1   Concentration-versus-time profile (Procedure 5.1)

4. Plot the data.
The assumed model (first-order in both reactants and irreversible) predicts that if the data are plotted as $(1/[\text{NaOH}]_0 - 1/[\text{NaOH}])$ versus time, a straight line passing through the origin should be obtained, and the slope of this line will be the reaction rate constant $-k$. A plot of the experimental data according to this model is shown in Fig. 5.1. As may be seen, the data fit the assumed model quite well. The reaction rate constant, as determined by measuring the slope, is found to be

$$k = 6.42 \ \text{L/(g} \cdot \text{mol)(min)}$$

Thus the rate of saponification of ethyl acetate with sodium hydroxide may be adequately modeled by a rate expression of the form

$$-r_{\text{NaOH}} = 6.42[\text{NaOH}]^{1.0}[\text{CH}_3\text{COOC}_2\text{H}_5]^{1.0}$$

Related Calculations.   (1) Integral analysis may be used on data from any reactor from which integral reaction rate data have been obtained. The preceding procedure applies equally well to data from an integral tubular-flow reactor, if the tube-flow material balance

$$+r_i = \frac{dC_i}{d\tau}$$

is used rather than the batch-reactor material balance. Here C_i is the concentration of species i, in moles per volume, and τ is the residence time. (2) The form of the material balance to be tested depends on the reacting system under investigation and the data available for testing. The preceding analysis was applied because the system was of constant density and the data were in the form of concentration versus time. Data are often presented as fractional conversion (X_A) versus time, where

$$X_A = \frac{C_{A0} - C_A}{C_{A0}}$$

5.2   DETERMINING A RATE EXPRESSION BY DIFFERENTIAL ANALYSIS OF BATCH-REACTOR DATA

Determine an appropriate rate expression for the gas-phase reaction $A \rightarrow 2B$ utilizing the following data from a constant-volume batch reactor:

Time, h	Total pressure, kPa (atm)
0	132.74 (1.31)
0.5	151.99 (1.5)
1	167.19 (1.65)
1.5	178.33 (1.76)
2	186.44 (1.84)
2.5	192.52 (1.90)
3	197.58 (1.95)
3.5	201.64 (1.99)
4	205.18 (2.025)
5	210.76 (2.08)
6	214.81 (2.12)
7	217.85 (2.15)
8	220.38 (2.175)

The reaction mixture consists of 76.94%. A with 23.06% inerts at 101.325 kPa (1 atm) and 287 K (57.2°F). The reaction is initiated by dropping the reactor into a constant-temperature bath at 373 K (212°F). Equilibrium calculations have shown the reaction to be essentially irreversible in this temperature range.

Calculation Procedure

1. _Propose a generalized rate expression for testing the data._
Analysis of rate data by the differential method involves utilizing the entire reaction-rate expression to find reaction order and the rate constant. Since the data have been obtained from a batch reactor, a general rate expression of the following form may be used:

$$\frac{dC_A}{dt} = -kC_A^\alpha$$

where k and α are the reaction rate constant and reaction order to be determined respectively.

2. _Convert the rate expression to units of pressure._
Since the data are in the form of total pressure versus time, the rate expression to be tested must also be in the form of total pressure versus time. Assuming ideal-gas behavior, $PV = nRT$, and

therefore,

$$C_i = \frac{n_i}{V} = \frac{P_i}{RT}$$

where P_i = partial pressure of species i. Thus the rate expression becomes

$$\frac{1}{RT}\frac{dP_A}{dt} = -\left(\frac{1}{RT}\right)^{\alpha} k\,P_A^{\alpha}$$

Now, partial pressure of species A must be related to total-system pressure. This may be done easily by a general mole balance on the system, resulting in the following relationships:

a. *For any reactant,*

$$P_R = P_{R0} - \frac{r}{\Delta n}(\pi - \pi_0)$$

b. *For any product,*

$$P_S = P_{S0} + \frac{s}{\Delta n}(\pi - \pi_0)$$

where P_{R0} and P_{S0} are initial partial pressures of reactant R and product S, r and s are molar stoichiometric coefficients on R and P, π is total pressure, π_0 is initial total pressure, and Δn is net change in number of moles, equaling total moles of products minus total moles of reactants.

In the present case, r for the reactant A equals 1, s for the product B equals 2, and Δn equals $(2 - 1) = 1$. Using the data, and the relationship between partial pressure and total pressure for a reactant, the form of the rate expression to be tested may be derived:

$$P_A = P_{A0} - \frac{1}{2-1}(\pi - \pi_0)$$

From the data, $\pi_0 = 132.74$ kPa (1.31 atm), so $P_{A0} = (132.74)(0.7694) = 102.13$ kPa (1.0 atm). Therefore, $P_A = 102.13 - (1/1)(\pi - 132.74) = 234.87 - \pi$, with P_A and π in kilopascals, and $dP_A/dt = -d\pi/dt$. Thus the rate expression becomes $d\pi/dt = k'(234.87 - \pi)^{\alpha}$, where $k' = k(RT)^{1-\alpha}$.

3. *Linearize the rate expression by taking logs, and plot the data.*
The proposed rate expression may be linearized by taking logs, resulting in the following expression:

$$\ln\left(\frac{d\pi}{dt}\right) = \ln k' + \alpha \ln(234.87 - \pi)$$

This expression indicates that if $\ln(d\pi/dt)$ is plotted against $\ln(234.87 - \pi)$, a straight line should result with slope α and y intercept $\ln k'$. Thus, to complete the rate-data analysis, the derivative $d\pi/dt$ must be evaluated.

Three methods are commonly used to estimate this quantity: (1) slopes from a plot of π versus t, (2) equal-area graphic differentiation, or (3) Taylor series expansion. For details on these, see a mathematics handbook. The derivatives as found by equal-area graphic differentiation and other pertinent data are shown in the following table:

Time, h	$234.87 - \pi$, kPa	$d\pi/dt$
0	102.13	44.5
0.5	82.88	34
1	67.68	26
1.5	56.54	19.5
2	48.43	15
2.5	42.35	11
3	37.29	9
3.5	33.23	7.5
4	29.69	6.5
5	24.11	4.5
6	20.06	3.5
7	17.02	2.5
8	14.49	1.5

Plotting $\ln(d\pi/dt)$ versus $\ln(234.87 - \pi)$ yields an essentially straight line with a slope of 1.7 and an intercept of 0.0165. Thus, an appropriate rate expression for this reaction is given by

$$d\pi/dt = 0.0165(234.87 - \pi)^{1.7}$$

or

$$-dP_A/dt = 0.0165P_A^{1.7}$$

Related Calculations.   The rate expression derived above may be converted back to concentration units by noting that $C_A = P_A / RT$ and using $T = 373$ K, $R = 0.0821$ (L)(atm)/(g · mol)(K).

5.3   FINDING REQUIRED VOLUME FOR AN ADIABATIC CONTINUOUS-FLOW STIRRED-TANK REACTOR

Determine the volume required for an adiabatic mixed-flow reactor processing 56.64 L/min (2 ft^3 min or 0.05664 m^3/min) of a liquid feed containing reactant R and inerts I flowing at a rate of 0.67 g · mol/min and 0.33 g · mol/min, respectively. In the reactor, R is isomerized to S and T (90 percent fractional conversion of R) by the following elementary reaction: $R \xrightarrow{k_1} S + T$. Feed enters the reactor at 300 K (80.6°F). Data on the system are as follows:

Heat Capacities

$$R = 7 \text{ cal/(g · mol)(°C)}$$
$$S = T = 4 \text{ cal/(g · mol)(°C)}$$
$$I = 8 \text{ cal/(g · mol)(°C)}$$

Reaction Rate Constant at 298 K

$$k_1 = 0.12 \text{ h}^{-1}$$

Activation Energy

$$25{,}000 \text{ cal/(g} \cdot \text{mol)}$$

Heat of Reaction at 273 K

$$\Delta H_R = -333 \text{ cal/(g} \cdot \text{mol) of } R$$

Calculation Procedure

1. *Write the material- and energy-balance expressions for the reactor.*
This problem must be solved by simultaneous solution of the material- and energy-balance relationships that describe the reacting system. Since the reactor is well insulated and an exothermic reaction is taking place, the fluid in the reactor will heat up, causing the reaction to take place at some temperature other than where the reaction rate constant and heat of reaction are known.

Assuming a constant-density reacting system, a constant volumetric flow rate through the reactor, and steady-state operation, a material balance on species R gives the expression

$$v C_{R0} - v C_R + V r_R = 0$$

where C_{R0} = concentration of R in feed
 C_R = concentration of R in products
 v = volumetric flow rate
 V = reactor volume
 r_R = net rate of formation of R

This equation can be rearranged into

$$\frac{C_{R0} - C_R}{-r_R} = \tau$$

where τ = residence time (V/v).

A first-law energy balance on the continuous-flow stirred-tank reactor gives the expression

$$Q = \sum_{i=1}^{n} F_{i0} \hat{C}_{p,i}(T - T_{i0}) + X[\Delta H° + \Delta \hat{C}_p(T - T°)]$$

where Q = rate of heat exchange with surroundings
 F_i = outlet molar flow rate of species i
 F_{i0} = inlet molar flow rate of species i
 $\hat{C}_{p,i}$ = mean heat capacity of species i
 T_{i0} = inlet (feed) temperature of species i
 T = reactor operating temperature
 X = molar conversion rate of species R $(= F_{R0} - F_R)$
 $\Delta H°$ = heat of reaction at $T°$
 $T°$ = reference temperature for heat-of-reaction data
 $\Delta \hat{C}_p = \Sigma v_i C_{p,i \text{ products}} - \Sigma v_i C_{p,i \text{ reactants}}$
 v_i = molar stoichiometric coefficient

2. *Calculate the operating temperature in the reactor.*
Application of the energy balance shown above allows the reaction mass temperature to be calculated, since all quantities in the expression are known except T. Pertinent calculations and parameters

are as follows:

$$Q = 0 \text{ (adiabatic)}$$
$$F_{R0} = 0.67 \text{ mol/min}$$
$$F_{I0} = 0.33 \text{ mol/min}$$
$$T_{R0} = T_{I0} = 300 \text{ K}$$
$$X = X_R F_{R0} = (0.90)(0.67) = 0.603 \text{ g} \cdot \text{mol } R \text{ per minute}$$
$$\Delta \hat{C}_p = 4 + 4 - 7 = 1 \text{ cal/(g} \cdot \text{mol)(°C)}$$

Substituting into the energy balance,

$$0 = (0.67)(7)(T - 300) + (0.33)(8)(T - 300) + 0.603[-333 + (1)(T - 273)]$$

Solving for reactor temperature,

$$T = 323.4 \text{ K (122.7°F)}$$

3. Calculate the reaction rate constant at the reactor operating temperature.
Since the temperature in the reactor is not 25°C (where the value for the reaction rate constant is known), the rate constant must be estimated at the reactor temperature. The Arrhenius form of the rate constant may be used to obtain this estimate:

$$k = A \exp[-E_A/(RT)]$$

where A = preexponential factor
E_A = activation energy
R = universal gas constant
T = absolute temperature

Dividing this Arrhenius equation for $T = T$ by the Arrhenius equation for $T = 298$ K (25°C) and noting that $(1/T) - (1/298) = (298 - T)/298T$, the following expression can be derived:

$$k_T = k_{298} \exp\left[\frac{-25,000}{1.987}\left(\frac{298 - T}{298T}\right)\right]$$

Now $k_{298} = 0.12$ h^{-1}, so when $T = 323.4$ K (the reactor operating temperature), the reaction rate constant becomes $k_{323.4} = 3.31$ h^{-1}.

4. Solve for reactor volume using the material-balance expression.
The material balance for the continuous-flow stirred-tank reactor may now be used to calculate the reactor volume required for the isomerization. Inserting the first-order rate expression into the material balance,

$$\tau = \frac{C_{R0} - C_R}{kC_R} = \frac{V}{v}$$

To apply this material balance, it is first necessary to calculate the inlet and outlet concentrations of species R. This may be easily accomplished from the given data and the relationships

$$C_i = \frac{F_i}{v} \qquad F_i = F_{i0}(1 - X_i)$$

Since v = constant, $C_i = C_{i0}(1 - X_i)$. In these relationships, C_i is the concentration of species i, F_i is the molar flow rate of species i, v is volumetric flow rate, and X_i is fractional conversion of species i.

For this example,

$$C_{R0} = \frac{F_{R0}}{v} = \frac{0.67}{56.64} = 0.012 g \cdot \text{mol/}L$$

and

$$C_R = 0.012(1 - 0.9) = 0.0012 \text{ g} \cdot \text{mol/}L$$

Thus the reactor volume may now be directly calculated after converting the volumetric flow rate to an hourly basis:

$$V = \frac{0.012 - 0.0012}{(3.31)(0.0012)}(56.64)(60) = 9240 \text{ L } (326 \text{ft}^3 \text{ or } 9.24 \text{ m}^3)$$

Related Calculations. (1) Since the reaction is irreversible, equilibrium considerations do not enter into the calculations. For reversible reactions, the ultimate extent of the reaction should always be checked first, using the procedures outlined in Section 4. If equilibrium calculations show that the required conversion cannot be attained, then either the conditions of the reaction (e.g., temperature) must be changed or the design is not feasible. Higher temperatures should be investigated to increase ultimate conversions for endothermic reactions, while lower temperatures will favor higher conversions for exothermic reactions.

(2) The simultaneous solving of the material- and energy-balance expressions may yield more than one solution. This is especially true for exothermic reactions occurring in continuous stirred-tank reactors. The existence of other feasible solutions may be determined by plotting the energy-balance and material-balance expressions on molar conversion rate versus temperature coordinates, as shown in Fig. 5.2. In the figure, points A and C represent stable operating points for the reactor, while point B is the metastable, or "ignition," point, where stable operation is difficult. This is due to the relative slopes of the material- and energy-balance lines at B. Small positive temperature excursions away from B result in "ignition" of the reaction mass because the rate of heat generation is greater than that of heat removal, and the reactor restabilizes at point A. Similarly, a small negative temperature deviation from B causes the reaction mass to "quench," and the reactor restabilizes at point C.

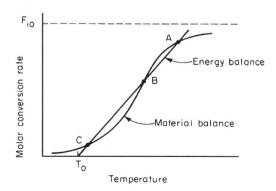

FIGURE 5.2 Molar conversion rate versus temperature (Procedure 5.3).

5.4 CALCULATING THE SIZE OF AN ISOTHERMAL PLUG-FLOW REACTOR

Laboratory experiments on the irreversible, homogeneous gas-phase reaction $2A + B = 2C$ have shown the reaction rate constant to be 1×10^5 (g · mol/L)$^{-2}$ s^{-1} at 500°C (932°F). Analysis of isothermal data from this reaction has indicated that a rate expression of the form $-r_A = kC_A C_B^2$ provides an adequate representation for the data at 500°C and 101.325 kPa (1 atm) total pressure. Calculate the volume of an isothermal, isobaric plug-flow reactor that would be required to process 6 L/s (0.212 ft^3/s) of a feed gas containing 25% A, 25% B, and 50% inserts by volume if a fractional conversion of 90% is required for component A.

Calculation Procedure

1. Develop a plug-flow-reactor design equation from the material balance.

To properly size a reactor for this reaction and feedstock, a relationship between reactor volume, conversion rate of feed, and rate of reaction is needed. This relationship is provided by the material balance on the plug-flow reactor.

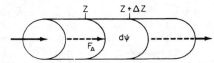

For a single ideal reactor, a component material balance on a differential volume element $d\psi$ (see Fig. 5.3) becomes, for species A

$$F_{AZ} - F_{AZ+\Delta Z} + r_A \Delta \psi = 0 \quad \text{or} \quad \frac{dF_A}{d\psi} = r_A$$

FIGURE 5.3 Differential volume element for plug-flow reactor (Procedure 5.4).

where ψ is a reactor-size parameter (volume, mass of catalyst, etc.), F_A is molar flow rate of A, and r_A is the rate of generation of A per unit of volume. Upon rearranging and integrating,

$$\int_{F_{A0}}^{F_A} \frac{dF_A}{r_A} \int_0^\psi d\psi = \psi$$

In terms of total molar conversion rate of species A, designated X, $F_A = F_{A0} - X$, and $dF_A = -dX$, where F_{A0} is molar flow rate of A in feed. Thus the material balance for this homogeneous reaction becomes

$$\psi = V = -\int_0^X \frac{dX}{r_A}$$

2. Relate molar flow rates of products and reactants to conversion rate of A.

To use the mass balance to solve for reactor size V, it is necessary to relate the rate of generation $+r_A$ of A to the molar conversion rate X of A. This is easily done through the rate expression. For this reaction

$$+r_A = -kC_A C_B^2$$

where C_A and C_B are molar concentrations of A and B. For this gas-phase reaction, the concentrations of A and B may be calculated from the ideal-gas law, so

$$C_A = n_A/V = \frac{p_A}{RT} = \frac{p_T y_A}{RT}$$

where p_A = partial pressure of A
V = volume
n_A = number of moles of A
p_T = total system pressure
R = universal gas constant
T = absolute temperature
y_A = mole fraction of A
The rate expression then becomes

$$+r_A = -k\left(\frac{p_T}{RT}\right)^3 y_A y_B^2$$

Mole fractions for components A and B are computed from a component-by-component mole balance on the reactor, noting that $1/2$ mol B and 1 mol C are involved per mole of A reacted. Thus,

$$F_A = F_{A0} - X$$
$$F_B = F_{B0} - X/2$$
$$F_C = F_{C0} + X$$
$$F_I = F_{I0}$$
$$\overline{F_T = F_{T0} - X/2}$$

In this balance, F_A, F_B, and F_C are the molar flow rates of species A, B, and C at any position in the reactor, F_{A0}, F_{B0}, and F_{C0} are the molar flow rates of A, B, and C in the feed, and F_I and F_{I0} are the molar flow rates of inerts in product and in the feed.

Now, by definition,

$$y_A = \frac{F_A}{F_T} = \frac{F_{A0} - X}{F_{T0} - X/2} \qquad \text{and} \qquad y_B = \frac{F_B}{F_T} = \frac{F_{B0} - X/2}{F_{T0} - X/2}$$

and the rate expression in terms of molar conversion rate of A therefore becomes:

$$+r_A = -k\left(\frac{pT}{RT}\right)^3 \left[\frac{(F_{A0} - X)(F_{B0} - X/2)^2}{(F_{T0} - X/2)^3}\right]$$

3. *Evaluate molar flow rates and integrate the mass balance.*
Once the known data on molar flow rates are substituted into the mass balance, it can be solved for reactor volume. From step 1, the equation to be solved is

$$V = \frac{1}{k\left(\dfrac{pT}{RT}\right)^3} \int_0^x \frac{dX}{\dfrac{(F_{A0} - X)(F_{B0} - X/2)^2}{(F_{T0} - X/2)^3}}$$

From the problem statement,

$$C_{A0} = \frac{p_{A0}}{RT} = \frac{0.25(1)}{0.082(773)} = 0.004 \ g \cdot \text{mol/L}$$

and

$$F_{A0} = v_0 C_{A0} = 6(0.004) = 0.024 \text{ g} \cdot \text{mol/L}$$

and

$$F_{B0} = F_{A0}$$

Further,

$$F_{I0} = F_I = 0.048 \text{ g} \cdot \text{mol/s}, \quad \text{and} \quad F_{T0} = F_{A0} + F_{B0} + F_{I0} = 0.096 \text{ g} \cdot \text{mol/s}$$

For 90 percent fractional conversion of A, the upper limit on the integral may be evaluated as

$$x = X_A F_{A0} = 0.9(0.024) = 0.022 \text{ g} \cdot \text{mol/s}$$

The integral may now be solved for reactor volume. Since analytical solution is difficult because of the complexity of the integrand, graphic integration (not shown here) may be used to find the required reactor volume.

By graphic integration, the value of the integral is evaluated and reactor volume is calculated as follows: The value of the integral $f(X)dX = 91$; then

$$k\left(\frac{P_T}{RT}\right)^3 = 10^5 \left[\frac{1}{0.082(773)}\right]^3 = 0.393$$

Therefore,

$$V = \frac{91}{0.393} = 232 \text{ L } (8.2 \text{ ft}^3 \text{ or } 0.23 \text{ m}^3)$$

Related Calculations.   (1) The integral may be evaluated more precisely using any of the various numerical integration routines (Euler, Runge-Kutta, for example), which may be readily programmed for computer solution.

(2) If the reaction is reversible instead of irreversible, its ultimate extent must be checked (using methods outlined in the preceding section) to make sure that the required 90% fractional conversion can be attained. If it cannot, either the problem specification must be altered to allow a lower conversion, or the reactor conditions (i.e., reaction temperature or total pressure) must be changed.

5.5   CALCULATING THE REQUIRED REACTION TIME IN A BATCH REACTOR

The heterogeneous gas-phase hydrogenation of acetylene to ethane at 1000 K (1341°F), that is,

$$C_2H_2 + 2H_2 \rightarrow C_2H_6$$

has been found to proceed with a reaction rate that may be adequately represented by the rate expression

$$-r_{C_2H_2} = kC_{H_2}^2 C_{C_2H_2}$$

Laboratory experiments at 1000 K indicate that the reaction rate constant is 1×10^5 g · mol/(L)(min). If a mixture of 75 mol % hydrogen and 25% acetylene is charged to a 1-L (0.035-ft^3 or 0.001-m^3) batch reactor and 0.001 g · mol acetylene is present initially in the reactor, calculate the holding time at 1000 K necessary to achieve 90% fractional conversion of acetylene to ethane. The reactor is to be operated at a constant temperature of 1000 K.

Calculation Procedure

1. ***Develop a batch-reactor design equation from the mass balance.***
To find the required holding time, a relationship between reaction time and the rate of conversion of acetylene must be developed. This may be developed from a mass balance on the batch reactor. Since the molar density of the reacting mixture is not constant (there is a net change in the number of moles due to reaction), the pressure of the reactor will have to change accordingly.

A material balance on component A in a batch reactor gives

$$\left(\frac{1}{V} \right) \frac{dN_A}{dt} = r_A$$

where r_A = net rate of generation of species A
V = volume
N_A = moles of A

Since concentration equals N/V, the rate of reaction of acetylene is given by

$$-r_{C_2H_2} = k \left[\frac{N_{H_2}}{V} \right]^2 \left[\frac{N_{C_2H_2}}{V} \right] \quad \text{or} \quad -r_{C_2H_2} = \frac{k}{V^3} N_{H_2}^2 N_{C_2H_2}$$

The design equation for the batch reactor thus becomes

$$\frac{dN_{C_2H_2}}{dt} = -\frac{k}{V^2} N_{H_2}^2 N_{C_2H_2}$$

This equation may be recast in terms of total moles X of acetylene converted, where X is a function of elapsed time. Since the reaction of 1 mol acetylene involves 2 mol hydrogen and produces 1 mol ethane, the batch reactor will contain at any given time t the following mixture (with subscript 0 referring to the amount initially present):

$$N_{C_2H_2} = N_{C_2H_{2,0}} - X$$
$$N_{H_2} = N_{H_{2,0}} - 2X$$
$$N_{C_2H_6} = X$$

The total number of moles is thus the sum of these three equations, or $N_{T,0} - 2X$. The design equation thus becomes

$$\frac{-dX}{dt} = \frac{-k}{V^2} (N_{H_{2,0}} - 2X)^2 (N_{C_2H_{2,0}} - X)$$

2. Separate and integrate the batch-reactor design equation.
The reaction time required for any given initial composition and conversion of reactants may now be calculated directly from the preceding expression, once this expression has been integrated and solved for $t = f(X)$. The appropriate solution method is as follows:

a. *Separate the variables.*

$$\int_0^x \frac{dX}{(N_{H_{2,0}} - 2X)^2 (N_{C_2H_{2,0}} - X)} = \frac{k}{V^2} \int_0^t dt$$

b. *Integrate.*

$$\int_0^x \frac{dX}{(N_{H_{2,0}} - 2X)^2 (N_{C_2H_{2,0}} - X)} = \frac{1}{(-N_{H_{2,0}} + 2N_{C_2H_{2,0}})} \frac{1}{(N_{H_{2,0}} - 2X)}$$

$$+ \frac{-1}{(-N_{H_{2,0}} + 2N_{C_2H_{2,0}})} \ln\left(\frac{N_{C_2H_{2,0}} - X}{N_{H_{2,0}} - 2X}\right) = \frac{k}{V^2} t$$

c. *Evaluate between limits.*
Evaluation of the integral between 0 and x for the left-hand side and between 0 and t for the right-hand side gives the following expression for reaction time as a function of total molar conversion:

$$\frac{1}{(N_{H_{2,0}} - 2X)(2N_{C_2H_{2,0}} - N_{H_{2,0}})} - \frac{1}{N_{H_{2,0}}(2N_{C_2H_{2,0}} - N_{H_{2,0}})}$$

$$+ \frac{1}{(2N_{C_2H_{2,0}} - N_{H_{2,0}})^2} \ln\left[\frac{N_{C_2H_{2,0}}/N_{H_{2,0}}}{(N_{C_2H_{2,0}} - X)/(N_{H_{2,0}} - 2X)}\right] = \frac{k}{V^2} t$$

3. Solve for reaction time.
The total molar conversion of acetylene at 90% fractional conversion may be found by direct application of the definitions for fractional and total molar conversion. Thus, let $X_{C_2H_2}$ stand for the fractional conversion of acetylene, which is defined as

$$\frac{N_{C_2H_{2,0}} - N_{C_2H_2}}{N_{C_2H_{2,0}}}$$

According to the problem statement, this equals 0.9. Since $N_{C_2H_{2,0}} = 0.001$ g $\cdot$ mol, $N_{C_2H_2} = 0.001 - (0.9)(0.001) = 0.0001$ g $\cdot$ mol from step 1, X equals total molar conversion of acetylene, defined as $N_{C_2H_{2,0}} - N_{C_2H_2}$. It is therefore equal to $(0.001 - 0.0001) = 0.0009$ g $\cdot$ mol.

This value for total molar conversion, along with the initial moles of C_2H_2 and H_2, now allows reaction time to be calculated. Since the initial mixture is 75% hydrogen and 25% acetylene, $N_{H_{2,0}} = 3(0.001) = 0.003$. Substituting into the design equation from step 2,

$$\frac{1}{(0.003 - 0.0018)(0.002 - 0.003)} - \frac{1}{0.003(0.002 - 0.003)}$$

$$+ \frac{1}{(0.002 - 0.003)^2} \ln\left[\frac{(0.001/0.003)}{(0.001 - 0.0009)/(0.003 - 0.0018)}\right] = \frac{10^5}{1^2} t$$

Solving, $t = 8.86$ min. Thus approximately 9 min is needed to convert 90% of the acetylene originally charged to the 1-L reactor to ethane.

Related Calculations. (1) The stoichiometry of the chemical reaction strongly influences the final form of the integrated design expression. Different rate expressions will lead to different functional relationships between time and total molar conversion. The preceding example was specific for an irreversible reaction, second-order in hydrogen and first-order in acetylene. If the rate expression had been simply first-order in acetylene (as might be the case with excess hydrogen), the integration of the design expression would yield

$$-\ln\left(\frac{N_{C_2H_2} - X}{N_{C_2H_{2,0}}}\right) = kt$$

Similarly, other rate expressions yield different forms for the time, molar conversion relationship.

(2) In many cases, analytical integration of the design equation is difficult. The integral may still be evaluated, however, by (a) numerical integration methods, such as Euler or Runge-Kutta, or (b) graphic evaluation of $\int_0^x f(X)dX$ by plotting $f(X)$ versus X and finding the area under the curve.

5.6 CALCULATING REACTION RATES FROM CONTINUOUS-FLOW STIRRED-TANK REACTOR DATA

In a study of the nitration of toluene by mixed acids, the following data were obtained in a continuous-flow stirred-tank reactor. It had been previously determined that the reactor was well mixed; the composition within the reactor and in the exit stream can be considered equal. In addition, it had been determined that mass-transfer effects were not limiting the process rate. Thus the rate measured is the true kinetic rate of reaction. Calculate that rate.

Reactor data:	
Mixed-acid feed rate, g/h	325.3
Toluene feed rate, g/h	91.3
Acid-phase leaving, g/h	301.4
Organic-phase leaving, g/h	117.3
Temperature, °C (°F)	36.1 (97.0)
Stirrer speed, r/min	1520
Reactor volume, cm^3	635
Volume fraction of acid phase in reactor	0.67
Organic-phase composition:	
Mononitrotoluene, mol %	68.1
Toluene, mol %	31.1
Sulfuric acid, mol %	0.8
Density at 25°C, g/cm^3	1.0710
Feed acid-phase composition:	
H_2SO_4, mol %	29.68
HNO_3, mol %	9.45
H_2O, mol %	60.87
Density at 25°C, g/cm^3	1.639
Spent acid composition:	
H_2SO_4, mol %	29.3
HNO_3, mol%	0.3
H_2O, mol %	70.4
Density, g/cm^3	1.603

1. Check the elemental material balances.
For reference, the molecular weights involved are as follows:

Component	Molecular weight
Toluene	92
Mononitrotoluene	137
Sulfuric acid	98
Nitric acid	63
Water	18

The feed consists of 0.0913 kg/h of toluene and 0.3253 kg/h of mixed acid. The latter stream can be considered as follows:

Acid component	Mol % (given)	Kilograms per 100 mol feed	Wt %	kg/h
H_2SO_4	29.68	2908.6	63.24	0.2057
HNO_3	9.45	595.4	12.94	0.0421
H_2O	60.87	1095.7	23.82	0.0775
		4599.7	100.00	0.3253

The output consists of 0.1173 kg/h of the organic phase and 0.3014 kg/h of spent acid. These two streams can be considered as follows:

Organic-phase component	Mol % (given)	Kilograms per 100 mol product	Wt %	kg/h
Toluene	31.1	2,861.2	23.32	0.0274
Mononitrotoluene	68.1	9,329.7	76.04	0.0892
Sulfuric acid	0.8	78.4	0.64	0.0007
		12,269.3	100.00	0.1173

Spent-acid-phase component	Mol % (given)	Kilograms per 100 mol acid	Wt %	kg/h
H_2SO_4	29.3	2871.4	69.07	0.2082
HNO_3	0.3	18.9	0.45	0.0014
H_2O	70.4	1267.2	30.48	0.0918
	100.0	4157.5	100.00	0.3014

The elemental material balances can then be checked. For carbon:

Component	kg/h of C in	kg/h of C out
Toluene	0.0834	0.0250
Mononitrotoluene	0.0000	0.0547
	0.0834	0.0797

Percent difference in C = (100)(0.0834 − 0.0797)/0.0834 = 4.44%

For hydrogen:

Component	kg/h of H in	kg/h of H out
Mononitrotoluene	0.0000	0.0046
Toluene	0.0079	0.0024
H_2SO_4	0.0042	0.0043
HNO_3	0.0007	0.0000
H_2O	0.0086	0.0102
	0.0214	0.0215

Percent difference in H = (100)(0.0215 − 0.0214)/0.0214 = 0.5%

For oxygen:

Component	kg/h of O in	kg/h of O out
Mononitrotoluene	0.0000	0.0208
H_2SO_4	0.1343	0.1364
HNO_3	0.0321	0.0011
H_2O	0.0689	0.0816
	0.2353	0.2399

Percent difference in O = (100)(0.2399 − 0.2353)/0.2353 = 2%

The elemental balances for C, H, and O suggest that the run is reasonably consistent (because the percent differences between feed and product are small), so there is no need to make material balances for the other elements.

2. Employ the material-balance data to determine the reaction rate.

In a continuous-flow stirred-tank reactor, the material balance for component A is as follows at steady state:

Rate of input of A − rate of output of A + rate of generation of $A = 0$

Thus,

$$(F_{A0} - F_A) + r_A V = 0$$

where F_{A0}, F_A = inlet and outlet molar flow rates
r_A = rate of reaction per unit volume of organic phase
V = volume of the organic phase

Rearranging the material balance,

$$-r_A = \frac{F_{A0} - F_A}{V} = \frac{W_{A0} - W_A}{MV}$$

where W_{A0} and W_A are mass flow rates, and M is the molecular weight of A.

The reaction stoichiometry is as follows:

$$C_6H_5CH_3 + HNO_3 \xrightarrow[H_2O]{H_2SO_4} C_6H_4(CH_3)(NO_2) + H_2O$$

Since the reaction involves 1 mol of each component, r(toluene) = $r(HNO_3)$ = $-r$(mononitrotoluene) = $-r(H_2O)$.

From the reactor data given in the statement of the problem, the organic-phase volume is
0.33(0.635 L) = 0.210 L. Thus, by material balance, the computed rates are

Component	r, kg · mol/(h)(L)
Toluene	0.00331
HNO_3	0.00308
Mononitrotoluene	0.00310
Water	0.00370

These can be averaged to give the mean rate of reaction: 0.00330 ± 0.00021 kg · mol/(h)(L). For a
reactant (e.g., toluene), a minus sign should be placed in front of it.

Related Calculations.   In this example, it was stated at the outset that mass-transfer effects were
not limiting the process rate. In the general case, however, it is important to calculate the effect of
mass-transfer resistance on the reaction rate.

Consider, for instance, the gasification of porous carbon pellets in a fixed-bed reactor using
steam and oxygen; the reaction rate could be affected both by external-film mass transfer and by
pore-diffusion mass transfer.

The first of these pertains to the stagnant film separating the particle surface from the bulk gas.
At steady state, the rate of transport to the surface is given by the standard mass-transfer expression

$$W = k_m A_p C (Y_B - Y_S) = k_m A_p (C_B - C_S)$$

where W = transfer rate, in moles per time per weight of solid
k_m = mass-transfer coefficient, in length per time
A_p = external surface area, per weight of solid
Y_B = bulk-gas concentration, in mole fraction units
Y_S = concentration of gas adjacent to surface, in mole fraction units
C = total gas concentration, in moles per volume
C_B = concentration of component in the bulk, in moles per volume
C_S = concentration of component adjacent to surface, in moles per volume.

The mass-transfer coefficient k_m is a weak function of absolute temperature and velocity. The
total concentration C is given approximately by the ideal-gas law $C = P/(RT)$, where P is absolute
pressure, R is the gas constant, and T is absolute temperature.

In fixed-bed operation, Satterfield [1] recommends correlations for mass (and heat) transfer coefficients based on the Colburn j factor, defined as follows:

$$j = \frac{k_m}{(\rho^* V)} Sc^{2/3}$$

where j = Colburn j factor, dimensionless
k_m = mass-transfer coefficient, in moles per unit of time per unit area of particle surface
ρ^* = molar density, in moles per volume
V = superficial velocity, based on empty reactor tube
Sc = Schmidt number, $\mu/\rho D$, dimensionless
μ = viscosity
ρ = mass density
D = diffusivity through the film

The j factor depends on the external bed porosity ϵ and the Reynolds number, $Re = D_p V\rho/\mu$, where D_p is the particle diameter, as follows:

$$\epsilon j = \frac{0.357}{Re^{0.359}} \quad 3 \leq Re \leq 2000$$

The appropriate particle diameter is given as

$$D_p = \frac{6V_{ex}}{S_{ex}}$$

where V_{ex} = volume of particle
S_{ex} = surface area of particle

External mass transfer reduces the concentration of reactant gas close to the particle surface and thus reduces the overall process rate. Thus, consider gasification to be a first-order reaction. Then at steady state, the rate of gasification equals the rate of mass transfer. For a nonporous solid, the surface reaction (whose rate constant is k^*) consumes the diffusing reactant:

$$k^* C_S = k_m A_p (C_B - C_S)$$

Solving for the surface concentration yields

$$C_S = \frac{k_m A_p C_B}{k^* + k_m A_p}$$

Now, the process rate is given by

$$-r_c = k^* C_S = \frac{k^* k_m A_p C_B}{k^* + k_m A_p}$$

So, if the mass-transfer rate constant k_m is large in comparison to k^*, the rate reduces to $-r_c = k^* C_B$; that is, the true kinetic rate is based directly on the bulk concentration.

Next, consider the effect of pore diffusion on the reaction rate. The gasification reaction occurs principally within the particle. Except at very high temperatures, reactants must diffuse into the pore to the reacting surface. The average reaction rate within the particle may be related to the rate based on the surface concentration in terms of an effectiveness factor η defined as

$$\eta = \frac{r_{avg}}{r_{surface}}$$

The effectiveness factor is a function of a dimensionless group termed the "Thiele modulus," which depends on the diffusivity in the pore, the rate constant for reaction, pore dimension, and external surface concentration C_S.

The effectiveness factor for a wide range of reaction kinetic models differs little from that of the first-order case. For an isothermal particle, the first-order reaction effectiveness factor is given as

$$\eta = \frac{\tanh\phi}{\phi}$$

where ϕ is the Thiele modulus, that is,

$$\phi = L_p \left(\frac{k C_S^{m-1}}{V_p D} \right)^{1/2}$$

where L_p = effective pore length, cm = $R/3$ for spheres (R = particle radius)
  k = reaction rate constant
  C_S = external surface concentration, in moles per cubic centimeter
  m = reaction order (equal to 1 for first-order)
  V_p = pore volume, in cubic centimeters per gram
  D = diffusivity

When porous solids are being used as catalysts or as reactants, the rate constant k^* in the global equation is replaced by ηk^*. Consequently, this equation applies to porous solids as well as nonporous solids.

When diffusion is fast relative to surface kinetics, $\phi \to 0$, $\eta \to 1$, and $r_{avg} = r_{surface}$. Under these conditions, all the pore area is accessible and effective for reaction. When $\phi \to \infty$, that is, when diffusion is slow relative to kinetics, the reaction occurs exclusively at the particle external surface; reactant gas does not penetrate into the pores.

External mass transport generally becomes dominant at temperatures higher than that at which pore diffusion limits the gasification rate. For small particles, smaller than 20 mesh, mass-transfer limitations generally are not important because these particles have external surface areas that are large compared with their unit volume. Furthermore, mass-transfer coefficients are greater in fluid-bed operations owing to the motion of the solid particles. Thus in fluid-bed operations, external mass-transfer limitation in the temperature region below about 900 to 1100°C is never important. For fixed-bed operation, however, mass transfer to large particles can be important.

5.7 TOTAL SURFACE AREA, ACTIVE SURFACE AREA, POROSITY, AND MEAN PORE RADIUS OF A CATALYST

A catalyst composed of 10 wt % nickel on γ-alumina is used to promote the catalytic methanation reaction

$$CO + 3H_2 \rightarrow CH_4 + H_2O$$

The important properties of the catalyst for characterization purposes are the total surface area, dispersion of metallic nickel, pore volume, and mean pore radius. Determine each of these, using the experimental data given in the respective calculation steps.

Calculation Procedure

1. Calculate the total surface area.
The weight gain of the catalyst due to the physical adsorption of nitrogen under various nitrogen pressures is a function of, and thus an indicator of, total surface area. The first three columns of Table 5.1 show the weight of adsorbed nitrogen and the corresponding pressure for experimental runs conducted at the atmospheric boiling point of liquid nitrogen.

The most common way of analyzing such data is by using the so-called BET equation. For multilayer adsorption, this equation can be set out in the form

$$\frac{P}{W(P^* - P)} = \frac{1}{W_m C} + \frac{C - 1}{C W_m}\frac{P}{P^*}$$

where W is weight adsorbed per gram of catalyst at pressure P, P^* is the vapor pressure of the adsorbent, C is a parameter related to the heat of adsorption, and W_m is the weight for monolayer coverage of the solid.

The last-named variable is the one of interest in the present case, because it represents the weight of adsorbed nitrogen that just covers the entire surface of the catalyst, internal and external. (Because the catalyst is highly porous, most of the area is pore wall and is internal to the solid.)

TABLE 5.1 BET Calculations for Prototype Catalyst (Procedure 5.7)

W, mg/g	P, mmHg	P, kPa	$\dfrac{P}{(P^* - P)W} \times 10^3$	$P/P^* \times 10^2$
13	6.25	0.83	0.637	0.82
17	15.6	2.08	1.233	2.05
20	25.0	3.33	1.700	3.29
22	34.4	4.59	2.155	4.53
25	56.3	7.50	3.200	7.41
28	84.4	11.2	4.462	11.11
32	163.0	21.7	8.532	21.45

Note: Let $y = P/(P^* - P)W$, $X = P/P^*$, $y = Sx + I$. By least squares,

$$D = (\Sigma x)^2 - n\Sigma x^2$$
$$S = (\Sigma y \Sigma x - n\Sigma xy)/D$$
$$I = (\Sigma x \Sigma xy - \Sigma y \Sigma x^2)/D$$
$$\Sigma y = 21.92 \times 10^{-3}$$
$$\Sigma xy = 2.747 \times 10^{-3}$$
$$\Sigma x^2 = 0.06747$$
$$\Sigma x = 0.5066$$
$$n = 7$$
$$S = 0.037 \text{ mg}^{-1}$$
$$I = 4.05 \times 10^{-4} \text{ mg}^{-1}$$

Now, from the form of the equation, a plot of $P/[W(P^* - P)]$ against P/P^* should yield a straight line. Let S and I, respectively, stand for the slope $(C - 1)/(CW_m)$ and the intercept $1/(W_m C)$ of that line. Then, by algebraic rearrangement, $W_m = 1/(S + I)$. Since the tests were conducted at the atmospheric boiling point, P^* was essentially 760 mmHg (101.3 kPa).

Values for $P/[W(P^* - P)]$ and P/P^* appear in the fourth and fifth columns of Table 5.1. Application of ordinary least-squares regression analysis to the resulting plot (not shown) shows S to be 0.0377 mg^{-1} and I to be 4.05×10^{-4} mg^{-1}. Therefore, $W_m = 26.26$ mg nitrogen per gram of catalyst. This equals $0.02626/28 = 9.38 \times 10^{-4}$ g $\cdot$ mol nitrogen. Employing Avogadro's number, this is $(9.38 \times 10^{-4})(6.023 \times 10^{23}) = 5.65 \times 10^{20}$ nitrogen molecules. Finally, the nitrogen molecule can be taken to have a surface area of 15.7×10^{-20} m^2. Therefore, the surface area of the catalyst (in intimate contact with the nitrogen monolayer) can be estimated to be $(5.65 \times 10^{20})(15.7 \times 10^{-20})$, or about 89 m^2/g.

2. *Estimate the dispersion of the nickel.*

Hydrogen dissociatively adsorbs on nickel whereas it does not interact with the catalyst support and is not significantly adsorbed within the nickel crystal lattice. Therefore, the amount of uptake of hydrogen by the catalyst is a measure of how well the nickel has been dispersed when deposited on the support.

At several pressures up to atmospheric, uptake of hydrogen proved to be constant, at 0.256 mg/g of catalyst, suggesting that the exposed nickel sites were saturated.

Thus, per gram of catalyst, the atoms of hydrogen adsorbed equals $(0.256 \times 10^{-3}\text{g})$ ($\frac{1}{2}$ mol/g)(2 atoms/ molecule)(6.023×10^{23} molecules/mol) = 1.542×10^{20} atoms. This can also be taken as the number of surface nickel atoms. Now, since the catalyst consists of 10% nickel, the total number of nickel atoms per gram of catalyst equals $(0.1 \text{ g})(1/58.71 \text{ mol/g})(6.023 \times 10^{23}$ atoms/mol) = 1.0259×10^{21} atoms. Therefore, the degree of dispersion of the nickel equals (surface nickel atoms)/(total nickel atoms) = $(1.542 \times 10^{20})/(1.0259 \times 10^{21})$ = 0.15. Thus only 15 percent of the nickel deposited has been dispersed and is available for catalysis.

3. *Calculate the porosity and the mean pore radius.*
The particle porosity may be readily determined by a helium pycnometer and a mercury porosimeter. In the pycnometer, the solid skeletal volume V_S is obtained. The skeletal density ρ_S is found from the sample weight W_S:

$$\rho_S = \frac{W}{V_S}$$

The total sample volume V_A, including pores, can be determined by mercury displacement at atmospheric pressure, since mercury will not enter the pores under these conditions. The apparent density ρ_A, then, is

$$\rho_A = \frac{W}{V_A}$$

And the porosity ϵ of the solid is given by

$$\epsilon = 1 - \rho_A/\rho_S$$

For the catalyst in question, the apparent density is 1.3 kg/dm^3 and the skeletal density is 3.0 kg/dm^3. The porosity is therefore $\epsilon = 1 - 1.3/3 = 0.57$.

The pore volume V_p equals the porosity divided by the apparent density: $V_p = \epsilon/\rho_A = 0.57/1.3 = 0.44$ dm^3/kg $= 0.44$ cm^3/g. Assuming cylindrical pores of uniform length and radius.

$$\frac{V_p}{A_p} = \frac{n\pi R_p^2 L_p}{n 2\pi R_p L_p}$$

where n = number of pores
   A_p = pore surface area (calculated in step 1)
   R_p = pore radius
   L_p = pore length

Therefore,

$$R_p = \frac{2V_p}{A_p} = \frac{2(0.44 \text{ cm}^3/\text{g})}{89 \text{ m}^2/\text{g}} = 99 \times 10^{-10} \text{ m} (99 \text{ Å})$$

5.8   SIZING AND DESIGN OF A SYSTEM OF STIRRED-TANK REACTORS

It is proposed to process 3 m^3/h of a reaction mixture in either one or two (in series) continuous-flow stirred-tank reactors. The reaction is $A + 2B \rightarrow C$. At 50°C, the kinetic rate expression is as follows:

$$-r_A = k_1 C_A C_B /(1 + k_2 C_A)$$

where $k_1 = 0.1$ and $k_2 = 0.6$, with concentrations in kilogram-moles per cubic meter and rates in kilogram-moles per cubic meter per hour.

The mixture specific gravity is constant and equal to 1.2 kg/dm^3. The molecular weight of the feed is 40. The feed contains 10 mol % A, 20% B, and 70% inert solvent S. The liquid viscosity is 0.8 mPa · s (cp) at reaction temperature.

Determine the reactor volume required for one reactor and that for two equal-sized reactors in series for 80 percent conversion of A. And if the capital cost of a continuous-flow stirred-tank reactor unit is given by $200,000(V/100)^{0.6}$ (where V is reactor volume in m^3), the life is 20 years with no salvage value, and power costs 3 cents per kilowatt-hour, determine which system has the economic advantage. Assume that overhead, personnel, and other operating costs, except agitation, are constant. The operating year is 340 days. Each reactor is baffled (with a baffle width to tank diameter of 1/12) and equipped with an impeller whose diameter is one-third the tank diameter. The impeller is a six-bladed turbine having a width-to-diameter ratio of 1/5. The impeller is located at one-third the liquid depth from the bottom. The tank liquid-depth-to-diameter ratio is unity.

Calculation Procedure

1. Develop the necessary material-balance expressions for a single reactor, and find its volume.
For a single reactor, the mass-balance design equation is

$$V = \frac{F_{A,0} - F_A}{-r_A} = \frac{X}{-r_A}$$

where V is the volume of material within the reactor, $F_{A,0}$ is inlet molar flow rate of species A, F_A is its exit molar flow rate, and X is moles of A reacted per unit of time. As noted, $F_A = (F_{A,0} - X)$, and since the reaction of 1 mol A involves 2 mol B and yields 1 mol C, and since the inlet is 10% A and 20% B, $F_B = F_{B,0} - 2X = 2(F_{A,0} - X)$ and $F_C = X$. Because the solvent is inert, $F_S = F_{S,0}$.
Now,

$$-r_A = \frac{K_1 C_A C_B}{1 + K_2 C_A} = \frac{K_1 (F_A/v)(F_B/v)}{1 + K_2 (F_A/v)} = \frac{2K_1 (F_A/v)^2}{1 + K_2 (F_A/v)}$$

where v is the volumetric flow rate at the outlet (which equals the inlet volumetric rate, since the system is of constant density).
Since 80% conversion of A is specified, $X/F_{A,0} = 0.8$. And since the total inlet molar flow rate $F_{T,0}$ is 10% A,

$$F_{A,0} = 0.1 F_{T,0} = 0.1 \left(\frac{3\,m^3}{h} \times \frac{1.2\,kg}{dm^3} \times \left(\frac{10\,dm}{m} \right)^3 \times \frac{1\,kg \cdot mol}{40\,kg} \right) = 9\,kg \cdot mol/h$$

Therefore, the molar conversion rate X is $0.8(9) = 7.2\,kg \cdot mol/h$. The outlet concentration of A is given by

$$C_A = F_A/v = (F_{A,0} - X)/v = \frac{9 - 7.2}{3} \frac{(kg \cdot mol/h)}{(m^3/h)} = 0.600\,kg \cdot mol/m^3$$

Thus the rate at the outlet conditions is given by

$$-r_A = \frac{(2)(0.1)(0.6)^2}{1 + (0.6)(0.6)} = 0.053\,kg \cdot mol/(m^3)(h)$$

Finally, the volume is

$$V = X/-r_A = \frac{7.2\,kg \cdot mol/h}{0.053\,kg \cdot mol/(m^3)(h)} = 136\,m^3 (4803\,ft^3)$$

2. Develop the equations for two reactors in series, and find their volume.
For a pair of reactors in series, define X_1 and X_2 as the moles of A reacted per unit of time in reactors 1 and 2, respectively.
Then,

$$V = X_1/-r_{A1} = X_2/-r_{A2} \quad \text{and} \quad X_1 + X_2 = 7.2\,kg \cdot mol/h$$

Therefore,

$$\frac{X_1}{7.2 - X_1} = \frac{-r_{A1}}{-r_{A2}}$$

By material balance,

$$-r_{A1} = \frac{2K_1(F_{A,0} - X_1)^2/v^2}{1 + K_2(F_{A,0} - X_1)/v} \quad \text{and} \quad -r_{A2} = \frac{2K_1(F_{A,0} - X_1 - X_2)^2/v^2}{1 + K_2(F_{A,0} - X_1 - X_2)/v}$$

Now, the rate $-r_{A2}$ is the same as the rate for the single reactor, since $X_1 + X_2$ = overall conversion; therefore, $-r_{A2} = 0.053$ kg · mol/(m³)(h). Accordingly,

$$\frac{0.053X_1}{7.2 - X_1} = \frac{2K_1(F_{A,0} - X_1)^2/v^2}{1 + K_2(F_{A,0} - X_1)/v} = \frac{0.022(9 - X_1)^2}{1 + 0.2(9 - X_1)}$$

Solution of this cubic equation (by, for example, Newton's method) gives $X_1 = 5.44$. Therefore, the volume for each reactor is as follows:

$$V = \frac{X_2}{-r_{A2}} = \frac{7.2 - X_1}{-r_{A2}} = \frac{1.76}{0.053} = 33.2 \text{ m}^3 \ (1173 \text{ ft}^3)$$

3. Conduct an economic analysis and decide between the one- and two-reactor systems.
The two costs to be considered are depreciation of capital and power cost for agitation. Using the volumes, 20-year life with no salvage, and the straight-line depreciation method:

System	Capital cost	Annual depreciation expense
One reactor, 136 m³	$240,520	$12,026
Two reactors, each 33.2 m³	206,415	10,321

For normal mixing, the Pfaudler agitation-index (γ) number for this low-viscosity fluid is 2 ft²/s³. Most stirrers are designed for impeller Reynolds numbers of 1000 or greater. For the impeller specified, the power number ψ_n is 0.6 at high Reynolds numbers [2].

The required impeller diameter D_i may be calculated from the given data. With the liquid height H_L equal to the tank diameter D_t,

$$V = \frac{\pi D_t^2}{4}(D_t) \quad D_i = \tfrac{1}{3}D_t$$

For the two cases, the impeller diameters are thus found to be 1.86 m (6.10 ft) and 1.16 m (3.81 ft), respectively. In terms of the Pfaudler index, for low-viscosity liquids,

$$3\gamma = \frac{4n^3 D_i^2}{\pi} \psi_n \left(\frac{D_i}{D_t}\right)^2 \frac{D_i}{H_L}$$

where n is the mixer revolutions per minute. Solving this equation for n, and noting that $\gamma = 2$ ft²/s³ = 0.186 m²/s³, the mixer revolutions per minute is given as follows:

$$n = 60 \text{ s/m} \left[\frac{3\gamma\pi}{4D_i^2} \psi_n \left(\frac{D_t}{D_i}\right)^3\right]^{1/3} = \frac{162.1}{(D_i)^{2/3}} \text{ r/min}$$

Substituting the impeller diameter in meters gives 107.2 r/min for the large tank and 146.8 r/min for each small tank. The Reynolds numbers

$$Re = \frac{nD_i^2 \rho_L}{\mu_L}$$

are 9.27×10^6 for the large and 4.95×10^6 for the small tank, respectively, so the assumption of 0.6 power number is satisfactory.

From the definition of the power number,

$$\psi_n = \frac{P}{\rho_L n^3 D_i^5}$$

Therefore,

$$P = \psi_n \rho_L n^3 D_i^5 (\text{kg})(\text{m}^5)/(\text{dm}^3)(\text{s}^3)$$

For the two cases, the power consumption is calculated to be $91,417$ N · m/s and $22,150$ N · m/s for a large and small tank, respectively (1 watt = 1 N · m/s). The power consumption on an annual basis is 745,960 kW for the large tank and 361,490 kW for two small tanks. Therefore, the annual cost advantage is as follows:

System	Depreciation	Power	Total cost
1 tank	$12,026	$22,379	$34,405
2 tanks	10,321	10,845	21,166

The benefit for the two-tank system is $13,239 per year.

For more detail on cost engineering, see Section 18. And for more on mixers, see Section 12.

5.9 DETERMINATION OF REACTION-RATE EXPRESSIONS FROM PLUG-FLOW-REACTOR DATA

A 25-cm-long by 1-cm-diameter plug-flow reactor was used to investigate the homogeneous kinetics of benzene dehydrogenation. The stoichiometric equations are as follows:

Reaction 1: $2C_6H_6 \rightleftarrows C_{12}H_{10} + H_2$ [2 Bz $\rightleftarrows$ Bi + H$_2$]

Reaction 2: $C_6H_6 + C_{12}H_{10} \rightleftarrows C_{18}H_{14} + H_2$ [Bz + Bi $\rightleftarrows$ Tri + H$_2$]

At 760°C (1400°F) and 101.325 kPa (1 atm), the data in Table 5.2 were collected. Find rate equations for the production of biphenyl (Bi) and triphenyl (Tri) at 760°C by the differential method.

TABLE 5.2   Kinetics of Benzene Dehydrogenation (Procedure 5.9)

Residence time		Mole fraction in product			
$\dfrac{(\text{ft}^3)(\text{h})}{\text{lb} \cdot \text{mol}}$	$\dfrac{(\text{dm}^3)(\text{h})}{\text{kg} \cdot \text{mol}}$	Benzene	Biphenyl	Triphenyl	Hydrogen
0	0	1.0	0	0	0
0.01	0.129	0.941	0.0288	0.00051	0.0298
0.02	0.257	0.888	0.0534	0.00184	0.0571
0.06	0.772	0.724	0.1201	0.0119	0.1439
0.12	1.543	0.583	0.163	0.0302	0.224
0.22	2.829	0.477	0.179	0.0549	0.289
0.30	3.858	0.448	0.175	0.0673	0.310
∞	∞	0.413	0.157	0.091	0.339

Calculation Procedure

1. Calculate equilibrium constants.
From the data, the equilibrium constants for the two reactions, respectively, may be determined as follows (see Section 4):

$$K_1 = \frac{a_{Bi}a_{H_2}}{a_{Bz}^2} \quad K_2 = \frac{a_{Tri}a_{H_2}}{a_{Bz}a_{Bi}}$$

where a_i is the activity of species i. For the standard state being ideal gases at 101.325 kPa and 760°C, the activity for the Bi component, for example, is

$$a_{Bi} = \frac{\hat{f}_{Bi}}{f_{Bi}^\circ} = \frac{\hat{f}_{Bi}}{101.325} = \frac{\hat{\phi}_{Bi}y_{Bi}P}{101.325}$$

where f is fugacity, ϕ is fugacity coefficient, y is mole fraction, and P is pressure. Assuming that the gases are ideal, $\phi_{Bi} = 1$, so $a_{Bi} = y_{Bi}$, and in general, $a_i = y_i$.

The mole fractions at infinite residence time can be taken as the equilibrium mole fractions. Then, from the preceding data, $K_1 = a_{Bi}a_{H_2}/a_{Bz}^2 = (0.157)(0.339)/(0.413)^2 = 0.312$, and $K_2 = a_{Tri}a_{H_2}/a_{Bz}a_{Bi} = (0.091)(0.339)/(0.413)(0.157) = 0.476$.

2. Develop the reactor mass balance.
Let x = mol/h of benzene reacted by reaction 1 and y = mol/h of benzene reacted by reaction 2, and let $F_{i,0}$ and F_i be the moles per hour of species i in the inflow and outflow, respectively. Then, noting that each mole of benzene in reaction 1 involves one-half mole each of biphenyl and hydrogen,

$$F_{Bz} = F_{Bz,0} - x - y$$
$$F_{Bi} = F_{Bi,0} + x/2 - y$$
$$F_{H_2} = F_{H_2,0} + x/2 + y$$
$$F_{Tri} = F_{Tri,0} + y$$

Adding these four equations and letting subscript t stand for moles per hour,

$$F_t = F_{t,0} = F_{Bz,0} \quad \text{and} \quad F_{Bi,0} = F_{H_2,0} = F_{Tri,0} = 0$$

Now, for a plug-flow reactor, the material balance is as follows for a differential volume at steady state:

(Rate of A input by flow) − (rate of A output by flow) + (rate of A generated) = 0

or

$$F_{A,0} - (F_{A,0} + dF_A) + r_A dV = 0$$

where r_A is the reaction rate for A and V is the reaction volume. Therefore, $r_A = dF_A/dV$. In this case, $F_A = y_A F_{t,0}$; thus, $dF_A = F_{t,0}dy_A$, and the material balance becomes

$$r_A = F_{t,0}\frac{dy_A}{dV} = \frac{dy_A}{d(V/F_{t,0})}$$

where $V/F_{t,0}$ is residence time in the reactor. Thus the reaction rate is the slope of the concentration-versus-residence-time plot.

3. Calculate reaction rates.

For the homogeneous plug-flow reactor, the conversion is a function of residence time. By carrying out a series of experiments at various residence times ($V/F_{t,0}$) in a reactor of fixed volume for a constant feed composition, one obtains the same concentration-versus-residence-time plot as if an infinitely long reactor had been used and the composition had been sampled along the reactor length. Thus the data given can be plotted to give a continuous concentration-versus-residence-time plot that may be differentiated according to the mass-balance equation to give rates of reaction. The concentrations corresponding to those rates are obtained from the data plot at the time for which the rate is evaluated. The differentiation may be accomplished by drawing tangents on the graph at various times to the concentration curves.

The concentration data given can be used to determine net rates of reaction by the material-balance expressions. These rates must be analyzed in terms of the stoichiometry to get the individual rates of reaction. Thus, for the trimer,

$$r_{\text{Tri}} = r_{\text{Tri by reaction 2}}$$

$$r_{\text{Bi}} = r_{\text{Bi by reaction 1}} + r_{\text{Bi by reaction 2}}$$

$$r_{\text{Tri by reaction 2}} = -r_{\text{Bi by reaction 2}}$$

Therefore,

$$r_{\text{Bi by reaction 1}} = r_{\text{Bi}} + r_{\text{Tri by reaction 2}}$$

The net triphenyl and biphenyl rates r_{Tri} and r_{Bi} can be found by plotting the mole fraction of triphenyl and biphenyl versus residence time and taking tangents. Figure 5.4 shows such a plot and Table 5.3 presents the results.

Assume the reactions are elementary, as a first guess. Then, letting P_t be total pressure and k_i be the specific reaction-rate constant for reaction 1, the two required equations are

$$r_{\text{Bi by reaction 1}} = k_1 P_t^2 (y_{\text{Bz}}^2 - y_{\text{H}_2} y_{\text{Bi}}/K_1)$$

and

$$r_{\text{Tri by reaction 2}} = k_2 P_t^2 (y_{\text{Bz}} y_{\text{Bi}} - y_{\text{Tri}} y_{\text{H}_2}/K_2)$$

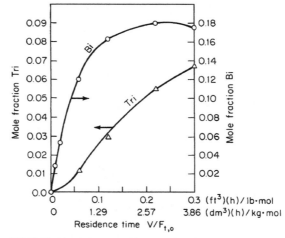

FIGURE 5.4 Exit composition as a function of residence time (Procedure 5.9).

TABLE 5.3   Rates of Reaction for Triphenyl and Biphenyl (Procedure 5.9)

Residence time $V/F_{t,0}$, $(dm^3)(h)/(kg \cdot mol)$	Rate of reaction, $kg \cdot mol/(dm^3)(h)$	
	r_{Tri}	r_{Bi} (by reaction 1)
0.129	0.00715	0.2148
0.257	0.01166	0.1893
0.772	0.02916	0.1322
1.543	0.02138	0.0467
2.829	0.01361	0.0194

Note: To obtain reaction rate in pound-moles per cubic foot per hour, multiply the preceding rates by 12.86.

In order to determine k_1 and k_2, and to check the suitability of the assumption, plot $r_{Bi \text{ by reaction 1}}$ against $(y_{Bz}^2 - y_{H_2} y_{Bi}/K_1)$ and plot r_{Tri} (which equals $r_{Tri \text{ by reaction 2}}$) against $(y_{Bz} y_{Bi} - y_{Tri} y_{H_2}/K_2)$. Check to see that the lines are indeed straight, and measure their slope (see Fig. 5.5). Define the slope as $k_i^* = k_i P_t^2$. Then $k_1 = k_1^* = 0.2496$ (by measurement of the slope), and $k_2 = k_2^* = 0.3079$.

With the numerical values for the reaction-rate constants and the equilibrium constants inserted, then,

$$r_{Bi \text{ by reaction 1}} = 0.2496(y_{Bz}^2 - y_{H_2} y_{Bi}/0.312)$$

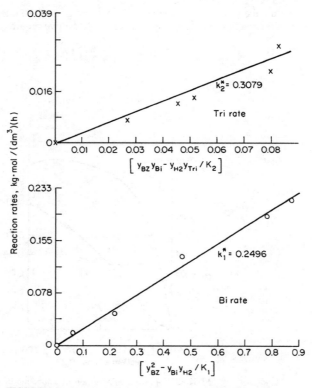

FIGURE 5.5   Rate plots for biphenyl and triphenyl (Procedure 5.9).

and

$$r_{\text{Tri by reaction 2}} = 0.3079(y_{Bz}y_{Bi} - y_{Tri}y_{H_2}/0.476)$$

Related Calculations.   For a batch reactor, the material balance is rate of accumulation of species A = rate of generation of species A, or $dN_A/dt = r_A$, where N is number of moles at time t and r is rate of reaction (which can be, for example, per unit of catalyst mass in the reactor, in which case it must be multiplied by the number of such units present). The rate at any given time can be found by plotting N_A against residence time and measuring the slope, but this technique can lead to large errors. A better approach is to use the Taylor-series interpolation formula (see mathematics handbooks for details).

5.10 REACTION-SEPARATION PROCESSES: OPTIMAL RECYCLE ALTERNATIVES AND MINIMUM REQUIRED SELECTIVITIES

Monochlorodecane (MCD) is to be produced from decane (DEC) and chlorine via the reaction

$$C_{10}H_{22} + Cl_2 \rightarrow C_{10}H_{21}Cl + HCl$$

DEC   chlorine   MCD   hydrogen chloride

A side reaction occurs in which dichlorodecane (DCD) is produced:

$$C_{10}H_{21}Cl + Cl_2 \rightarrow C_{10}H_{20}Cl_2 + HCl$$

MCD   chlorine   DCD   hydrogen chloride

The by-product, DCD, is not required for this project. Hydrogen chloride can be sold to a neighboring plant. Assume at this stage that all separations can be carried out by distillation. The normal boiling points are given in the table.

1. Determine alternative recycle structures for the process by assuming different levels of conversion of raw materials and different excesses of reactants.
2. Which structure is most effective in suppressing the side reaction?
3. What is the minimum selectivity of decane that must be achieved for profitable operation? The values of the materials involved together with their molecular weights are given in the table.

Material	Molecular weight	Normal boiling point (K)	Value ($ kg^{-1})
Hydrogen chloride	36	188	0.35
Chlorine	71	239	0.21
Decane	142	447	0.27
Monochlorodecane	176	488	0.45
Dichlorodecane	211	514	0

Calculation Procedure

1. Summarize and assess the alternative recycle structures that are possible.
Four possible arrangements can be considered:

> **I.** *Complete conversion of both feeds.* Figure I shows the most desirable arrangement: complete conversion of the decane and chlorine in the reactor. The absence of reactants in the reactor effluent means that no recycles are needed.

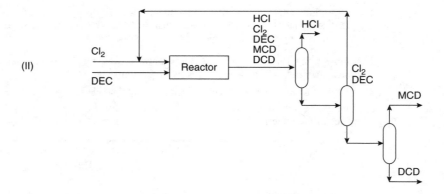

(I)

Although the flowsheet shown in Fig. I is very attractive, it is not practical. This would require careful control of the stoichiometric ratio of decane to chlorine, taking into account both the requirements of the primary and byproduct reactions. Even if it were possible to balance out the reactants exactly, a small upset in process conditions would create an excess of either decane or chlorine, and these would then appear as components in the reactor effluent. If these components appear in the reactor effluent of the flowsheet in Fig. I, there are no separators to deal with their presence and no means of recycling unconverted raw materials.

Also, although there are no selectivity data for the reaction, the selectivity losses would be expected to increase with increasing conversion. Complete conversion would tend to produce unacceptable selectivity losses. Finally, the reactor volume required to give a complete conversion would be extremely large.

II. *Incomplete conversion of both feeds.* If complete conversion is not practical, let us consider incomplete conversion. This is shown in Fig. II. However, in this case, all components are present in the reactor effluent, and one additional separator and a recycle are required. Thus the complexity is somewhat increased compared with complete conversion.

(II)

Note that no attempt has been made to separate the chlorine and decane, since they are remixed after recycling to the reactor.

III. *Excess chlorine.* Use of excess chlorine in the reactor can force the decane to effectively complete conversion (see Fig. III). Now there is effectively no decane in the reactor effluent, and again, three separators and a recycle are required.

In practice, there is likely to be a trace of decane in the reactor effluent. However, this should not be a problem, since it can either be recycled with the unreacted chlorine or leave with the product, monochlorodecane (providing it can still meet product specifications).

At this stage, how great the excess of chlorine should be for Fig. III to be feasible cannot be specified. Experimental work on the reaction chemistry would be required to establish this. However, the size of the excess does not change the basic structure.

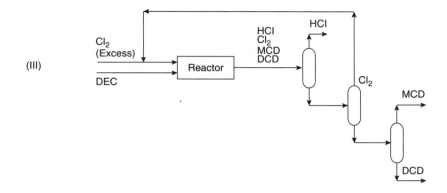

IV. *Excess decane.* Use of excess decane in the reactor forces the chlorine to effectively complete conversion (see Fig. IV). Now there is effectively no chlorine in the reactor effluent. Again, three separators and a recycle of unconverted raw material are required.

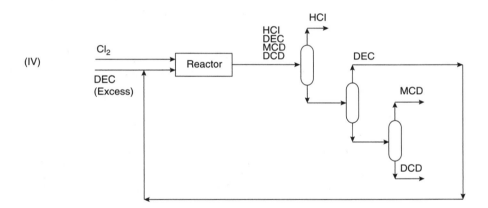

Again, in practice, there is likely to be a trace of chlorine in the reactor effluent. This can be recycled to the reactor with the unreactcd decane or allowed to leave with the hydrogen chloride byproduct (providing this meets with the by-product specification).

It cannot be said at this stage exactly how great an excess of decane would be required to make Fig. IV feasible. This would have to be established experimentally, but the size of the excess does not change the basic structure.

An arrangement is to be chosen to inhibit the side reaction (i.e., give low selectivity losses). The side reaction is suppressed by starving the reactor of either monochlorodecane or chlorine. Since the reactor is designed to produce monochlorodecane, the former option is not practical. However, it is practical to use an excess of decane.

The last of the four flowsheet options generated, which features excess decane in the reactor, is therefore preferred (see Fig. IV).

2. *Determine the minimum selectivity of decane needed for profitable operation.*
The selectivity S is defined by

$$S = \frac{\text{(MCD produced in the reactor)}}{\text{(DEC consumed in the reactor)}} \times \text{stoichiometric factor}$$

For more on selectivities, see Section 4 in this book.

In this case, the stoichiometric factor is 1. This is a measure of the MCD obtained from the DEC consumed. To assess the selectivity losses, the MCD produced in the primary reaction is split into that fraction which will become final product and that which will become the byproduct. Thus the reaction stoichiometry is

$$C_{10}H_{22} + Cl_2 \rightarrow SC_{10}H_{12}Cl + (1 - S)C_{10}H_{21}Cl + HCl$$

and for the byproduct reaction the stoichiometry is

$$(1 - S)C_{10}H_{21}Cl + (1 - S)Cl_2 \rightarrow (1 - S)C_{10}H_{20}Cl_2 + (1 - S)HCl$$

Adding the two reactions gives overall

$$C_{10}H_{22} + (2 - S)Cl_2 \rightarrow SC_{10}H_{21}Cl + (1 - S)C_{10}H_{20}Cl_2 + (2 - S)HCl$$

Considering raw materials costs only, the economic potential (EP) of the process is defined as

$$EP = \text{value of products} - \text{raw materials cost}$$
$$= [176 \times S \times 0.45 + 36 \times (2 - S) \times 0.35]$$
$$- [142 \times 1 \times 0.27 + 71 \times (2 - S) \times 0.21]$$
$$= 79.2S - 2.31(2 - S) - 38.34 \ (\$ \ \text{kmol}^{-1} \ \text{decane reacted})$$

The minimum selectivity that can be tolerated is given when the economic potential is just zero:

$$0 = 79.2S - 2.31(2 - S) - 38.34$$
$$S = 0.53$$

In other words, the process must convert at least 53% of the decane that reacts to monochlorodecane rather than to dichlorodecane for the process to be economic. This figure assumes selling the hydrogen chloride to a neighboring process. If this is not the case, there is no value associated with the hydrogen chloride. Assuming that there are no treatment and disposal costs for the now waste hydrogen chloride, the minimum economic potential is given by

$$0 = (176 \times S \times 0.45) - [142 \times 1 \times 0.27 + 71 \times (2 - S) \times 0.21]$$
$$= 79.2S - 14.91(2 - S) - 38.34$$
$$S = 0.72$$

Now the process must convert at least 72 percent of the decane to monochlorodecane.

If the hydrogen chloride cannot be sold, it must be disposed of somehow. Alternatively, it could be converted back to chlorine via the reaction

$$2HCl + \tfrac{1}{2}O_2 \rightleftarrows Cl_2 + H_2O$$

and then recycled to the MCD reactor. Now the overall stoichiometry changes, since the $(2 - S)$ moles of HCl that were being produced as byproduct are now being recycled to substitute fresh chlorine feed:

$$(2 - S)HCl + \tfrac{1}{4}(2 - S)O_2 \rightarrow \tfrac{1}{2}(2 - S)Cl_2 + \tfrac{1}{2}(2 - S)H_2O$$

Thus the overall reaction now becomes

$$C_{10}H_{22} + \tfrac{1}{2}(2 - S)Cl_2 + \tfrac{1}{4}(2 - S)O_2 \rightarrow SC_{10}H_{21}Cl + (1 - S)C_{10}H_{20}Cl_2 + \tfrac{1}{2}(2 - S)H_2O$$

The economic potential is now given by

$$0 = (176 \times S \times 0.45) - [142 \times 1 \times 0.27 + 71 \times \tfrac{1}{2}(2 - S) \times 0.21]$$
$$= 79.2S - 7.455(2 - S) - 38.34$$
$$S = 0.61$$

The minimum selectivity that can now be tolerated becomes 61 percent.

This example is adapted from Smith, *Chemical Process Design*, McGraw-Hill. For more on calculations and decisions involving recycle streams, see Section 2 in this book.

*5.11 THERMODYNAMIC ANALYSIS OF A LINDE SYSTEM

Make a thermodynamic analysis of a Linde system for the separation of gaseous oxygen and nitrogen, as shown in Fig. 5.6. Table 5.4 lists a set of operating conditions for the numbered points in Fig. 5.6. Heat leaks into the column of 147 J/mol of entering air and into the exchanger of 63 J/mol entering air are assumed. This calculation will be made on the basis of 1 mol of entering air, which is assumed to contain 79 mol % N_2 and 21% O_2.

Calculation Procedure

1. Set the basis for the thermodynamic analysis of the system.
The object of a thermodynamic analysis of a real process is the determination of the efficiency of the process from the standpoint of energy utilization. Further, it is useful to calculate the influence of each irreversibility individually on the overall efficiency of the process.

The present calculation will be limited to consideration of steady-flow processes, for which the energy equation resulting from the first law of thermodynamics is

$$\Delta H + \Delta E_P + \Delta E_K = Q - W_s$$

or
$$W_s = Q - \Delta H - \Delta E_P - \Delta E_K \tag{5.1}$$

where H = enthalpy
E_P = potential energy
E_K = kinetic energy
Q = heat
W_s = shaft work

and Δ signifies a difference in values between outlet and inlet streams.

Figure 5.7 is a schematic representation of the general process considered, which may be simple or complex. We presume that the process exists in surroundings which constitute a heat reservoir at the constant temperature T_0. Heat exchange between process and surroundings causes entropy changes in the surroundings in the amount

$$\Delta S_0 = Q_0/T_0$$

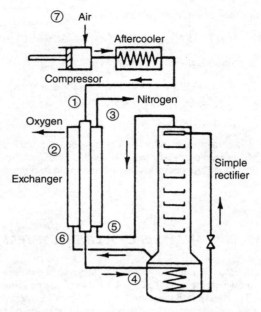

FIGURE 5.6 Diagram of a simple gaseous oxygen process.

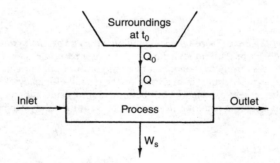

FIGURE 5.7 Schematic diagram of a steady-flow process.

TABLE 5.4 States and Values of Properties for the Process of Fig. 5.6[†]

Point	P, bar	T, K	Composition	State	H, J/mol	S, J/(mol · K)
1	55.22	300	Air	Superheated	12,046	82.98
2	1.01	295	Pure O_2	Superheated	13,460	118.48
3	1.01	295	91.48% N_2	Superheated	12,074	114.34
4	55.22	147.2	Air	Superheated	5850	52 08
5	1.01	79.4	91.48% N_2	Saturated vapor	5773	75.82
6	1.01	90	Pure O_2	Saturated vapor	7485	83.69
7	1.01	300	Air	Superheated	12,407	117.35

[†]Properties on the basis of Miller and Sullivan, U.S. Bur. Mines Tech. Pap. 424 (1928).

Since $Q_0 = -Q$, this equation may also be written as

$$Q = -T_0 \Delta S_0 \tag{5.2}$$

Combination of Eqs. (5.1) and (5.2) and rearrangement give

$$W_s = -T_0 \Delta S_0 - \Delta H - \Delta E_P - \Delta E_k \tag{5.3}$$

As it stands, this equation is of little use because the entropy change of the surroundings ΔS_0 is rarely known. However, for the special case of a completely reversible process, the second law of thermodynamics provides the equation

$$\Delta S + \Delta S_0 = 0$$

or

$$\Delta S_0 = -\Delta S \tag{5.4}$$

where ΔS is the entropy change of the flowing stream between inlet and outlet. Substitution of Eq. (5.4) into Eq. (5.3) provides

$$W_{\text{ideal}} = T_0 \Delta S - \Delta H - \Delta E_P - \Delta E_K \tag{5.5}$$

where the work is now represented by W_{ideal} so as to indicate clearly that it is the work associated with a completely reversible process for which the change of state is implied by the property changes ΔS, ΔH, ΔE_P, and ΔE_K. When these property changes are values taken for a real process, then Eq. (5.5) yields the work required to bring about *the same change of state* in a completely reversible process. It is the minimum work requirement or the maximum work obtainable, depending on whether the process requires or produces work. The stipulation of *complete* reversibility requires not only that the process be internally reversible but also that heat transfer between system and surroundings also be reversible. Such a process is taken as the standard or ideal against which to measure the efficiencies of real processes that accomplish the same change of state. Thus the thermodynamic efficiency η is given by

$$\eta(\text{work produced}) = W_s / W_{\text{ideal}} \tag{5.6a}$$

$$\eta(\text{work required}) = W_{\text{ideal}} / W_s \tag{5.6b}$$

The difference between the ideal work for a given change of state and the real work of a process that brings about the same change is called the lost work. Thus by definition,

$$W_{\text{lost}} = W_{\text{ideal}} - W_s \tag{5.7}$$

and is given as the difference between Eqs. (5.5) and (5.1), both written for the same change of state:

$$W_{\text{lost}} = T_0 \Delta S - Q \tag{5.8}$$

Alternatively,

$$W_{\text{lost}} = T_0 \Delta S + Q_0$$

Since

$$Q_0 = T_0 \Delta S_0$$

then

$$W_{\text{lost}} = T_0 \Delta S + T_0 \Delta S_0 = T_0 (\Delta S + \Delta S_0)$$

or

$$W_{\text{lost}} = T_0 \Delta S_{\text{total}} \tag{5.9}$$

By the second law of thermodynamics, $\Delta S_{\text{total}} \geq 0$. Thus,

$$W_{\text{lost}} \geq 0 \tag{5.10}$$

The engineering significance of this result is clear. The greater the irreversibility of a process, the greater the increase in total entropy accompanying it and the greater the amount of energy that becomes unavailable as work. Thus every irreversibility in a process carries with it a price.

For processes of more than one step it is advantageous to calculate W_{lost} for each step separately. Then Eq. (5.7) becomes

$$\sum W_{\text{lost}} = W_{\text{ideal}} - W_s \tag{5.11}$$

For processes that require work this equation is written

$$W_s = W_{\text{ideal}} - \sum W_{\text{lost}} \tag{5.11a}$$

The terms on the right side represent an analysis of the actual work, showing the part ideally required to bring about the change of state and the parts that are required as a result of the irreversibilities in the various steps of the process.

For processes that produce work, Eq. (5.11) is written

$$W_{\text{ideal}} = W_s + \sum W_{\text{lost}} \tag{5.11b}$$

Here the terms on the right side represent an analysis of the ideal work showing the part actually produced and the parts that become unavailable because of irreversibilities in the various steps of the process.

A material balance on the nitrogen gives

$$0.79 = 0.9148x$$

or

$$x = 0.8636 \text{ mol of nitrogen product}$$

Therefore, the N_2 product stream contains 0.8636 mol, and the O_2 product stream contains 0.1364 mol.

2. Calculate the ideal work.
If changes in kinetic and potential energies are neglected, Eq. (5.5) becomes

$$W_{\text{ideal}} = T_0 \Delta S - \Delta H$$

The surroundings temperature T_0 will be taken as 300 K. With the material quantities as calculated previously and the property values as given in engineering handbooks, ΔH and ΔS for the overall process are determined as follows:

$$\Delta H = (13,460)(0.1364) + (12,074)(0.8636) - (12,407)(1)$$
$$= -144 \text{ J}$$
$$\Delta S = (118.48)(0.1364) + (114.34)(0.8636) - (117.35)(1)$$
$$= -2.445 \text{ J/K}$$

Therefore

$$W_{\text{ideal}} = (300)(-2.445) + 144 = -590 \text{ J}$$

3. *Calculate the actual work of compression.*
In order not to complicate this example, the work of compression will be calculated by means of the equation for an ideal gas in a three-stage reciprocating compressor with complete intercooling. This equation is based on isentropic compression in each stage, and it will be assumed that the work so calculated represents 80 percent of the actual work. This equation may be found in any number of standard textbooks on thermodynamics:

$$W_s = \frac{-n\gamma RT_1}{(0.8)(\gamma - 1)}\left[\left(\frac{P_2}{P_1}\right)^{(\gamma-1)/n\gamma} - 1\right]$$

where n = number of stages, here taken as 3
γ = ratio of heat capacities, here taken as 1.4
T_1 = initial absolute temperature, equal to 300 K
P_2/P_1 = overall pressure ratio, equal to 54.5
R = universal gas constant, equal to 8.314 J/(mol · K)

The efficiency factor of 0.8 has already been incorporated in the equation. Substitution of values gives

$$W_s = \frac{-(3)(1.4)(8.314)(300)}{(0.8)(0.4)}[(54.5)^{0.4/(3)(1.4)} - 1]$$
$$= 15,171 \text{ J}$$

The heat transferred to the surroundings during compression as a result of intercooling and after-cooling is determined from the first law:

$$Q = \Delta H + W_s = (12,046 - 12,407) - 15,171$$
$$= -15,532 \text{ J}$$

4. *Determine the lost work in the system components.*
The equation used is Eq. (5.3),

$$W_{\text{lost}} = T_0 \Delta S - Q$$

and it remains only to evaluate ΔS and Q for the various steps of the process.

a.*   *Compression

$$\Delta S = 82.98 - 117.35 = -34.37 \text{ J/K}$$
$$Q = -15,532 \text{ J}$$
$$W_{\text{lost}} = (300)(-34.37) + 15,532$$
$$= 5221 \text{ J}$$

b.*   *Heat exchanger

$$\Delta S = (0.8636)(114.34 - 75.82) + (0.1364)(118.48 - 83.69) + (1)(52.08 - 82.98)$$
$$= 7.11 \text{ J/K}$$
$$Q = 63 \text{ J}$$
$$W_{\text{lost}} = (300)(7.11) - 63$$
$$= 2070 \text{ J}$$

>   *c.   Column*

$$\Delta S = (0.8636)(75.82) + (0.1364)(83.69) + (1)(52.08)$$
$$= 24.81 \text{ J/K}$$
$$Q = 147 \text{ J}$$
$$W_{\text{lost}} = (300)(24.81) - 147$$
$$= 7296 \text{ J}$$

5. Summarize the calculation results.
Since the process requires work, Eq. (5.11a) is appropriate for the thermodynamic analysis:

$$W_s = W_{\text{ideal}} - \Sigma W_{\text{lost}}$$

The various terms on the right appear as entries in the following summary of results:

	J	% of W_s
W_{ideal}	−591	3.9
$-W_{\text{lost}}$		
Compression	−5221	34.4
Heat exchanger	−2070	13.6
Column	−7296	48.1
$W_s = W_{\text{ideal}} - \Sigma W_{\text{lost}}$	−15,178	100.0

The value of W_s determined by summing the individual terms should be the same as that calculated by the compressor formula (−15,171 J). The slight discrepancy is the result of the accumulation of round-off errors. The thermodynamic efficiency of the process is 3.9 percent, as given in the first row of the summary table. The largest lost-work term results from irreversibilities in the column.

Related Calculations: Use this basic method to perform thermodynamic analyses of various operating systems. This procedure is the work of Hendrick C. Van Ness, and Michael M. Abbott, as given in Perry—*Chemical Engineers' Handbook,* 5th ed., McGraw-Hill 1983.

*5.12 SIZING REACTOR DESUPERHEATER CONDENSERS ECONOMICALLY

A reactor exhausts 27,958 lb/h (3.52 kg/s) of isobutane with a small amount of *n*-butane at 200°F (93.3°C) and 85 lb/in² (gage) (586.0 kPa). This gas becomes saturated at 130°F (54.4°C) and condenses completely at 125°F (51.7°C). The gas is to be cooled and condensed by a horizontal counterflow heat exchanger like that in Fig. 5.8 using well water at 65°F (18.3°C) inlet temperature and an outlet temperature of 100°F (37.8°C). How much heat-transfer area is required in this exchanger?

Calculation Procedure

1. Check for condensation at the hot end.
Desuperheater-condensers are widely used in the process, petrochemical, chemical, and power industries. Figure 5.8 shows a horizontal in-shell design which might be used as a high-pressure

feed heater, an inter- or aftercondenser in a steam-jet ejector system, or a gas cooler in a compressor system.

Conventional design practice splits the heat load and sizes the desuperheating and condensing zones separately. This assumes that the superheated vapor cools as if it were a dry gas, which is true only when the tube-wall temperature T_w in the desuperheating zone is greater than the vapor's saturation temperature T_{sat}.

When T_w is less than T_{sat}, the superheated vapor condenses directly, in the same way as saturated vapor. Because the heat flux in condensation is much greater than in desuperheating, this situation requires less heat-transfer area than the conventional design would prescribe.

For a counterflow desuperheater-condenser with vapor on the shell side like that in Fig. 5.8, the energy balance for the desuperheating zone is $h_d(T_1 - T_w) = U_d(T_1 - t_2)$, where h_d is the desuperheating heat-transfer coefficient; U_d is the overall heat-transfer coefficient; T_1, is the vapor inlet temperature; t_2 is the cooling-medium outlet temperature. Using this energy-balance equation and

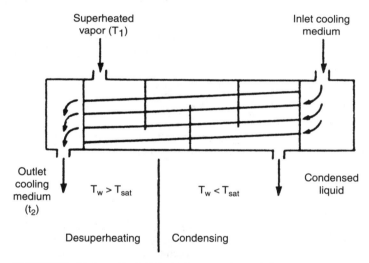

FIGURE 5.8   Horizontal in-shell desuperheater-condenser. (*Chemical Engineering.*)

the data from Fig. 5.9, we find the tube-wall temperature at the hot end of the unit is $45.2(200 - T_w) = 42.3(200 - 100)$; $T_w = 106.4°F$ (41.3°C). Since T_w is less than T_{sat}, condensation does take place as soon as the superheated vapor enters the shell.

2. Determine the area required for the condensing load.

Since condensation does take place, the entire desuperheating load, 860,000 Btu/h (251.9 W), should be treated as a condensing load, by using the condensing heat-transfer coefficient shown in Fig. 5.9. The heat-transfer area required for this load is $(Btu/h)/U_c(LMTD)$; or $A = 860,000/158(33) = 165$ ft^2 (15.3 m^2).

3. Compare the conventional approach to this approach.

Using the conventional approach, such as that in Kern—*Process Heat Transfer*, McGraw-Hill, and in the heat-exchanger calculation procedures given elsewhere in this handbook (see index), shows that the heat duty for the exchanger is split into two zones—condensing and desuperheating. The area

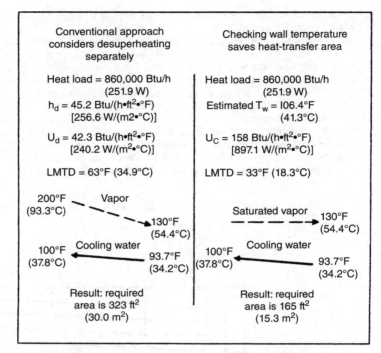

Conventional approach considers desuperheating separately

Heat load = 860,000 Btu/h
(251.9 W)

h_d = 45.2 Btu/(h•ft^2•°F)
[256.6 W/(m2•°C)]

U_d = 42.3 Btu/(h•ft^2•°F)
[240.2 W/(m^2•°C)]

LMTD = 63°F (34.9°C)

200°F (93.3°C) — Vapor → 130°F (54.4°C)

100°F (37.8°C) ← Cooling water — 93.7°F (34.2°C)

Result: required area is 323 ft^2 (30.0 m^2)

Checking wall temperature saves heat-transfer area

Heat load = 860,000 Btu/h
(251.9 W)

Estimated T_w = 106.4°F
(41.3°C)

U_C = 158 Btu/(h•ft^2•°F)
[897.1 W/(m^2•°C)]

LMTD = 33°F (18.3°C)

Saturated vapor → 130°F (54.4°C)

100°F (37.8°C) ← Cooling water — 93.7°F (34.2°C)

Result: required area is 165 ft^2 (15.3 m^2)

FIGURE 5.9 Example shows importance of checking wall temperature. (*Chemical Engineering.*)

required for the condensing zone is 523 ft^2 (48.6 m^2); for the desuperheating zone it is 323 ft^2 (30 m^2). Important data for the desuperheating zone are shown in Fig. 5.9.

The total heat-transfer area, when the approach given here is used, will be the condensing-zone area from the conventional approach + desuperheating area computed in step 2. Or, A_{total} = 523 + 165 = 688 ft^2 (63.9 m^2). This compares with 523 + 323 = 846 ft^2 (78.6 m^2) for the conventional approach, or 23 percent greater area.

Note that the desuperheat was a sizable fraction (over 20 percent) of the total heat load in this exchanger, which is cooling an organic vapor. With steam, the importance of desuperheating is generally less.

Related Calculations. Energy conservation studies often show a longer breakeven period and a smaller payout because equipment costs are excessive. Higher equipment costs lead to greater cost of money for the heat-recovery unit or units. Hence, it is important that any equipment chosen to conserve heat be sized properly. The procedure given here shows how a saving of nearly 25 percent can be made in the area of certain types of heat exchangers. Such savings can significantly reduce the required investment, leading to an earlier breakeven and higher payout. Thus, energy conservation will be easier to justify when this procedure is used.

Since this procedure is relatively simple, it should be applied in selecting a heat exchanger which involves desuperheating. The method is applicable in land, marine, chemical, petrochemical, and process heat-exchanger selection. This procedure is the work of P. S. V. Kurmarao, Ph.D., EDC (Heat Exchangers) Bharat Heavy Electricals Ltd., as reported in *Chemical Engineering* magazine.

*5.13   DESIGN OF A COMPLETE-MIX ACTIVATED SLUDGE REACTOR

Domestic wastewater with an average daily flow of 4.0 Mgd (15,140 m³/d) has a 5-day biochemical oxygen demand (BOD₅) of 240 mg/L after primary settling. The effluent is to have a BOD₅ of 10 mg/L or less. Design a complete-mix activated sludge reactor to treat the wastewater including reactor volume, hydraulic retention time, quantity of sludge wasted, oxygen requirements, food to microorganism ratio, volumetric loading, and waste activated sludge (WAS) and return activated sludge (RAS) requirements.

Calculation Procedure

1. Compute the reactor volume.
The volume of the reactor can be determined using the following equation derived from Monod kinetics:

$$V_r = \frac{\theta_c QY(S_o - S)}{X_a(1 + k_d\theta_c)}$$

where V_r = reactor volume (Mgal) (m³)

θ_c = mean cell residence time, or the average time that the sludge remains in the reactor (sludge age). For a complete-mix activated sludge process, θ_c ranges from 5 to 15 days. The design of the reactor is based on θ_c on the assumption that substantially all the substrate (BOD) conversion occurs in the reactor. A θ_c of 8 days will be assumed.

Q = average daily influent flow rate (Mgd) = 4.0 Mgd (15,140 m³/d)

Y = maximum yield coefficient (mg VSS/mg BOD₅). For the activated sludge process for domestic wastewater Y ranges from 0.4 to 0.8. A Y of 0.6 mg VSS/mg BOD₅ will be assumed. Essentially, Y represents the maximum mg of cells produced per mg organic matter removed.

S_O = influent substrate (BOD₅) concentration (mg/L) = 240 mg/L

S = effluent substrate (BOD₅) concentration (mg/L) = 10 mg/L

X_a = concentration of microorganisms in reactor = mixed liquor volatile suspended solids (MLVSS) in mg/L. It is generally accepted that the ratio MLVSS/MLSS ≈ 0.8, where MLSS is the mixed liquor suspended solids concentration in the reactor. MLSS represents the sum of volatile suspended solids (organics) and fixed suspended solids (in-organics). For a complete-mix activated sludge process, MLSS ranges from 1000 to 6500 mg/L. An MLSS of 4500 mg/L will be assumed. Thus MLVSS = (0.8)(4500 mg/L) = 3600 mg/L.

k_d = endogenous decay coefficient (d^{-1}) which is a coefficient representing the decrease of cell mass in the MLVSS. For the activated sludge process for domestic wastewater k_d ranges from 0.025 to 0.075 d^{-1}. A value of 0.06 d^{-1} will be assumed.

Therefore:

$$V_r = \frac{(8\ d)(4.0\ \text{Mgd})(0.6\ \text{mg VSS/mg BOD}_5)(240 - 10)\text{mg/L}}{(3600\ \text{mg/L})(1 + (0.06\ d^{-1})\,(8\ d))}$$

$$= 0.83\ \text{Mgal}\,(110{,}955\ \text{ft}^3)\,(3140\ \text{m}^3)$$

2. Compute the hydraulic retention time.
The hydraulic retention time (θ) in the reactor is the reactor volume divided by the influent flow rate: V_r/Q. Therefore, θ = (0.83 Mgal)/(4.0 Mgd) = 0.208 days = 5.0 hours. For a complete-mix activated sludge process, θ is generally 3 to 5 hours. Therefore, the hydraulic retention time is acceptable.

3. Compute the quantity of sludge wasted.
The observed cell yield, $Y_{obs} = Y/1 + k_d\theta_c = 0.6/(1 + (0.06 \ d^{-1})(8 \ d)) = 0.41$ mg/mg represents the actual cell yield that would be observed. The observed cell yield is always less than the maximum cell yield (Y).

The increase in MLVSS is computed using the following equation:

$$P_x = Y_{obs}Q(S_O - S)(8.34 \ \text{lb/Mgal/mg/L})$$

where P_x is the net waste activated sludge produced each day in (lb VSS/d).
Using values defined above

$$P_x = \left(0.41 \frac{\text{mg VSS}}{\text{mg BOD}_5}\right)(4.0 \ \text{Mgd})\left(240 \frac{\text{mg}}{\text{L}} - 10 \frac{\text{mg}}{\text{L}}\right)\left(8.34 \frac{\text{lb/Mgal}}{\text{mg/L}}\right)$$

$$= 3146 \ \text{lb VSS/d} \ (1428.3 \ \text{kg VSS/d})$$

This represents the increase of volatile suspended solids (organics) in the reactor. Of course the total increase in sludge mass will include fixed suspended solids (inorganics) as well. Therefore, the increase in the total mass of mixed liquor suspended solids (MLSS) = $P_{x(ss)}$ = (3146 lb VSS/d)/(0.8) = 3933 lb SS/d (1785.6 kg SS/d). This represents the total mass of sludge that must be wasted from the system each day.

4. Compute the oxygen requirements based on ultimate carbonaceous oxygen demand (BOD_L).
The theoretical oxygen requirements are calculated using the BOD_5 of the wastewater and the amount of organisms (P_x) wasted from the system each day. If all BOD_5 were converted to end products, the total oxygen demand would be computed by converting BOD_5 to ultimate BOD (BOD_L), using an appropriate conversion factor. The "Quantity of Sludge Wasted" calculation illustrated that a portion of the incoming waste is converted to new cells which are subsequently wasted from the system. Therefore, if the BOD_L of the wasted cells is subtracted from the total, the remaining amount represents the amount of oxygen that must be supplied to the system. From stoichiometry, it is known that the BOD_L of 1 M of cells is equal to 1.42 times the concentration of cells. Therefore, the theoretical oxygen requirements for the removal of the carbonaceous organic matter in wastewater for an activated-sludge system can be computed using the following equation:

$$\text{lb O}_2/\text{d} = (\text{total mass of BOD}_L \text{ utilized, lb/d}) - 1.42 \ (\text{mass of organisms wasted, lb/d})$$

Using terms that have been defined previously where f = conversion factor for converting BOD_5 to BOD_L (0.68 is commonly used):

$$\text{lb O}_2/\text{d} = \frac{Q(S_O - S)\left(8.34 \ \dfrac{\text{lb/Mgal}}{\text{mg/L}}\right)}{f} - (1.42)(P_x)$$

Using the above quantities

$$\text{lb O}_2/\text{d} = \frac{(4.0 \ \text{Mgd})(240 \ \text{mg/L} - 10 \ \text{mg/L})(8.34)}{0.68} - (1.42)(3146 \ \text{lb/d})$$

$$= 6816 \ \text{lb O}_2/\text{d} \ (3094.5 \ \text{kg O}_2/\text{d})$$

This represents the theoretical oxygen requirement for removal of the influent BOD_5. However, to meet sustained peak organic loadings, it is recommended that aeration equipment be designed with a safety factor of at least 2. Therefore, in sizing aeration equipment a value of (2)(6816 lb O$_2$/d) = 13,632 lb O$_2$/d (6188.9 kg O$_2$/d) is used.

5. *Compute the food to microorganism ratio (F:M) and the volumetric loading (V_L).*
In order to maintain control over the activated sludge process, two commonly used parameters are
(1) the food to microorganism ratio (F:M) and, (2) the mean cell residence time (θ_c). The mean cell
residence time was assumed in step 1 to be 8 days.
   The food to microorganism ratio is defined as:

$$F:M = S_O/\theta X_a$$

where F:M is the food to microorganism ratio in d^{-1}.
   F:M is simply a ratio of the *food* or BOD_5 of the incoming waste, to the concentration of *micro-organisms* in the aeration tank or MLVSS. Therefore, using values defined previously

$$F:M = \frac{240 \text{ mg/L}}{(0.208 \text{ d})(3600 \text{ mg/L})} = 0.321 \; d^{-1}$$

   Typical values for F:M reported in literature vary from 0.05 d^{-1} to 1.0 d^{-1} depending on the type
of treatment process used.
   A low value of F:M can result in the growth of filamentous organisms and is the most common
operational problem in the activated sludge process. A proliferation of filamentous organisms in the
mixed liquor results in a poorly settling sludge, commonly referred to as "bulking sludge."
   One method of controlling the growth of filamentous organisms is through the use of a separate
compartment as the initial contact zone of a biological reactor where primary effluent and return
activated sludge are combined. This concept provides a high F:M at controlled oxygen levels which
provides selective growth of floc forming organisms at the initial stage of the biological process. An
F:M ratio of at least 2.27 d^{-1} in this compartment is suggested in the literature. However, initial F:M
ratios ranging from 20 to 25 d^{-1} have also been reported.
   The volumetric (organic) loading (V_L) is defined as

$$V_L = S_O Q/V_r = S_O/\theta$$

V_L is a measure of the pounds of BOD_5 applied daily per thousand cubic feet of aeration tank
volume. Using values defined previously

$$V_L = (240 \text{ mg/L})/(0.208 \text{ d}) = 1154 \text{ mg/L} \cdot \text{d} = 72 \text{ lb}/10^3 \text{ft}^3 \cdot \text{d} \, (1.15 \text{ kg/Mm}^3 \cdot \text{d})$$

   Volumetric loading can vary from 20 to more than 200 $\text{lb}/10^3\text{ft}^3 \cdot \text{d}$ (0.32–3.2 $\text{kg/Mm}^3 \cdot$ d), and
may be used as an alternate (although crude) method of sizing aeration tanks.

6. *Compute the waste activated sludge (WAS) and return activated sludge (RAS) requirements.*
Control of the activated sludge process is important to maintain high levels of treatment performance
under a wide range of operating conditions. The principle factors used in process control are (1)
maintaining dissolved-oxygen levels in the aeration tanks, (2) regulating the amount of return activated sludge (RAS), and (3) controlling the waste activated sludge (WAS). As outlined previously in
step 5 the most commonly used parameters for controlling the activated sludge process are the F:M
ratio and the mean cell residence time (θ_c). The mixed liquor volatile suspended solids (MLVSS)
concentration may also be used as a control parameter. RAS is important in maintaining the MLVSS
concentration and the WAS is important in controlling the mean cell residence time (θ_c).
   The excess waste activated sludge produced each day (see step 3) is wasted from the system to
maintain a given F:M or mean cell residence time. Generally, sludge is wasted from the return sludge
line because it is more concentrated than the mixed liquor in the aeration tank, hence smaller waste
sludge pumps are required. The waste sludge is generally discharged to sludge thickening and digestion facilities. The alternative method of sludge wasting is to withdraw mixed liquor directly from
the aeration tank where the concentration of solids is uniform. Both methods of calculating the waste
sludge flow rate are illustrated below.

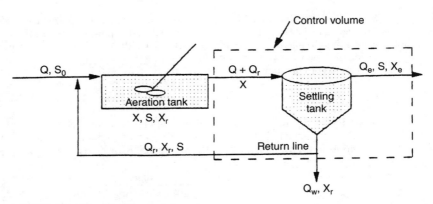

FIGURE 5.10   Settling tank mass balance.

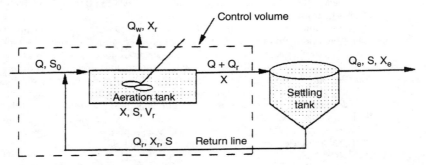

FIGURE 5.11   Aeration tank mass balance.

Use Figs. 5.10 and 5.11 when performing mass balances for the determination of RAS and WAS.

X = mixed liquor suspended solids (MLSS) - see Step 1.
Q_r = return activated sludge pumping rate (Mgd).
X_r = concentration of sludge in the return line (mg/L). When lacking site specific operational data, a value commonly assumed is 8000 mg/L.
Q_e = effluent flow rate (Mgd).
X_e = concentration of solids in effluent (mg/L). When lacking site-specific operational data, this value is commonly assumed to be zero.
Q_w = wasted activated sludge (WAS) pumping rate from the reactor (Mgd).
$Q_{w'}$ = waste activated sludge (WAS) pumping rate from the return line (Mgd).

Other variables are as defined previously.

The actual amount of liquid that must be pumped to achieve process control depends on the method used and the location from which the wasting is to be accomplished. Also note that because the solids capture of the sludge processing facilities (i.e., thickeners, digesters, etc.) is not 100 percent and some solids are returned, the actual wasting rate will be higher than the theoretically determined value.

(a) Waste activated sludge (WAS) pumping rate from the return line.   If the mean cell residence time is used for process control and the wasting is from the sludge return line (Fig. 5.10), the wasting rate is computed using the following:

$$\theta_c = \frac{V_r X}{(Q_{w'} X_r + Q_e X_e)}$$

Assuming that the concentration of solids in the effluent from the settling tank (X_e) is low, then the above equation reduces to:

$$\theta_c \approx \frac{V_r X}{Q_{w'} X_r} \Rightarrow Q_{w'} = \frac{V_r X}{\theta_c X_r}$$

Using values defined previously

$$Q_{w'} = \frac{(0.83 \text{ Mgal})(4500 \text{ mg/L})}{(8 \text{ d}) (8000 \text{ mg/L})} = 0.0584 \text{ Mgd} = 58{,}400 \text{ gal/day} (221 \text{ m}^3/\text{d})$$

To determine the WAS pumping rate using this method, the solids concentration in both the aeration tank and the return line must be known.

If the food to microorganism ratio (F:M) method of control is used, the WAS pumping rate from the return line is determined using the following:

$$P_{x(ss)} = Q_{w'} X_r (8.34 \text{ lb/Mgal/mg/L})$$

Therefore:

$$Q_{w'} = \frac{3933 \text{ lb/d}}{(8000 \text{ mg/L})(8.34)} = 0.059 \text{ Mgd} = 59{,}000 \text{ gal/day} (223.3 \text{ m}^3/\text{d})$$

In this case, the concentration of solids in the sludge return line must be known. Note that regardless of the method used for calculation, if wasting occurs from the return line, the WAS pumping rate is approximately the same.

(b) Waste activated sludge (WAS) pumping rate from the aeration tank. If the mean cell residence time is used for process control, wasting is from the aeration tank (Fig. 5.11), and the solids in the plant effluent (X_e) are again neglected, then the WAS pumping rate is estimated using the following:

$$\theta_c \approx \frac{V_r}{Q_w} \Rightarrow Q_w \approx \frac{V_r}{\theta_c}$$

Using values defined previously

$$Q_w = \frac{0.83 \text{ Mgal}}{8 \text{ day}} = 0.104 \text{ Mgd} = 104{,}000 \text{ gal/day} (393.6 \text{ m}^3/\text{d})$$

Note that in case (a) or (b) above, the weight of sludge wasted is the same (3933 lb SS/d) (1785.6 kg SS/d), and that either wasting method will achieve a θ_c of 8 days. As can be seen, wasting from the aeration tank produces a much higher waste flow rate. This is because the concentration of solids in the bottom of the settling tank (and hence the return line) is higher than in the aeration tank. Consequently, wasting a given mass of solids per day is going to require a larger WAS pumping rate (and larger WAS pumps) if done from the aeration tank as opposed to the return line. The return activated sludge (RAS) pumping rate is determined by performing a mass balance analysis around either the settling tank or the aeration tank. The appropriate control volume for either mass balance analysis is illustrated in Fig. 5.10 and 5.11 respectively. Assuming that the sludge blanket level in the settling tank remains constant and that the solids in the effluent from the settling tank (X_e) are negligible, a mass balance around the settling tank (Fig. 5.10) yields the following equation for RAS pumping rate:

$$Q_r = \frac{XQ - X_r Q_{w'}}{X_r - X}$$

Using values defined previously, the RAS pumping rate is computed to be:

$$Q_r = \frac{(4500 \text{ mg/L})(4.0 \text{ Mgd}) - (8000 \text{ mg/L})(0.0584 \text{ Mgd})}{8000 \text{ mg/L} - 4500 \text{ mg/L}}$$

$$= 5.0 \text{ Mgd} \,(18,925 \text{ m}^3/\text{d})$$

As outlined above, the required RAS pumping rate can also be estimated by performing a mass balance around the aeration tank (Fig. 5.11). If new cell growth is considered negligible, then the solids entering the tank will equal the solids leaving the tank. Under conditions such as high organic loadings, this assumption may be incorrect. Solids enter the aeration tank in the return sludge and in the influent flow to the secondary process. However, because the influent solids are negligible compared to the MLSS in the return sludge, the mass balance around the aeration tank yields the following equation for RAS pumping rate:

$$Q_r = \frac{X(Q - Q_w)}{X_r - X}$$

Using values defined previously, the RAS pumping rate is computed to be:

$$Q_r = \frac{(4500 \text{ mg/L})(4.0 \text{ Mgd} - 0.104 \text{ Mgd})}{8000 \text{ mg/L} - 4500 \text{ mg/L}} = 5.0 \text{ Mgd} \,(18,925 \text{ m}^3/\text{d})$$

The ratio of RAS pumping rate to influent flow rate, or recirculation ratio (α), may now be calculated:

$$\alpha = \frac{Q_r}{Q} = \frac{5.0 \text{ Mgd}}{4.0 \text{ Mgd}} = 1.25$$

Recirculation ratio can vary from 0.25 to 1.50 depending upon the type of activated sludge process used. Common design practice is to size the RAS pumps so that they are capable of providing a recirculation ratio ranging from 0.50 to 1.50.

It should be noted that if the control volume were placed around the aeration tank in Fig. 5.10 and a mass balance performed, or the control volume placed around the settling tank in Fig. 5.11 and a mass balance performed, that a slightly higher RAS pumping rate would result. However, the difference between these RAS pumping rates and the ones calculated above is negligible.

This procedure is the work of Kevin D. Wills, M.S.E., P.E., who, at the time of its preparation, was consulting engineer, Stanley Consultants, Inc.

REFERENCES

1. Satterfield—*Mass Transfer in Heterogeneous Catalysis*, MIT Press (1970).
2. Barona—*Hydrocarbon Proc.* 59(7), 1979.
3. Smith—*Chemical Process Design*, McGraw-Hill.

SECTION 6
FLOW OF FLUIDS AND SOLIDS†

---

†Procedure 6.10 is from *Chemical Engineering* magazine. All the other examples are from T. G. Hicks. *Standard Handbook of Engineering Calculations.* McGraw-Hill.

6.1   BERNOULLI'S THEOREM, AND EQUATION OF CONTINUITY

A piping system is conveying 10 ft³/s (0.28 m³/s) of ethanol. At a particular cross section of the system, section 1, the pipe diameter is 12 in (0.30 m), the pressure is 18 lb/in² (124 kPa), and the elevation is 140 ft (42.7 m). At another cross section further downstream, section 2, the pipe diameter is 8 in (0.20 m), and the elevation is 106 ft (32.3 m). If there is a head loss of 9 ft (2.74 m) between these sections due to pipe friction, what is the pressure at section 2? Assume that the specific gravity of the ethanol is 0.79.

Calculation Procedure

1. Compute the velocity at each section.
Use the equation of continuity

$$Q = A_1 V_1 = A_2 V_2$$

where Q is volumetric rate of flow, A is cross-sectional area, V is velocity, and the subscripts refer to sections 1 and 2. Now, $A = (\pi/4)d^2$, where d is (inside) pipe diameter, so $A_1 = (\pi/4)(1\ \text{ft})^2 = 0.785\ \text{ft}^2$ and $A_2 = (\pi/4)(8/12\ \text{ft})^2 = 0.349\ \text{ft}^2$; and $Q = 10\ \text{ft}^3$ s. Therefore, $V_1 = 10/0.785 = 12.7$ ft/s, and $V_2 = 10/0.349 = 28.7$ ft/s.

2. Compute the pressure at section 2.
Use Bernoulli's theorem, which in one form can be written as

$$\frac{V_1^2}{2g} + \frac{p_1}{\rho} + z_1 = \frac{V_2^2}{2g} + \frac{p_2}{\rho} + z_2 + h_L$$

where g is the acceleration due to gravity, 32.2 ft/s²; p is pressure; ρ is density; z is elevation; and h_L is loss of head between two sections. In this case, $\rho = 0.79(62.4\ \text{lb/ft}^3) = 49.3\ \text{lb/ft}^3$. Upon rearranging the equation for Bernoulli's theorem,

$$\frac{p_2 - p_1}{\rho} = \frac{V_1^2 - V_2^2}{2g} + z_1 - z_2 - h_L$$

or $(p_2 - p_1)/49.3 = (12.7^2 - 28.7^2)/64.4 + 140 - 106 - 9$, so $(p^2 - p^1) = 725.2\ \text{lb/ft}^2$, or $725.2/144 = 5.0\ \text{lb/in}^2$. Therefore, $p_2 = 18 + 5 = 23\ \text{lb/in}^2$ (159 kPa).

6.2   SPECIFIC GRAVITY AND VISCOSITY OF LIQUIDS

An oil has a specific gravity of 0.8000 and a viscosity of 200 SSU (Saybolt Seconds Universal) at 60°F (289 K). Determine the API gravity and Bé gravity of this oil at 70°F (294 K) and its weight in pounds per gallon. What is the kinematic viscosity in centistokes? What is the absolute viscosity in centipoise?

Calculation Procedure

1. Determine the API gravity of the liquid.
For any oil at 60°F (15.6 C), its specific gravity S, in relation to water at 60°F, is $S = 141.5/(131.5 + °\text{API})$; or API $= (141.5 - 131.5S)/S$. For this oil, $°\text{API} = [141.5 - 131.5(0.80)]/0.80 = 45.4°$ API.

2. Determine the Bé gravity of the liquid.

For any liquid lighter than water, $S = 140/(130 + Bé)$; or $Bé = (140 - 130S)/S$. For this oil, Bé = $[140 - 130(0.80)]/0.80 = 45$ Bé.

3. Compute the weight per gallon of liquid.

With a specific gravity of S, the weight of 1 ft^3 oil equals (S) (weight of 1 ft^3 fresh water at 60°F) = $(0.80)(62.4) = 49.92$ lb/ft^3. Since 1 gal liquid occupies 0.13368 ft^3, the weight of this oil per gal is $49.92(0.13368) = 6.66$ lb/gal (800 kg/m^3).

4. Compute the kinematic viscosity of the liquid.

For any liquid having a viscosity between 32 and 99 SSU, the kinematic viscosity $k = 0.226$ SSU − 195/SSU Cst. For this oil, $k = 0.226(200) - 195/200 = 44.225$ Cst.

5. Convert the kinematic viscosity to absolute viscosity.

For any liquid, the absolute viscosity, cP, equals (kinematic viscosity, Cst)(specific gravity). Thus, for this oil, the absolute viscosity = $(44.225)(0.80) = 35.38$ cP.

Related Calculations.   For liquids *heavier* than water, $S = 145/(145 - Bé)$. When the SSU viscosity is greater than 100 s, $k = 0.220$ SSU − 135/SSU. Use these relations for any liquid—brine, gasoline, crude oil, kerosene, Bunker C, diesel oil, etc. Consult the *Pipe Friction Manual* and King and Crocker—*Piping Handbook* for tabulations of typical viscosities and specific gravities of various liquids.

6.3   *PRESSURE LOSS IN PIPING WITH LAMINAR FLOW*

Fuel oil at 300°F (422 K) and having a specific gravity of 0.850 is pumped through a 30,000-ft-long 24-in pipe at the rate of 500 gal/min (0.032 m^3/s). What is the pressure loss if the viscosity of the oil is 75 cP?

Calculation Procedure

1. Determine the type of flow that exists.

Flow is laminar (also termed viscous) if the Reynolds number Re for the liquid in the pipe is less than about 2000. Turbulent flow exists if the Reynolds number is greater than about 4000. Between these values is a zone in which either condition may exist, depending on the roughness of the pipe wall, entrance conditions, and other factors. Avoid sizing a pipe for flow in this critical zone because excessive pressure drops result without a corresponding increase in the pipe discharge.

Compute the Reynolds number from Re = $3.162G/kd$, where G = flow rate, gal/min; k = kinematic viscosity of liquid, Cst = viscosity z cP/specific gravity of the liquid S; d = inside diameter of pipe, in. From a table of pipe properties, $d = 22.626$ in. Also, $k = z/S = 75/0.85 = 88.2$ Cst. Then, Re = $3162(500)/[88.2(22.626)] = 792$. Since Re < 2000, laminar flow exists in this pipe.

2. Compute the pressure loss using the Poiseuille formula.

The Poiseuille formula gives the pressure drop p_d lb/in^2 = $2.73(10^{-4})luG/d^4$, where l = total length of pipe, including equivalent length of fittings, ft; u = absolute viscosity of liquid, cP; G = flow rate, gal/min; d = inside diameter of pipe, in. For this pipe, $p_d = 2.73(10^{-4})(30,000)(75)(500)/262,078 = 1.17$ lb/in^2 (8.07 kPa).

Related Calculations.   Use this procedure for any pipe in which there is laminar flow of liquid. Table 6.1 gives a quick summary of various ways in which the Reynolds number can be expressed. The symbols in Table 6.1, in the order of their appearance, are D = inside diameter of pipe, ft; v = liquid velocity, ft/s; ρ = liquid density, lb/ft^3; μ = absolute viscosity of liquid, lb mass/ft · s;

TABLE 6.1 Reynolds Number

Reynolds number Re	Numerator				Denominator	
	Coefficient	First symbol	Second symbol	Third symbol	Fourth symbol	Fifth symbol
Dvp/μ	—	ft	ft/s	lb/ft^3	lb mass/ft · s	—
$124dv\rho/z$	124	in	ft/s	lb/ft^3	cP	—
$50.7G\rho/dz$	50.7	gal/min	lb/ft^3	—	in	cP
$6.32W/dz$	6.32	lb/h	—	—	in	cP
$35.5B\rho/dz$	35.5	bbl/h	lb/ft^3	—	in	cP
$7{,}742dv/k$	7742	in	ft/s	—	—	cP
$3{,}162G/dk$	3162	gal/min	—	—	in	cP
$2{,}214\ B/dk$	2214	bbl/h	—	—	in	cP
$22{,}735q\rho/dz$	22,735	ft^3/s	lb/ft^3	—	in	cP
$378.9\ Q\rho/dz$	378.9	ft^3/min	lb/ft^3	—	in	cP

d = inside diameter of pipe, in. From a table of pipe properties, $d = 22.626$ in. Also, $k = z/S$ liquid flow rate, lb/h; B = liquid flow rate, bbl/h; k = kinematic viscosity of the liquid, Cst; q = liquid flow rate, ft³/s; Q = liquid flow rate, ft³/min. Use Table 6.1 to find the Reynolds number for any liquid flowing through a pipe.

6.4 *DETERMINING THE PRESSURE LOSS IN PIPES*

What is the pressure drop in a 5000-ft-long 6-in oil pipe conveying 500 bbl/h (0.022 m³/s) kerosene having a specific gravity of 0.813 at 65°F, which is the temperature of the liquid in the pipe? The pipe is schedule 40 steel.

Calculation Procedure

1. *Determine the kinematic viscosity of the oil.*
Use Fig. 6.1 and Table 6.2 or the Hydraulic Institute—*Pipe Friction Manual* kinematic viscosity and Reynolds number chart to determine the kinematic viscosity of the liquid. Enter Table 6.2 at kerosene and find the coordinates as $X = 10.2$, $Y = 16.9$. Using these coordinates, enter Fig. 6.1 and find the absolute viscosity of kerosene at 65°F as 2.4 cP. Using the method of Procedure 6.2, the kinematic viscosity, in cSt, equals absolute viscosity, cP/specific gravity of the liquid = 2.4/0.813 = 2.95 cSt. This value agrees closely with that given in the *Pipe Friction Manual*.

2. *Determine the Reynolds number of the liquid.*
The Reynolds number can be found from the *Pipe Friction Manual* chart mentioned in step 1 or computed from Re = 2214 B/dk = 2214(500)/[(6.065)(2.95)] = 61,900.

To use the *Pipe Friction Manual* chart, compute the velocity of the liquid in the pipe by converting the flow rate to cubic feet per second. Since there are 42 gal/bbl and 1 gal = 0.13368 ft³, 1 bbl = (42)(0.13368) = 5.6 ft³. With a flow rate of 500 bbl/h, the equivalent flow in ft³ = (500)(5.6) = 2800 ft³/h, or 2800/3600 s/h = 0.778 ft³/s. Since 6-in schedule 40 pipe has a cross-sectional area of 0.2006 ft² internally, the liquid velocity, in ft/s, equals 0.778/0.2006 = 3.88 ft/s. Then, the product (velocity, ft/s)(internal diameter, in) = (3.88)(6.065) = 23.75. In the *Pipe Friction Manual*, project horizontally from the kerosene specific-gravity curve to the *vd* product of 23.75 and read the Reynolds number as 61,900, as before. In general, the Reynolds number can be found faster by computing it using the appropriate relation given in Table 6.1, unless the flow velocity is already known.

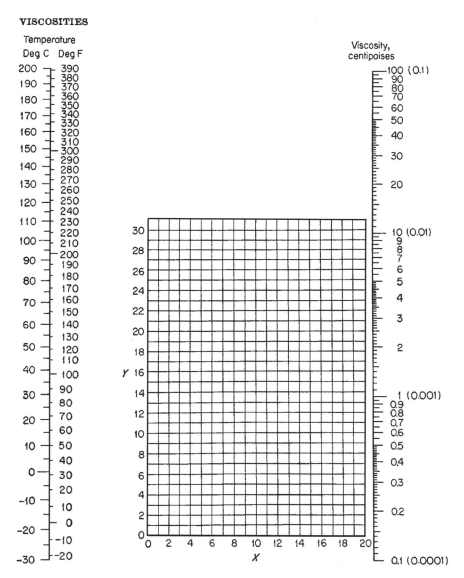

VISCOSITIES

FIGURE 6.1   Viscosities of liquids al 1 atm (101.3 kPa). For coordinates, see Table 6.2.

3. *Determine the friction factor of this pipe.*

Enter Fig. 6.2 at the Reynolds number value of 61,900 and project to the curve 4 as indicated by Table 6.3. Read the friction factor as 0.0212 at the left. Alternatively, the *Pipe Friction Manual* friction-factor chart could be used, if desired.

4. *Compute the pressure loss in the pipe.*

Use the Fanning formula $p_d = 1.06(10^{-4})fpl\ B^2/d^5$. In this formula, ρ = density of the liquid, lb/ft^3. For kerosene, p = (density of water, lb/ft^3) × (specific gravity of the kerosene) = (62.4)(0.813) = 50.6 lb/ft^3. Then, $p_d = 1.06(10^{-4})\ (0.0212)(50.6)(5000)(500)^2/8206 = 17.3$ lb/in^2 (119 kPa).

TABLE 6.2 Viscosities of Liquids (coordinates for use with Fig. 6.1)

No.	Liquid	X	Y	No.	Liquid	X	Y
1	Acetaldehyde	15.2	4.8	56	Freon-22	17.2	4.7
	Acetic acid:			57	Freon-13	12.5	11.4
2	100%	12.1	14.2		Glycerol:		
3	70%	9.5	17.0	58	100%	2.0	30.0
4	Acetic anhydride	12.7	12.8	59	50%	6.9	19.6
	Acetone:			60	Heptene	14 1	8.4
5	100%	14.5	7.2	61	Hexane	14.7	7.0
6	35%	7.9	15.0	62	Hydrochloric acid, 31.5%	13.0	16.6
7	Allyl alcohol	10.2	14.3	63	Isobutyl alcohol	7.1	18.0
	Ammonia:			64	Isobutyric acid	12.2	14.4
8	100%	12.6	2.0	65	Isopropyl alcohol	8.2	16.0
9	26%	10.1	13.9	66	Kerosene	10.2	16.9
10	Amyl acetate	11.8	12.5	67	Linseed oil. raw	7.5	27.2
11	Amyl alcohol	7.5	18.4	68	Mercury	18.4	16.4
12	Aniline	8.1	18.7		Methanol:		
13	Anisole	12.3	13.5	69	100%	12.4	10.5
14	Arsenic trichloride	13.9	14.5	70	90%	12.3	11.8
15	Benzene	12.5	10.9	71	40%	7.8	15.5
	Brine:			72	Methyl acetate	14.2	8.2
16	CaCl$_2$, 25%	6.6	15.9	73	Methyl chloride	15.0	3.8
17	NaCl, 25%	10.2	16.6	74	Methyl ethyl ketone	13.9	8.6
18	Bromine	14.2	13.2	75	Naphthalene	7.9	18.1
19	Bromotoluene	20.0	15.9		Nitric acid:		
20	Butyl acetate	12.3	11.0	76	95%	12.8	13.8
21	Butyl alcohol	8.6	17.2	77	60%	10.8	17.0
22	Butyric acid	12.1	15.3	78	Nitrobenzene	10.6	16.2
23	Carbon dioxide	11.6	0.3	79	Nitrotoluene	11.0	17.0
24	Carbon disulfide	16.1	7.5	80	Octane	13.7	10.0
25	Carbon tetrachloride	12.7	13.1	81	Octyl alcohol	6.6	21.1
26	Chlorobenzene	12.3	12.4	82	Pentachloroethane	10.9	17.3
27	Chloroform	14.4	10.2	83	Pentane	14.9	5.2
28	Chlorosulfonic acid	11.2	18.1	84	Phenol	6.9	20.8
	Chlorotoluene:			85	Phosphorus tribromide	13.8	16.7
29	Ortho	13.0	13.3	86	Phosphorus trichloride	16.2	10.9
30	Meta	13.3	12.5	87	Propionic acid	12.8	13.8
31	Para	13.3	12.5	88	Propyl alcohol	9.1	16.5
32	Cresol. meta	2.5	20.8	89	Propyl bromide	14.5	9.6
33	Cyclohexanol	2.9	24.3	90	Propyl chloride	14.4	7.5
34	Dibromoethane	12.7	15.8	91	Propyl iodide	14.1	11.6
35	Dichloroethane	13.2	12.2	92	Sodium	16.4	13.9
36	Dichloromethane	14.6	8.9	93	Sodium hydroxide, 50%	3.3	25.8
37	Diethyl oxalate	11.0	16.4	94	Stannic chloride	13.5	12.8
38	Dimethyl oxalate	12.3	15.8	95	Sulfur dioxide	15.2	7.1
39	Diphenyl	12.0	18.3		Sulfuric acid:		
40	Dipropyl oxalate	10.3	17.7	96	110%	7.2	27.4
41	Ethyl acetate	13.7	9.1	97	98%	7.0	24.8
	Ethyl alcohol:			98	60%	10.2	21.3
42	100%	10.5	13.8	99	Sulfuryl chloride	15.2	12.4
43	95%	9.8	14.3	100	Tetrachloroethane	11.9	15.7
44	40%	6.5	16.6	101	Tetrachloroethylene	14.2	12.7
45	Ethyl benzene	13.2	11.5	102	Titanium tetrachloride	14.4	12.3
46	Ethyl bromide	14.5	8.1	103	Toluene	13.7	10.4
47	Ethyl chloride	14.8	6.0	104	Trichloroethylene	14 8	10.5
48	Ethyl ether	14.5	5.3	105	Turpentine	11.5	14.9
49	Ethyl formate			106	Water	10.2	13.0
50	Ethyl iodide	14.7	10.3		Xylene:		
51	Ethylene glycol	6.0	23.6	107	Ortho	13.5	12.1
52	Formic acid	10.7	15.8	108	Meta	13.9	10.6
53	Freon-11	14.4	9.0	109	Para	13.9	10.9
54	Freon-12	16.8	5.6				
55	Freon-21	15.7	7.5				

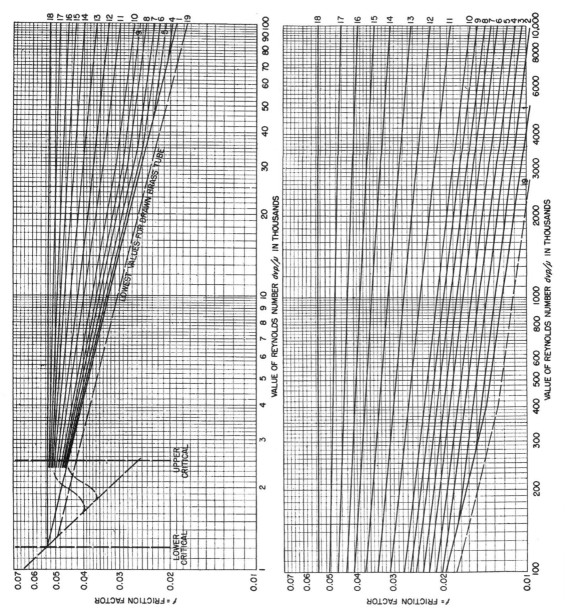

FIGURE 6.2 Friction-factor curves. (*Mechanical Engineering.*)

6.7

TABLE 6.3 Data for Fig. 6.2

Percent roughness	For value of f see curve	Drawn tubing, brass, tin, lead, glass		Clean steel, wrought iron		Clean, galvanized		Best cast iron		Average cast iron		Heavy riveted, spiral riveted	
		in	mm	in	mm	in	mm	in	mm	in	mm	in	mm
0.2	1	0.35 up	8.89 up	72	1829	—	—	—	—	—	—	—	—
1.35	4	—	—	6–12	152–305	10–24	254–610	20–48	508–1219	42–96	1067–2438	84–204	2134–5182
2.1	5	—	—	4–5	102–127	6–8	152–203	12–16	305–406	24–36	610–914	48–72	1219–1829
3.0	6	—	—	2–3	51–76	3–5	76–127	5–10	127–254	10–20	254–508	20–42	508–1067
3.8	7	—	—	1½	38	2½	64	3–4	76–102	6–8	152–203	16–18	406–457
4.8	8	—	—	1–1½	25–32	1½–2	38–51	2–2½	51–64	4–5	102–127	10–14	254–356
6.0	9	—	—	¾	19	1¼	32	1½	38	3	76	8	203
7.2	10	—	—	½	13	1	25	1¼	32	—	—	5	127
10.5	11	—	—	⅜	9.5	¾	19	1	35	—	—	4	102
14.5	12	—	—	¼	6.4	½	13	—	—	—	—	3	76
24.0	14	0.125	3.18	—	—	⅜	9.5	—	—	—	—	—	—
31.5	16	—	—	—	—	¼	6.4	—	—	—	—	—	—
37.5	18	0.0625	1.588	—	—	⅛	3.2	—	—	—	—	—	—

Diameter (actual of drawn tubing, nominal of standard-weight pipe)

Related Calculations.   The Fanning formula is popular with oil-pipe designers and can be stated in various ways: (1) with velocity v, in ft/s, $p_d = 1.29(10^{-3})f\rho v^2 l/d$; (2) with velocity V, in ft/min. $p_d = 3.6(10^{-7})f\rho V^2 l/d$; (3) with flow rate G, in gal/min, $p_d = 2.15(10^{-4})f\rho l G^2/d^2$; (4) with the flow rate W, in lb/h, $p_d = 3.36(10^{-6}) f_l W^2/d^5 \rho$.

Use this procedure for any fluid—crude oil, kerosene, benzene, gasoline, naptha, fuel oil, Bunker C, diesel oil toluene, etc. The tables and charts presented here and in the *Pipe Friction Manual* save computation time.

6.5   EQUIVALENT LENGTH OF A COMPLEX-SERIES PIPELINE

Figure 6.3 shows a complex-series pipeline made up of four lengths of different size pipe. Determine the equivalent length of this pipe if each size of pipe has the same friction factor.

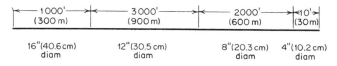

FIGURE 6.3   Complex-series pipeline.

Calculation Procedure

1. *Select the pipe size for expressing the equivalent length.*
The usual procedure when analyzing complex pipelines is to express the equivalent length in terms of the smallest, or next-to-smallest, diameter pipe. Choose the 8-in size as being suitable for expressing the equivalent length.

2. *Find the equivalent length of each pipe.*
For any complex-series pipeline having equal friction factors in all the pipes, L_e = equivalent length, ft, of a section of constant diameter = (actual length of section, ft) (inside diameter, in, of pipe used to express the equivalent length/inside diameter, in, of section under consideration)^5.

For the 16-in pipe, $L_e = (1000)(7.981/15.000)^5 = 42.6$ ft. The 12-in pipe is next; for it, $L_e = (3000)(7.981/12.00)^5 = 390$ ft. For the 8-in pipe, the equivalent length = actual length = 2000 ft. For the 4-in pipe, $L_e = (10)(7.981/4.026)^5 = 306$ ft. Then, the total equivalent length of 8-in pipe = sum of the equivalent lengths = 42.6 + 390 + 2000 + 306 = 2738.6 ft, or rounding off, 2740 ft of 8-in pipe (835 m of 0.2-m pipe) will have a frictional resistance equal to the complex-series pipeline shown in Fig. 6.3. To compute the actual frictional resistance, use the methods given in previous Calculation Procedures.

Related Calculations.   Use this general procedure for any complex-series pipeline conveying water, oil, gas, steam, etc. See King and Crocker—*Piping Handbook* for derivation of the flow equations. Use the tables in King and Crocker to simplify finding the fifth power of the inside diameter of a pipe.

Choosing a flow rate of 1000 gal/min and using the tables in the Hydraulic Institute *Pipe Friction Manual* gives an equivalent length of 2770 ft for the 8-in pipe. This compares favorably with the 2740 ft computed above. The difference of 30 ft is negligible.

The equivalent length is found by summing the friction-head loss for 1000 gal/min flow for each length of the four pipes—16, 12, 8, and 4 in—and dividing this by the friction-head loss for 1000 gal/min flowing through an 8-in pipe. Be careful to observe the units in which the friction-head loss is stated because errors are easy to make if the units are ignored.

6.6 HYDRAULIC RADIUS AND LIQUID VELOCITY IN PIPES

What is the velocity of 1000 gal/min (0.064 m³/s) of water flowing through a 10-in inside-diameter cast-iron water-main pipe? What is the hydraulic radius of this pipe when it is full of water? When the water depth is 8 in (0.203 m)?

Calculation Procedure

1. Compute the water velocity in the pipe.
For any pipe conveying liquid, the liquid velocity, in ft/s, is $v = $ (gal/min)/($2.448d^2$), where $d = $ internal pipe diameter, in. For this pipe, $v = 1000/[2.448(100)] = 4.08$ ft/s, or $(60)(4.08) = 244.8$ ft/min.

2. Compute the hydraulic radius for a full pipe.
For any pipe, the hydraulic radius is the ratio of the cross-sectional area of the pipe to the wetted perimeter, or $d/4$. For this pipe, when full of liquid, the hydraulic radius $= 10/4 = 2.5$.

3. Compute the hydraulic radius for a partially full pipe.
Use the hydraulic radius tables in King and Brater—*Handbook of Hydraulics* or compute the wetted perimeter using the geometric properties of the pipe, as in step 2. Using the King and Brater table, the hydraulic radius $= Fd$, where $F = $ table factor for the ratio of the depth of liquid, in/diameter of channel, in $= 8/10 = 0.8$. For this ratio, $F = 0.304$. Then, hydraulic radius $= (0.304)(10) = 3.04$ in.

6.7 FRICTION-HEAD LOSS IN WATER PIPING OF VARIOUS MATERIALS

Determine the friction-head loss in 2500 ft of clean 10-in new tar-dipped cast-iron pipe when 2000 gal/min (0.126 m³/s) of cold water is flowing. What is the friction-head loss 20 years later? Use the Hazen-Williams and Manning formulas and compare the results.

Calculation Procedure

1. Compute the friction-head loss using the Hazen-Williams formula.
The Hazen-Williams formula is $h_f = (v/1.318CR_h^{0.63})^{1.85}$, where $h_f = $ friction-head loss per foot of pipe, in feet of water; $v = $ water velocity, in ft/s; $C = $ a constant depending on the condition and kind of pipe; and $R_h = $ hydraulic radius of pipe, in ft.

For a water pipe, $v = $ (gal/min)/($2.44d^2$); for this pipe, $v = 2000/[2.448(10)^2] = 8.18$ ft/s. From Table 6.4 or King and Crocker—*Piping Handbook, C* for new pipe $= 120$; for 20-year-old pipe,

TABLE 6.4 Values of C in Hazen-Williams Formula

Type of pipe	$C^†$	Type of pipe	$C^†$
Cement-asbestos	140	Cast iron or wrought iron	100
Asphalt-lined iron or steel	140	Welded or seamless steel	100
Copper or brass	130	Concrete	100
Lead, tin, or glass	130	Corrugated steel	60
Wood stave	110		

†Values of C commonly used for design. The value of C for pipes made of corrosive materials decreases as the age of the pipe increases: the values given are those which apply at an age of 15 to 20 years. For example, the value of C for cast-iron pipes 30 in in diameter or greater at various ages is approximately as follows: new, 130; 5 years old, 120; 10 years old, 115; 20 years old, 100; 30 years old, 90; 40 years old, 80; and 50 years old, 75. The value of C for smaller-size pipes decreases at a more rapid rate.

$C = 90$; $R_h = d/4$ for a full-flow pipe $= 10/4 = 2.5$ in, or $2.5/12 = 0.208$ ft. Then, $h_f = (8.18/1.318 \times 120 \times 0.208^{0.63})^{1.85} = 0.0263$ ft of water per foot of pipe. For 2500 ft of pipe, the total friction-head loss $= 2500(0.0263) = 65.9$ ft (20.1 m) of water for the new pipe.

For 20-year-old pipe using the same formula, except with $C = 90$, $h_f = 0.0451$ ft of water per foot of pipe. For 2500 ft of pipe, the total friction-head loss $= 2500(0.0451) = 112.9$ ft (34.4 m) of water. Thus the friction-head loss nearly doubles (from 65.9 to 112.9 ft) in 20 years. This shows that it is wise to design for future friction losses; otherwise, pumping equipment may become overloaded.

2. Compute the friction-head loss using the Manning formula.
The Manning formula is $h_f = n^2 v^2/(2.208\, R_h^{4/3})$, where $n = $ a constant depending on the condition and kind of pipe; other symbols as before.

Using $n = 0.011$ for new coated cast-iron pipe from Table 6.5 or King and Crocker—*Piping Handbook*, $h_f = (0.011)^2(8.18)^2/[2.208(0.208)^{4/3}] = 0.0295$ ft of water per foot of pipe. For 2500 ft of pipe, the total friction-head loss $= 2500(0.0295) = 73.8$ ft (22.5 m) of water, as compared with 65.9 ft of water computed with the Hazen-Williams formula.

For coated cast-iron pipe in fair condition, $n = 0.013$, and $h_f = 0.0411$ ft of water. For 2500 ft of pipe, the total friction-head loss $= 2500(0.0411) = 102.8$ ft (31.4 m) of water, as compared with 112.9 ft of water computed with the Hazen-Williams formula. Thus the Manning formula gives results higher than the Hazen-Williams in one case and lower in another. However, the differences in each case are not excessive; $(73.8 - 65.9)/65.9 = 0.12$, or 12 percent higher, and $(112.9 - 102.8)/102.8 = 0.0983$, or 9.83 percent lower. Both these differences are within the normal range of accuracy expected in pipe friction-head calculations.

TABLE 6.5   Roughness Coefficients (Manning's n) for Closed Conduits

Type of conduit			Manning's n	
			Good construction[†]	Fair construction[†]
Concrete pipe			0.013	0.015
Corrugated metal pipe or pipe arch, $2\frac{2}{3} \times \frac{1}{2}$ in corrugation, riveted:				
Plain			0.024	—
Paved invert:				
Percent of circumference paved	25	50		
Depth of flow:				
Full	0.021	0.018		
0.8D	0.021	0.016		
0.6D	0.019	0.013		
Vitrified clay pipe			0.012	0.014
Cast-iron pipe, uncoated			0.013	—
Steel pipe			0.011	—
Brick			0.014	0.017
Monolithic concrete:				
Wood forms, rough			0.015	0.017
Wood forms, smooth			0.012	0.014
Steel forms			0.012	0.013
Cemented-rubble masonry walls:				
Concrete floor and top			0.017	0.022
Natural floor			0.019	0.025
Laminated treated wood			0.015	0.017
Vitrified-clay liner plates			0.015	—

[†]For poor-quality construction, use larger values of n.

Related Calculations. The Hazen-Williams and Manning formulas are popular with many piping designers for computing pressure losses in cold-water piping. To simplify calculations, most designers use the precomputed tabulated solutions available in King and Crocker—*Piping Handbook*, King and Brater—*Handbook of Hydraulics*, and similar publications. In the rush of daily work these precomputed solutions are also preferred over the more complex Darcy-Weisbach equation used in conjunction with the friction factor f, the Reynolds number Re, and the roughness-diameter ratio.

Use the method given here for sewer lines, water-supply pipes for commercial, industrial, or process plants, and all similar applications where cold water at temperatures of 33 to 90°F flows through a pipe made of cast iron, riveted steel, welded steel, galvanized iron, brass, glass, wood-stove, concrete, vitrified, common clay, corrugated metal, unlined rock, or enameled steel. Thus either of these formulas, used in conjunction with a suitable constant, gives the friction-head loss for a variety of piping materials. Suitable constants are given in Tables 6.4 and 6.5 and in the preceding references. For the Hazen-Williams formula, the constant C varies from about 70 to 140, while n in the Manning formula varies from about 0.017 for $C = 70$ to $n = 0.010$ for $C = 140$. Values obtained with these formulas have been used for years with satisfactory results.

6.8   RELATIVE CARRYING CAPACITY OF PIPES

What is the equivalent steam-carrying capacity of a 24-in-inside-diameter pipe in terms of a 10-in-inside-diameter pipe? What is the equivalent water-carrying capacity of a 23-in-inside-diameter pipe in terms of a 13.25-in-inside-diameter pipe?

Calculation Procedure

1. *Compute the relative carrying capacity of the steam pipes.*
For steam, air, or gas pipes, the number N of small pipes of inside diameter d_2 in equal to one pipe of larger inside diameter d_1 in is $N = (d_1^3 \sqrt{d_2} + 3.6)/(d_2^3 + \sqrt{d_1} + 3.6)$. For this piping system, $N = (24^3 + \sqrt{10} + 3.6)/(10^3 + \sqrt{24} + 3.6) = 9.69$, say 9.7. Thus a 24-in-inside-diameter steam pipe has a carrying capacity equivalent to 9.7 pipes having a 10-in inside diameter.

2. *Compute the relative carrying capacity of the water pipes.*
For water, $N = (d_2/d_1)^{2.5} = (23/13.25)^{2.5} = 3.97$. Thus one 23-in-inside-diameter pipe can carry as much water as 3.97 pipes of 13.25 in inside diameter.

Related Calculations. King and Crocker—*Piping Handbook* and certain piping catalogs contain tabulations of relative carrying capacities of pipes of various sizes. Most piping designers use these tables. However, the equations given here are useful for ranges not covered by the tables and when the tables are unavailable.

6.9   FLOW RATE AND PRESSURE LOSS IN COMPRESSED-AIR AND GAS PIPING

Dry air at 80°F (300 K) and 150 psia (1034 kPa) flows at the rate of 500 ft³/min (0.24 m³/s) through a 4-in schedule 40 pipe from the discharge of an air compressor. What is the flow rate in pounds per hour and the air velocity in feet per second? Using the Fanning formula, determine the pressure loss if the total equivalent length of the pipe is 500 ft (153 m).

Calculation Procedure

1. Determine the density of the air or gas in the pipe.

For air or a gas, $pV = MRT$, where p = absolute pressure of the gas, in lb/ft^2; V = volume of M lb of gas, in ft^3; M = weight of gas, in lb; R = gas constant, in ft · lb/(lb)(°F); T = absolute temperature of the gas, in R. For this installation, using 1 ft^3 of air, $M = pV/RT$, $M = (150)(144)/$ [$(53.33)(80 + 459.7)$] = 0.754 lb/ft^3. The value of R in this equation was obtained from Table 6.6.

2. Compute the flow rate of the air or gas.

For air or a gas, the flow rate W_h, in lb/h, = (60)(density, lb/ft^3)(flow rate, ft^3/min); or W_h = (60)(0.754)(500) = 22,620 lb/h.

3. Compute the velocity of the air or gas in the pipe.

For any air or gas pipe, velocity of the moving fluid v, in ft/s, = $183.4\ W_h/(3600\ d^2\rho)$, where d = internal diameter of pipe, in; ρ = density of fluid, lb/ft^3. For this system, $v = (183.4)(22,620)/$ [$(3600)(4.026)^2(0.754)$] = 95.7 ft/s.

4. Compute the Reynolds number of the air or gas.

The viscosity of air at 80°F is 0.0186 cP, obtained from King and Crocker—*Piping Handbook*, Perry et al.—*Chemical Engineers' Handbook*, or a similar reference. Then, using the Reynolds number relation given in Table 6.1, Re = $6.32W/dz$ = 6.32(22,620)/[4.026(0.0186)] = 1,910,000.

5. Compute the pressure loss in the pipe.

Using Fig. 6.2 or the Hydraulic Institute *Pipe Friction Manual*, f = 0.02 for a 4-in schedule 40 pipe when the Reynolds number = 1,910,000. Using the Fanning formula from Procedure 6.4, $p_d = 3.36(10^{-6})f\ lW^2/(d^5\rho)$, or $p_d = 3.36(10^{-6})$ (0.02)(500)(22,620)^2/[(4.026)^5)(0.754)] = 21.63 lb/in^2 (121 kPa).

Related Calculations.   Use this procedure to compute the pressure loss, velocity, and flow rate in compressed-air and gas lines of any length. Gases for which this procedure can be used include ammonia, carbon dioxide, carbon monoxide, ethane, ethylene, hydrogen, hydrogen sulfide, isobutane, methane, nitrogen, *n*-butane, oxygen, propane, propylene, and sulfur dioxide.

TABLE 6.6 Gas Constants

Gas	R		C for critical-velocity equation
	ft · lb/(lb)(°F)	J/(kg)(K)	
Air	53.33	286.9	2870
Ammonia	89.42	481.1	2080
Carbon dioxide	34.87	187.6	3330
Carbon monoxide	55.14	296.7	2820
Ethane	50.82	273.4	
Ethylene	54.70	294.3	2480
Hydrogen	767.04	4126.9	750
Hydrogen sulfide	44.79	240.9	
Isobutane	25.79	138.8	
Methane	96.18	517.5	2030
Natural gas	—	—	2070–2670
Nitrogen	55.13	296.6	2800
n-butane	25.57	137.6	
Oxygen	48.24	259.5	2990
Propane	34.13	183.6	
Propylene	36.01	193.7	
Sulfur dioxide	23.53	126.6	3870

Alternate relations for computing the velocity of air or gas in a pipe are $v = 144 W_s/(a\rho)$; $v = 183.4 W_s/(d^2\rho)$; $v = 0.0509 W_s v_g/d^2$, where W_s = flow rate, in lb/s; a = cross-sectional area of pipe, in in²; v_g = specific volume of the air or gas at the operating pressure and temperature, in ft³/lb.

6.10 CALCULATIONS FOR PARTIALLY FILLED PIPES DURING LIQUID FLOW

A horizontal run of 3-in pipe handles 40 gal/min of water. Determine whether the flow is below or above the minimum needed to keep the pipe sealed (full of liquid). If the pipe instead is only partly filled, determine the height of liquid in it, the velocity of the fluid, and the equivalent diameter of the flooded section.

Calculation Procedure

1. Calculate the minimum flow required to seal this pipe, and compare it with the flow rate stated for this problem.
Use the equation $Q = 10.2D^{2.5}$, where Q is minimum liquid flow rate required for seal flow in gallons per minute and D is the pipe diameter in inches, Thus, $Q = 10.2(3)^{2.5} = 159$ gal/min.

Since the stated flow rate is only 40 gal/min, the pipe is not sealed.

2. Determine the height of liquid in the pipe, H.
First, calculate the quantity, $Q/D^{2.5}$. It equals $40/(3)^{2.5}$, or 2.56. With this value, enter the figure below along its ordinate and, from the abscissa, read H/D as 0.7. Therefore, $H = 0.7 D = 0.7(3) = 2.1$ in.

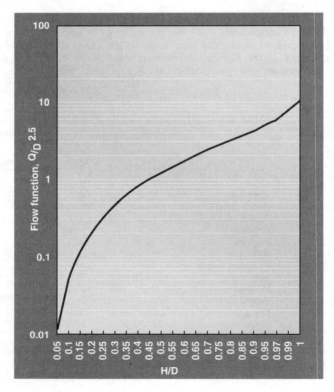

3. Determine the cross-sectional area A of the flow.
Enter the figure below along its abscissa with an H/D value of 0.7 and, from the ordinate along the right side of the graph, read A/D^2 as about 0.59. Therefore, $A = 0.59(3)^2 = 5.31$ in^2, or 0.369 ft^2.

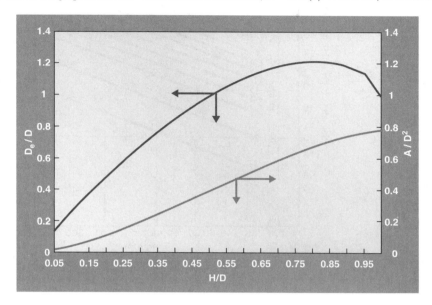

4. Calculate the velocity V of the fluid flow.
The calculation is straightforward:

$$V = (40 \text{ gal/min})(1 \text{ ft}^3/7.48 \text{ gal})(1 \text{ min}/60 \text{ s})(1/0.0369 \text{ ft}^2) = 2.42 \text{ ft/s}$$

5. Determine the equivalent diameter D_e for this flow.
Enter the figure in step 3 along its abscissa with an H/D value of 0.7 and, from the ordinate along the left side of the graph, read D_e/D as about 1.18. Thus, $D_e = 1.18(3) = 3.54$ in.

6.11 FRICTION LOSS IN PIPES HANDLING SOLIDS IN SUSPENSION

What is the friction loss in 800 ft of 6-in schedule 40 pipe when 400 gal/min (0.025 m^3/s) of sulfate paper stock is flowing? The consistency of the sulfate stock is 6 percent.

Calculation Procedure

1. Determine the friction loss in the pipe.
There are few general equations for friction loss in pipes conveying liquids having solids in suspension. Therefore, most practicing engineers use plots of friction loss available in engineering handbooks, *Cameron Hydraulic Data, Standards of the Hydraulic Institute,* and from pump engineering data. Figure 6.4 shows one set of typical friction-loss curves based on work done at the University of Maine on the data of Brecht and Heller of the Technical College, Darmstadt, Germany, and

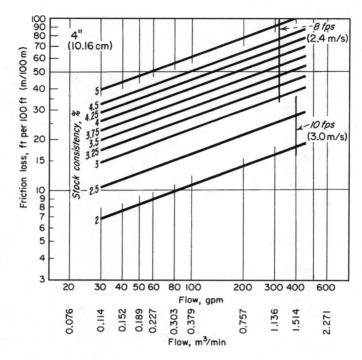

FIGURE 6.4   Friction loss of paper stock in 4-in (0.10-m) steel pipe. (*Goulds Pumps, Inc.*)

published by Goulds Pumps, Inc. There is a similar series of curves for commonly used pipe sizes from 2 through 36 in.

Enter Fig. 6.4 at the pipe flow rate, 400 gal/min, and project vertically upward to the 6 percent consistency curve. From the intersection, project horizontally to the left to read the friction loss as 60 ft of liquid per 100 ft of pipe. Since this pipe is 800 ft long, the total friction-head loss in the pipe is (800/100)(60) = 480 ft (146 m) of liquid flowing.

2. Correct the friction loss for the liquid consistency.

Friction-loss factors are usually plotted for one type of liquid, and correction factors are applied to determine the loss for similar, but different, liquids. Thus, with the Goulds charts, a factor of 0.9 is used for soda, sulfate, bleached sulfite, and reclaimed paper stocks. For ground wood, the factor is 1.40.

When the stock consistency is less than 1.5 percent, water-friction values are used. Below a consistency of 3 percent, the velocity of flow should not exceed 10 ft/s. For suspensions of 3 percent and above, limit the maximum velocity in the pipe to 8 ft/s.

Since the liquid flowing in this pipe is sulfate stock, use the 0.9 correction factor, or the actual total friction head = (0.9)(480) = 432 ft (132 m) of sulfate liquid. Note that Fig. 6.4 shows that the liquid velocity is less than 8 ft/s.

Related Calculations.   Use this procedure for soda, sulfate, bleached sulfite, and reclaimed and ground-wood paper stock. The values obtained are valid for both suction and discharge piping. The same general procedure can be used for sand mixtures, sewage, trash, sludge, foods in suspension in a liquid, and other slurries.

6.12 DETERMINING THE PRESSURE LOSS IN STEAM PIPING

Use a suitable pressure-loss chart to determine the pressure loss in 510 ft of 4-in flanged steel pipe containing two 90° elbows and four 45° bends. The schedule 40 piping conveys 13,000 lb/h (1.64 kg/s) of 40-psig 350°F superheated steam. List other methods of determining the pressure loss in steam piping.

Calculation Procedure

1. *Determine the equivalent length of the piping.*
The equivalent length of a pipe L_e, in ft, equals length of straight pipe, ft + equivalent length of fitting, ft. Using data from the Hydraulic Institute, King and Crocker—*Piping Handbook,* or Fig. 6.5, find the equivalent of a 90° 4-in elbow as 10 ft of straight pipe. Likewise, the equivalent length of a 45° bend is 5 ft of straight pipe. Substituting in the preceding relation and using the straight lengths and the number of fittings of each type, $L_e = 510 + (2)(10) + 4(5) = 550$ ft of straight pipe.

2. *Compute the pressure loss using a suitable chart.*
Figure 6.6 presents a typical pressure-loss chart for steam piping. Enter the chart at the top left at the superheated steam temperature of 350°F and project vertically downward until the 40-psig superheated steam pressure curve is intersected. From here, project horizontally to the right until the outer border of the chart is intersected. Next, project through the steam flow rate, 13,000 lb/h on scale *B* of Fig. 6.6 to the pivot scale *C*. From this point, project through 4-in (101.6-mm) schedule 40 pipe on scale *D* of Fig. 6.6. Extend this line to intersect the pressure-drop scale and read the pressure loss as 7.25 lb/in^2 (5. kPa) per 100 ft (30.4 m) of pipe.
  Since the equivalent length of this pipe is 550 ft (167.6 m), the total pressure loss in the pipe is $(550/100)(7.25) = 39.875$ lb/m^2 (274.9 kPa), say 40 lb/in^2 (275.8 kPa).

3. *List the other methods of computing pressure loss.*
Numerous pressure-loss equations have been developed to compute the pressure drop in steam piping. Among the better-known equations are those of Unwin, Fritzche, Spitz-glass, Babcock, Gutermuth, and others. These equations are discussed in some detail in King and Crocker—*Piping Handbook* and in the engineering data published by valve and piping manufacturers.
  Most piping designers use a chart to determine the pressure loss in steam piping because a chart saves time and reduces the effort involved. Further, the accuracy obtained is sufficient for all usual design practice.
  Figure 6.7 is a popular flowchart for determining steam flow rate, pipe size, steam pressure, or steam velocity in a given pipe. Using this chart, the designer can determine any one of the four variables listed above when the other three are known. In solving a problem on the chart in Fig. 6.7, use the steam-quantity lines to intersect pipe sizes and the steam-pressure lines to intersect steam velocities. Here are two typical applications of this chart.
  Example: What size schedule 40 pipe is needed to deliver 8000 lb/h (3600 kg/h) of 120-psig (827.3-kPa) steam at a velocity of 5000 ft/min (1524 m/min)?
  Solution: Enter Fig. 6.7 at the upper left at a velocity of 5000 ft/min and project along this velocity line until the 120-psig pressure line is intersected. From this intersection, project horizontally until the 8000 lb/h (3600 kg/h) vertical line is intersected. Read the *nearest* pipe size as 4 in (101.6 mm) on the *nearest* pipe-diameter curve.
  Example: What is the steam velocity in a 6-in (152.4-mm) pipe delivering 20,000 lb/h (9000 kg/h) of steam at 85 psig (586 kPa)?
  Solution: Enter the bottom of Fig. 6.7 at the flow rate, 20,000 lb/h, and project vertically upward until the 6-in pipe (152.4-mm) curve is intersected. From this point, project horizontally to the 85-psig (586-kPa) curve. At the intersection, read the velocity as 7350 ft/min (2240.3 m/min).

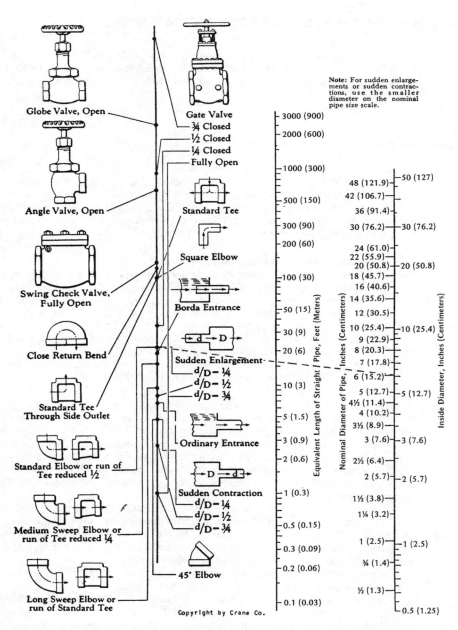

Note: For sudden enlargements or sudden contractions, use the smaller diameter on the nominal pipe size scale.

Globe Valve, Open

Angle Valve, Open

Swing Check Valve, Fully Open

Close Return Bend

Standard Tee Through Side Outlet

Standard Elbow or run of Tee reduced ½

Medium Sweep Elbow or run of Tee reduced ¼

Long Sweep Elbow or run of Standard Tee

Gate Valve
¾ Closed
½ Closed
¼ Closed
Fully Open

Standard Tee

Square Elbow

Borda Entrance

Sudden Enlargement
d/D – ¼
d/D – ½
d/D – ¾

Ordinary Entrance

Sudden Contraction
d/D – ¼
d/D – ½
d/D – ¾

45° Elbow

3000 (900)
2000 (600)
1000 (300)
500 (150)
300 (90)
200 (60)
100 (30)
50 (15)
30 (9)
20 (6)
10 (3)
5 (1.5)
3 (0.9)
2 (0.6)
1 (0.3)
0.5 (0.15)
0.3 (0.09)
0.2 (0.06)
0.1 (0.03)

Equivalent Length of Straight / Pipe, Feet (Meters)

Nominal Diameter of Pipe, Inches (Centimeters)

48 (121.9)
42 (106.7)
36 (91.4)
30 (76.2)
24 (61.0)
22 (55.9)
20 (50.8)
18 (45.7)
16 (40.6)
14 (35.6)
12 (30.5)
10 (25.4)
9 (22.9)
8 (20.3)
7 (17.8)
6 (15.2)
5 (12.7)
4½ (11.4)
4 (10.2)
3½ (8.9)
3 (7.6)
2½ (6.4)
2 (5.7)
1½ (3.8)
1¼ (3.2)
1 (2.5)
¾ (1.4)
½ (1.3)

Inside Diameter, Inches (Centimeters)

50 (127)
30 (76.2)
20 (50.8)
10 (25.4)
5 (12.7)
3 (7.6)
2 (5.7)
1 (2.5)
0.5 (1.25)

Copyright by Crane Co.

FIGURE 6.5 Equivalent length of pipe fittings and valves. (*Crane Co.*)

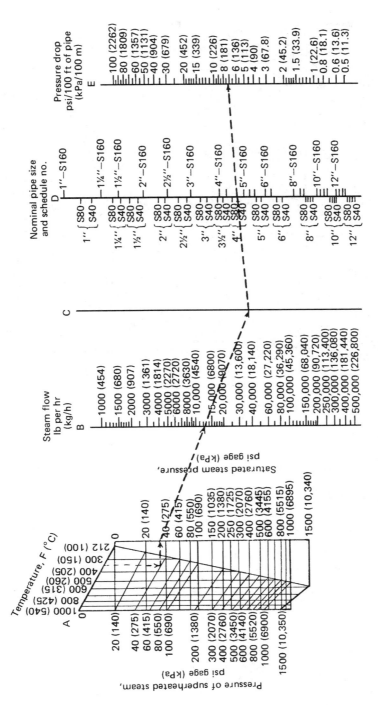

FIGURE 6.6 Pressure loss in steam pipes based on the Fritzche formula. (*Power.*)

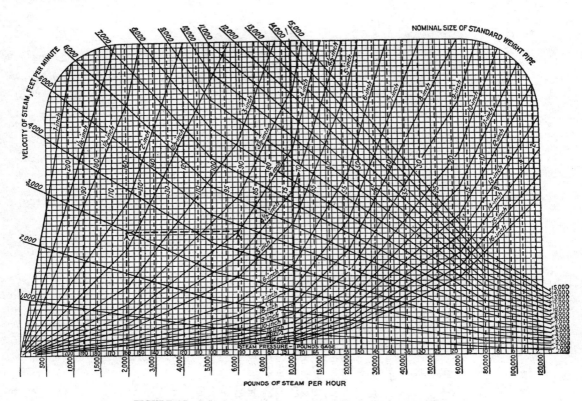

FIGURE 6.7   Spitzglass chart for saturated steam flowing in schedule 40 pipe.

Table 6.7 shows typical steam velocities for various industrial and commercial applications. Use the given values as guides when sizing steam piping.

6.13   STEAM-TRAP SELECTION FOR PROCESS APPLICATIONS

Select steam traps for the following five types of equipment: (1) where the steam directly heats solid materials, as in autoclaves, retorts, and sterilizers; (2) where the steam indirectly heats a liquid through a metallic surface, as in heat exchangers and kettles where the quantity of liquid heated is known and unknown; (3) where the steam indirectly heats a solid through a metallic surface, as in dryers using cylinders or chambers and platen presses; and (4) where the steam indirectly heats air through metallic surfaces, as in unit heaters, pipe coils, and radiators.

Calculation Procedure

1.   *Determine the condensate load.*

The first step in selecting a steam trap for any type of equipment is determination of the condensate load. Use the following general procedure:

a. Solid materials in autoclaves, retorts, and sterilizers. How much condensate is formed when 2000 lb of solid material with a specific heat of 1.0 is processed in 15 min at 240°F by 25-psig steam from an initial temperature of 60°F in an insulated steel retort?

TABLE 6.7   Steam Velocities Used in Pipe Design

Steam condition	Steam pressure		Steam use	Steam velocity	
	lb/in²	kPa		ft/min	m/min
Saturated	0–15	0–103.4	Heating	4000–6000	1219.2–1828.8
Saturated	50–150	344.7–1034.1	Process	6000–10,000	1828.8–3048.0
Superheated	200 and higher	1378.8 and higher	Boiler leads	10,000–15,000	3048.0–4572.0

For this type of equipment, use $C = WsP$, where C = condensate formed, in lb/h; W = weight of material heated, in lb; s = specific heat, in Btu/(lb)(°F); P = factor from Table 6.8. Thus, for this application, $C = (2000)(1.0)(0.193) = 386$ lb of condensate. Note that P is based on a temperature rise of $240 - 60 = 180$°F and a steam pressure of 25 psig. For the retort, using the specific heat of steel from Table 6.9, $C = (4000)(0.12)(0.193) = 92.6$ lb of condensate, say 93 lb (41.9 kg). The total weight of condensate formed in 15 min = $386 + 93 = 479$ lb (215.6 kg). In 1 h, $479(60/15) = 1916$ lb (862.2 kg) of condensate is formed.

TABLE 6.8   Factors, $P = (T - t)/L$, to Find Condensate Load

Pressure		Temperature		
psia	kPa	160°F(71.1°C)	180°F (82.2°C)	200°F (93.3°C)
20	137.8	0.170	0.192	0.213
25	172.4	0.172	0.193	0.214
30	206.8	0.172	0.194	0.215

TABLE 6.9   Use These Specific Heats When Calculating Condensate Load

Solids	Btu /(lb)(°F)	kJ/(kg)(°C)	Liquids	Btu/(lb)(°F)	kJ/(kg)(°C)
Aluminum	0.23	0.96	Alcohol	0.65	2.7
Brass	0.10	0.42	Carbon tetrachloride	0.20	0.84
Copper	0.10	0.42	Gasoline	0.53	2.22
Glass	0.20	0.84	Glycerin	0.58	2.43
Iron	0.13	0.54	Kerosene	0.47	1.97
Steel	0.12	0.50	Oils	0.40–0.50	1.67–2.09

A safety factor must be applied to compensate for radiation and other losses. Typical safety factors used in selecting steam traps are

Steam mains and headers	2–3
Steam heating pipes	2–6
Purifiers and separators	2–3
Retorts for process	2–4
Unit heaters	3
Submerged pipe coils	2–4
Cylinder dryers	4–10

Using a safety factor of 4 for this process retort, the trap capacity = $(4)(1916) = 7664$ lb/h (3449 kg/h), say 7700 lb/h (3465 kg/h).

TABLE 6.10 Ordinary Ranges of Overall Coefficients of Heat Transfer

Type of heat exchanger	State of controlling resistance		Typical fluid	Typical apparatus
	Free convection, U	Forced convection, U		
Liquid to liquid	25–60 [141.9–340.7]	150–300 [851.7–1703.4]	Water	Liquid-to-liquid heat exchangers
Liquid to liquid	5–10 [28.4–56.8]	20–50 [113.6–283.9]	Oil	
Liquid to gas†	1–3 [5.7–17.0]	2–10 [11.4–56.8]	—	Hot-water radiators
Liquid to boiling liquid	20–60 [113.6–340.7]	50–150 [283.9–851.7]	Water	Brine coolers
Liquid to boiling liquid	5–20 [28.4–113.6]	25–60 [141.9–340.7]	Oil	
Gas† to liquid	1–3 [5.7–17.0]	2–10 [11.4–56.8]	—	Air coolers, economizers
Gas† to gas	0.6–2 [3.4–11.4]	2–6 [11.4–34.1]	—	Steam superheaters
Gas† to boiling liquid	1–3 [5.7–17.0]	2–10 [11.4–56.8]	—	Steam boilers
Condensing vapor to liquid	50–200 [283.9–1136]	150–800 [851.7–4542.4]	Steam to water	Liquid heaters and condensers
Condensing vapor to liquid	10–30 [56.8–170.3]	20–60 [113.6–340.7]	Steam to oil	
Condensing vapor to liquid	40–80 [227.1–454.2]	60–150 [340.7–851.7]	Organic vapor to water	
Condensing vapor to liquid	—	15–300 [85.2–1703.4]	Steam–gas mixture	
Condensing vapor to gas†	1–2 [5.7–11.4]	2–10 [11.4–56.8]	—	Steam pipes in air, air heaters
Condensing vapor to boiling liquid	40–100 [227.1–567.8]	—		Scale-forming evaporators
Condensing vapor to boiling liquid	300–800 [1703.4–4542.4]	—	Steam to water	
Condensing vapor to boiling liquid	50–150 [283.9–851.7]	—	Steam to oil	

† At atmospheric pressure.

Note: U = Btu/(h)(ft²)(°F)[W/(m²)(°C)]. Under many conditions, either higher or lower values may be realized.

b(1). Submerged heating surface and a known quantity of liquid. How much condensate forms in the jacket of a kettle when 500 gal (1892.5 L) of water is heated in 30 min from 72 to 212°F (22.2 to 100°C) with 50-psig (344.7-kPa) steam?

For this type of equipment, $C = GwsP$, where G = gallons of liquid heated; w = weight of liquid, in lb/gal. Substitute the appropriate values as follows: $C = (500)(8.33)(1.0) \times (0.154) = 641$ lb, or $(641)(60/30) = 1282$ lb/h. Using a safety factor of 3, the trap capacity = $(3)(1282) = 3846$ lb/h; say 3900 lb/h.

b(2). Submerged heating surface and an unknown quantity of liquid. How much condensate is formed in a coil submerged in oil when the oil is heated as quickly as possible from 50 to 250°F by 25-psig steam if the coil has an area of 50 ft² and the oil is free to circulate around the coil?

For this condition, $C = UAP$, where U = overall coefficient of heat transfer, in Btu/(h)(ft²)(°F), from Table 6.10: A = area of heating surface, in ft². With free convection and a condensing-vapor-to-liquid type of heat exchanger, $U = 10$ to 30. Using an average value of $U = 20$, $C = (20)(50)(0.214) = 214$ lb/h of condensate. Choosing a safety factor 3, the trap capacity = $(3)(214) = 642$ lb/h; say 650 lb/h.

b(3). Submerged surfaces having more area then needed to heat a specified quantity of liquid in a given time with condensate withdrawn as rapidly as formed. Use Table 6.11 instead of steps $b(1)$ or $b(2)$. Find the condensation rate by multiplying the submerged area by the appropriate factor from Table 6.11. Use this method for heating water, chemical solutions, oils, and other liquids. Thus, with steam at 100 psig and a temperature of 338°F and heating oil from 50 to 226°F with a submerged surface having an area of 500 ft², the mean temperature difference equals steam temperature minus the average liquid temperature = $Mtd = 338 - (50 + 226/2) = 200°F$. The factor from Table 6.11 for 100-psig steam and a 200°F Mtd is 56.75. Thus the condensation rate = $(56.75) \times (500) = 28,375$ lb/h. With a safety factor of 2, the trap capacity = $(2)(28,375) = 56,750$ lb/h.

c. Solids indirectly heated through a metallic surface. How much condensate is formed in a chamber dryer when 1000 lb of cereal is dried to 750 lb by 10-psig steam? The initial temperature of the cereal is 60°F and the final temperature equals that of the steam.

For this condition, $C = 970(W - D)/h_{fg} + WP$, where D = dry weight of the material, in lb; h_{fg} = enthalpy of vaporization of the steam at the trap pressure, in Btu/lb. Using the steam tables and Table 6.8, $C = 970(1000 - 750)/952 + (1000)(0.189) = 443.5$ lb/h of condensate. With a safety factor of 4, the trap capacity = $(4)(443.5) = 1774$ lb/h.

d. Indirect heating of air through a metallic surface. How much condensate is formed in a unit heater using 10-psig steam if the entering-air temperature is 30°F and the leaving-air temperature is 130°F? Airflow is 10,000 ft³/min.

Use Table 6.12, entering at a temperature difference of 100°F and projecting to a steam pressure of 10 psig. Read the condensate formed as 122 lb/h per 1000 ft³/min. Since 10,000 ft³/min of air is being heated, the condensation rate = $(10,000/1000)(122) = 1220$ lb/h. With a safety factor of 3, the trap capacity = $(3)(1220) = 3660$ lb/h, say 3700 lb/h.

Table 6.13 shows the condensate formed by radiation from bare iron and steel pipes in still air and with forced-air circulation. Thus, with a steam pressure of 100 psig and an initial air temperature of 75°F, 1.05 lb/h of condensate will be formed per square foot of heating surface in still air. With forced-air circulation, the condensate rate is $(5)(1.05) = 5.25$ lb/h per square foot of heating surface.

TABLE 6.11   Condensate Formed in Submerged Steel[†] Heating Elements, lb/(ft²)(h) [kg/(m²)(min)]

Mtd[‡]		Steam pressure				
°F	°C	75 psia (517.1 kPa)	100 psia (689.4 kPa)	150 psia (1034.1 kPa)	Btu/(ft²)(h)	kW/m²
175	97.2	44.3(3.6)	45.4 (3.7)	46.7 (3.8)	40,000	126.2
200	111.1	54.8(4.5)	56.8 (4.6)	58.3 (4.7)	50,000	157.7
250	138.9	90.0(7.3)	93.1 (7.6)	95.7 (7.8)	82,000	258.6

[†]For copper, multiply table data by 2.0. For brass, multiply table data by 1.6.

[‡]Mean temperature difference, °F or °C, equals temperature of steam minus average liquid temperature. Heat-transfer data for calculating this table obtained from and used by permission of the American Radiator & Standard Sanitary Corp.

TABLE 6.12 Steam Condensed by Air, lb/h at 1000 ft³/min (kg/h at 28.3 m³/min)[†]

Temperature difference		Pressure		
°F	°C	5 psig (34.5 kPa)	10 psig (68.9 kPa)	50 psig (344.7 kPa)
50	27.8	61 (27.5)	61 (27.5)	63 (28.4)
100	55.6	120 (54.0)	122 (54.9)	126 (56.7)
150	83.3	180 (81.0)	183 (82.4)	189 (85.1)

[†]Based on 0.0192 Btu (0.02 kJ) absorbed per cubic foot (0.028 m³) of saturated air per °F (0.556°C) at 32°F (0°C). For 0°F (−17.8°C), multiply by 1.1.

TABLE 6.13 Condensate Formed by Radiation from Bare Iron and Steel[†], lb/(ft²)(h) [kg/(m²)(h)]

Air temperature		Steam pressure			
°F	°C	50 psig (344.7 kPa)	75 psig (517.1 kPa)	100 psig (689.5 kPa)	150 psig (1034 kPa)
65	18.3	0.82 (3.97)	1.00 (5.84)	1.08 (5.23)	1.32 (6.39)
70	21.2	0.80 (3.87)	0.98 (4.74)	1.06 (5.13)	1.21 (5.86)
75	23.9	0.77 (3.73)	0.88 (4.26)	1.05 (5.08)	1.19 (5.76)

[†]Based on still air; for forced-air circulation, multiply by 5.

Unit heaters have a *standard rating* based on 2-psig steam with entering air at 60°F. If the steam pressure or air temperature is different from these standard conditions, multiply the heater Btu/h capacity rating by the appropriate correction factor from Table 6.14. Thus a heater rated at 10,000 Btu/h with 2-psig steam and 60°F air would have an output of (1.290)(10,000) = 12,900 Btu/h with 40°F inlet air and 10-psig steam. Trap manufacturers usually list heater Btu ratings and recommend trap model numbers and sizes in their trap engineering data. This allows easier selection of the correct trap.

2. *Select the trap size based on the load and steam pressure.*
Obtain a chart or tabulation of trap capacities published by the manufacturer whose trap will be used. Figure 6.8 is a capacity chart for one type of bucket trap manufactured by Armstrong Machine Works. Table 6.15 shows typical capacities of impulse traps manufactured by the Yarway Company. Be sure to use up-to-date vendor data.

To select a trap from Fig. 6.8, when the condensation rate is uniform and the pressure across the trap is constant, enter at the left at the condensation rate—say 8000 lb/h (3600 kg/h) (as obtained from step 1)—and project horizontally to the right to the vertical ordinate representing the pressure across the trap (= Δp = steam-line pressure, in psig—return-line pressure with trap valve closed, in psig). Assume $\Delta p = 20$ psig (138 kPa) for this trap. The intersection of the horizontal 8000 lb/h (3600 kg/h) projection and the vertical 20-psig (137.9-kPa) projection is on the sawtooth capacity curve for a trap having a ⁹⁄₁₆-in (14.3-mm) orifice. If these projections intersected beneath this curve, a ⁹⁄₁₆-in (14.3-mm) orifice would still be used if the point was between the verticals for this size orifice.

TABLE 6.14 Unit-Heater Correction Factors

Steam pressure		Temperature of entering air		
		20°F (−6.7°C)	40°F (4.4°C)	60°F (15.6°C)
psig	kPa			
5	34.5	1.370	1.206	1.050
10	68.9	1.460	1.290	1.131
15	103.4	1.525	1.335	1.194

Source: Yarway Corporation; SI values added by *Handbook* editor.

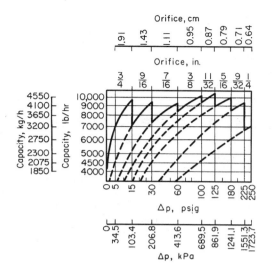

FIGURE 6.8   Capacities of one type of bucket steam trap. (*Armstrong Machine Works.*)

The dashed lines extending downward from the sawtooth curves show the capacity of a trap at reduced Δp. Thus the capacity of a trap with a ⅜-in (9.53-mm) orifice at $\Delta p = 30$ psig (207 kPa) is 6200 lb/h (2790 kg/h), read at the intersection of the 30-psig (207-kPa) ordinate and the dashed curve extended from the ⅜-in (9.53-mm) solid curve.

To select an impulse trap from Table 6.15, enter the table at the trap inlet pressure—say 125 psig (862 kPa)—and project to the desired capacity—say 8000 lb/h (3600 kg/h), determined from step 1. Table 6.15 shows that a 2-in (50.8-mm) trap having an 8530 lb/h (3839 kg/h) capacity must be used because the next smallest size has a capacity of 5165 lb/h (2324 kg/h). This capacity is less than that required.

Some trap manufacturers publish capacity tables relating various trap models to specific types of equipment. Such tables simplify trap selection, but the condensation rate must still be computed as given here.

Related Calculations.   Use the procedure given here to determine the trap capacity required for any industrial, commercial, or domestic application.

When using a trap-capacity diagram or table, be sure to determine the basis on which it was prepared. Apply any necessary correction factors. Thus *cold-water capacity ratings* must be corrected for traps operating at higher condensate temperatures. Correction factors are published in trap engineering

TABLE 6.15   Capacities of Impulse Traps, lb/h (kg/h)
[maximum continuous discharge of condensate, based on condensate at 30°F (16.7°C) below steam temperature]

Pressure at trap inlet		Trap nominal size	
psig	kPa	1.25 in (38.1 mm)	2.0 in (50.8 mm)
125	861.8	6165 (2774)	8530 (3839)
150	1034.1	6630 (2984)	9075 (4084)
200	1378.8	7410 (3335)	9950 (4478)

Source: Yarway Corporation.

data. The capacity of a trap is greater at condensate temperatures less than 212°F (100°C), because at or above this temperature condensate forms flash steam when it flows into a pipe or vessel at atmospheric [14.7 psia (101.3 kPa)] pressure. At altitudes above sea level, condensate flashes into steam at a lower temperature, depending on the altitude.

The method presented here is the work of L. C. Campbell. Yarway Corporation, as reported in *Chemical Engineering*.

6.14   ORIFICE-METER SELECTION FOR A STEAM PIPE

Steam is metered with an orifice meter in a 10-in boiler lead having an internal diameter of $d_p = 9.760$ in. Determine the maximum rate of steam flow that can be measured with a steel orifice plate having a diameter of $d_o = 5.855$ in at 70°F (294 K). The upstream pressure tap is $1D$ ahead of the orifice, and the downstream tap is $0.5D$ past the orifice. Steam pressure at the orifice inlet $p_p = 250$ psig (1825 kPa); temperature is 640°F (611 K). A differential gage fitted across the orifice has a maximum range of 120 in of water. What is the steam flow rate when the observed differential pressure is 40 in of water? Use the ASME Research Committee on Fluid Meters method in analyzing the meter. Atmospheric pressure is 14.696 psia.

Calculation Procedure

1. *Determine the diameter ratio and steam density.*
For any orifice meter, diameter ratio = β = meter orifice diameter, in/pipe internal diameter, in = 5.855/9.760 = 0.5999.

Determine the density of the steam by entering the superheated steam table at $250 + 14.696 = 264.696$ psia and 640°F and reading the specific volume as 2.387 ft³/lb. For steam, the density = 1/specific volume = $d_s = 1/2.387 = 0.4193$ lb/ft³.

2. *Determine the steam viscosity and meter flow coefficient.*
From the ASME publication *Fluid Meters—Their Theory and Application*, the steam viscosity gu_1 for a steam system operating at 640°F is $gu_1 = 0.0000141$ in · lb/(°F)(s)(ft²).

Find the flow coefficient K from the same ASME source by entering the 10-in nominal pipe diameter table at $\beta = 0.5999$ and projecting to the appropriate Reynolds number column. Assume that the Reynolds number = 10^7, approximately, for the flow conditions in this pipe. Then, $K = 0.6486$. Since the Reynolds number for steam pressures above 100 lb/in² ranges from 10^6 to 10^7, this assumption is safe because the value of K does not vary appreciably in this Reynolds number range. Also, the Reynolds number cannot be computed yet because the flow rate is unknown. Therefore, assumption of the Reynolds number is necessary. The assumption will be checked later.

3. *Determine the expansion factor and the meter area factor.*
Since steam is a compressible fluid, the expansion factor Y_1 must be determined. For superheated steam, the ratio of the specific heat at constant pressure c_p to the specific heat at constant volume c_v is $k = c_p/c_v = 1.3$. Also, the ratio of the differential maximum pressure reading h_w, in in of water, to the maximum pressure in the pipe, in psia, equals $120/246.7 = 0.454$. Using the expansion-factor curve in the ASME *Fluid Meters*, $Y_1 = 0.994$ for $\beta = 0.5999$, and the pressure ratio = 0.454. And, from the same reference, the meter area factor $F_a = 1.0084$ for a steel meter operating at 640°F.

4. *Compute the rate of steam flow.*
For square-edged orifices, the flow rate, in lb/s, is $w = 0.0997\ F_a K_d{}^2 Y_1 (h_w d_s)^{0.5} = (0.0997)(1.0084)$ $(0.6468)(5.855)^2(0.994)(120 \times 0.4188)^{0.5} = 15.75$ lb/s.

5. *Compute the Reynolds number for the actual flow rate.*
For any steam pipe, the Reynolds number Re = $48w/d_p gu_1 = 48(15.75)/[3.1416(0.760)(0.0000141)] = 1,750,000$.

6. *Adjust the flow coefficient for the actual Reynolds number.*
In step 2, Re = 10^7 was assumed and $K = 0.6486$. For Re = 1,750,000, $K = 0.6489$, from ASME *Fluid Meters*, by interpolation. Then, the actual flow rate w_h = (computed flow rate) (ratio of flow coefficients based on assumed and actual Reynolds numbers) = (15.75)(0.6489/0.6486)(3600) = 56,700 lb/h, closely, where the value 3600 is a conversion factor for changing lb/s to lb/h.

7. *Compute the flow rate for a specific differential gage deflection.*
For a 40-in H_2O deflection, F_a is unchanged and equals 1.0084. The expansion factor changes because $h_w/p_p = 40/264.7 = 0.151$. Using the ASME *Fluid Meters*, $Y_1 = 0.998$. Assuming again that Re = 10^7, $K = 0.6486$, as before; then $w = 0.0997(1.0084)(0.6486)(5.855)^2 (0.998)(40 \times 0.4188)^{0.5} =$ 9.132 lb/s. Computing the Reynolds number as before, Re = 40(0.132)/[3.1416(0.76)(0.0000141)] = 1,014,000. The value of K corresponding to this value, as before, is from ASME *Fluid Meters;* $K = 0.6497$. Therefore, the flow rate for a 40-in H_2O reading, in lb/h, is w_h = 0.132(0.6497/0.6486) (3600) = 32,940 lb/h (4.15 kg/s).

Related Calculations.   Use these steps and the ASME *Fluid Meters* or comprehensive meter engineering tables giving similar data to select or check an orifice meter used in any type of steam pipe—main, auxiliary, process, industrial, marine, heating, or commercial—conveying wet, saturated, or superheated steam.

6.15   SELECTION OF A PRESSURE-REGULATING VALVE FOR STEAM SERVICE

Select a single-seat spring-loaded diaphragm-actuated pressure-reducing valve to deliver 350 lb/h (0.044 kg/s) of steam at 50 psig when the initial pressure is 225 psig. Also select an integral pilot-controlled piston-operated single-seat pressure-regulating valve to deliver 30,000 lb/h (3.78 kg/s) of steam at 40 psig with an initial pressure of 225 psig saturated. What size pipe must be used on the downstream side of the valve to produce a velocity of 10,000 ft/min (50.8 m/s)? How large should the pressure-regulating valve be if the steam entering the valve is at 225 psig and 600°F (589 K)?

Calculation Procedure

1. *Compute the maximum flow for the diaphragm-actuated valve.*
For best results in service, pressure-reducing valves are selected so that they operate 60 to 70 percent open at normal load. To obtain a valve sized for this opening, divide the desired delivery, in lb/h, by 0.7 to obtain the maximum flow expected. For this valve, then, the maximum flow is 350/0.7 = 500 lb/h.

2. *Select the diaphragm-actuated valve size.*
Using a manufacturer's engineering data for an acceptable valve, enter the appropriate valve-capacity table at the valve inlet steam pressure, 225 psig, and project to a capacity of 500 lb/h, as in Table 6.16. Read the valve size as ¾ in at the top of the capacity column.

TABLE 6.16   Pressure-Reducing-Valve Capacity, lb/h (kg/h)

Inlet pressure		Valve size		
psig	kPa	½ in (12.7 mm)	¾ in (19.1 mm)	1 in (25.4 mm)
200	1379	420 (189)	460 (207)	560 (252)
225	1551	450 (203)	500 (225)	600 (270)
250	1724	485 (218)	560 (252)	650 (293)

Source: Clark-Reliance Corporation.

3. Select the size of the pilot-controlled pressure-regulating valve.
Enter the capacity table in the engineering data of an acceptable pilot-controlled pressure-regulating valve, similar to Table 6.17, at the required capacity, 30,000 lb/h, and project across until the correct inlet steam pressure column, 225 psig, is intercepted, and read the required valve size as 4 in.

Note that it is not necessary to compute the maximum capacity before entering the table, as in step 1, for the pressure-reducing valve. Also note that a capacity table such as Table 6.17 can be used only for valves conveying saturated steam, unless the table notes state that the values listed are valid for other steam conditions.

4. Determine the size of the downstream pipe.
Enter Table 6.17 at the required capacity, 30,000 lb/h (13,500 kg/h), and project across to the valve *outlet pressure*, 40 psig (275.8 kPa), and read the required pipe size as 8 in (203.2 mm) for a velocity of 10,000 ft/min (3048 m/min). Thus the pipe immediately downstream from the valve must be enlarged from the valve size, 4 in (101.6 mm), to the required pipe size, 8 in (203.2 mm), to obtain the desired steam velocity.

5. Determine the size of the valve handling superheated steam.
To determine the correct size of a pilot-controlled pressure regulating valve handling superheated steam, a correction must be applied. Either a factor may be used or a tabulation of corrected pressures, such as Table 6.18. To use Table 6.18, enter at the valve inlet pressure, 225 psig (1551.2 kPa), and project across to the total temperature, 600°F (316°C), to read the corrected pressure, 165 psig (1137.5 kPa). Enter Table 6.17 at the *next highest* saturated steam pressure, 175 psig, and project down to the required capacity, 30,000 lb/h (13,500 kg/h), and read the required valve size as 5 in (127 mm).

Related Calculations. To simplify pressure-reducing and pressure-regulating valve selection, become familiar with two or three acceptable valve manufacturers' engineering data. Use the procedures given in the engineering data or those given here to select valves for industrial, marine, utility, heating. process, laundry, kitchen, or hospital service with a saturated or superheated steam supply.

Do not oversize reducing or regulating valves. Oversizing causes chatter and excessive wear.

When an anticipated load on the downstream side will not develop for several months after installation of a valve, fit to the valve a reduced-area disk sized to handle the present load. When the load increases, install a full-size disk. Size the valve for the ultimate load, not the reduced load.

TABLE 6.17 Pressure-Regulating-Valve Capacity

Steam capacity		Initital steam pressure, saturated			
lb/h	kg/h	40 psig (276 kPa)	175 psig (1206 kPa)	225 psig (1551 kPa)	300 psig (2068 kPa)
20,000	9,000	6[†] (152.4)	4 (101.6)	4 (101.6)	3 (76.2)
30,000	13,500	8 203.2)	5 (127.0)	4 (101.6)	4 (101.6)
40,000	18,000	— —	5 (127.0)	5 (127.0)	4 (101.6)

[†]Value diameter measured in inches (millimeters).
Source: Clark-Reliance Corporation.

TABLE 6.18 Equivalent Saturated Steam Values for Superheated Steam at Various Pressures and Temperatures

Steam pressure		Steam temp.		Total temperature					
				500°F	600°F	700°F	260.0°C	315.6°C	371.1°C
psig	kPa	°F	°C	Steam values, psig			Steam values, kPa		
205	1413.3	389	198	171	149	133	1178.9	1027.2	916.9
225	1551.2	397	203	190	165	147	1309.9	1137.5	1013.4
265	1826.9	411	211	227	200	177	1564.9	1378.8	1220.2

Source: Clark-Reliance Corporation.

Where there is a wide variation in demand for steam at the reduced pressure, consider installing two regulators piped in parallel. Size the smaller regulator to handle light loads and the larger regulator to handle the difference between 60 percent of the light load and the maximum heavy load. Set the larger regulator to open when the minimum allowable reduced pressure is reached. Then both regulators will be open to handle the heavy load. Be certain to use the actual regulator inlet pressure and not the boiler pressure when sizing the valve if this is different from the inlet pressure. Data in this calculation procedure are based on valves built by the Clark-Reliance Corporation, Cleveland, Ohio.

Some valve manufacturers use the valve-flow coefficient C_v for valve sizing. This coefficient is defined as the flow rate, in lb/h, through a valve of given size when the pressure loss across the valve is 1 lb/in^2. Tabulations such as Tables 6.16 and 6.17 incorporate this flow coefficient and are somewhat, easier to use. These tables make the necessary allowances for downstream pressures less than the critical pressure (= 0.55 × absolute upstream pressure, in lb/in^2, for superheated steam and hydrocarbon vapors, and 0.58 × absolute upstream pressure, in lb/in^2, for saturated steam). The accuracy of these tabulations equals that of valve size determined by using the flow coefficient.

6.16 PRESSURE-REDUCING-VALVE SELECTION FOR WATER PIPING

What size pressure-reducing valve should be used to deliver 1200 gal/h (1.26 L/s) of water at 40 lb/in^2 (275.8 kPa) if the inlet pressure is 140 lb/in^2 (965.2 kPa)?

Calculation Procedure

1. Determine the valve capacity required.
Pressure-reducing valves in water systems operate best when the nominal load is 60 to 70 percent of the maximum load. Using 60 percent, the maximum load for this valve = 1200/0.6 = 2000 gal/h (2.1 L/s).

2. Determine the valve size required.
Enter a valve-capacity table in suitable valve engineering data at the valve inlet pressure and project to the exact, or next higher, valve capacity. Thus, enter Table 6.19 at 140 lb/in^2 (965.2 kPa) and project to the next higher capacity, 2200 gal/h (2.3 L/s), since a capacity of 2000 gal/h (2.1 L/s) is not tabulated. Read at the top of the column the required valve size as 1 in (25.4 mm).

Some valve manufacturers present the capacity of their valves in graphic instead of tabular form. One popular chart, Fig. 6.9, is entered at the difference between the inlet and outlet pressures on the abscissa, or 140 − 40 = 100 lb/in^2 (689.4 kPa). Project vertically to the flow rate of 2000/60 = 33.3 gal/min (2.1 L/s). Read the valve size on the intersecting valve-capacity curve, or on the next curve if there is no intersection with the curve. Figure 6.9 shows that a 1-in valve should be used. This agrees with the tabulated capacity.

TABLE 6.19 Maximum Capacities of Water Pressure-Reducing Valves, gal/h (L/h)

Inlet pressure		Valve size		
psig	kPa	¾ in (19.1 mm)	1 in (25.4 mm)	1¼ in (31.8 mm)
120	827.3	1550 (5867)	2000 (7570)	4500 (17,033)
140	965.2	1700 (6435)	2200 (8327)	5000 (18,925)
160	1103.0	1850 (7002)	2400 (9084)	5500 (20,818)

Source: Clark-Reliance Corporation.

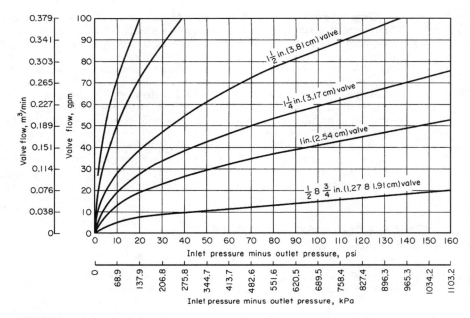

FIGURE 6.9 Pressure-reducing valve flow capacity. (*Foster Engineering Co.*)

Related Calculations. Use this method for pressure-reducing valves in any type of water piping—process, domestic, commercial—where the water temperature is 100°F (37.8°C) or less. Table 6.19 is from data prepared by the Clark-Reliance Corporation; Figure 6.9 is from Foster Engineering Company data. Be sure to use up-to-date vendor data.

Some valve manufacturers use the valve-flow coefficient C_v for valve sizing. This coefficient is defined as the flow rate, in gal/min, through a valve of given size when the pressure loss across the valve is 1 lb/in². Tabulations such as Table 6.19 and flowcharts such as Fig. 6.9 incorporate this flow coefficient and are somewhat easier to use. Their accuracy equals that of the flow-coefficient method.

6.17 SIMILARITY OR AFFINITY LAWS FOR CENTRIFUGAL PUMPS

A centrifugal pump designed for an 1800-r/min operation and a head of 200 ft (61 m) has a capacity of 3000 gal/min (0.19 m³/s) with a power input of 175 hp. What effect will a speed reduction to 1200 r/min have on the head, capacity, and power input of the pump? What will be the change in these variables if the impeller diameter is reduced from 12 to 10 in while the speed is held constant at 1800 r/min?

Calculation Procedure

1. Compute the effect of a change in pump speed.
For any centrifugal pump in which the effects of fluid viscosity are negligible or are neglected, the similarity or affinity laws can be used to determine the effect of a speed, power, or head change. For a *constant impeller diameter*, these laws are $Q_1/Q_2 = N_1/N_2$; $H_1/H_2 = (N_1/N_2)^2$; $P_1/P_2 = (N_1/N_2)^3$. For a *constant speed*, $Q_1/Q_2 = D_1/D_2$; $H_1/H_2 = (D_1/D_2)^2$; $P_1/P_2 = (D_1/D_2)^3$. In both sets of laws, Q = capacity,

in gal/min: N = impeller r/min; D = impeller diameter, in in; H = total head, in ft of liquid; P = bhp input. The subscripts 1 and 2 refer to the initial and changed conditions, respectively.

For this pump, with a constant impeller diameter, $Q_1/Q_2 = N_1/N_2$; $3000/Q_2 = 1800/1200$; $Q_2 = 2000$ gal/min (0.13 m³/s). And, $H_1/H_2 = (N_1/N_2)^2 = 200/H_2 = (1800/1200)^2$; $H_2 = 88.9$ ft (27.1 m). Also, $P_1/P_2 = (N_1/N_2)^3 = 175/P_2 = (1800/1200)^3$; $P_2 = 51.8$ bhp.

2. Compute the effect of a change in impeller diameter.
With the speed constant, use the second set of laws. Or for this pump, $Q_1/Q_2 = D_1/D_2$; $3000/Q_2 = {}^{12}\!/_{10}$; $Q_2 = 2500$ gal/min (0.016 m³/s). And, $H_1/H_2 = (D_1/D_2)^2$; $200/H_2 = ({}^{12}\!/_{10})^2$; $H_2 = 138.8$ ft (42.3 m). Also, $P_1/P_2 = (D_1/D_2)^3$; $175/P_2 = ({}^{12}\!/_{10})^3$; $P_2 = 101.2$ bhp.

Related Calculations.   Use the similarity laws to extend or change the data obtained from centrifugal-pump characteristic curves. These laws are also useful in field calculations when the pump head, capacity, speed, or impeller diameter is changed.

The similarity laws are most accurate when the efficiency of the pump remains nearly constant. Results obtained when the laws are applied to a pump having a constant impeller diameter are somewhat more accurate than for a pump at constant speed with a changed impeller diameter. The latter laws are more accurate when applied to pumps having a low specific speed.

If the similarity laws are applied to a pump whose impeller diameter is increased, be certain to consider the effect of the higher velocity in the pump suction line. Use the similarity laws for any liquid whose viscosity remains constant during passage through the pump. However, the accuracy of the similarity laws decreases as the liquid viscosity increases.

6.18 SIMILARITY OR AFFINITY LAWS IN CENTRIFUGAL-PUMP SELECTION

A test-model pump delivers, at its best efficiency point, 500 gal/min (0.03 m³/s) at a 350-ft (107-m) head with a required net positive suction head (NPSH) of 10 ft (3.05 m) and a power input of 55 hp (41 kW) at 3500 r/min, when using a 10.5-in-diameter impeller. Determine the performance of the model at 1750 r/min. What is the performance of a full-scale prototype pump with a 20-in impeller operating at 1170 r/min? What are the specific speeds and the suction specific speeds of the test-model and prototype pumps?

Calculation Procedure

1. Compute the pump performance at the new speed.
The similarity or affinity laws can be stated in general terms, with subscripts p and m for prototype and model, respectively, as $Q_p = K_d^3 K_n Q_m$; $H_p = K_d^2 K_n^2 H_m$; $\text{NPSH}_p = K_d^2 K_n^2 \text{NPSH}_m$; $P_p = K_d^5 K_n^5 P_m$, where K_d = size factor = prototype dimension/model dimension. The usual dimension used for the size factor is the impeller diameter. Both dimensions should be in the same units of measure. Also, K_n = prototype speed, r/min/model speed, r/min. Other symbols are the same as in the previous example.

When the model speed is reduced from 3500 to 1750 r/min, the pump dimensions remain the same and $K_d = 1.0$; $K_n = 1750/3500 = 0.5$. Then $Q = (1.0)(0.5)(500) = 250$ r/min; $H = (1.0)^2(0.5)^2(350) = 87.5$ ft (26.7 m); $\text{NPSH} = (1.0)^2(0.5)^2(10) = 2.5$ ft (0.76 m); $P = (1.0)^5(0.5)^3(55) = 6.9$ hp. In this computation, the subscripts were omitted from the equation because the same pump, the test model, was being considered.

2. Compute performance of the prototype pump.
First, K_d and K_n must be found. $K_d = 20/10.5 = 1.905$; $K_n = 1170/3500 = 0.335$. Then, $Q_p = (1.905)^3(0.335)(500) = 1158$ gal/min (0.073 m³/s); $H_p = (1.905)^2(0.335)^2(350) = 142.5$ ft (43.4m); $\text{NPSH}_p = (1.905)^2(0.335)^2(10) = 4.06$ ft; $P_p = (1.905)^5(0.335)^3(55) = 51.8$ hp.

3. Compute the specific speed and suction specific speed.

The specific speed or, as Horwitz [2] says, "more correctly, discharge specific speed," $N_s = N(Q)^{0.5}/(H)^{0.75}$, while the suction specific speed $S = N(Q)^{0.5}/NPSH^{0.75}$, where all values are taken at the best efficiency point of the pump.

For the model, $N_s = 3500(500)^{0.5}/350^{0.75} = 965$; $S = 3500(500)^{0.5}/10^{0.75} = 13,900$. For the prototype. $N_s = 1170(1158)^{0.5}/142.5^{0.75} = 965$; $S = 1170(1156)^{0.5}/4.06^{0.75} = 13,900$. The specific speed and suction specific speed of the model and prototype are equal because these units are geometrically similar or homologous pumps and both speeds are mathematically derived from the similarity laws.

Related Calculations.   Use the procedure given here for any type of centrifugal pump where the similarity laws apply. When the term "model" is used, it can apply to a production test pump or to a standard unit ready for installation. The procedure presented here is the work of R. P. Horwitz, as reported in *Power* magazine [2].

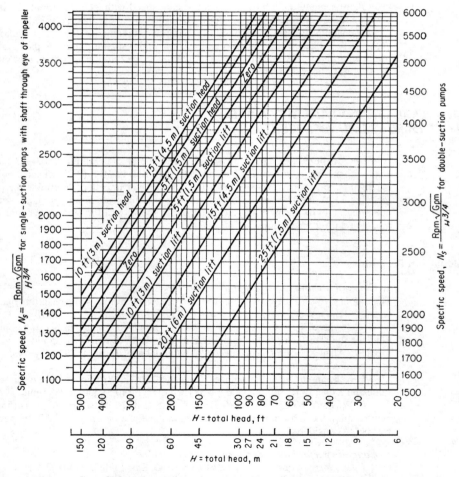

FIGURE 6.10   Upper limits of specific speeds of single-stage single- and double-suction centrifugal pumps handling clear water at 85°F (29°C) at sea level. (*Hydraulic Institute.*)

6.19 SPECIFIC-SPEED CONSIDERATIONS
IN CENTRIFUGAL-PUMP SELECTION

What is the upper limit of specific speed and capacity of a 1750-r/min single-stage double-suction centrifugal pump having a shaft that passes through the impeller eye if it handles clear water at 85°F (302 K) at sea level at a total head of 280 ft with a 10-ft suction lift? What is the efficiency of the pump and its approximate impeller shape?

Calculation Procedure

1. Determine the upper limit of specific speed.
Use the Hydraulic Institute upper-specific-speed curve, Fig. 6.10, for centrifugal pumps or a similar curve, Fig. 6.11, for mixed- and axial-flow pumps. Enter Fig. 6.10 at the bottom at 280-ft total head and project vertically upward until the 10-ft suction-lift curve is intersected. From here, project horizontally to the right to read the specific speed $N_S = 2000$. Figure 6.11 is used in a similar manner.

2. Compute the maximum pump capacity.
For any centrifugal, mixed- or axial-flow pump. $N_s = (\text{gal/min})^{0.5} (\text{r/min})/H_t^{0.75}$, where H_t = total head on the pump, in ft of liquid. Solving for the maximum capacity, gal/min = $[N_s H_t^{0.75}/(\text{r/min})]^2 = (2000 \times 280^{0.75}/1750)^2 = 6040$ gal/min.

3. Determine the pump efficiency and impeller shape.
Figure 6.12 shows the general relation between impeller shape, specific speed, pump capacity, efficiency, and characteristic curves. At $N_s = 2000$, efficiency = 87 percent. The impeller, as shown in Fig. 6.12, is moderately short and has a relatively large discharge area. A cross section of the impeller appears directly under the $N_s = 2000$ ordinate.

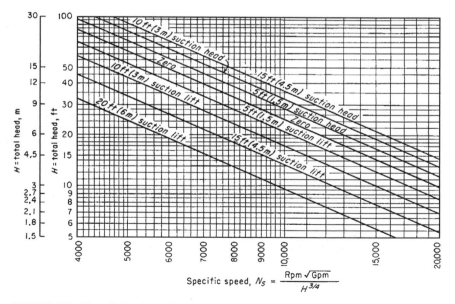

FIGURE 6.11 Upper limits of specific speeds of single-suction mixed-flow and axial-flow pumps. (*Hydraulic Institute.*)

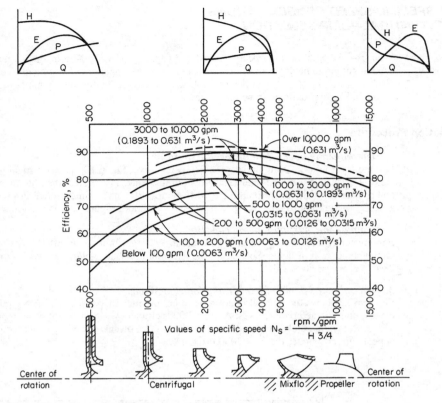

FIGURE 6.12 Approximate relative impeller shapes and efficiency variations for various specific speeds of centrifugal pumps. (*Worthington Corp.*)

Related Calculations. Use the method given here for any type of pump whose variables are included in the Hydraulic Institute curves (Figs. 6.10 and 6.11) and in similar curves available from the same source. *Operating specific speed*, computed as above, is sometimes plotted on the performance curve of a centrifugal pump so that the characteristics of the unit can be better understood. *Type specific speed* is the operating specific speed giving maximum efficiency for a given pump and is a number used to identify a pump. Specific speed is important in cavitation and suction-lift studies. The Hydraulic Institute curves (Figs. 6.10 and 6.11) give upper limits of speed, head, capacity, and suction lift for cavitation-free operation. When making actual pump analyses, be certain to use the curves (Figs. 6.10 and 6.11 herewith) in the latest edition of the *Standards of the Hydraulic Institute.*

6.20   SELECTING THE BEST OPERATING SPEED FOR A CENTRIFUGAL PUMP

A single-suction centrifugal pump is driven by a 60-Hz ac motor. The pump delivers 10,000 gal/min (0.63 m³/s) of water at a 100-ft (30.5-m) head. The available net positive suction head is 32 ft(9.75 m) of water. What is the best operating speed for this pump if the pump operates at its best efficiency point?

Calculation Procedure

1. Determine the specific speed and suction specific speed.

Alternating-current motors can operate at a variety of speeds, depending on the number of poles. Assume that the motor driving this pump might operate at 870, 1160, 1750, or 3500 r/min. Compute the specific speed $N_s = N(Q)^{0.5}/H^{0.75} = N(10,000)^{0.5}/100^{0.75} = 3.14N$, and the suction specific speed $S = N(Q)^{0.5}/\text{NPSH}^{0.75} = N(10,000)^{0.5}/32^{0.75} = 7.43N$ for each of the assumed speeds, and tabulate the results as follows:

Operating speed, r/min	Required specific speed	Required suction specific speed
870	2740	6460
1160	3640	8620
1750	5500	13,000
3500	11,000	26,000

2. Choose the best speed for the pump.

Analyze the specific speed and suction specific speed at each of the various operating speeds using the data in Tables 6.20 and 6.21. These tables show that at 870 and 1160 r/min, the suction specific-speed rating is poor. At 1750 r/min, the suction specific-speed rating is excellent, and a turbine or mixed-flow type of pump will be suitable. Operation at 3500 r/min is unfeasible because a suction specific speed of 26,000 is beyond the range of conventional pumps.

Related Calculations. Use this procedure for any type of centrifugal pump handling water for plant services, cooling, process, fire protection, and similar requirements. This procedure is the work of R. P. Horwitz, Hydrodynamics Division, Peerless Pump, FMC Corporation, as reported in *Power* magazine.

TABLE 6.20 Pump Types Listed by Specific Speed

Specific speed range	Type of pump
Below 2000	Volute, diffuser
2000–5000	Turbine
4000–10,000	Mixed-flow
9000–15,000	Axial-flow

Source: Peerless Pump Division, FMC Corporation.

TABLE 6.21 Suction Specific-Speed Ratings

Single-suction pump	Double-suction pump	Rating
Above 11,000	Above 14,000	Excellent
9000–11,000	11,000–14,000	Good
7000–9000	9000–11,000	Average
5000–7000	7000–9000	Poor
Below 5000	Below 7000	Very poor

Source: Peerless Pump Division, FMC Corporation.

6.21 TOTAL HEAD ON A PUMP HANDLING VAPOR-FREE LIQUID

Sketch three typical pump piping arrangements with static suction lift and submerged, free, and varying discharge head. Prepare similar sketches for the same pump with static suction head. Label the various heads. Compute the total head on each pump if the elevations are as shown in Fig. 6.13 and the pump discharges a maximum of 2000 gal/min (0.126 m³/s) of water through 8-in schedule 40 pipe. What horsepower is required to drive the pump? A swing check valve is used on the pump suction line and a gate valve on the discharge line.

Calculation Procedure

1. Sketch the possible piping arrangements.
Figure 6.13 shows the six possible piping arrangements for the stated conditions of the installation. Label the total static head—i.e., the *vertical* distance from the surface of the source of the liquid supply to the free surface of the liquid in the discharge receiver, or to the point of free discharge from the discharge

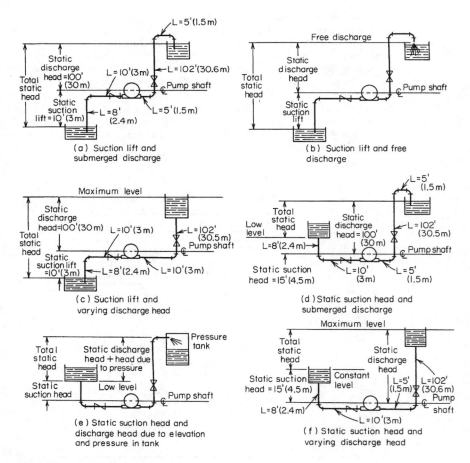

FIGURE 6.13   Typical pump suction and discharge piping arrangements.

pipe. When both the suction and discharge surfaces are open to the atmosphere, the total static head equals the vertical difference in elevation. Use the free-surface elevations that cause the maximum suction lift and discharge head—i.e., the *lowest* possible level in the supply tank and the *highest* possible level in the discharge tank or pipe. When the supply source is *below* the pump centerline, the vertical distance is called the "static suction lift." With the supply *above* the pump centerline, the vertical distance is called "static suction head." With variable static suction head, use the lowest liquid level in the supply tank when computing total static head. Label the diagrams as shown in Fig. 6.13.

2. Compute the total static head on the pump.

The total static head, in feet, is H_{ts} = static suction lift, in feet, h_{sl} + static discharge head, in feet, h_{sd}, where the pump has a suction lift, s in Fig. 6.13a, b, and c. In these installations, $H_{ts} = 10 + 100 = 110$ ft. Note that the static discharge head is computed between the pump centerline and the water level with an underwater discharge (Fig. 6.13a), to the pipe outlet with a free discharge (Fig. 6.13b), and to the maximum water level in the discharge tank (Fig. 6.13c). When a pump is discharging into a closed compression tank, the total discharge head equals the static discharge head plus the head equivalent, in feet of liquid, of the internal pressure in the tank, or 2.31 × tank pressure, in lb/in^2.

Where the pump has a static suction head, as in Fig. 6.13d, e, and f, the total static head, in feet, is $H_{ts} = h_{sd}$ − static suction head, in feet, h_{sh}. In these installations, $H_t = 100 - 15 = 85$ ft.

The total static head, as computed above, refers to the head on the pump without liquid flow. To determine the total head on the pump, the friction losses in the piping system during liquid flow must also be determined.

3. Compute the piping friction losses.

Mark the length of each piece of straight pipe on the piping drawing. Thus, in Fig. 6.13a, the total length of straight pipe L_r in feet, is $8 + 10 + 5 + 102 + 5 = 130$ ft, starting at the suction tank and adding each length until the discharge tank is reached. To the total length of straight pipe must be added the *equivalent* length of the pipe fittings. In Fig. 6.13a there are four long-radius elbows, one swing check valve, and one globe valve. In addition, there is a minor head loss at the pipe inlet and at the pipe outlet.

The equivalent length of one 8-in-long-radius elbow is 14 ft of pipe, from Table 6.22. Since the pipe contains four elbows, the total equivalent length is $4(14) = 56$ ft of straight pipe. The open gate valve has an equivalent resistance of 4.5 ft, and the open swing check valve has an equivalent resistance of 53 ft.

The entrance loss h_e, in feet, assuming a basket-type strainer is used at the suction-pipe inlet, is $Kv^2/(2g)$, where K = a constant from Fig. 6.14; v = liquid velocity, in ft/s; and $g = 32.2$ ft/s^2. The exit loss occurs when the liquid passes through a sudden enlargement, as from a pipe to a tank. Where the area of the tank is large, causing a final velocity that is zero, $h_{ex} = v^2/2g$.

The velocity v, in feet per second, in a pipe is $(gal/min)/(2.448d^2)$. For this pipe, $v = 2000/[2.448(7.98)^2] = 12.82$ ft/s. Then, $h_e = 0.74(12.82)^2/[2(32.2)] = 1.89$ ft, and $h_{ex} = (12.82)^2/[2(32.2)] = 2.56$ ft (0.78 m). Hence the total length of the piping system in Fig. 6.13a is $130 + 56 + 4.5 + 53 + 1.89 + 2.56 = 247.95$ ft (75.6 m), say 248 ft (75.6 m).

Use a suitable head-loss equation, or Table 6.23, to compute the head loss for the pipe and fittings. Enter Table 6.23 at an 8-in (203.2-mm) pipe size and project horizontally across to 2000 gal/min (126.2 L/s) and read the head loss as 5.86 ft of water per 100 ft (1.8 m/30.5 m) of pipe.

The total length of pipe and fittings computed above is 288 ft (87.8 m). Then, total friction-head loss with a 2000-gal/min (126.2 L/s) flow is $H_f = (5.86)(248/100) = 14.53$ ft (4.5 m).

4. Compute the total head on the pump.

The total head on the pump $H_t = H_{ts} + H_f$. For the pump in Fig. 6.13a, $H_t = 110 + 14.53 = 124.53$ ft (38.0 m), say 125 ft (38.0 m). The total head on the pump in Fig. 6.13b and c would be the same. Some engineers term the total head on a pump the "total dyamic head" to distinguish between static head (no-flow vertical head) and operating head (rated flow through the pump).

The total head on the pumps in Fig. 6.13d, c, and f is computed in the same way as described above, except that the total static head is less because the pump has a static suction head—that is, the elevation of the liquid on the suction side reduces the total distance through which the pump must

TABLE 6.22 Resistance of Fittings and Valves (length of straight pipe giving equivalent resistance)

Pipe size		Standard ell		Medium-radius ell		Long-radius ell		45° Ell		Tee		Gate valve, open		Globe valve, open		Swing check, open	
in	mm	ft	m	ft	m	ft	m	ft	m	ft	m	ft	m	ft	m	ft	m
6	152.4	16	4.9	14	4.3	11	3.4	7.7	2.3	33	10.1	3.5	1.1	160	48.8	40	12.2
8	203.2	21	6.4	18	5.5	14	4.3	10	3.0	43	13.1	4.5	1.4	220	67.0	53	16.2
10	254.0	26	7.9	22	6.7	17	5.2	13	3.9	56	17.1	5.7	1.7	290	88.4	67	20.4
12	304.8	32	9.8	26	7.9	20	6.1	15	4.6	66	20.1	6.7	2.0	340	103.6	80	24.4

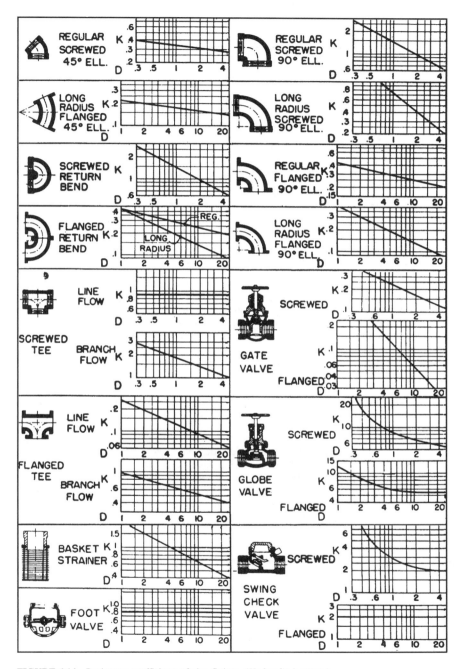

FIGURE 6.14 Resistance coefficients of pipe fittings. (*Hydraulic Institute.*)

TABLE 6.23   Pipe Friction Loss for Water (wrought-iron or steel schedule 40 pipe in good condition)

Diameter		Flow		Velocity		Velocity head		Friction loss/100 ft (30.5 m) pipe	
in	mm	gal/min	L/s	ft/s	m/s	ft water	m water	ft water	m water
6	152.4	1000	63.1	11.1	3.4	1.92	0.59	6.17	1.88
6	152.4	2000	126.2	22.2	6.8	7.67	2.3	23.8	7.25
6	152.4	4000	252.4	44.4	13.5	30.7	9.4	93.1	28.4
8	203.2	1000	63.1	6.41	1.9	0.639	0.195	1.56	0.475
8	203.2	2000	126.2	12.8	3.9	2.56	0.78	5.86	1.786
8	203.2	4000	252.4	25.7	7.8	10.2	3.1	22.6	6.888
10	254.0	1000	63.1	3.93	1.2	0.240	0.07	0.497	0.151
10	254.0	3000	189.3	11.8	3.6	2.16	0.658	4.00	1.219
10	254.0	5000	315.5	19.6	5.9	5.99	1.82	10.8	3.292

discharge liquid; thus the total static head is less. The static suction head is *subtracted* from the static discharge head to determine the total static head on the pump.

5. *Compute the horsepower required to drive the pump.*
The brake horsepower input to a pump equals (gal/min)(H_t)(s)/3960e, where s = specific gravity of the liquid handled, and e = hydraulic efficiency of the pump, expressed as a decimal. The usual hydraulic efficiency of a centrifugal pump is 60 to 80 percent; reciprocating pumps, 55 to 90 percent; rotary pumps, 50 to 90 percent. For each class of pump, the hydraulic efficiency decreases as the liquid viscosity increases.

Assume that the hydraulic efficiency of the pump in this system is 70 percent and the specific gravity of the liquid handled is 1.0. Then, input brake horsepower equals (2000)(125)(1.0)/[3960(0.70)] = 90.2 hp (67.4 kW).

The theoretical or *hydraulic horsepower* equals (gal/min)(H_t)(s)/3960 = (2000)(125)(1.0)/3900 = 64.1 hp (47.8 kW).

Related Calculations.   Use this procedure for any liquid-water, oil, chemical, sludge, etc.—whose specific gravity is known. When liquids other than water are being pumped, the specific gravity and viscosity of the liquid must be taken into consideration. The procedure given here can be used for any class of pump—centrifugal, rotary, or reciprocating.

Note that Fig. 6.14 can be used to determine the equivalent length of a variety of pipe fittings. To use Fig. 6.14, simply substitute the appropriate K value in the relation $h = Kv^2/2g$, where h = equivalent length of straight pipe; other symbols as before.

6.22   *PUMP SELECTION FOR ANY PUMPING SYSTEM*

Give a step-by-step procedure for choosing the class, type, capacity, drive, and materials for a pump that will be used in an industrial pumping system.

Calculation Procedure

1. *Sketch the proposed piping layout.*
Use a single-line diagram (Fig. 6.15) of the piping system. Base the sketch on the actual job conditions. Show all the piping, fittings, valves, equipment, and other units in the system. Mark the *actual* and *equivalent* pipe length (see the previous example) on the sketch. Be certain to include

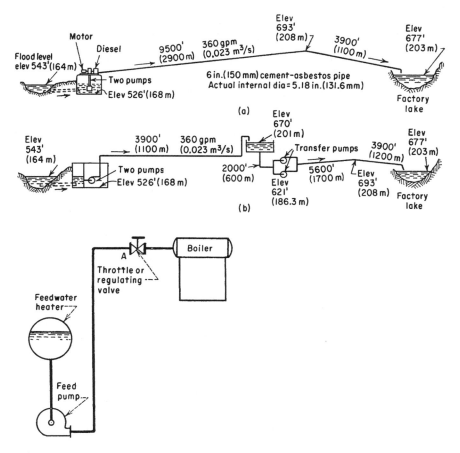

FIGURE 6.15 (*a*) Single-line diagrams for an industrial pipeline; (*b*) single-line diagram of a boiler feed system. (*Worthington Corp.*)

all vertical lifts, sharp bends, sudden enlargements, storage tanks, and similar equipment in the proposed system.

2. *Determine the required capacity of the pump.*

The required capacity is the flow rate that must be handled in gal/min, million gal/day, ft³/s, gal/h, bbl/day, lb/h, acre-ft/day, mil/h, or some similar measure. Obtain the required flow rate from the process conditions—for example, boiler feed rate, cooling-water flow rate, chemical feed rate, etc. The required flow rate for any process unit is usually given by the manufacturer.

Once the required flow rate is determined, apply a suitable factor of safety. The value of this factor of safety can vary from a low of 5 percent of the required flow to a high of 50 percent or more, depending on the application. Typical safety factors are in the 10 percent range. With flow rates up to 1000 gal/min, and in the selection of process pumps, it is common practice to round off a computed required flow rate to the next highest round-number capacity. Thus, with a required flow rate of 450 gal/min and a 10 percent safety factor, the flow of 450 + 0.10(450) = 495 gal/min would be rounded off to 500 gal/min *before* selecting the pump. A pump of 500 gal/min, or larger, capacity would be selected.

3. *Compute the total head on the pump.*

Use the steps given in the previous example to compute the total head on the pump. Express the result in feet of water—this is the most common way of expressing the head on a pump. Be certain to use the exact specific gravity of the liquid handled when expressing the head in feet of water. A specific gravity less than 1.00 *reduces* the total head when expressed in feet of water, whereas a specific gravity greater than 1.00 *increases* the total head when expressed in feet of water. Note that variations in the suction and discharge conditions can affect the total head on the pump.

Summary of Essential Data Required in Selection of Centrifugal Pumps

1. Number of Units Required

2. Nature of the Liquid to Be Pumped
   Is the liquid:
   a. Fresh or salt water, acid or alkali, oil, gasoline, slurry, or paper stock?
   b. Cold or hot and if hot, at what temperature? What is the vapor pressure of the liquid at the pumping temperature?
   c. What is its specific gravity?
   d. Is it viscous or nonviscous?
   e. Clear and free from suspended foreign matter or dirty and gritty? If the latter, what is the size and nature of the solids, and are they abrasive? If the liquid is of a pulpy nature, what is the consistency expressed either in percentage or in lb per cu ft of liquid? What is the suspended material?
   f. What is the chemical analysis, pH value, etc.? What are the expected variations of this analysis? If corrosive, what has been the past experience, both with successful materials and with unsatisfactory materials?

3. Capacity
   What is the required capacity as well as the minimum and maximum amount of liquid the pump will ever be called upon to deliver?

4. Suction Conditions
   Is there:
   a. A suction lift?
   b. Or a suction head?
   c. What are the length and diameter of the suction pipe?

5. Discharge Conditions
   a. What is the static head? Is it constant or variable?
   b. What is the friction head?
   c. What is the maximum discharge pressure against which the pump must deliver the liquid?

6. Total Head
   Variations in items 4 and 5 will cause variations in the total head.

7. Is the service continuous or intermittent?

8. Is the pump to be installed in a horizontal or vertical position? If the latter,
   a. In a wet pit?
   b. In a dry pit?

9. What type of power is available to drive the pump and what are the characteristics of this power?

10. What space, weight, or transportation limitations are involved?

11. Location of installation
    a. Geographical location
    b. Elevation above sea level
    c. Indoor or outdoor installation
    d. Range of ambient temperatures

12. Are there any special requirements or marked preferences with respect to the design, construction, or performance of the pump?

FIGURE 6.16 Typical selection chart for centrifugal pumps. (*Worthington Corp.*)

4. *Analyze the liquid conditions.*

Obtain complete data on the liquid pumped. These data should include the name and chemical formula of the liquid, maximum and minimum pumping temperature, corresponding vapor pressure at these temperatures, specific gravity, viscosity at the pumping temperature, pH, flash point, ignition temperature, unusual characteristics (such as tendency to foam, curd, crystallize, become gelatinous or tacky), solids content, type of solids and their size, and variation in the chemical analysis of the liquid.

Enter the liquid conditions on a pump-selection form such as that in Fig. 6.16. Such forms are available from many pump manufacturers or can be prepared to meet special job conditions.

5. *Select the class and type of pump.*

Three *classes* of pumps are used today—centrifugal, rotary, and reciprocating (Fig. 6.17). Note that these terms apply only to the mechanics of moving the liquid—not to the service for which the pump was designed. Each class of pump is further subdivided into a number of *types* (Fig. 6.17).

Use Table 6.24 as a general guide to the class and type of pump to be used. For example, when a large capacity at moderate pressure is required, table 6.24 shows that a centrifugal pump would probably be best. Table 6.24 also shows the typical characteristics of various classes and types of pumps used in industrial process work.

Consider the liquid properties when choosing the class and type of pump, because exceptionally severe conditions may rule out one or another class of pump at the start. Thus, screw- and gear-type rotary pumps are suitable for handling viscous, nonabrasive liquid (Table 6.24). When an abrasive liquid must be handled, either another class of pump or another type of rotary pump must be used.

Also consider all the operating factors related to the particular pump. These factors include the type of service (continuous or intermittent), operating-speed preferences, future load expected and its effect on pump head and capacity, maintenance facilities available, possibility of parallel or series hookup, and other conditions peculiar to a given job.

Once the class and type of pump are selected, consult a rating table (Table 6.25) or rating chart (Fig. 6.18) to determine if a suitable pump is available from the manufacturer whose unit will be used. When the hydraulic requirements fall between two standard pump models, it is usual practice to choose the next larger size of pump, unless there is some reason why an exact head and capacity

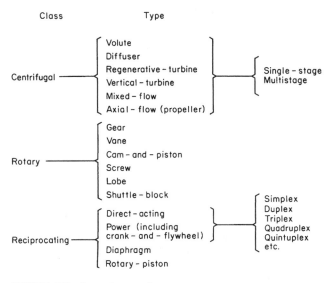

FIGURE 6.17 Pump classes and types.

TABLE 6.24 Characteristics of Modern Pumps

	Centrifugal		Rotary	Reciprocating		
	Volute and diffuser	Axial flow	Screw and gear	Direct acting steam	Double acting power	Triplex
Discharge flow Usual maximum suction lift, ft (m)	Steady 15 (4.6)	Steady 15 (4.6)	Steady 22 (6.7)	Pulsating 22 (6.7)	Pulsating 22 (6.7)	Pulsating 22 (6.7)
Liquids handled	Clean, clear; dirty, abrasive; liquids with high solids content		Viscous, nonabrasive	Clean and clear		
Discharge pressure range	Low to high		Medium	Low to highest produced		
Usual capacity range	Small to largest available		Small to medium	Relatively small		
How increased head affects: Capacity Power input	Decrease Depends on specific speed		None Increase	Decrease Increase	None Increase	None Increase
How decreased head affects: Capacity Power input	Increase Depends on specific speed		None Decrease	Small increase Decrease	None Decrease	None Decrease

TABLE 6.25 Typical Centrifugal-Pump Rating Table

Size		Total head			
gal/min	L/s	20 ft, r/min—hp	6.1 m, r/min—kW	25 ft, r/min—hp	7.6 m, r/min—kW
3 CL:					
200	12.6	910–1.3	910–0.97	1010–1.6	1010–1.19
300	18.9	1000–1.9	1000–1.41	1100–2.4	1100–1.79
400	25.2	1200–3.1	1200–2.31	1230–3.7	1230–2.76
500	31.5	—	—	—	—
4 C:					
400	25.2	940–2.4	940–1.79	1040–3	1040–2.24
600	37.9	1080–4	1080–2.98	1170–4.6	1170–3.43
800	50.5	—	—	—	—

Example: 1080–4 indicates pump speed is 1080 r/min; actual input required to operate the pump is 4 hp (2.98 kW).
Source: Condensed from data of Goulds Pumps, Inc.; SI values added by *Handbook* editor.

are required for the unit. When one manufacturer does not have the desired unit, refer to the engineering data of other manufacturers. Also keep in mind that some pumps are custom-built for a given job when precise head and capacity requirements must be met.

Other pump data included in manufacturer's engineering information include characteristic curves for various diameter impellers in the same casing (Fig. 6.19) and variable-speed

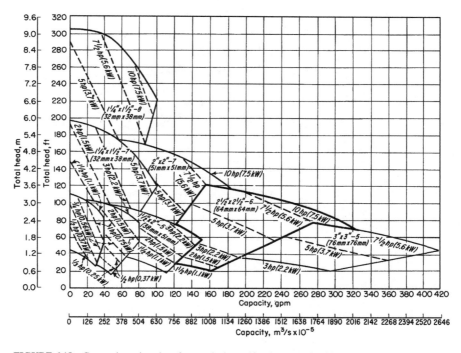

FIGURE 6.18   Composite rating chart for a typical centrifugal pump. (*Goulds Pumps, Inc.*)

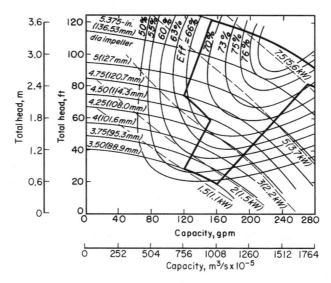

FIGURE 6.19   Pump characteristics when impeller diameter is varied within the same casing.

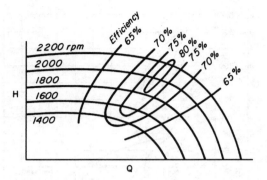

FIGURE 6.20 Variable-speed head-capacity curves for a centrifugal pump.

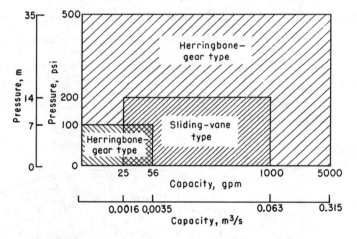

FIGURE 6.21 Capacity ranges of some rotary pumps. (*Worthington Corp.*)

head-capacity curves for an impeller of given diameter (Fig. 6.20). Note that the required power input is given in Figs. 6.18 and 6.19 and may also be given in Fig. 6.20. Use of Table 6.25 is explained in the table.

Performance data for rotary pumps is given in several forms. Figure 6.21 shows a typical plot of the head and capacity ranges of different types of rotary pumps. Reciprocating-pump capacity data are often tabulated, as in Table 6.26.

6. Evaluate the pump chosen for the installation.

Check the specific speed of a centrifugal pump using the method given in Procedure 6.20. Once the specific speed is known, the impeller type and approximate operating efficiency can be found from Fig. 6.12.

Check the piping system, using the method of Procedure 6.20, to see if the available net positive suction head equals, or is greater than, the required net positive suction head of the pump.

TABLE 6.26 Capacities of Typical Horizontal Duplex Plunger pumps

Size		Cold-water pressure service			
				Piston speed	
in	cm	gal/min	L/s	ft/min	m/min
6 × 3½ × 6	15.2 × 8.9 × 15.2	60	3.8	60	18.3
7½ × 4½ × 10	19.1 × 11.4 × 25.4	124	7.8	75	22.9
9 × 5 × 10	22.9 × 12.7 × 25.4	153	9.7	75	22.9
10 × 6 × 12	25.4 × 15.2 × 30.5	235	14.8	80	24.4
12 × 7 × 12	30.5 × 17.8 × 30.5	320	20.2	80	24.4

Size		Boiler-feed service					
				Boiler		Piston	speed
in	cm	gal/min	L/s	hp	kW	ft/min	m/min
6 × 3½ × 6	15. 2 × 8.9 × 15.2	36	2.3	475	354.4	36	10.9
7½ × 4½ × 10	19.1 × 11.4 × 25.4	74	4.7	975	727.4	45	13.7
9 × 5 × 10	22.9 × 12.7 × 25.4	92	5.8	1210	902.7	45	13.7
10 × 6 × 12	25.4 × 15.2 × 30.5	141	8.9	1860	1387.6	48	14.6
12 × 7 × 12	30.5 × 17.8 × 30.5	192	12.1	2530	1887.4	48	14.6

Source: Courtesy of Worthington Corporation.

Determine whether a vertical or horizontal pump is more desirable. From the standpoint of floor space occupied, required NPSH, priming, and flexibility in changing the pump use, vertical pumps may be preferable to horizontal designs in some installations. But where headroom, corrosion, abrasion, and ease of maintenance are important factors, horizontal pumps may be preferable.

As a general guide, single-suction centrifugal pumps handle up to 50 gal/min (0.0032 m³/s) at total heads up to 50 ft (15 m); either single- or double-suction pumps are used for the flow rates to 1000 gal/min (0.063 m³/s) and total heads to 300 ft (91 m); beyond these capacities and heads, double-suction or multistage pumps are generally used.

Mechanical seals have fully established themselves for all types of centrifugal pumps in a variety of services. Though more costly than packing, the mechanical seal reduces pump maintenance costs.

Related Calculations.   Use the procedure given here to select any class of pump—centrifugal, rotary, or reciprocating—for any type of service—power plant, atomic energy, petroleum processing, chemical manufacture, paper mills, textile mills, rubber factories, food processing, water supply, sewage and sump service, air conditioning and heating, irrigation and flood control, mining and construction, marine services, industrial hydraulics, iron and steel manufacture, etc.

6.23 ANALYSIS OF PUMP AND SYSTEM CHARACTERISTIC CURVES

Analyze a set of pump and system characteristic curves for the following conditions: friction losses without static head, friction losses with static head, pump without lift, system with little friction and much static head, system with gravity head, system with different pipe sizes, system with two discharge heads, system with diverted flow, and effect of pump wear on characteristic curve.

Calculation Procedure

1. *Plot the system-friction curve.*
Without static head, the system friction curve passes through the origin (0, 0) (Fig. 6.22), because
when no head is developed by the pump, flow through the piping is zero. For most piping systems,

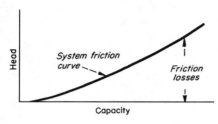

the friction-head loss varies as the square of the liquid
flow rate in the system. Hence, a system-friction curve,
also called a "friction-head curve," is parabolic—the
friction head increasing as the flow rate or capacity
of the system increases. Draw the curve as shown in
Fig. 6.22.

2. *Plot the piping system and system-head curve.*
Figure 6.23*a* shows a typical piping system with
a pump operating against a static discharge head.
Indicate the total static head (Fig. 6.23*b*) by a dashed
line—in this installation $H_{ts} = 110$ ft. Since static head

FIGURE 6.22   Typical system-friction curve.

is a physical dimension, it does not vary with flow rate and is a constant for all flow rates. Draw the
dashed line parallel to the abscissa (Fig. 6.23*b*).

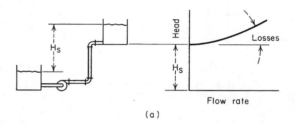

(a)

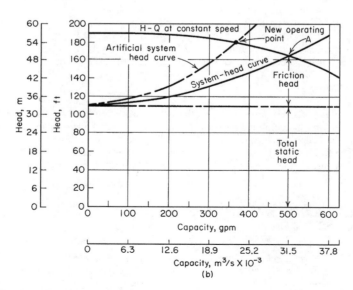

(b)

FIGURE 6.23   (*a*) Significant friction loss and lift; (*b*) system-head curve
superimposed on pump head-capacity curve. (*Peerless Pumps.*)

From the point of no flow—zero capacity—plot the friction-head loss at various flow rates—100, 200, 300 gal/min, etc. Determine the friction-head loss by computing it as shown in Procedure 6.7. Draw a curve through the points obtained. This is called the "system-head curve."

Plot the pump head-capacity (H-Q) curve of the pump on Fig. 6.23b. The H-Q curve can be obtained from the pump manufacturer or from a tabulation of H and Q values for the pump being considered. The point of intersection, A, between the H-Q and system-head curves is the operating point of the pump.

Changing the resistance of a given piping system by partially closing a valve or making some other change in the friction alters the position of the system-head curve and pump operating point. Compute the frictional resistance as before and plot the artificial system-head curve as shown. Where this curve intersects the H-Q curve is the new operating point of the pump. System-head curves are valuable for analyzing the suitability of a given pump for a particular application.

3. Plot the no-lift system-head curve and compute the losses.

With no static head or lift, the system-head curve passes through the origin (0, 0) (Fig. 6.24). For a flow of 900 gal/min (56.8 L/s), in this system, compute the friction loss as follows using the Hydraulic Institute—*Pipe Friction Manual* tables or the method of Procedure 6.7:

	ft	m
Entrance loss from tank into 10-in (254-mm) suction pipe, $0.5v^2/2g$	0.10	0.03
Friction loss in 2 ft (0.61 m) of suction pipe	0.02	0.01
Loss in 10-in (254-mm) 90° elbow at pump	0.20	0.06
Friction loss in 3000 ft (914.4 m) of 8-in (203.2-mm) discharge pipe	74.50	22.71
Loss in fully open 8-in (203.2-mm) gate valve	0.12	0.04
Exit loss from 8-in (203.2-mm) pipe into tank, $v^2/2g$	0.52	0.16
Total friction loss	75.46	23.01

Compute the friction loss at other flow rates in a similar manner and plot the system-head curve (Fig. 6.24). Note that if all losses in this system except the friction in the discharge pipe are ignored, the total head would not change appreciably. However, for the purposes of accuracy, all losses should always be computed.

4. Plot the low-friction, high-head system-head curve.

The system-head curve for the vertical pump installation in Fig. 6.25 starts at the total static head, 15 ft (4.6 m), and zero flow. Compute the friction head for 15,000 gal/min (946.4 L/s) as follows:

	ft	m
Friction in 20 ft (6.1 m) of 24-in pipe	0.40	0.12
Exit loss from 24-in pipe into tank, $v^2/2g$	1.60	0.49
Total friction loss	2.00	0.61

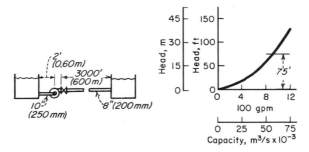

FIGURE 6.24   No lift; all friction head. (*Peerless Pumps.*)

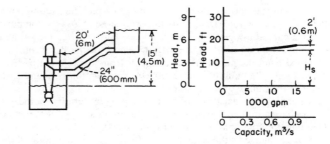

FIGURE 6.25   Mostly lift; little friction head. (*Peerless Pumps.*)

Hence, almost 90 percent of the total head of $15 + 2 = 17$ ft at 15,000 gal/min (946.4-L/s) flow is static head. But neglect of the pipe friction and exit losses could cause appreciable error during selection of a pump for the job.

5. *Plot the gravity-head system-head curve.*

In a system with gravity head (also called "negative lift"), fluid flow will continue until the system friction loss equals the available gravity head. In Fig. 6.26 the available gravity head is 50 ft (15.2 m). Flows up to 7200 gal/min (454.3 L/s) are obtained by gravity head alone. To obtain larger flow rates, a pump is needed to overcome the friction in the piping between the tanks. Compute the friction loss for several flow rates as follows:

	ft	m
At 5000 gal/min (315.5 L/s) friction loss in 1000 ft (305 m) of 16-in pipe	25	7.6
At 7200 gal/min (454.3 L/s) friction loss = available gravity head	50	15.2
At 13,000 gal/min (820.2 L/s) friction loss	150	45.7

Using these three flow rates, plot the system-head curve (Fig. 6.26).

6. *Plot the system-head curves for different pipe sizes.*

When different diameter pipes are used, the friction-loss-vs.-flow rate is plotted independently for the two pipe sizes. At a given flow rate, the total friction loss for the system is the sum of the loss for the two pipes. Thus the combined system-head curve represents the sum of the static head and the friction losses for all portions of the pipe.

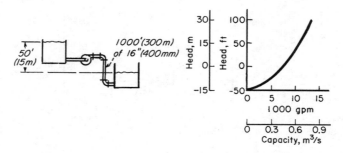

FIGURE 6.26   Negative lift (gravity head). (*Peerless Pumps.*)

Figure 6.27 shows a system with two different pipe sizes. Compute the friction losses as follows:

	ft	m
At 150 gal/min (9.5 L/s), friction loss in 200 ft (60.9 m) of 4-in (102-mm) pipe	5	1.52
At 150 gal/min (9.5 L/s), friction loss in 200 ft (60.9 m) of 3-in (76.2-mm) pipe	19	5.79
Total static head for 3- (76.2-mm) and 4-in (102-mm) pipes	10	3.05
Total head at 150-gal/min (9.5 L/s) flow	34	10.36

Compute the total head at other flow rates and plot the system-head curve as shown in Fig. 6.27.

7. *Plot the system-head curve for two discharge heads.*
Figure 6.28 shows a typical pumping system having two different discharge heads. Plot separate system-head curves when the discharge heads are different. Add the flow rates for the two pipes at the same head to find points on the combined system-head curve (Fig. 6.28). Thus,

	ft	m
At 550 gal/min (34.7 L/s), friction loss in 1000 ft (305 m) of 8-in pipe	= 10	3.05
At 1150 gal/min (72.6 L/s), friction	= 38	11.6
At 1150 gal/min (72.6 L/s), friction + lift in pipe 1 = 38 + 50	= 88	26.8
At 550 gal/min (34.7 L/s), friction + lift in pipe 2 = 10 + 78	= 88	26.8

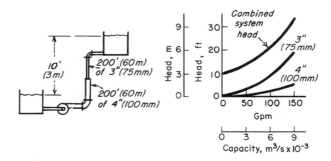

FIGURE 6.27   System with two different pipe sizes. (*Peerless Pumps.*)

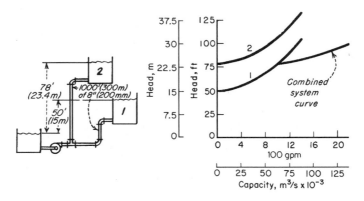

FIGURE 6.28   System with two different discharge heads. (*Peerless Pumps.*)

The flow rate for the combined system at a head of 88 ft is $1150 + 550 = 1700$ gal/min (107.3 L/s). To produce a flow of 1700 gal/min (107.3 L/s) through this system, a pump capable of developing an 88-ft (26.8-m) head is required.

8. *Plot the system-head curve for diverted flow.*

To analyze a system with diverted flow, assume that a constant quantity of liquid is tapped off at the intermediate point. Plot the friction-loss-vs.-flow rate in the normal manner for pipe 1 (Fig. 6.29). Move the curve for pipe 3 to the right at zero head by an amount equal to Q_2, since this represents the quantity passing through pipes 1 and 2 but not through pipe 3. Plot the combined system-head curve by adding, at a given flow rate, the head losses for pipes 1 and 3. With $Q = 300$ gal/min (18.9 L/s), pipe 1 = 500 ft (152.4 m) of 10-in (254-mm) pipe, and pipe 3 = 50 ft (15.2 m) of 6-in (152.4-mm) pipe:

	ft	m
At 1500 gal/min (94.6 L/s) through pipe 1, friction loss	= 11	3.35
Friction loss for pipe 3 (1500 − 300 = 1200 gal/min) (75.7 L/s)	= 8	2.44
Total friction loss at 1500-gal/min (94.6-L/s) delivery	= 19	5.79

9. *Plot the effect of pump wear.*

When a pump wears, there is a loss in capacity and efficiency. The amount of loss depends, however, on the shape of the system-head curve. For a centrifugal pump (Fig. 6.30), the capacity loss is greater for a given amount of wear if the system-head curve is flat, as compared with a steep system-head curve.

Determine the capacity loss for a worn pump by plotting its *H-Q* curve. Find this curve by testing the pump at different capacities and plotting the corresponding head. On the same chart, plot the

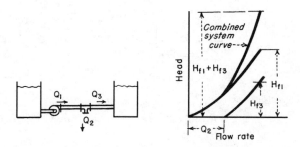

FIGURE 6.29   Part of the fluid flow diverted from the main pipe. (*Peerless Pumps.*)

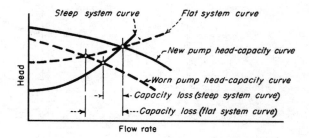

FIGURE 6.30   Effect of pump wear on pump capacity. (*Peerless Pumps.*)

H-Q curve for a new pump of the same size (Fig. 6.30). Plot the system-head curve and determine the capacity loss as shown in Fig. 6.30.

Related Calculations. Use the techniques given here for any type of pump—centrifugal, reciprocating, or rotary—handling any type of liquid—oil, water, chemicals, etc. The methods given here are the work of Melvin Mann, as reported in *Chemical Engineering*, and Peerless Pump Div. of FMC Corp.

6.24 NET POSITIVE SUCTION HEAD FOR HOT-LIQUID PUMPS

What is the maximum capacity of a double-suction condensate pump operating at 1750 r/min if it handles 100°F (311 K) water from a hot well in a condenser having an absolute pressure of 2.0 in Hg (6.8 kPa) if the pump centerline is 10 ft (3.05 m) below the hot-well liquid level and the friction-head loss in the suction piping and fitting is 5 ft (1.5 m) of water?

Calculation Procedure

1. *Compute the net positive suction head on the pump.*
The net positive suction head h_n on a pump when the liquid supply is *above* the pump inlet equals pressure on liquid surface + static suction head–friction-head loss in suction piping and pump inlet–vapor pressure of the liquid, all expressed in feet absolute of liquid handled. When the liquid supply is *below* the pump centerline—i.e., there is a static suction lift—the vertical distance of the lift is *subtracted* from the pressure on the liquid surface instead of added as in the preceding relation.

The density of 100°F water is 62.0 lb/ft^3. The pressure on the liquid surface, in absolute feet of liquid, is (2.0 inHg)(1.133)(62.4/62.0) = 2.24 ft. In this calculation, 1.133 = ft of 39.2°F water = 1 inHg; 62.4 = lb/ft^3 of 39.2°F water. The temperature of 39.2°F is used because at this temperature water has its maximum density. Thus, to convert inches of mercury to feet of absolute of water, find the product of (inHg)(1.133)(water density at 39.2°F)/(water density at operating temperature). Express both density values in the same unit, usually lb/ft^3.

The static suction head is a physical dimension that is measured in feet of liquid at the operating temperature. In this installation, h_{sh} = 10 ft absolute.

The friction-head loss is 5 ft of water at maximum density. To convert to feet absolute, multiply by the ratio of water densities at 39.2°F and the operating temperature or (5)(62.4/62.0) = 5.03 ft.

The vapor pressure of water at 100°F is 0.949 psia, from the steam tables. Convert any vapor pressure to feet absolute by finding the result of (vapor pressure, psia)(144 in^2/ft^2)/liquid density at operating temperature, or (0.949)(144)/62.0 = 2.204 ft absolute.

With all the heads known, the net positive suction head is h_n = 2.24 + 10 − 5.03 − 2.204 = 5.01 ft (1.53 m) absolute.

2. *Determine the capacity of the condensate pump.*
Use the Hydraulic Institute curve (Fig. 6.31) to determine the maximum capacity of the pump. Enter at the left of Fig. 6.31 at a net positive suction head of 5.01 ft and project horizontally to the right until the 3500-r/min curve is intersected. At the top, read the capacity as 278 gal/min (0.0175 m^3/s).

Related Calculations. Use this procedure for any condensate or boiler-feed pump handling water at an elevated temperature. Consult the *Standards of the Hydraulic Institute* for capacity curves of pumps having different types of construction. In general, pump manufacturers who are members of the Hydraulic Institute rate their pumps in accordance with the *Standards,* and a pump chosen from a catalog capacity table or curve will deliver the stated capacity. A similar procedure is used for computing the capacity of pumps handling volatile petroleum liquids. When using this procedure, be certain to refer to the latest edition of the *Standards*.

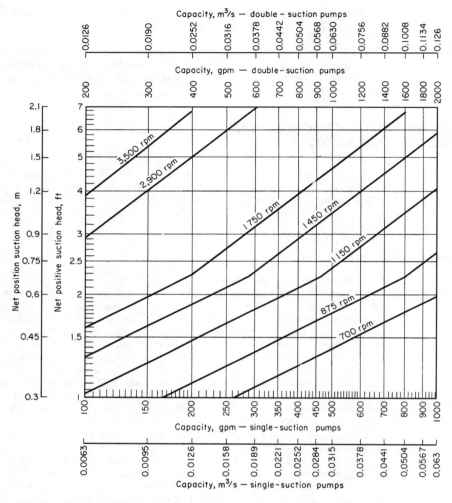

FIGURE 6.31 Capacity and speed limitations of condensate pumps with the shaft through the impeller eye. (*Hydraulic Institute.*)

6.25 *MINIMUM SAFE FLOW FOR A CENTRIFUGAL PUMP*

A centrifugal pump handles 220°F (377 K) water and has a shutoff head (with closed discharge valve) of 3200 ft (975 m). At shutoff, the pump efficiency is 17 percent and the input brake horsepower is 210. What is the minimum safe flow through this pump to prevent overheating at shutoff? Determine the minimum safe flow if the NPSH is 18.8 ft (5.73 m) of water and the liquid specific gravity is 0.995. If the pump contains 500 lb (227 kg) of water, determine the rate of the temperature rise at shutoff.

Calculation Procedure

1. Compute the temperature rise in the pump.

With the discharge valve closed, the power input to the pump is converted to heat in the casing and causes the liquid temperature to rise. The temperature rise $t = (1 - e) \times H_s/778e$, where t is temperature rise during shutoff, °F; e is pump efficiency, expressed as a decimal; H_s is shutoff head, ft. For this pump, $t = (1 - 0.17)(3200)/[778(0.17)] = 20.4°F$ (11.3°C).

2. Compute the minimum safe liquid flow.

For general-service pumps, the minimum safe flow M, in gal/min, is 6.0(bhp input at shutoff)/t. Or, $M = 6.0(210)/20.4 = 62.7$ gal/min (0.00396 m³/s). This equation includes a 20 percent safety factor.

Centrifugal boiler-feed pumps usually have a maximum allowable temperature rise of 15°F. The minimum allowable flow through the pump to prevent the water temperature from rising more than 15°F is 30 gal/min for each 100 bhp input at shutoff.

3. Compute the temperature rise for the operating NPSH.

An NPSH of 18.8 ft is equivalent to a pressure of 18.8(0.433)(0.995) = 7.78 psia at 220°F, where the factor 0.433 converts feet of water to pounds per square inch. At 220°F, the vapor pressure of the water is 17.19 psia, from the steam tables. Thus the total vapor pressure the water can develop before flashing occurs equals NPSH pressure + vapor pressure at operating temperature = 7.78 + 17.19 = 24.97 psia. Enter the steam tables at this pressure and read the corresponding temperature as 240°F. The allowable temperature rise of the water is then 240 − 220 = 20°F. Using the safe-flow relation of step 2, the minimum safe flow is 62.9 gal/min (0.00397 m³/s).

4. Compute the rate of temperature rise.

In any centrifugal pump, the rate of temperature rise t_r, in °F per minute, is 42.4(bhp input at shutoff)/wc, where w is weight of liquid in the pump, lb; c is specific heat of the liquid in the pump, Btu/(lb)(°F). For this pump containing 500 lb of water with a specific heat c of 1.0, $t_r = 42.4(210)/[500(1.0)] = 17.8°F/$min (0.16 K/s). This is a very rapid temperature rise and could lead to overheating in a few minutes.

Related Calculations.   Use this procedure for any centrifugal pump handling any liquid in any service—power, process, marine, industrial, or commercial. Pump manufacturers can supply a temperature-rise curve for a given model pump if it is requested. This curve is superimposed on the pump characteristic curve and shows the temperature rise accompanying a specific flow through the pump.

6.26 SELECTING A CENTRIFUGAL PUMP TO HANDLE A VISCOUS LIQUID

Select a centrifugal pump to deliver 750 gal/min (0.047 m³/s) of 1000-SSU oil at a total head of 100 ft (30.5 m). The oil has a specific gravity of 0.90 at the pumping temperature. Show how to plot the characteristic curves when the pump is handling the viscous liquid.

Calculation Procedure

1. Determine the required correction factors.

A centrifugal pump handling a viscous liquid usually must develop a greater capacity and head, and it requires a larger power input than the same pump handling water. With the water performance of the pump known—either from the pump characteristic curves or a tabulation of pump performance parameters—Fig. 6.32, prepared by the Hydraulic Institute, can be used to find suitable correction factors. Use this chart only within its scale limits; do not extrapolate. Do not use the chart for mixed-flow or axial-flow pumps or for pumps of special design. Use the chart only for pumps handling uniform liquids; slurries, gels, paper stock, etc. may cause incorrect results. In using the chart, the available net positive suction head is assumed adequate for the pump.

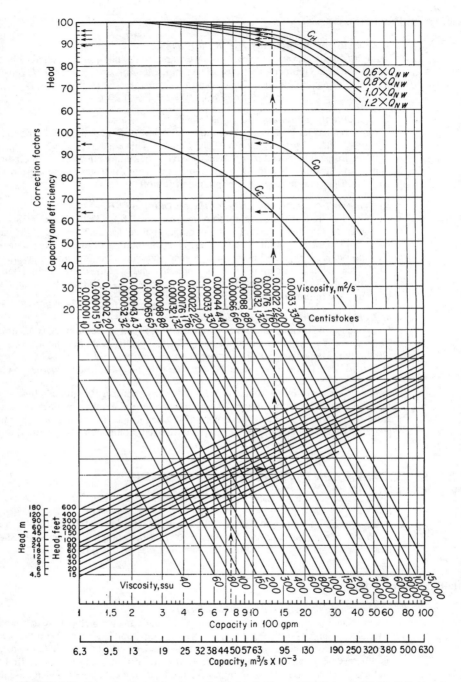

FIGURE 6.32 Correction factors for viscous liquids handled by centrifugal pumps. (*Hydraulic Institute.*)

To use Fig. 6.32, enter at the bottom at the required capacity, 750 gal/min, and project vertically to intersect the 100-ft head curve, the required head. From here project horizontally to the 1000-SSU viscosity curve and then vertically upward to the correction-factor curves. Read $C_E = 0.635$; $C_Q = 0.95$; $C_H = 0.92$ for $1.0Q_{NW}$. The subscripts E, Q, and H refer to correction factors for efficiency, capacity, and head, respectively, and NW refers to the water capacity at a particular efficiency. At maximum efficiency, the water capacity is given as $1.0Q_{NW}$; other efficiencies, expressed by numbers equal to or less than unity, give different capacities.

2. Compute the water characteristics required.
The water capacity required for the pump $Q_w = Q_v/C_Q$, where Q_v is viscous capacity, gal/min. For this pump, $Q_w = 750/0.95 = 790$ gal/min. Likewise, water head $H_w = H_v/C_H$ where H_v is viscous head. Or, $H_w = 100/0.92 = 108.8$, say 109 ft water.
   Choose a pump to deliver 790 gal/min of water at 109-ft head of water and the required viscous head and capacity will be obtained. Pick the pump so that it is operating at or near its maximum efficiency on water. If the water efficiency $E_w = 81$ percent at 790 gal/min for this pump, the efficiency when handling the viscous liquid $E_w = E_w C_E$. Or, $E_v = 0.81(0.635) = 0.515$, or 51.5 percent.
   The power input to the pump when handling viscous liquids is given by $P_v = Q_v H_v s/(3960 E_v)$, where s is specific gravity of the viscous liquid. For this pump, $P_v = (750)(100)(0.90)/[3960(0.515)]$ = 33.1 hp (24.7 kW).

3. Plot the characteristic curves for viscous-liquid pumping.
Follow these eight steps to plot the complete characteristic curves of a centrifugal pump handling a viscous liquid when the water characteristics are known: (*a*) Secure a complete set of characteristic curves (*H, Q, P, E*) for the pump to be used. (*b*) Locate the point of maximum efficiency for the pump when handling water. (*c*) Read the pump capacity, Q gal/min, at this point, (*d*) Compute the values of $0.6Q$, $0.8Q$, and $1.2Q$ at the maximum efficiency, (*e*) Using Fig. 6.32, determine the correction factors at the capacities in steps *c* and *d*. Where a multistage pump is being considered, use the head per stage (= total pump head, ft/number of stages), when entering Fig. 6.32. (*f*) Correct the head, capacity, and efficiency for each of the flow rates in *c* and *d* using the correction factors from Fig. 6.32. (*g*) Plot the corrected head and efficiency against the corrected capacity, as in Fig. 6.33. (*h*) Compute the power input at each flow rate and plot. Draw smooth curves through the points obtained (Fig. 6.33).

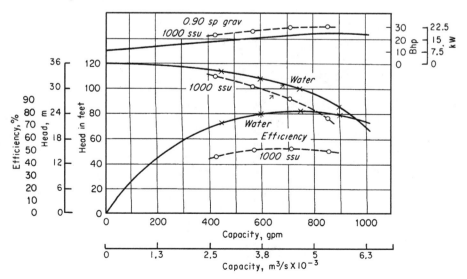

FIGURE 6.33 Characteristic curves for water (*solid line*) and oil (*dashed line*). (*Hydraulic Institute.*)

Related Calculations. Use the method given here for any uniform viscous liquid—oil, gasoline, kerosene, mercury, etc.—handled by a centrifugal pump. Be careful to use Fig. 6.32 only within its scale limits; *do not extrapolate*. The method presented here is that developed by the Hydraulic Institute. For new developments in the method, be certain to consult the latest edition of the Hydraulic Institute—*Standards*.

6.27 EFFECT OF LIQUID VISCOSITY ON REGENERATIVE PUMP PERFORMANCE

A regenerative (turbine) pump has the water head-capacity and power-input characteristics shown in Fig. 6.34. Determine the head-capacity and power-input characteristics for four different viscosity oils to be handled by the pump—400, 600, 900, and 1000 SSU. What effect does increased viscosity have on the performance of the pump?

Calculation Procedure

1. Plot the water characteristics of the pump.
Obtain a tabulation or plot of the water characteristics of the pump from the manufacturer or from the engineering data. With a tabulation of the characteristics, enter the various capacity and power points given and draw a smooth curve through them (Fig. 6.34).

2. Plot the viscous-liquid characteristics of the pump.
The viscous-liquid characteristics of regenerative-type pumps are obtained by test of the actual unit. Hence the only source of this information is the pump manufacturer. Obtain these characteristics from the pump manufacturer or the test data and plot them on Fig. 6.34, as shown, for each oil or other liquid handled.

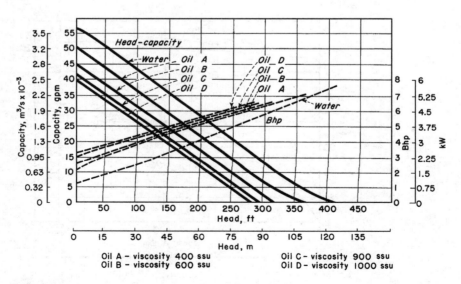

Oil A – viscosity 400 ssu   Oil C – viscosity 900 ssu
Oil B – viscosity 600 ssu   Oil D – viscosity 1000 ssu

FIGURE 6.34 Regenerative pump performance when handling water and oil. (*Aurora Pump Division, The New York Air Brake Co.*)

3. *Evaluate the effect of viscosity on pump performance.*
Study Fig. 6.34 to determine the effect of increased liquid viscosity on the performance of the pump. Thus at a given head—say 100 ft—the capacity of the pump decreases as the liquid viscosity increases. At 100-ft head, this pump has a water capacity of 43.5 gal/min (Fig. 6.34). The pump capacity for the various oils at 100-ft head is 36 gal/min for 400 SSU, 32 gal/min for 600 SSU, 28 gal/min for 900 SSU, and 26 gal/min for 1000 SSU, respectively. There is a similar reduction in capacity of the pump at the other heads plotted in Fig. 6.34. Thus, as a general rule, it can be stated that the capacity of a regenerative pump decreases with an increase in liquid viscosity at constant head. Or conversely, at constant capacity, the head developed decreases as the liquid viscosity increases.

Plots of the power input to this pump show that the input power increases as the liquid viscosity increases.

Related Calculations.   Use this procedure for a regenerative-type pump handling any liquid—water, oil, kerosene, gasoline, etc. A decrease in the viscosity of a liquid—as compared with the viscosity of water—will produce the opposite effect from that of increased viscosity.

6.28 EFFECT OF LIQUID VISCOSITY ON RECIPROCATING-PUMP PERFORMANCE

A direct-acting steam-driven reciprocating pump delivers 100 gal/min (0.0063 m³/s) of 70°F (294 K) water when operating at 50 strokes/min. How much 2000-SSU crude oil will this pump deliver? How much 125°F (325 K) water will this pump deliver?

Calculation Procedure

1. *Determine the recommended change in the pump performance.*
Reciprocating pumps of any type—direct-acting or power—having any number of liquid-handling cylinders—1 to 5, or more—are usually rated for maximum delivery when handling 250-SSU liquids or 70°F (21°C) water. At higher liquid viscosities or water temperatures, the speed—strokes or revolutions per minute—is reduced. Table 6.27 shows typical recommended speed-correction factors for reciprocating pumps for various liquid viscosities and water temperatures. This table shows that with a liquid viscosity of 2000 SSU, the pump speed should be reduced 20 percent. When handling 125°F (51.7°C) water, the pump speed should be reduced 25 percent, as shown in Table 6.27.

2. *Compute the delivery of the pump.*
The delivery capacity of any reciprocating pump is directly proportional to the number of strokes per minute it makes or to its revolutions per minute.

When handling 2000-SSU oil, the pump strokes/min must be reduced 20 percent, or $(50)(0.20) = 10$ strokes/min. Hence the pump speed will be $50 - 10 = 40$ strokes/min. Since the delivery is directly proportional to speed, the delivery of 2000-SSU oil is $(40/50)(100) = 80$ gal/min (5.1 L/s).

TABLE 6.27   Speed-Correction Factors

Liquid viscosity, SSU	Speed reduction, %	Water Temperature		Speed reduction, %
		°F	°C	
250	0	70	21.1	0
500	4	80	26.2	9
1000	11	100	37.8	18
2000	20	125	51.7	25
3000	26	150	65.6	29
4000	30	200	97.3	34
5000	35	250	121.1	38

When handling 125°F (51.7°C) water, the pump strokes per minute must be reduced 25 percent, or (50)(0.5) = 12.5 strokes/min. Hence the pump speed will be 50.0 − 12.5 = 37.5 strokes/min. Since the delivery is directly proportional to speed, the delivery of 125°F (51.7°C) water is (37.5/50)(100) = 75 gal/min (4.7 L/s).

Related Calculations. Use this procedure for any type of reciprocating pump handling liquids falling within the range of Table 6.27. Such liquids include oil, kerosene, gasoline, brine, water, etc.

6.29 EFFECT OF VISCOSITY AND DISSOLVED GAS ON ROTARY PUMPS

A rotary pump handles 8000-SSU liquid containing 5% entrained gas and 10% dissolved gas at a 20-in(508-mm)Hg pump inlet vacuum. The pump is rated at 1000 gal/min (63.1 L/s) when handling gas-free liquids at viscosities less than 600 SSU. What is the output of this pump without slip? With 10 percent slip?

Calculation Procedure

1. Compute the required speed reduction of the pump.
When the liquid viscosity exceeds 600 SSU, many pump manufacturers recommend that the speed of a rotary pump be reduced to permit operation without excessive noise or vibration. The speed reduction ususally recommended is shown in Table 6.28.

With this pump handling 8000-SSU liquid, a speed reduction of 40 percent is necessary, as shown in Table 6.28. Since the capacity of a rotary pump varies directly with its speed, the output of this pump when handling 8000-SSU liquid is (1000 gal/min) × (1.0 − 0.40) = 600 gal/min (37.9 L/s).

2. Compute the effect of gas on the pump output.
Entrained or dissolved gas reduces the output of a rotary pump, as shown in Table 6.29. The gas in the liquid expands when the inlet pressure of the pump is below atmospheric and the gas occupies part of the pump chamber, reducing the liquid capacity.

With a 20-in (508-mm)Hg inlet vacuum, 5% entrained gas, and 10% dissolved gas. Table 6.29 shows that the liquid displacement is 74 percent of the rated displacement. Thus, the output of the pump when handling this viscous, gas-containing liquid will be (600 gal/min)(0.74) = 444 gal/min (28.0 L/s) without slip.

TABLE 6.28 Rotary-Pump Speed Reduction for Various Liquid Viscosities

Liquid viscosity, SSU	Speed reduction, percent of rated pump speed
600	2
800	6
1000	10
1500	12
2000	14
4000	20
6000	30
8000	40
10,000	50
20,000	55
30,000	57
40,000	60

TABLE 6.29 Effect of Entrained or Dissolved Gas on the Liquid Displacement of Rotary Pumps (liquid displacement: percent of displacement)

Vacuum at pump inlet, inHg (mmHg)	Gas entrainment					Gas solubility					Gas entrainment and gas solubility combined				
	1%	2%	3%	4%	5%	2%	4%	6%	8%	10%	1% 2%	2% 4%	3% 6%	4% 8%	5% 10%
5 (127)	99	97½	96½	95	93½	99½	99	98½	97	97½	98½	96½	96	92	91
10 (254)	98½	97¼	95½	94	92	99	97½	97	95	95	97½	95	90	90	88¼
15 (381)	98	96½	94½	92½	90½	97	96	94	92	90½	96	93	89½	86½	83¼
20 (508)	97½	94½	92	89	86½	96	92	89	86	83	94	88	83	78	74
25 (635)	94	89	84	79	75½	90	83	76½	71	66	85½	75½	68	61	55

For example: with 5 percent gas entrainment at 15 in Hg (381 mmHg) vacuum, the liquid displacement will be 90½ percent of the pump displacement neglecting slip, or with 10 percent dissolved gas, liquid displacement will be 90½ percent of pump displacement, and with 5 percent entrained gas combined with 10 percent dissolved gas, the liquid displacement will be 83¼ percent of pump displacement.

Source: Courtesy of Kinney Mfg. Div., The New York Air Brake Co.

3. *Compute the effect of slip on the pump output.*
Slip reduces rotary-pump output in direct proportion to the slip. Thus, with 10 percent slip, the output of this pump is (444 gal/min)(1.0 − 0.10) = 399.6 gal/min (25 L/s).

Related Calculations. Use this procedure for any type of rotary pump—gear, lobe, screw, swinging-vane, sliding-vane, or shuttle-block—handling any clear, viscous liquid. Where the liquid is gas-free, apply only the viscosity correction. Where the liquid viscosity is less than 600 SSU but the liquid contains gas or air, apply the entrained or dissolved gas correction, or both corrections.

6.30 SELECTING FORCED- AND INDUCED-DRAFT FANS

Combustion calculations show that an oil-fired watertube boiler requires 200,000 lb/h (25.2 kg/s) for air of combustion at maximum load. Select forced- and induced-draft fans for this boiler if the average temperature of the inlet air is 75°F (297 K) and the average temperature of the combustion gas leaving the air heater is 350°F (450 K) with an ambient barometric pressure of 29.9 inHg. Pressure losses on the air-inlet side are, in inH_2O: air heater, 1.5; air supply ducts, 0.75; boiler windbox, 1.75; burners, 1.25. Draft losses in the boiler and related equipment are, in inH_2O: furnace pressure, 0.20; boiler, 3.0; superheater, 1.0; economizer, 1.50; air heater, 2.00; uptake ducts and dampers, 1.25. Determine the fan discharge pressure and horsepower input. The boiler burns 18,000 lb/h (2.27 kg/s) of oil at full load.

Calculation Procedure

1. *Compute the quantity of air required for combustion.*
The combustion calculations show that 200,000 lb/h of air is theoretically required for combustion in this boiler. To this theoretical requirement must be added allowances for excess air at the burner and leakage out of the air heater and furnace. Allow 25 percent excess air for this boiler. The exact allowance for a given installation depends on the type of fuel burned. However, a 25 percent excess-air allowance is an average used by power-plant designers for coal, oil, and gas firing. Using this allowance, the required excess air is 200,000(0.25) = 50,000 lb/h.

Air-heater air leakage varies from about 1 to 2 percent of the theoretically required airflow. Using 2 percent, the air-heater leakage allowance is 200,000(0.02) = 4,000 lb/h.

Furnace air leakage ranges from 5 to 10 percent of the theoretically required airflow. Using 7.5 percent, the furnace leakage allowance is 200,000(0.075) = 15,000 lb/h.

The total airflow required is the sum of the theoretical requirement, excess air, and leakage, or 200,000 + 50,000 + 4000 + 15,000 = 269,000 lb/h. The forced-draft fan must supply at least this quantity of air to the boiler. Usual practice is to allow a 10 to 20 percent safety factor for fan capacity to ensure an adequate air supply at all operating conditions. This factor of safety is applied to the total airflow required. Using a 10 percent factor of safety, fan capacity is 269,000 + 269,000(0.1) = 295,000 lb/h. Round this off to 296,000 lb/h (37.3 kg/s) fan capacity.

2. *Express the required airflow in cubic feet per minute.*
Convert the required flow in pounds per hour to cubic feet per minute. To do this, apply a factor of safety to the ambient air temperature to ensure an adequate air supply during times of high ambient temperature. At such times, the density of the air is lower and the fan discharges less air to the boiler. The usual practice is to apply a factor of safety of 20 to 25 percent to the known ambient air temperature. Using 20 percent, the ambient temperature for fan selection is 75 + 75(0.20) = 90°F. The density of air at 90°F is 0.0717 lb/ft^3, found in Baumeister—*Standard Handbook for Mechanical Engineers.* Converting, ft^3/min = lb/h/60(lb/ft^3) = 296,000/60(0.0717) = 69,400 ft^3/min. This is the minimum capacity the forced-draft fan may have.

3. Determine the forced-draft discharge pressure.

The total resistance between the forced-draft fan outlet and furnace is the sum of the losses in the air heater, air-supply ducts, boiler windbox, and burners. For this boiler, the total resistance, in inH_2O, is $1.5 + 0.75 + 1.75 + 1.25 = 5.25$ inH_2O. Apply a 15 to 30 percent factor of safety to the required discharge pressure to ensure adequate airflow at all times. Or fan discharge pressure, using a 20 percent factor of safety, is $5.25 + 5.25(0.20) = 6.30$ inH_2O. The fan must therefore deliver at least 69,400 ft^3/min (32.7 m^3/s) at 6.30 inH_2O.

4. Compute the power required to drive the forced-draft fan.

The air horsepower for any fan is $0.0001753\ H_f C$, where H_f is total head developed by fan, in inH_2O; C is airflow, in ft^3/min. For this fan, air hp = $0.0001753(6.3)\ (69,400)$, = 76.5 hp. Assume or obtain the fan and fan-driver efficiencies at the rated capacity (69,400 ft^3/min) and pressure (6.30 inH_2O). With a fan efficiency of 75 percent and assuming the fan is driven by an electric motor having an efficiency of 90 percent, the overall efficiency of the fan-motor combination is $(0.75)(0.90) = 0.675$, or 67.5 percent. Then the motor horsepower required equals air horsepower/overall efficiency = $76.5/0.675 = 113.2$ hp (84.4 kW). A 125-hp motor would be chosen because it is the nearest, next larger, unit readily available. Usual practice is to choose a *larger* driver capacity when the computed capacity is lower than a standard capacity. The next larger standard capacity is generally chosen, except for extremely large fans where a special motor may be ordered.

5. Compute the quantity of flue gas handled.

The quantity of gas reaching the induced draft fan is the sum of the actual air required for combustion from step 1, air leakage in the boiler and furnace, and the weight of fuel burned. With an air leakage of 10 percent in the boiler and furnace (this is a typical leakage factor applied in practice), the gas flow is as follows:

	lb/h	kg/s
Actual airflow required	296,000	37.3
Air leakage in boiler and furnace	29,600	3.7
Weight of oil burned	18,000	2.3
Total	343,600	43.3

Determine from combustion calculations for the boiler the density of the flue gas. Assume that the combustion calculations for this boiler show that the flue-gas density is 0.045 lb/ft^3 (0.72 kg/m^3) at the exit-gas temperature. To determine the exit-gas temperature, apply a 10 percent factor of safety to the given exit temperature, 350°F (176.6°C). Hence exit-gas temperature is $350 + 350(0.10) = 385$°F (196.1°C). Then, flue-gas flow, in ft^3/min, is (flue-gas flow, lb/h)/(60)(flue-gas density, lb/ft^3) = $343,600/[(60)(0.045)] = 127,000$ ft^3/min (59.9 m^3/s). Apply a 10 to 25 percent factor of safety to the flue-gas quantity to allow for increased gas flow. Using a 20 percent factor of safety, the actual flue-gas flow the fan must handle is $127,000 + 127,000(0.20) = 152,400$ ft^3/min (71.8 m^3/s), say 152,500 ft^3/min for fan-selection purposes.

6. Compute the induced-draft fan discharge pressure.

Find the sum of the draft losses from the burner outlet to the induced-draft inlet. These losses are, for this boiler:

	inH$_2$O	kPa
Furnace draft loss	0.20	0.05
Boiler draft loss	3.00	0.75
Superheater draft loss	1.00	0.25
Economizer draft loss	1.50	0.37
Air heater draft loss	2.00	0.50
Uptake ducts and damper draft loss	1.25	0.31
Total draft loss	8.95	2.23

Allow a 10 to 25 percent factor of safety to ensure adequate pressure during all boiler loads and furnace conditions. Using a 20 percent factor of safety for this fan, the total actual pressure loss is $8.95 + 8.95(0.20) = 10.74$ inH$_2$O (2.7 kPa). Round this off to 11.0 inH$_2$O (2.7 kPa) for fan-selection purposes.

7. Compute the power required to drive the induced-draft fan.
As with the forced-draft fan, air horsepower is $0.0001753\, H_f C = 0.0001753(11.0)(127,000) = 245$ hp (182.7 kW). If the combined efficiency of the fan and its driver, assumed to be an electric motor, is 68 percent, the motor horsepower required is $245/0.68 = 360.5$ hp (268.8 kW). A 375-hp (279.6-kW) motor would be chosen for the fan driver.

8. Choose the fans from a manufacturer's engineering data.
Use Procedure 6.31 to select the fans from the engineering data of an acceptable manufacturer. For larger boiler units, the forced-draft fan is usually a backward-curved blade centrifugal-type unit. Where two fans are chosen to operate in parallel, the pressure curve of each fan should decrease at the same rate near shutoff so that the fans divide the load equally. Be certain that forced-draft fans are heavy-duty units designed for continuous operation with well-balanced rotors. Choose high-efficiency units with self-limiting power characteristics to prevent overloading the driving motor. Airflow is usually controlled by dampers on the fan discharge.

Induced-draft fans handle hot, dusty combustion products. For this reason, extreme care must be used to choose units specifically designed for induced-draft service. The usual choice for large boilers is a centrifugal-type unit with forward- or backward-curved, or flat blades, depending on the type of gas handled. Flat blades are popular when the flue gas contains large quantities of dust. Fan bearings are generally water-cooled.

Related Calculations.   Use this procedure for selecting draft fans for all types of boilers—firetube, packaged, portable, marine, and stationary. Obtain draft losses from the boiler manufacturer. Compute duct pressure losses using the methods given in later procedures in this handbook.

6.31.   POWER-PLANT FAN SELECTION FROM CAPACITY TABLES

Choose a forced-draft fan to handle 69,400 ft^3/min (32.7 m^3/s) of 90°F (305 K) air at 6.30 inH$_2$O static pressure and an induced-draft fan to handle 152,500 ft^3/min (71.9 m^3/s) of 385°F (469 K) gas at 11.0 inH$_2$O static pressure. The boiler that these fans serve is installed at an elevation of 5000 ft (1524 m) above sea level. Use commercially available capacity tables for making the fan choice. The flue-gas density is 0.045 lb/ft^3 (0.72 kg/m^3) at 385°F (469 K).

Calculation Procedure

1. Compute the correction factors for the forced-draft fan.
Commercial fan-capacity tables are based on fans handling standard air at 70°F at a barometric pressure of 29.92 inHg and having a density of 0.075 lb/ft^3. Where different conditions exist, the fan flow rate must be corrected for temperature and altitude.

Obtain the engineering data for commercially available forced-draft fans and turn to the temperature and altitude correction-factor tables. Pick the appropriate correction factors from these tables for the prevailing temperature and altitude of the installation. Thus, in Table 6.30, select the correction factors for 90°F air and 5000-ft altitude. These correction factors are $C_T = 1.018$ for 90°F air and $C_A = 1.095$ for 5000-ft altitude.

Find the composite correction factor CCF by taking the product of the temperature and altitude correction factors, or CCF $= C_T C_A = 1.018(1.095) = 1.1147$. Now divide the given ft^3/min by the composite correction factor to find the capacity-table ft^3/min. Or, capacity-table ft^3/min is $69,400/1.1147 = 62,250$ ft^3/min.

TABLE 6.30   Fan Correction Factors

Temperature		Correction factor	Altitude		Correction factor
°F	°C		ft	m	
80	26.7	1.009	4500	1371.6	1.086
90	32.2	1.018	5000	1524.0	1.095
100	37.8	1.028	5500	1676.4	1.106
375	190.6	1.255			
400	204.4	1.273			
450	232.2	1.310			

2. Choose the fan size from the capacity table.
Turn to the fan-capacity table in the engineering data and look for a fan delivering 62,250 ft^3/min at 6.3 inH$_2$O static pressure. Inspection of the table shows that the capacities are tabulated for pressures of 6.0 and 6.5 inH$_2$O static pressure. There is no tabulation for 6.3 inH$_2$O. The fan must therefore be selected for 6.5 inH$_2$O static pressure.

Enter the table at the nearest capacity to that required, 62,250 ft^3/min, as shown in Table 6.31. This table, excerpted with permission from the American Standard Inc. engineering data, shows that the nearest capacity of this particular type of fan is 62,595 ft^3/min. The difference, or 62,595 − 62,250 = 345 ft^3/min, is only 345/62,250 = 0.0055, or 0.55 percent. This is a negligible difference, and the 62,595-ft^3/min fan is well suited for its intended use. The extra static pressure, 6.5 − 6.3 = 0.2 inH$_2$O, is desirable in a forced-draft fan because furnace or duct resistance may increase during the life of the boiler. Also, the extra static pressure is so small that it will not markedly increase the fan power consumption.

3. Compute the fan speed and power input.
Multiply the capacity-table r/min and bhp by the composite correction factor to determine the actual r/min and bhp. Thus, using data from Table 6.31, the actual r/min is (1096)(1.1147) = 1221.7 r/min. Actual bhp is (99.08)(1.1147) = 110.5 hp. This is the horsepower input required to drive the fan and is close to the 113.2 hp computed in the previous example. The actual motor horsepower would be the same in each case because a standard-size motor would be chosen. The difference of 113.2 − 110.5 = 2.7 hp results from the assumed efficiencies that depart from the actual values. Also, a sea-level altitude was assumed in the previous example. However, the two methods used show how accurately fan capacity and horsepower input can be estimated by judicious evaluation of variables.

4. Compute the correction factors for the induced-draft fan.
The flue-gas density is 0.045 lb/ft^3 at 385°F. Interpolate in the temperature correction-factor table because a value of 385°F is not tabulated. Find the correction factor for 385°F thus: (Actual temperature − lower temperature)/(higher temperature − lower temperature) × (higher temperature-correction factor − lower temperature-correction factor) + lower-temperature-correction factor. Or, [(385 − 375)/(400 − 375)](1.273 − 1.255) + 1.255 = 1.262.

TABLE 6.31   Typical Fan Capacities

Capacity		Outlet velocity		Outlet velocity pressure		Ratings at 6.5 inH$_2$O (1.6 kPa) static pressure		
ft^3/min	m^3/s	ft/min	m/s	inH$_2$O	kPa	r/min	bhp	kW
61,204	28.9	4400	22.4	1.210	0.3011	1083	95.45	71.2
62,595	29.5	4500	22.9	1.266	0.3150	1096	99.08	73.9
63,975	30.2	4600	23.4	1.323	0.3212	1109	103.0	76.8

The altitude-correction factor is 1.095 for an elevation of 5000 ft, as shown in Table 6.30. As for the forced-draft fan, CCF = $C_T C_A$ = (1.262)(1.095) = 1.3819. Use the CCF to find the capacity-table ft³/min in the same manner as for the forced-draft fan. Or, capacity-table ft³/min is (given ft³/min)/ CCF = 152,500/1.3819 = 110,355 ft³/min.

5. *Choose the fan size from the capacity table.*
Check the capacity table to be sure that it lists fans suitable for induced-draft (elevated temperature) service. Turn to the 11-in static-pressure-capacity table and find a capacity equal to 110,355 ft³/ min. In the engineering data used for this fan, the nearest capacity at 11-in static pressure is 110,467 ft³/min, with an outlet velocity of 4400 ft³/min, an outlet velocity pressure of 1.210 inH₂O, a speed of 1222 r/min, and an input horsepower of 255.5 bhp. The tabulation of these quantities is of the same form as that given for the forced-draft fan (step 2). The selected capacity of 110,467 ft³/min is entirely satisfactory because it is only (110,467 − 110,355)/110,355 = 0.00101, or 0.1 percent, higher than the desired capacity.

6. *Compute the fan speed and power input.*
Multiply the capacity-table r/min and brake horsepower by the CCF to determine the actual r/min and brake horsepower. Thus, the actual r/min is (1222)(1.3819) = 1690 r/min. Actual brake horsepower is (255.5)(1.3819) = 353.5 bhp (263.7 kW). This is the horsepower input required to drive the fan and is close to the 360.5 hp computed in the previous example. The actual motor horsepower would be the same in each case because a standard-size motor would be chosen. The difference in horsepower of 360.5 − 353.5 = 7.0 hp results from the same factors discussed in step 3.

Note: The static pressure is normally used in most fan-selection procedures because this is the pressure value used in computing pressure and draft losses in boilers, economizers, air heaters, and ducts. In any fan system, the total air pressure equals static pressure + velocity pressure. However, the velocity pressure at the fan discharge is not considered in draft calculations unless there are factors requiring its evaluation. These requirements are generally related to pressure losses in the fan-control devices.

6.32 *DETERMINATION OF THE MOST ECONOMICAL FAN CONTROL*

Determine the most economical fan control for a forced- or induced-draft fan designed to deliver 140,000 ft³/min (66.03 m³/s) at 14 inH₂O (3.5 kPa) at full load. Plot the power-consumption curve for each type of control device considered.

Calculation Procedure

1. *Determine the types of controls to consider.*
There are five types of controls used for forced- and induced-draft fans: (*a*) a damper in the duct with constant-speed fan drive, (*b*) two-speed fan driver, (*c*) inlet vanes or inlet louvres with a constant-speed fan drive, (*d*) multiple-step variable-speed fan drive, and (*e*) hydraulic or electric coupling with constant-speed driver giving wide control over fan speed.

2. *Evaluate each type of fan control.*
Tabulate the selection factors influencing the control decision as follows, using the control letters in step 1:

Control type	Control cost	Required power input	Advantages (A), and disadvantages (D)
a	Low	High	(A) Simplicity; (D) High power input
b	Moderate	Moderate	(A) Lower input power; (D) Higher cost
c	Low	Moderate	(A) Simplicity; (D) ID fan erosion
d	Moderate	Moderate	(D) Complex; also needs dampers
e	High	Low	(A) Simple; no dampers needed

3. Plot the control characteristics for the fans.
Draw the fan head-capacity curve for the airflow or gasflow range considered (Fig. 6.35). This plot shows the maximum capacity of 140,000 ft³/min and required static head of 14 inH₂O, point *P*.

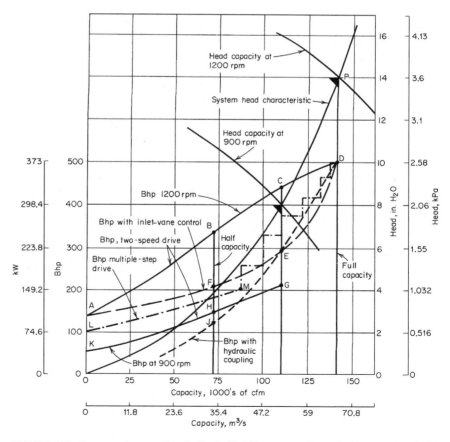

FIGURE 6.35 Power requirements for a fan fitted with different types of controls. (*American Standard Inc.*)

Plot the power-input curve *ABCD* for a constant-speed motor or turbine drive with damper control—type *a*, listed above—after obtaining from the fan manufacturer or damper builder the input power required at various static pressures and capacities. Plotting these values gives curve *ABCD*. Fan speed is 1200 r/min.

Plot the power-input curve *GHK* for a two-speed drive, type *b*. This drive might be a motor with an additional winding, or it might be a second motor for use at reduced boiler capacities. With either arrangement, the fan speed at lower boiler capacities is 900 r/min.

Plot the power-input curve *AFED* for inlet-vane control on the forced-draft fan or inlet-louvre control on induced-draft fans. The data for plotting this curve can be obtained from the fan manufacturer.

Multiple-step variable-speed fan control, type *d*, is best applied with steam-turbine drives. In a plant with ac auxiliary motor drives, slip-ring motors with damper integration must be used between steps, making the installation expensive. Although dc motor drives would be less costly, few power plants other than marine propulsion plants have direct current available. And since marine units normally operate at full load 90 percent or more of the time, part-load operating economics are unimportant. If steam-turbine drive will be used for the fans, plot the power-input curve *LMD*, using data from the fan manufacturer.

A hydraulic coupling or electric magnetic coupling, type *e*, with a constant-speed motor drive would have the power-input curve *DEJ*.

Study of the power-input curves shows that the hydraulic and electric couplings have the smallest power input. Their first cost, however, is usually greater than any other types of power-saving devices. To determine the return on any extra investment in power-saving devices, an economic study, including a load-duration analysis of the boiler load, must be made.

4. Compare the return on the extra investment.

Compute and tabulate the total cost of each type of control system. Then determine the extra investment for each of the more costly systems by subtracting the cost of type *a* from the cost of each of the other types. With the extra investment known, compute the lifetime savings in power input for each of the more efficient control methods. With the extra investment and savings resulting from it known, compute the percentage return on the extra investment. Tabulate the findings as in Table 6.32.

In Table 6.32, considering control type *c*, the extra cost of type *c* over type *b* is $75,000 − 50,000 = $25,000. The total power saving of $6500 is computed on the basis of the cost of energy in the plant for the life of the control. The return on the extra investment then is $6500/$25,000 = 0.26, or 26 percent. Type *e* control provides the highest percentage return on the extra investment. It would probably be chosen if the only measure of investment desirability is the return on the extra investment. However, if other criteria are used—such as a minimum rate of return on the extra investment—one of the other control types might be chosen. This is easily determined by studying the tabulation in conjunction with the investment requirement. For more on investment decisions, see Section 18.

Related Calculations.   The procedure used here can be applied to heating, power, marine, and portable boilers of all types. Follow the same steps given above, changing the values to suit the existing conditions. Work closely with the fan and drive manufacturer when analyzing drive power input and costs.

TABLE 6.32   Fan Control Comparison

	Type of control used				
	a	*b*	*c*	*d*	*e*
Total cost, $	30,000	50,000	75,000	89,500	98,000
Extra cost, $	—	20,000	25,000	14,500	8,500
Total power saving, $	—	8,000	6,500	3,000	6,300
Return on extra investment, %	—	40	26	20.7	74.2

6.33 VACUUM-PUMP SELECTION FOR HIGH-VACUUM SYSTEMS

Choose a mechanical vacuum pump for use in a laboratory fitted with a vacuum system having a total volume, including the piping, of 12,000 ft³ (340 m³). The operating pressure of the system is 0.10 torr, and the optimum pump-down time is 150 min. (*Note:* 1 torr = 1 mmHg.)

Calculation Procedure

1. Make a tentative choice of pump type.
Mechanical vacuum pumps of the reciprocating type are well suited for system pressures in the 0.0001- to 760-torr range. Hence, this type of pump will be considered first to see if it meets the desired pump-down time.

2. Obtain the pump characteristic curves.
Many manufacturers publish pump-down factor curves such as those in Fig. 6.36a and b. These curves are usually published as part of the engineering data for a given line of pumps. Obtain the curves from the manufacturers whose pumps are being considered.

3. Compute the pump-down time for the pumps being considered.
Three reciprocating pumps can serve this system: (a) a single-stage pump, (b) a compound or two-stage pump, or (c) a combination of a mechanical booster and a single-stage backing or roughing-down pump. Figure 6.36 gives the pump-down factor for each type of pump.

To use the pump-down factor, apply this relation: $t = V F/d$, where t is pump-down time, min; V is system volume, ft³; F is pump-down factor for the pump; d is pump displacement, ft³/min.

Thus, for a single-stage pump, Fig. 6.36a shows that $F = 10.8$ for a pressure of 0.10 torr. Assuming a pump displacement of 1000 ft³/min, $t = 12,000(10.8)/1000 = 129.6$ min; say 130 min.

For a compound pump, $F = 9.5$ from Fig. 6.36a. Hence, a compound pump having the same displacement, or 1000 ft³/min, will require $t = 12,000(9.5)/1000 = 114.0$ min.

With a combination arrangement, the backing or roughing pump, a 130-ft³/min unit, reduces the system pressure from atmospheric, 760 torr, to the economical transition pressure, 15 torr (Fig. 636b). Then the single-stage mechanical booster pump, a 1200-ft³/min unit, takes over and in combination with the backing pump reduces the pressure to the desired level, or 0.10 torr. During this part of the cycle, the unit operates as a two-stage pump. Hence the total pump-down time consists of the sum of the backing-pump and booster-pump times. The pump-down factors are, respectively, 4.2 for the backing pump at 15 torr and 6.9 for the booster pump at 0.10 torr. Hence the respective pump-down times are $t_1 = 12,000(4.2)/130 = 388$ min; $t_2 = 12,000(6.9)/1200 = 69$ min. The total time is thus $388 + 69 = 457$ min.

The pump-down time with the combination arrangement is greater than the optimum 150 min. Where a future lower operating pressure is anticipated, making the combination arrangement desirable, an additional large-capacity single-stage roughing pump can be used to assist the 130-ft³/min unit. This large-capacity unit is operated until the transition pressure is reached and roughing down is finished. The pump is then shut off and the balance of the pumping down is carried on by the combination unit. This keeps the power consumption at a minimum.

Thus, if a 1200-ft³/min single-stage roughing pump were used to reduce the pressure to 15 torr, its pump-down time would be $t = 12,000(4.0)/1200 = 40$ min. The total pump-down time for the combination would then be $40 + 69 = 109$ min, using the time computed above for the two pumps in combination.

4. Apply the respective system factors.
Studies and experience show that the calculated pump-down time for a vacuum system must be corrected by an appropriate system factor. This factor makes allowance for the normal outgassing of surfaces exposed to atmospheric air. It also provides a basis for judging whether a system is pumping

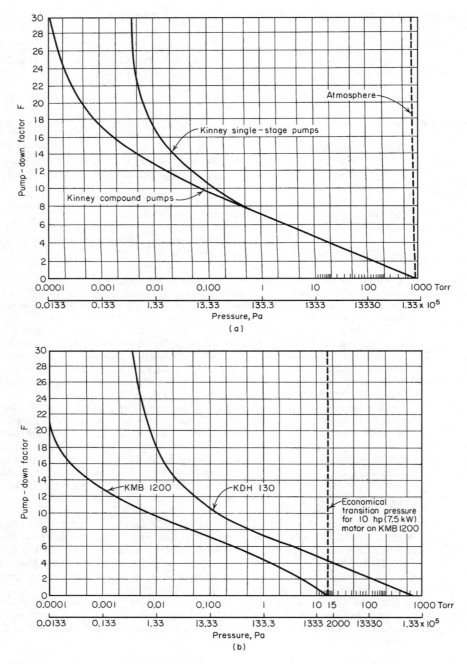

FIGURE 6.36 (*a*) Pump-down factor for single-stage and compound vacuum pumps; (*b*) pump-down factor for mechanical booster and backing pump. (*After Kinney Vacuum Division, The New York Air Brake Co., and Van Atta.*)

TABLE 6.33 Recommended System Factors

Pressure range		System factors		
torr	Pa	Single-stage mechanical pump	Compound mechanical pump	Mechanical booster pump[†]
760–20	115.6 kPa–3000	1.0	1.0	—
20–1	3000–150	1.1	1.1	1.15
1–0.5	150–76	1.25	1.25	1.15
0.5–0.1	76–15	1.5	1.25	1.35
0.1–0.02	15–3	—	1.25	1.35
0.02–0.001	3–0.15	—	—	2.0

[†]Based on bypass operation until the booster pump is put into operation. Larger system factors apply if rough pumping flow must pass through the idling mechanical booster. Any time needed for operating valves and getting the mechanical booster pump up to speed must also be added.

Source: From Van Atta—*Vacuum Science and Engineering,* McGraw-Hill, New York, 1965.

down normally or whether some problem exists that must be corrected. Table 6.33 lists typical system factors that have proven reliable in many tests. To use the system factor for any pump, apply it this way: $t_a = tS$, where t_a is actual pump-down time, in min; t is computed pump-down time from step 3, in min; S is system factor for the type of pump being considered.

Thus, using the appropriate system factor for each pump, the actual pump-down time for the single-stage mechanical pump is $t_a = 130(1.5) = 195$ min. For the compound mechanical pump, $t_a = 114(1.25) = 142.5$ min. For the combination mechanical booster pump, $t_a = 109(1.35) = 147$ min.

5. *Choose the pump to use.*
Based on the actual pump-down time, either the compound mechanical pump or the combination mechanical booster pump can be used. The final choice of the pump should take other factors into consideration—first cost, operating cost, maintenance cost, reliability, and probable future pressure requirements in the system. Where future lower pressure requirements are not expected, the compound mechanical pump would be a good choice. However, if lower operating pressures are anticipated in the future, the combination mechanical booster pump would probably be a better choice.

Van Atta [4] gives the following typical examples of pumps chosen for vacuum systems:

Pressure range, torr	Typical pump choice
Down to 50 (7.6 kPa)	Single-stage oil-sealed rotary; large water or vapor load may require use of refrigerated traps
0.05 to 0.01 (7.6 to 1.5 Pa)	Single-stage or compound oil-sealed pump plus refrigerated traps, particularly at the lower pressure limit
0.01 to 0.005 (1.5 to 0.76 Pa)	Compound oil-sealed plus refrigerated traps, or single-stage pumps backing diffusion pumps if a continuous large evolution of gas is expected
1 to 0.0001 (152.1 to 0.015 Pa)	Mechanical booster and backing pump combination with interstage refrigerated condenser and cooled vapor trap at the high-vacuum inlet for extreme freedom from vapor contamination
0.0005 and lower (0.076 Pa and lower)	Single-stage pumps backing diffusion pumps, with refrigerated traps on the high-vacuum side of the diffusion pumps and possibly between the single-stage and diffusion pumps if evolution of condensable vapor is expected

6.34 *VACUUM-SYSTEM PUMPING SPEED AND PIPE SIZE*

A laboratory vacuum system has a volume of 500 ft^3 (14 m^3). Leakage into the system is expected at the rate of 0.00035 ft^3/min. What backing pump speed, i.e., displacement, should an oil-sealed vacuum pump serving this system have if the pump blocking pressure is 0.150 mmHg and the desired operating pressure is 0.0002 mmHg? What should be the speed of the diffusion pump? What pump size is needed for the connecting pipe of the backing pump if it has a displacement or pumping speed of 388 ft^3/min (0.18 m^3/s) at 0.150 mmHg and a length of 15 ft (4.6 m)?

Calculation Procedure

1. *Compute the required backing pump speed.*
Use the relation $d_b = G/P_b$, where d_b is backing pump speed or pump displacement, in ft^3/min; G is gas leakage or flow rate, in mm/(ft^3/min). To convert the gas or leakage flow rate to mm/(ft^3/min), multiply the ft^3/min by 760 mm, the standard atmospheric pressure, in mmHg. Thus, $d_b = 760(0.00035)/0.150 = 1.775$ ft^3/min.

2. *Select the actual backing pump speed.*
For practical purposes, since gas leakage and outgassing are impossible to calculate accurately, a backing pump speed or displacement of at least twice the computed value, or $2(1.775) = 3.550$ ft^3/min—say 4 ft^3/min (0.002m^3/s)—would probably be used.

   If this backing pump is to be used for pumping down the system, compute the pump-down time as shown in the previous example. Should the pump-down time be excessive, increase the pump displacement until a suitable pump-down time is obtained.

3. *Compute the diffusion pump speed.*
The diffusion pump reduces the system pressure from the blocking point, 0.150 mmHg, to the system operating pressure of 0.0002 mmHg. (*Note:* 1 torr = 1 mmHg.) Compute the diffusion pump speed from $d_d = G/P_d$, where d is diffusion pump speed, in ft^3/min; P_d is diffusion-pump operating pressure, mmHg. Or $d_d = 760(0.00035)/0.0002 = 1330$ ft^3/min (0.627 m^3/s). To allow for excessive leaks, outgassing, and manifold pressure loss, a 3000- or 4000-ft^3/min diffusion pump would be chosen. To ensure reliability of service, two diffusion pumps would be chosen so that one could operate while the other was being overhauled.

4. *Compute the size of the connection pipe.*
In usual vacuum-pump practice, the pressure drop in pipes serving mechanical pumps is not allowed to exceed 20 percent of the inlet pressure prevailing under steady operating conditions. A correctly designed vacuum system, where this pressure loss is not exceeded, will have a pump-down time which closely approximates that obtained under ideal conditions.

   Compute the pressure drop in the high-pressure region of vacuum pumps from $p_d = 1.9d_b L/d^4$, where p_d is pipe pressure drop, in μ; d_b is backing pump displacement or speed, in ft^3/min; L is pipe length, in ft; d is inside diameter of pipe, in in. Since the pressure drop should not exceed 20 percent of the inlet or system operating pressure, the drop for a backing pump is based on its blocking pressure, or 0.150 mmHg, or 150 μ. Hence $p_d = 0.20(150) = 30 \mu$. Then, $30 = 1.9(380)(15)/d^4$, and $d = 4.35$ in (0.110 m). Use a 5-in-diameter pipe.

   In the low-pressure region, the diameter of the converting pipe should equal, or be larger than, the pump inlet connection. Whenever the size of a pump is increased, the diameter of the pipe should also be increased to conform with the above guide.

Related Calculations. Use the general procedures given here for laboratory- and production-type high-vacuum systems.

6.35 *BULK-MATERIAL ELEVATOR AND CONVEYOR SELECTION*

Choose a bucket elevator to handle 150 tons/h (136.1 tonnes/h) of abrasive material weighing 50 lb/ft^3 (800.5 kg/m^3) through a vertical distance of 75 ft (22.9 m) at a speed of 100 ft/min (30.5 m/min). What horsepower input is required to drive the elevator? The bucket elevator discharges onto a horizontal conveyor which must transport the material 1400 ft (426.7 m). Choose the type of conveyor to use and determine the required power input needed to drive it.

Calculation Procedure

1. *Select the type of elevator to use.*
Table 6.34 summarizes the various characteristics of bucket elevators used to transport bulk materials vertically. This table shows that a continuous bucket elevator would be a good choice, because it is a recommended type for abrasive materials. The second choice would be a pivoted bucket elevator. However, the continuous bucket type is popular and will be chosen for this application.

2. *Compute the elevator height.*
To allow for satisfactory loading of the bulk material, the elevator length is usually increased by about 5 ft (1.5 m) more than the vertical lift. Hence the elevator height is 75 + 5 = 80 ft (24.4 m).

3. *Compute the required power input to the elevator.*
Use the relation $hp = 2CH/1000$, where C is elevator capacity, in tons/h; H is elevator height, in ft. Thus, for this elevator, $hp = 2(150)(80)/1000 = 24.9$ hp (17.9 kW).

The power input relation given above is valid for continuous bucket, centrifugal-discharge, perfect- discharge, and supercapacity elevators. A 25-hp (18.7-kW) motor would probably be chosen for this elevator.

4. *Select the type of conveyor to use.*
Since the elevator discharges onto the conveyor, the capacity of the conveyor should be the same, per unit time, as the elevator. Table 6.35 lists the characteristics of various types of conveyors. Study of the tabulation shows that a belt conveyor would probably be best for this application, based on the speed, capacity, and type of material it can handle, hence, it will be chosen for this installation.

TABLE 6.34 Bucket Elevators

	Centrifugal discharge	Perfect discharge	Continuous bucket	Gravity discharge	Pivoted bucket
Carrying paths	Vertical	Vertical to inclination 15° from vertical	Vertical to inclination 15° from vertical	Vertical and horizontal	Vertical and horizontal
Capacity range, tons/h (tonnes/h), material weighing 50 lb/ft^3 (800.5 kg/m^3)	78 (70.8)	34 (30.8)	345(312.9)	191 (173.3)	255 (231.3)
Speed range, ft/min (m/min)	306 (93.3)	120 (36.6)	100 (30.5)	100 (30.5)	80 (24.4)
Location of loading point	Boot	Boot	Boot	On lower horizontal run	On lower horizontal run
Location of discharge point	Over head wheel	Over head wheel	Over head wheel	On horizontal run	On horizontal run
Handling abrasive materials	Not preferred	Not preferred	Recommended	Not recommended	Recommended

Source: Link-Belt Div. of FMC Corp.

TABLE 6.35 Conveyor Characteristics

	Belt conveyor	Apron conveyor	Flight conveyor	Drag chain	En masse conveyor	Screw conveyor	Vibratory conveyor
Carrying paths	Horizontal to 18°	Horizontal to 25°	Horizontal to 45°	Horizontal or slight incline, 10°	Horizontal to 90°	Horizontal to 15°; may be used up to 90° but capacity falls off rapidly	Horizontal or slight incline, 5° above or below horizontal
Capacity range, tons/h (tonnes/h) material weighing 50 lb/ft^3	2160 (1959.5)	100 (90.7)	360 (326.6)	20 (18.1)	100 (90.7)	150 (136.1)	100 (90.7)
Speed range, ft/min	600 (182.9 m/min)	100 (30.5 m/min)	150 (45.7 m/min)	20 (6.1 m/min)	80 (24.4 m/min)	100 (30.5 m/min)	40 (12.2 m/min)
Location of loading point	Any point	Any point	Any point	Any point	On horizontal runs	Any point	Any point
Location of discharge point	Over end wheel and intermediate points by tripper or plow	Over end wheel	At end of trough and intermediate points by gates	At end of trough	Any point on horizontal runs by gate	At end of trough and intermediate points by gates	At end of trough
Handling abrasive materials	Recommended	Recommended	Not recommended	Recommended with special steels	Not recommended	Not preferred	Recommended

Source: Link-Belt Div. of FMC Corp.

5. *Compute the required power input to the conveyor.*
The power input to a conveyor is composed of two portions: (*a*) the power required to move the empty belt conveyor, and (*b*) the power required to move the load horizontally.

Determine from Fig. 6.37 the power required to move the empty belt conveyor, after choosing the required belt width. Determine the belt width from Table 6.36.

Thus, for this conveyor, Table 6.36 shows that a belt width of 42 in (106.7 cm) is required to transport up to 150 tons/h (136.1 tonnes/h) at a belt speed of 100 ft/min (30.5 m/min). [Note that the next *larger* capacity, 162 tons/h (146.9 tonnes/h), is used when the exact capacity required is not tabulated.] Find the horsepower required to drive the empty belt by entering Fig. 6.37 at the belt distance between centers, 1400 ft (426.7 m), and projecting vertically upward to the belt width, 42 in (106.7 cm). At the left, read the required power input as 7.2 hp (5.4 kW).

Compute the power required to move the load horizontally from $hp = (C/100)(0.4 + 0.00345 \, L)$, where L is distance between conveyor centers, in ft; other symbols as before. For this conveyor, $hp = (150/100)(0.4 + 0.00325 \times 1400) = 6.83$ hp (5.1 kW). Hence the total horsepower to drive this horizontal conveyor is $7.2 + 6.83 = 14.03$ hp (10.5 kW).

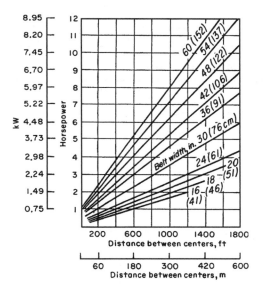

FIGURE 6.37 Power required to move an empty conveyor belt at 100 ft/min (0.508 m/s).

TABLE 6.36 Capacities of Troughed Rest [tons/h (tonnes/h) with belt speed of 100 ft/min (30.5 m/min)]

Belt width, in (cm)	Weight of material, lb/ft³ (kg/m³)			
	30 (480.3)	50 (800.5)	100 (1601)	150 (2402)
30 (9.1)	47 (42.6)	79 (71.7)	158 (143.3)	237 (214.9)
36 (10.9)	69 (62.6)	114 (103.4)	228 (206.8)	342 (310.2)
42 (12.8)	97 (87.9)	162 (146.9)	324 (293.9)	486 (440.9)
48 (14.6)	130 (117.9)	215 (195.0)	430 (390.1)	645 (585.1)
60 (18.3)	207 (187.8)	345 (312.9)	690 (625.9)	1035 (938.9)

Source: United States Rubber Co.

The total horsepower input to this conveyor installation is the sum of the elevator and conveyor belt horsepowers, or 14.03 + 24.0 = 38.03 hp (28.4 kW).

Related Calculations. This procedure is valid for conveyors using rubber belts reinforced with cotton duck, open-mesh fabric, cords, or steel wires. It is also valid for stitched-canvas belts, balata belts, and flat-steel belts. The required horsepower input includes any power absorbed by idler pulleys.

Table 6.37 shows the minimum recommended belt widths for lumpy materials of various sizes. Maximum recommended belt speeds for various materials are shown in Table 6.38.

When a conveyor belt is equipped with a tripper, the belt must rise about 5 ft (1.5 m) above its horizontal plane of travel.

This rise must be included in the vertical lift power input computation. When the tripper is driven by the belt, allow 1 hp (0.75 kW) for a 16-in (406.4-mm) belt, 3 hp (2.2 kW) for a 36-in (914.4-mm) belt, and 7 hp (5.2 kW) for a 60-in (1524-mm) belt. Where a rotary cleaning brush is driven by the conveyor shaft, allow about the same power input to the brush for belts of various widths.

TABLE 6.37 Minimum Belt Width for Lumps

Belt width, in (mm)	24 (609.6)	36 (914.4)	42 (1066.8)	48 (1219.2)
Sized materials, in (mm)	4½ (114.3)	8 (203.2)	10 (254)	12 (304.9)
Unsized material, in (mm)	8 (203.2)	14 (355.6)	20 (508)	35 (889)

TABLE 6.38 Maximum Belt Speeds for Various Materials

Width of belt		Light or free-flowing materials, grains dry sand, etc.		Moderately free-flowing sand, gravel, fine stone, etc.		Lump coal, coarse stone, crushed ore		Heavy sharp lumpy materials, heavy ores, lump coke	
in	mm	ft/min	m/min	ft/min	m/min	ft/min	m/min	ft/min	m/min
12–14	305–356	400	122	250	76	—	—	—	—
16–18	406–457	500	152	300	91	250	76	—	—
20–24	508–610	600	183	400	122	350	107	250	76
30–36	762–914	750	229	500	152	400	122	300	91

6.36 SCREW-CONVEYOR POWER INPUT AND CAPACITY

What is the required input for a 100-ft (30.5-m) long screw conveyor handling dry coal ashes having a maximum density of 40 lb/ft³ if the conveyor capacity is 30 tons/h (27.2 tonnes/h)?

Calculation Procedure

1. *Select the conveyor diameter and speed.*
Refer to a manufacturer's engineering data or Table 6.39 for a listing of recommended screw-conveyor diameters and speeds for various types of materials. Dry coal ashes are commonly rated as group 3 materials (Table 6.40)—i.e., materials with small mixed lumps with fines.

To determine a suitable screw diameter, assume two typical values and obtain the recommended r/min from the sources listed above or Table 6.39. Thus the maximum r/min recommended for a 6-in (152.4-mm) screw when handling group 3 material is 90, as shown in Table 6.39; for a 20-in (508.0-mm) screw, 60 r/min. Assume a 6-in (152.4-mm) screw as a trial diameter.

2. *Determine the material factor for the conveyor.*
A material factor is used in the screw conveyor power input computation to allow for the character of the substance handled. Table 6.40 lists the material factor for dry ashes as $F = 4.0$. Standard references show that the average weight of dry coal ashes is 35 to 40 lb/ft³ (640.4 kg/m³).

TABLE 6.39   Screw-Conveyor Capacities and Speeds

Material group	Max material density, lb/ft³ (kg/m³)	Max r/min for diameters of	
		6 in (152 mm)	20 in (508 mm)
1	50 (801)	170	110
2	50 (801)	120	75
3	75 (1201)	90	60
4	100 (1601)	70	50
5	125 (2001)	30	25

TABLE 6.40   Material Factors for Screw Conveyors

Material group	Material type	Material factor
1	Lightweight:	
	Barley, beans, flour, oats, pulverized coal, etc.	0.5
2	Fines and granular:	
	Coal—slack or fines	0.9
	Sawdust, soda ash	0.7
	Flyash	0.4
3	Small lumps and fines:	
	Ashes, dry alum	4.0
	Salt	1.4
4	Semiabrasives; small lumps:	
	Phosphate, cement	1.4
	Clay, limestone	2.0
	Sugar, white lead	1.0
5	Abrasive lumps:	
	Wet ashes	5.0
	Sewage sludge	6.0
	Flue dust	4.0

3. *Determine the conveyor size factor.*

A size factor that is a function of the conveyor diameter is also used in the power input computation. Table 6.41 shows that for a 6-in diameter conveyor the size factor $A = 54$.

4. *Compute the required power input to the conveyor.*

Use the relation $hp = 10^{-6}(ALN + CWLF)$, where hp is hp input to the screw conveyor head shaft; A is size factor from step 3; L is conveyor length, in ft; N is conveyor r/min; C is quantity of material handled, in ft³/h; W is density of material, in lb/ft³; F is material factor from step 2. For this conveyor, using the data listed above, $hp = 10^{-6}(54 \times 100 \times 60 + 1500 \times 40 \times 100 \times 4.0) = 24.3$ hp (18.1 kW). With a 90 percent motor efficiency, the required motor rating would be 24.3/0.90 = 27 hp (20.1 kW). A 30-hp (22.4-kW) motor would be chosen to drive this conveyor. Since this is not an excessive power input, the 6-in (152.4-mm) conveyor is suitable for this application.

TABLE 6.41   Screw Conveyor Size Factors

Conveyor diameter, in (mm)	Size factor	Conveyor diameter, in (mm)	Size factor
6 (152.4)	54	16 (406.4)	336
9 (228.6)	96	18 (457.2)	414
10 (254)	114	20 (508)	510
12 (304.8)	171	24 (609.6)	690

If the calculation indicates that an excessively large power input—say 50 hp (37.3 kW) or more—is required, the larger-diameter conveyor should be analyzed. In general, a higher initial investment in conveyor size that reduces the power input will be more than recovered by the savings in power costs.

Related Calculations.   Use this procedure for screw or spiral conveyors and feeders handling any material that will flow. The usual screw or spiral conveyor is suitable for conveying materials for distances up to about 200 ft (60 m), although special designs can be built for greater distances. Conveyors of this type can be sloped upward to angles of 35° with the horizontal. However, the capacity of the conveyor decreases as the angle of inclination is increased. Thus the reduction in capacity at a 10° inclination is 10 percent over the horizontal capacity; at 35° the reduction is 78 percent.

The capacities of screw and spiral conveyors are generally stated in ft^3/h of various classes of materials at the maximum recommended shaft r/min. As the size of the lumps in the material conveyed increases, the recommended shaft r/min decreases. The capacity of a screw or spiral conveyor at a lower speed is found from (capacity at given speed, in ft^3/h)(lower speed, r/min/higher speed, r/min). Table 6.39 shows typical screw conveyor capacities at usual operating speeds.

Various types of screws are used for modern conveyors. These include short-pitch, variable-pitch, cut flights, ribbon, and paddle screws. The procedure given above also applies to these screws.

*6.37   PUMP SELECTION FOR CHEMICAL PLANTS

Choose a pump to handle 26,000 gal/min (1640 L/s) of water at 60°F (15.6°C) in a chemical plant when the total dynamic head is 37 ft (11.3 m) of water. What is the required hp input to the pump if the pump efficiency is 85 percent? What type of pump should be used if the rotational speed is limited to 880 r/min?

Calculation Procedure

1. *Determine the required power input to the pump.*
A quick way to determine the power input to a pump handling water at normal atmospheric temperatures is to use Fig. 6.38. Enter on the left at the total dynamic head, 37 ft (11.3 m), and project to the right to the required pump capacity, 26,000 gal/min (1640 L/s). At the intersection with the hp stem, read the required power input as 285 hp (212.6 kW).

2. *Select the type of pump to use.*
From the rotational speed, 880 r/min on the bottom stem, draw a straight line at right angles to the first construction line, as shown. At the intersection with the top stem, read the type of pump as a propeller pump having a specific speed of 9500 r/min.

Related Calculations.   Note that this pump application chart applies to rotating-type centrifugal pumps. Where a reciprocating pump is desired, use the methods given in this section. The chart in Fig. 6.38 was developed by H. W. Hamm and was first presented in *Power* magazine.

Pending environmental regulations will strictly limit pump leakage in chemical and process plants of all kinds. Today's laws require plant operators to report leakage of toxic substances of 0.0001 percent of the pump's capacity.

There are both national (EPA) and state laws controlling pump leakage. For example, the state of New Jersey has a Toxic Catastrophe Prevention Act (TCPA) which strictly controls pump leakage. This and similar state environmental laws controlling pump seal leakage of toxic materials will probably become stricter in the future. For this reason, careful selection of pump shaft seals is important to every engineer working with toxic materials.

Typical toxic materials whose leakage must be prevented from pumps are sulfuric and nitric acid. Where water-flushed seals are used to contain leakage of such materials, the acidic flush water must be treated before disposal. Leakage of toxic materials must be prevented both while a pump is operating

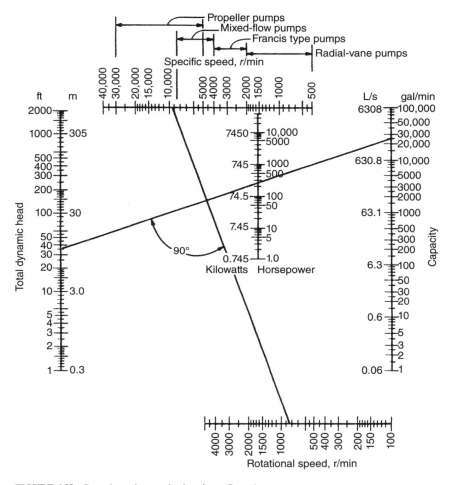

FIGURE 6.38 Pump hp and type selection chart. (*Power.*)

and while it is idle. Often, a seal that prevents leakage while the pump is operating will allow leakage when the pump is shut down. Such leakage is just as unlawful as leakage while the pump is operating.

For these reasons, engineers must carefully specify leak-free seals when choosing pumps handling toxic materials. The best seals can only be chosen after thorough study and consultation with both the pump and seal manufacturers.

*6.38 DETERMINING THE FRICTION FACTOR FOR FLOW OF BINGHAM PLASTICS

A coal slurry is being pumped through a 0.4413-m (18-in) diameter schedule 20 pipeline at a flow rate of 400 m³/h. The slurry behaves as a Bingham plastic, with the following properties (at the relevant temperature): $\tau_0 = 2$ N/m² (0.0418 lbf/ft²); $\mu_\infty = 0.03$ Pa·s (30 cP); $\rho = 1500$ kg/m³ (93.6 lbm/ft³). What is the Fanning friction factor for this system?

Calculation Procedure

1. *Determine the Bingham Reynolds number and the Hedstrom number.*
Engineers today often must size pipe or estimate pressure drops for fluids that are nonnewtonian in nature—coal suspensions, latex paint, or printer's ink, for example. This procedure shows how to find the friction factors needed in such calculations for the many fluids that can be described by the Bingham-plastic flow mode. The method is convenient to use and applies to all regimes of pipe flow.

A Bingham plastic is a fluid that exhibits a yield stress; that is, the fluid at rest will not flow unless some minimum stress τ_0 is applied. Newtonian fluids, in contrast, exhibit no yield stress, as Fig. 6.39 shows.

The Bingham-plastic flow model can be expressed in terms of either shear stress τ versus shear rate $\dot{\gamma}$, as in Fig. 6.39, or apparent viscosity η versus shear rate:

$$\tau = \tau_0 + \mu_\infty \dot{\gamma} \tag{6.1}$$

$$\eta = \frac{\tau}{\dot{\gamma}} = \frac{\tau_0}{\dot{\gamma}} + \mu_\infty \tag{6.2}$$

Equation 6.2 means that the apparent viscosity of a Bingham plastic depends on the shear rate. The parameter μ_∞ is sometimes called the coefficient of rigidity, but it is really a limiting viscosity. As Eq. 6.2 shows, apparent viscosity approaches μ_∞ as shear rate increases indefinitely. Thus, the Bingham plastic behaves almost like a newtonian at sufficiently high shear rates, exhibiting a viscosity of μ_∞ at such conditions. Table 6.42 shows values of τ_0 and μ_∞ for several actual fluids.

For any incompressible fluid flowing through a pipe, the friction loss per unit mass F can be expressed in terms of a Fanning friction factor f:

$$F = \frac{2fLv^2}{D} \tag{6.3}$$

where L is the lengths of the pipe section, D is its diameter, and v is the fluid velocity.

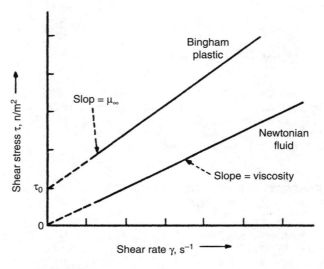

FIGURE 6.39   Bingham plastics exhibit a yield stress. (*Chemical Engineering.*)

TABLE 6.42   Values of τ_0 and μ_∞

Fluid	τ_0, N/m^2	μ_∞, Pa·s	Ref.
Blood (45% hematocrit)	0.005	0.0028	[10]
Printing-ink pigment in varnish (10% by wt.)	0.4	0.25	[11]
Coal suspension in methanol (35% by vol.)	1.6	0.04	[12]
Finely divided galena in water (37% by vol.)	4.0	0.057	[13]
Molten chocolate (100°F)	20	2.0	[14]
Thorium oxide in water (50% by vol.)	300	0.403	[15]

An exact description of friction loss for Bingham plastics in fully developed laminar pipe flow was first published by Buckingham [5]. His expression can be rewritten in dimensionless form as follows:

$$f_L = \frac{16}{N_{Re}} \left(1 + \frac{N_{He}}{6N_{Re}} - \frac{N_{He}^4}{3f_L^3 N_{Re}^7} \right) \tag{6.4}$$

where N_{Re} is the Bingham Reynolds number (Dvp/μ_∞) and N_{He} is the Hedstrom number ($D^2 p\tau_0/\mu^2$). Equation 6.4 is implicit in f_L, the laminar friction factor, but can be readily solved either by Newton's method or by iteration. Since the last term in Eq. 6.4 is normally small, the value of f obtained by omitting this term is usually a good starting point for iterative solution.

For this pipeline

$$N_{Re} = \frac{4Q\rho}{\pi D\mu_\infty} = \frac{4(400)(1/3)(600)(1500)}{\pi(0.4413)(0.03)} = 16{,}030$$

$$N_{He} = \frac{D^2 \rho\tau_0}{\mu_\infty^2} = \frac{(0.4413)^2(1500)(2)}{(0.03)^2} = 649{,}200$$

2. Find the friction factor f_L for the laminar-flow regime.
Substituting the values for N_{Re} and N_{He} into Eq. 6.4, we find $f_L = 0.007138$.

3. Determine the friction factor f_T for the turbulent-flow regime.
Equation 6.4 describes the laminar-flow sections. An empirical expression that fits the turbulent-flow regime is

$$f_T = 10^a N_{Re}^{-0.193} \tag{6.5}$$

where

$$a = -1.378 \, [1 + 0.146 \, \exp(-2.9 \times 10^{-5} N_{He})] \tag{6.6}$$

We now have friction-factor expressions for both laminar and turbulent flow. Equation 6.6 does not apply when N_{He} is less than 1000, but this is not a practical constraint for most Bingham plastics with a measurable yield stress.

When N_{He} is above 300,000, the exponential term in Eq. 6.6 is essentially zero. Thus, $a = -1{,}378$ here, and Eq. 6.5 becomes

$$f_T = 10^{-1.378}(16{,}030)^{-0.193}$$

$$= 0.006463$$

4. *Find the friction factor f.*
Combine the f_L and f_T expressions to get a single friction factor valid for all flow regimes:

$$f = (f_L^m + f_T^m)\frac{1}{m} \tag{6.7}$$

where f_L and f_T are obtained from Eqs. 6.4 and 6.5, respectively, and the power m depends on the Bingham Reynolds number:

$$m = 1.7 + \frac{40,000}{N_{\text{Re}}} \tag{6.8}$$

The values of f predicted by Eq. 6.7 coincide with Hanks's values in most places, and the general agreement is excellent. Relative roughness is not a parameter in any of the equations because the friction factor for nonnewtonian fluids, and particularly plastics, is not sensitive to pipe roughness.
   Substituting yields $m = 1.7 + 40,000/16,030 = 4.20$, and $f = [(0.007138)^{4.20} + (0.006463)^{4.20}]^{1/4.20} = 0.00805$.
   If m had been very large, the bracketed term above would have approached zero. Generally, when N_{Re} is below 4000, Eq. 6.8 should be solved by taking f equal to the greater of f_L and f_T.

Related Calculations.   This procedure is valid for a variety of fluids met in many different industrial and commercial applications. The procedure is the work of Ron Darby, Professor of Chemical Engineering. Texas A & M University, College of Engineering, and Jeff Melson, Undergraduate Fellow, Texas A & M, as reported in *Chemical Engineering* magazine. In their report they cite works by Hanks and Pratt [6], Hanks and Dadia [7], Churchill [8], and Churchill and Usagi [9] as important in the procedure described and presented here.

REFERENCES

1. Hicks—*Standard Handbook of Engineering Calculations*, McGraw-Hill, New York.
2. Horowitz—"Affinity Laws and Specific Speed Can Simplify Centrifugal Pump Selection." *Power,* November 1964.
3. *Chemical Engineering,* March 1998, pp. 129ff.
4. Van Atta—*Vacuum Science and Engineering*, McGraw-Hill, New York, 1965.
5. Buckingham—"On Plastic Flow through Capillary Tubes," *ASTM Proc. 21*: 1154, 1921.
6. Hanks and Pratt—"On the Flow of Bingham Plastic Slurries in Pipes and Between Parallel Plates," *Soc. Petrol. Eng. J. 1*:342, 1967.
7. Hanks and Dadia—"Theoretical Analysis of the Turbulent Flow of Non-Newtonian Slurries in Pipes," *AIChE Journal 17*:554, 1971.
8. Churchill—"Friction-Factor Equation Spans All Fluid-Flow Regimes," *Chem. Eng.*, 91–92, Nov. 7, 1977.
9. Churchill and Usagi—"A General Expression for the Correlation of Rates of Transfer and Other Phenomena," *AIChE Journal 18(6)*: 1121–1128, 1972.
10. Whitmore—*Rheology of the Circulation,* Pergamon Press, Oxford, 1968.
11. Casson—"A Flow Equation for Pigment-Oil Dispersions of the Printing Ink Type," Ch. 5 in *Rheology of Disperse Systems*, C. C. Mill (ed.), Perganion Press, Oxford, 1959.
12. Darby and Rogers—"Non-Newtonian Viscous Properties of Methacoal Suspensions," *AIChE Journal 26*:310, 1980.
13. Govier and Aziz—*The Flow of Complex Mixtures in Pipes*, Van Nostrand Reinhold, New York, 1972.
14. Steiner—The Rheology of Molten Chocolate. Ch. 9 in C. C. Mill (ed.), *op. cit.*
15. Bird, Stewart, and Lightfoot—*Transport Phenomena*, John Wiley & Sons, New York, 1960.

SECTION 7
HEAT TRANSFER[†]

Paul E. Minton
Principal Engineer
Union Carbide Corporation
South Charleston, WV

Edward S. S. Morrison
Senior Staff Engineer
Union Carbide Corporation
Houston, TX

[†]Procedure 7.36. Heat Exchanger Networking, has been adapted from Smith, *Chemical Process Design*, McGraw-Hill.

7.1 SPECIFYING RADIATION SHIELDING FOR CHEMICAL-PLANT EQUIPMENT

A furnace is to be located next to a dense complex of cryogenic propane piping. To protect the cold equipment from excessive heat loads, reflective aluminum radiation shielding sheets are to be placed between the piping and the 400°F (477 K) furnace wall. The space between the furnace wall and the cold surface is 2 ft (0.61 m). The facing surfaces of the furnace and the piping array are each 25 × 40 ft (7.6 × 12.2 m). With the ice-covered surface of the cold equipment at an average temperature of 35°F (275 K), how many 25 × 40 ft aluminum sheets must be installed between the two faces to keep the last sheet at or below 90°F (305 K)? Emittances of the furnace wall and cryogenic equipment are 0.90 and 0.65, respectively, and that of the aluminum shields is 0.1.

Calculation Procedure

1. Analyze the arrangement to assess the type(s) of heat transfer involved.
The distance separating the hot and cold surfaces is small compared with the size of the surfaces. The approximation can thus be made that the furnace wall, the dense network of cryogenic piping, and the radiation shields are all infinitely extended parallel planes. This is a conservative assumption, since the effect of proximity to an edge is to introduce a source of moderate temperature, thus allowing the hot wall to cool off. Convection is omitted with the same justification. So the problem can be treated as pure radiation.

Radiant heat transfer between two parallel, infinite plates is given by

$$\frac{q_{1-2}}{A} = \frac{\sigma(T_1^4 - T_2^4)}{(1/\epsilon_1) + (1/\epsilon_2) - 1}$$

where q_{1-2}/A is the heat flux between hotter surface 1 and colder surface 2, σ is the Stefan-Boltzmann constant, 0.1713×10^{-8} Btu/(h)(ft^2)(°R^4), T_1 and T_2 are absolute temperatures of the hotter and colder surfaces, and ϵ_1 and ϵ_2 are the emittances of the two surfaces. Emittance is the ratio of radiant energy emitted by a given real surface to the radiant energy that would be emitted by a theoretical, perfectly radiating black surface. Emittances of several materials are given in Table 7.1. The variations indicated between values of ϵ for apparently similar surfaces are not unusual, and they indicate the advisability of using measured data whenever available.

TABLE 7.1   Emittances of Some Real Surfaces

Material	Theoretical emittance ϵ_{th}
Aluminum foil, bright, foil, at 700°F (644 K)	0.04
Aluminum alloy 6061T6, H_2CrO_4 anodized, 300°F (422 K)	0.17
Aluminum alloy 6061T6, H_2SO_4 anodized, 300°F (422 K)	0.80
Cast iron, polished, at 392°F (473 K)	0.21
Cast iron, oxidized, at 390°F (472 K)	0.64
Black lacquer on iron at 76°F (298 K)	0.88
White enamel on iron at 66°F (292 K)	0.90
Lampblack on iron at 68°F (293 K)	0.97
Firebrick at 1832°F(1273 K)	0.75
Roofing paper at 69°F (294 K)	0.91

Source: Chapman [7].

Note that the numerator expresses a potential and the denominator a resistance. If a series of n radiation shields with emittance ϵ_s is interspersed between the two infinite parallel plates, the equivalent resistance can be shown to be

$$\left[\frac{1}{\epsilon_1} + \frac{1}{\epsilon_2} - 1\right] + n\left[\frac{2}{\epsilon_s} - 1\right]$$

2. Set up an equation for heat flux from the furnace to the cold equipment, expressed in terms of n.
Substituting into the equation in step 1,

$$\frac{q}{A} = \frac{(0.1713 \times 10^{-8})[(400 + 460)^4 - (35 + 460)^4]}{[(1/0.9) + (1/0.65) - 1] + n[(2/0.1) - 1]} = 834.19/(1.65 + 19n)$$

3. Set up an equation for heat flux from the furnace to the last shield, expressed in terms of n.
In this case, the heat flows through $n - 1$ shields en route to the last shield, whose temperature is to be kept at or below 90°F and whose emittance (like that of the other shields) is 0.1. Substituting again into the equation in step 1,

$$\frac{q}{A} = \frac{(0.1713 \times 10^{-8})[(400 + 460)^4 - (90 + 460)^4]}{[(1/0.9) + (1/0.1) - 1] + (n - 1)[(2/0.1) - 1]}$$

$$= \frac{780.27}{10.11 + (n - 1)(19)}$$

$$= \frac{780.27}{(-8.89 + 19n)}$$

4. Solve for n.
Since all the heat must pass through all shields, the expressions of steps 2 and 3 are equal. Setting them equal to each other and solving for n, it is found to be 8.499. Rounding off, we specify 9 shields.

5. Check the answer by back-calculating the temperature of the last shield.
Substituting 9 for n in the equation in step 2, the heat transferred is

$$\frac{q}{A} = \frac{(0.1713 \times 10^{-8})(860^4 - 435^4)}{[(1/0.9) + 1/(0.65) - 1] + 9[(2/0.1) - 1]}$$

$$= 4.83 \text{ Btu/(h)(ft}^2)$$

Next, let T_2 in the equation in step 3, that is, $(90 + 460)$, become an unknown, while substituting 9 for n and 4.83 for q/A. Thus, $4.83 = (0.1713 \times 10^8)(860^4 - T^4)/\{[(1/0.9) + (1/0.1) - 1] + [9 - 1] [(2/0.1) - 1]\}$. Solving for T_2, we find it to be 547.6°R (or 87.6°F), thus satisfying the requirement of the problem.

7.2   EFFECT OF SOLAR HEAT ON A STORAGE TANK

A flat-topped, nitrogen-blanketed atmospheric-pressure tank in a plant at Texas City, Texas, has a diameter of 30 ft and a height of 20 ft (9.1 m diameter and 6.1 m high) and is half full of ethanol at 85°F (302 K). As a first step in calculating nitrogen flow rates into and out of the tank during operations, calculate the solar heating of the tank and the tank skin temperature in the ullage space at a maximum-temperature condition. The tank has a coating of white zinc oxide paint, whose solar absorptance is 0.18. The latitude of Texas City is about N29°20′. For the maximum-temperature condition, select noon on June 20, the summer solstice, when the solar declination is 23.5°. Assume that the solar constant (the solar flux on a surface perpendicular to the solar vector) is 343 Btu/(h) (ft²) (1080 W/m²), the air temperature is 90°F (305 K), and the effective sky temperature is 5°F (258 K). Also assume that surrounding structural and other elements (such as hot pipes) are at 105°F (314 K) and have a radiant interchange factor of 0.2 with the tank and that the effective film coefficients for convection heat transfer between (1) the air and the outside of the tank and (2) the inside of the tank and the contained material are 0.72 and 0.75 Btu/(h)(ft²)(°F) [4.08 and 4.25 W/(m²)(K)], respectively.

Calculation Procedure

1. Calculate the solar-heat input to the tank.
Since the sun, although an extremely powerful emitter, subtends a very small solid angle, it has an only minute radiant interchange factor with objects on earth. The earth's orbital distance from the sun is nearly constant throughout the year. Therefore, it is a valid simplification to consider solar radiation simply as a heat source independent of the radiation environment and governed solely by the solar absorptance of each surface and the angular relationship of the surface to the solar vector.

The magnitude of the solar heating is indicated by the so-called solar constant. In space, at the radius of the earth's orbit, the solar constant is about 443 Btu/(h)(ft²) (1396 W/m²). However, solar radiation is attenuated by passage through the atmosphere; it is also reflected diffusely by the atmosphere, which itself varies greatly in composition. Table 7.2 provides representative values of the solar constant for use at ground level, as well as of the apparent daytime temperature of the sky for radiation purposes.

Since the solar constant G_n is a measure of total solar radiation *perpendicular to* the solar vector, it is necessary to also factor in the actual angle which the solar vector makes with the surface(s) being heated. This takes into account the geographic location, the time of year, the time of day, and the geometry of the surface and gives a corrected solar constant G_i.

For a horizontal surface, such as the tank roof, $G_i = G_n \cos Z$, where $Z = \cos^{-1}(\sin \phi \sin \delta_s + \cos \phi \cos \delta_s \cos h)$, ϕ is the latitude, δ_s is the solar declination, and h the hour angle, measured from 0° at high noon. In the present case, $Z = \cos^{-1}(\sin 29°20′ \sin 23.5° + \cos 29°20′ \cos 23.5° \cos 0°) = \cos^{-1} 0.995 = 5°50′$, and $G_i = 343(0.995) = 341.2$ Btu/(h)(ft²).

For a surface that is tilted $\psi°$ from horizontal and whose surface normal has an azimuth of $\alpha°$ from due south (westward being positive), $G_i = G_n(\cos |Z - \psi| - \sin Z \sin \psi + \sin Z \sin \psi \cos |A - \alpha|)$, where $A = \sin^{-1} \{\cos\delta_s \sin h/[\cos(90 - Z)]\}$. In the present case, $A = 0°$ because $\sin h = 0$.

Because the solar effect is distributed around the vertical surface of the tank to a varying degree (the effect being strongest from the south, since the sun is in the south), select wall segments 30° apart and calculate each separately. In the G_i equation, α will thus assume values ranging from −90° (facing due east) to +90° (due west); the northern half of the tank will be in shadow. Because the wall is vertical, $\psi = 90°$.

TABLE 7.2   Representative Values of Solar Constant and Sky Temperature

Conditions for total normal incident solar radiation	Solar constant		Effective sky temp., °F	Effective sky temp., K
	Btu/(h)(ft²)	kJ/(s)(m²)		
Southwestern United States, June, 6:00 A.M., extreme	252.3	0.797	−30 to −22	239 to 243
Southwestern United States, June, 12:00 M., extreme	385.1	1.216	−30 to −22	239 to 243
Southwestern United States, December, 9:00 A.M., extreme	307.7	0.972	−30 to −22	239 to 243
Southwestern United States, December, 12:00 M., extreme	396.2	1.251	−30 to −22	239 to 243
NASA recommended high design value, 12:00 M.	363.0	1.146	—	—
NASA recommended low design value, 12:00 M.	75.0	0.237	—	—
Southern United States, maximum for extremely bad weather	111.0	0.350	—	—
Southern United States desert, maximum for extremely bad weather	177.0	0.559	—	—

Source: Daniels [13].

It is also necessary to take sky radiation into account, that is, sunlight scattered by the atmosphere and reflected diffusely and which reaches all surfaces of the tank, including those not hit by direct sunlight because they are in shadow. This diffuse radiation G_s varies greatly but is generally small, between about 2.2 Btu/(h)(ft²) (6.93 W/m²) on a clear day and 44.2 Btu/(h)(ft²) (139 W/m²) on a cloudy day. For the day as described, assume that G_s is 25 Btu/(h)(ft²). This value must be added to all surfaces, including those in shadow.

Finally, take the solar absorptance of the paint α_s into account. Thus, the solar heat absorbed q_s equals $\alpha_s(G_i + G_s)$. The calculations can be summarized as follows:

	Roof	1	2	3	4	5	6	7
Segment (azimuth)		−90	−60	−30	0	30	60	90
G_i	341.2	0	17.4	30.2	34.9	30.2	17.4	0
$G_i + G_s$	366.2	25	42.4	55.2	59.9	55.2	42.4	25
q_s/A, Btu/(h)(ft²)	65.92	4.50	7.63	9.94	10.8	9.94	7.63	4.50
q_s/A, W/m²	208	14.2	24.0	31.3	34.0	31.3	24.0	14.2

Since this calculation procedure treats the tank as if it had 12 ftat sides, the calculated G_i for segments 1 and 7 is zero. Of course, G_i is also zero for the shaded half of the tank [and $q_s/A = 4.5$ Btu/(h)(ft²)].

2. Calculate the equilibrium temperature of each of the tank surfaces.
It can be shown that conduction between the segments is negligible. Then, at equilibrium, each segment must satisfy the heat-balance equation, that is, solar-heat absorption + net heat input by radiation + heat transferred in by outside convection + heat transferred in by inside convection = 0, or

$$q_s/A + \sigma \mathcal{J}_o\left(T_o^4 - T_w^4\right) + \sigma \mathcal{J}_R\left(T_R^4 - -T_w^4\right) + h_a(T_a - T_w) + h_i(T_i - T_w) = 0$$

where T_w is the tank-wall temperature; subscript o refers to surrounding structural and other elements having a radiant interchange factor $\mathcal{J}_o$ with the segments; subscript R refers to the atmosphere, having an equivalent radiation temperature T_R and a radiant interchange factor $\mathcal{J}_R$; subscript a refers to the air surrounding the tank, and subscript i refers to the gas inside the tank. The heat balance for the roof is solved as follows (similar calculations can be made for each segment of the tank wall):

Now, $q_s/A = 65.92$ Btu/(h)(ft^2), $T_o = 105°F = 565°R$ (due to hot pipes and other equipment in the vicinity), $\mathcal{J}_o = 0.2$, $T_R = 5°F = 465°R$, a good assumption for $\mathcal{J}_R$ is 0.75, $h_a = 0.72$, and $h_i = 0.75$. Therefore,

$$65.92 + (0.1713 \times 10^{-8})(0.2)\left(565^4 - T_w^4\right) + (0.1713 \times 10^{-8})(0.75)\left(465^4 - T_w^4\right)$$

$$+ \; 0.72(90 + 460 - T_w) + 0.75(85 + 460 - T_w) = 0$$

$$65.92 + 34.91 - (3.426 \times 10^{-10} + 12.848 \times 10^{-10})T_w^4 + 60.07 + 396$$

$$- \; (0.72 + 0.75)T_w + 408 = 0$$

$$1.6274 \times 10^{-9}T_w^4 + 1.47T_w = 965$$

This is solved by trial and error, to yield $T_w = 553°R = 93°F$ (307 K). Note that if the paint had been black, α_s might have been 0.97 instead of 0.18. In that case, the temperature would have been about 200°F (366 K).

The same procedure is then applied to each of the other tank segments.

7.3 HEAT LOSS FROM AN UNINSULATED SURFACE TO AIR

A steam line with a diameter of 3.5 in (0.089 m) and a length of 50 ft (15.2 m) transports steam at 320°F (433 K). The carbon steel pipe [thermal conductivity of 25 Btu/(h)(ft)(°F) or 142 W/(m^2)(K)] is not insulated. Its emissivity is 0.8. Calculate the heat loss for calm air and also for a wind velocity of 15 mi/h (24 km/h), if the air temperature is 68°F (293 K).

Calculation Procedure

1. Calculate the heat loss due to radiation.
Because the coefficient for heat transfer from the outside of the pipe as a result of radiation and convection is much less than all other heat-transfer coefficients involved in this example, the surface temperature of the pipe can be assumed to be that of the steam. To calculate the heat loss, use the straightforward radiation formula

$$\frac{Q}{A} = 0.1713\epsilon\left[\left(\frac{T_s}{100}\right)^4 - \left(\frac{T_s}{100}\right)^4\right]$$

where Q is heat loss in British thermal units per hour, A is heat-transfer area in square feet, T_s is absolute temperature of the surface in degrees Rankine, T_a is absolute temperature of the air, and ϵ is the emissivity of the pipe. (Note that in this version of the formula, the 10^{-8} portion of the Stefan-Boltzmann constant is built into the temperature terms.)

Thus,

$$\frac{Q}{A} = 0.1713(0.8)\left[\left(\frac{460 + 320}{100}\right)^4 - \left(\frac{460 + 68}{100}\right)^4\right] = 401 \text{ Btu/(h)(ft)}^2 \; (1264 \text{ W/m}^2)$$

2. Calculate the heat loss as a result of natural convection in calm air.
Use the formula

$$\frac{Q}{A} = \frac{0.27\,\Delta T^{1.25}}{D^{0.25}}$$

where $\Delta T = T_s - T_a$ in degrees Fahrenheit and D is pipe diameter in feet. Thus,

$$\frac{Q}{A} = \frac{0.27(320 - 68)^{1.25}}{(3.5/12)^{0.25}} = 369 \text{ Btu/(h)(ft}^2) \text{ (1164 W/m}^2)$$

3. Calculate the total heat loss for the pipe in calm air.
Now, $Q = (Q/A)\,A$, and $A = \pi(3.5/12) \times 50 = 45.81$ ft^2 (4.26 m^2), so $Q = (401 + 369)(45.81) = 35{,}270$ Btu/h (10,330 W).

4. Calculate the heat loss by convection for a wind velocity of 15 mi/h.
First, determine the mass velocity G of the air: $G = \rho v$, where ρ is density and v is linear velocity. For air $\rho = 0.075$ lb/ft^3 (1.20 kg/m^3). In this problem, $v = 15$ mi/h (5280 ft/mi) = 79,200 ft/h (24,140 m/h), so $G = 0.075(79{,}200) = 5940$ lb/(h)(ft^2) [29,000 kg/(h)(m^2)].

Next, determine the heat-transfer coefficient, using the formula $h = 0.11cG^{0.6}/D^{0.4}$, where h is heat-transfer coefficient in British thermal units per hour per square foot per degree Fahrenheit, c is specific heat in British thermal units per pound per degree Fahrenheit (0.24 for air), G is mass velocity in pounds per hour per square foot, and D is diameter in feet. Thus, $h = 0.11\,(0.24)(5940)^{0.6}/(3.5/12)^{0.4} = 7.94$ Btu/(h)(ft^2)(°F) [45.05 W/(m^2)(K)].

Finally, use this coefficient to determine the heat loss due to convection via the formula $Q/A = h(T_s - T_a)$. Thus, $Q/A = 7.94(320 - 68) = 2000.9$ Btu/(h)(ft^2)(6307 W/m^2).

5. Calculate the total heat loss for the pipe when the wind velocity is 15 mi/h.
As in step 3, $Q = (401 + 2000.9)(45.81) = 110{,}030$ Btu/h (32,240 W).

7.4 HEAT LOSS FROM AN INSULATED SURFACE TO AIR

Calculate the heat loss from the steam line in Procedure 7.3 if it has insulation 2 in (0.050 m) thick having a thermal conductivity of 0.05 Btu/(h)(ft)(°F) [0.086 W/(m)(K)]. The inside diameter of the pipe is 3 in (0.076 m), and the heat-transfer coefficient from the condensing steam to the pipe wall is 1500 Btu/(h)(ft^2)(°F) [8500 W/(m^2)(K)]. Assume that the wind velocity is 15 mi/h (24 km/h). The pipe is illustrated in Fig. 7.1.

Calculation Procedure

1. Set out the appropriate overall heat-transfer equation.
The heat loss can be calculated from the equation $Q = UA_o(T_s - T_a)$, where Q is heat loss, U is overall heat-transfer coefficient, A_o is area of the outside surface of the insulation ($= 2\pi Lr_3$; see Fig. 7.1), T_s is steam temperature, and T_a is air temperature. The overall heat-transfer coefficient can be calculated with the equation

$$U = \frac{1}{\dfrac{r_3}{r_1 h_i} + \dfrac{r_3 \ln(r_2/r_1)}{k_1} + \dfrac{r_3 \ln(r_3/r_2)}{k_2} + \dfrac{1}{h_o}}$$

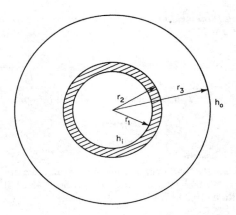

r_1 = 3.0/2 in = 1.5 in = 0.125 ft (0.0381 m)
r_2 = 3.5/2 in = 1.75 in = 0.1458 ft (0.0445 m)
r_3 = 7.5/2 in = 3.75 in = 0.3125 ft (0.0953 m)

FIGURE 7.1 Cross section of insulated pipe in Procedure 7.4.

where the r_i are as described in Fig. 7.1, k_1 is the thermal conductivity of the pipe, k_2 is that of the insulation, and h_o and h_i are the outside and inside heat-transfer coefficients, respectively.

2. Calculate h_o.
The outside heat-transfer coefficient h_o is the sum of the heat-transfer coefficient for convection to the wind h_c and the transfer of heat as a result of radiation h_r. The latter is approximately 1.0 Btu/(h)(ft^2)(°F) [5.68 W/(m^2)(K)], and we will use that value here. From step 4 of Procedure 7.3, $h_c = 7.94$. Thus, $h_o = h_c + h_r = 7.94 + 1.0 = 8.94$ Btu/(h)(ft^2)(°F) [50.78 W/(m^2)(K)].

3. Calculate U.
Substituting into the equation in step 1.

$$U = \cfrac{1}{\cfrac{0.3125}{(0.125)(1500)} + \cfrac{0.3125 \ln (1.75/1.5)}{25} + \cfrac{0.3125 \ln (3.75/1.75)}{0.05} + \cfrac{1}{8.94}}$$

$$= 0.205 \text{ Btu/(h)(ft}^2\text{)(°F) } [1.16 \text{ W/(m}^2\text{)(K)}]$$

4. Calculate the heat loss.
From step 1, $Q = 0.205(2\pi)(50)(0.3125)(320 - 68) = 5070$ Btu/h (1486 W).

7.5 HEAT LOSS FROM A BURIED LINE

A steam line with a diameter of 12.75 in (0.3239 m) is buried with its center 6 ft (1.829 m) below the surface in soil having an average thermal conductivity of 0.3 Btu/(h)(ft)(°F) [0.52 W/(m)(K)]. Calculate the heat loss if the pipe is 200 ft (60.96 m) long, the steam is saturated at a temperature of 320°F (433 K), and the surface of the soil is at a temperature of 40°F (277 K).

Calculation Procedure

1. *Select the appropriate heat-transfer equation.*
The heat loss can be calculated from the equation

$$Q = Sk(T_1 - T_2)$$

where Q is heat loss, k is thermal conductivity of the soil, T_1 is surface temperature of the pipe, T_2 is surface temperature of the soil, and S is a shape factor.

2. *Determine the shape factor.*
Use the equation $S = 2\pi L/\cosh^{-1}(2z/D)$ (when z is much less than L), where L is length, z is distance from ground surface to the center of the pipe, and D is diameter. Now,

$$\cosh^{-1}(2z/D) = \ln\left\{(2z/D) + [(2z/D)^2 - 1]^{1/2}\right\}$$

For this example, $2z/D = 2(6)/(12.75/12) = 11.294$, and $\cosh^{-1}(2z/D) = \ln[11.294 + (11.294^2 - 1)^{1/2}] = 3.116$. Thus, $S = 2\pi 200/3.116 = 403.29$ ft (122.92 m).

3. *Calculate the heat loss.*
Because the heat-transfer resistance through the soil is much greater than all other resistances to heat transfer, the surface of the pipe can be assumed to be at the temperature of the steam. Thus, $Q = 403.29(0.3)(320 - 40) = 33,876$ Btu/h (9926 W).

Related Calculations. This approach can be used for any geometry for which shape factors can be evaluated. Shape factors for many other geometries are presented in several of the references, including Krieth [2] and Holman [4].

7.6 CONDUCTION OF HEAT FROM A BELT COOLER

A stainless steel belt cooler with a width of 3 ft (0.914 m), a thickness of ⅛ in (3.175 mm), and a length of 100 ft (30.48 m) is to be used to cool 10,000 lb/h (4535.9 kg/h) of material with the physical properties listed below. The bottom of the belt is sprayed with water at 86°F (303 K), and the heat-transfer coefficient h from the belt to the water is 500 Btu/(h)(ft²)(°F) [2835 W/(m²)(K)]. Calculate the surface temperature of the material as it leaves the belt if the material is placed on the belt at a temperature of 400°F (477 K) and the belt moves at a speed of 150 ft/min (45.72 m/min).

Physical Properties of the Material

ρ = density = 60 lb/ft³ (961.1 kg/m³)
k = thermal conductivity = 0.10 Btu/(h)(ft)(°F) [0.17 W/(m)(K)]
α = thermal diffusivity = 0.0042 ft²/h (3.871 cm³/h)
c = specific heat = 0.4 Btu/(lb)(°F) [1.7 kJ/(kg)(K)]

Physical Properties of the Belt

ρ = density = 500 lb/ft³ (8009.2 kg/m³)
k = thermal conductivity = 10 Btu/(h)(ft)(°F)[17 W/(m)(K)]
c = specific heat = 0.11 Btu/(lb)(°F) [0.46 kJ/(kg)(K)]
α = thermal diffusivity = 0.182 ft²/h (0.169 m²/h)

Calculation Procedure

1. Calculate the thickness of the material on the belt.
By dimensional analysis, (lb/h)(ft³/lb)(min/ft belt)(h/min) (1/ft belt) = ft, so (10,000)(1/60)(1/150)(1/60)(1/3) = 0.00617 ft (1.881 mm).

2. Calculate the time the material is in contact with the cooled belt.
Here, (ft)(min/ft)(h/min) = h, so 100(1/150)(1/60) = 0.0111 h.

3. Calculate $\alpha\tau/L^2$ for the belt.
Use the curves in Fig. 7.2. Assume that the water flow rate is high enough to neglect the temperature rise as the water removes heat from the material. Then, $\alpha\tau/L^2 = 0.182(0.0111)/(0.125/12)^2 = 18.62$.

4. Calculate $k/(hL)$ for the belt.
Now, $k/(hL) = 10/[500(0.125/12)] = 1.92$.

5. Determine θ_o/θ_i for the belt.
From Fig. 7.2, for $\alpha\tau/L^2 = 18.62$ and $k/(hL) = 1.92$, θ_o/θ_i is less than 0.001. Therefore, pending the outcome of steps 6 through 8, the conduction of heat through the belt will presumably be negligible compared with the conduction of heat through the material to be cooled.

6. Calculate $\alpha\tau/L^2$ for the material.
Now, $\alpha\tau/L^2 = 0.0042(0.0111)/0.00617^2 = 1.225$.

7. Calculate $k/(hL)$ for the material.
Now, $k/(hL) = 0.10/[500(0.00617)] = 0.0324$.

8. Determine θ_o/θ_i.
From Fig. 7.2, for $\alpha\tau/L^2 = 1.225$ and $k/(hL) = 0.0324$, $\theta_o/\theta_i = 0.075$.

9. Determine the final surface temperature of the material.
Now, $\theta_o/\theta_i = (T - T_\infty)/(T_i - T_\infty)$, where T is the temperature at the end of the belt, T_i is the initial material temperature, and T_∞ is the water temperature. So, $\theta_o/\theta_i = (T - 86)/(400 - 86) = 0.075$, and $T = 109.6°F$ (316.3 K).

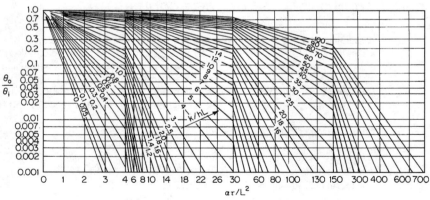

L = half slab thickness
h = heat-transfer coefficient
k = thermal conductivity
α = thermal diffusivity
τ = time
$\theta_o = T - T_\infty$ = temperature at time τ minus surrounding temperature
$\theta_i = T_i - T_\infty$ = initial temperature minus surrounding temperature

FIGURE 7.2   Midplane temperature for a plate of thickness 2L.

7.7   SIZING A BELT COOLER

For the conditions in the preceding procedure, calculate the length of belt required if the belt speed is reduced to 100 ft/min (30.48 m/min) and an outlet temperature of 125°F (324.8 K) is acceptable.

Calculation Procedure

The curves in Fig. 7.2 will be used.

1. Calculate the thickness of the material on the belt.
By dimensional analysis, (lb/h)(ft^3/lb)(min/ft belt)(h/min)(1/ft belt) = ft, so (10,000)(1/60)(1/100)(1/60)(1/3) = 0.0093 ft (2.82 mm).

2. Calculate k/(hL) for the material.
Now, $k/(hL) = 0.10/[500(0.0093)] = 0.0215$.

3. Calculate θ_o/θ_i.
Thus, $\theta_o/\theta_i = (125 - 86)/(400 - 86) = 0.1242$.

4. Determine $\alpha\tau/L^2$.
From Fig. 7.2, for $\theta_o/\theta_i = 0.1242$ and $k/(hL) = 0.0215$, $\alpha\tau/L^2 = 1.0$.

5. Determine the time required for contact with the water.
Since $\alpha\tau/L^2 = 1.0 = 0.0042\,\tau/0.0093^2$, $\tau = 0.0206$ h.

6. Calculate the length of belt required.
Now, $\tau = $ h $=$ ft(min/ft)(h/min) $= 0.0206$, which is feet of belt(1/100)(1/60), so the length of belt is (0.0206)(100)(60) = 123.6 ft (37.67 m).

7.8   BATCH HEATING: INTERNAL COIL, ISOTHERMAL HEATING MEDIUM

A tank containing 50,000 lb (22,679.5 kg) of material with a specific heat of 0.5 Btu/(lb)(°F) [2.1 kJ/(kg)(K)] is to be heated from 68°F (293 K) to 257°F (398 K). The tank contains a heating coil with a heat-transfer surface of 100 ft^2 (9.29 m^2), and the overall heat-transfer coefficient from the coil to the tank contents is 150 Btu/(h)(ft^2)(°F) [850 W/(m^2)(K)]. Calculate the time required to heat the tank contents with steam condensing at 320°F (433 K).

Calculation Procedure

1. Select and apply the appropriate heat-transfer formula.
When heating a batch with an internal coil with an isothermal heating medium, the following equation applies:

$$\ln\left(\frac{T_1 - t_1}{T_1 - t_2}\right) = \left(\frac{UA}{Mc}\right)\theta$$

where T_1 = heating-medium temperature
t_1 = initial batch temperature
t_2 = final batch temperature
U = overall heat-transfer coefficient
A = heat-transfer surface

M = weight of batch
c = specific heat of batch
θ = time

For this procedure, $\ln[(320 - 68)/(320 - 257)] = \{150(100)/[50,000(0.5)]\}\,\theta$, so $\theta = 2.31$ h.

Related Calculations. This procedure can also be used for batch cooling with internal coils and isothermal cooling media. The equation in such cases is

$$\ln\left(\frac{T_1 - t_1}{T_2 - t_1}\right) = \left(\frac{UA}{Mc}\right)\theta$$

where T_1 = initial batch temperature
T_2 = final batch temperature
t_1 = cooling-medium temperature

7.9 BATCH COOLING: INTERNAL COIL, NONISOTHERMAL COOLING MEDIUM

For the tank described in the preceding procedure, calculate the time required to cool the batch from 257°F (398 K) to 104°F (313 K) if cooling water is available at a temperature of 86°F (303 K) and a flow rate of 10,000 lb/h (4535.9 kg/h).

Calculation Procedure

1. Select and apply the appropriate heat-transfer formula.
When cooling a batch with an internal coil and a nonisothermal cooling medium, the following equation applies:

$$\ln\left(\frac{T_1 - t_1}{T_2 - t_1}\right) = \frac{w_c c_c}{Mc}(K_2 - 1)\left(\frac{1}{K_2}\right)\theta$$

where $K_2 = e^{UA/w_c c_c}$
T_1 = initial batch temperature
T_2 = final batch temperature
t_1 = initial coolant temperature
w_c = coolant flow rate
c_c = coolant specific heat
U = overall heat-transfer coefficient
A = heat-transfer surface
M = weight of batch
c = specific heat of batch
θ = time

For this procedure, $K_2 = \exp 150(100)/[10,000(1.0)] = 4.4817$, so

$$\ln\frac{257 - 86}{104 - 86} = \frac{(10,000)1.0}{(50,000)0.5}\frac{4.4817 - 1}{4.4817}\theta$$

Therefore, $\theta = 7.245$ h.

Related Calculations. This procedure can also be used for batch heating with internal coils and isothermal heating media. The equations in such cases are

$$\ln\frac{T_1 - t_1}{T_1 - t_2} = \frac{W_h c_h}{Mc}(K_3 - 1)\frac{1}{K_3}\theta \quad \text{and} \quad K_3 = \exp\frac{UA}{W_h c_h}$$

where T_1 = heating-medium temperature
$\quad t_1$ = initial batch temperature
$\quad t_2$ = final batch temperature
$\quad W_h$ = heating-medium flow rate
$\quad c_h$ = heating-medium specific heat

7.10 BATCH COOLING: EXTERNAL HEAT EXCHANGER (COUNTERFLOW), NONISOTHERMAL COOLING MEDIUM

Calculate the time required to cool the batch described in the preceding procedure if an external heat exchanger with a heat-transfer surface of 200 ft^2 (18.58 m^2) is available. The batch material is circulated through the exchanger at the rate of 25,000 lb/h (11,339.8 kg/h). The overall heat-transfer coefficient in the heat exchanger is 200 Btu/(h)(ft^2)(°F) [1134 W/(m^2)(K)].

Calculation Procedure

1. Select and apply the appropriate heat-transfer formula.
When cooling a batch with an external heat exchanger and a nonisothermal cooling medium, the following equations apply:

$$\ln\frac{T_1 - t_1}{T_2 - t_1} = \frac{K_4 - 1}{M}\frac{W_b w_c c_c}{K_4 w_c c_c - W_b c}\theta \quad \text{and} \quad K_4 = \exp U\,A\left(\frac{1}{W_b c} - \frac{1}{w_c c_c}\right)$$

where T_1 = initial batch temperature
$\quad T_2$ = final batch temperature
$\quad t_1$ = initial coolant temperature
$\quad w_c$ = coolant flow rate
$\quad c_c$ = coolant specific heat
$\quad W_b$ = batch flow rate
$\quad c$ = batch specific heat
$\quad M$ = weight of batch
$\quad U$ = overall heat-transfer coefficient
$\quad A$ = heat transfer surface
$\quad \theta$ = time

For this problem,

$$K_4 = (200)(200)\left[\frac{1}{(25,000)0.5} - \frac{1}{(10,000)1.0}\right] = 0.4493$$

So,

$$\ln\frac{257 - 86}{104 - 86} = \frac{0.4493 - 1}{50,000}\frac{25,000(10,000)(1.0)}{[0.4493(10,000)(1.0) - 25,000(0.5)]}\theta$$

Therefore, $\theta = 6.547$ h.

Related Calculations. This procedure can also be used for batch heating with external heat exchangers and nonisothermal heating media. The equations in such cases are

$$\ln\frac{T_1 - t_1}{T_1 - t_2} = \frac{K_5 - 1}{M}\frac{W_b W_h c_h}{K_5 W_h c_h - W_b c}\theta \quad \text{and} \quad K_5 = \exp U A\left(\frac{1}{W_b c} - \frac{1}{W_h c_h}\right)$$

where T_1 = heating-medium initial temperature
t_1 = initial batch temperature
t_2 = final batch temperature
W_h = heating-medium flow rate
c_h = specific heat of heating medium

7.11 BATCH COOLING: EXTERNAL HEAT EXCHANGER (1–2 MULTIPASS), NONISOTHERMAL COOLING MEDIUM

Calculate the time required to cool the batch described in the preceding procedure if the external heat exchanger is a 1–2 multipass unit rather than counterflow.

Calculation Procedure

1. Select and apply the appropriate heat-transfer formula.
When cooling a batch with an external 1–2 multipass heat exchanger and a nonisothermal cooling medium, the following equations apply:

$$\ln\frac{T_1 - t_1}{T_2 - t_1} = S\frac{w_c c_c}{Mc}\theta$$

and

$$S = \frac{2(K_7 - 1)}{K_7\left[R + 1 + (R^2 + 1)^{1/2}\right] - \left[R + 1 - (R^2 + 1)^{1/2}\right]}$$

where $K_7 = \exp\dfrac{UA}{w_c c_c}(R^2 + 1)^{1/2}$

$R = \dfrac{w_c c_c}{W_b c}$

T_1 = initial batch temperature
T_2 = final batch temperature
t_1 = initial coolant temperature
w_c = coolant flow rate
c_c = coolant specific heat
M = weight of batch
W_b = batch flow rate
c = specific heat of batch
θ = time required to cool the batch
U = overall heat-transfer coefficient
A = heat-transfer surface

For the problem here, $R = 10,000(1.0)/[25,000(0.5)] = 0.80$, so

$$K_7 = \exp\frac{200(200)}{10,000(1.0)}(0.80^2 + 1)^{1/2}$$
$$= 167.75$$

and

$$S = \frac{2(167.75 - 1)}{167.75[0.80 + 1 + (0.8^2 + 1)^{1/2}] - [0.8 + 1 - (0.8^2 + 1)^{1/2}]}$$
$$= 0.646$$

Therefore,

$$\ln\frac{257 - 86}{104 - 86} = \frac{0.646(10,000)(1.0)}{50,000(0.5)}\theta$$

and $\theta = 8.713$ h

Related Calculations.   This procedure can also be used for batch heating with external 1–2 multipass heat exchangers and nonisothermal heating media. The equation in such a case is

$$\ln\frac{T_1 - t_1}{T_1 - t_2} = \frac{SW_h}{M}\theta$$

where S is defined by the preceding equation, and

$$R = \frac{W_b c}{W_h c_h}$$

$$K_7 = \exp\frac{UA}{W_b c}(R^2 + 1)^{1/2}$$

T_1 = initial temperature of heating medium
t_1 = initial temperature of batch
t_2 = final batch temperature
W_h = heating-medium flow rate
c_h = specific heat of heating medium

7.12   *HEAT TRANSFER IN AGITATED VESSELS*

Calculate the heat-transfer coefficient from a coil immersed in an agitated vessel with a diameter of 8 ft (2.44 m). The agitator is a turbine 3 ft (0.91 m) in diameter and turns at 150 r/min. The fluid has these properties:

ρ = density = 45 lb/ft^3 (720.8 kg/m^3)
μ = viscosity = 10 lb/(ft)(h) (4.13 cP)
c = specific heat = 0.7 Btu/(lb)(°F) [2.9 kJ/(kg)(K)]
k = thermal conductivity = 0.10 Btu/(h)(ft)(°F) [0.17 W/(m)(K)]

The viscosity may be assumed to be constant with temperature.

Calculation Procedure

1. Select and apply the appropriate heat-transfer formula.

The following equation can be used to predict heat-transfer coefficients from coils or tank walls in agitated tanks:

$$\frac{hD_j}{k} = a\left(\frac{L^2 N\rho}{\mu}\right)^{2/3}\left(\frac{c\mu}{k}\right)^{1/3}\left(\frac{\mu_b}{\mu_w}\right)^{0.14}$$

The term a has these values:

Agitator	Surface	a
Turbine	Jacket	0.62
Turbine	Coil	1.50
Paddle	Jacket	0.36
Paddle	Coil	0.87
Anchor	Jacket	0.46
Propeller	Jacket	0.54
Propeller	Coil	0.83

The other variables in the equation are

h = heat-transfer coefficient
D_j = diameter of vessel
k = thermal conductivity
L = diameter of agitator
N = speed of agitator in revolutions per hour
ρ = density
μ = viscosity
c = specific heat
μ_b = viscosity at bulk fluid temperature
μ_w = viscosity at surface temperature

Therefore, $h = 1.50(k/D_j)(L^2 N\rho/\mu)^{2/3}(c\mu/k)^{1/3}(\mu_b/\mu_w)^{0.14}$. As noted above, assume that $(\mu_b/\mu_w)^{0.14} = 1.0$. Then,

$$h = 1.50(0.10/8)\left[\frac{3^2(150)(60)(45)}{10}\right]^{2/3}\left[\frac{0.7(10)}{0.1}\right]^{1.3}(1.0)$$

$$= 394 \text{ Btu/(h)(ft}^2)(°\text{F)}[2238 \text{ W/(m}^2)(\text{K})]$$

7.13   NATURAL-CONVECTION HEAT TRANSFER

Calculate the heat-transfer coefficient from a coil immersed in water with the physical properties listed below. The coil has a diameter of 1 in (0.025 m), and the temperature difference between the surface of the coil and the fluid is 10°F (5.56 K). The properties of the water are

c = specific heat = 1.0 Btu/(lb)(°F) [4.19 kJ/(kg)(K)]
ρ = liquid density = 60 lb/ft^3 (961.1 kg/m^3)
k = thermal conductivity = 0.395 Btu/(h)(ft)(°F)[0.683 W/(m)(K)]
μ = viscosity = 0.72 lb/(ft)(h) (0.298 cP)
β = coefficient of expansion = 0.0004°F^{-1} (0.00022 K^{-1})

Calculation Procedure

1. Consider the natural-convection equations available.
Heat-transfer coefficients for natural convection may be calculated using the equations presented below. These equations are also valid for horizontal plates or discs. For horizontal plates facing upward which are heated or for plates facing downward which are cooled, the equations are applicable directly. For heated plates facing downward or cooled plates facing upward, the heat-transfer coefficients obtained should be multiplied by 0.5.

Vertical surfaces	Horizontal cylinders
For Reynolds numbers greater than 10,000:	
$[h/(cG)](c\mu/k)^{2/3} = 0.13/(LG/\mu)^{1/3}$	$= 0.13/(DG/\mu)^{1/3}$
For Reynolds numbers from 100 to 10,000:	
$[h/(cG)](c\mu/k)^{3/4} = 0.59/(LG/\mu)^{1/2}$	$= 0.53/(DG/\mu)^{1/2}$
For Reynolds numbers less than 100:	
$[h/(cG)](c\mu/k)^{5/6} = 1.36/(LG/\mu)^{2/3}$	$= 1.09/(DG/\mu)^{2/3}$

In these equations, G is mass velocity, that is, $(g\beta\Delta T\rho^2 L)^{1/2}$ for vertical surfaces or $(g\beta\Delta\rho^2 D)^{1/2}$ for horizontal cylinders, and,

h = heat-transfer coefficient
c = specific heat
L = length
D = diameter
μ = viscosity
g = acceleration of gravity $(4.18 \times 10^8 \text{ ft/h}^2)$
β = coefficient of expansion
ΔT = temperature difference between surface and fluid
ρ = density
k = thermal conductivity

2. Calculate mass velocity.
Thus, $G = (g\beta\Delta Tp^2 D)^{1/2} = [(4.18 \times 10^8)(0.0004)(10)60^2(1/12)]^{1/2} = 22{,}396.4 \text{ lb/(ft}^2)(\text{h})$.

3. Calculate the Reynolds number DG/μ.
Thus, $DG/\mu = (1/12)(22{,}396.4)/0.72 = 2592.2$.

4. Calculate the heat-transfer coefficient h.
Thus, $h = 0.53cG/[(c\mu/k)^{3/4}(DG/\mu)^{1/2}] = 0.53(1.0)(22{,}396.4)/\{[1.0(0.72)/0.395]^{3/4}(2592.2)^{1/2}\} = 148.6 \text{ Btu/(h)(ft}^2)(°F) [844 \text{ W/(m}^2)(K)]$.

Related Calculations. The equations presented above can be simplified for natural-convection heat transfer in air. The usual cases yield the following equations: For vertical surfaces,

$$h = 0.29(\Delta T/L)^{1/4}$$

for horizontal cylinders,

$$h = 0.27(\Delta T/D)^{1/4}$$

In these, h is in Btu/(h)(ft^2)(°F), ΔT is in °F, and L and D are in feet.

7.14 HEAT-TRANSFER COEFFICIENTS FOR FLUIDS FLOWING INSIDE TUBES: FORCED CONVECTION, SENSIBLE HEAT

Calculate the heat-transfer coefficient for a fluid with the properties listed below flowing through a tube 20 ft (6.1 m) long and of 0.62-in (0.016-m) inside diameter. The bulk fluid temperature is 212°F (373 K), and the tube surface temperature is 122°F (323 K). Calculate the heat-transfer coefficient if the fluid is flowing at a rate of 2000 lb/h (907.2 kg/h). Also calculate the heat-transfer coefficient if the flow rate is reduced to 100 lb/h (45.36 kg/h).

Physical Properties of the Fluid

c = specific heat = 0.65 Btu/(lb)(°F) [2.72 kJ/(kg)(K)]
k = thermal conductivity = 0.085 Btu/(h)(ft)(°F) [0.147 W/(m)(K)]
μ_w = viscosity at 122°F = 4.0 lb/(ft)(h) (1.65 cP)
μ_b = viscosity at 212°F = 1.95 lb/(ft)(h) (0.806 cP)

Calculation Procedure

1. Select the appropriate heat-transfer coefficient equation.
Heat-transfer coefficients for fluids flowing inside tubes or ducts can be calculated using these equations:

a. For Reynolds numbers (DG/μ) greater than 8000,

$$\frac{h}{cG} = \frac{0.023}{(c\mu/k)^{2/3}(D_iG/\mu)^{0.2}(\mu_w/\mu_b)^{0.14}}$$

b. For Reynolds numbers (DG/μ) less than 2100,

$$\frac{h}{cG} = \frac{1.86}{(c\mu/k)^{2/3}(D_iG/\mu)^{2/3}(L/D_i)^{1/3}(\mu_w/\mu_b)^{0.14}}$$

In these equations,

h = heat-transfer coefficient
c = specific heat
G = mass velocity (mass flow rate divided by cross-sectional area)
μ = viscosity
μ_w = viscosity at the surface temperature
μ_b = viscosity at the bulk fluid temperature
k = thermal conductivity
D_i = inside diameter
L = length

2. Calculate D_iG/μ for a 2000 lb/h flow rate.

$$\frac{D_iG}{\mu} = \frac{0.62}{12}\frac{2000}{(0.62/12)^2(\pi/4)}\frac{1}{1.95}$$

$$= 25,275$$

3. Calculate h for the 2000 lb/h flow rate.
Because DG/μ is greater than 8000,

$$\frac{h}{cG} = \frac{0.023}{(c\mu/k)^{2/3}(D_iG/\mu)^{0.2}(\mu_w/\mu_b)^{0.14}}$$

$$= \frac{0.023(0.65)[2000/(0.62/12)^2(\pi/4)]}{[0.65(1.95)/0.085]^{2/3}\,25,275^{0.2}(4.0/1.95)^{0.14}}$$

$$= 280.3 \text{ Btu/(h)(ft}^2\text{)(K)}[1592 \text{ W/(m}^2\text{)(K)}]$$

4. Calculate DG/μ for a 100 lb/h flow rate.

$$D_iG/\mu = 25,275(100/2000) = 1263.8$$

5. Calculate h for the 100 lb/h flow rate.
Because DG/μ is less than 2100,

$$\frac{h}{cG} = \frac{1.86}{(c\mu/k)^{2/3}(DG/\mu)^{2/3}(L/D)^{1/3}(\mu_w/\mu_b)^{0.14}}$$

$$= \frac{1.86(0.65)\{100/[(0.62/12)^2(\pi/4)]\}}{\left[\frac{0.65(1.95)}{0.085}\right]^{2/3}1263.8^{2/3}\left[\frac{20}{(0.62/12)}\right]^{1/3}\left(\frac{4.0}{1.95}\right)^{0.14}}$$

$$= 10.1 \text{ Btu/(h)(ft}^2\text{)(°F)}[57.4 \text{ W/(m}^2\text{)(K)}].$$

Related Calculations. Heat transfer for fluids with Reynolds numbers between 2100 and 8000 is not stable, and the heat-transfer coefficients in this region cannot be predicted with certainty. Equations have been presented in many of the references. The heat-transfer coefficients in this region can be bracketed by calculating the values using both the preceding equations for the Reynolds number in question.

The equations presented here can also be used to predict heat-transfer coefficients for the shell side of shell-and-tube heat exchangers in which the baffles have been designed to produce flow parallel to the axis of the tube. For such cases, the diameter that should be used is the equivalent diameter

$$D_e = \frac{4a}{P}$$

where a = flow area
P = wetted perimeter

Here, $a = (D_s^2 - nD_o^2)(\pi/4)$, where D_s is the shell inside diameter, D_o is the tube outside diameter, and n is the number of tubes; and $P = \pi(D_s + nD_o)$.

For shells with triple or double segmental baffles, the heat-transfer coefficient calculated for turbulent flow (DG/μ greater than 8000) should be multiplied by a value of 1.3.

For gases, the equation for heat transfer in the turbulent region (DG/μ greater than 8000) can be simplified because the Prandtl number ($c\mu/k$) and the viscosity for most gases are approximately constant. Assigning the values $c\mu/k = 0.78$ and $\mu = 0.0426$ lb/(h)(ft) (0.0176 cP) results in the following equation for gases:

$$h = 0.0144\frac{cG^{0.8}}{D_i^{0.2}}$$

with the variables defined in English units.

7.15 HEAT-TRANSFER COEFFICIENTS FOR FLUIDS FLOWING INSIDE HELICAL COILS

Calculate the heat-transfer coefficient for a fluid with a flow rate of 100 lb/h (45.36 kg/h) and the physical properties outlined in Procedure 7.14. The inside diameter of the tube is 0.62 in (0.016 m), and the tube is fabricated into a helical coil with a helix diameter of 24 in (0.61 m).

Calculation Procedure

1. Select the appropriate heat-transfer coefficient equation.
Heat-transfer coefficients for fluids flowing inside helical coils can be calculated with modifications of the equations for straight tubes. The equations presented in Procedure 7.14 should be multiplied by the factor $1 + 3.5\,D_i/D_c$, where D_i is the inside diameter and D_c is the diameter of the helix or coil. In addition, for laminar flow, the term $(D_c/D_i)^{1/6}$ should be substituted for the term $(L/D)^{1/3}$. The Reynolds number required for turbulent flow is $2100[1 + 12(D_i/D_c)^{1/2}]$.

2. Calculate the minimum Reynolds number for turbulent flow.
Now,

$$
\begin{aligned}
(DG/\mu)_{\min} &= 2100\left[1 + 12(D_i/D_c)^{1/2}\right] \\
&= 2100\left[1 + 12(0.62/24)^{1/2}\right] \\
&= 6150
\end{aligned}
$$

3. Calculate h.
From the preceding calculations, $DG/\mu = 1263.8$ at a flow rate of 100 lb/h. Therefore,

$$
\begin{aligned}
h &= \frac{1.86cG(1 + 3.5D_i/D_c)}{(c\mu/k)^{2/3}(DG/\mu)^{2/3}(D_c/D_i)^{1/6}(\mu_w/\mu_b)^{0.14}} \\[2mm]
&= \frac{1.86(0.65)\{100/[(0.62/12)^2(\pi/4)]\}[1 + 3.5(0.62/24)]}{[0.65(1.95)/0.085]^{2/3}1263.8^{2/3}(24/0.62)^{1/6}(4.0/1.95)^{0.14}} \\[2mm]
&= 43.7\,\text{Btu/(h)(ft}^2)(°F)[248\ \text{W/(M}^2)(\text{K})]
\end{aligned}
$$

7.16 HEAT-TRANSFER COEFFICIENTS: FLUIDS FLOWING ACROSS BANKS OF TUBES; FORCED CONVECTION, SENSIBLE HEAT

Calculate the heat-transfer coefficient for a fluid with the properties listed in Procedure 7.14 if the fluid is flowing across a tube bundle with the following geometry. The fluid flows at a rate of 50,000 lb/h (22,679.5 kg/h). Calculate the heat-transfer coefficient for both clean and fouled conditions.

Tube Bundle Geometry

D_s = shell diameter = 25 in (2.08 ft or 0.635 m)
B = baffle spacing = 9.5 in (0.79 ft or 0.241 m)
D_o = outside tube diameter = 0.75 in (0.019 m)
s = tube center-to-center spacing = 0.9375 in (0.0238 m)

The tubes are spaced on a triangular pattern.

Calculation Procedure

1. *Select the appropriate heat-transfer coefficient equation.*
Heat-transfer coefficients for fluids flowing across ideal-tube banks may be calculated using the equation

$$\frac{h}{cG} = \frac{a}{(c\mu/k)^{2/3}(D_o G/\mu)^m (\mu_m/\mu_b)^{0.14}}$$

The values of a and m are as follows:

Reynolds number	Tube pattern	m	a
Greater than 200,000	Staggered	0.300	0.166
Greater than 200,000	In-line	0.300	0.124
300 to 200,000	Staggered	0.365	0.273
300 to 200,000	In-line	0.349	0.211
Less than 300	Staggered	0.640	1.309
Less than 300	In-line	0.569	0.742

In these equations,

h = heat-transfer coefficient
c = specific heat
G = mass velocity = W/a_c
a_c = flow area
W = flow rate
k = thermal conductivity
D_o = outside tube diameter
μ = viscosity
μ_w = viscosity at wall temperature
μ_b = viscosity at bulk fluid temperature

For triangular and square tube patterns,

$$a_c = \frac{BD_s(s - D_o)}{s}$$

For rotated square tube patterns,

$$a_c = \frac{1.5 BD_s(s - D_o)}{s}$$

where B = baffle spacing
D_s = shell diameter
s = tube center-to-center spacing
D_o = tube outside diameter

The values of a in the preceding table are based on heat exchangers that have no bypassing of the bundle by the fluid. For heat exchangers built to the standards of the Tubular Exchangers Manufacturers Association [11] and with an adequate number of sealing devices, the heat-transfer coefficient calculated with the preceding equation must be corrected as below to reflect the bypassing of the fluid:

$$h_o = hF_1 F_r$$

where $F_1 = 0.8(B/D_s)^{1/6}$ for bundles with typical fouling
$\quad F_1 = 0.8(B/D_s)^{1/4}$ for bundles with no fouling
$\quad F_r = 1.0$ for D_oG/μ greater than 100
$\quad F_r = 0.2(D_oG/\mu)^{1/3}$ for D_oG/μ less than 100

2. Calculate D_oG/μ.

$$a_c = \frac{BD_s(s - D_o)}{s}$$

$$= \frac{(9.5/12)(25/12)(0.9375 - 0.75)}{0.9375}$$

$$= 0.3299 \text{ ft}^2 \ (0.0306 \text{ m}^2)$$

$$G = \frac{W}{a_c} = \frac{50,000}{0.3299} = 151,578.9 \text{ lb/(h)(ft}^2)$$

$$\frac{D_oG}{\mu} = \frac{(0.75/12)151,578.9}{1.95} = 4858.9$$

3. Calculate h.
Now,

$$h = \frac{0.273cG}{(c\mu/k)^{2/3}(D_oG/\mu)^{0.365}(\mu_w/\mu_b)^{0.14}}$$

Thus,

$$h = \frac{0.273(0.65)(151,578.9)}{[0.65(1.95)/0.085]^{2/3}(4858.3)^{0.365}(4.0/1.95)^{0.14}}$$

$$= 181.2 \text{ Btu/(h)(ft}^2)(°F)[1029 \text{ W/(m}^2)(K)]$$

4. Calculate h_o for the fouled condition.
Now, $h_o = hF_1F_r$, where $F_r = 1.0$ and $F_1 = 0.8 \times (B/D_s)^{1/6} = 0.8(9.5/25)^{1/6} = 0.6809$, so $h_o = 181.5(0.6809)(1.0) = 123.6$ Btu/(h)(ft²)(°F) [702 W/(m²)(K)].

5. Calculate h_o for the clean condition.
Here, $F_1 = 0.8(B/D_s)^{1/4} = 0.8(9.5/25)^{1/4} = 0.6281$, so $h_o = 181.5(0.6281)(1.0) = 114.0$ Btu/(h)(ft²)(°F) [647.5 W/(m²)(K)].

Related Calculations. The preceding equations for F_1 and F_r are based on heat exchangers that have been fabricated to minimize bypassing of the bundle by the fluid. The assumption has also been made that the baffle cut for segmental baffles is 20 percent of the shell diameter and that the layout includes tubes in the baffle window areas. For conditions removed from these assumptions, the effects of fluid bypassing should be evaluated. Several methods have been presented; a widely used one is that of Bell [15].

For gases, the preceding equations for the turbulent regime can be simplified because the Prandtl number ($c\mu/k$) and viscosity for most gases are approximately constant. Assigning the values $c\mu/k = 0.78$ and $\mu = 0.0426$ lb/(ft)(h) (0.0176 cP) results in the equation

$$h = \frac{bcG^{1-m}}{D_o^m}$$

with the variables defined in English units and with b having these values:

Reynolds number	Tube pattern	b
300 to 200,000	Staggered	0.102
300 to 200,000	In-line	0.083
Above 200,000	Staggered	0.076
Above 200,000	In-line	0.057

7.17 MEAN TEMPERATURE DIFFERENCE FOR HEAT EXCHANGERS

A warm stream enters a heat exchanger at 200°C (392°F), T_h, and is cooled to 100°C (212°F), T_c. The cooling stream enters the exchanger at 20°C (68°F), t_c, and is heated to 95°C (203°F), t_h. Calculate the mean temperature difference for the following cases:

1. Countercurrent flow
2. Cocurrent flow (also called "parallel flow")
3. 1–2 Multipass (one pass on the shell side, two passes on the tube side)
4. 2–4 Multipass (two passes on the shell side, four passes on the tube side)
5. 1–1 Cross flow (one cross-flow pass on the shell side, one pass on the tube side)
6. 1–2 Cross flow (one cross-flow pass on the shell side, two passes on the tube side)

These flow arrangements are illustrated in Fig. 7.3.

Calculation Procedure

1. Use equation for countercurrent flow.
The mean temperature difference for countercurrent flow is the log mean temperature difference as calculated from the equation

$$\Delta T_{LM} = \frac{(T_h - t_h) - (T_c - t_c)}{\ln[(T_h - t_h)/(T_c - t_c)]}$$

Thus,

$$\Delta T_{LM} = \frac{(200 - 95) - (100 - 20)}{\ln[(200 - 95)/(100 - 20)]}$$
$$= 91.93°C(165.47°F)$$

2. Use the equation for cocurrent flow.
The mean temperature difference for cocurrent flow is the log mean temperature difference as calculated from the equation

$$\Delta T_{LM} = \frac{(T_h - t_c) - (T_c - t_h)}{\ln[(T_h - t_c)/(T_c - t_h)]}$$

Thus,

$$\Delta T_{LM} = \frac{(200 - 20) - (100 - 95)}{\ln[(200 - 20)/(100 - 95)]}$$
$$= 48.83°C(87.89°F)$$

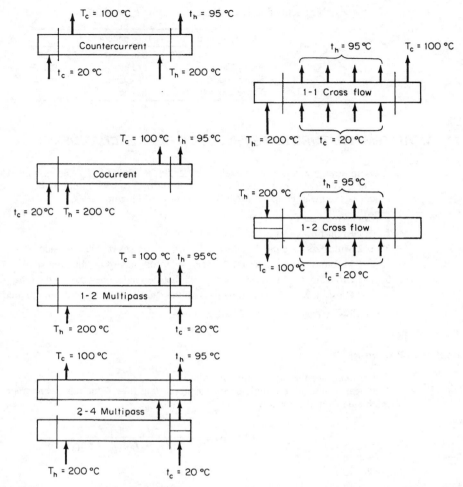

FIGURE 7.3 Flow arrangements through heat exchangers. Note: T refers to the stream being cooled; t refers to the stream being heated.

3. Use equation for 1–2 multipass.

The mean temperature difference for a 1–2 multipass heat exchanger can be calculated from the equation

$$\Delta T_m = \frac{M}{\ln[(P + M)/(P - M)]}$$

where $P = (T_h - t_h) + (T_c - t_c)$ and $M = [(T_h - T_c)^2 + (t_h - t_c)^2]^{1/2}$. Thus, $P = (200 - 95) + (100 - 20) = 185$, $M = [(200 - 100)^2 + (95 - 20)^2]^{1/2} = 125$, and

$$\Delta T_m = \frac{125}{\ln[(185 + 125)/(185 - 125)]}$$

$$= 76.12°C\,(137.02°F)$$

4. Use equation for 2–4 multipass.

The mean temperature difference for a 2–4 multipass heat exchanger can be calculated from the equation

$$\Delta T_m = \frac{M/2}{\ln[(Q + M)/(Q - M)]}$$

where M is defined above and

$$Q = \left[(T_h - t_h)^{1/2} + (T_c - t_c)^{1/2} \right]^2$$

Thus, $Q = [(200 - 95)^{1/2} + (100 - 20)^{1/2}]^2 = 368.30$, and

$$\Delta T_m = \frac{125/2}{\ln[(368.30 + 125)/(368.30 - 125)]}$$

$$= 88.42°C \ (125.16°F)$$

5. Use the correction factor for a 1–1 cross flow.

The mean temperature difference for a 1–1 cross-flow heat exchanger can be calculated by using the correction factor determined from Fig. 7.4. The mean temperature difference will be the product of this factor and the log mean temperature difference for countercurrent flow. To obtain the correction factor F, calculate the value of two parameters P and R:

$$P = \frac{t_h - t_c}{T_h - t_c}$$

$$= \frac{95 - 20}{200 - 20} = 0.42$$

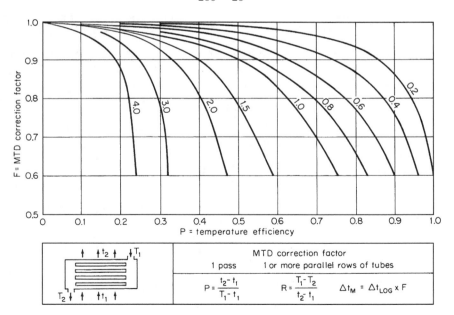

FIGURE 7.4 Correction factor for one-pass crossflow. (*From "Engineering Data Book, 1966," Natural Gas Processors Suppliers Association.*)

$$R = \frac{T_h - T_c}{t_h - t_c}$$

$$= \frac{200 - 100}{95 - 20} = 1.33$$

From Fig. 7.4, with these values of P and R, F is found to be 0.91. Then,

$$\Delta T_m = F\Delta T_{LM} = 0.91(91.93) = 83.66°C (150.58°F)$$

6. Use the correction factor for a 1–2 cross flow.
The mean temperature difference for a 1–2 cross-flow heat exchanger can be calculated using the same procedure as used for a 1–1 cross-flow heat exchanger, except that the value of F is determined from Fig.7.5.
   For $P = 0.42$ and $R = 1.33$, $F = 0.98$ (from Fig. 7.5), so

$$\Delta T_m = F\Delta T_{LM} = 0.98(91.93) = 90.09°C (162.16°F)$$

Related Calculations.   Mean temperature differences for multipass heat exchangers may also be calculated by using appropriate correction factors for the log mean temperature difference for countercurrent flow

$$\Delta T_m = F\Delta T_{LM}$$

Curves for determining values of F are presented in many of the references [1–4,6,10,11].
   For shell-and-tube heat exchangers with cross-flow baffles, the preceding methods assume that an adequate number of baffles has been provided. If the shell-side fluid makes less than eight passes across the tube bundle, the mean temperature difference may need to be corrected for this cross-flow

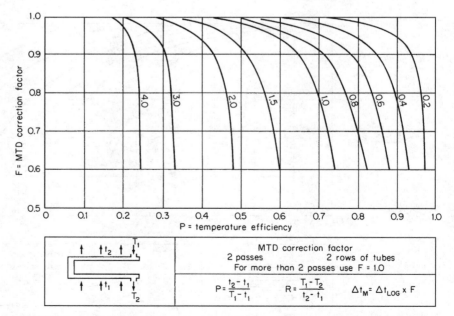

FIGURE 7.5   Correction factor for two-pass crossflow. (*From "Engineering Data Book, 1966," Natural Gas Processors Suppliers Association.*)

condition. Appropriate curves are presented in Caglayan and Buthod [20]. The curves in this reference may also be used to determine correction factors for cross-flow exchangers with one shell pass and more than two tube passes.

The methods presented above are applicable only for conditions in which the heat transferred is a straight-line function of temperature. For systems that do not meet this condition, the total heat-release curve can be treated in sections, each section of which closely approximates the straight-line requirement. A log mean temperature difference can then be calculated for each section. Common examples in which this approach is encountered include (1) total condensers in which the condensate is subcooled after condensation, and (2) vaporizers in which the fluid enters as a subcooled liquid, the liquid is heated to the saturation temperature, the fluid is vaporized, and the vapor is heated and leaves in a superheated state.

Other types of heat-exchanger configurations not illustrated here are sometimes used. Curves for determining the value of F for these are presented in Ref. 11. These configurations include divided-flow and split-flow exchangers.

7.18 OVERALL HEAT-TRANSFER COEFFICIENT FOR SHELL-AND-TUBE HEAT EXCHANGER

A shell-and-tube exchanger with the following geometry is available:

D_s = shell diameter = 25 in (0.635 m)
n = number of tubes = 532
D_o = tube outside diameter = 0.75 in (0.019 m)
D_i = tube inside diameter = 0.62 in (0.016 m)
L = tube length = 16 ft (4.88 m)
s = tube spacing = 0.9375 in, triangular (0.024 m)
B = baffle spacing = 9.5 in (0.241 m)

There is one tube-side pass and one shell-side pass, and the tube material is stainless steel with a thermal conductivity of 10 Btu/(h)(ft)(°F) [17 W/(m)(K)].

Calculate the overall heat-transfer coefficient for this heat exchanger under the following service conditions:

Tube Side

Liquid undergoing sensible-heat transfer:

W_t = flow rate = 500,000 lb/h (226,795 kg/h)
c = specific heat = 0.5 Btu/(lb)(°F) [2.1 kJ/(kg)(K)]
μ = viscosity = 1.21 lb/(ft)(h) (0.5 cP)
    specific gravity = 0.8
k = thermal conductivity = 0.075 Btu/(h)(ft)(°F) [0.13 W/(m)(K)]

Shell Side

Liquid undergoing sensible-heat transfer:

W_s = flow rate = 200,000 lb/h (90,718 kg/h)
c = specific heat = 1.0 Btu/(lb)(°F) [4.19 kJ/(kg)(K)]
μ = viscosity = 2.0 lb/(ft)(h) (0.83 cP)
    specific gravity = 1.0
k = thermal conductivity = 0.36 Btu/(h)(ft)(°F) [0.62 W/(m)(K)]

In addition, the fouling heat-transfer coefficient is 1000 Btu/(h)(ft^2)(°F) [5670 W/(m^2)(K)]. Assume that the change in viscosity with temperature is negligible, that is, $\mu_w/\mu_b = 1.0$.

Calculation Procedure

1. Calculate the Reynolds number inside the tubes.
The Reynolds number is DG/μ, where G is the mass flow rate.

$$G = \frac{500,000}{532\,\text{tubes}\,(0.62/12)^2(\pi/4)} = 448,278 \ \text{lb/(h)(ft)}^2$$

$$DG/\mu = \frac{(0.62/12)448,278}{1.21} = 19,141$$

2. Calculate h_i, the heat-transfer coefficient inside the tubes.
For Reynolds numbers greater than 8000,

$$h_i = \frac{0.023cG}{(c\mu/k)^{2/3}(DG/\mu)^{0.2}}$$

Thus,

$$h_i = \frac{0.023(0.5)(448,278)}{[0.5(1.21)/0.075]^{2/3}19,141^{0.2}}$$

$$= 178.6 \ \text{Btu/(h)(ft}^2)(°F)[1013 \ \text{W/(m}^2)(K)]$$

3. Calculate the Reynolds number for the shell side.
For the shell side, $G = W_s/a_c$, where the flow area $a_c = BD_s(s - D_o)/s$. Then,

$$a_c = \frac{(9.5/12)(25/12)(0.9375 - 0.75)}{0.9375}$$

$$= 0.33\,\text{ft}^2(0.0307\,\text{m}^2)$$

So,

$$D_oG/\mu = \frac{(0.75/12)(200,000/0.33)}{2.0}$$

$$= 18,940$$

4. Calculate h_o, the heat-transfer coefficient outside the tubes.
Now,

$$h = \frac{0.273cG}{(c\mu/k)^{2/3}(D_oG/\mu)^{0.365}}$$

$$= \frac{0.273(1.0)(200,000/0.33)}{[1.0(2.0)/0.36]^{2/3}18,940^{0.365}}$$

$$= 1450.7\,\text{Btu/(h)(ft}^2)(°F)$$

Then, correcting for flow bypassing (see Procedure 7.16),

$$h_o = hF_1F_r = h(0.8)(B/D_s)^{1/6}(1.0)$$

So,

$$h_o = 1450.7(0.8)(9.5/25)^{1/6}(1.0)$$

$$= 987.7\,\text{Btu/(h)(ft}^2)(°F)[5608 \ \text{W/(m}^2)(K)]$$

5. Calculate h_w, the heat-transfer coefficient across the tube wall.
The heat-transfer coefficient across the tube wall h_w can be calculated from $h_w = 2k/(D_o - D_i)$. Thus,

$$h_w = \frac{2(10)}{(0.75/12) - (0.62/12)}$$

$$= 1846 \, \text{Btu}/(\text{h})(\text{ft}^2)(°\text{F})[10.480 \, \text{W}/(\text{m}^2)(\text{K})]$$

6. Calculate the overall heat-transfer coefficient.
The formula is

$$\frac{1}{U} = \frac{1}{h_o} + \frac{1}{h_i(D_i/D_o)} + \frac{1}{h_w} + \frac{1}{h_s}$$

where U = overall heat-transfer coefficient
h_o = outside heat-transfer coefficient
h_i = inside heat-transfer coefficient
h_w = heat transfer across tube wall
h_s = fouling heat-transfer coefficient
D_i = inside diameter
D_o = outside diameter

Then,

$$\frac{1}{U} = \frac{1}{987.7} + \frac{1}{178.6(0.62/0.75)} + \frac{1}{1846} + \frac{1}{1000}$$

So, $U = 107.2 \, \text{Btu}/(\text{h})(\text{ft}^2)(°\text{F}) \, [608.8 \, \text{W}/(\text{m}^2)(\text{K})]$.

Related Calculations. The method described for calculating the overall heat-transfer coefficient is also used to calculate the overall resistance to conduction of heat through a composite wall containing materials in series that have different thicknesses and thermal conductivities. For this case, each individual heat-transfer coefficient is equal to the thermal conductivity of a particular material divided by its thickness. The amount of heat transferred by conduction can then be determined from the formula

$$Q = (t_1 - t_2)\frac{1}{R}A$$

where Q = heat transferred
t_1 = temperature of the hot surface
t_2 = temperature of the cold surface
A = area of wall
R = overall resistance, which equals

$$\frac{x_1}{k_1} + \frac{x_2}{k_2} + \frac{x_3}{k_3} + \cdots$$

where x = thickness
k = thermal conductivity

It is seen from inspection that $U = 1/R$.

7.19   OUTLET TEMPERATURES FOR COUNTERCURRENT HEAT EXCHANGER

For the heat exchanger described in Procedure 7.18, calculate the outlet temperatures and the amount of heat transferred if the tube-side fluid enters at 68°F (293 K) and the shell-side fluid enters at 500°F (533 K).

Calculation Procedure

1. Determine the heat-transfer surface.
The surface $A = \pi n D_o L = \pi(532)(0.75/12)(16)$, so $A = 1670$ ft^2 (155.2 m^2).

2. Determine the thermal effectiveness.
The thermal effectiveness of a countercurrent heat exchanger can be determined from Fig. 7.6. Now,

$$\frac{U\,A}{wc} = \frac{107.2(1670)}{500,000(0.5)}$$
$$= 0.716$$

and,

$$R = \frac{500,000(0.5)}{200,000(1.0)}$$
$$= 1.25$$

So, from Fig. 7.6, the thermal effectiveness $P = 0.39$.

3. Determine outlet temperatures.
Because $P = (t_2 - t_1)/(T_1 - t_1) = 0.39$,

$$t_2 = t_1 + P(T_1 - t_1) = 68 + (0.39)(500 - 68)$$
$$= 236.5°F$$

Now, by definition, $wc(t_2 - t_1) = WC(T_1 - T_2)$, and $R = wc/(WC)$, so $T_2 = T_1 - R(t_2 - t_1)$. Thus,

$$T_2 = 500 - 1.25(236.5 - 68)$$
$$= 289.4°F(416 \text{ K})$$

4. Determine the amount of heat transferred.
Use the formula,

$$Q = wc(t_2 - t_1)$$

where Q = heat rate
$\quad w$ = flow rate for fluid being heated
$\quad c$ = specific heat for fluid being heated

Then,

$$Q = 500,000(0.5)(236.5 - 68)$$
$$= 42,120,000 \text{ Btu/h} (12,343,000 \text{ W})$$

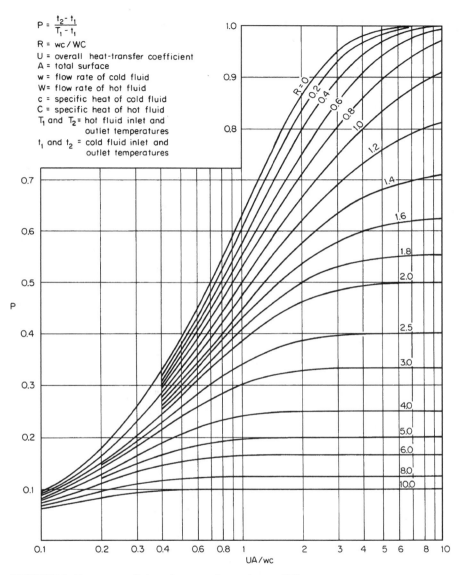

$$P = \frac{t_2 - t_1}{T_1 - t_1}$$

$R = wc/WC$
U = overall heat-transfer coefficient
A = total surface
w = flow rate of cold fluid
W = flow rate of hot fluid
c = specific heat of cold fluid
C = specific heat of hot fluid
T_1 and T_2 = hot fluid inlet and
            outlet temperatures
t_1 and t_2 = cold fluid inlet and
            outlet temperatures

FIGURE 7.6   Temperature efficiency for counterflow exchangers [11].

(The calculation can instead be based on the temperatures, flow rate, and specific heat for the fluid being cooled.)

Related Calculations.   Figure 7.6 can also be used to determine the thermal effectiveness for exchangers in which one fluid is isothermal. For this case, $R = 0$.

7.20   OUTLET TEMPERATURES FOR 1–2 MULTIPASS HEAT EXCHANGER

Calculate the outlet temperatures and the amount of heat transferred for the exchanger described in Procedure 7.19 if the tube side is converted from single-pass to two-pass.

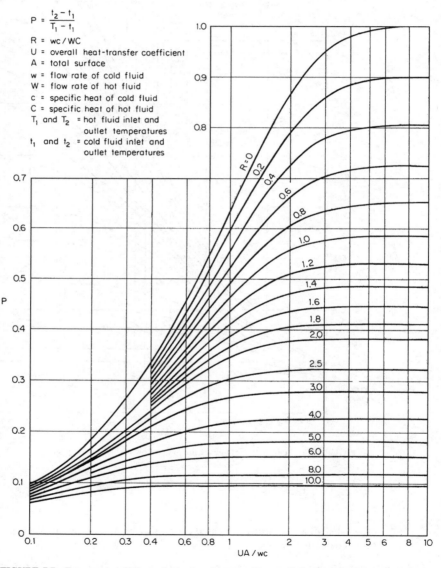

$$P = \frac{t_2 - t_1}{T_1 - t_1}$$

R = wc / WC
U = overall heat-transfer coefficient
A = total surface
w = flow rate of cold fluid
W = flow rate of hot fluid
c = specific heat of cold fluid
C = specific heat of hot fluid
T_1 and T_2 = hot fluid inlet and outlet temperatures
t_1 and t_2 = cold fluid inlet and outlet temperatures

FIGURE 7.7   Temperature efficiency for heat exchangers with one shell pass and even number of tube passes [11].

Calculation Procedure

1. Determine the overall heat-transfer coefficient.

The heat-transfer coefficient on the tube side is proportional to $G^{0.8}$. The mass velocity G will be doubled because the exchanger is to be converted to two passes on the tube side. Therefore, $h_i = 178.6(2)^{0.8} = 310.9$ Btu/(h)(ft^2)(°F) [1764 W/(m^2)(K)]. The new overall heat-transfer coefficient can now be calculated:

$$\frac{1}{U} = \frac{1}{987.7} + \frac{1}{(310.9)(0.62/0.75)} + \frac{1}{1846} + \frac{1}{1000}$$

$$= 155.2 \text{ Btu/(h)(ft}^2\text{)(°F)}[880.4 \text{ W/(m}^2\text{)(K)}]$$

2. Determine the thermal effectiveness.

The thermal effectiveness of a 1–2 multipass exchanger can be determined from Fig. 7.7. Now,

$$\frac{U A}{wc} = \frac{155.2(1670)}{500,000(0.5)}$$

$$= 1.037$$

and $R = 1.25$, so from Fig. 7.7, the thermal effectiveness $P = 0.43$.

3. Determine the outlet temperatures.

See Procedure 7.19, step 3. Now, $t_2 - t_1 = 0.43(500 - 68) = 185.8$°F, so $t_2 = 185.8 + 68 = 253.8$°F; and $T_2 - T_1 = R(t_2 - t_1) = 1.25(185.8) = 232.2$°F, so $T_2 = 500 - 232.2 = 267.8$°F (404 K).

4. Calculate the amount of heat transferred.

See Procedure 7.19, step 4. Now, $Q = 500,000 \times (0.5)(185.8) = 46,450,000$ Btu/h (13,610,000 W).

Related Calculations. The thermal effectiveness of a 2–4 multipass heat exchanger can be determined from Fig. 7.8. The value of R for exchangers that have one isothermal fluid is zero, for both 1–2 and 2–4 multipass heat exchangers.

7.21 CONDENSATION FOR VERTICAL TUBES

Calculate the condensing coefficient for a vertical tube with an inside diameter of 0.62 in (0.016 m) if steam with these properties is condensing on the inside of the tube at a rate of 50 lb/h (22.68 kg/h):

ρ_L = liquid density = 60 lb/ft^3 (961.1 kg/m^3)
ρ_v = vapor density = 0.0372 lb/ft^3 (0.60 kg/m^3)
μ_L = liquid viscosity = 0.72 lb/(ft)(h) (0.298 cP)
μ_v = vapor viscosity = 0.0313 lb/(ft)(h) (0.0129 cP)
c = liquid specific heat = 1.0 Btu/(lb)(°F) [4.19 kJ/(kg)(K)]
k = liquid thermal conductivity = 0.395 Btu/(h)(ft)(°F) [0.683 W/(m)(K)]

Calculation Procedure

1. Select the calculation method to be used.

Use the Dukler theory [18], which assumes that three fixed factors must be known to establish the value of the average heat-transfer coefficient for condensing inside vertical tubes. These are the

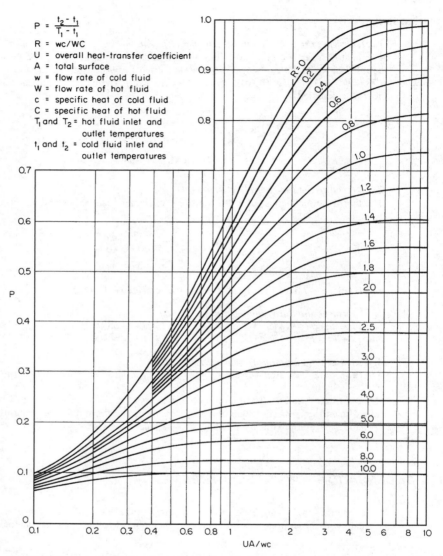

$$P = \frac{t_2 - t_1}{T_1 - t_1}$$

R = wc/WC
U = overall heat-transfer coefficient
A = total surface
w = flow rate of cold fluid
W = flow rate of hot fluid
c = specific heat of cold fluid
C = specific heat of hot fluid
T_1 and T_2 = hot fluid inlet and
    outlet temperatures
t_1 and t_2 = cold fluid inlet and
    outlet temperatures

FIGURE 7.8 Temperature efficiency for heat exchangers with two shell passes and with four or a multiple of four tube passes [11].

terminal Reynolds number $(4\Gamma/\mu)$, the Prandtl number $(c\mu/k)$ of the condensed phase, and a dimensionless group designated A_d and defined as follows:

$$A_d = \frac{0.250\mu_L^{1.173}\mu_v^{0.16}}{g^{2/3}D_i^2\rho_L^{0.553}\rho_v^{0.78}}$$

In these equations,

$$\Gamma = W/(n\pi D_i)$$

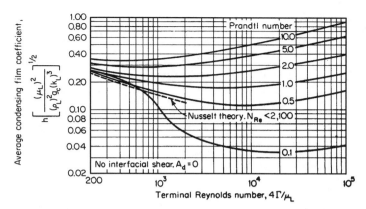

FIGURE 7.9 Dukler plot for condensing-film coefficients.

where W = mass flow rate
n = number of tubes
D_i = inside tube diameter
ρ_L = liquid density
ρ_v = vapor density
g = gravitational constant
μ_L = liquid viscosity
μ_o = vapor viscosity

The Reynolds and Prandtl numbers are related to the condensing coefficient in Fig. 7.9. However, that figure is based on $A_d = 0$, which assumes no interfacial shear. The following factors can be used to evaluate the effects of any interfacial shear (h = coefficient with interfacial shear; h_0 = coefficient with no interfacial shear):

Terminal Reynolds number $4\Gamma/\mu$	h/h_0			
	$A_d = 0$	$A_d = 10^{-5}$	$A_d = 10^{-4}$	$A_d = 2 \times 10^{-4}$
200	1.0	1.0	1.03	1.05
500	1.0	1.03	1.15	1.28
1000	1.0	1.07	1.40	1.68
3000	1.0	1.25	2.25	2.80
10,000	1.0	1.90	4.35	6.00
30,000	1.0	3.30	8.75	13.00

2. Calculate the terminal Reynolds number $4\Gamma/\mu$.
Since $\Gamma = W/(n\pi D_i) = 50/[1.0\pi(0.62/12)] = 308$ lb/(h)(ft), $4\Gamma/\mu = 4(308)/0.72 = 1711.3$.

3. Calculate the Prandtl number.
Thus, $c\mu/k = 1.0(0.72)/0.395 = 1.82$.

4. Calculate h_0.
Refer to Fig. 7.9. For $4\Gamma/\mu = 1711.3$ and $c\mu/k = 1.82$, $(h\mu_L^2/\rho_L^2 gk^3)^{1/3} = 0.21$. Therefore,

$$h = 0.21\left[\frac{60^2(4.18 \times 10^8)0.395^3}{0.72^2}\right]^{1/3}$$

$$= 1183 \text{ Btu/(h)(ft}^2)(°F)[6720 \text{ W/(m}^2)(K)]$$

5. Calculate A_d.
Thus,

$$A_d = \frac{0.250\mu_L^{1.173}\mu_v^{0.16}}{g^{2/3}D_i^2\rho_L^{0.553}\rho_v^{0.78}}$$

$$= \frac{(0.250)(0.72)^{1.173}(0.0313)^{0.16}}{(4.18\times10^8)^{2/3}(0.62/12)^2(60)^{0.553}(0.0374)^{0.78}}$$

$$= 8.876\times10^{-5}$$

6. Calculate h.
For $A_d = 8.876\times10^{-5}$ and $4\Gamma/\mu = 1711.3$, h/h_0 is approximately 1.65 from step 1. Then, $h = 1.65\,h_0 = 1.65(1183) = 1952$ Btu/(h)(ft²)(°F) [11,088 W/(m²)(K)].

Related Calculations. For low values of Reynolds number ($4\Gamma/\mu$), the Nusselt equation can be used to predict condensing heat-transfer coefficients for vertical tubes:

$$h = 0.925k\left(\frac{\rho^2 g}{\mu\Gamma}\right)^{1/3}$$

where h = heat-transfer coefficient
k = liquid thermal conductivity
ρ = liquid density
g = gravitational constant
μ = liquid viscosity
Γ = condensate rate per unit periphery

It can be seen from Fig. 7.9 that the condensing heat-transfer coefficient for a fluid with a Prandtl number of approximately 2 (for instance, steam) is not strongly dependent on flow rate or Reynolds number. For this reason, heat-transfer coefficients for steam condensing on vertical tubes are frequently not calculated, but are assigned a value of 1500 to 2000 Btu/(h)(ft²)(°F) [8500 to 11,340 W/(m²)(K)].

This approach can be used for condensing on the outside of vertical tubes. The equivalent diameter should be used in evaluating the value of A_d, and the outside tube diameter should be used in calculating the terminal Reynolds number.

Heat-transfer coefficients for falling films can also be predicted from Fig. 7.9. The coefficient obtained should be multiplied by 0.75 to obtain heat-transfer coefficients for falling-film mechanisms.

A minimum flow rate is required to produce a falling film on vertical tubes when not condensing. The minimum flow rate can be predicted with the equation

$$\Gamma_{min} = 19.5(\mu s\sigma^3)^{1/5}$$

where Γ_{min} = tube loading, in lb/(h)(ft)
μ = liquid viscosity, in centipoise
s = liquid specific gravity
σ = surface tension, in dyn/cm

If this minimum is not achieved, the perimeter of the tube will not be uniformly wetted. Once a falling film has been induced, however, a lower terminal flow rate can be realized without destroying the film. This minimum terminal rate can be predicted with the equation

$$\Gamma_T = 2.4(\mu s\sigma^3)^{1/5}$$

where Γ_T = terminal tube loading, in lb/(h)(ft).

This criterion establishes the maximum amount of material that can be vaporized with a falling-film vaporizer. Approximately 85 percent of the entering material can be vaporized in a single pass without destroying the film. If tube loadings below the terminal loading are attempted, the film will break and form rivulets. Part of the tube surface will not be wetted, and the result will be reduced heat transfer with possible increased fouling of the heat-transfer surface.

For vertical condensers, condensate can be readily subcooled if required. The subcooling occurs as falling-film heat transfer, so the procedure discussed for falling-film heat exchangers can be used to calculate heat-transfer coefficients.

Condensation of mixed vapors of immiscible liquids is not well understood. The conservative approach is to assume that two condensate films are present and all the heat must be transferred through both films in series. Another approach is to use a mass fraction average thermal conductivity and calculate the heat-transfer coefficient using the viscosity of the film-forming component (the organic component for water-organic mixtures).

The recommended approach is to use a shared-surface model and calculate the effective heat-transfer coefficient as

$$h_L = V_A h_A + (1 - V_A)h_B$$

where h_L = the effective heat-transfer coefficient
h_A = the heat-transfer coefficient for liquid A assuming it only is present
h_B = the heat-transfer coefficient for liquid B assuming it only is present
V_A = the volume fraction of liquid A in the condensate

7.22   CONDENSATION INSIDE HORIZONTAL TUBES

Calculate the effective condensing coefficient for a horizontal tube with an inside diameter of 0.62 in (0.016 m) and a length of 9 ft (2.74 m) for a fluid with the following properties condensing at a rate of 126 lb/h (57.15 kg/h):

ρ_L = liquid density = 50 lb/ft^3 (800.9 kg/m^3)
ρ_v = vapor density = 1.0 lb/ft^3 (16.02 kg/m^3)
μ = liquid viscosity = 0.25 lb/(ft)(h) (0.104 cP)
c = liquid specific heat = 0.55 Btu/(lb)(°F) [2.30 kJ/(kg)(K)]
k = liquid thermal conductivity = 0.08 Btu/(h)(ft)(°F) [0.14 W/(m)(K)]

Calculation Procedure

1. Select the calculation method to be used.
Condensation inside horizontal tubes can be predicted assuming two mechanisms. The first assumes stratified flow, with laminar film condensation. The second assumes annular flow and is approximated with single-phase heat transfer using an equivalent mass velocity to reflect the two-phase flow. For the stratified-flow assumption, the further assumption is made that the rate of condensation on the stratified layer of liquid running along the bottom of the tube is negligible. Consequently, this layer of liquid must not exceed values assumed without being appropriately accounted for.

The following equations can be used to predict heat-transfer coefficients for condensation inside horizontal tubes: For stratified flow,

$$h_c = 0.767k\left(\frac{\rho_L^2 g L}{n\mu W}\right)^{1/3}$$

For annular flows,

a. $h_a D_i/k = 0.0265(c\mu/k)^{1/3}(DG_E/\mu)^{0.8}$
   when Re_L is greater than 5000 and Re_v greater than 20,000, or

b. $h_a D_i/k = 5.03(c\mu/k)^{1/3}(DG_E/\mu)^{1/3}$
   when Re_L is less than 5000 or Re_v less than 20,000.

In these equations, $Re_L = DG_L/\mu$, $Re_v = (DG_v/\mu)(\rho_L/\rho_v)^{1/2}$, and $G_E = G_L + G_v(\rho_L/\rho_v)^{1/2}$,

where h = heat-transfer coefficient
 ρ_L = liquid density
 ρ_v = vapor density
 g = gravitational constant
 k = liquid thermal conductivity
 μ = liquid viscosity
 L = tube length
 n = number of tubes
 W = condensate flow rate
 G_E = equivalent mass velocity
 G_L = liquid mass velocity assuming only liquid is flowing
 G_v = vapor mass velocity assuming only vapor is flowing
 D_i = inside tube diameter
 c = liquid specific heat
 Re_L = liquid Reynolds number
 Re_v = vapor Reynolds number, defined above

Calculate the heat-transfer coefficient using both mechanisms and select the higher value calculated as the effective heat-transfer coefficient h_L. The annular-flow assumption results in heat-transfer coefficients that vary along the tube length. The condenser should be broken into increments, with the average vapor and liquid flow rates for each increment used to calculate heat-transfer coefficients. The total is the integrated value of all the increments.

For this problem, we calculate the heat-transfer coefficient for nine increments, each 1 ft (0.305 m) long and each condensing the same amount. The actual calculations are shown for only three increments—the first, the middle, and the last. All nine increments, however, are shown in the summary.

2. Calculate h_c, the condensing coefficient if stratified flow is assumed.
Because an equal amount condenses in each increment, h_c will be the same in each increment. Thus, $W_{c,av} = 126/9 = 14$ lb/(h)(ft). Therefore,

$$
\begin{aligned}
h_c &= 0.767k\left(\frac{\rho_L^2 gL}{n\mu W_{c,av}}\right)^{1/3} \\
&= 0.767(0.08)\left[\frac{50^2(4.18 \times 10^8)(1.0)}{1(0.25)(14)}\right]^{1/3} \\
&= 410.1 \text{ Btu/(h)(ft}^2)(°F)[2330 \text{ W/(m}^2)(K)]
\end{aligned}
$$

3. Evaluate the effect of condensate loading on h_c.
The preceding equation for h_c assumes a certain condensate level on the bottom of the tube. This should be evaluated, which can be done by comparing the value of $W/(n\rho D_i^{2.56})$ the values shown

TABLE 7.3 Effect of Condensate Loading on Condensing Coefficient

$W/\left(n\rho D_i^{2.56}\right)$	h/h_c	$W/\left(n\rho D_i^{2.56}\right)$	h/h_c
0	1.30	7000	0.85
50	1.25	9000	0.80
200	1.20	10,000	0.75
500	1.15	12,000	0.70
750	1.10	17,000	0.60
1500	1.05	20,000	0.50
2650	1.00	25,000	0.40
4000	0.95	30,000	0.25
6000	0.90	35,000	0.00

in Table 7.3, W being the condensate flow rate at the end of the tube, in pounds per hour, ρ is the density, in pounds per cubic foot, and D_i is the inside diameter, in feet.

Now $W/(n\rho D_i^{2.56}) = 126/[1(50)(0.62/12)^{2.56}] = 4961$. From Table 7.3, $h/h_c = 0.92$, so $h = 0.92(410.1) = 377.3$ Btu/(h)(ft²)(°F) [2144 W/(m²)(K)].

4. Calculate h_a, the condensing coefficient if annular flow is assumed.
As noted earlier, this coefficient must be calculated separately for each increment of tube length. Here are the calculations for three of the nine increments:

First Increment

a. *Determine average liquid and vapor flow rates.* Of the total vapor input, 126 lb/h, there is 14 lb/h condensed in this increment. Therefore, $W_{L1} = (0 + 14)/2 = 7$ lb/h of liquid, and $W_{v1} = 126 - 7 = 119$ lb/h of vapor.

b. *Calculate G_E and $D_i G_{E/\mu}$.* Thus,

$$G_L = \frac{W_{L1}}{a} = \frac{7}{(0.62/12)^2(\pi/4)} = 3338.8 \,\text{lb/(h)(ft}^2)$$

$$G_v = \frac{W_{v1}}{a} = \frac{119}{(0.62/12)^2(\pi/4)} = 56,759.2$$

$$G_E = G_L + G_v(\rho_L/\rho_v)^{1/2} = 3338.8 + (56,759.2)(50/1.0)^{1/2}$$
$$= 3338.8 + 401,348.1 = 404,686.9 \,\text{lb/(h)(ft}^2)$$

$$\frac{D_i G_E}{\mu} = \frac{(0.62/12)404,686.9}{0.25} = 83.635.3$$

c. *Determine Re_v and Re_L.* Thus,

$$Re_v = \frac{D_i G_v(\rho_L/\rho_v)^{1/2}}{\mu} = \frac{(0.62/12)401,348.1}{0.25}$$
$$= 82,945.3$$

$$Re_L = \frac{D_i G_L}{\mu} = \frac{(0.62/12)3338.8}{0.25}$$
$$= 690.0$$

d. *Calculate h_a.* Since Re_v is greater than 20,000 but Re_L less than 5000, $h_a = 5.03(k/D_i)$ $(c\mu/k)^{1/3}(D_iG_E/\mu)^{1/3}$. Now, $(c\mu/k)^{1/3} = [0.55(0.25)/0.08^{1/3}] = 1.1979$, so

$$h_a = 5.03\frac{0.08}{(0.62/12)}1.1979(83,635.3)^{1/3}$$

$$= 408.0 \text{ Btu/(h)(ft}^2)(°F)[2318 \text{ W/(m}^2)(K)]$$

Middle (Fifth) Increment

a. *Determine average liquid and vapor flow rates.* Since the four upstream increments each condensed 14 lb/h, $W_{L5} = 14(4 + 5)/2 = 63$ lb/h and $W_{v5} = 126 - 63 = 63$ lb/h.

b. *Calculate G_E and D_iG_E/μ.* Now, $G_L = G_v = W_{L5}/a = 63/[(0.62/12)^2(\pi/4)] = 30,049.0$ lb/(h)(ft^2); $G_E = G_L + G_v(\rho_L/\rho_v)^{1/2} = 30,049.0 + 30,049.0(50/1.0)^{1/2} = 30,049.0 + 212,478.4 = 242,527.4$ lb/(h)(ft^2); and $D_iG_E/\mu = (0.62/12)242,527.4/0.25 = 50,122.3$.

c. *Determine Re_v and R_{eL}.* Now,

$$Re_v = \frac{D_iG_v(\rho_L/\rho_v)^{1/2}}{\mu} = \frac{(0.62/12)212,478.4}{0.25}$$

$$= 43,912.2$$

$$Re_L = \frac{D_iG_L}{\mu} = \frac{(0.62/12)30,049.0}{0.25}$$

$$= 6210.1$$

d. *Calculate h_a.* Since Re_v is greater than 20,000 and Re_L is greater than 5000,

$$h_a = 0.0265(k/D_i)(c\mu/k)^{1/3}(D_iG_E/\mu)^{0.8}$$

$$= 0.0265[0.08/(0.62/12)]1.1979(50,122.3)^{0.8}$$

$$= 282.8 \text{ Btu/(h)(ft}^2)(°F)[1607 \text{ W/(m}^2)(K)]$$

Last Increment

a. *Determine average liquid and vapor flow rates.* Now, $W_{L9} = 14(8 + 9)/2 = 119$ lb/h and $W_{v9} = 126 - 119 = 7$ lb/h.

b. *Calculate G_E and D_iG_E/μ.* Now, $G_L = 119/[(0.62/12)^2(\pi/4)] = 56,759.2$ lb/(h)(ft^2); $G_v = 7/[(0.62/12)^2(\pi/4)] = 3338.8$ lb/(h)(ft^2); $G_E = G_L + G_v(\rho_L/\rho_v)^{1/2} = 56,759.2 + 3338.8(50/1.0)^{1/2} = 56,759.2 + 23,608.7 = 80,367.9$ lb/(h)(ft^2); and $D_iG_E/\mu = (0.62/12)80,367.9/0.25 = 16,609.4$.

c. *Determine Re_v and Re_L.* Thus, $Re_L = D_iG_L/\mu = (0.62/12)56,759.2/0.25 = 11,730.2$; $Re_v = D_iG_v(\rho_L/\rho_v)^{1/2}/\mu = (0.62/12)23.608.7/0.25 = 4879.1$.

d. *Calculate h_a.* Since Re_v is less than 20,000 and Re_L greater than 5000,

$$h_a = 5.03(k/D_i)(c\mu/k)^{1/3}(D_iG_E/\mu)^{1/3}$$

$$= 5.03[0.08/(0.62/12)]1.1979(16,609.4)^{1/3}$$

$$= 238.0 \text{ Btu/(h)(ft}^2)(°F)[1353 \text{ W/(m}^2)(K)]$$

5. *Compare heat-transfer coefficients.*
The effective heat-transfer coefficient h_L is the larger value of h_c and h_a, as indicated here:

Increment	h_c	h_a	h_L
1	377.3	408.0	408.0
2	377.3	392.7	392.7
3	377.3	378.7	378.7
4	377.3	362.2	377.3
5	377.3	282.8	377.3
6	377.3	244.3	377.3
7	377.3	204.3	377.3
8	377.3	272.8	377.3
9	377.3	238.0	377.3

The average value of h_L is 382.6 Btu/(h)(ft^2)(°F) [2174 W/(m^2)(K)].

Related Calculations. Condensation of mixed vapors of immiscible liquids can be treated in the same manner as outlined in Procedure 7.21.

No good methods are available for calculating heat-transfer coefficients when appreciable subcooling of the condensate is required. A conservative approach is to calculate a superficial mass velocity assuming the condensate fills the entire tube and use the equations presented above for single-phase heat transfer inside tubes. This method is less conservative for higher condensate loads.

7.23 CONDENSATION OUTSIDE HORIZONTAL TUBES

For the following conditions, calculate the heat-transfer coefficient when condensing at a rate of 54,000 lb/h (24,493.9 kg/h) on the outside of a tube bundle with a diameter of 25 in (0.635 m) with nine baffle sections each 12 in (0.305 m) long. The bundle contains 532 tubes with an outside diameter of 0.75 in (0.019 m). The tubes are on a triangular pitch and are spaced 0.9375 in (0.02381 m) center to center. Assume equal amounts condense in each baffle section.

Physical Properties

ρ_L = liquid density = 87.5 lb/ft^3 (1401.6 kg/m^3)
ρ_v = vapor density = 1.03 lb/ft^3 (16.5 kg/m^3)
μ = liquid viscosity = 0.7 lb/(ft)(h) (0.29 cP)
c = liquid specific heat = 0.22 Btu/(lb)(°F) [0.92 kJ/(kg)(K)]
k = liquid thermal conductivity = 0.05 Btu/(h)(ft)(°F) [0.086 W/(m)(K)]

Calculation Procedure

1. *Select the calculation method to be used.*
Condensation on the outside of banks of horizontal tubes can be predicted assuming two mechanisms. The first assumes laminar condensate flow; the second assumes that vapor shear dominates the heat transfer. The following equations can be used to predict heat-transfer coefficients for condensation on banks of horizontal tubes: For laminar-film condensation,

$$h_c = ak\left(\frac{\rho_L^2 gnL}{\mu W}\right)^{1/3}\left(\frac{1}{N_r}\right)^{1/6}$$

where a = 0.951 for triangular tube patterns, 0.904 for rotated square tube patterns, or 0.856 for square tube patterns.

For vapor-shear-dominated condensation,

$$\frac{h_s D_o}{k} = b \left(\frac{D_o \rho_L v_G}{\mu} \right)^{1/2} \left(\frac{1}{N_r} \right)^{1/6}$$

where $b = 0.42$ for triangular tube patterns, 0.39 for square tube patterns, or 0.43 for rotated square tube patterns. In these equations, h_c is the laminar-film heat-transfer coefficient, h_s is the vapor-shear-dominated heat-transfer coefficient, k is liquid thermal conductivity, ρ_L is liquid density, g is the gravitational constant, μ is liquid viscosity, D_o is outside tube diameter, L is tube length, W is condensate flow rate, n is the number of tubes, v_G is maximum vapor velocity (defined below), and N_r is the number of vertical tube rows (defined below). Specifically,

$$N_r = \frac{m D_s}{s}$$

where D_s = shell diameter
 s = tube center-to-center spacing
 m = 1.0 for square tube patterns, 1.155 for triangular tube patterns, or 0.707 for rotated square tube patterns

and,

$$v_G = \frac{W_v}{\rho_v a_c}$$

where W_v = vapor flow rate
 ρ_v = vapor density
 a_c = minimum flow area

In this equation, $a_c = B D_s (s - D_o)/s$ for square and triangular tube patterns or $a_c = 1.5 B D_s (s - D_o)/s$ for rotated square tube patterns, B being the baffle spacing.

Calculate the heat-transfer coefficient using both mechanisms, and select the higher value calculated as the effective heat-transfer coefficient h_L. The vapor-shear effects vary for each typical baffle section. The condenser should be calculated in increments, with the average vapor velocity for each increment used to calculate vapor-shear heat-transfer coefficients.

When the heat-transfer coefficients for laminar flow and for vapor shear are nearly equal, the effective heat-transfer coefficient is increased above the higher of the two values. The following table permits the increase to be approximated:

h_s/h_c	h_L/h_c
0.5	1.05
0.75	1.125
1.0	1.20
1.25	1.125
1.5	1.05

For this problem, we calculate the heat-transfer coefficient for the first increment, the middle increment, and the last increment. The summary will show the results of all increments. The increment chosen for illustration is one baffle section.

2. Calculate h_c, the condensing coefficient if laminar-film condensation is assumed.
Because an equal amount condenses in each increment, h_c will be the same in each increment. The average condensate flow rate in each increment will be $W_i = W_c/9 = 54{,}000/9 = 6000$ lb/h.

Now,

$$h_c = 0.951k\left(\frac{\rho_L^2 g^n L}{\mu W_i}\right)^{1/3}\left(\frac{1}{N_r}\right)^{1/6}$$

where $N_r = 1.55D_s/s = 1.155(25)/0.9375 = 30.8$. So,

$$h_c = 0.951(0.05)\left[\frac{87.5^2(4.18\times10^8)(532)1.0}{0.7(6000)}\right]^{1/3}\left(\frac{1}{30.8}\right)^{1/6}$$

$$= 198.7\,\text{Btu/(h)(ft}^2\text{)(}°\text{F)}[1129\,\text{W/(m}^2\text{)(K)}]$$

3. Calculate h_s, the coefficient if vapor-shear domination is assumed.

As noted earlier, this must be calculated separately for each increment. Here are the calculations for three of the nine increments:

First Increment

a. *Calculate average vapor velocity.* Now, $W_{v1} = (54,000 + 54,000 - 6000)/2 = 51,000$ lb/h. And,

$$a_c = \frac{BD_s(s - D_o)}{s}$$

$$= \frac{(12/12)(25/12)(0.9375 - 0.75)}{0.9375}$$

$$= 0.4167\,\text{ft}^2\,(0.0387\,\text{m}^2)$$

Therefore,

$$v_G = \frac{W_{v1}}{\rho a_c}$$

$$= \frac{51,000}{1.03(0.4167)}$$

$$= 118,825.4\,\text{ft/h}$$

b. *Calculate h_s.* Now,

$$h_s = 0.42\left(\frac{k}{D_o}\right)\left(\frac{D_o\rho_L v_G}{\mu}\right)^{1/2}\left(\frac{1}{N_r}\right)^{1/6}$$

$$= 0.42\frac{0.05}{(0.75/12)}\left[\frac{(0.75/12)(87.5)118,825.4}{0.7}\right]^{1/2}\left(\frac{1}{30.8}\right)^{1/6}$$

$$= 182.9\,\text{Btu/(h)(ft}^2\text{)(}°\text{F)}[1039\,\text{W/(m}^2\text{)(K)}]$$

c. *Calculate h_L.* Now, $h_s/h_c = 182.9/198.7 = 0.921$. From the preceding table, $h_L/h_c = 1.18$. So, $h_L = 1.18(198.7) = 234.5$ Btu/(h)(ft^2)(°F) [1332 W/(m^2)(K)].

Middle (Fifth) Increment

a. *Determine average vapor flow rate.* Thus, $W_{v5} = 54,000 - 4(6000) - 0.5(6000) = 27,000$ lb/h.

b. *Calculate* h_s. Refer to the calculation for the first increment. Then, $h_s = 182.9(27,000/51,000)^{1/2} = 133.1$ Btu/(h)(ft²)(°F).

c. *Calculate* h_L. Thus, $h_s/h_c = 133.1/198.7 = 0.67$. From the preceding table, $h_L/h_c = 1.105$. So, $h_L = 1.105(198.7) = 219.6$ Btu/(h)(ft²)(°F) [1248 W/(m²)(K)].

Last Increment

a. *Determine average vapor flow rate.* Thus, $W_{v9} = 54,000 - 8(6000) - 0.5(6000) = 3000$ lb/h.

b. *Calculate* h_s. Refer to the calculation for the first increment. Then, $h_s = 182.9(3000/51,000)^{1/2} = 44.4$ Btu/(h)(ft²)(°F).

c. *Calculate* h_L. Now, $h_s/h_c = 44.4/198.7 = 0.223$. For this low value, assume that $h_L = h_c = 198.7$ Btu/(h)(ft²)(°F) [1129 W/(m²)(K)].

4. Summarize the results.
The following table summarizes the calculations for all nine increments:

Increment	h_c	h_s	h_L
1	198.7	182.7	234.5
2	198.7	171.8	231.5
3	198.7	159.9	227.5
4	198.7	147.1	223.5
5	198.7	133.1	219.6
6	198.7	117.4	213.6
7	198.7	99.2	198.7
8	198.7	76.8	198.7
9	198.7	44.4	198.7

The average value of h_L is 216.3 Btu/(h)(ft²)(°F) [1229 W/(m²)(K)].

Related Calculations. For most cases, vapor-shear condensation is not important, and the condensing heat-transfer coefficient can be calculated simply as the laminar-film coefficient.

It is important that the shell side of horizontal shell-side condensers be designed to avoid excessive condensate holdup caused by the baffle or nozzle types selected by the designer.

For bundles with slight slopes, the following correction should be applied:

$$h_t = h_c(\cos \alpha)^{1/3} \quad \text{for } L/D_o > 1.8 \tan \alpha$$

where h_t = heat-transfer coefficient for sloped tube
h_c = heat-transfer coefficient for horizontal tube
α = angle from horizontal, in degrees
L = tube length
D_o = outside tube diameter

Condensation of mixed vapors of immiscible fluids can be treated in the same manner as outlined in Procedure 7.21.

Condensate subcooling when condensing on banks of horizontal tubes can be accomplished in two ways. The first method requires holding a condensate level on the shell side; heat transfer can then be calculated using the appropriate single-phase correlation. The second method requires that the vapor make a single pass across the bundle in a vertical downflow direction. Subcooling heat transfer can then be calculated using falling-film correlations.

For low-fin tubes, the laminar condensing coefficient can be calculated by applying an appropriate correction factor F to the value calculated using the preceding equation for laminar-film condensation. The factor F is defined thus:

$$F = \left(\eta \frac{A_t}{A_i} \frac{D_i}{D_r} \right)^{1/4}$$

where η = weighted fin efficiency
    A_t = total outside surface
    A_i = inside surface
    D_i = inside tube diameter
    D_r = diameter at root of fins

7.24   *CONDENSATION IN THE PRESENCE OF NONCONDENSABLES*

A mixture of vapor and noncondensable gases flows through a vertical tube bundle containing 150 tubes with an inside diameter of 0.62 in (0.016 m) and an outside diameter of 0.75 in (0.019 m) at a rate of 25,000 lb/h (11,339.8 kg/h). The mixture enters at a temperature of 212°F (373 K) and is cooled to a temperature of 140°F (333 K). As cooling occurs, 15,000 lb/h (6803.9 kg/h) of the mixture condenses. The condensation may be assumed to be straight-line condensation. Calculate the tube length required if the outside heat-transfer coefficient is 300 Btu/(h)(ft^2)(°F) [1700 W/(m^2)(K)] and the temperature is isothermal at 104°F (313 K). Assume the fouling heat-transfer coefficient is 1000 Btu/(h)(ft^2)(°F) [5680 W/(m^2)(K)]. The tube-wall thermal conductivity is 10 Btu/(h)(ft)(°F) [117.28 W/(m)(K)]. The physical properties of the system are as follows:

Vapor specific heat = 0.35 Btu/(lb)(°F) [1.47 kJ/(kg)(K)]
Liquid specific heat = 0.7 Btu/(lb)(°F) [2.93 kJ/(kg)(K)]
Vapor viscosity = 0.048 lb/(ft)(h) (0.020 cP)
Liquid viscosity = 0.24 lb/(ft)(h) (0.10 cP)
Vapor thermal conductivity = 0.021 Btu/(h)(ft)(°F) [0.036 W/(m)(K)]
Liquid thermal conductivity = 0.075 Btu/(h)(ft)(°F) [0.13 W/(m)(K)]
Heat of vaporization = 200 Btu/lb (465 kJ/kg)
Liquid density = 30 lb/ft^3 (481 kg/m^3)
Vapor density = 0.9 lb/ft^3 (14.4 kg/m^3)

Calculation Procedure

1. *Select the calculation method to be used.*
A heat-transfer coefficient that predicts heat transfer when both sensible heat and latent heat are being transferred can be calculated using the equation

$$h_{c,g} = \frac{1}{\dfrac{Q_g}{Q_T} \dfrac{1}{h_g} + \dfrac{1}{h_c}}$$

where $h_{c,g}$ = combined cooling-condensing heat-transfer coefficient
    h_g = heat-transfer coefficient for gas cooling only
    h_c = condensing heat-transfer coefficient
    Q_g = heat-transfer rate for cooling the gas only
    Q_T = total heat-transfer rate (includes sensible heat for gas and condensate cooling and latent heat for condensing)

2. Calculate the sensible and latent heat loads.
Since straight-line condensing occurs, the mean condensing temperature may be taken as the arithmetic mean (176°F; 353 K). The heat loads are calculated assuming that all gas and vapor are cooled to the mean condensing temperature, that all condensing then occurs at this temperature, and that the uncondensed mixture plus the condensate are further cooled to the outlet temperature.

Inlet Vapor Cooling

$$Q_i = 25,000(0.35)(212 - 176) = 315,000 \text{ Btu/h}$$

Latent Heat of Condensation

$$Q_c = 15,000(200) = 3,000,000 \text{ Btu/h}$$

Outlet Vapor Cooling

$$Q_o = 10,000(0.35)(176 - 140) = 126,000 \text{ Btu/h}$$

Condensate Cooling

$$Q_s = 15,000(0.7)(176 - 140) = 378,000 \text{ Btu/h}$$

Therefore, the total heat load Q_T equals $Q_i + Q_c, + Q_o + Q_s = 3,819,000$ Btu/h. And $Q_g = Q_i + Q_o = 441,000$ Btu/h, so $Q_T/Q_g = 3,819,000/441,000 = 8.66$.

3. Calculate the gas cooling heat-transfer coefficient.
Use the equation $h = 0.023cG/[(c\mu/k)^{2/3}(DG/\mu)^{0.2}]$. Base the mass velocity on the average vapor flow rate, that is, $(25,000 + 10,000)/2 = 17,500$ lb/h. Then $G = 17,500/[150 (0.62/12)^2(\pi/4)] = 55,646.3$ lb/(h)(ft^2), and

$$h_g = \frac{0.023(0.35)(55,646.3)}{[0.35(0.048)/0.021]^{2/3}[(0.62/12)55.646,3/0.48]^{0.2}}$$
$$= 57.58 \text{ Btu/(h)(ft}^2)(°F) [327 \text{ W/(m}^2)(K)]$$

4. Calculate the condensing coefficient.
See Procedure 7.21. Now, $4\Gamma/\mu = 4[15,000/150\pi(0.62/12)]/0.24 = 10,268$, and $c\mu/k = 0.7(0.24)/0.075 = 2.24$. From Fig. 7.9, for the parameters calculated above, $h_c[\mu^2/(\rho^2 gk^3)]^{1/3} = 0.28$. Therefore,

$$h_c = 0.28[30^2(4.18 \times 10^8)0.075^3/0.24^2]^{1/3}$$
$$= 392.5 \text{ Btu/(h)(ft}^2)(°F) [2229 \text{ W/(m}^2)(K)]$$

5. Calculate $h_{c,g}$.
Now,

$$h_{c,g} = 1/[(Q_g/Q_T)(1/h_g) + (1/h_c)]$$
$$= 1/[(1/8.66)(1/57.58) + (1/392.5)]$$
$$= 219.6 \text{ Btu/(h)(ft}^2)(°F) [1247 \text{ W/(m2)(K)]}$$

6. Calculate the overall heat-transfer coefficient.
Use the equation

$$\frac{1}{U} = \frac{1}{h_o} + \frac{1}{h_s} + \frac{1}{h_w} + \frac{1}{h_{c,g}(d_i/d_o)}$$

where U = overall heat-transfer coefficient
  h_o = outside heat-transfer coefficient
  h_s = fouling heat-transfer coefficient
  h_w = heat-transfer coefficient through the wall (thermal conductivity of the wall divided by its thickness)
  $h_{c.g}$ = inside heat-transfer coefficient
  d_i = inside tube diameter
  d_o = outside tube diameter

Then,

$$\frac{1}{U} = \frac{1}{300} + \frac{1}{1000} + \frac{1}{10/[(0.75 - 0.62)/12(2)]} + \frac{1}{219.6(0.62/0.75)}$$
$$= 96.3 \ \text{Btu/(h)(ft}^2\text{)(°F)[547 W/(m}^2\text{)(K)]}$$

7. Calculate the log mean temperature difference.
See Procedure 7.17 and use the equation for countercurrent flow. Thus,

$$\Delta T_{LM} = \frac{(212 - 104) - (140 - 104)}{\ln[(212 - 104)/(140 - 104)]}$$
$$= 65.5°F \ (36.4 \ \text{K})$$

8. Calculate the heat-transfer surface required.
The equation is

$$A = \frac{Q_T}{U \Delta T_{LM}} = \frac{3{,}819.00}{96.3(65.5)}$$
$$= 605.4 \ \text{ft}^2 (56.24 \ \text{m}^2)$$

9. Calculate the tube length required.
Since $A = n\pi D_o L$, $L = 605.4/[150\pi(0.75/12)] = 20.6$ ft (6.27 m).

Related Calculations. For condensers that do not exhibit straight-line condensing, this procedure can be used by treating the problem as a series of condensing zones that approximate straight-line segments. The vapor and condensate flow rates, the heat load, the overall heat-transfer coefficient, and the log mean temperature difference will vary with each zone. The total answer is obtained by integrating all the segments. It is usually not necessary to use a large number of zones. The accuracy lost by using less than 10 zones is not significant in most cases.

For horizontal shell-side condensers, the condensate falls to the bottom of the shell, and vapor and liquid do not coexist, as assumed by the preceding method. The effect this has on the heat transfer must be considered. It is recommended that shell-side condensers with noncondensable gases present be somewhat overdesigned: perhaps 20 percent excess surface should be provided.

7.25 MAXIMUM VAPOR VELOCITY FOR CONDENSERS WITH UPFLOW VAPOR

Calculate the maximum velocity to avoid flooding for the vapor conditions of Procedure 7.24. Flooding occurs in upflow condensers when the vapor velocity is too high to permit the condensate to drain. Unstable conditions exist when flooding occurs.

Calculation Procedure

1. Select the appropriate equation.
The following equation can be used to establish the condition for flooding of vertical-upflow vapor condensers:

$$v_v^{1/2}\rho_v^{1/4} + v_L^{1/2}\rho_L^{1/4} \leq 0.6[gD_i(\rho_L - \rho_v)]^{1/4}$$

where v_v = vapor velocity assuming only vapor is flowing in the tube
v_L = liquid velocity of the condensate assuming only condensate is flowing in the tube
ρ_v = vapor density
ρ_L = liquid density
D_i = inside tube diameter
g = acceleration of gravity

2. Calculate the maximum allowable vapor mass-flow rate G_v.
Now, $v_L = G_L/\rho_L$, and $v_v = G_v/\rho_v$, where G is mass flow rate and the subscripts L and v refer to liquid and vapor. For the problem at hand, $G_L = (15,000/25,000)G_v = 0.6G_v$; therefore, $v_L = 0.6G_v/\rho_L$. Then, substituting into the expression in step 1, $(G_v/\rho_v)^{1/2}\rho_v^{1/4} + (0.6G_v/\rho_L)^{1/2}\rho_L^{1/4} = 0.6[gD_i(\rho_L - \rho_v)]^{1/4}$ or $G_v^{1/2}/\rho_v^{1/4} + 0.6^{1/2}G_v^{1/2}/\rho_L^{1/4} = 0.6[gD_i(\rho_L - \rho_v)]^{1/4}$. Then $G_v^{1/2}/0.9^{1/4} + 0.6^{1/2}G_v^{1/2}/30^{1/4} = 0.6[32.2(0.62/12)(30 - 0.9)]^{1/4}$, and $G_v = 1.3589$ lb/(ft^2)(s).

3. Calculate the maximum allowable velocity.
Now, $G_v = \rho_v v_v$, so $v_v = G_v/\rho_v = 1.3589/0.9 = 1.51$ ft/s (0.46 m/s).

Related Calculations.   The velocity calculated above is the threshold for flooding of the condenser. Flooding causes unstable operation. A safety factor of 0.85 is commonly used when designing upflow condensers. The maximum velocity at which flooding occurs can be increased by cutting the bottom of the tube at an angle. Improvements that can be achieved are as follows:

Angle of cut (measured from horizontal)	Increase in maximum flooding velocity
30°	5%
60°	25%
75°	55%

7.26   MAXIMUM HEAT FLUX FOR KETTLE-TYPE REBOILER

A kettle-type reboiler with a shell diameter of 30 in (0.76 m) contains a tube bundle with a diameter of 15 in (0.38 m). The bundle contains 80 tubes, each with a diameter of 1 in (0.025 m) and a length of 12 ft (3.66 m). Determine the maximum heat flux for a fluid with the following physical properties: heat of vaporization of 895 Btu/lb (2082 kJ/kg), surface tension of 0.00308 lb/ft (0.045 N/m), liquid density of 56.5 lb/ft^3 (905 kg/m^3), and vapor density of 0.2 lb/ft^3 (3.204 kg/m^3).

Calculation Procedure

1. Determine the maximum superficial vapor velocity.
The maximum superficial vapor velocity to avoid film boiling can be calculated using the equation

$$v_c = \frac{3(\rho_L - \rho_v)^{1/4}\sigma^{1/4}}{\rho_v^{1/2}}$$

where v_c = superficial vapor velocity, in ft/s
ρ_L = liquid density, in lb/ft^3
ρ_v = vapor density, in lb/ft^3
σ = surface tension, in lb/ft

For this problem,

$$v_c = \frac{3(56.5 - 0.2)^{1/4} 0.00308^{1/4}}{0.2^{1/2}}$$
$$= 4.33 \text{ ft/s (1.329 m/s)}$$

2. Determine the maximum heat rate.

The superficial vapor velocity is based on the projected area of the tube bundle

$$a = D_b L$$

where a = projected area of the bundle
D_b = bundle diameter
L = bundle length

In this case, $a = (15/12)12 = 15$ ft^2 (1.39 m^2).

The maximum heat transferred is determined from the equation

$$Q = 3600 a v_c \rho_v \lambda$$

where Q = heat transferred, in Btu/h
a = projected area, in ft^2
v_c = superficial vapor velocity, in ft/s
ρ_v = vapor density, in lb/ft^3
λ = heat of vaporization, in Btu/lb

In this case, $Q = 3600(15)(4.33)(0.2)(895) = 41,854,000$ Btu/h (12,266 kW).

3. Calculate the maximum heat flux.

Maximum heat flux = Q/A, where Q is maximum heat transferred and A is total heat-transfer surface for all n tubes. In this case, $A = n\pi DL = 80\pi(1/12)12 = 251$ ft^2(23.35 m^2), so maximum heat flux = 41,854,000/251 = 166,750 Btu/(h)(ft^2) (525,600 W/m^2).

Related Calculations. The preceding equation for maximum superficial vapor velocity is applicable only to kettle-type reboilers having a shell diameter 1.3 to 2.0 times greater than the diameter of the tube bundle. (For small-diameter bundles, the ratio required is greater than that for large bundles.) This ratio is generally sufficient to permit liquid circulation adequate to obtain the superficial vapor velocity predicted by the equation presented here. For single tubes or for tube bundles with geometries that do not permit adequate liquid circulation, the superficial vapor velocity calculated using this equation should be multiplied by 0.3.

For vertical tubes, the superficial vapor velocity (based on the total heat-transfer surface) can be obtained by multiplying the value calculated from the preceding equation by 0.22. This assumes that there is adequate liquid circulating past the surface to satisfy the mass balance. For thermosiphon reboilers, a detailed analysis must be made to establish circulation rate, boiling pressure, sensible heat-transfer zone, boiling heat-transfer zone, and mean temperature difference. If liquid circulation rates are not adequate, all liquid will be vaporized and superheating of the vapor will occur with a resultant decrease in heat-transfer rates.

The procedure for design of thermosiphon reboilers presented by Fair [23] has been widely used. Special surfaces are available commercially that permit much higher superficial vapor velocities than calculated by the method presented here: see Gottzmann, O'Neill, and Minton [32].

7.27   *DOUBLE-PIPE HEAT EXCHANGERS WITH BARE TUBES*

Calculate the outlet temperature for air entering the annulus of a double-pipe exchanger at 68°F (293 K) at a flow rate of 500 lb/h (226.8 kg/h) if steam is condensing inside the tube at 320°F (433 K). The shell of the double-pipe exchanger is 3.068 in (0.0779 m) in diameter. The steel tube is 1.9 in (0.048 m) in outside diameter and 1.61 in (0.0409 m) in inside diameter and has a thermal conductivity of 25 Btu/(h)(ft)(°F) [43 W/(m)(K)]. The tube length is 20 ft (6.1 m). Air has a viscosity of 0.0426 lb/(ft)(h) and a specific heat of 0.24 Btu/(lb)(°F) [1.01 kJ/(kg)(K)]. The fouling coefficient is 1000 Btu/(h)(ft²)(°F)[5680 W/(m²)(K)].

The physical properties for the steam condensate are:

μ = liquid viscosity = 0.5 lb/(ft)(h) (0.207 cP)
ρ = liquid density = 55.5 lb/ft³ (889 kg/m³)
λ = heat of vaporization = 895 Btu/lb (2,081, 770 J/kg)
k = liquid thermal conductivity = 0.395 Btu/(h)(ft)(°F) [0.683 W/(m)(K)]

Calculation Procedure

1. Calculate condensing coefficient h_i.
The steam-condensing coefficient will be much larger than the air-side coefficient; this permits us to approximate the condensing coefficient by assuming the maximum steam condensate loading. The maximum heat transferred would occur if the air were heated to the steam temperature. Thus, Q_{max} = $W_{air}c(T_{steam} - t_{air})$, where W is mass flow rate and c is specific heat. Or, Q_{max} = 500(0.24)(320 − 68) = 30,240 Btu/h. Then, maximum condensate flow $W_{c,max}$ = Q_{max}/λ = 30,240/895 = 33.8 lb/h.

Now, with reference to Procedure 7.22, $W_c/n\rho D_i^{2.56}$ = 33.8/[1(55.5)(1.61/12)$^{2.56}$] = 104. This is a relatively low value, so stratified flow can be assumed. Then,

$$h_c = 0.767k\left(\frac{\rho^2 gL}{n\mu W}\right)^{1/3}$$

$$= 0.767(0.39)\left(\frac{55.5^2(4.18 \times 10^8)(20)}{1(0.5)(33.8)}\right)^{1/3}$$

$$= 3486 \text{ Btu/(h)(ft}^2)(°F)$$

Now, for a $W_c/(n\rho D_i^{2.56})$ of 104, h_i = 1.2/h_c, as indicated in Procedure 7.22. Thus, h_i = 1.2(3486) = 4183 Btu/(h)(ft²)(°F), and $h_i(D_i/D_o)$ = 4183(1.61/1.9) = 3545 Btu/(h)(ft²)(°F)[20,136 W/(m²)(K)].

2. Calculate h_w, the heat-transfer coefficient through the tube wall.
By definition, h_w = $2k/(D_o - D_i)$ = 2(25)/(1.9 − 1.61)(1/12) = 2068 Btu/(h)(ft²)(°F) [11,750 W/(m²)(K)].

3. Calculate h_o, the heat-transfer coefficient for the outside of the tube.
Because the fluid flowing is air, we can use simplified equations for heat-transfer coefficients. Cross-sectional area of the annulus a_c = $(\pi/4)(D_s^2 - D_o^2)$ = $(\pi/4)(3.068^2 - 1.9^2)(1/12)^2$ = 0.0316 ft². Then, mass flow rate through the annulus G = W_{air}/a_c = 500/0.0316 = 15,800 lb/(h)(ft²), and equivalent diameter D_e = $4a_c/P$, where P = $\pi(D_s + D_o)$. Thus, by algebraic simplification, D_e = $D_s - D_o$ = (3.068/12) − (1.9/12) = 0.0973 ft (0.0297 m).

Since the Reynolds number, that is, $D_e G/\mu$ = 0.0973(15,800)/0.0426 = 36,088, is greater than 8000 (see Procedure 7.18),

$$h_o = 0.0144cG^{0.8}/D_e^{0.2}$$

$$= 0.0144(0.24)(15,800)^{0.8}/0.0973^{0.2}$$

$$= 12.58 \text{ Btu/(h)(ft}^2)(°F)[71.45 \text{ W/(m}^2)(K)]$$

4. Calculate U, the overall heat-transfer coefficient.
Now,

$$\frac{1}{U} = \frac{1}{h_o} + \frac{1}{h_i(D_i/D_o)} + \frac{1}{h_w} + \frac{1}{h_s}$$

$$= \frac{1}{12.58} + \frac{1}{3545} + \frac{1}{2068} + \frac{1}{1000}$$

So, $U = 12.32$ Btu/(h)(ft^2)(°F) [69.97 W/(m^2)(K)].

5. Calculate outlet air temperature.
Use Fig. 7.6. Now, $A = \pi D_o L = \pi(1.9/12)(20) = 9.95$ ft^2, so $UA/(wc) = 12.32(9.95)/[500(0.24)] = 1.02$. In addition, $R = 0$ (isothermal on steam side).
   For these values of $UA/(wc)$ and R, $P = 0.64$ from Fig. 7.6. Then $t_2 = t_1 + P(T_s - t_1) = 68 + 0.64(320 - 68) = 229.2$°F (382.8 K).

7.28   DOUBLE-PIPE HEAT EXCHANGERS WITH LONGITUDINALLY FINNED TUBES

Calculate the outlet air temperature for the double-pipe heat exchanger under the conditions of Procedure 7.27 if the tube has 24 steel fins 0.5 in (0.013 m) high and 0.03125 in (0.794 mm) thick.

Calculation Procedure

1. Calculate the relevant fin areas.
The relevant areas are cross-sectional flow area a_c, fin surface A_f, outside bare surface A_o, and inside surface A_i. Now,

$$a_c = (\pi/4)\left(D_s^2 - D_o^2\right) - nlb$$

where n = number of fins
   l = fin height
   b = fin thickness

Thus, $a_c = 0.0316 - 24(0.5/12)(0.03125/12) = 0.029$ ft^2 (0.0028 m^2).
   Further, A_f is the heat-transfer surface on both sides and the tip of the fins, and it equals $2nl + nb = 2(24)(0.5/12) + 24(0.03125/12) = 2.0625$ ft^2 per foot of tube length. And, A_o is the outside bare surface exclusive of the area beneath the fins, and it equals $\pi D_o - nb = \pi(1.9/12) - 24(0.03125/12) = 0.4349$ ft^2 per foot of tube length. Finally, $A_i = \pi D_i = \pi(1.61/12) = 0.4215$ ft^2 per foot of tube length.

2. Calculate D_e for the fin tube.
Use the formula $D_e = 4a_c/P$, where $P = \pi(D_s + D_o) + 2nl$. Now, $P = \pi(3.068 + 1.9)(1/12) + 2(24)(0.5/12) = 3.3006$ ft. Then, $D_e = 4(0.029)/3.3006 = 0.0351$ ft (0.0107 m).

3. Calculate h_o.
Mass flow rate through the annulus $G = W_{air}/a_c = 500/0.029 = 17,241.4$ lb/(h)(ft^2). Since the Reynolds number, that is, $D_e G/\mu = 0.0351(17,241.4)/0.0426 = 14,206$, is greater than 8000,

$$h_o = 0.0144cG^{0.8}/D_e^{0.2}$$

$$= 0.0144(0.24)(17,241.4)^{0.8}/0.0351^{0.2}$$

$$= 16.55 \text{ Btu/(h)(ft}^2)(°F)[93.99 \text{ W/(m}^2)(K)]$$

4. *Calculate fin efficiency.*
Because the heat must be transferred through the fin by conduction, the fin is not as effective as a bare tube with the same heat-transfer surface and heat-transfer coefficient. The fin efficiency is a measure of the actual heat transferred compared with the amount that could be transferred if the fin were uniformly at the temperature of the base of the fin.

For a fin with rectangular cross section,

$$\eta = \frac{\tanh ml}{ml}$$

where η = fin efficiency
l = fin height
$m = (2h_o/kb)^{1/2}$, where k is fin thermal conductivity, and b is fin thickness

For this problem, $m = \{2(16.55)/[25(0.03125/12)]\}^{1/2} = 22.55$, so $ml = 22.55(0.5/12) = 0.9395$ and $\eta = (\tanh 0.9395)/0.9395$. Now, $\tanh x = (e^x - e^{-x})/(e^x + e^{-x})$, so

$$\tanh 0.9395 = \frac{e^{0.9395} - e^{-0.9395}}{e^{0.9395} + e^{-0.9395}}$$
$$= 0.735$$

Therefore, $\eta = 0.735/0.9395 = 0.782$.

5. *Calculate $h_{f,i}$, pertaining to the outside of the fin tube.*
It is convenient to base the overall heat-transfer coefficient on the inside area of a fin tube. Then the relevant outside coefficient is

$$h_{f,i} = (\eta A_f + A_o)(h_o/A_i)$$
$$= [0.782(2.0625) + 0.4349]\frac{16.55}{0.4215}$$
$$= 80.40 \text{ Btu/(h)(ft}^2\text{)(}^\circ\text{F)}[456.6 \text{ W/(m}^2\text{)(K)]}$$

6. *Calculate U_i.*
The formula is

$$\frac{1}{U_i} = \frac{1}{h_{f,i}} + \frac{1}{h_i} + \frac{1}{h_w} + \frac{1}{h_s}$$

Thus, $1/U_i = 1/80.4 + 1/4183 + 1/2068 + 1/1000$, so $U_i = 70.6$ Btu/(h)(ft^2)($^\circ$F) [401] W/(m^2)(K)].

7. *Calculate outlet air temperature.*
From step 1, total area $A_i = 0.4215(20) = 8.43$ ft^2 (0.783 m^2). Then, $U_i A_i/(wc) = 70.6(8.43)/[500(0.24)] = 4.96$. From Fig. 7.6, for $R = 0$, $P = 0.99$. Then $t_2 = t_1 + P(T_s - t_1) = 68 + 0.99(320 - 68) = 317.5^\circ$F (431.8 K).

Related Calculations.   This procedure can also be used for fins with cross sections other than rectangular. Fin-efficiency curves for some of these shapes are presented in Refs. 2 through 4.

7.29   HEAT TRANSFER FOR LOW-FIN TUBES

An existing heat exchanger with the following geometry must be retubed. Bare copper-alloy tubes [$k = 65$ Btu/(h)(ft)($^\circ$F) or 112 W/(m)(K)] cost \$1 per foot; low-fin copper-alloy tubes cost \$1.75 per foot. Heat is to be exchanged between two process streams operating under the following conditions.

The cool stream must be further heated downstream with steam that has a heat of vaporization of 895 Btu/lb (2082 kJ/kg). The warm stream must be further cooled downstream with cooling water that can accept a maximum temperature rise of 30°F (16.67 K). Pressure drop for the tube side is not a penalty because the tube-side fluid must be throttled downstream. Is it economical to retube the exchanger with low-fin tubes if the evaluated cost of the steam is $50 per pound per hour and the evaluated cost of the cooling water is $25 per gallon per minute?

Tube-Side Fluid

Condition = liquid, sensible-heat transfer
Flow rate = 50,000 lb/h (22,679.5 kg/h)
c = specific heat, 1.0 Btu/(lb)(°F) [4.2 kJ/(kg)(K)]
μ = viscosity, 1.21 lb/(ft)(h) (0.5 cP)
k = thermal conductivity, 0.38 Btu/(h)(ft)(°F) [0.66 W/(m)(K)]
Inlet temperature = 320°F (433 K)
Assume that $(\mu_w/\mu_b)^{0.14} = 1.0$

Shell-Side Fluid

Condition = liquid, sensible-heat transfer
Flow rate = 30,000 lb/h (13,607.7 kg/h)
c = specific heat, 0.7 Btu/(lb)(°F) [2.9 kJ/(kg)(K)]
μ = viscosity, 10 lb/(h)(ft) (4.13 cP)
k = thermal conductivity, 0.12 Btu/(h)(ft)(°F) [0.21 W/(m)(K)]
Inlet temperature =41°F (278 K)
Assume that $(\mu_w/\mu_b)^{0.14} = 0.9$

Heat-Exchanger Geometry

D_s = shell diameter, 8.071 in (0.205 m)
n = number of tubes, 36
D_o = tube outside diameter, 0.75 in (0.019 m)
D_i = tube inside diameter, 0.62 in (0.016 m) for bare tube and 0.495 in (0.0126 m) for low-fin tube
L = tube length, 16 ft (4.88 m)
B = baffle spacing, 4 in (0.102 m)
s = tube spacing, 0.9375 in (0.02381 m) with a triangular pitch
D_r = root diameter of fin tube, 0.625 in (0.0159 m)
A_o/A_i = 3.84

There is one tube pass, one shell pass. Assume an overall fouling coefficient of 1000 Btu/(h)(ft²)(°F) [5680 W/(m²)(K)].

Calculation Procedure

1. Calculate h_i, the inside film coefficient, for bare tubes.
See Procedure 7.18. Here, $G = W/a_c = 50,000/[36(0.62/12)^2(\pi/4)] = 662,455.5$ lb/(h)(ft²). Then $D_iG/\mu = (0.62/12)662,455.5/1.21 = 28,286.7$, and

$$h_i = \frac{0.023cG}{(c\mu/k)^{2/3}(D_iG/\mu)^{0.2}}$$

$$= \frac{0.023(1.0)(662,455.5)}{[1.0(1.21)/0.38]^{2/3}28,286.7^{0.2}}$$

$$= 906.9 \text{ Btu/(h)(ft}^2)(°F)[5151\,W/(m^2)(K)]$$

2. Calculate h_i for low-fin tubes.

Use the relationship $h_{i,\text{fin}} = h_{i,\text{bare}}(D_{i,\text{bare}}/D_{i,\text{fin}})^{1.8}$. Thus, $h_{i,\text{fin}} = 906.9\ (0.62/0.495)^{1.8} = 1360.1$ Btu/(h)(ft^2)(°F) [7725 W/(m^2)(K)].

3. Calculate h_o, the outside film coefficient, for bare tubes.

See Procedure 7.18. Here, $a_c = D_s B(s - D_o)/s = (8.071/12)(4/12)(0.9375 - 0.75)/0.9375 = 0.0448$ ft^2. Then, $G = W/a_c = 30{,}000/0.0448 = 669{,}062$ lb/(h)(ft^2), and the Reynolds number $D_o G/\mu = (0.75/12)669{,}062/10 = 4181.6$. Therefore, $h = 0.273cG/[(c\mu/k)^{2/3}(D_o G/\mu)^{0.365}(\mu_w/\mu_o)^{0.14}]$. Now, $(c\mu/k)^{2/3} = [0.7(10)/0.12]^{2/3} = 15.04$, and $h = 0.273(0.7)(669{,}062)/[15.04(4181.6)^{0.365}0.9] = 450.2$ Btu/(h)(ft^2)(°F). Finally, $h_o = hF_1 F_r$, $F_r = 1.0$, and $F_1 = 0.8(B/D_s)^{1/6}$. Thus, $h_o = 450.2(0.8)(4/8.071)^{1/6}(1.0) = 320.4$ Btu/(h)(ft^2)(°F) [1819 W/(m^2)(K)].

4. Calculate h_o for low-fin tubes.

For low-fin tubes, the shell-side mass velocity is reduced because of the space between the fins. This reduction can be closely approximated with the expression $(s - D_o)/(s - D_o + 0.09)$, each term being expressed in inches.

For this problem, the expression has the value $(0.9375 - 0.75)/(0.9375 - 0.75 + 0.09) = 0.676$. The diameter that should be used for calculating the Reynolds number is the root diameter of the fin tube. By applying the diameter ratio and the velocity reduction to the Reynolds number from step 3, the result is a Reynolds number for this case of $D_r G/\mu = 4181.6(0.625/0.75)(0.676) = 2355.6$. Then,

$$
\begin{aligned}
h &= \frac{0.273cG}{(c\mu/k)^{2/3}(D_r G/\mu)^{0.365}(\mu_w/\mu_b)^{0.14}} \\
&= \frac{0.273(0.7)(669{,}062)(0.676)}{15.04(2355.6)^{0.365}0.9} \\
&= 375.3 \text{ Btu/(h)(ft}^2\text{)(°F)}
\end{aligned}
$$

Then, as in step 3, $h_o = 375.3(0.8)(4/8.071)^{1/6}\ 1.0 = 267.1$ Btu/(h)(ft^2)(°F) [1517 W/(m^2)(K)].

5. Determine weighted fin efficiency.

Weighted fin efficiencies for low-fin tubes are functions of the outside heat-transfer coefficient. Weighted fin efficiencies η can be determined from curves provided by various manufacturers. Table 7.4 permits approximation of weighted fin efficiencies. For this problem, the weighted efficiency η is 0.94.

6. Calculate $h_{f,i}$, pertaining to the outside of the fin tubes.

For fin tubes, it is convenient to base the overall heat-transfer coefficient on the inside surface of the tube. Then the relevant outside coefficient is $h_{f,i} = \eta(A_o/A_i)h_o = 0.94(3.84)(267.1) = 964.1$ Btu/(h)(ft^2)(°F) [5476.2 W/(m^2)(K)].

TABLE 7.4   Weighted Fin Efficiency for Low-Fin Tubes

h_o	Weighted fin efficiency η				
	$k = 10[17.3]$	$k = 25[43.3]$	$k = 65[112]$	$k = 125[216]$	$k = 225[389]$
10 [56.7]	0.97	0.98	0.99	1.00	1.00
50 [284]	0.94	0.97	0.98	0.99	1.00
100 [567]	0.89	0.94	0.97	0.98	0.99
200 [1134]	0.81	0.90	0.95	0.97	0.98
500 [2840]	0.68	0.80	0.90	0.93	0.96

Note: h_o, the outside film coefficient, is in Btu/(h)(ft^2)(°F) [W/(m^2)(K)]; k, the thermal conductivity, is in Btu/(h)(ft)(°F)[W/(m)(K)].

7. Calculate h_w, the coefficient for heat transfer through the tube wall.
The formulas are $h_w = 2k/(D_o - D_i)$ for bare tubes, and $h_w = 2k/(D_r - D_i)$ for low-fin tubes. For this problem, $D_o - D_i = D_r - D_i = 0.13/12$, so $h_w = 2(65)/(0.13/12) = 12,000$ Btu/(h)(ft^2)(°F) [68,160 W/(m^2)(K)].

8. Calculate U_o for bare tubes.
The formula is

$$\frac{1}{U_o} = \frac{1}{h_i(D_i/D_o)} + \frac{1}{h_o} + \frac{1}{h_w} + \frac{1}{h_s}$$

Thus, $1/U_o = 1/[906.9(0.62/0.75)] + 1/320.4 + 1/12,000 + 1/1000$, so $U_o = 180.6$ Btu/(h) (ft^2)(°F) [1026 W/(m^2)(K)].

9. Calculate U_i for low-fin tubes.
Again, $1/U_i = 1/h_i + 1/h_{f,i} + 1/h_w + 1/h_s$. Thus, $1/U_i = 1/1360.1 + 1/964.1 + 1/12,000 + 1/1000$, so $U_i = 350.2$ Btu/(h)(ft^2)(°F) [1989 W/(m^2)(K)].

10. Determine outlet temperatures for bare tubes.
Use Fig. 7.6. Now, $A_o = n\pi D_o L = 36\pi(0.75/12)16 = 113.1$ ft^2, so $UA/(wc) = 180.6(113.1)/[30,000(0.7)] = 0.973$, and $R = wc/(WC) = 30,000(0.7)/[50,000(1.0)] = 0.420$. For these values, $P = 0.57$. Therefore, $t_2 = t_1 + P(T_1 - t_1) = 41 + 0.57(320 - 41) = 200.0$°F (366.5 K).

11. Determine outlet temperatures for low-fin tubes.
Use Fig. 7.6 again. In this case, the calculation is based on the inside diameter. So, $A_i = n\pi D_i L = 36\pi(0.495/12)16 = 74.6$ ft^2. And $UA/(wc) = 350.2(74.6)/(30,000(0.7)] = 1.244$. For this value and $R = 0.42$, $P = 0.65$. Therefore, $t_2 = t_1 + P(T_1 - t_1) = 41 + 0.65(320 - 41) = 222.4$°F (378.9 K).

12. Calculate water savings using low-fin tubes.
The tube-side fluid must be further cooled; the water savings is represented by the difference in heat recovery between bare and low-fin tubes. Thus,

$$Q_{saved} = wc\Delta t = 30,000(0.7)(222.4 - 200.0)$$
$$= 470,400 \text{ Btu/h}$$

Since the cooling water can accept a temperature rise of 30°C, the water rate = $Q_{saved}/(c\Delta T)$ = 470,400/[1(30)] = 15.680 lb/h. Then, dollars saved (lb water/h)[(gal/min)/(lb/h)][$/(gal/min)] = 15,680(1/500)(25) = $784.

13. Calculate steam savings using low-fin tubes.
The shell-side fluid must be further heated; the steam savings is represented by the difference in heat recovery between bare and low-fin tubes:

$$\text{Steam rate} = Q_{saved}/\lambda = 470,400/895 = 525.6 \text{ lb/h}$$

Then dollars saved = (lb steam/h)[$/(lb/h)] = equivalent savings = 525.6(50) = $26,280.

14. Compare energy savings with additional tubing cost.
Additional cost for retubing is $nL(\$1.75 - \$1.00) = 36(16)(1.75 - 1.00) = \$432$. The equivalent energy savings is $784 + $26,280 = $27,064.

Related Calculations.   Low-fin tubes are tubes with extended surfaces that have the same outside diameter as bare tubes. They can therefore be used interchangeably with bare tubes in tubular exchangers. Various geometries and materials of construction are available from several manufacturers.
   Low-fin tubes find wide application when the heat-transfer coefficient on the inside of the tube is much greater than the coefficient on the outside of the tube. A guideline suggests that low-fin tubes should be considered when the outside coefficient is less than one-third that on the inside.

Low-fin tubes also find application in some fouling services because the fin tubes are more easily cleaned by hydroblasting than are bare tubes. In addition, low-fin tubes are used when boiling at low temperature differences, because the minimum temperature difference required for nucleate boiling is reduced with the use of low-fin tubes. Low-fin tubes are also used for condensing, primarily for materials with low surface tension.

The procedure outlined in this example can also be used when designing equipment using low-fin tubes. The same approach is used when condensing or boiling on the outside of low-fin tubes.

7.30   HEAT TRANSFER FOR BANKS OF FINNED TUBES

Calculate the outlet temperature from a duct cooler if hydrogen is flowing at a rate of 1000 lb/h (453.6 kg/h) with a duct velocity of 500 ft/min (152.4 m/min). The duct cooler contains a bank of finned tubes described below. Hydrogen enters the cooler at 200°F (366.5 K) and has a specific heat of 3.4 Btu/(lb)(°F) [14.2 kJ/(kg)(K)], a viscosity of 0.225 lb/(ft)(h), (0.0093 cP), a thermal conductivity of 0.11 Btu/(h)(ft)(°F) [0.19 W/(m)(K)], and a density of 0.0049 lb/ft³ (0.0785 kg/m³). The coolant is water entering the tubes at 86°F (303 K) and a rate of 10,000 lb/h (4535.0 kg/h). The water-side heat-transfer coefficient h_i is 1200 Btu/(h)(ft²)(°F) [6800 W/(m²)(K)], the heat-transfer coefficient through the tube wall h_w is 77,140 Btu/(h)(ft²)(°F) [437,380 W/(m²)(K)], and the overall fouling coefficient is 1000 Btu/(h)(ft²)(°F) [5680 W/(m²)(K)].

Duct Cooler (Finned-Tube Bank)

D_r = root diameter (bare-tube outside diameter) = 0.625 in (0.0159 m)
  l = fin height = 0.5 in (0.013 m)
  b = fin thickness = 0.012 in (0.305 mm)
  n = number of fins per inch = 12
  s = tube spacing = 1.75 in (0.0445 m)
  x = tube-wall thickness = 0.035 in (0.889 mm)
A_o = bare heat-transfer surface = 50 ft² (4.64 m²)

The tube pattern is triangular, the fin material is aluminum, and the tube material is copper.

Calculation Procedure

1. Select the appropriate equation for the outside heat-transfer coefficient.
The heat-transfer coefficient for finned tubes can be calculated using the equation

$$\frac{hD_r}{k} = a\left(\frac{c\mu}{k}\right)^{1/3}\left(\frac{D_r G}{\mu}\right)^{0.681}\left(\frac{y^3}{l^2 b}\right)^{0.1}$$

where a = 0.134 for triangular tube patterns or $0.128(y/l)^{0.15}/[1 + (s - D_f)/(s - D_r)]$ for inline tube patterns
  h = outside heat-transfer coefficient
  D_r = root diameter (diameter at base of fins)
  D_f = diameter of fins (equal to $D_r + 2l$)
  k = thermal conductivity
  c = specific heat
  μ = viscosity
  G = mass velocity, based on minimum flow area
  l = fin height
  b = fin thickness
  y = distance between fins = $(1/n) - b$

n = number of fins per unit length
s = tube spacing perpendicular to flow
p = tube spacing parallel to flow

The equation is valid for triangular spacings with s/p ranging from 0.7 to 1.1.

2. Calculate the relevant fin areas.

These are heat-transfer surface of fins A_f, outside heat-transfer surface of bare tube A_o, total outside heat-transfer surface A_t (equal to $A_f + A_o$), and ratio of maximum flow area to minimum flow area a_r. For this problem

$$A_f = \left(D_f^2 - D_r^2\right)(\pi/4)2n$$

$$= (1.625^2 - 0.625^2)(1/12)^2(\pi/4)(2)(12)(12)$$

$$= 3.5343 \text{ ft}^2 \text{ per foot of tube length}$$

And $A_o = \pi D_o = \pi(0.625/12) = 0.1636$ ft^2 per foot of tube length. Then, $A_t = A_f + A_o = 3.5343 + 0.1636 = 3.6979$ ft^2 per foot of tube length, and $A_t/A_o = 3.6979/0.1636 = 22.6$. Finally, $a_r = s/(s - D_o - 2nlb) = 1.75/[1.75 - 0.625 - 2(12)(0.5)(0.012)] = 1.7839$.

3. Calculate h_o, the outside film coefficient.

From step 1, $h_o = 0.134(k/D_r)(c\pi/k)^{1/3} \times (D_r G/\mu)^{0.681} (y^3/l^2 b)^{0.1}$. Now, $G = 60\rho FVa_r$, where ρ is the density in pounds per cubic foot and FV is the face velocity (duct velocity) in feet per minute. Then, $G = 60(0.0049)(500)(1.7839) = 262.2$ lb/(h)(ft^2), and $D_r G/\mu = (0.625/12)262.2/0.0225 = 607.0$. Now, $y = (1/n) - b = (1/12) - 0.012 = 0.0713$ in (1.812 mm).

The Prandtl number for hydrogen ($c\mu/k$) = 0.7, so

$$h_o = 0.134\left(\frac{0.11}{0.625/12}\right)0.7^{1/3}607.0^{0.681}\left[\frac{0.0713^3}{0.5^2(0.012)}\right]^{0.1}$$

$$= 15.99 \text{ Btu/(h)(ft}^2)(°F)[90.82 \text{ W/(m}^2)(K)]$$

4. Calculate $h_{f,o}$, pertaining to the outside of the finned tubes.

For transverse fins fabricated by finning a bare tube, it is convenient to base the overall heat-transfer coefficient on the outside surface of the bare tube. Thus

$$h_{f,o} = h_o(A_t/A_o)\eta$$

where η is the weighted fin efficiency.

The weighted fin efficiency may be approximated from Table 7.5. For the present problem, use a value of 0.80. Then, $h_{f,o} = 15.99(22.6)(0.80) = 289.1$ Btu/(h)(ft^2)(°F) [1642 W/(m^2)(K)].

TABLE 7.5   Weighted Fin Efficiency for Transverse Fins

h_o	Weighted fin efficiency η			
	Copper, $k = 220$ [381]	Aluminum, $k = 125$ [216]	Steel, $k = 25$ [43.3]	Stainless steel, $k = 10$ [17.3]
5 [28.4]	0.95	0.90	0.70	0.50
10 [56.7]	0.90	0.85	0.60	0.35
25 [142]	0.85	0.70	0.40	0.20
50 [284]	0.70	0.55	0.25	0.15
100 [567]	0.55	0.40	0.15	0.10

Note: h_o, the outside film coefficient, is in Btu/(h)(ft^2)(°F) [W/(m^2)(K)]: k, the thermal conductivity, is in Btu/(h)(ft)(°F) [W/(m)(K)]. This table is valid for fin heights in the vicinity of 0.5 in and fin thicknesses around 0.012 in.

5. Calculate U_o.
The formula is

$$\frac{1}{U_o} = \frac{1}{h_{f,o}} + \frac{1}{h_i(D_i/D_o)} + \frac{1}{h_w} + \frac{1}{h_s}$$

$$= \frac{1}{289.1} + \frac{1}{1200(0.555/0.625)} + \frac{1}{77{,}140} + \frac{1}{1000}$$

$$= 185 \text{ Btu/(h)(ft}^2)(°\text{F}) \text{ [1050 W/(m}^2)(\text{K})]}$$

6. Calculate the outlet temperature.
Assume that there are several tube-side passes. The outlet temperature can then be calculated using Fig. 7.6. This figure can be employed to directly calculate the outlet temperature of the hot fluid (which is of interest in this example); in such a case, the abscissa becomes $UA/(WC)$, $R = WC/(wc)$, and $P = (T_2 - T_1)/(t_1 - T_1)$ Then, for the present example, $UA/(WC) = 185(50)/[1000(3.4)] = 2.72$, and $R = 1000(3.4)/[10{,}000(1.0)] = 0.34$. For these parameters, Fig. 7.6 shows P to be 0.88. Therefore, $T_2 = T_1 + P(t_1 - T_1) = 200 + 0.88(86 - 200) = 99.7°\text{F}$ (310.6 K).

7.31 AIR-COOLED HEAT EXCHANGERS

Design an air-cooled heat exchanger to cool water under the following conditions. The design ambient air temperature is 35°C (95°F). The tubes to be used are steel tubes [thermal conductivity = 25 Btu/(h)(ft^2)(°F), or 43 W/(m)(K)] with aluminum fins. The steel tube is 1 in (0.0254 m) outside diameter and 0.834 in (0.0212 m) inside diameter. The inside heat-transfer coefficient h_i and the fouling coefficient h_s are each 1000 Btu/(h)(ft^2)(°F) [5680 W/(m^2)(K)]. Heat-exchanger purchase cost is $22 per square foot for four-tube-row units, $20 per square foot for five-tube-row units, and $18 per square foot for six-tube-row units.

Water Conditions

Flow rate = 500,000 lb/h (226,795 kg/h), multipass
Inlet temperature = 150°C (302°F)
Outlet temperature = 50°C (122°F)

Calculation Procedure

1. Decide on the design approach.
A four-tube-row unit, a five-tube-row unit, and a six-tube-row unit will be designed and the cost compared. Standard tube geometries for air-cooled exchangers are 1 in (0.0254 m) outside diameter with ⅝ in (0.0159 m) high aluminum fins spaced 10 fins per inch. Also available are 8 fins per inch. Standard spacings for tubes are listed in Table 7.6.

Typical face velocities (*FVs*) used for design are also tabulated in Table 7.6. These values result in air-cooled heat exchangers that approach an optimum cost, taking into account the purchase cost, the cost for installation, and the cost of power to drive the fans. Each designer may wish to establish his or her own values of typical design face velocities; these should not vary greatly from those tabulated.

For air-cooled equipment, the tube spacing is normally determined by the relative values of the inside heat-transfer coefficient and the air-side heat-transfer coefficient. For inside coefficients much greater than the air-side coefficient, tubes spaced on 2.5-in (0.064-m) centers are normally justified. For low values of the inside coefficient, tubes spaced on 2.375-in (0.060-m) centers are normally

TABLE 7.6 Design Face Velocities for Air-Cooled Exchangers

	Face velocity, ft/min(m/s)		
Number of tube rows	8 fins/in (315 fins/m), 2.375-in (0.0603-m) pitch	10 fins/in (394 fins/m), 2.375-in (0.0603-m) pitch	10 fins/in (394 fins/m), 2.5-in(0.0635-m) pitch
3	650 (3.30)	625 (3.18)	700 (3.56)
4	615 (3.12)	600 (3.05)	660 (3.35)
5	585 (2.97)	575 (2.92)	625 (3.18)
6	560 (2.84)	550 (2.79)	600 (3.05)

justified. In the present case, because water is being cooled, the inside coefficient is indeed likely to be much greater, so specify the use of tubes on 2.5-in centers.

2. Determine h_a, the air-side heat-transfer coefficient.
The air-side coefficient is frequently calculated on the basis of the outside surface of a bare tube. The equations for air can be simplified as follows:

$$h_a = 8(FV)^{1/2} \quad \text{for 10 fins per inch}$$

$$h_a = 6.75(FV)^{1/2} \quad \text{for 8 fins per inch}$$

where h_a is in Btu's per hour per square foot per degree Fahrenheit and FV is in feet per minute.

For this problem, FV for 5 rows = 625 ft/min, so $h_a = 8(625)^{1/2} = 200$ Btu/(h)(ft^2)(°F) [1136 W/(m^2)(K)]. FV for 4 rows = 660 ft/min, so $h_a = 8(660)^{1/2} = 205$ Btu/(h)(ft^2)(°F) [1164 W/(m^2)(K)]. FV for 6 rows = 600 ft/min, so $h_a = 8(601)^{1/2} = 196$ Btu/(h)(ft^2)(°F) [1113 W/(m^2)(K)].

3. Calculate h_w, the coefficient of heat transfer through the tube wall.
The formula is $h_w = 2k/(D_o - D_i) = 2(25)/[(1 - 0.834)(1/12)] = 3614$ Btu/(h)(ft^2)(°F) [20,520 W/(m^2)(K)].

4. Calculate U, the overall heat-transfer coefficient.
The formula is

$$\frac{1}{U} = \frac{1}{h_a} + \frac{1}{h_i(D_i/D_o)} + \frac{1}{h_w} + \frac{1}{h_s}$$

For five-tube rows, $1/U = 1/200 + 1/[1000(0.834/1)] + 1/3614 + 1/1000$, so $U = 134$ Btu/(h)(ft^2)(°F). For four-tube rows, $1/U = 1/205 + 1/[1000(0.834/1)] + 1/3614 + 1/1000$, so $U = 136$ Btu/(h)(ft^2)(°F). For six-tube rows, $1/U = 1/196 + 1/[1000(0.834/1)] + 1/3614 + 1/1000$, so $U = 132$ Btu/(h)(ft^2)(°F). Since the values are so close, use the same value of U for four, five, and six rows, namely, $U = 135$ Btu/(h)(ft^2)(°F)[767 W/(m^2)(K)].

5. Design a five-row unit.
Air-cooled equipment is fabricated in standard modules. The standards begin with fin-tube bundles 48 in (1.22 m) wide and increase in 6-in (0.152-m) increments up to a 144-in (3.66-m) maximum. These modules can then be placed in parallel to obtain any size exchanger needed. The maximum tube length is 48 ft (14.63 m). In general, long tubes result in economical heat exchangers. In the present case, assume that the plant layout allows a maximum tube length of 40 ft (12.2 m).

The design of heat-transfer equipment is a trial-and-error procedure because various design standards are followed in order to reduce equipment cost. To design an air-cooled exchanger, an outlet air temperature and a tube length are assumed, which establishes the amount of air to be pumped by the fan. The amount of air to be pumped establishes a face area, because we have assumed a face velocity. The face area fixes the heat-transfer area for a given tube length and number of tube rows.

TABLE 7.7   Estimated Outlet Air Temperatures for Air-Cooled Exchangers

Process inlet temperature, °C	Outlet air temperature, °C		
	$U = 50$	$U = 100$	$U = 150$
175	90	95	100
150	75	80	85
125	70	75	80
100	60	65	70
90	55	60	65
80	50	55	60
70	48	50	55
60	45	48	50
50	40	41	42

Note: U is the overall heat-transfer coefficient in Btu/(h)(ft²)(°F) [1 Btu/(h)(ft²)(°F) = 5.67 W/(m²)(K)].

Table 7.7 permits an estimate of the outlet air temperature, based on 90 to 95°F (305 to 308 K) design ambient air temperature.

For this problem, $U = 135$ and inlet process temperature = 150°C, so assume an outlet air temperature of 83°C. Then, Q = heat load = $wc(t_2 - t_1)$ = 500,000(1.0)(150 − 50)(1.8°F/°C) = 90,000,000 Btu/h. The face area FA can be estimated from the equation

$$FA = \frac{Q}{FV(T_2 - T_1)(1.95)}$$

where Q is the heat load in Btu/h, FA is the face area in square feet, FV is the face velocity in feet per minute, and T_1 and T_2 are inlet and outlet air temperatures in degrees Celsius. Thus, FA = 90,000,000/[625(83 − 35)(1.95)] = 1540 ft².

The exchanger width can now be determined:

$$Y = \text{width} = \text{face area/tube length} = FA/L = 1540/40 = 38.5 \text{ ft}$$

However, standard widths have 6-in increments. Therefore, assume four 9.5-ft-wide bundles. Then, $FA = 4(9.5) = 38$ ft.

For this face area, calculate the outlet air temperature. First, air temperature rise (in °C) is

$$\Delta T_a = \frac{Q}{Y(FV)(L)1.95}$$
$$= \frac{90,000}{36(625)(40)1.95}$$
$$= 48.5°C$$

Then, $T_2 = T_1 + \Delta T_a = 35 + 48.5 = 83.5°C$. Next, calculate mean temperature difference. The formula is

$$\Delta T_m = \frac{(t_1 - T_2) - (t_2 - T_1)}{\ln[(t_1 - T_2)/(t_2 - T_1)]}$$

Thus, ΔT_m = [(150 − 83.5) − (50 − 35)]/ln [(150 − 83.5)/(50 − 35)] = 34.6°C (62.3°F).

Now, calculate the area required if this is the available temperature difference. Thus, $A = Q/(U\Delta T_m)$ = 90,000,000/[135(34.6)(1.8)] = 10,705 ft² (994.4 m²). Next, calculate the area actually available. The number of tubes per row N_t can be approximated by dividing the bundle width by the tube spacing, that is, $N_t = Y/s = 38/(2.5/12) = 182$ per row. Then, letting N_r equal the number of rows,

$A = N_r N_t \pi D_o L = 5(182)\pi(1/12)(40) = 9530$ ft². Therefore, the area available is less than that required. Accordingly, increase the bundle width from 9.5 to 10 ft and calculate the new ΔT_a:

$$\Delta T_a = 48.5(9.5/10) = 46.1°C$$

Therefore, $T_2 = 35 + 46.1 = 81.1°C$.
  Next, calculate the new ΔT_m:

$$\Delta T_m = [(150 - 81.1) - (50 - 35)]/\ln[(150 - 81.1)/(50 - 35)]$$
$$= 35.4°C \ (63.7°F)$$

And calculate the new area required

$$A = 90,000,000/[135(35.4)(1.8)] = 10.460 \text{ ft}^2$$

Finally, calculate the new area available: $N_t = 40/(2.5/12) = 192$, so $A = 5(192)\pi(1/12)(40) = 10,050$ ft². Or more precisely, the actual tube count provided by manufacturers' standards for a bundle 10 ft wide with five tube rows is 243. Letting N_b equal the number of bundles, $A = N_b N \pi D_o L = 4(243)\pi(1/12)(40) = 10.180$ ft² (945.7 m²). This area would normally be accepted because the relatively high design air temperature occurs for relatively short periods and the fouling coefficient is usually arbitrarily assigned. If the design air temperature must be met at all times, then a wider bundle or greater face velocity would be required.
  With the dimensions of the cooler established, the next step is to calculate the air-side pressure drop. For air, the following relations can be used:

$$\Delta P_a = 0.0047 N_r (FV/100)^{1.8} \quad \text{for 10 fins per inch, 2.375-in spacing}$$
$$\Delta P_a = 0.0044 N_r (FV/100)^{1.8} \quad \text{for 8 fins per inch, 2.375-in spacing}$$
$$\Delta P_a = 0.0037 N_r (FV/100)^{1.8} \quad \text{for 10 fins per inch, 2.5-in spacing}$$

In these equations, ΔP_a is air-side pressure drop, in inches of water, N_r is the number of tube rows, and FV is face velocity, in feet per minute. For this problem, $\Delta P_a = 0.0037(5)(625/100)^{1.8} = 0.501$ in water.
  Now, calculate the power required to pump the air. Use the formula

$$bhp = (FV)(FA)(T_2 + 273)(\Delta P_a + 0.1)/(1.15 \times 10^6)$$

where bhp = brake horsepower
  FV = face velocity, in ft/min
  FA = face area, in ft² (equals $N_b YL$)
  T_2 = outlet air temperature, in °C
  ΔP_a = air-side pressure drop, in inches of water

In this case, $FA = N_b YL = 4(10)(40) = 1600$ ft², so $bhp = 625(1600)(81.1 + 273)(0.501 + 0.1)/(1.15 \times 10^6) = 185$ hp (138 kW).

6. *Compare costs.*
The same procedure can be followed to design a six-row unit and a four-row unit. The following comparison can then be made:

	Number of tube rows		
	4	5	6
Number of bundles	4	4	4
Bundle width, ft	12	10	9.5
Heat transfer area, ft²	9800	10,180	11,510
Equipment cost, dollars	215,600	203,600	207,180
Power required, bhp	195	185	189

From this comparison, a five-row unit would be the most economical.

7.32 PRESSURE DROP FOR FLOW INSIDE TUBES: SINGLE-PHASE FLUIDS

Calculate the pressure drop for the water flowing through the air-cooled heat exchanger designed in Procedure 7.31 if the number of tube-side passes is 10. The density of the water is 60 lb/ft^3 (961.1 kg/m^3), and the viscosity is 0.74 lb/(ft)(h) (0.31 cP). Assume that the velocity in the nozzles is 10 ft/s (3.05 m/s) and that the viscosity change with temperature is negligible.

Calculation Procedure

1. Consider the causes of the pressure drop and the equations to find each.
The total pressure drop for fluids flowing through tubes results from factional pressure drop as the fluid flows along the tube, from pressure drop as the fluid enters and leaves the tube-side heads or channels, and from pressure drop as the fluid enters and leaves the tubes from the heads or channels.
The frictional pressure drop can be calculated from the equation

$$\Delta P_f = \frac{4 f G^2 L N_p}{2(144) g \rho D}$$

where ΔP_f = pressure drop, in lb/in^2
   f = friction factor
   G = mass velocity, in lb/(h)(ft^2)
   L = tube length, in ft
   N_p = number of tube-side passes
   g = the gravitational constant
   ρ = density, in lb/ft^3
   D = tube inside diameter, in ft

For fluids with temperature-dependent viscosities, the pressure drop must be corrected by the ratio of

$$(\mu_w / \mu_b)^n$$

where μ_w = viscosity at the surface temperature
   μ_b = viscosity at the bulk fluid temperature
   n = 0.14 for turbulent flow or 0.25 for laminar flow

The friction factor f can be calculated from the equations

$$f = 16/(DG/\mu) \quad \text{for } DG/\mu \text{ less than } 2100$$

$$f = 0.054/(DG/\mu)^{0.2} \quad \text{for } DG/\mu \text{ greater than } 2100$$

where μ is viscosity, in lb/(ft)(h).
The pressure drop as the fluid enters and leaves a radial nozzle at the heads or channels can be calculated from

$$\Delta P_n = k \rho v_n^2 / 9266$$

where ΔP_n = pressure drop, in lb/in^2
   ρ = density, in lb/ft^3
   v_n = velocity in the nozzle, in ft/s
   k = 0 for inlet nozzles or 1.25 for outlet nozzles

The pressure drop associated with inlet and outlet nozzles can be reduced by selection of other channel types, but the expense is not warranted except for situations in which pressure drop is critical or costly.

The pressure drop as a result of entry and exit from the tubes can be calculated from

$$\Delta P_e = k N_p \rho v_t^2 / 9266$$

where ΔP_e = pressure drop, in lb/in^2
     N_p = number of tube-side passes
     v_t = velocity in tube, in ft/s
     ρ = density, in lb/ft^3
     $k = 1.8$

The total pressure drop is the sum of all these components:

$$\Delta P_t = \Delta P_f + \Delta P_n + \Delta P_e$$

2. Calculate frictional pressure drop.
The total number of tubes from the previous design is $4(243) = 972$. The number of tubes per pass is $972/10 = 97.2$. Letting a_c be flow area,

$$G = \frac{W}{a_c} = \frac{500{,}000}{97.2(0.834/12)^2(\pi/4)}$$

$$= 1{,}355{,}952 \text{ lb/(h)(ft}^2\text{)[6,619,758 kg/(h)(m}^2\text{)]}$$

Then, $DG/\mu = (0.834/12)1{,}355{,}952/0.74 = 127{,}350$. Therefore, $f = 0.054/(DG/\mu)^{0.2} = 0.054/127{,}350^{0.2} = 0.0051$, and

$$\Delta P_f = \frac{4 f G^2 L N_p}{2(144) g \rho D}$$

$$= \frac{4(0.0051)(1{,}355{,}952)^2(40)10}{2(144)(4.18 \times 10^8)(60)(0.834/12)}$$

$$= 29.89 \text{ lb/in}^2$$

3. Calculate nozzle pressure drop.
From step 1, $\Delta P_n = 1.25 \rho v_n^2 / 9266 = 1.25(60)(10)^2 / 9266 = 0.81 \text{ lb/in}^2$.

4. Calculate tube entry/exit pressure drop.
From step 1, $\Delta P_e = 1.8 N_p \rho v_t^2 / 9266$. Now, by definition, $v_t = G/3600\rho = 1{,}355{,}952/[3600(60)] = 6.28$ ft/s (1.91 m/s). So, $\Delta P_e = 1.8(10)(60)(6.28)^2/9266 = 4.59 \text{ lb/in}^2$.

5. Calculate total pressure drop.
Thus, $\Delta P_t = \Delta P_f + \Delta P_n + \Delta P_e = 29.89 + 0.81 + 4.59 = 35.29 \text{ lb/in}^2$ (243.32 kPa).

Related Calculations. *Helical Coils.* The same procedure can be used to calculate the pressure drop in helical coils. For turbulent flow, a friction factor for curved flow is substituted for the friction factor for straight tubes. For laminar flow, the friction loss for a curved tube is expressed as an equivalent length of straight tube and the friction factor for straight tubes is used. The Reynolds number required for turbulent flow is $2100[1 + 12(D_i/D_c)^{1/2}]$, where D_i is the inside diameter of the tube and D_c is the coil diameter.

The friction factor for turbulent flow is calculated from the equation $f_c(D_c/D_i)^{1/2} = 0.0073 + 0.076[(DG/\mu)(D_i/D_c)^2]^{-1/4}$, for $(DG/\mu)(D_i/D_c)^2$ between 0.034 and 300, where f_c is the friction factor for curved flow. For values of $(DG/\mu)(D_i/D_c)^2$ below 0.034, the friction factor for curved flow is practically the same as that for straight pipes.

For laminar flow, the equivalent length L_e can be predicted as follows: For $(DG/\mu)(D_i/D_c)^{1/2}$ between 150 and 2000,

$$L_e/L = 0.23\left[(DG/\mu)(D_i/D_c)^{1/2}\right]^{0.4}$$

For $(DG/\mu)(D_i/D_c)^{1/2}$ between 10 and 150,

$$L_e/L = 0.63\left[(DG/\mu)(D_i/D_c)^{1/2}\right]^{0.2}$$

For $(DG/\mu)(D_i/D_c)^{1/2}$ less than 10,

$$L_e/L = 1$$

In these equations, L is straight length, and L_e is equivalent length of a curved tube.

Longitudinal fin tubes. The same procedure can be used for longitudinally finned tubes. The equivalent diameter D_e is substituted for D_i. The friction factor can be determined from these equations: For Reynolds numbers below 2100,

$$f_{lf} = \frac{16}{(D_e G/\mu)}$$

For Reynolds numbers greater than 2100,

$$f_{lf} = \frac{0.103}{(D_e G/\mu)^{0.25}}$$

In these equations, f_{lf} is the friction factor for longitudinally finned tubes, and D_e is equivalent diameter ($= 4a_c/P$, where a_c is the cross-sectional flow area and P is the wetted perimeter).

7.33   PRESSURE DROP FOR FLOW INSIDE TUBES: TWO-PHASE FLUIDS

Calculate the frictional pressure drop for a two-phase fluid flowing through a tube 0.62 in (0.0158 m) in inside diameter D and 20 ft (6.1 m) in length L at a rate of 100 lb/h (45.4 kg/h). The mixture is 50 percent gas by weight and 50 percent liquid by weight, having the following properties:

Liquid Properties

ρ = density = 50 lb/ft^3 (800 kg/m^3)
μ = viscosity = 2.0 lb/(ft)(h) (0.84 cP)

Gas Properties

ρ = density = 0.1 lb/ft^3 (1.6 kg/m^3)
μ = viscosity = 0.045 lb/(ft)(h) (0.019 cP)

Calculation Procedure

1. *Select the method to be used.*
For two-phase flow, the friction pressure drop inside a tube can be calculated using the equation [31]

$$\Delta P_{tp} = \left[\Delta P_L^{1/n} + \Delta P_G^{1/n}\right]^n$$

where ΔP_{tp} = two-phase pressure drop

ΔP_L = pressure drop for the liquid phase, assuming only the liquid phase is present

ΔP_G = pressure drop for the gas phase, assuming only the gas phase is present

n = 4.0 when both phases are in turbulent flow or 3.5 when one or both phases are in laminar flow

2. Calculate the liquid-phase pressure drop.

Let W_L be the mass flow rate of the liquid, equal to 50 percent of 100 lb/h, or 50 lb/h, and let a_c be the cross-sectional area of the tube. Then G_L = mass velocity of liquid phase = $W_L/a_c = 50/[(0.62/12)^2(\pi/4)]$ = 23,848.4 lb/(h)(ft^2). Then the Reynolds number $DG_L/\mu = (0.62/12)23,848.4/2.0 = 616.1$. Then, as in step 1 of Procedure 7.32, f = friction factor = $16/(DG_L/\mu) = 16/616.1 = 0.026$, and

$$\Delta P_L = \frac{4fG_L^2LN_p}{2(144)g\rho D}$$

$$= \frac{4(0.026)(23,848.4)^2(20)(1)}{2(144)(4.18 \times 10^8)(50)(0.62/12)}$$

$$= 0.0038 \text{ lb/in}^2$$

3. Calculate the gas-phase pressure drop.

The mixture is 50 percent each phase, so G_G = mass velocity of gas phase = G_L from step 2. Then $DG_G/\mu = (0.62/12)23,848.4/0.045 = 27,381.5$. Again referring to step 1 of Procedure 7.32, f = friction factor = $0.054/(DG_G/\mu)^{0.2} = 0.054/27,381.5^{0.2} = 0.0070$, and

$$\Delta P_G = \frac{4fG_G^2LN_p}{2(144)g\rho D}$$

$$= \frac{4(0.0070)(23,848.4)^2(20)(1)}{2(144)(4.18 \times 10^8)(0.1)(0.62/12)}$$

$$= 0.5120 \text{ lb/in}^2$$

4. Calculate the two-phase pressure drop.

The liquid phase is in laminar flow; the gas phase is in turbulent flow. Therefore,

$$\Delta P_{tp} = \left[\Delta P_L^{1/3.5} + \Delta P_G^{1/3.5} \right]^{3.5}$$

$$= \left(0.0038^{1/3.5} + 0.5120^{1/3.5} \right)^{3.5}$$

$$= 1.107 \text{ lb/in}^3 (7.633 \text{ kPa})$$

Related Calculations.   *Homogeneous flow method.* Two-phase flow pressure drop can also be calculated using a homogeneous-flow model that assumes that gas and liquid flow at the same velocity (no slip) and that the physical properties of the fluids can be suitably averaged. The correct averages are

$$\rho_{ns} = \rho_L\lambda_L + \rho_G(1 - \lambda_L) \quad \text{and} \quad \mu_{ns} = \mu_L\lambda_L + \mu_G(1 - \lambda_L)$$

where $\lambda_L = Q_L/(Q_L + Q_G)$

ρ_{ns} = no-slip density

μ_{ns} = no-slip viscosity

ρ_L = liquid density

ρ_G = gas density

μ_L = liquid viscosity

μ_G = gas viscosity
Q_L = liquid volumetric flow rate
Q_G = gas volumetric flow rate

Once the average density and viscosity are calculated, an average no-slip Reynolds number can be calculated:

$$\text{No-slip Reynolds number} = DG_T/\mu_{ns}$$

where G_T = total mass velocity = $G_L + G_G$
D = tube diameter

The frictional pressure drop is then

$$\Delta P_f = \frac{4fG_T^2 \mathrm{L} \mathrm{N}_p}{2(144)g\rho_{ns}D}$$

where f is the friction factor calculated using the method presented for single-phase fluids but based on the no-slip Reynolds number.

In addition to the frictional pressure drop, there will be a pressure loss associated with the expansion of the gas, termed an "acceleration loss" and calculated from

$$\Delta P_a = \frac{G_T^2}{2(144)g}\left(\frac{1}{\rho_{G2}} - \frac{1}{\rho_{G1}}\right)$$

where ΔP_a = acceleration pressure drop, in lb/in^2
G_T = total mass velocity = $G_L + G_G$, in lb/(h)(ft^2)
ρ_{G2} = gas density at outlet, in lb/ft^3
ρ_{G1} = gas density at inlet, in lb/ft^3
g = gravitational constant, in ft/h^2

The total pressure drop ΔP_{tp} is, then,

$$\Delta P_{tp} = \Delta P_f + \Delta P_a$$

For flashing flow, the pressure drop is

$$\Delta P_{tp} = \frac{G_T^2}{2(144)g}\left[2\left(\frac{x_2}{\rho_{G2}} - \frac{x_1}{\rho_{G1}}\right) + \frac{4f\,\mathrm{L}\mathrm{N}_p}{D\rho_{ns,avg}}\right]$$

where x_1 = mass fraction vapor at inlet
x_2 = mass fraction vapor at outlet
$\rho_{ns,avg}$ = average no-slip density

Pressure drop for condensers. For condensers, the frictional pressure drop can be estimated using the relation

$$\Delta P_c = 1/2\frac{1 + v_2}{v_1}\Delta P_1$$

where ΔP_c = condensing pressure drop
ΔP_1 = pressure drop based on the inlet conditions of flow rate, density, and viscosity
v_2 = vapor velocity at the outlet
v_1 = vapor velocity at the inlet

For a total condenser, this becomes

$$\Delta P_c = \frac{\Delta P_1}{2}$$

7.34   PRESSURE DROP FOR FLOW ACROSS TUBE BANKS: SINGLE-PHASE FLUIDS

Calculate the pressure drop for the conditions of Procedure 7.16. Assume that the nozzle velocities are 5 ft/s (1.52 m/s), that the fluid density is 55 lb/ft³ (881 kg/m³), and that there are 24 baffles. Assume that there is also an impingement plate at the inlet nozzle. Calculate the pressure drop for both fouled and clean conditions.

Calculation Procedure

1. Consider the causes of the pressure drop, and select equations to calculate each.
The pressure drop for fluids flowing across tube banks may be determined by calculating the following components:

a. Inlet-nozzle pressure drop

b. Outlet-nozzle pressure drop

c. Frictional pressure drop for inlet and outlet baffle sections

d. Frictional pressure drop for intermediate baffle sections

e. Pressure drop for flow through the baffle windows

Heat exchangers with well-constructed shell sides will have a certain amount of bypassing that will reduce the pressure drop experienced with an ideal-tube bundle (one with no fluid bypassing or leakage). The amount of bypassing for a clean heat exchanger is more than that for a fouled heat exchanger. The following leakage factors are based on data from operating heat exchangers and include the typical effects of fouling on pressure drop.

Pressure drop for nozzles may be calculated from

$$\Delta P_n = \frac{k\rho v_n^2}{9266}$$

where $k = 0$ for inlet nozzles with no impingement plate, 1.0 for inlet nozzles with impingement plates, or 1.25 for outlet nozzles

ΔP_n = pressure drop, in lb/in²

ρ = density, in lb/ft³

v_n = velocity in the nozzle, in ft/s

Frictional pressure drop for tube banks may be calculated as follows: For intermediate baffle sections.

$$\Delta P_f = \frac{4f G^2 N_r (N_b - 1) R_1 R_b \phi}{2(144)g\rho}$$

and for inlet and outlet baffle sections combined,

$$\Delta P_{fi} = \frac{4(2.66)f G^2 N_r R_b \phi}{2(144)g\rho}$$

where $R_1 = 0.6(B/D_s)^{1/2}$ for clean bundles or $0.75(B/D_s)^{1/3}$ for bundles with assumed fouling
$\quad R_b = 0.80(D_s)^{0.08}$ for clean bundles or $0.85(D_s)^{0.08}$ for bundles with assumed fouling
$\quad \phi = (\mu_w/\mu_b)^n$, where $n = 0.14$ for DG/μ greater than 300 or 0.25 for DG/μ less than 300
$\quad N_r = bD_s/s$, where $b = 0.7$ for triangular tube spacing, 0.6 for square tube spacing, or 0.85 for rotated square tube spacing

The friction factor f can be calculated. For $D_g G/\mu$ greater than 100,

$$f = \frac{z}{(D_g G/\mu)^{0.25}}$$

where $z = 1.0$ for square and triangular tube patterns or 0.75 for rotated square tube patterns. For $D_g G/\mu$ less than 100,

$$f = \frac{r}{(D_g G/\mu)^{0.725}}$$

where $r = 10$ for triangular tube patterns or 5.7 for square and rotated square tube patterns. In these equations, D_g is defined as the gap between the tubes, that is, $D_g = s - D_o$.

Pressure drop for baffle windows may be calculated as follows: For $D_o G/\mu$ greater than 100,

$$\Delta P_w = \frac{G^2}{2(144)g\rho} \frac{a_c}{a_w} (2 + 0.6N_w)N_b R_1$$

in which the factor $(2 + 0.6N_w)$ can be approximated with the term $2 + 0.6N_w = m(D_s)^{5/8}$, where $m = 3.5$ for triangular tube patterns, 3.2 for square tube patterns, or 3.9 for rotated square tube patterns. For $D_o G/\mu$ less than 100,

$$\Delta P_w = \frac{26\mu G}{g\rho}\left(\frac{a_c}{a_w}\right)^{1/2}\left(\frac{N_w}{s - D_o} + \frac{B}{D_e^2}\right) + \frac{2G^2}{2g\rho}\frac{a_c}{a_w}$$

Baffles are normally cut on the centerline of a row of tubes. Baffles should be cut on the centerline of a row whose location is nearest the value of 20 percent of the shell diameter. For baffle cuts meeting this criterion, the following values can be used:

Tube geometry	$N_w/(s - D_o)$	D_e, ft
$D_o = \frac{5}{8}$ in(0.016 m); $s = \frac{13}{16}$ in(0.020 m); triangular pitch	$173D_s$	0.059
$D_o = \frac{5}{8}$ in(0.016 m); $s = \frac{7}{8}$ in(0.022 m); square pitch	$105D_s$	0.090
$D_o = \frac{3}{4}$ in(0.019 m); $s = \frac{15}{16}$ in(0.024 m); triangular pitch	$151D_s$	0.063
$D_o = \frac{3}{4}$ in(0.019 m); $s = 1$ in(0.025 m); square pitch	$92D_s$	0.097
$D_o = 1$ in(0.025 m); $s = 1\frac{1}{4}$ in(0.032 m); triangular pitch	$85D_s$	0.083
$D_o = 1$ in(0.025 m); $s = 1\frac{1}{4}$ in(0.032 m); square pitch	$73.5D_s$	0.104

In these equations, $a_c = B D_s(s - D_o)/s$ for triangular and square tube patterns or $1.5 B D_s(s - D_o)/s$ for rotated square tube patterns, and $a_w = 0.055D_s^2$ for triangular tube patterns or $0.066D_s^2$ for square and rotated square tube patterns. Moreover, the units for the various terms are as follows:

$\Delta P_f =$ total pressure drop for intermediate baffle sections, in lb/in^2
$\Delta P_{fi} =$ total pressure drop for inlet and outlet baffle sections, in lb/in^2
$\Delta P_w =$ total pressure drop for baffle windows, in lb/in^2
$\quad f =$ friction factor (dimensionless)
$\quad G =$ mass velocity, in lb/(h)(ft^2) ($= W/a_c$, where $W =$ flow rate)

g = gravitational constant, in ft/h^2
ρ = density, in lb/ft^3
N_r = number of tube rows crossed
N_b = number of baffles
a_c = minimum flow area for cross flow, in ft^2
a_w = flow area in baffle window, in ft^2
D_s = shell diameter, in ft
s = tube center-to-center spacing, in ft
B = baffle spacing, in ft
D, D_o = outside tube diameter, in ft
D_g = gap between tubes, in ft (= $s - D_o$)
μ = viscosity, in lb/(ft)(h)
μ_w = viscosity at wall temperature, in lb/(ft)(h)
μ_b = viscosity at bulk fluid temperature, in lb/(ft)(h)
R_1 = correction factor for battle leakage
R_b = correction factor for bundle bypass
D_e = equivalent diameter of baffle window, in ft

The total pressure drop is the sum of all these components:

$$\Delta P = \Delta P_n + \Delta P_f + \Delta P_{fi} + \Delta P_w$$

2. Calculate the total pressure drop assuming fouled conditions.

a. *Calculate nozzle pressure drops.* The equation is

$$\Delta P_n = \frac{k\rho v_n^2}{9266}$$

$$= \frac{2.25\rho v_n^2}{9266}$$

Thus, with k equal to 1.0 plus 1.25,

$$\Delta P_n = \frac{2.25(55)(5)^2}{9266}$$

$$= 0.334 \text{ lb/in}^2 (2.303 \text{ kPa})$$

b. *Calculate frictional pressure for intermediate baffle sections.* Now, $a_c = BD_s(s - D_o)/s = (9.5/12)$ $(25/12)(0.9375 - 0.75)/0.9375 = 0.3299$ ft^2 (0.0306 m^2). Therefore, $G = W/a_c = 50,000/0.3299 =$ 151,561.1 lb/(h)(ft^2), and $D_g G/\mu = (0.9375 - 0.75)(1/12)151,561.1/1.95 = 1214.4$. Accordingly, $f = 1/(D_g G/\mu)^{0.25} = 1/1214.4^{0.25} = 0.1694$. And in the equation $\Delta P_f = 4fG^2N_r(N_b - 1)R_1R_b\phi/$ $[2(144)g\rho]$, $R_1 = 0.75(B/D_s)^{1/3} = 0.75(9.5/25)^{1/3} = 0.5432$, $R_b = 0.85(D_s)^{0.08} = 0.85(25/12)^{0.08} =$ 0.9014, $\phi = (4.0/1.95)^{0.14} = 1.1058$, and $N_r = 0.7 D_s/s = 0.7(25)/0.9375 = 18.7$. Therefore,

$$\Delta P_f = \frac{4(0.1694)(151,561.1)^2(18.7)(24 - 1)(0.5432)(0.9014)1.1058}{2(144)(4.18 \times 10^8)55}$$

$$= 0.547 \text{ lb/in}^2 \ (3.774 \text{ kPa})$$

c. *Calculate fictional pressure drop for inlet and outlet baffle sections.* The equation is

$$\Delta P_{fi} = \frac{4(2.66)f G^2 N_r R_b \phi}{2(144)g\rho}$$

Thus,

$$\Delta P_{fi} = \frac{4(2.66)(0.1694)(151,561.1)^2(18.7)(0.9014)1.1058}{2(144)(4.18 \times 10^8)55}$$

$$= 0.117 \text{ lb/in}^2 (0.803 \text{ kPa})$$

d. *Calculate pressure drop for window sections.* The equation is

$$\Delta P_w = \frac{3.5G^2}{2(144)g\rho} \frac{a_c}{a_w} D_s^{5/8} N_b R_1$$

Now, $a_w = 0.055D_s^2 = 0.055(25/12)^2 = 0.2387 \text{ ft}^2$. Therefore,

$$\Delta P_w = \frac{3.5(151,561.1)^2(0.3299/0.2387)(25/12)^{5/8}(24)0.5432}{2(144)(4.18 \times 10^8)55}$$

$$= 0.346 \text{ lb/in}^2 \ (2.38 \text{ kPa})$$

e. *Calculate total pressure drop.* The equation is

$$\Delta P = \Delta P_n + \Delta P_f + \Delta P_{fi} + \Delta P_w$$

Thus, $\Delta P = 0.334 + 0.547 + 0.117 + 0.346 = 1.344 \text{ lb/in}^2(9.267 \text{ kPa})$.

3. Calculate the total pressure drop assuming clean conditions.

a. *Calculate R_1 and R_b for clean conditions.* Here, $R_1 = 0.6(B/D_s)^{1/2} = 0.6(9.5/25)^{1/2} = 0.3699$, and $R_b = 0.80(D_s)^{0.08} = 0.80(25/12)^{0.08} = 0.848$.

b. *Calculate total pressure drop.* The clean-pressure-drop components will be the fouled-pressure-drop components multiplied by the appropriate ratios of R_1 and R_b for the clean and fouled conditions. Thus,

$$\Delta P = \Delta P_n + \Delta P_f \frac{R_1}{R_1} \frac{R_b}{R_b} + \Delta P_{fi} \frac{R_b}{R_b} + \Delta P_w \frac{R_1}{R_1}$$

$$= 0.334 + 0.547 + \frac{0.3699}{0.5432} \frac{0.848}{0.9014} + 0.117\frac{0.848}{0.9014} + 0.346\frac{0.3699}{0.5432}$$

$$= 0.334 + 0.350 + 0.110 + 0.236$$

$$= 1.03 \text{ lb/in}^2(7.102 \text{ kPa})$$

Related Calculations. The preceding equations for pressure drop assume a well-constructed tube bundle with baffle cuts amounting to 20 percent of the shell diameter and tubes included in the baffle windows. The corrections for fluid bypassing (R_1 and R_b) are based on the standards of the Tubular Exchangers Manufacturers Association [11] and assume an adequate number of sealing devices. The values for fouling are based on plant data for typical services. Methods that evaluate the effects of the various bypass streams have been presented; the most widely used is that of Bell [15]. For poorly constructed tube bundles or for conditions greatly different from those assumed, this reference should be used for evaluating the effects on the calculated pressure drop.

Bundles with tubes omitted from baffle windows. Frequently, tubes are omitted from the baffle-window areas. For this configuration, maldistribution of the fluid as it flows across the bank of tubes may occur as a result of the momentum of the fluid as it flows through the baffle window. For this reason, baffle cuts less than 20 percent of the shell diameter should only be used with caution. Maldistribution will normally be minimized if the fluid velocity in the baffle window is equal to or less than the fluid velocity in crossflow across the bundle. This frequently requires baffle cuts greater than 20 percent of the shell diameter. For such cases, the number of tubes in cross-flow will be less than that assumed in the preceding methods, so a correction is required. In addition, the pressure drop for the first and last baffle sections assumes tubes in the baffle windows; the factor 2.66 should be reduced to 2.0 for the case of 20 percent baffle cuts.

For baffles with no tubes in the baffle windows, the pressure drop for the window section can be calculated from

$$\Delta P_w = \frac{1.8 \rho v_w^2 N_b}{9266}$$

where ΔP_w = pressure drop, in lb/in^2
$\quad \rho$ = density, in lb/ft^3
$\quad v_w$ = velocity in the baffle window, in ft/s
$\quad N_b$ = number of baffles

The flow area may be calculated from $a_w = 0.11 D_s^2$ for 20 percent baffle cuts or $0.15 D_s^2$ for 25 percent baffle cuts, where D_s is the shell diameter.

Low-fin tubes. The method presented above can be used to predict pressure drop for banks of low-fin tubes. For low-fin tubes, the pressure drop is calculated assuming that the tubes are bare. The mass velocity and tube diameter used for calculation are those for a bare tube with the same diameter as the fins of the low-fin tube.

Banks of finned tubes. For fin tubes other than low-fin tubes, the pressure drop for flowing across banks of transverse fin tubes can be calculated from

$$\Delta P_f = \frac{4 f_r G_m^2 N_r \phi}{2(144) g \rho}$$

In this equation, for values of $D_r G_m / \mu$ from 2000 to 50,000,

$$f_r = \frac{a}{(D_r G_m / \mu)^{0.316} (s/D_r)^{0.927}}$$

where $a = 10.5(a/P)^{1/2}$ for triangular tube patterns or $6.5(D_f/s)^{1/2}$ for square tube patterns. The units for the various terms are as follows:

ΔP_f = frictional pressure drop, in lb/in^2
$\quad f_r$ = friction factor (dimensionless)
$\quad G_m$ = mass velocity based on minimum flow area ($= W/a_c$) in $lb/(h)(ft^2)$
$\quad N_r$ = number of tube rows crossed by the fluid
$\quad \rho$ = density, in lb/ft^3
$\quad g$ = gravitational constant, in ft/h^2
$\quad D_r$ = root diameter of fin, in ft
$\quad \mu$ = viscosity, in $lb/(ft)(h)$
$\quad s$ = tube spacing perpendicular to flow, in ft
$\quad p$ = tube spacing parallel to flow, in ft
$\quad D_f$ = diameter of fins, in ft
$\quad a_c$ = minimum flow area, in ft^2

And,

$$\phi = (\mu_w / \mu_b)^{0.14}$$

where μ_w is the viscosity at the wall temperature and μ_b is the viscosity at the bulk fluid temperature, and

$$a_c = s - D_r - 2nlb$$

where n = number of fins per foot
$\quad\quad l$ = fin height, in ft
$\quad\quad b$ = fin thickness, in ft

7.35 PRESSURE DROP FOR FLOW ACROSS TUBE BUNDLES: TWO-PHASE FLOW

Calculate the pressure drop for the conditions of Procedure 7.34 if 10,000 lb/h (4535.9 kg/h) of the total fluid is a gas with a density of 0.5 lb/ft^3 (8.01 kg/m^3) and a viscosity of 0.05 lb/(ft)(h) (0.0207 cP). Assume the nozzle velocities are 50 ft/s (15.24 m/s). The flow is vertical up-and-down flow.

Calculation Procedure

1. Select the appropriate equation.
Two-phase-flow pressure drop for flow across tube bundles may be calculated using the equation

$$\frac{\Delta P_{tp}}{\Delta P_{LO}} = 1 + (K^2 - 1)\left[Bx^{(2-n)/n}(1 - x)^{(2-n)/n} + x^{2-n} \right]$$

where ΔP_{tp} = the two-phase pressure drop
$\quad\quad \Delta P_{LO}$ = the pressure drop for the total mass flowing as liquid
$\quad\quad x$ = the mass fraction vapor
$\quad\quad K = (\Delta P_{GO}/\Delta P_{LO})^{1/2}$, where ΔP_{GO} is the pressure drop for the total mass flowing as vapor
$\quad\quad B$ = for vertical up-and-down flow, 0.75 for horizontal flow other than stratified flow, or 0.25 for horizontal stratified flow

The value of n can be calculated from the relation

$$K = (\Delta P_{GO}/\Delta P_{LO})^{1/2} = (\rho_L/\rho_G)^{1/2}(\mu_G/\mu_L)^{n/2}$$

where ρ_G = gas density
$\quad\quad \rho_L$ = liquid density
$\quad\quad \mu_G$ = gas viscosity
$\quad\quad \mu_L$ = liquid viscosity

For the baffle windows, $n = 0$, and the preceding equation becomes

$$\frac{\Delta P_{tp}}{\Delta P_{LO}} = 1 + (K^2 - 1)[Bx(1 - x) + x^2]$$

where $B = (\rho_{ns}/\rho_L)^{1/4}$ for vertical up-and-down flow or $2/(K + 1)$ for horizontal flow, and where ρ_L is liquid density and ρ_{ns} is no-slip density (homogeneous density).

2. Determine ΔP_{LO} for cross flow.
The pressure drop for the total mass flowing as liquid was calculated in the preceding example. The total friction pressure drop across the bundle is $\Delta P_f + \Delta P_{fi}$. Thus, $\Delta P_f + \Delta P_{fi} = 0.547 + 0.117 = 0.664$ lb/in^2 (4.578 kPa).

3. Calculate ρ_{nr}.
The formula is

$$\rho_{ns} = \rho_L \lambda_L + \rho_G(1 - \lambda_L)$$

where $\lambda_L = Q_L/(Q_L + Q_G)$
$\quad Q_L$ = liquid volumetric flow rate = 40,000/55 = 727.3 ft^3/h
$\quad Q_G$ = gas volumetric flow rate = 10,000/0.5 = 20,000 ft^3/h

Thus λ_L = 727.3/(727.3 + 20,000) = 0.0351, so ρ_{ns} = 55(0.0351) + 0.5(1 − 0.0351) = 2.412 lb/ft^3 (38.64 kg/m^3).

4. Calculate ΔP_{GO} for cross flow.
Analogously to step 2, $\Delta P_{GO} = \Delta P_f + \Delta P_{fi}$. From the preceding example, G = 151,561.1 lb/(h)(ft^2). Then, $D_g G/\mu$ = (0.9375 − 0.75)(1/12)151,561.1/0.05 = 47,362.8, and $f = 1/(D_g G/\mu)^{0.25}$ = $1/47,362.8^{0.25}$ = 0.0678. Now,

$$\Delta P_f = \frac{4fG^2 N_r(N_b - 1)R_1 R_b \phi}{2(144)g\rho}$$

From the preceding example,

N_r = 18.7
$N_b - 1$ = 23
R_1 = 0.5432
R_b = 0.9014

For the gas, ϕ = 1.0. Then,

$$\Delta P_f = \frac{4(0.0678)(151,516.1)^2(18.7)(23)(0.5432)(0.9014)1.0}{2(144)(4.18 \times 10^8)0.5}$$
$$= 21.796 \text{ lb/in}^2 (150.28 \text{ kPa})$$

Also,

$$\Delta P_{fi} = \frac{4(2.66)fG^2 N_r R_b \phi}{2(144)g\rho}$$
$$= \frac{4(2.66)(0.0678)(151,561.1)^2(18.7)(0.9014)1.0}{2(144)(4.18 \times 10^8)0.5}$$
$$= 4.641 \text{ lb/in}^2 (32.00 \text{ kPa})$$

Thus, ΔP_{GO} = 21.796 + 4.641 = 26.437 lb/in^2 (182.28 kPa).

5. Calculate K for cross flow.
As noted above, $K = (\Delta P_{GO}/\Delta P_{LO})^{1/2}$ = (26.437/0.664)$^{1/2}$ = 6.3099.

6. Calculate n for cross flow.
As noted above, $K = (\rho_L/\rho_G)^{1/2}(\mu_G/\mu_L)^{n/2}$. Then, K = 6.3099 = (55/0.5)$^{1/2}$(0.05/1.95)$^{n/2}$, so n = 0.2774.

7. Calculate x, the mass friction of vapor.
Straightforwardly, x = 10,000/50,000 = 0.2.

8. Calculate P_{tp} for cross flow.
As noted above, $\Delta P_{tp}/\Delta P_{LO} = 1 + (K^2 - 1)[Bx^{(2-n)/2}(1-x)^{(2-n)/2} + x^{2-n}]$. Thus,

$$\frac{\Delta P_{tp}}{\Delta P_{LO}} = 1 + (6.3099^2 - 1)\left[1.0(0.2)^{(2-0.2774)/2}(1-0.2)^{(2-0.2774)/2} + (0.2)^{(2-0.2774)}\right]$$

$$= 11.434$$

Therefore, $\Delta P_{tp} = 11.434\Delta P_{LO} = 11.434(0.664) = 7.592$ lb/in^2 (52.35 kPa).

9. Calculate ΔP_w, the pressure drop for the baffle windows, for the total mass flowing as gas.
From the preceding example, ΔP_w for the total mass flowing as a liquid is 0.346 lb/in^2. Now, ΔP_w for the gas $= \Delta P_w$ for the liquid (ρ_L/ρ_G). So, $\Delta P_w = 0.346(55/0.5) = 38.06$ lb/in^2 (262.42 kPa).

10. Calculate K for the flow through the baffle windows.
Here, $K = (\Delta P_{GO}/\Delta P_{LO})^{1/2} = (38.06/0.346)^{1/2} = 10.4881$.

11. Calculate ΔP_{tp} for the baffle windows.
As noted in step 1, $\Delta P_{tp}/\Delta P_{LO} = 1 + (K^2 - 1)[Bx(1-x) + x^2]$. Now, $B = (\rho_{ns}/\rho_L)^{1/4} = (2.412/55)^{1/4} = 0.4576$. So, $\Delta P_{tp}/\Delta P_{LO} = 1 + (10.4881^2 - 1)[0.4576(0.2)(1-0.2) + 0.2^2] = 13.34$. Therefore, $\Delta P_{tp} = 13.34\Delta P_{LO} = 13.34 \times (0.346) = 4.616$ lb/in^2 (31.82 kPa).

12. Calculate ΔP_n, the pressure drop through the nozzles.
Use the same approach as in step 2a of the previous example, employing the no-slip density. Thus, $\Delta P_n = 2.25\rho_{ns}U_n^2/9266 = 2.25(2.412)(50)^2/9266 = 1.464$ lb/in^2 (10.09 kPa).

13. Calculate total two-phase pressure drop.
The total is the sum of the pressure drops for cross flow, window flow, and nozzle flow. Therefore $\Delta P_{tp} = 7.592 + 4.616 + 1.464 = 13.672$ lb/in^2 (94.27 kPa).

Related Calculations.   For situations where $\Delta P_{tp}/\Delta P_{GO}$ is much less than $1/K^2$ and $1/K^2$ is much less than 1, the equation for two-phase pressure drop can be written as

$$\frac{\Delta P_{tp}}{\Delta P_{GO}} = Bx^{(2-n)/2}(1-x)^{(2-n)/2} + x^{(2-n)}$$

This form may be more convenient for condensers.
   The method presented here is taken from Grant and Chisholm [16]. This reference should be consulted to determine the flow regime for a given two-phase system.
   Acceleration pressure drop. When the gas density or the vapor mass fraction changes, there is an acceleration pressure drop calculated from

$$\Delta P_a = \frac{G_T^2}{144g}\left(\frac{x_2}{\rho_{G2}} - \frac{x_1}{\rho_{G1}}\right)$$

where ΔP_a = acceleration pressure drop, in lb/in^2
   G_T = total mass velocity, in lb/(h)(ft^2)
   x_2 = mass fraction gas at outlet
   x_1 = mass fraction gas at inlet
   ρ_{G2} = gas density at outlet, in lb/ft^3
   ρ_{G1} = gas density at inlet, in lb/ft^3
   g = gravitational constant, in ft/(h)^2

Shell-side condensation. The frictional pressure drop for shell-side condensation can be calculated from the equation

$$\Delta P_c = 1/2 \left(1 + \frac{v_2}{v_1} \right) \Delta P_1$$

where ΔP_c = condensing pressure drop
  ΔP_1 = pressure drop based on the inlet conditions of flow rate, density, and viscosity
  v_1 = vapor velocity at the inlet
  v_2 = vapor velocity at the outlet

For a total condenser, this becomes

$$\Delta P_c = \frac{\Delta P_1}{2}$$

7.36   HEAT EXCHANGER NETWORKING

For the low-temperature distillation process shown in the flow diagram in Fig. 7.10, calculate the minimum hot-utility requirement and the location of the heat recovery pinch. Assume that the minimum acceptable temperature difference, ΔT_{min}, equals 5°C. The specifics of the process streams passing through the seven heat exchangers appear in the first five columns of Table 7.8, the mass flow rates of the streams being represented within the enthalpy (ΔH) values.

Calculation Procedure

1. For each stream, calculate its heat capacity flow rate.
For a given stream, the heat capacity flow rate, CP, is defined as ΔH divided by the absolute value of the difference between the supply temperature and the target temperature. For instance, for Stream 1,

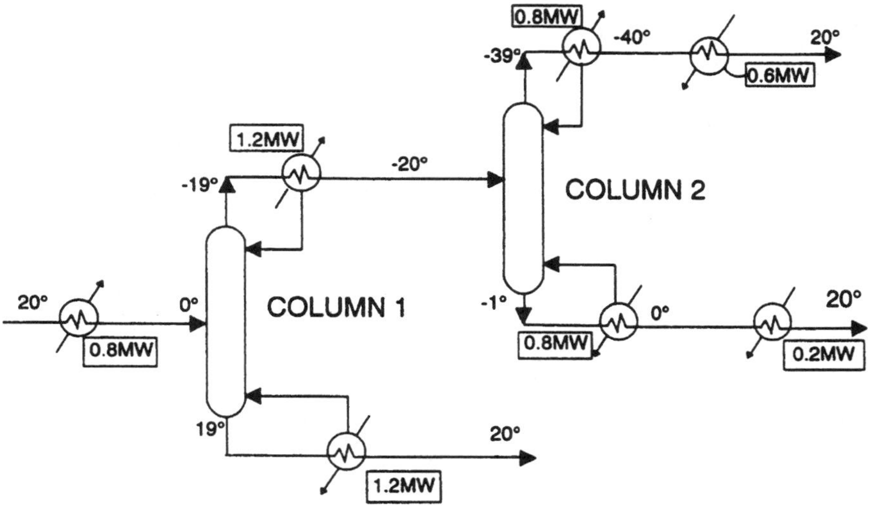

FIGURE 7.10   A low-temperature distillation process.

TABLE 7.8   Stream Data for Low-Temperature Distillation Process

Stream	Type	Supply temp. $T_S(°C)$	Target temp. $T_T(°C)$	ΔH (MW)	Heat capacity flow rate CP (MW $°C^{-1}$)
1. Feed to column 1	Hot	−20	0	0.8	0.04
2. Column 1 condenser	Hot	−19	−20	1.2	1.2
3. Column 2 condenser	Hot	−39	−40	0.8	0.8
4. Column 1 reboiler	Cold	19	20	1.2	1.2
5. Column 2 reboiler	Cold	−1	0	0.8	0.8
6. Column 2 bottoms	Cold	0	20	0.2	0.01
7. Column 2 overheads	Cold	−40	20	0.6	0.01

the feed to Column 1, $CP = 0.8/(20 − 0) = 0.04$. The heat capacity flow rates for the seven streams appears as the last column of Table 7.8.

2. For each stream, modify the supply and target temperatures to assure that the minimum-temperature-difference requirement is met.

This step consists of lowering the supply and target temperatures of each hot stream by $\Delta T_{min}/2$ and raising the supply and target temperatures of each cold stream by $\Delta T_{min}/2$. For Stream 1 (a hot stream), for instance, the supply temperature drops from 20 to 17.5°C, and the target temperature drops from 0 to −2.5°C. The shifted temperatures for the seven streams appear in Table 7.9.

3. Carry out a heat balance within each interval between the shifted temperatures.

To display the temperature intervals, take all the 14 shifted supply and target temperatures (two for each of the seven streams) and list them in descending order (noting that a few temperatures, such as 22.5°C, appear more than once). The resulting list is the first column in Fig. 7.11. Draw horizontal lines extending leftward through each temperature value on the list. The regions between the lines are the temperature intervals, whose numerical values appear in the third column of Fig. 7.11. For example, the region between 21.5 and 17.5°C represents a temperature interval of $(21.5 − 17.5)$, or 4°C.

Then, represent each of the seven streams by drawing vertical lines, starting at the shifted supply temperature and extending downward (for hot streams being cooled) or upward (for cold streams being heated) until reaching the shifted target temperature. For Stream 1, for instance, the line begins at the line for 17.5°C and extends downward to −2.5°C. Label each line with its CP value (for instance, $CP = 0.04$ for Stream 1). The resulting seven vertical stream lines appear in the second column of Fig. 7.11, labeled "Stream population."

For each of the 10 temperature intervals, sum the heat capacity flow rates for the cold streams being heated that fall within that interval, then subtract the heat capacity flow rate of each hot stream being cooled that falls within the interval. The resulting algebraic sums are in the fourth column of Fig. 7.11, For example, there are three streams within the interval between 17.5 and 2.5°C, namely, Stream 1, being cooled, with $CP = 0.04$; Stream 6, being heated, with $CP = 0.01$; and Stream 7, being heated, with $CP = 0.01$. The resulting algebraic sum is $(0.01 + 0.01 − 0.04)$, or −0.02.

TABLE 7.9   Shifted Temperatures for the Data in Table 7.8

Stream	Type	T_S	T_T	$T_S^\dagger$	$T_T^\dagger$
1	Hot	20	0	17.5	−2.5
2	Hot	−19	−20	−21.5	−22.5
3	Hot	−39	−40	−41.5	−42.5
4	Cold	19	20	21.5	22.5
5	Cold	−1	0	1.5	2.5
6	Cold	0	20	2.5	22.5
7	Cold	−40	20	−37.5	22.5

Interval Temperature	Stream Population	$\Delta T_{INTERVAL}$	$\Sigma CP_C - \Sigma CP_H$	$\Delta H_{INTERVAL}$	Surplus/Deficit
22.5°C					
		1	1.22	1.22	Deficit
21.5°C					
		4	0.02	0.08	Deficit
17.5°C					
		15	-0.02	-0.30	Surplus
2.5°C					
		1	0.77	0.77	Deficit
1.5°C					
		4	-0.03	-0.12	Surplus
-2.5°C					
		19	0.01	0.19	Deficit
-21.5°C					
		1	-1.19	-1.19	Surplus
-22.5°C					
		15	0.01	0.15	Deficit
-37.5°C					
		4	0	0	—
-41.5°C					
		1	-0.8	-0.8	Surplus
-42.5°C					

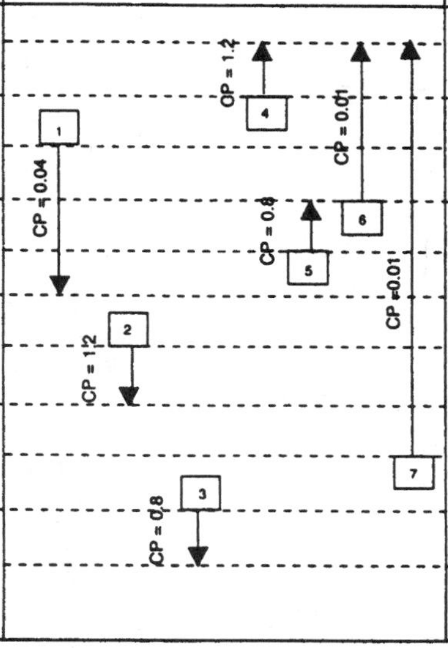

FIGURE 7.11   Temperature-interval heat balances for Procedure 7.36.

Finally, multiply each temperature difference in Column 3 by the corresponding algebraic sum of heat capacity flow rates in Column 4. The result is the heat balance, or enthalpy change, ΔH, in megawatts for that interval. The intervals having a negative ΔH, such as the interval between 17.5 and 2.5°C, are designated as being in surplus; regions with a positive ΔH are designated as being in deficit. The results are in the fifth and sixth columns of Fig. 7.11.

4. Cascade the surplus and deficit heat down the temperature scale from interval to interval, and note the largest net deficit that results. That number is the minimum hot-utility requirement.
Arrange the ΔH values as shown in Fig. 7.12a, with the temperatures taken from the first column of Fig. 7.13. Working downward, calculate at each temperature the cumulative sum of the ΔH values above it. Assign a positive number if the cumulative value is surplus or a negative number if it is instead in deficit.

For example: working downward to 17.5°C, we add up two successive deficits of 1.22 and 0.08 MW, so the cumulative sum on the 17.5°C line is −1.30 MW; but, in the next downward step to 2.5°C, we pick up a surplus of 0.30 MW, so the cumulative sum on the 2.5°C line is −1.00 MW.

Upon completing the task, we note that the largest net deficit is 1.84 MW, on the line for −21.5°C. Thus, the minimum hot-utility requirement is 1.84 MW.

To confirm this, repeat the downward cascading but begin with an assumed input of 1.84 MW at the top of the cascade, as shown in Fig. 7.12b. Note that, as a result, no deficit results.

5 Find the heat-recovery pinch point.
The pinch arises at the point where, after adding the minimum heat utility, the cumulative sum of ΔH values is zero. In this case, it occurs at −21.5°C. Assuming that the heat exchangers become networked appropriately, and keeping in mind that the pinch result of −21.5°C emerges after the temperatures were shifted, in step 2, we note that, at the pinch, the hot stream being cooled is at a temperature of $(-21.5 + \Delta T_{min}/2)$, or $(-21.5 + 2.5)$, or −19°C, and the cold stream being heated is at $(-21.5 - \Delta T_{min}/2)$, or −24°C.

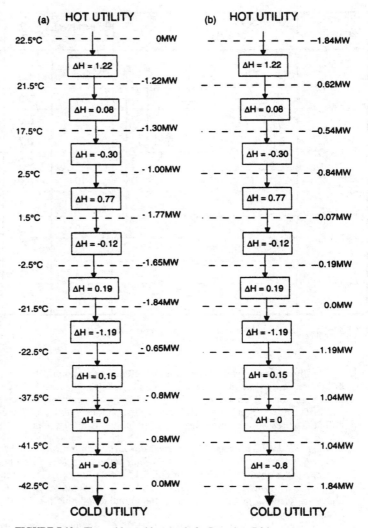

FIGURE 7.12   The problem table cascade for Procedure 7.36.

Related Calculations.   Heat-exchanger networking problems can instead be solved by the more-familiar composite diagrams, consisting of graphs in which enthalpy is the horizontal axis, temperature is the vertical axis, individual cold streams are combined into a composite cold stream, and individual hot streams are combined into a composite hot stream. However, the table algorithm used in the example here is more convenient for calculations.

With respect to economizing on thermal energy, the networking and pinch technology principle can be extended beyond heat exchangers and condensers to also embrace other operations within the plant that involve heat generation or transfer, such as reactors and heat pumps. Beyond that, networking and pinch technology principles also have relevance for optimal use of other plant resources, such as water or hydrogen.

*7.37 HEAT-EXCHANGER CHOICE FOR SPECIFIC CHEMICAL-PLANT APPLICATIONS

Determine the type of heat exchanger to use for each of the following applications: (1) heating oil with steam; (2) cooling internal combustion engine liquid coolant; (3) evaporating a hot liquid in a chemical process. For each heater chosen, specify the typical pressure range for which the heater is usually built and the typical range of the overall coefficient of heat transfer U.

Calculation Procedure:

1. Determine the heat-transfer process involved.
In a heat exchanger, one or more of four processes may occur: heating, cooling, boiling, or condensing Table 7.10 lists each of these four processes and shows the usual heat-transfer fluids involved. Thus, the heat exchangers being considered here involve (a) oil heater—heating—vapor-liquid; (b) internal-combustion engine coolant—cooling—gas-liquid; (c) hot-liquid evaporation—boiling—liquid-liquid.

2. Specify the heater action and the usual type selected.
Using the same identifying letters for the heaters being selected, Table 7.10 shows the action and usual type of heater chosen. Thus,

	Action	Type
a.	Steam condensed; oil heated	Shell-and-tube
b.	Air heated; water cooled	Tubes in open air
c.	Waste liquid cooled; water boiled	Shell-and-tube

3. Specify the usual pressure range and typical U.
Using the same identifying letters for the heaters being selected, Table 7.10 shows the action and usual type of heater chosen. Thus,

	Usual pressure range	Typical U range	
		Btu/(h · °F · ft²)	W/(m² · °C)
a.	0–500 lb/in² (abs)(0 to 3447 kPa)	20–60	113.6–340.7
b.	0–100 lb/in² (abs)(0 to 689.4 kPa)	2–10	11.4–56.8
c.	0–500 lb/in² (abs)(0 to 3447 kPa)	40–150	227.1–851.7

4. Select the heater for each service.
Where the heat-transfer conditions are normal for the type of service met, the type of heater listed in step 2 can be safely used. When the heat-transfer conditions are unusual, a special type of heater may be needed. To select such a heater, study the data in Table 7.10 and make a tentative selection. Check the selection by using the methods given in the following calculation procedures in this section.

Related Calculations.   Use Table 7.10 as a general guide to heat-exchanger selection in any industry— petroleum, chemical, power, marine, textile, lumber, etc. Once the general type of heater and its typical U value are known, compute the required size, using the procedure given later in this section.

TABLE 7.10 Heat-Exchanger Selection Guide[†]

	Heat-transfer fluids	Equipment	Action	Type[‡]	Pressure range[§]	Typical range of U[‖]
	Liquid-liquid	Boiler-water blowdown exchanger	Blowdown cooled, feedwater heated	S	M, H	50–300 (0.28–1.7)
		Laundry-water heat reclaimer	Waste water cooled, feed heated	S	L	30–200 (0.17–1.1)
		Service-water heater	Waste liquid cooled, water heated	S	L, H	50–300 (0.28–1.7)
	Vapor-liquid	Bleeder heater	Steam condensed, feedwater heated	S	L, H	200–800 (1.1–4/5)
		Deaerating feed heater	Steam condensed, feedwater heated	M	L, M	DC
		Jet heater	Steam condensed, water heated	M	L	DC
		Process kettle	Steam condensed, liquid heated	S	L, M	100–500 (0.57–2.8)
		Oil heater	Steam condensed, oil heated	S	L, M	20–60 (0.11–0.34)
Heating		Service-water heater	Steam condensed, water heated	S	L, M	200–800 (1.1–4.5)
		Open flow-through heater	Steam condensed, water heated	M	L	DC
		Liquid-sodium steam superheater	Sodium cooled, steam superheated	S	M, H	50–200 (0.28–1.1)
	Gas-liquid	Waste-heat water heater	Waste gas cooled, water heated	T	L	2–10 (0.011–0.05)
		Boiler economizer	Flue gas cooled, feedwater heated	T	M, H	2–10 (0.011–0.05)
		Hot-water radiator	Water cooled, air heated	T	L	1–10 (0.0057–0.05)
	Gas-gas	Boiler air heater	Flue gas cooled, combustion air heated	T, R	L	2–10 (0.011–0.05)
		Gas-turbine regenerator	Flue gas cooled, combustion air heated	T	L	2–10 (0.011–0.05)
	Vapor-gas	Boiler superheater	Combustion gas cooled, steam superheated	T	M, H	2–20 (0.011–0.11)
		Steam pipe coils	Steam condensed, air heated	T	L, M	2–10 (0.011–0.05)
		Steam radiator	Steam condensed, air heated	T	L	2–10 (0.011–0.05)
Cooling	Liquid-liquid	Oil cooler	Water heated, oil cooled	S, D	L, M	20–200 (0.11–1.1)
		Water chiller	Refrigerant boiled, water cooled	S	L, M	30–151 (0.17–0.86)
		Brine cooler	Refrigerant boiled, brine cooled	S	L, M	30–150 (0.17–0.86)
		Transformer-oil cooler	Water heated, oil cooled	S	L, M	20–50 (0.11–0.88)
	Vapor-liquid	Boiler desuperheater	Boiler water heated, steam desuperheated	S, M	M, H	150–800 (0.85–4.5)

			Exchanger[‡]	Fluid[§]	U[¶]	
Cooling	Gas-liquid	Compressor intercoolers and aftercoolers	Water heated, compressed air cooled	S	L, H	10–20	(0.057–0.11)
		Internal-combustion-engine radiator	Air heated, water cooled	T	L	2–10	(0.011–0.05)
		Generator hydrogen, air coolers	Water heated, hydrogen or air cooled	S	L	2–10	(0.011–0.05)
		Air-conditioning cooler	Water heated, air cooled	T	L	2–10	(0.011–0.05)
		Refrigeration heat exchanger	Brine heated, air cooled	T	L, M	2–10	(0.011–0.05)
		Refrigeration evaporator	Refrigerant boiled, air cooled	T	L, M	2–10	(0.011–0.05)
	Vapor-gas	Boiler desuperheater	Flue gas heated, steam desuperheated	T	L, H	2–8	(0.011–0.04)
	Liquid-liquid	Hot-liquid evaporator	Waste liquid cooled, water boiled	S	L, H	40–150	(0.23–0.85)
		Liquid-sodium steam generator	Sodium cooled, water boiled	S	M, H	500–	(2.8–5.7)
Boiling	Vapor-liquid	Evaporator (vacuum)	Steam condensed, water boiled	S	L	1000	(2.3–3.4)
		Evaporator (high pressure)	Steam condensed, water boiled	S	L, M	400–600	(2.3–3.4)
		Mercury condenser-boiler	Mercury condensed, water boiled	S	M, H	400–600	(2.8–4.0)
	Gas-liquid	Waste-heat steam boiler	Flue gas cooled, water boiled	T	L, H	500–700	(0.011–0.05)
		Direct-fired steam boiler	Combustion gas cooled, water boiled	T	L, H	2–10	(0.011–0.05)
Condensing	Vapor-liquid	Refrigeration condenser	Water heated, refrigerant condensed	S, D	L, M	2–10	(0.45–1.4)
		Steam surface condenser	Water heated, steam condensed	S	L	80–250	(1.7–4.5)
		Steam mixing condenser	Water heated, steam condensed	M	L	300–800	
	Vapor-gas	Intercondenser and aftercondenser	Condensate heated, steam condensed	S	L	DC	(0.085–1.7)
		Air-cooled surface condenser	Air heated, steam condensed	T	L	15–300	(0.011–0.09)
						2–16	

†Power

‡S—shell-and-tube exchanger, M—direct contact mixing exchanger, T—tubes in path of moving fluid, or exchanger open to surrounding air, R—regenerative plate-type or simple plate-type exchanger; D—double-tube exchanger.

§L—highest pressure ranges from 0 to 100 lb/in² (abs) (0 to 689.4 kPa), M—highest pressure from 100 to 500 lb/in² (abs) (689.4 to 3447 kPa), H-500 lb/in² (abs) (3447 kPa) up.

¶Values of U represent range of overall heat-transfer coefficients that might be expected in various exchangers. Coefficients are stated in Btu/(h · °F · ft) [W/(m² · °C)] of heating surface. Total heat transferred in exchanger, in Btu/h, is obtained by multiplying a specific value of U for that type of exchanger by the surface and the log mean temperature difference. DC indicates direct exchange of heat.

*7.38 SIZING SHELL-AND-TUBE HEAT EXCHANGERS FOR CHEMICAL PLANTS

What is the required heat-transfer area for a parallel-flow shell-and-tube heat exchanger used to heat oil if the entering oil temperature is 60°F (15.6°C), the leaving oil temperature is 120°F (48.9°C), and the heating medium is steam at 200 lb/in² (abs) (1378.8 kPa)? There is no subcooling of condensate in the heat exchanger. The overall coefficient of heat transfer $U = 25$ Btu/(h · °F · ft²) [141.9 W/(m² · °C)]. How much heating steam is required if the oil flow rate through the heater is 100 gal/min (6.3 L/s), the specific gravity of the oil is 0.9, and the specific heat of the oil is 0.5 Btu/ (lb · °F) [2.84 W/(m² · °C)]?

Calculation Procedure:

1. Compute the heat-transfer rate of the heater.
With a flow rate of 100 gal/min (6.3 L/s) or (100 gal/min)(60 min/h) = 6000 gal/h (22,710 L/h), the weight flow rate of the oil, using the weight of water of specific gravity 1.0 as 8.33 lb/gal, is (6000 gal/h) (0.9 specific gravity)(8.33 lb/gal) = 45,000 lb/h (20,250 kg/h), closely.

Since the temperature of the oil rises 120 − 60 = 60°F (33.3°C) during passage through the heat exchanger and the oil has a specific heat of 0.50, find the heat-transfer rate of the heater from the general relation $Q = wc \, \Delta t$, where Q = heat-transfer rate, Btu/h; w = oil flow rate, lb/h; c = specific heat of the oil, Btu/(lb · °F); Δt = temperature rise of the oil during passage through the heater. Thus, $Q = (45,000)(0.5)(60) = 1,350,000$ Btu/h (0.4 MW).

2. Compute the heater logarithmic mean temperature difference.
The LMTD is found from LMTD = $(G − L)/\ln (G/L)$, where G = greater terminal temperature difference of the heater, °F; L = lower terminal temperature difference of the heater, °F; ln = logarithm to the base e. This relation is valid for heat exchangers in which the number of shell passes equals the number of tube passes.

In general, for parallel flow of the fluid streams, $G = T_1 − t_1$ and $L = T_2 − t_2$, where T_1 = heating fluid inlet temperature, °F; T_2 = heating fluid outlet temperature, °F; t_1 = heated fluid inlet temperature, °F; t_2 = heated fluid outlet temperature, °F. Figure 7.13 shows the maximum and minimum terminal temperature differences for various fluid flow paths.

For this parallel-flow exchanger, $G = T_1 − t_1, = 382 − 60 = 322$°F (179°C), where 382°F (194°C) = the temperature of 200-lb/in² (abs) (1379-kPa) saturated steam, from a table of steam properties. Also, $L = T_2 − t_2 = 382 − 120 = 262$°F (145.6°C), where the condensate temperature = the saturated steam temperature because there is no subcooling of the condensate. Then LMTD = $G − L/\ln (G/L)$ = (322 − 262)/ln (322/262) = 290°F (16°C).

3. Compute the required heat-transfer area.
Use the relation $A = Q/U × \text{LMTD}$, where A = required heat-transfer area, ft²; U = overall coefficient of heat transfer, Btu/(ft² · h · °F). Thus, $A = 1,350,000/[(25)(290)] = 186.4$ ft² (17.3 m²), say 200 ft² (18.6 m²).

4. Compute the required quantity of heating steam.
The heat added to the oil = Q = 1,350,000 Btu/h, from step 1. The enthalpy of vaporization of 200-lb/in² (abs) (1379-kPa) saturated steam is, from the steam tables, 843.0 Btu/lb (1960.8 kJ/kg). Use the relation $W = Q/h_{fg}$, where W = flow rate of heating steam, lb/h; h_{fg} = enthalpy of vaporization of the heating steam, Btu/lb. Hence, $W = 1,350,000/843.0 = 1600$ lb/h (720 kg/h).

Related Calculations. Use this general procedure to find the heat-transfer area, fluid outlet temperature, and required heating-fluid flow rate when true parallel flow or counterflow of the fluids occurs in the heat exchanger. When such a true flow does *not* exist, use a suitable correction factor, as shown in the next calculation procedure.

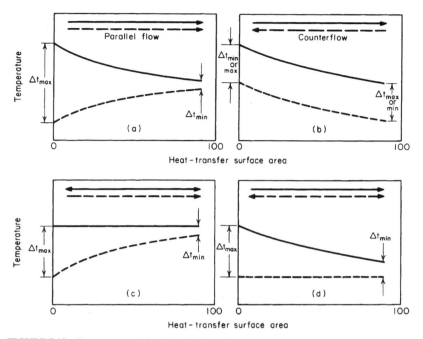

FIGURE 7.13   Temperature relations in typical parallel-flow and counterflow heat exchangers.

The procedure described here can be used for heat exchangers in power plants, heating systems, marine propulsion, air-conditioning systems, etc. Any heating or cooling fluid—steam, gas, chilled water, etc.—can be used.

To select a heat exchanger by using the results of this calculation procedure, enter the engineering data tables available from manufacturers at the computed heat-transfer area. Read the heater dimensions directly from the table. Be sure to use the next *larger* heat-transfer area when the exact required area is not available.

When there is little movement of the fluid on either side of the heat-transfer area, such as occurs during heat transmission through a building wall, the arithmetic mean (average) temperature difference can be used instead of the LMTD. Use the LMTD when there is rapid movement of the fluids on either side of the heat-transfer area and a rapid change in temperature in one, or both, fluids. When one of the two fluids is partially, but not totally, evaporated or condensed, the true mean temperature difference is different from the arithmetic mean and the LMTD. Special methods, such as those presented in Perry—*Chemical Engineers' Handbook,* McGraw-Hill, 2007, must be used to compute the actual temperature difference under these conditions.

When two liquids or gases with constant specific heats are exchanging heat in a heat exchanger, the area between their temperature curves, Fig. 7.14 is a measure of the total heat being transferred. Figure 7.14 shows how the temperature curves vary with the amount of heat-transfer area for counterflow and parallel-flow exchangers when the fluid inlet temperatures are kept constant. As Fig. 7.14 shows, the counterflow arrangement is superior.

If enough heating surface is provided, in a counterflow exchanger, the leaving cold-fluid temperature can be raised above the leaving hot-fluid temperature. This cannot be done in a parallel-flow exchanger, where the temperatures can only approach each other regardless of how much surface is used. The counterflow arrangement transfers more heat for given conditions and usually proves more economical to use.

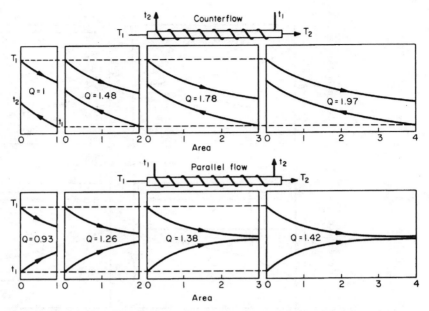

FIGURE 7.14   For certain conditions, the area between the temperature curves measures the amount of heat being transferred.

*7.39 TEMPERATURE DETERMINATION IN HEAT-EXCHANGER OPERATION

A counterflow shell-and-tube heat exchanger has one shell pass for the heating fluid and two shell passes for the fluid being heated. What is the actual LMTD for this exchanger if $T_1 = 300°F$ (148.9°C), $T_2 = 250°F$ (121°C), $t_1 = 100°F$ (37.8°C), and $t_2 = 230°F$ (110°C)?

Calculation Procedure

1. Determine how the LMTD should be computed.
When the numbers of shelf and tube passes are unequal, true counterflow does not exist in the heat exchanger. To allow for this deviation from true counterflow, a correction factor must be applied to the logarithmic mean temperature difference (LMTD). Figure 7.15 gives the correction factor to use.

2. Compute the variables for the correction factor.
The two variables that determine the correction factor are shown in Fig.7.15 as $P = (t_2 - t_1)/(T_1 - t_1)$ and $R = (T_1 - T_2)/(t_2 - t_1)$. Thus, $P = (230 - 100)/(300 - 100) = 0.65$, and $R = (300 - 250)/(230 - 100) = 0.385$. From Fig. 7.15, the correction factor is $F = 0.90$ for these values of P and R.

3. Compute the theoretical LMTD.
Use the relation LMTD $= (G - L)/\ln(G/L)$, where the symbols for counterflow heat exchange are $G = T_2 - t_1$; $L = T_1 - t_2$; ln = logarithm to the base e. All temperatures in this equation are expressed in °F. Thus, $G = 250 - 100 = 150°F$ (83.3°C); $L = 300 - 230 = 70°F$ (38.9°C). Then LMTD $= (150 - 70)/\ln (150/70) = 105°F$ (58.3°C).

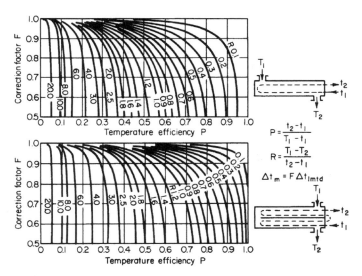

FIGURE 7.15   Correction factors for LMTD when the heater flow path differs from the counterflow. (*Power.*)

4. *Compute the actual LMTD for this exchanger.*
The actual LMTD for this or any other heat exchanger is LMTD $_{actual}$ = F(LMTD $_{computed}$) = 0.9(105) = 94.5°F (52.5°C). Use the actual LMTD to compute the required exchanger heat-transfer area.

Related Calculations.   Once the corrected LMTD is known, compute the required heat-exchanger size in the manner shown in the previous calculation procedure. The method given here is valid for both two- and four-pass shell-and-tube heat exchangers. Figure 7.16 simplifies the computation of the uncorrected LMTD for temperature differences ranging from 1 to 1000°F (−17 to 537.8°C). It gives LMTD with sufficient accuracy for all normal industrial and commercial heat-exchanger applications. Correction-factor charts for three shell passes, six or more tube passes, four shell passes, and eight or more tube passes are published in the *Standards of the Tubular Exchanger Manufacturers Association.*

*7.40   SELECTING AND SIZING HEAT EXCHANGERS BASED ON FOULING FACTORS

A heat exchanger having an overall coefficient of heat transfer of U = 100 Btu/(ft^2 · h · °F) [567.8 W/(m^2 · °C)] is used to cool lean oil. What effect will the tube fouling have on the value of U for this exchanger?

Calculation Procedure

1. *Determine the heat exchange fouling factor.*
Use Table 7.11 to determine the fouling factor for this exchanger. Thus, the fouling factor for lean oil = 0.0020.

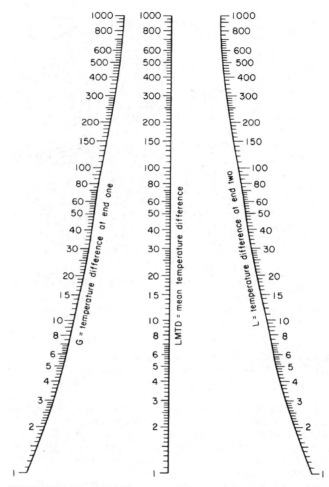

FIGURE 7.16 Logarithmic mean temperature for a variety of heat-transfer applications.

2. Determine the actual U for the heat exchanger.

Enter Fig. 7.17 at the bottom with the clean heat-transfer coefficient of $U = 100$ Btu/(h · ft² · °F) [567.8 W/(m² · °C)] and project vertically upward to the 0.002 fouling-factor curve. From the intersection with this curve, project horizontally to the left to read the design or actual heat-transfer coefficient as $U_a = 78$ Btu/(h · ft² · °F) [442.9 W/(m² · °C)]. Thus, the fouling of the tubes causes a reduction of the U value of $100 - 78 = 22$ Btu/(h · ft² · °F) [124.9 W/(m² · °C)]. This means that the required heat-transfer area must be increased by nearly 25 percent to compensate for the reduction in heat transfer caused by fouling.

Related Calculations. Table 7.11 gives fouling factors for a wide variety of service conditions in applications of many types. Use these factors as described above, or add the fouling factor to the film resistance for the heat exchanger to obtain the total resistance to heat transfer. Then $U =$ the reciprocal of the total resistance. Use the actual value U_a of the heat-transfer coefficient when sizing a heat exchanger. The method given here is that used by Condenser Service and Engineering Company, Inc.

TABLE 7.11 Heat-Exchanger Fouling Factors[†]

Fluid heated or cooled	Fouling factor
Fuel oil	0.0055
Lean oil	0.0020
Clean recirculated oil	0.0010
Quench oils	0.0042
Refrigerants (liquid)	0.0011
Gasoline	0.0006
Steam-clean and oil-free	0.0001
Refrigerant vapors	0.0023
Diesel exhaust	0.013
Compressed air	0.0022
Clean air	0.0011
Seawater under 130°F (54°C)	0.0006
Seawater over 130°F (54°C)	0.0011
City or well water under 130°F (54°C)	0.0011
City or well water over 130°F (54°C)	0.0021
Treated boiler feedwater under 130°F, 3 ft/s (54°C, 0.9 m/s)	0.0008
Treated boiler feedwater over 130°F, 3 ft/s (54°C, 0.9 m/s)	0.0009
Boiler blow-down	0.0022

[†]Condenser Service and Engineering Company, Inc.

*7.41 CHEMICAL-PLANT ELECTRIC PROCESS HEATER SELECTION AND APPLICATION

Choose the heating capacity of an electric heater to heat a pot containing 600 lb (272.2 kg) of lead from the charging temperature of 70°F (21.1°C) to a temperature of 750°F (398.9°C) if 600 lb (272.2 kg) of the lead is to be melted and heated per hour. The pot is 30 in (76.2 cm) in diameter and 18 in (45.7 cm) deep.

Calculation Procedure

1. Compute the heat needed to reach the melting point.

When a solid is melted, first it must be raised from its ambient or room temperature to the melting temperature. The quantity of heat required is $H = $ (weight of solid, lb) [specific heat of solid, Btu/(lb · °F)] $(t_m - t_t)$, where $H = $ Btu required to raise the temperature of the solid, °F; $t_t = $ room, charging, or initial temperature of the solid, °F; $t_m = $ melting temperature of the solid, °F.

For this pot with lead having a melting temperature of 620°F (326.7°C) and an average specific heat of 0.031 Btu/(lb · °F) [0.13 kJ/(kg · °C)], $H = (600)(0.031)(620 - 70) = 10,240$ Btu/h (3.0 kW), or (10,240 Btu/h)/(3412 Btu/kWh) = 2.98 kWh.

2. Compute the heat required to melt the solid.

The heat H_m Btu required to melt a solid is $H_m = $ (weight of solid melted, lb)(heat of fusion of the solid, Btu/lb). Since the heat of fusion of lead is 10 Btu/lb (23.2 kJ/kg), $H_m = (600)(10) = 6000$ Btu/h, or 6000/3412 = 1.752 kWh.

3. Compute the heat required to reach the working temperature.

Use the same relation as in step 1, except that the temperature range is expressed as $t_w - t_m$, where $t_w = $ working temperature of the melted solid. Thus, for this pot, $H = (600)(0.031)(750 - 620) = 2420$ Btu/h (709.3 W), or 2420/3412 = 0.709 kWh.

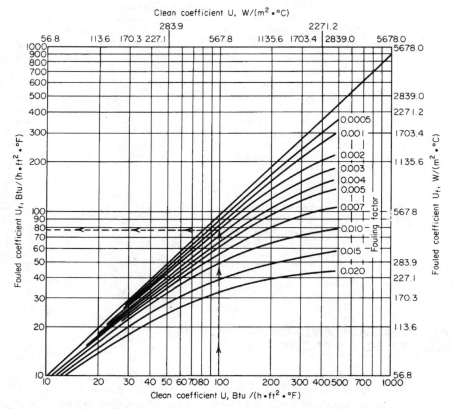

FIGURE 7.17   Effect of heat-exchanger fouling on the overall coefficient of heat transfer. (*Condenser Service and Engineering Co., Inc.*)

4. Determine the heat loss from the pot.

Use Fig. 7.18 to determine the heat loss from the pot. Enter at the bottom of Fig. 7.18 at 750°F (398.9°C), and project vertically upward to the 30-in (76.2-cm) diameter pot curve. At the left, read the heat loss at 7.3 kWh/h.

5. Compute the total heating capacity required.

The total heating capacity required is the sum of the individual capacities, or 2.98 + 1.752 + 0.708 + 7.30 = 12.74 kWh. A 15-kW electric heater would be chosen because this is a standard size and it provides a moderate extra capacity for overloads.

Related Calculations.   Use this general procedure to compute the capacity required for an electric heater used to melt a solid of any kind—lead, tin, type metal, solder, etc. When the substance being heated is a liquid—water, dye, paint, varnish, oil, etc.—use the relation H = (weight of liquid heated, lb) [specific heat of liquid, Btu/(lb · °F)] (temperature rise desired, °F), when the liquid is heated to approximately its boiling temperature, or a lower temperature.

For space heating of commercial and residential buildings, two methods used for computing the approximate wattage required are the W/ft^3 and the "35" method. These are summarized in Table 7.12. In many cases, the results given by these methods agree closely with more involved calculations. When the desired room temperature is different from 70°F (21.1°C), increase or decrease the required kilowatt capacity proportionally, depending on whether the desired temperature is higher than or lower than 70°F (21.1°C).

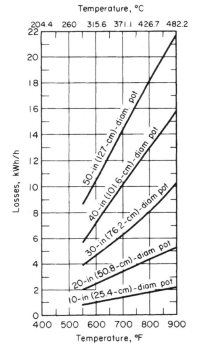

FIGURE 7.18 Heat losses from melting pots. (*General Electric Co.*)

For heating pipes with electric heaters, use a heater capacity of 0.8 W/ft^2 (8.6 W/m^2) of uninsulated exterior pipe surface per °F temperature difference between the pipe and the surrounding air. If the pipe is insulated with 1 in (2.5 cm) of insulation, use 30 percent of this value, or [0.24 (W/(ft^2 · °F)] [4.7 W/(m^2 · °C)].

The types of electric heaters used today include immersion (for water, oil, plating, liquids, etc.), strip, cartridge, tubular, vane, fin, unit, and edgewound resistor heaters. These heaters are used in a wide variety of applications including liquid heating, gas and air heating, oven warming, deicing, humidifying, plastics heating, pipe heating, etc.

For pipe heating, a tubular heating element can be fastened to the bottom of the pipe and run parallel with it. For large-wattage applications, the heater can be spiraled around the pipe. For

TABLE 7.12 Two Methods for Determining Wattage for Heating Buildings Electrically[†]

	W/ft^3 method	W/m^3 method
1. Interior rooms with no or little outside exposure	0.75–1.25	25.6–44.1
2. Average rooms with moderate windows and doors	1.25–1.75	44.1–61.8
3. Rooms with severe exposure and great window and door space	1.0–4.0	35.3–141.3
4. Isolated rooms, cabins, watch houses, and similar buildings	3.0–6.0	105.9–211.9
	The "35" method	
1. Volume in ft^3 for one air change × 0.35 =	0.01 W	
2. Exposed net wall, roof, or ceiling and floor in ft^2 × 3.5 =	0.1 W	
3. Area of exposed glass and doors in ft^2 × 35.0 =	1 W	

[†]General Electric Company.

temperatures below 165°F (73.9°C), heating cable can be used. Electric heating is often used in place of steam tracing of outdoor pipes.

The procedure presented above is the work of General Electric Company.

REFERENCES

1. McAdams—*Heat Transmission*, 3d ed., McGraw-Hill.
2. Krieth—*Principles Heat Transfer*, 3d ed., Intext Educational Publishers.
3. Kern—*Process Heat Transfer*, McGraw-Hill.
4. Holman— *Heat Transfer*, 2d ed., McGraw-Hill.
5. Collier—*Convective Boiling and Condensation*, McGraw Hill.
6. Rohsenow and Hartnett (eds.)—*Handbook of Heat Transfer,* McGraw-Hill.
7. Chapman—*Heat Transfer*, 3d ed., McMillan.
8. Oppenheim—"Radiation Analysis by the Network Method," in Hartnett et al.—*Recent Advances in Heat and Mass Transfer*, McGraw-Hill.
9. Siegel and Howell—*Thermal Radiation Heat Transfer,* NASA SP-164 vols. I, II, and III, Lewis Research Center, 1968, 1969, and 1970. Office of Technology Utilization.
10. Butterworth—*Introduction to Heat Transfer,* Oxford University Press.
11. *Standards of Tubular Exchangers Manufacturers Association,* 6th ed., Tubular Exchangers Manufacturers Association.
12. Blevins—*Flow-Induced Vibration,* Van Nostrand-Reinhold.
13. Daniels—*Terrestrial Environment (Climatic) Criteria Guidelines for Use in Aerospace Vehicle Development*, 1973 Revision, NASA TMS-64757, NASA/MSFC.
14. *The American Ephemeris and Nautical Almanac*, U.S. Government Printing Office.
15. Bell—*University of Delaware Experimental Station Bulletin* 5, also *Petro/Chem Engineer,* October 1960, p. C-26.
16. Grant and Chisholm— ASME paper 77-WA/HT-22.
17. Butterworth—ASME paper 77-WA/HT-24.
18. Dukler—*Chemical Engineering Progress Symposium Series No. 30,* vol. 56, 1, 1960.
19. Akers and Rosson—*Chemical Engineering Progress Symposium Series,* "Heat Transfer—Storrs," vol. 56, 3, 1960.
20. Caglayan and Buthod— *Oil Gas J,* Sept. 6, 1976, p. 91.
21. Bergles—*Sixth International Heat Transfer Conference,* Toronto, 1978, vol. 6, p. 89.
22. Briggs and Young—*Chemical Engineering Progress Symposium Series,* vol. 59, no. 41, pp. 1–10.
23. Fair—*Petroleum Refiner,* February 1960, p. 105.
24. Gilmour—*Chem. Engineer,* October 1952, p. 144; March 1953, p. 226; April 1953, p. 214; October 1953, p. 203; February 1954, p. 190; March 1954, p. 209; August 1954, p. 199.
25. Lord, Minton, and Slusser—*Chem. Engineer,* Jan. 26, 1970, p. 96.
26. Minton—*Chem. Engineer,* May 4, 1970, p. 103.
27. Minton—*Chem. Engineer,* May 18, 1970, p. 145.
28. Lord, Minton, and Slusser—*Chem. Engineer,* March 23, 1970, p. 127.
29. Lord, Minton, and Slusser—*Chem. Engineer,* June 1, 1970, p. 153.
30. Michiyoshi—*Sixth International Heat Transfer Conference,* Toronto, 1978. vol 6, p. 219.
31. Wallis—*One-Dimensional Two-Phase Flow,* McGraw-Hill.
32. Gottzmann, O'Neill, and Minton—*Chem. Engineer,* July 1973, p. 69.
33. Smith—*Chemical Process Design,* McGraw-Hill, 1995.

SECTION 8
DISTILLATION[†]

Otto Frank

Consultant (retired)
Air Products and Chemicals Inc.
Allentown, PA

8.1 CALCULATION OF EQUILIBRIUM STAGES

Design a distillation column to separate benzene, toluene, and xylene, using (1) the McCabe-Thiele *xy* diagram and (2) the Fenske-Underwood-Gilliland (FUG) method. Compare the results with each other. Assume that the system is ideal.

Feed consists of 60 mol/h benzene, 30 mol/h toluene, and 10 mol/h xylene. There must be no xylene in the overhead, and the concentration of toluene must be no greater than 0.2 percent. Benzene concentration in the bottoms should be minimized; however, recognizing that only one end of a column can be closely controlled (in this case, the overhead is being so controlled), a 2% concentration is specified. Feed temperature is 40°C (104°F); the reflux is not subcooled. The column will be operated at atmospheric conditions, with an average internal back pressure of 850 mmHg (113.343 kPa or 1.13 bar).

Calculation Procedure

1. Assess the problem to make sure that a hand-calculation approach is adequate.
The separation of multicomponent and nonideal mixtures is difficult to calculate by hand and is evaluated largely by the application of proper computer programs. However, under certain conditions, hand calculation can be justified, and it will be necessary for the designer to have access to suitable shortcut procedures. This may be the case when (1) making preliminary cost estimates, (2) performing parametric evaluations of operating variables, (3) dealing with situations that call for separations having only coarse purity requirements, or (4) dealing with a system that is thermodynamically ideal or nearly so. For guidelines as to ideality, see Table 8.1.

[†]Portions of the remarks under "Related calculations" at the conclusions of Procedures 8.1, 8.3, 8.4, and 8.7 were compiled and appended by the handbook editor.

TABLE 8.1 Rules of Thumb on Equilibrium Properties of Vapor–Liquid Mixtures

Declining ideality ↓	Mixtures of isomers usually form ideal solutions. Mixtures of close-boiling aliphatic hydrocarbons are nearly ideal below a pressure of 10 atm. Mixtures of compounds close in molecular weight and structure frequently do not deviate greatly from ideality (e.g., ring compounds, unsaturated compounds, naphthenes, etc.). Mixtures of simple aliphatics with aromatic compounds deviate modestly from ideality. "Inserts," such as CO_2, H_2S, H_2, N_2, etc., that are present in mixtures of heavier components tend to behave nonideally with respect to the other compounds. Mixtures of polar and nonpolar compounds are always strongly nonideal. (Look for polarity in molecules containing oxygen, chlorine, fluorine, or nitrogen, in which electrons in bonds between these atoms and hydrogen are not equally shared.) Azeotropes and phase separation represent the ultimate in nonideality, and their occurrence should always be confirmed before detailed distillation studies are undertaken.

Source: Otto Frank, "Shortcuts for Distillation Design," *Chemical Engineering*. March 14, 1977.

In the present case, hand calculation and shortcut procedures are adequate because the benzene-toluene-xylene system is close to ideal.

2. *Establish the equilibrium relationship among the constituents.*
The equilibrium data are based on vapor pressures. Therefore, this step consists of plotting the vapor pressure-temperature curves for benzene, toluene, and xylene. The vapor pressures can be determined by methods such as those discussed in Section 1 or can be found in the literature. In any case, the results are shown in Fig. 8.1.

3. *Convert the given ternary system to a pseudobinary system.*
This step greatly reduces the complexity of calculation, although at some expense in accuracy. (Based on limited data, a rough estimate as to the loss in accuracy is 10 percent, i.e., a 10 percent error). Benzene is the lightest-boiling component in the bottoms stream, so it is designated the "light key" and taken as one component of the binary system. Toluene is the higher-boiling of the two constituents in the overhead stream, so it is designated the "heavy key" and taken as the other component of the binary system.

Since there are 60 mol/h benzene and 30 mol/h toluene in the feed stream, the percentage composition of the pseudobinary feed is 100[60/(30 + 60)], i.e., 66.7% benzene and 100[30/(30 + 60)] = 33.3% toluene.

The overhead composition is stipulated to be 99.8% benzene and 0.2% toluene. By material-balance calculation, the bottoms is found to be 2.6% benzene and 97.4% toluene.

4. *Apply the McCabe-Thiele method of column design, based on an xy diagram.*

a. Develop the equilibrium line or xy curve. This can be done using the procedure described in Procedure 3.12. The result appears as the upper curve in Fig. 8.2 (including both the full figure and the portion that depicts in expanded form the enriching section). This curve shows the composition of vapor *y* that is in equilibrium with any given liquid of composition *x* in the benzene-toluene binary system.
Another way of plotting the curve is to employ the equation

$$y = \frac{\alpha x}{1 + (\alpha - 1)x}$$

where α is the relative volatility of benzene with respect to toluene (see Section 3). Relative volatility may vary with temperature from point to point along the column; if the values at the top and bottom of the column are within 15 percent of each other, an average can be used to establish the equilibrium line. The present example meets this criterion, and its average volatility is 2.58.

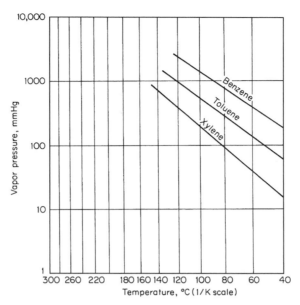

FIGURE 8.1   Vapor pressures of benzene, toluene, and xylene
(Procedure 8.1). Note: 1 mmHg = 0.133 kPa.

b. *Calculate the bubble point of the feed.* This is done via procedures outlined in Section 3. In the
present case, where both the vapor and liquid phases can be considered ideal, the vapor-liquid
equilibrium ratio K_i equals vapor pressure of ith component divided by system pressure. The
bubble point is found to be 94°C. At 40°C, then, the feed is subcooled 54°C.

c. *Calculate the q-line slope, to compensate for the feed not entering at its bubble point (40°C versus
94°C).* The thermal condition of the feed is taken into account by the slope of a q line, where

$$q = \frac{\text{heat needed to convert 1 mol feed to saturated vapor}}{\text{molar heat of vaporization}}$$

The enthalpy to raise the temperature of the benzene to the bubble point and vaporize it
is sensible heat required + latent heat of vaporization, or (94 − 40°C)[32.8 cal/(mol)(°C)] +
7566 cal/mol = 9337 cal/mol. The enthalpy to raise the temperature of the toluene to the bubble
point and vaporize it is (94 − 40°C)[40.5 cal/(mol)(°C)] + 7912 cal/mol = 10,100 cal/mol.
    Then

$$q = \frac{(0.667 \text{ mol benzene}) \, 9337 + (0.333 \text{ mol toluene}) \, 10,100}{(0.667 \text{ mol benzene}) \, 7566 + (0.333 \text{ mol toluene}) \, 7912}$$

$$= 1.25$$

And by definition of the q line, its slope is $q/(q - 1)$, or 1.25/0.25 = 5.

d. *Plot the q line on the xy diagram.* This line, also known as the "feed line," has the slope calculated
in the previous step, and it passes through the point on the xy diagram diagonal that has its abscissa
(and ordinate) corresponding to the feed composition x_F. In Fig. 8.2, the diagonal is *AB* and the q
line is *CD*. The operating lines, calculated in the next step, intersect with each other along the q line.

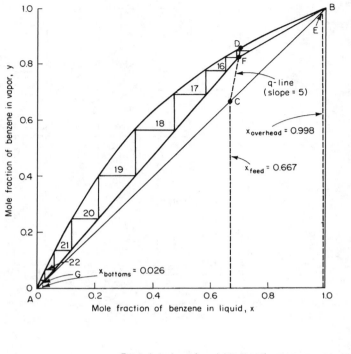

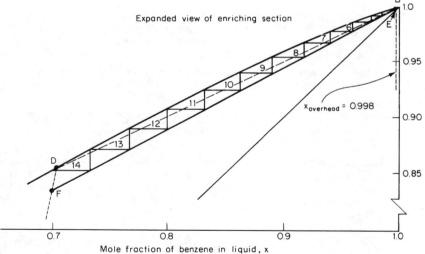

FIGURE 8.2 Application of McCabe-Thiele method of column design (Procedure 8.1).

e. *Construct the operating lines.* There are two operating lines, one for the enriching (upper) section of the column and the other for the stripping (lower) section, the feed plate marking the point of separation between the two. The upper line intersects the diagonal at the abscissa corresponding to the overhead-product composition (in Fig. 8.2, at point *E*); the lower line intersects it at the abscissa corresponding to the bottoms composition. The slope of each is the ratio of (descending) liquid flow to (ascending) vapor flow, or L/V, in that particular section of the column. The two lines intersect along the q line, as noted earlier.

The ratio L/V depends on the amount of reflux in the column. At minimum reflux, the operating lines intersect each other and the q line at the point where the q line intersects the xy curve (in Fig. 8.2, at point D).

In this example, the approach used for constructing the operating lines is based on the relationship that typically holds, in practice, between the minimum reflux and the amount of reflux actually employed. First, draw the operating line that would pertain for minimum reflux in the enriching section. As noted earlier, this line is defined by points D and E. Next, measure its slope $(L/V)_{\min}$ on the diagram. This is found to be 0.5.

For water- or air-cooled columns, the actual reflux ratio R_{actual} is normally 1.1 to 1.3 times the minimum reflux ratio $R_{\min}$. The optimal relationship between R_{actual} and $R_{\min}$ can be established by an economic analysis that compares the cost of energy (which rises with rising reflux) with number of column trays (which declines with rising reflux). For the present example, assume that $R_{\text{actual}}/R_{\min} = 1.2$.

In the enriching section of the column, the descending liquid consists of the reflux. So, $R = L/(V - L)$. By algebra, $1/R = (V - L)/L = V/L - 1$. Since $(L/V)_{\min}$ in the present example is 0.5, $1/R_{\min} = 1/0.5 - 1 = 1$. Therefore, $R_{\min} = 1$, and $R_{\text{actual}} = 1.2R_{\min} = 1.2(1) = 1.2$. Therefore, $1/1.2 = V/L - 1$, so $V/L = 1.83$, and L/V, the slope of the actual operating line for the enriching section, is 0.55. This line, then, is constructed by passing a line through point E having a slope of 0.55. This intersects the q line at F in Fig. 8.2. Finally, the operating line for the stripping section is constructed by drawing a line connecting point F with the point on the diagonal that corresponds to the bottoms composition, namely, point G. The two operating lines for the actual column, then, are EF and FG.

f. *Step off the actual number of theoretical stages.* Draw a horizontal line through point E, intersecting the xy curve. Through that intersection, draw a vertical line intersecting the operating line EF. Through the latter intersection, draw a horizontal line intersecting the xy curve. Continue the process until a horizontal line extends to the left of point G (as for the steps after the one in which a horizontal line crosses the q line, the vertical lines, of course, intersect operating line FG rather than EF). Count the number of horizontal lines; this number is found to be 22.2 (taking into account the shortness of the last line). That is the number of theoretical stages required for the column. The number of lines through or above the q line, in this case 14, is the number of stages in the enriching section, i.e., above the feed plate.

g. *Consider the advantages and disadvantages of the McCabe-Thiele method.* This example illustrates both the advantages and the disadvantages. On the positive side, it is easy to alter the slope of the operating line and thus the reflux ratio, changes introduced by varying the feed quality and feed location can be judged visually, and the effect of the top and bottom concentrations on number of equilibrium stages can be readily established by moving the operating lines to new points of origin. On the negative side, the example indicates the limit to which the xy diagram can be reasonably applied: When the number of stages exceeds 25, the accuracy of stage construction drops off drastically. This is usually the case with a small relative volatility or high purity requirements or when a very low reflux ratio is called for.

5. *As an alternative to the McCabe-Thiele method, design the column by using the Fenske-Underwood-Gilliland correlations*

a. *Calculate the minimum number of stages needed (implying total reflux).* Apply the Fenske correlation:

$$N_{\min} = \frac{\log[(x_{LK}/x_{HK})_D (x_{HK}/x_{LK})_B]}{\log \alpha}$$

where $N_{\min}$ is the minimum number of stages, x is mole fraction, mole percent, or actual number of moles, α is relative volatility of the light key (benzene) with respect to the heavy key (toluene), and the subscripts LK, HK, D, and B refer to light key, heavy key, overhead (or distillate) product, and bottoms product, respectively. In this example, α is 2.58, so

$$N_{min} = \frac{\log(99.8/0.2)(97.4/2.6)}{\log 2.58} = 10.4 \text{ stages}$$

b. *Estimate the feed tray.* Again, apply the Fenske correlation, but this time replace the bottoms-related ratio in the numerator with one that relates to the feed conditions, namely, $(x_{HK}/x_{LK})_F$. Thus,

$$N_{min} = \frac{\log(99.8/0.2)(33.3/66.7)}{\log 2.58} = 5.8 \text{ stages}$$

Therefore, the ratio of feed stages to total stages is $5.8/10.4 = 0.56$, so 56 percent of all trays should be located above the feed point.

c. *Calculate the minimum reflux required for this specific separation.* Apply the Underwood correlation: $(L/D)_{min} + 1 = \Sigma[\alpha x_D/(\alpha - \theta)]$ and $1 - q = \Sigma[\alpha x_F/(\alpha - \theta)]$. The two summations are over each component in the distillate and feed, respectively (thus, the system is not treated as pseudo-binary). Each relative volatility α is with respect to the heavy key (toluene), and $(L/D)_{min}$ is minimum reflux, x is mole fraction, q is the heat needed to convert one mole of feed to saturated vapor divided by the molar heat of vaporization, and θ is called the "Underwood constant." The value of q is 1.25, as found in step 4c. The relative volatilities, handled as discussed in step 4a, are $\alpha_{\text{benzene-toluene}} = 2.58$, $\alpha_{\text{toluene-toluene}} = 1.00$, and $\alpha_{\text{xylene-toluene}} = 0.36$.

The calculation procedure is as follows: Estimate the value of θ by trial and error using the second of the two preceding equations; then employ this value in the first equation to calculate $(L/D)_{min}$.

Thus, it is first necessary to assume values of θ in the equation

$$1 - q = \frac{\alpha_{B-T}x_B}{\alpha_{B-T} - \theta} + \frac{\alpha_{T-T}x_T}{\alpha_{T-T} - \theta} + \frac{\alpha_{X-T}x_X}{\alpha_{X-T} - \theta}$$

where the mole fractions pertain to the feed and where B, T, and X refer to benzene, toluene, and xylene, respectively. Trial and error shows the value of θ to be 1.22; in other words, $2.58(0.60)/(2.58 - 1.22) + 1(0.30)/(1 - 1.22) + 0.36(0.1)/(0.36 - 1.22) = -0.26$, which is close enough to $(1 - 1.25)$.

Now, substitute $\theta = 1.22$ into

$$(L/D)_{min} + 1 = \frac{\alpha_{B-T}x_B}{\alpha_{B-T} - \theta} + \frac{\alpha_{T-T}x_T}{\alpha_{T-T} - \theta}$$

where the mole fractions pertain to the overhead product (there is no xylene-related term because there is no xylene in the overhead). Thus,

$$(L/D)_{min} + 1 = \frac{2.58(0.998)}{2.58 - 1.22} + \frac{1(0.002)}{1 - 1.22}$$
$$= 0.88$$

[Note that because $(L/D)_{min} = R_{min}$ and $1/R = V/L - 1$, this $(L/D)_{min}$ of 0.88 corresponds to an $(L/V)_{min}$ of 0.47, versus the value of 0.5 that was found in the McCabe-Thiele method, step 4e above.]

A direct solution for θ can be obtained by using a graph developed by Van Winkle and Todd (Fig. 8.3) in which the abscissa is based on the system being treated as pseudobinary. Strictly speaking, this graph applies only to a liquid at its bubble point, but in the present case, with $(x_{LK}/x_{HK})_F = (x_B/x_T)_F = 66.7/33.3 = 2$, the graph shows for an α of 2.58 that $\theta = 1.26$. Then, from the Underwood equation, $(L/D)_{min} = 0.94$ and $(L/V)_{min} = 0.48$, so for typical feed conditions (only a limited amount of subcooling or flashing), the simplification introduced by the graph is acceptable.

d. *Estimate the actual number of theoretical stages using the Gilliland correlation.* This step employs Fig. 8.4. From step 5c, $(L/D)_{min}$ or R_{min} is 0.88. Assume (as was done in the McCabe-Thiele procedure, step 4e) that the actual reflux ratio R is 1.2 times R_{min}. Then $R = 1.2(0.88) = 1.06$, and the abscissa of Fig. 8.4 $(R - R_{min})/(R + 1) = (1.06 - 0.88)/(1.06 + 1) = 0.087$. From the graph, $(N - N_{min})/(N + 1) = 0.57$, where N is the actual number of stages and N_{min} is the minimum number. From step 5a, $N_{min} = 10.4$. Then $(N - 10.4)/(N + 1) = 0.57$, so $N = 25.5$ theoretical stages. (Note that if the R_{min} of 0.94, based on the graphically derived Underwood constant, is employed, the resulting number of theoretical stages is nearly the same, at 24.9.)

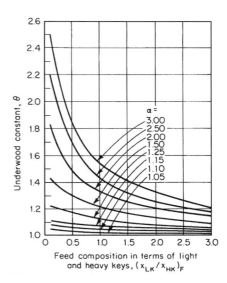

FIGURE 8.3 Underwood constant as a function of feed composition and relative volatility.

e. *Comment on the usage of the Fenske-Underwood-Gilliland method.* A reasonably good estimate of N can be obtained by simply doubling the N_{min} that emerges from the Fenske correlation. The Underwood correlation can handle multicomponent systems, whereas the McCabe-Thiele *xy* diagram is confined to binary (or pseudobinary) systems. However, if there is a component that has a vapor pressure between that of the light and heavy keys, then the Underwood-calculation procedure becomes more complicated than the one outlined above.

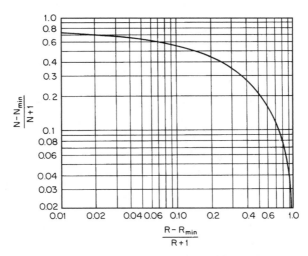

FIGURE 8.4 Gilliland correlation relating number of stages to reflux ratio. (*From Chemical Engineering, McGraw-Hill, 1977.*)

6. *Compare the results of the McCabe-Thiele and the Fenske-Underwood-Gilliland methods.*
Summarizing the results found in this example, the comparison is as follows:

| | Results | | |
Method	Reflux ratio	Number of theoretical stages	Optimal feed stage
McCabe-Thiele	1.2	22.2	14
Fenske-Underwood-Gilliland:			
Using calculated θ	1.06	25.5	14
Using θ from graph	1.13	24.9	14
Using nomograph	—	22	—
Using approximation, $2N_{min}$	—	20.8	12

Related Calculations.   Special McCabe-Thiele graph paper that expands the top and bottom of the *xy* diagram is available. This makes stepping off stages easier and more precise. However, it does not materially improve the accuracy of the procedure.

Variations of the basic McCabe-Thiele method, such as incorporating tray efficiencies, noncon-stant molal overflow, side streams, or a partial condenser, are often outlined in standard texts. These modifications usually are not justified; modern computer programs can calculate complex column arrangements and nonideal systems considerably faster and more accurately than use of any hand-drawn diagram can.

A nomograph for the overall Fenske-Underwood-Gilliland method has been derived that consid-erably reduces the required calculation effort without undue loss of accuracy. It is based on Fig. 8.5,

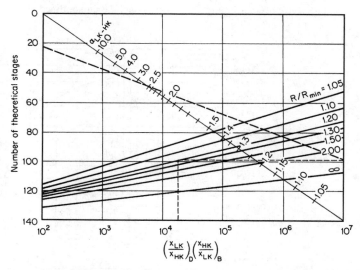

FIGURE 8.5   Graphical representation of Fenske-Underwood-Gilliland procedure. (*From Chemical Engineering, McGraw-Hill, 1977.*)

where the subscripts D and B in the abscissa refer to overhead and bottoms product streams, respectively, and the other symbols are as used in step 5. Find the relevant abscissa; in this case, (99.8/0.2)(97.4/2.6) = 18,700. Erect a vertical line and let it intersect the positively sloped line that represents the assumed ratio of actual to minimum reflux ratios, in this case, 1.2. Through this intersection, draw a horizontal line that intersects the right-hand vertical border of the diagram. Finally, draw a line through the latter intersection and the relevant α_{LK-HK} point (in this case, 2.58) on the negatively sloped line that represents relative volatilities, and extend this last-constructed line until it intersects the left-hand border of the diagram. The point of intersection (in this case, it turns out to be 22) represents the number of theoretical stages.

More recently, a different approach for using the McCabe-Thiele diagram has been presented in the literature: Instead of the vapor mole fraction, y, being plotted on the vertical axis, the difference between vapor and liquid mole fraction $(y - x)$ is plotted. The horizontal axis remains the liquid mole fraction, x. An example appears in the figure.[†] This transformation is claimed to be especially useful for situations involving relative volatilities less than 1.25; when such cases are plotted on the conventional McCabe-Thiele diagram, the stage-equilibria data and the operating-line data alike lie very close to the diagonal line, which complicates the usage of the diagram. Plotting $(y - x)$ instead of y spreads out the conventionally close-to-the-diagonal region over the entire height of the plot, because the diagonal coincides with the abscissa.

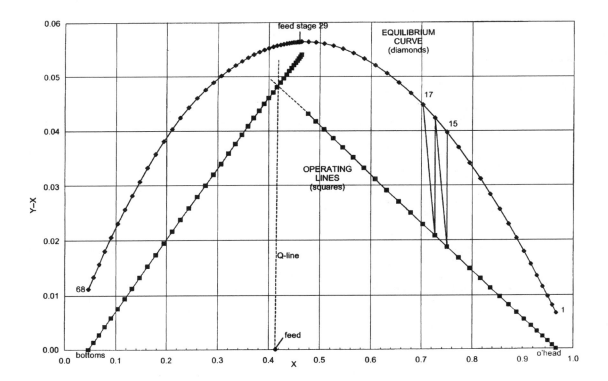

[†]*Source:* Ryan. J., Replotting the McCabe–Thiele Diagram, *Chemical Engineering*, May 2001, pp. 109–113. With permission.

8.2 DISTILLATION-TRAY SELECTION AND DESIGN

Design the trays for a distillation column separating dichlorobenzene (DCB) from a high-boiling reaction product. Include designs for sieve trays and valve trays, and discuss the applications of each. The product is temperature-sensitive, so sump pressure should be held at about 100 mmHg (3.9 inHg or 0.13 bar). The separation requires 20 actual trays.

Column conditions in the top section are as follows:

Vapor flow rate = 69,000 lb/h (31,300 kg/h)

Liquid flow rate = 20,100 lb/h (9120 kg/h)

Liquid viscosity = 0.35 cP (3.5×10^{-4} Pa · s)

Liquid density = 76.7 lb/ft^3 (1230 kg/m^3)

Surface tension of DCB liquid = 16.5 dyn/cm (1.65×10^{-4} N/cm)

Temperature (average for top section) = 200°F (366 K)

Pressure (average for top section) = 50 mmHg (6.67 kPa)

Molecular weight of DCB = 147

Note: Pressure drop for the vapor stream is assumed to be 3 mmHg (0.4 kPa) per tray; therefore, the pressure at the top of the column is set at 40 mmHg (5.3 kPa), given that the sump pressure is to be no greater than 100 mmHg (13.3 kPa).

Calculation Procedure

1. Set tray spacing.

Tray spacing is selected to minimize entrainment. A large distance between trays is needed in vacuum columns, where vapor velocities are high and excessive liquid carryover can drastically reduce tray efficiencies. A tradeoff between column diameter (affecting vapor velocity) and tray spacing (affecting the disengaging height) is often possible.

Trays are normally 12 to 30 in (0.305 to 0.762 m) apart. A close spacing of 12 to 15 in (0.305 to 0.381 m) is usually quite suitable for moderate- or high-pressure columns. Frequently, 18 in (0.457 m) is selected for atmospheric columns. High-vacuum systems may require a tray spacing of 24 to 30 in (0.61 to 0.762 m). A spacing as low as 9 in (0.229 m) is sometimes found in high-pressure systems. Spacings greater than 30 in are seldom if ever justified. As a starting point for the column in question, select a 24-in (0.61-m) tray spacing.

2. Estimate column diameter.

In high-pressure columns, liquid flow is the dominant design consideration for calculating column diameter. However, at moderate and low pressures (generally below 150 psig or 10.34 bar), as is the case for the column in this example, vapor flow governs the diameter.

In vacuum columns, the calculations furthermore are based on conditions at the column top, where vapor density is lowest and vapor velocity highest. The F-factor method used here is quite satisfactory for single- and two-pass trays; it has an accuracy well within ±15 percent. The F factor F_c is *defined* as follows:

$$F_c = v\rho_v^{1/2}$$

where v is the superficial vapor velocity in the tower (in ft/s), and ρ_v is the vapor density (in lb/ft^3). In the calculation procedure, first the appropriate value of F_c is determined from Fig. 8.6, which gives F_c as a function of column pressure (in this case, 0.97 psia) and tray spacing (in this case, 24 in, from step 1). The figure shows F_c to be 1.6. (Note that if the straight-line portion of the curve for 24-in tray spacing were continued at pressures below about 15 psia, the value of F_c would instead be about 1.87. The lower value, 1.6, takes into account the efficiency loss at low pressures that is due to entrainment.)

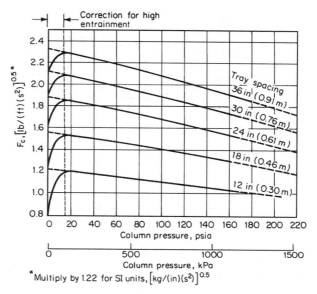

FIGURE 8.6   *F* factor as a function of column pressure and tray spacing.

Next, the F_c value is used in calculating required free area for vapor flow, via the relation

$$A_F = \frac{W}{F_c \rho_v^{1/2}}$$

where A_F is free area (in ft^2), and W is vapor mass flow rate (in lb/s). (Note that this is merely a rearrangement of the preceding definition for F_c.) In this case, $W = (69,000 \text{ lb/h})/(3600 \text{ s/h}) = 19.17$ lb/s, and (from the ideal-gas law, which is acceptable at up to moderate pressures) $\rho = (0.97 \text{ psia})$ $(147 \text{ lb/lb} \cdot \text{mol})/(10.73 \text{ psia} \cdot \text{ft}^3/R)[200 + 460)R] = 0.02$ lb/ft^3. Therefore, $A_F = 19.17/1.6(0.02)^{1/2} = 84.72$ ft^2.

Actual column cross-sectional area A_T is the sum of the free area A_F plus the downcomer area A_D. Downcomer cross-sectional area must be adequate to permit proper separation of vapor and liquid. Although downcomer size is a direct function of liquid flow, total downcomer area usually ranges between 3 and 20 percent of total column area, and for vacuum columns it is usually 3 to 5 percent. In this example, assume for now that it is 5 percent; then $A_T = A_F /0.95 = 84.72/0.95 = 89.18$ ft^2. Then, by geometry, column diameter $= [A_T /(\pi/4)]^{1/2} = (89.18/0.785)^{1/2} = 10.66$ ft; say, 11 ft (3.4 m).

3. Calculate the flooding velocity.
Use the Fair calculation. Flooding velocity $U_f = C(\sigma/20)^{0.2}[(\rho_L - \rho_V)/\rho_V]^{0.5}$, where σ is surface tension (in dyn/cm), ρ is density, the subscripts L and V refer to liquid and vapor, respectively, and C is a parameter that is related to tray spacing, liquid flow rate L, vapor flow rate G, and liquid and vapor densities by the graph in Fig. 8.7. In this case, the abscissa is $(20,100/69,000)[0.02/(76.7 - 0.02)]^{0.5} = 0.005$, and tray spacing is 24 in from step 1, so C is found from the graph to be 0.3. Then,

$$U_f = 0.3 \left(\frac{16.5}{20} \right)^{0.2} \left(\frac{76.7 - .02}{0.02} \right)^{0.5}$$

$$= 17.9 \text{ ft/s}$$

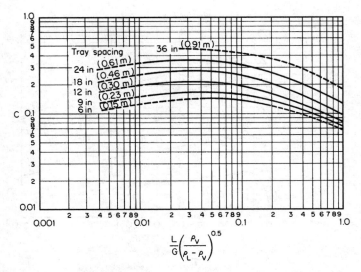

$$\frac{L}{G}\left(\frac{\rho_V}{\rho_L - \rho_V}\right)^{0.5}$$

FIGURE 8.7   Determination of parameter C for Fair calculation of flooding velocity.

The following adjustments, however, must be incorporated into the calculated flooding velocity to assure a reasonable design safety factor:

Open area (cap slots or holes)	Multiplying factor
10 percent of active tray area	1.0
8 percent of active tray area	0.95
6 percent of active tray area	0.90

System properties	Multiplying factor
Known nonfoaming system at atmospheric or moderate pressures	0.9
Nonfoaming systems; no prior experience	0.85
Systems thought to foam	0.75
Severely foaming systems	0.70
Vacuum systems (based on entrainment curves; see Fig. 8.17)	0.60 to 0.80
For downcomerless trays if the open area is less than 20 percent of active tray area (Note: it should never be less than 15 percent)	0.85

In this example, no adjustment is necessary for percent holes per active tray area, because in low-pressure systems, the hole area is usually greater than 10 percent. However, because we are dealing with a vacuum system, the flooding velocity must be downgraded. Referring to Fig. 8.17, take as the maximum acceptable fractional entrainment a ψ of 0.15. Then for an abscissa of $(L/G)(\rho_v/\rho_L)^{0.5} = (20,100/69,000)(0.02/76.7)^{0.5} = 0.005$, establish at what percent of flooding the column can be operated. The graph shows this to be 60 percent. Therefore, the column should be designed for a vapor velocity of 0.6(17.9) = 10.74 ft/s (3.27 m/s).

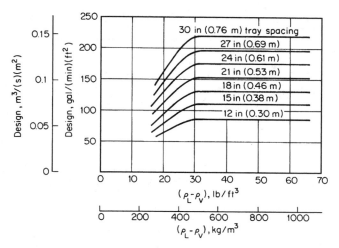

FIGURE 8.8 Downcomer area as a function of densities and tray spacing.
Note: Multiply clear liquid rate (ordinate) by 0.7 for foaming systems.

4. *Reestimate the required free cross-sectional area, this time to accommodate the maximum allowable vapor velocity.*

Since mass flow rate equals density times cross-sectional area times velocity, free column area $A_F =$ (69,000 lb/h)/[(0.02 lb/ft^3)(3600 s/h)(10.74 ft/s)] = 89.23 ft^2. Note that this turns out to be slightly larger than the area calculated in step 2.

5. *Set the downcomer configuration.*

A downcomer area can be selected from Fig. 8.8. In this case, the abscissa is (76.7 − 0.02) and the tray spacing is 24 in, so Design gal/(min)(ft^2) is read to be 175 (by extrapolating the horizontal lines on the graph to the right). Since 76.7 lb/ft^3 = 10.25 lb/gal, the gal/min rate is (20,100 lb/h)/ [(60 min/h)(10.25 lb/gal)] = 32.68 gal/min. Therefore, the minimum required downcomer area is 32.68/175 = 0.187 ft^2.

However, flow distribution across the tray must also be taken into account. To optimize this distribution, it is usually recommended that weir length be no less than 50 percent of the column diameter. Thus, from Fig. 8.9, the downcomer area subtended by a weir having half the length of the column diameter should be at least 3 percent of the column area (5 percent was assumed in step 2 during the initial estimate of column diameter).

Allowing 3 percent for downcomer area, then, total cross-sectional area for the column must be 89.23/0.97 = 91.99 ft^2 (8.55 m^2), and the downcomer area required for good flow distribution is (91.99 − 89.23) = 2.76 ft^2 (0.26 m^2). (Note that this is considerably larger than the minimum 0.187 ft^2 required for flow in the downcomer itself). Finally, the required column diameter is (91.99/0.785)$^{1/2}$ = 10.83 ft; say, 11 ft (3.4 m). (This turns out the same as the diameter estimated in step 2.)

Because of the low liquid rates in vacuum systems, downcomers will usually be oversized, and specific flow rates across the weir will be low. However, liquid rates in high-pressure columns may exceed values recommended for optimum tray performance across a single weir. The maximum specific flow is 70 gal/(min · ft)[53 m^3/(h · m)] for a straight segmental weir and 80 gal/(min · ft) [60 m^3/(h · m)] for a weir with relief wings. Above 80 gal/(min · ft), a multiple downcomer arrangement should be considered.

When dealing with foaming or high-pressure systems, a frequent recommendation is the installation of sloped downcomers. This provides for adequate liquid-vapor disengaging volume at the top, while leaving a maximum active area on the tray below.

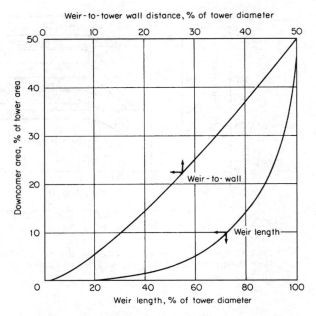

FIGURE 8.9 Design chart for segmental downcomers. (*Glitsch, Inc.*)

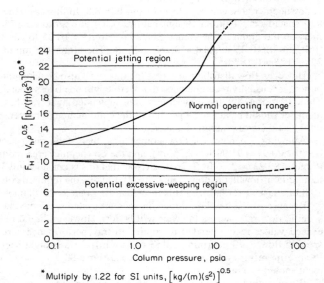

*Multiply by 1.22 for SI units, $\left[kg/(m)(s^2) \right]^{0.5}$

FIGURE 8.10 Approximate operating range of sieve trays. Note: 1 psia = 6.895 kPa.

6. *Design suitable sieve trays*

a. *Determine the required hole area.* A workable hole-area calculation can be based on the empirically developed relationship shown in Fig. 8.10. In this F factor (the ordinate), v_h is the velocity of vapor passing through the holes. The graph shows that columns operating at low pressures have less flexibility (lower turndown ratios) than those operating at higher pressures.

Assume that F_H is 12. Then $v_h\ F_H/\rho_v^{0.5} = 12./0.02^{0.5} = 85$ ft/s (26 m/s). Therefore, total hole area required is $(69,000\ \text{lb/h})/[(0.02\ \text{lb/ft}^3)(3600\ \text{s/h})(85\ \text{ft/s})] = 11.27\ \text{ft}^2$. Note that since each tray is fed by one downcomer and drained by another and each downcomer occupies 3 percent of the total tray area according to step 5, the hole area as a percent of active area is

$$\frac{(11.27\ \text{ft}^2)(100)}{[1.00 - (2)(0.03)](11\ \text{ft})^2(\pi/4)} = 12.62\ \text{percent}$$

b. *Specify hole size, weir height, and downcomer.* Guidelines for this step are as follows:

Hole pitch-to-diameter ratio	2 to 4.5
Hole size	Use ½ (0.0127 m) for normal service, ⅜ or ¼ in (0.0095 or 0.0064 m) for clean vacuum systems, ¾ in (0.0191 m) for fouling service.
Open area/active area	Use 4 to 16 percent, depending on system pressure and vapor velocity.
Weir height	Use 1 to 4 in (0.025 to 0.102 m) (but no higher than 15 percent of tray spacing) as follows: 1 to 2 in (0.025 to 0.051 m) for vacuum and atmospheric columns or 1½ 3 in (0.038 to 0.076 m) for moderate to high pressures.
Downcomer seal	Use one-half of weir height ¾ in (0.0191 m), whichever is greater. Clearance between downcomer and tray deck should never be less than ½ in (0.0127 m), and the velocity through the clearance should be under 1 ft/s (0.30 m/s).

Select ¼-in (6.4-mm) holes and a 1-in (2.54-mm) high weir, with a ¾-in downcomer clearance.

c. *Calculate pressure drop to confirm suitability of design.* The total pressure drop h_t is the sum of the pressure drop across the holes h_{dry} and the drop through the aerated material above the holes h_{liq}. Now,

$$h_{\text{dry}} = 0.186\frac{\rho_V}{\rho_L}\left(\frac{v_h}{C}\right)^2$$

where C is a discharge coefficient obtainable from Fig. 8.11. From step 6b, hole area divided by active area is 0.1262, and for a 14-gauge tray (0.078 in or 2 mm, a reasonable thickness for this service), the tray thickness divided by hole diameter ratio is 0.078/0.25 = 0.31. Therefore, C is read as 0.74. Then, $h_{\text{dry}} = 0.186(0.02/76.7)(85/0.74)^2 = 0.64$ in of liquid. However, this must be adjusted for entrainment:

$$h_{\text{dry}}(\text{corrected}) = h_{\text{dry}}\left(1 + \frac{\psi}{1 - \psi}\right)$$

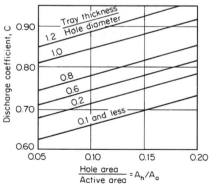

FIGURE 8.11 Discharge coefficient for sieve-tray performance. Note: Most perforated trays are fabricated from 14-gauge stainless steel or 12-gauge carbon-steel plates.

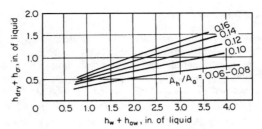

A_h/A_a = ratio of hole area to active tray area
h_w = height of weir, in.
h_{ow} = height over weir, in.
h_{dry} = dry pressure drop, in. liquid
h_σ = resistance due to liquid surface tension = $\dfrac{0.04\sigma}{\rho_L d_h}$
d_h = hole dia., in.
σ = surface tension, dynes/cm
ρ_L = liquid density, lb/ft^3

FIGURE 8.12 Estimate of excessive weeping. Note: Operating points above the respective lines represent "salf" designs. A point below the line may indicate uncertainty, but not necessarily a dumping station. (*From Chemical Engineering, McGraw-Hill, 1977.*)

where ψ is the entrainment function. Since this is 0.15 (see step 3),

$$h_{dry}(\text{corrected}) = 0.64\left(1 + \frac{0.15}{0.85}\right) = 0.75 \text{ in liquid}$$

Dry pressure drop for sieve trays should fall between 0.75 and 3 in of liquid (19 and 76 mm) to prevent excessive weeping on the one hand or jetting on the other. Since, in this case, the dry pressure drop is barely within the recommended range, it is desirable that the tray be checked for weeping using Fig. 8.12.

Now, $A_h/A_A = 0.1262$, from step 6a. And

$$h_\sigma = \frac{0.04\sigma}{\rho_L(\text{hold diameter})} = \frac{0.04(16.5)}{76.7(0.25)} = 0.034 \text{ in liquid}$$

Therefore, the ordinate in Fig. 8.12, $(h_{dry} + h_\sigma)$, is $0.75 + 0.03 = 0.78$. As for the abscissa, h_w (weir height) was chosen to be 1 in, and $h_{ow} = 0.5[(\text{liquid flow, gal/min})/(\text{weir length, in})]^{0.67} = 0.5\{32.68/[\frac{1}{2}(11 \text{ ft})(12 \text{ in/ft})]\}^{0.67} = 0.31$ in liquid, so $h_w + h_{ow} = 1.00 + 0.31 = 1.31$. In Fig. 8.12, the point (1.31, 0.78) falls above the line $A_H/A_A = 0.12$; therefore, weeping will not be excessive.

As for the pressure drop through the aerated material,

$$h_{liq} = \beta(h_w + h_{ow})$$

where β is an aeration factor obtainable from Fig. 8.13. In the abscissa, the vapor-velocity term refers to velocity through the column (10.74 ft/s from step 3) rather than to velocity through the holes, so $F_c = 10.74(0.02)^{1/2} = 1.52$. Thus β is read to be 0.6. So, $h_{liq} = 0.6(1.31) = 0.79$ in liquid. Finally, the total pressure drop $h_t = 0.75 + 0.79 = 1.54$ in liquid. Liquid density is 76.7 lb/ft^3, so the equivalent pressure drop is $1.54(76.7/62.4) = 1.89$ in water or 3.5 mmHg (0.47 kPa) per tray.

A calculated pressure drop of 3.5 mmHg, although somewhat higher than the assumed 3.0 mmHg, falls within the range of accuracy for the outlined procedure. However, if there is an overriding concern to avoid exceeding the specified 100 mmHg pressure in the column sump, the following options could be considered: (1) redesign the trays using an F_H factor (in step 6a) of

11.0 or 11.5, or (2) lower the design top pressure to 20 to 30 mmHg and redesign the trays for a pressure drop of 4 or 3.5 mmHg, respectively.

Note that the 3 mmHg pressure drop for this column can in fact be attained if turndown requirements are not excessive. A turndown to 80 percent of the design vapor load is probably the lower limit at which this column can be operated without loss of efficiency. At higher pressures, when pressure drop is not critical, the base design should be for a pressure drop greater than 4.5 mmHg; this will usually permit operation at turndown rates as high as 50 percent.

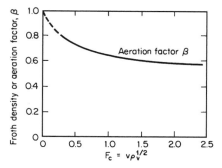

FIGURE 8.13   Aeration factor for pressure-drop calculations.

d. Discuss the applicability of sieve trays. The sieve tray is probably the most versatile contacting device. It should be considered first for the design of a tray column. It has the lowest installed cost of any equilibrium-stage-type device, its fouling tendencies are low, and it offers good efficiency when properly designed.

However, sieve trays are not recommended for the following conditions: (1) when very low pressure drop (less than 2.5 mmHg, or 0.39 kPa) is required; (2) when high turndown ratios are required at low pressure drop; or (3) when very low liquid rates are required: below either 0.25 gal/(min)(ft^2) [0.6 m^3/(h)(m^2)] of active tray area, or 1 gal/(min)(ft) [0.75 m^3/(h)(m)] of average flow-path width.

7. Design suitable valve trays (as an alternative to sieve trays)

a. Select the valve layout. Although valve trays come in several configurations, all have the same basic operational principle: Vapor passing through orifices in the tray lifts small metal disks or strips, thereby producing a variable opening that is proportional to the flow rate.

Because of their proprietary nature, valve trays are usually designed by their respective vendors based on process specifications supplied by the customer. However, most fabricators publish technical manuals that make it possible to estimate some of the design parameters. The procedure for calculating valve-tray pressure drop outlined here has been adapted from the *Koch Design Manual.* As for the other column specifications required, they can be obtained via the same calculation procedures outlined above for the sieve-tray design.

The number of valve caps that can be fitted on a tray is at best an estimate unless a detailed tray layout is prepared. However, a standard has evolved for low- and moderate-pressure operations: a 3 × ½ in pattern that is the tightest arrangement available, accommodating about 14 caps/ft^2 (150 caps/m^2). The active area does not take into account liquid-distribution areas at the inlet and outlet, nor edge losses due to support rings, nor unavailable space over tray-support beams. In smaller columns, it is possible that as much as 25 percent of the active tray area may not be available for functioning valves. For this column, which operates at low pressure, select the standard 3 × ½ in pitch.

b. Calculate the pressure drop per tray. The vapor velocity at the top of the column is (69,000 lb/h)/ [(60 min/h)(0.02 lb/ft^3)] = 57,500 ft^3/min. Assume that 15 percent of the tray is not available for functioning valves. Then, since the active tray area (subtracting both the descending and ascending downcomers; see step 6*a*) is 0.94(11)^2(π/4) = 89.3 ft^2, the number of valves per tray is 0.85(89.3 ft^2)(14 caps/ft^2) = 1063 valves. Then, in entering Fig. 8.14 so as to find the pressure drop h_{dry} through the valves, the abscissa (ft^3/min) air/cap is [(57,500 ft^3/min)/1063 valves] (0.02/0.0735)$^{0.5}$ = 28.2. From Fig. 8.14, using the curve for the Venturi orifice valve, ΔP = 0.8 in water, or 0.65 in liquid, or 1.5 mmHg, which when corrected for entrainment (same as for sieve trays; see step 6*c*) becomes 0.65(1 + 0.15/0.85) = 0.76 in liquid.

The pressure drop h_{liq} through the aerated liquid can be obtained directly from Fig. 8.15. For a liquid rate of 32.68 gal/min (see step 5) flowing over a 5.5-ft-long, 1-in weir (see steps 5 and 6*b*),

The figure axes (Figure 8.13): y-axis labeled "Froth density or aeration factor, β" with values 0, 0.2, 0.4, 0.6, 0.8, 1.0; x-axis labeled "$F_c = v \rho_v^{1/2}$" with values 0, 0.5, 1.0, 1.5, 2.0, 2.5; curve labeled "Aeration factor β".

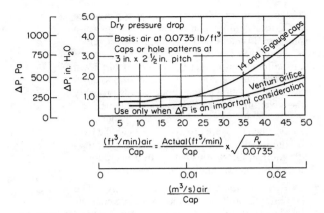

FIGURE 8.14   Dry pressure drop h_{dry} across valve trays. (*From Chemical Engineering, McGraw-Hill, 1977.*)

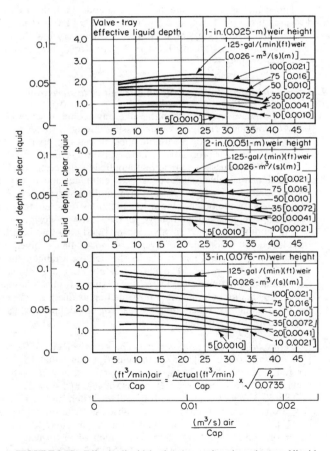

FIGURE 8.15   Effective liquid depth (pressure drop through aerated liquid, h_{liq}) on valve trays. (*From Chemical Engineering, McGraw-Hill, 1977.*)

or 6 gal/(min)(ft), h_{liq} is found to be 0.5 in liquid. Therefore, the total pressure drop across a valve tray is $0.76 + 0.5 = 1.26$ in liquid, or 1.55 in water, or 2.9 mmHg (0.39 kPa).

c. *Discuss the applicability of valve trays.* The relatively low pressure drop that can be maintained without undue loss of turndown in vacuum columns is probably the valve tray's greatest attribute. Although the accuracy of either the sieve or valve-tray calculation procedure is probably no better than ±20 percent, a lower pressure drop is likely to be achieved with Venturi orifice valves than with sieve trays if a reasonable turndown ratio (say 60 percent) is required. This is of little concern at column pressures above 400 mmHg (53 kPa), when pressure drop becomes a minor consideration.

A word of caution: When the valves are exposed to a corrosive environment, it is likely that their constant movement will induce fatigue stresses, which frequently lead to the rapid deterioration of the retaining lugs and valve caps. It is not unusual to find valves missing in that part of the column where corrosive constituents are concentrated.

8.3 COLUMN EFFICIENCY

Determine the efficiency in the upper portion of the DCB distillation column designed in the preceding example. Relative volatility for the system is 3.6.

Calculation Procedure

1. *Determine the effects of the physical properties of the system on column efficiency.*
Tray efficiency is a function of (1) physical properties of the system, such as viscosity, surface tension, relative volatility, and diffusivity; (2) tray hydraulics, such as liquid height, hole size, fraction of tray area open, length of liquid flow path, and weir configuration; and (3) degree of separation of the liquid and vapor streams leaving the tray. Overall column efficiency is based on the same factors, but will ordinarily be less than individual-tray efficiency.

The effect of physical properties on column efficiency can be roughly estimated from Fig. 8.16. For this system, viscosity is 0.35 cP (see statement of previous example) and relative volatility is 3.6, so the abscissa is 1.26. The ordinate, or column efficiency, is read to be 68 percent.

2. *Determine the effects of tray hydraulics on the efficiency.*
Tray hydraulics will affect efficiency adversely only if submergence, hole size, open tray area, and weir configuration are outside the recommended limits outlined in the previous example. Since that is not the case, no adverse effects need be expected.

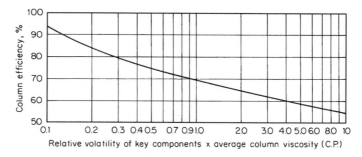

FIGURE 8.16 Column efficiency as a function of average column viscosity and relative volatility.

3. *Determine the effects of inadequate separation of liquid and vapor (e.g., entrainment) on the efficiency.*
The effects of entrainment on efficiency can be quite drastic, especially in vacuum columns, as the vapor rate in the column approaches flooding velocities. The corrected-for-entrainment efficiency can be calculated as follows:

$$E_c = \frac{E_i}{1 + E_i[\psi / (1 - \psi)]}$$

where E_c is the corrected efficiency, E_i is the efficiency that would prevail if entrainment were no problem (i.e., 68 percent, from step 1), and ψ is the fractional entrainment, whose relationship to other column parameters is defined in Fig. 8.17. In step 3 of the previous example, ψ was assumed to be 0.15. Therefore, $E_c = 0.68/\{1 + 0.68[0.15/(1 - 0.15)]\} = 0.607$; say, 60 percent.

Accordingly, an actual column efficiency of 60 percent for the top section of the column would be reasonable.

Related Calculations. Since efficiencies are likely to vary from column top to bottom, it is usually well to estimate them at various points (at least two) along the column.

"Normal" column efficiencies run between 60 and 85 percent. They will tend toward the lower part of this range in vacuum columns, where entrainment can be a major factor, and in systems where

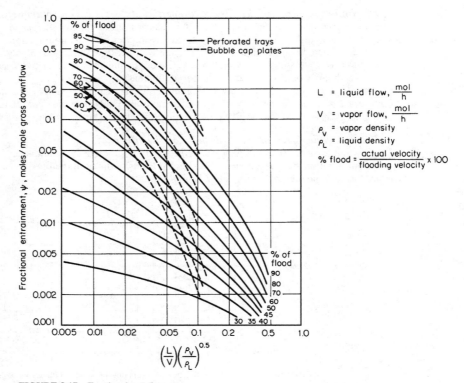

FIGURE 8.17 Fractional entrainment.

high purities are demanded (<100 ppm of a contaminant). Moderate-pressure systems frequently show higher efficiencies. For instance, the benzene-toluene-xylene system outlined in Procedure 8.1 has a viscosity of 0.3 cP and a 2.58 relative volatility. From Fig 8.16, this calls for an efficiency of 71 percent. Little if any correction for entrainment will be necessary, and a final assumed efficiency of 70 percent would not be unreasonable.

Apparent efficiencies for high-pressure systems have frequently been reported in the 90 to 100 percent range.

Weeping is considered excessive and will adversely affect efficiency when the major fraction of the liquid drops through the holes rather than flows over the weir. Figure 8.12 provides a good guide for selecting safe operating conditions of trays; no derating of the basic efficiency is necessary if the operating point falls above the appropriate area-ratio curve.

A correction for dynamic column instability must be made in order to adjust for the continuous shifting of the concentration profile that results from the interaction of controllers with changes in feed flow, cooling water rates, and the ambient temperature. To ensure that product quality will always stay within specifications, it is recommended that 10 percent, but not less than three trays, be added to the calculated number of trays. For instance, recall that the benzene-toluene-xylene system in Procedure 8.1 requires 25 theoretical stages. With a 70 percent column efficiency, this is now raised to 36 actual trays. Another 4 trays should be added to account for dynamic instability, making a total of 40 installed trays. (Although texts frequently suggest that reboilers and condensers be counted as a theoretical stage, this is, strictly speaking, true only of kettle-type reboilers and partial condensers. Generally, the safest approach in a column design is to ignore both the reboiler and condenser when counting equilibrium stages.)

There may be a drop in efficiency for very large diameter columns (>12 ft) due to their size, unless special jets or baffles are provided to ensure an even flow pattern.

As pointed out by Humphrey and Keller in *Separation Process Technology* (McGraw-Hill, 1997), a rate-based design method for distillation (and other staged processes) has been developed. Component material and energy balances for each phase, together with mass and energy transfer-rate equations, as well as equilibrium equations for the phase interface, are solved to determine the actual separation directly. Calculations take place on an incremental basis as the designer proceeds through the column. The uncertainties of computations that use average tray efficiencies, based on individual components, are entirely avoided.

8.4 PACKED-COLUMN DESIGN

Specify a packing and the column dimensions for a distillation column separating ethyl benzene and styrene at 1200 mmHg (23.21 psia). The separation requires 30 theoretical stages. Vapor flow is 12,000 lb/h (5455 kg/h), average vapor density is 0.3 lb/ft^3 (4.8 kg/m^3), liquid flow is 10,000 lb/h (4545 kg/h), and average liquid density is 52 lb/ft^3 (833 kg/m^3). Liquid kinematic viscosity is 0.48 cSt (4.8×10^{-7} m^2/s).

Calculation Procedure

1. *Select a type, arrangement, and size of packing.*
For this particular system, if one is dealing with a new column, the use of random (dumped) packing is probably a better economic choice than a systematically packed column. Although there is no clear line of demarcation between the two, the latter type is generally favored for very low pressure operations and for expanding the capacity of an existing column.

Somewhat arbitrarily, let us choose a metal slotted ring, say Hy-Pak, as the packing type to be used for this service. Hy-Pak is the Norton Company (Akron, Ohio) version of a slotted ring. There are other packing devices, such as metal saddles and half or pyramidal Pall rings, that would be equally suitable for this service.

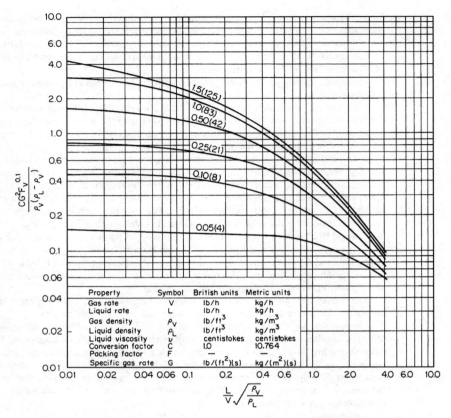

FIGURE 8.18 Generalized pressure-drop correlation for packed towers. Note: Parameter of curves is pressure drop in inches of water per foot. Numbers in parentheses are millimeters of water per meter of packed height.

A packing size should be selected so that the column-diameter-to-packing-size ratio is greater than 30 for Raschig rings, 15 for ceramic saddles, and 10 for slotted rings or plastic saddles. When dealing with distillation columns larger than 24 in (0.6 m) in diameter, a 2-in, or no. 2, packing should probably be given the first consideration. Assume (essentially based on trial and error) that this column will be larger than 24 in. Therefore, select a 2-in size for the packing.

2. Determine the column diameter.

The generally accepted design procedures for sizing randomly packed columns are modifications of the Sherwood correlation. A widely applied version is that developed by the Norton Company. It has been adapted slightly for this text to permit its application to low-pressure systems.

Tower diameter is primarily a function of throughput rate and packing configuration. A specific design gas rate G can be determined from Fig. 8.18 if liquid and vapor flows are known and a suitable packing and the proper pressure drop have been selected. Recommended design pressure drops are as follows: 0.4 to 0.75 in water per foot of packing (32 to 63 mm water per meter) for moderate- and high-pressure distillation; 0.1 to 0.2 in water per foot of packing (8 to 16 mm water per meter) for vacuum distillation; and 0.2 to 0.6 in water per foot of packing (16 to 48 mm per meter) for absorbers and strippers.

TABLE 8.2   Packing Factors for Column Packings (1 in = 0.0254 m)

Packing type	Material	Nominal packing size, in										
		$1/4$	$3/8$	$1/2$	$5/8$	$3/4$	1 or no. 1	$1^1/4$	$1^1/2$	2 or no.2	3	$3^1/2$ or no.3
Hy–Pak	Metal						43			18		15
Super Intalox saddles	Ceramic						60			30		
Super Intalox saddles	Plastic						33			21		16
Pall rings	Plastic				97		52		40	24		16
Pall rings	Metal				70		48		33	20		16
Intalox saddles	Ceramic	725	330	200		145	92		52	40	22	
Raschig rings	Ceramic	1600	1000	580	380	255	155	125	95	65	37	
Raschig rings	Metal, $1/32$ in	700	390	300	170	155	115					
Raschig rings	Metal, $1/16$ in			410	290	220	137	110	83	57	32	
Berl saddles	Ceramic	900		240		170	110		65	45		
Tellerettes	Plastic						38			19		
Mas Pac	Plastic									32		20
Quartz rock										160		
Cross partition	Ceramic										80	
Flexipac	Metal						33			22		16
Interlox	Metal						41			27		18
Chempak	Metal						29					

Note: Many of the values are those listed in vendors' literature and are frequently based solely on pilot tests. It may be prudent on occasion to assume slightly larger packing factors in order to represent more precisely the pressure drop of newly marketed packing in commercial columns. Consult with vendors for data on newer, proprietary packings.

First, evaluate the abscissa for Fig. 8.18. Thus, $(L/V)(\rho_V/\rho_L)^{0.5} = (10,000/12,000)(0.3/52)^{0.5} = 0.063$. (Note: ρ_V, especially for vacuum columns, should be determined for the top of the bed, because this is where the density is lowest and the vapor velocity highest.)

Next, select a pressure drop of 0.15 in/ft of packing, and on the figure read 0.55 as the ordinate. Thus, $0.55 = CG^2 Fv^{0.1}/[\rho_V(\rho_L - \rho_V)]$. From Table 8.2, F is 18. Solving for G, $G = [0.55(0.3)(52 - 0.3)/1(18)(0.48)^{0.1}]^{0.5} = 0.71$ lb/(ft^2)(s).

Accordingly, the required column cross-sectional area is $(12,000$ lb/h$)/[(3600$ s/h$)0.71$ lb/(ft^2)(s)$] = 4.7$ ft^2. Finally, column diameter is $[(4.7$ ft$^2)/(\pi/4)]^{0.5} = 2.45$ ft; say, 2.5 ft (0.75 m). (Thus the initial assumption in step 1 that column diameter would be greater than 24 in is valid.)

3. Determine the column height.

To translate the 30 required theoretical stages into an actual column height, use the height equivalent to a theoretical plate (HETP) concept. HETP values are remarkably constant for a large number of organic and inorganic systems.

As long as dumped packing in *commercial* columns is properly wetted [more than 1000 lb/(h)(ft^2) or 5000 kg/(h)(m^2)], the following HETP values will result in a workable column:

Nominal packing size, for slotted rings or Intalox saddles	HETP
1 in (or no. 1)	1.5 ft (0.46 m)
$1^1/2$	2.2 ft (0.67 m)
2 in (or no. 2)	3.0 ft (0.91 m)

Because the irrigation rate in vacuum columns often falls below 1000 lb/(h)(ft^2), it may be wise to add another 6 in to the listed HETP values as a safety factor. For the ethylbenzene/styrene system in this example, the specific irrigation rate is $(10{,}000 \text{ lb/h})/4.7 \text{ ft}^2 = 2128 \text{ lb/(h)(ft}^2)$, which is well above the minimum required for good wetting. Total packing height is $30(3.0) = 90$ ft.

Since liquid maldistribution may become a problem unless the flow is periodically redistributed in a tall column, the total packing height must be broken up into a number of individual beds:

Packing	Maximum bed height
Raschig rings	2.5 to 3.0 bed diameters
Ceramic saddles	5 to 8 bed diameters
Slotted rings and plastic saddles	5 to 10 bed diameters

With 2-in slotted ring packing in a 2.5-ft (0.75-m) column, each bed should be no higher than 25 ft (7.6 m). Therefore, for the given system, four 23-ft (7-m) beds would be appropriate.

Related Calculations.   With ceramic packing, the height of a single bed is, for structural reasons, frequently restricted to no more than 20 ft (≈ 6 m).

In step 1, the column-diameter-to-packing-size ratio for Raschig rings can be less than 30 for scrubbing applications in which the liquid-irrigation rate is high and the column is operated at above 70 percent of flooding.

If it is desirable to have systematically packed column internals, then the choice is from among various types of mesh pads, open-grid configurations, springs, spirals, and corrugated elements. Since no generalized design correlations can be applied to all configurations, it will be necessary to have the respective manufacturers develop the final packing design.

The question arises whether trays or packing should be selected for a given distillation task. Humphrey and Keller (see reference at end of Procedure 8.3) cite the following factors as favoring trays: high liquid rate (occurs when high column pressures are involved); large column diameter (because packing is prone to maldistribution); complex columns with multiple feeds and takeoffs; and variation in feed composition. Furthermore, scaleup is less risky with trays, and trayed columns weigh less than packed columns. Conversely, the following factors favor packings: vacuum conditions; a need for low pressure drop; corrosive systems; foaming systems; and systems with low liquid holdup.

8.5 BATCH DISTILLATION

Establish the separation capability of a single-stage (differential) batch still processing a mixture of two compounds having a relative volatility of 4.0. At the start of the batch separation, there are 600 mol of the more-volatile compound A and 400 mol of compound B in the kettle. When the remaining charge in the kettle is 80 percent B, how much total material has been boiled off, and what is the composition of the accumulated distillate?

Calculation Procedure

1. Calculate the amount of material boiled off.
Use the integrated form of the Rayleigh equation

$$\ln \frac{L_1}{L_2} = \frac{1}{\alpha - 1}\left(\ln \frac{x_1}{x_2} + \alpha \ln \frac{1 - x_2}{1 - x_1} \right)$$

where L_1 is the amount of initial liquid in the kettle (in moles), L_2 is the amount of final residual liquid (in moles), α is the relative volatility (4.0), x_1 is the initial mole fraction of more-volatile compound in the kettle (0.6), and x_2 is the final mole fraction of more-volatile component in the kettle (0.2). Then,

$$\ln\frac{L_1}{L_2} = \frac{1}{4-1}\left(\ln\frac{0.6}{0.2} + 4\ln\frac{1-0.2}{1-0.6}\right) = 1.29$$

So, $L_1/L_2 = 3.63$, or $L_2 = 1000/3.63 = 275$ mol. Therefore, the amount of material distilled over is $1000 - 275 = 725$ mol.

2. Calculate the distillate composition.
Compound A in the initial charge consisted of 600 mol. Compound A in the residue amounts to $275(0.2) = 55$ mol. Therefore, compound A in the distillate amounts to $600 - 55 = 545$ mol. Therefore, mole fraction A in the distillate is $^{545}/_{725} = 0.75$, and mole fraction B is $1 - 0.75 = 0.25$.

Related Calculations.   Since a simple batch kettle provides only a single theoretical stage, it is impossible to achieve any reasonable separation unless the magnitude of the relative volatility approaches infinity. This is the case with the removal of very light components of a mixture, particularly of heavy residues. In this example, even with a comfortable relative volatility of 4, it was only possible to increase the concentration of A from 60 to 75% and to strip the residue to 20%.

Obviously, the more distillate that is boiled over, the lower will be the separation efficiency. Conversely, higher overhead concentrations can be obtained at the expense of a larger loss of compound A in the residue. For instance, if distillation in this example is stopped after 50 percent of the charge has been boiled off, the final concentration of A in the kettle x_2 can be obtained as follows:

$$\ln\frac{1000}{500} = \frac{1}{4-1}\left(\ln\frac{0.6}{x_2} + 4\ln\frac{1-x_2}{1-0.6}\right)$$

from which $x_2 = 0.395$ (by trial-and-error calculation). The amount of A remaining in the residue is 198 mol, making a 40% concentration, and the amount of A in the distillate is 402 mol, making an 80% concentration.

8.6 *BATCH-COLUMN DESIGN*

Estimate the required size of a batch still, with vapor rectification, to recover a dye intermediate from its coproduct and some low- and high-boiling impurities. It has been specified that 13,000 lb (5900 kg), consisting of fresh reactor product and recycled "slop" cuts, must be processed per batch.

The initial composition in the kettle is as follows:

Low-boiling impurities: 500 lb (227 kg)

Dye intermediate: 5500 lb (2500 kg)

Coproduct: 5000 lb (2273 kg)

High-boiling impurities: 2000 lb (909 kg)

Tests in laboratory columns have indicated that to ensure adequate removal of the low-boiling impurities, 500 lb (227 kg) of the dye intermediate is lost in the low boiler's cut. Similarly, high boilers remaining in the kettle at the end of the distillation will retain 500 lb (227 kg) of the coproduct. This leaves 9500 lb (4320 kg) of the two recoverable products. The specification for the dye intermediate requires a concentration of less than 0.5 mol % of the coproduct. Of the two, the coproduct has the higher boiling point.

The conditions for the separation of the dye intermediate from the coproduct are as follows:

Relative volatility $\alpha = 2$

Molecular weight of dye intermediate = 80

Molecular weight of coproduct = 100

Average liquid density = 62.0 lb/ft^3 (993 kg/m^3)

Column pressure (top) = 350 mmHg (46.6 kPa)

Column temperature (top) = 185°F (358 K)

Calculation Procedure

1. *Assess the applicability of a design not based on computer analyses.*
The precise design of a batch still is extremely complex because of the transient behavior of the column. Not only do compositions change continuously during the rectification of a charge, but successive batches may start with varying compositions as "slop" cuts and heels are recycled. Only sophisticated software programs can optimize the size, collection time, and reflux ratios for each cut. In addition to the recycle streams, these programs must also take into account nonideal equilibrium (where applicable) and the effect of holdup on trays. The following is not a detailed design for batch rectification, but instead an outline of how to estimate a "workable" facility with reasonable assurance that it will do the desired job.

Before continuing with a step-by-step procedure, consider these rules of thumb for batch stills:

1. Too low a reflux ratio cannot produce the required product specification no matter how many trays are installed. Conversely, even infinite reflux will not be sufficient if an inadequate number of equilibrium stages has been provided.

2. For optimum separation efficiency, reflux holdup should be minimized by eliminating surge drums and using flow splitters that retain little or no liquid.

3. Too little or too much holdup in the column is detrimental to separation efficiency. A reasonable amount provides a flywheel effect that dampens the effects of equilibrium-condition fluctuations; too much, especially at higher reflux ratios, makes it difficult to achieve good purity levels. A holdup equivalent to 10 or 15 percent of the initial batch charge is recommended.

4. Since the column consists solely of a rectifying section, there is a limit to how many trays can be profitably installed. The system will "pinch" regardless of stages once the low-boiler concentration in the reboiler approaches the intersection of the operating line with the equilibrium curve.

5. Once a workable column has been installed, capacity to produce at a given rate and product specification is only minimally affected by changes in reflux ratio or length of a cut.

6. As the more-volatile component is being removed from the reboiler, separation becomes progressively more difficult.

7. It is impossible to recover in a single operation, at high purity, a low-boiling component that represents only a small fraction of the initial charge.

2. *Set up an estimated batch-processing time schedule.*
A reasonable time schedule is as follows:

Charge new batch into kettle; heat up charge: 3 h

Run column at total reflux to stabilize concentration; distill off the low-boilers cut: 6 h

Recover (i.e., distill off) the dye intermediate while increasing the reflux ratio one or two times: 6 h

Distill off center (slop) cut while further increasing reflux ratio one or two times: 6 h

Distill off coproduct from its mixture with the high boilers while again increasing reflux ratio once (if necessary): 4 h

Drain the high-boiling residue and get ready for next batch; recycle the center cut and prepare to charge fresh feed: 3 h

Total elapsed time: 28 h

3. Estimate the number of theoretical trays needed to recover the dye intermediate.
Enter Fig. 8.19 along its ordinate at a relative volatility of 2. For product purity of 99.5 percent, the graph shows that 11 stages are needed.

4. Draw the relevant xy diagram (equilibrium curve).
The relevant diagram is one that pertains to a relative volatility α of 2. As indicated in Procedure 8.1, it can be plotted from the equation $y = \alpha x/[1 + (\alpha - 1)x]$. This is the uppermost curve in Fig. 8.20

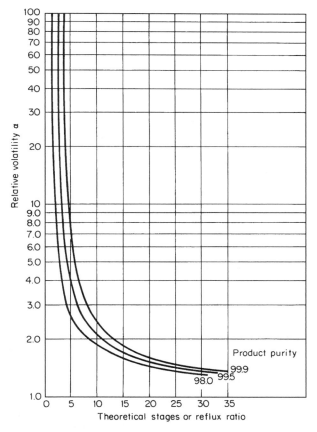

FIGURE 8.19   Estimate of theoretical stages, or of reflux ratio, for batch distillation.

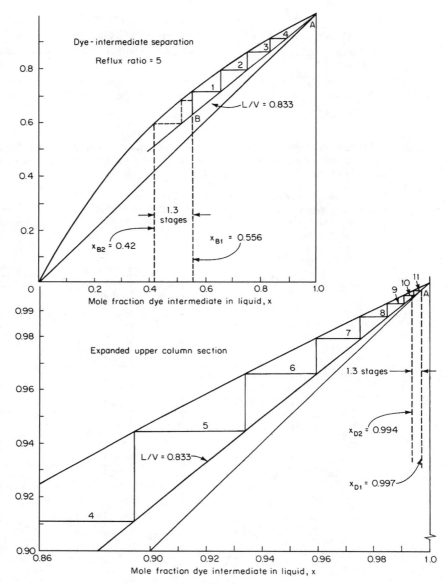

FIGURE 8.20 McCabe-Thiele diagram for Procedure 8.6.

(including both the full portion at the top of the diagram and the expanded upper-column section at the bottom).

5. *Applying the McCabe-Thiele principles, position an operating line on the xy diagram that will accommodate 11 stages between the initial kettle composition and the distillate composition that pertain to the time-schedule step in which the dye intermediate is recovered.*

The McCabe-Thiele principles applied here are outlined in more detail in Procedure 8.1. Using the dye intermediate and the coproduct as the two key components, the initial kettle composition after removal of the low boilers cut (based on the key components only) is $(5000/80)/[(5000/80) + (5000/100)] = 0.556$ mole fraction, or 55.6 mol %, dye intermediate and $(100 - 55.6) = 44.4$ mol % coproduct. The abscissa corresponding to this composition x_{B1} marks one end (the "feed" end) of the 11-stage separation; the other end (the overhead-product end) is marked by the abscissa x_{D1} that corresponds to the required dye-intermediate purity, 99.5%.

By trial-and-error positioning, it is found that an operating line having a slope of 0.833 will accommodate 11 stages (the overhead product corresponding to this line is very slightly purer than 99.5%, namely, 99.7%). In Fig. 8.20, the operating line is line AB. By definition (see Procedure 8.1), its 0.833 slope establishes (and is equal to) the ratio L/V of descending liquid to rising vapor. This in turn establishes the reflux ratio L/D, or $L/(V - L)$, where D is the amount of overhead product taken. Since $L/V = 0.833$, then $L/D = 0.833V/(V - 0.833V) = 5$. Thus, during this period, the column must be operated at a reflux ratio of 5.

6. Determine the effect of elapsed time on overhead-product concentration.

It is reasonable to assume that about 35 to 50 percent of the dye intermediate can be removed efficiently at the initial, relatively low reflux ratio of 5. In this case, assume that 2100 lb of the 5000 lb is thus removed. It is also reasonable to assume that this takes place during half of the 6 h allotted in step 2 for recovering this intermediate. (Note that the shorter the allotted time, the bigger the required column diameter.)

Therefore, $D = 2100/3 = 700$ lb/h. Because $L/D = 5$, $L = 3500$ lb/h. And since $D = V - L$, $V = 4200$ lb/h. After 3 h (assuming negligible column holdup), the kettle will contain $(5000 - 2100)/80 = 36.25$ mol dye intermediate. Since the overhead stream contains (initially, at least) only 0.3% coproduct, the amount of coproduct in the kettle after 3 h is about $[5000 - 0.005(2100)]/100 = 49.9$ mol. The mole percentages thus are 42% intermediate and 58% coproduct. If the abscissa corresponding to this new kettle composition x_{B2} is extended upward to the operating line and equilibrium curve, it can be seen that this "feed" composition has shifted downward by 1.3 stages. This will likewise lower the overhead (distillate) composition by 1.3 stages, which means that the concentration of dye intermediate drops to 99.4% after 2100 lb of distillate has boiled over and been collected. Average concentration during this period of operating at a reflux ratio of 5 is thus $(99.7 + 99.4)/2 = 99.55$ percent, just slightly above the specified purity.

7. Adjust the reflux ratio so as to maintain the required overhead-product composition.

It is necessary to raise the reflux ratio in order to keep the concentration of dye intermediate in the overhead product high enough. In this case, assume that an additional 1400 lb intermediate can be recovered in the remaining 3 h while maintaining the original vapor rate of 4200 lb/h. Then D for this latter 3 h is $1400/3 = 467$ lb/h; L is $(4200 - 467) = 3733$ lb/h; and the reflux ratio L/D for this portion of operation must be $3733/467 = 8$.

The slope L/V of the new operating line is $3733/4200 = 0.89$. Drawing this line on the replotted xy diagram in Fig. 8.21 and stepping off 11 stages, we find that the initial overhead concentration x_{D3} while operating at the new reflux ratio is 99.7% dye intermediate. (Note that x_{B3} in Fig. 8.21 is a deliberate repetition of x_{B2} in Fig. 8.20, whereas x_{D3} is not a deliberate repetition but instead a coincidental result that emerges from the stepping off of the stages.)

8. Again determine the effect of elapsed time on overhead-product composition and readjust the reflux ratio of step 7 if necessary.

It follows from the preceding two steps that after the second 3 h, the kettle will contain $(2900 - 1400)/80 = 18.75$ mol dye intermediate and about $[5000 - 0.005(3500)]/100 = 49.8$ mol coproduct. Thus the mole percentages will be 27% intermediate and 73% coproduct. The abscissa x_{B4} corresponding to this in Fig. 8.21 indicates that the composition will have shifted downward about 1.6 stages. This in turn means that the distillate composition will have dropped 1.6 stages during the 3 h, to a composition x_{D4} of 99.2% dye intermediate. Average concentration while operating at a reflux ratio of 8 is therefore $(99.7 + 99.2)/2 = 99.45$ percent.

Since this average concentration is slightly below the required purity of 99.5%, it would be wise to recalculate step 7 at a higher reflux ratio with a correspondingly longer time for removing the same 1400 lb. A 4-h removal time, corresponding to a reflux ratio of 11, should be more than adequate.

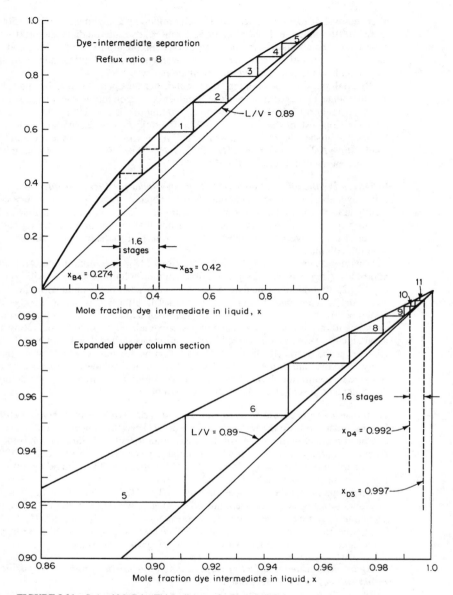

FIGURE 8.21 Second McCabe-Thiele diagram for Procedure 8.6.

9. *Determine the column diameter.*

The column diameter can be determined by taking the boil-up rate established for the separation of components having the smallest relative volatility (which, in this case, is given to be the dye-intermediate/coproduct separation) and using the design procedures outlined in Procedures 8.2 and 8.4.

For a 4200 lb/h boil-up rate (from step 6) of dye intermediate having a density ρ_V of (80 lb/mol) $(6.77 \text{ lb/in}^2)/[10.73(460 + 185R)] = 0.078 \text{ lb/ft}^3$ (where 10.73 is the gas constant), the total vapor flow rate is $4200/[0.078(3600)] = 14.9 \text{ ft}^3/\text{s}$. A tray spacing of 18 in seems appropriate. The column diameter required for an 18-in tray spacing can be obtained with the aid of Fig. 8.6, where F_c is found to be about 1.4 $[\text{lb}/(\text{ft} \cdot s^2)]^{0.5}$. Then free tray area $A_F = W/(F_c\rho_V^{0.5}) = (4200 \text{ lb/h})/[3600 \text{ s/h}]$ $(1.4)(0.078 \text{ lb/ft}^3)^{0.5}] = 2.98 \text{ ft}^2$. Allowing 5 percent of the tray area for segmental downcomers, total tray area is $2.98/0.95 = 3.14 \text{ ft}^2$. Finally, column diameter is $[3.14/(\pi/4)]^{0.5} = 2$ ft (0.41 m).

10. *Determine the column height.*
Efficiencies of batch columns vary greatly, since the concentration profile in the column shifts over a wide range. For appreciable intervals, separation may of necessity take place under pinched conditions, conducive to low efficiencies. So an overall column efficiency of 50 percent is not unreasonable. Thus the total number of trays provided is (11 stages)/ 0.5 = 22 trays. At a spacing of 18 in, the required column height is 33 ft (10 m). The trays should be cartridge-type trays, because normal trays cannot be readily installed in a 2-ft column.

Related Calculations. Alternatively, a packed column can be considered for this separation. Its design should follow the procedure outlined in Procedure 8.4.

If the coproduct must also be recovered at a reasonably high purity, then steps 5 through 8 should also be repeated for distilling off the slop cut (essentially a mixture of dye intermediate and coproduct) and for the coproduct cut (a mixture of coproduct and high-boiling residue).

In the final analysis, processing time is the main criterion of batch-still design. To achieve optimal cost-effective performance requires a large number of trial calculations, such that the best combination of equilibrium stages, reflux ratio, batch size, and batch-processing time can be established. It is extremely difficult to successfully carry out such a procedure by hand calculation.

8.7 OVERALL COLUMN SELECTION AND DESIGN

Select and specify an efficient distillation column to separate ethylbenzene (EB) and ethyl cyclohexane (ECH), and develop the appropriate heat and material balances. Feed rate is 10 lb · mol/h (4.54 kg · mol/h): 75 mol % ECH and 25 mol % EB. Concentration of EB in the overhead product must be less than 0.1%; concentration of ECH in the bottoms stream must be less than 5%. Feed is at ambient temperature (25°C). Specific heat of the distillate and bottoms streams can be taken as 0.39 and 0.45 Btu/(lb)(°F) [1.63 and 1.89 kJ/(kg)(K)], respectively. The normal boiling points for EB and ECH are 136.19°C and 131.78°C, respectively; their latent heats of vaporization can be taken as 153 and 147 Btu/lb (356 and 342 kJ/kg), respectively. It can be assumed that EB and ECH form an ideal mixture.

Calculation Procedure

1. *Make a rough estimate of the number of theoretical stages required.*
This step employs the procedures developed in Procedure 8.1. Use the Fenske-Underwood-Gilliland approach rather than the McCabe-Thiele, because the small boiling-point difference indicates that a large number of stages will be needed.

The appropriate vapor pressures can be obtained from the Antoine equation:

$$\log P = A - \frac{B}{t + C}$$

where P is vapor pressure (in mmHg), t is temperature (in °C), and the other letters are the Antoine constants, which in this case are

	A	B	C
Ethylbenzene (EB)	6.96	1424	213
Ethyl cyclohexane (ECH)	6.87	1384	215

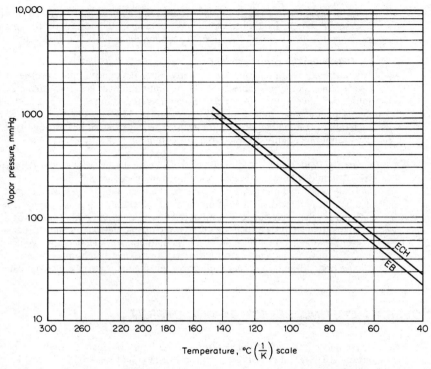

FIGURE 8.22 Vapor pressures of ethylbenzene and ethyl cyclohexane (Procedure 8.7). Note: 1 mmHg = 0.133 kPa.

The resulting vapor-pressure plots are shown in Fig. 8.22.

Since the system is ideal, relative volatility α can be determined from the vapor-pressure ratio P_{ECH}/P_{EB}. If this is determined at, say, 50, 100, and 135°C, the effect of temperature (and thus, of system pressure) on relative volatility can be gaged as follows:

Temperature, °C (°F)	P_{EB}, mmHg	P_{ECH}, mmHg	α
50 (122)	35.1	44.4	1.265
100 (212)	257	300	1.167
135 (275)	738	824	1.117

There is a noticeable increase in the relative volatility as the system temperature and pressure are lowered. In the case of two closely boiling liquids, as is the present case, this can mean a substantial lessening of required equilibrium stages. This can be seen by employing the Fenske-Underwood-Gilliland correlations as expressed graphically in Fig. 8.5. The abscissa is (0.999/0.001)/(0.95/0.05) = 19,000, and assuming an R/R_{min} of 1.3, an α of 1.265 calls for 86 theoretical stages (as read along the ordinate), whereas an α of 1.117 calls for 176 stages (by extrapolating the ordinate downward). However, this comparison assumes an isobaric column. In actual practice, the bottoms pressures will be considerably higher, so the difference in stages, although substantial, will not be quite so dramatic. Depending on column pressures, it seems reasonable to assume that roughly 100 stages will be needed.

2. Evaluate the effects of choosing trays, random packing, or systematic packing.
It is recommended that the top column pressure be set for the lowest reasonable overhead pressure consistent with the use of a water- or air-cooled condenser. In this case, pressure would be 50 mmHg, corresponding to a temperature of 53°C (127°F).

The bottoms pressure is usually selected to permit use of a readily available heating medium (steam or hot oil), as well as to stay below a temperature that could cause product degradation. In the ECH-EB system, degradation is not considered a problem, and column bottoms pressure is solely a function of the pressure drop across the tower internals. Because, as seen in step 1, relative volatility can vary appreciably with pressure, it is advantageous in this case to install low-pressure-drop, high-efficiency tower internals.

a. Evaluation of cross-flow trays. Assume as a first try that 110 equilibrium stages are needed. Assume further that a tray column in this service will operate at 70 percent efficiency. Then the actual number of trays needed is 110/0.7 = 157.

It is reasonable to allow a pressure drop of 3 mmHg per tray. Then the reboiler pressure will be 50 + 3(157) = 521 mmHg. At this pressure, the relative volatility (from Fig. 8.22) is 1.14, and the average relative volatility in the column is then (see step 1) (1.265 + 1.14)/2 = 1.2. From Fig. 8.5, the estimated number of equilibrium stages is 107, which confirms that the initially selected 110 trays was a reasonable assumption.

If the upper 25 percent of the trays (where the vapor velocity and therefore entrainment are highest) are spaced 24 in apart and the distance between the remaining trays is 18 in, then the total equipment height can be estimated as follows:

	ft (m)
Total height of trays: 40(2 ft) + 117(1.5 ft) =	256
Overhead disengaging area	4
Manways: 8(1.5 ft) =	12
Reboiler	6
Total column height	278
Skirt (minimum to ensure adequate pump suction head)	12
Total equipment height	290 (90)

b. Evaluation of random packing. This step draws on information brought out in Procedure 8.4. Again, recognizing that relative volatility decreases with pressure, estimate the actual number of equilibrium stages and check the assumption from Fig. 8.5. If no. 2 slotted rings are selected, the pressure drop should be set at 0.15 in water per foot (12 mm water per meter), and the height equivalent to a theoretical-plate (or stage) HETP may be as large as 3.5 ft (*ca.* 1 m). For dumped packing, assume that the specified separation requires 100 stages. Then the total height of packing needed is 100(3.5 ft) = 350 ft, requiring at least 10 separate beds.

The pressure in the column reboiler can now be developed:

	mmHg
Pressure drop through packing (no. 2 slotted rings): (350 ft)(0.15 in water per foot) = 52.5 in water =	98
Pressure drop through 10 support plates	10
Pressure drop through 10 distributors	10
Column top pressure	50
Total reboiler pressure	168

Again using Fig. 8.22, the relative volatility at the bottom is found to be 1.16, resulting in an average of 1.21. Again from Fig. 8.5 we find that the assumption of 100 stages is reasonable.

Now calculate the column height:

	ft (m)
Total height of packing	350
Spacing between beds: (9)(4 ft) =	36
Disengaging and distribution	6
Reboiler	6
Total column height	398
Skirt	12
Total equipment height	410(125)

c. Evaluation of a systematically packed column. One example of a high-efficiency packing (large number of stages per unit of pressure drop) is corrugated-wire-gauze elements, sections of which are assembled inside the tower. This type of packing has an HETP of 10 to 12 in (25 to 30 cm) and a pressure drop of 0.3 to 0.5 mmHg (53 Pa or 0.22 in water) per equilibrium stage. Assume that 95 stages are needed, each entailing 12 in of packing. Then the pressure at the bottom of 95 ft of packing can be determined as follows:

	mmHg
Pressure drop through wire-gauze packing: (95)(0.4 mmHg/stage) =	38
Pressure drop through four distributors and four packing supports	8
Column top pressure	50
Total reboiler pressure	96

The average relative volatility for the column is 1.22, and the assumption of 95 plates is reasonable.

The next step is to calculate the column height:

	ft (m)
Total height of packing	95
Spacing between beds: (3)(4 ft) =	12
Disengaging and distribution	6
Reboiler	6
Total column height	119
Skirt	12
Total equipment height	131(40)

3. *Choose between the tray column, the randomly packed column, and the systematically packed column.*

It is unlikely that a single 300-ft tray column or a 400-ft randomly packed column would be installed for this system. In addition, multiple-column operation, by its nature, is expensive. So it is reasonable to opt for the 131-ft column containing systematically packed corrugated-wire-gauze packing. However, this type of packing is quite expensive. So a detailed economic analysis should be performed before a final decision is made.

4. *Estimate the column diameter.*

This step draws on information brought out in Procedure 8.2. If corrugated-wire-gauze packing is the final choice, the column size can be estimated by applying an appropriate F_c factor. The usual range for this packing is 1.6 to 1.9 $[lb/(ft)(s^2)]^{0.5}$, with 1.7 being a good design point.

In order to establish the column vapor rate, it is necessary first to determine the required reflux ratio and then to set up a column material balance.

Using the Underwood equation (see Procedure 8.1) and noting from Fig. 8.3 that $\theta = 1.08$,

$$(L/D)_{min} = 1.22(0.999)/(1.22 - 1.08) + 1(0.001)/(1 - 1.08) - 1$$
$$= 7.69$$

With an arbitrarily selected R/R_{min} ratio of 1.3, then, the reflux ratio $L/D = 7.69(1.3) = 10$.

Material-balance calculations (see Section 2) indicate that the withdrawal rate for overhead product from the system is 7.38 mol/h; therefore, 73.8 mol/h must be refluxed, and the vapor flow from the top of the column must be $73.8 + 7.38 = 81.18$ mol/h. The vapor consists of almost pure ECH, whose molecular weight is 112, so mass flow from the top is $81.18(112) = 9092$ lb/h (4133 kg/h). The bottoms stream is 2.62 mol/h.

Take the average pressure in the upper part of the column to be 55 mmHg (1.06 psia). Assuming that this consists essentially of ECH and referring to Fig. 8.22, the corresponding temperature is 55°C (131°F). Then, by the ideal-gas law, vapor density $\rho_V = PM/RT = 1.06(112)/[10.73(460 + 131)] = 0.0187$ lb/ft^3, where P is pressure, M is molecular weight, R is the gas-law constant, and T is absolute temperature. (Note: For high-pressure systems of greater than 5 atm, correlations for vapor-phase nonideality should be used instead; see Section 3.) Then, as discussed in Procedure 8.2, design vapor velocity $V = F_c/\rho^{0.5} = 1.7/0.0187^{0.5} = 12.43$ ft/s. The volumetric flow rate is (9092 lb/h)/ $[(3600 \text{ s/h})(0.0187 \text{ lb/ft}^3)] = 135$ ft^3/s. Column cross-sectional area A, then, must be $135/12.43 = 10.9$ ft^2, and column diameter must be $[A/(\pi/4)]^{0.5} = (10.9/0.785)^{0.5} = 3.73$ ft(1.1 m).

5. *Calculate the heat duty of the reboiler and of the condenser.*

Assembly of a heat balance can be simplified by assuming constant molal overflow (not unreasonable in this case) and no subcooling in the condenser (usually, less than 5°C of cooling takes place in a well-designed heat exchanger). For constant molal overflow, the latent heats may be averaged, yielding 150 Btu/lb.

The reboiler duly is the sum of three parts:

1. Enthalpy to heat distillate (consisting essentially of ECH) from 25 to 53°C:

$$(7.38 \text{ lb} \cdot \text{mol/h})(112 \text{ lb/mol}) \times (53 - 25°C)(1.8°F/°C)[0.39 \text{ Btu}/(lb)(°F)] = 16,240 \text{ Btu/h}$$

2. Enthalpy to heat bottoms (essentially EB) from 25 to 72°C:

$$(2.62 \text{ lb} \cdot \text{mol/h})(106 \text{ lb/mol}) \times (72 - 25°C)(1.8°F/°C)[0.45 \text{ Btu}/(lb)(°F)] = 10,570 \text{ Btu/h}$$

3. Enthalpy to boil up the vapor:

$$(81.18 \text{ lb} \cdot \text{mol/h})(106 \text{ lb/mol})(153 \text{ Btu/lb}) = 1,316,580 \text{ Btu/h}$$

Therefore, total reboiler duty is $16,240 + 10,570 + 1,316,580 = 1,343,390$ Btu/h (393,310 W). As for the condenser duty, it is $(81.18 \text{ lb} \cdot \text{mol/h})(112 \text{ lb/mol})(147 \text{ Btu/lb}) = 1,336,550$ Btu/h (391,350 W).

Related Calculations. Sequences of distillation columns are often used for separating multicomponent mixtures. The question arises: In what order should the individual components be separated? For example, should a given component be separated in the first column, the second column, the *n*th column, or the last column? Should it be removed as the overhead or as the bottoms? Each case should be examined on its own merits, but a variety of heuristics have evolved. For example (based on Douglas, *Conceptual Design of Chemical Processes*, McGraw-Hill): Remove corrosive components as soon as possible; remove reactive components as soon as possible; remove desired products as distillates, not bottoms; remove the lightest components first; make the high-recovery separations last; make the difficult separations last; favor splits that are equimolar. Obviously, however, reliance on multiple heuristics can lead to contradictory results.

8.8 SIZING RUPTURE DISKS FOR GASES AND LIQUIDS

What diameter rupture disk is required to relieve 50,000 lb/h (6.3 kg/s) of hydrogen to the atmosphere from a pressure of 80 lb/in^2 (gage) (551.5 kPa)? Determine the diameter of a rupture disk required to relieve 100 gal/min (6.3 L/s) of a liquid having a specific gravity of 0.9 from 200 lb/in^2 (gage) (1378.8 kPa) to atmosphere.

Calculation Procedure

1. Determine the rupture disk diameter for the gas.
For a gas, use the relation $d = (W/146P)^{0.5} (1/Mw)^{0.25}$, where d = minimum rupture-disk diameter, in; W = relieving capacity, lb/h; P = relieving pressure, lb/in^2 (abs); Mw = molecular weight of gas being relieved. By substituting, $d = [50,000/146(94.7)]^{0.5} (1/2)^{0.25} = 1.60$ in (4.1 cm).

2. Find the rupture-disk diameter for the liquid.
Use the relation $d = 0.236(Q)^{0.5}(Sp)^{0.25}/P^{0.25}$, where the symbols are the same as in step 1 except that Q = relieving capacity, gal/min; Sp = liquid specific gravity. So $d = 0.236(100)^{0.5}(0.0)^{0.25}/(214.7)^{0.25} = 0.60$ in (1.52 cm).

Related Calculations. Rupture disks are used in a variety of applications—process, chemical, power, petrochemical, and marine plants. These disks protect pressure vessels from pressure surges and are used to separate safety and relief valves from process fluids of various types.

Pressure-vessel codes give precise rules for installing rupture disks. Most manufacturers will guarantee rupture disks as they size according to the capacities and operating conditions set forth in a purchase requisition or specification.

Designers, however, often must know the needed size of a rupture disk long before bids are received from a manufacturer so the designer can specify vessel nozzles, plan piping, etc.

The equations given in this procedure are based on standard disk sizing computations. They provide a quick way of making a preliminary estimate of rupture-disk diameter for any gas or liquid whose properties are known. The procedure is the work of V. Ganapathy, Bharat Heavy Electricals Ltd., as reported in *Chemical Engineering* magazine.

SECTION 9
EXTRACTION AND LEACHING

Frank H. Verhoff, Ph.D.
Vice President
USTech
Cincinnati, OH

9.1 MULTISTAGE COUNTERCURRENT LIQUID-LIQUID EXTRACTION

Alcohol is to be extracted from an aqueous solution by pure ether in an extraction column. The alcohol solution, containing 30% alcohol by weight, enters the top of the column at a rate of 370 kg/h. The ether is to be fed to the column bottom at 350 kg/h. About 90% of the alcohol is to be extracted; that is, alcohol concentration in the exiting aqueous stream should be about 3%. Experimental data on the compositions of pairs of water-rich and ether-rich phases in equilibrium are given in Table 9.1. Calculate the flow rates and compositions of the exiting raffinate (i.e., alcohol-depleted, aqueous) phase and the extract (i.e., alcohol-enriched, ether) phase. Also calculate the number of extraction stages needed.

Calculation Procedure

1. Plot the equilibrium-composition data on a right-triangular diagram.
The plot, shown in Fig. 9.1, is prepared as follows: Let the vertex labeled E represent 100% ether, let the one labeled A represent 100% alcohol, and let the one labeled W represent 100% water. Then the scale along the abscissa represents the weight fraction alcohol, and the scale along the ordinate represents the weight fraction ether. Take each pair of points in Table 9.1 and plot them, joining any given pair by a straight line, called a "tie line." (Note that for the first pair, the tie line coincides with part of the ordinate, since the alcohol concentration is 0.0 in each of the two phases.) Draw a curve through points at the lower ends of the tie lines and another curve through the points at the upper ends.

These two curves divide the diagram into three regions. Any composition falling within the uppermost region will consist solely of an ether-rich phase; any composition within the lowermost region will consist solely of an aqueous phase; any composition within the middle region will constitute a combination of those two liquid phases.

2. Calculate the mean concentrations of ether, alcohol, and water within the system.
This step is easier to visualize if a stage diagram is first drawn (see Fig. 9.2). Each box represents an extraction stage, with stage 1 being the stage at the top of the column. Stage N is at the bottom of the

TABLE 9.1   Equilibrium Data for Alcohol-Water-Ether System (Procedure 9.1)

Weight fraction in phase					
Water phase			Ether phase		
Alcohol	Ether	Water	Alcohol	Water	Ether
0.0	0.075	0.925	0.0	0.225	0.775
0.1	0.077	0.823	0.090	0.170	0.740
0.2	0.090	0.710	0.175	0.120	0.705
0.31	0.095	0.595	0.250	0.080	0.670
0.44	0.118	0.442	0.290	0.05	0.660
0.530	0.150	0.320	0.31	0.035	0.655
0.645	0.195	0.160	0.33	0.019	0.651
0.75	0.25	0.0	0.35	0.0	0.65

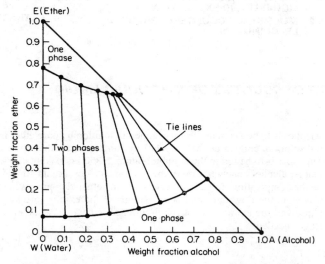

FIGURE 9.1   Equilibrium phase diagram for alcohol, water, ether system (Procedure 9.1).

column. The streams labeled L are the aqueous-phase streams; those labeled S are the ether-phase (i.e., solvent-phase) streams. Let the given letter stand for the actual flow rates.

Thus the alcohol-water feed mixture L_0 enters the top of the column, where the ether-solvent-phase stream S_1 leaves. At the bottom of the column, ether-solvent stream S_{N+1} enters, while the stripped water stream L_N leaves.

Now, the mean concentration of ether W_{EM} and alcohol W_{AM} in the system can be calculated from the equations

$$W_{EM} = \frac{L_0 W_{E,0} + S_{N+1} e_{E,N+1}}{L_0 + S_{N+1}}$$

$$W_{AM} = \frac{L_0 W_{A,0} + S_{N+1} e_{A,N+1}}{L_0 + S_{N+1}}$$

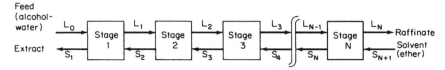

FIGURE 9.2 Stage diagram for countercurrent multistage liquid-liquid extraction (Procedure 9.1).

where L_0 is the feed rate of the water phase, S_{N+1} is the feed rate of the solvent, $W_{E,0}$ and $W_{A,0}$ are the concentrations of ether and alcohol, respectively, in the entering water phase, and $e_{E,N+1}$ and $e_{A,N+1}$ are the concentrations of ether and alcohol, respectively, in the entering ether (solvent) phase. Then

$$W_{EM} = \frac{370(0) + 350(1.0)}{370 + 350} = 0.49$$

$$W_{AM} = \frac{370(0.3) + 350(0)}{370 + 350} = 0.154$$

And by difference, the mean concentration of water is $(1.0 - 0.49 - 0.154) = 0.356$.

3. Calculate the compositions of the exiting raffinate and extract streams.
Replot (or trace) Fig. 9.1 without the tie lines (Fig. 9.3). On Fig. 9.3, plot the mean-concentration point M. Now from mass-balance considerations, the exit concentrations must lie on the two phase-boundary lines and on a straight line passing through the mean concentration point. We know we want the water phase to have an exit concentration of 3 wt % alcohol. Such a concentration corresponds to point L_N on the graph. At point L_N, the ether concentration is seen to be 7.6 wt % (this can be found more accurately in the present case by numerical extrapolation of the water-phase data in Table 9.1). Therefore, the composition of the raffinate stream is 3% alcohol, 7.6% ether, and (by difference) 89.4% water.

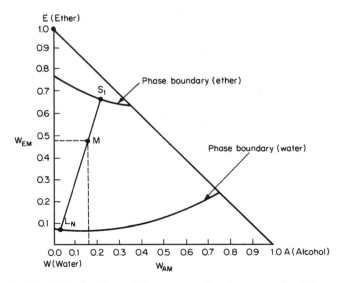

FIGURE 9.3 Overall mass balance on extraction column (Procedure 9.1).

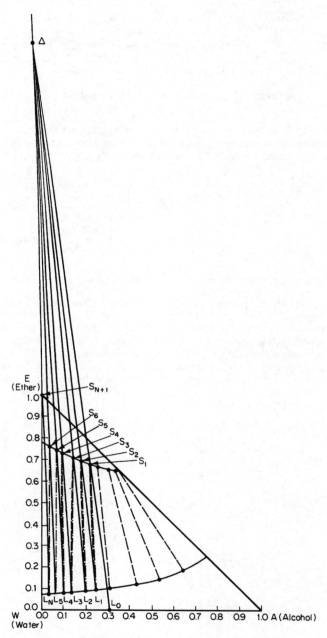

FIGURE 9.4 Determination of number of stages (Procedure 9.1).

To determine the composition of the exiting extract stream, extend line $L_N M$ until it intersects the ether-phase boundary, at point S_1, and read the composition graphically. It is found to be 69% ether, 22% alcohol, and (by difference) 9% water.

4. Calculate the flow rates of the exiting raffinate and extract streams.
Set up an overall mass balance:

$$S_1 + L_N = L_0 + S_{N+1} = 370 + 350 = 720$$

where (see Fig. 9.2) L_0 and S_{N+1} are the entering streams (see step 2), and S_1 and L_N are the exiting extract and raffinate streams, respectively.

Now set up a mass balance for the ether, using the compositions found in step 3:

$$S_1(0.69) + L_N(0.076) = 370(0) + 350(1.0) = 350$$

Solving the overall and the ether balances simultaneously shows the exiting raffinate stream L_N to be 239 kg/h and the exiting extract stream S_1 to be 481 kg/h.

5. Calculate the number of stages needed.
The number of stages can be found graphically on Fig. 9.1 (repeated for convenience as Fig. 9.4), including the tie lines. Plot on the graph the points corresponding to L_0 (30% alcohol, 0% ether) and S_1 (69% ether, 22% alcohol), draw a straight line through them, and extend the line upward. Similarly, plot L_N (3% alcohol, 7.6% ether) and S_{N+1} (100% ether), and draw a line through them, extending it upward. Designate the intersection of the two lines as Δ.

Now, begin to step off the stages by drawing an interpolated tie line from point S_1 down to the water-phase boundary. Designate their intersection as point L_1. Draw a line joining L_1 with Δ, and label the intersection of this line with the ether-phase boundary as S_2. Then repeat the procedure, drawing a tie line through S_2 and intersecting the water-phase boundary at L_2. This sequence is repeated until a point is reached on the water-phase boundary that has an alcohol content less than that of L_N. Count the number of steps involved. In this case, 6 steps are required. Thus the extraction operation requires 6 stages.

Related Calculations. This basic calculation procedure can be extended to the case of countercurrent multistage extraction with reflux. A schematic of the basic extractor is shown in Fig. 9.5. For this extractor there are N stages in the extracting section, $1E$ to NE, and there are M stages in the stripping section, $1S$ to MS.

The overall mass balance on the entire extractor as well as the mass balances on the solvent separator, feed stage, and solvent mixer are performed separately. The calculations on the extraction section and the stripping sections are then performed as described above.

9.2 MULTISTAGE COUNTERCURRENT LEACHING

Hot water is to be used to leach a protein out of seaweed in an isothermal multistage countercurrent system, as shown in Fig. 9.6. The seaweed slurry, consisting of 48.1% solids, 2.9% protein, and 49% water, enters at a rate of 400 kg/h. The hot water is fed at a rate of 500 kg/h. It is desired to have the outlet underflow (the spent seaweed) have a maximum residual concentration of 0.2% protein on a solids-free basis. Table 9.2 shows experimental data for the operation, taken by (1) contacting the seaweed with hot water for a period of time with mixing, (2) stopping the mixing and letting the seaweed settle, and (3) sampling the bottom slurry (underflow) phase and the upper extract (overflow) phase. Calculate the number of equilibrium leaching stages needed.

Calculation Procedure

1. *Calculate the weight ratio of solids to total liquid and the weight concentration of protein on a solids-free basis for each of the two phases in each run of the experimental data.*

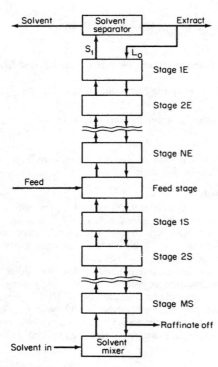

Let N_o and N_u designate the weight ratio of solids to total liquid in the overflow and underflow phases, respectively. Similarly, let x_o and x_u represent the weight concentration of protein on a solids-free basis in the overflow and underflow phases, respectively. For run 1, for instance,

$$N_o = \frac{0.002}{0.952 + 0.046} = 0.002$$

$$N_u = \frac{0.432}{0.542 + 0.026} = 0.760$$

$$x_o = \frac{0.046}{0.952 + 0.046} = 0.046$$

$$x_u = \frac{0.026}{0.542 + 0.026} = 0.046$$

The calculated results for all runs are as follows (the sequence being rearranged for convenience in step 2):

FIGURE 9.5   Reflux extractor.

Run number	x_u	N_u	x_o	N_o
1	0.046	0.760	0.046	0.002
2	0.032	0.715	0.032	0.001
3	0.022	0.669	0.021	0
4	0.011	0.661	0.011	0
5	0.006	0.658	0.006	0
6	0.002	0.656	0.002	0

2. *Plot the equilibrium data.*
See step 2 in Procedure 9.1. In the present case, however, a rectangular rather than a triangular diagram is employed. The abscissa is x, from above; similarly, the ordinate is N. For each run, plot the points x_o, N_o and x_u, N_u and join the pair by a tie line. Then connect the x_o, N_o points, thereby generating the overflow equilibrium line, and similarly, connect the x_u, N_u points to generate the underflow equilibrium line. The result is shown as Fig. 9.7.

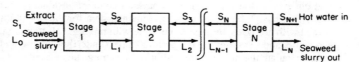

FIGURE 9.6   Stage diagram for multistage countercurrent leaching (Procedure 9.2).

TABLE 9.2   Equilibrium Data for Seaweed-Water System (Procedure 9.2)

| Run number | Extract (overflow) phase | | | Slurry (underflow) phase | | |
	Water	Solution protein	Solids	Water	Solution protein	Solids
1	0.952	0.046	0.002	0.542	0.026	0.432
2	0.967	0.032	0.001	0.564	0.019	0.417
3	0.979	0.021	0.00	0.586	0.013	0.401
4	0.989	0.011	0.0	0.5954	0.0066	0.398
5	0.994	0.006	0.0	0.5994	0.0036	0.397
6	0.998	0.002	0.0	0.6028	0.0012	0.396

(Weight fraction in phase)

3. Calculate the mean values for x and N.

Refer to step 2 in Procedure 9.1. In the present case, the mean values can be found from the equations

$$x_M = \frac{L_0 W_{P,0} + S_{N+1} e_{P,N+1}}{L_0(W_{P,0} + W_{W,0}) + S_{N+1}(e_{P,N+1} + e_{W,N+1})}$$

and

$$N_M = \frac{L_0 W_{S,0} + S_{N+1} e_{S,N+1}}{L_0(W_{P,0} + W_{W,0}) + S_{N+1}(e_{P,N+1} + e_{W,N+1})}$$

where x_M is the mean weight concentration of protein on a solids-free basis, N_M is the mean weight ratio of solids to total liquid, L_0 is the feed rate of the seawater slurry, $W_{P,0}$, $W_{W,0}$, and $W_{S,0}$ are the weight fractions of protein, water, and solids, respectively, in the seawater slurry, S_{N+1} is the feed rate of the hot water, and $e_{P,N+1}$, $e_{W,N+1}$, and $e_{S,N+1}$ are the weight fractions of protein, water, and solids, respectively, in the entering hot water. Then,

$$x_M = \frac{400(0.029) + 500(0)}{400(0.029 + 0.49) + 500(0 + 1)}$$

$$= 0.0164$$

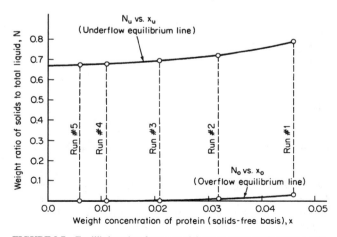

FIGURE 9.7   Equilibrium data for seaweed, hot water system (Procedure 9.2).

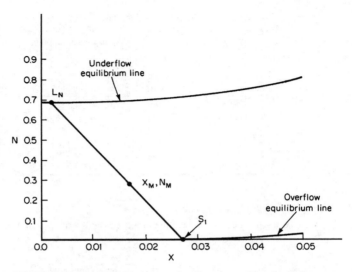

FIGURE 9.8   Overall mass balance on leaching system (Procedure 9.2).

and

$$N_M = \frac{400(0.481) + 500(0)}{400(0.029 + 0.49) + 500(0 + 1)}$$

$$= 0.272$$

4. Calculate the concentration of protein in the product extract (overflow) stream.
Replot (or trace) Fig. 9.7 without the tie lines (Fig. 9.8). On it, plot the point x_M, N_M. Locate along the underflow equilibrium line the point L_N that corresponds to the desired maximum residual concentration of protein, 0.2% or 0.002 weight fraction. Now, from mass-balance considerations, the exit concentrations must lie on the two equilibrium lines and on a straight line passing through x_M, N_M. So draw a line through L_N and x_M, N_M, and extend it to the overflow equilibrium line, labeling the intersection as S_1. The concentration of protein in the exiting overflow stream is read, then, as 0.027 weight fraction, or 2.7%. (Note also that there are virtually no solids in this stream; that is, the value for N at S_1 is virtually zero.)

5. Calculate the number of stages.
With reference to step 1, calculate the values of x and N for each of the two entering streams. Thus, for the seawater slurry,

$$N_{L,0} = \frac{0.481}{0.49 + 0.029} = 0.927$$

and

$$x_{L,0} = \frac{0.029}{0.49 + 0.029} = 0.0559$$

For the entering hot water, $N_{S,N+1}$ and $x_{S,N+1}$ both equal 0, because this stream contains neither solids nor protein.

Now, the number of stages can be found graphically on Fig. 9.7 (redrawn as Fig. 9.9), including the tie lines, in a manner analogous to that of step 5 in Procedure 9.1. Plot on Fig. 9.9 the points L_0 (that is, $x = 0.0559$, $N = 0.927$) and S_1 (from Fig. 9.8), draw a line through them, and extend it downward. Similarly, plot the points S_{N+1} (that is, $x = 0$, $N = 0$) and L_N, draw a line through them, and extend it downward. Designate the intersection of the two lines as Δ.

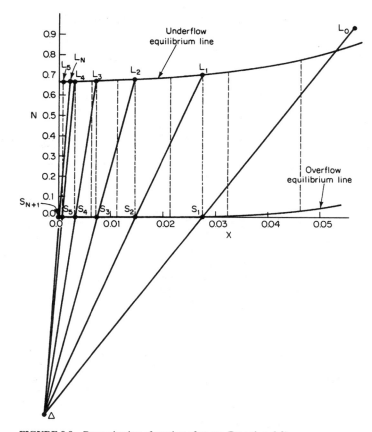

FIGURE 9.9   Determination of number of stages (Procedure 9.2).

Now begin to step off the stages by drawing an interpolated tie line through point S_1 up to the underflow equilibrium line. Designate their intersection as point L_1. Draw a line joining L_1 with Δ, and label the intersection of this line with the overflow equilibrium line as S_2. Then repeat the procedure, drawing a tie line through S_2 and intersecting the underflow equilibrium line at L_2. This sequence is repeated until a point is reached on the underflow equilibrium line that has a protein content less than that of L_N. Count the number of steps involved. In this case, 5 steps are required. Thus the leaching operation requires 5 equilibrium stages.

Related Calculations.   This example is a case of variable-underflow conditions. However, the same procedure can be applied to constant underflow.

*9.3 *DETERMINING THE NUMBER OF STAGES FOR A COUNTERCURRENT EXTRACTOR*

Oil is to be extracted from meal by means of benzene, using a continuous countercurrent extractor. The unit is to treat 2000 lb (908 kg) of meal (based on completely exhausted solid) per hour. The untreated meal contains 800 lb (364 kg) of oil and 50 lb (22.7 kg) of benzene. The fresh solvent

TABLE 9.3 Data for Calculation Procedure

Concentration, lb oil/lb solution	Solution retained, lb/lb solid
0.0	0.500
0.1	0.505
0.2	0.515
0.3	0.530
0.4	0.550
0.5	0.571
0.6	0.595
0.7	0.620

mixture contains 20 lb (9.1 kg) of oil and 1310 lb (595 kg) of benzene. The exhausted solids are to contain 120 lb (54.5 kg) of unextracted oil. Experiments carried out under conditions identical with those of the projected battery show that the solution retained depends on the concentration of the solution, as shown in Table 9.3.

Find: (a) the concentration of the strong solution, or extract, (b) the concentration of the solution adhering to the extracted solids, (c) the mass of solution leaving with the extracted meal, (d) the mass of extract, (e) the number of stages required.

Calculation Procedure

1. Compute, and plot, the Ponchon-Savarit diagram for this extraction.
The data in Table 9.3 provide the coordinates of the Y vs. X line for all underflows. Values of X are given in the first column of the table, and corresponding values of Y are the reciprocals of the numbers in the second column. The YX line for the underflow is plotted as curve ab in Fig. 9.10. From the conditions of the problem, the coordinates of points L_a and V_b are

Point L_a:
$$X = \frac{800}{800 + 50} = \frac{800}{850} = 0.941 \quad Y = \frac{2000}{850} = 2.35$$

Point V_b:
$$X = \frac{20}{1310 + 20} = \frac{20}{1330} = 0.015 \quad Y = 0$$

Points L_a and V_b are plotted on Fig. 9.10, and line L_aV_b drawn. Point J lies on this line, a distance $850/(1330 + 850) = 0.39$ times the distance between points V_b and L_a, measured from V_b. Point J is plotted. The ratio of Y to X for point L_b is the ratio of the solid to the solute in this stream, or $2000/120 = 16.67$. Point L_b lies on line ab. A straight line through the origin with a slope of 16.67 intersects line ab at point L_b, and another straight line through points L_b and J intersects the $X = 0$ line at point V_a. The X coordinate of point L_b is 0.12, and that of point V_a is 0.592.

2. Determine the solution components and the process results.
The total solution input is $1330 + 850 = 2180$ lb (988 kg), and this equals the total solution leaving in streams V_a and L_b. This flow is divided between the two streams in proportion to the line segments on line $V_a JL_b$. The X coordinate of point J is 0.372, and, by the center-of-gravity principle,

$$L_b = \frac{0.592 - 0.372}{0.592 - 0.120} 2180 = 1020 \text{ lb (464 kg)}$$
$$V_a = 2180 - 1020 = 1160 \text{ lb (527 kg)}$$

The answers to parts (a) to (d) are: part (a), 0.592; part (b), 0.12; part (c) 1020 lb (464 kg) part (d) 1160 lb (527 kg).

3. *Find the number of ideal stages for this extractor.*
To determine the number of ideal stages, point P is established as the intersection between straight lines through points L_a and V_a and L_b and V_b, as shown in Fig. 9.10. The Ponchon-Savarit construction is shown in the figure, and from this the number of ideal stages is found to be slightly less than four. The answer to part (*e*), then, is four stages.

Related Calculations.   A special case of leaching is encountered when the solute is of limited solubility and the concentrated solution reaches saturation. This situation can be treated by the above methods. The solvent input to stage N should be the maximum that is consistent with a saturated overflow from stage 1, and all liquids except that adhering to the underflow from stage 1 should be unsaturated. If saturation is attained in stages other than the first, all but one of the "saturated" stages are unnecessary, and the solute concentration in the underflow from stage N is higher than it needs to be.

This procedure is the work of Warren L. McCabe and Julian C. Smith in their book *Unit Operations of Chemical Engineering*, McGraw-Hill, 1956.

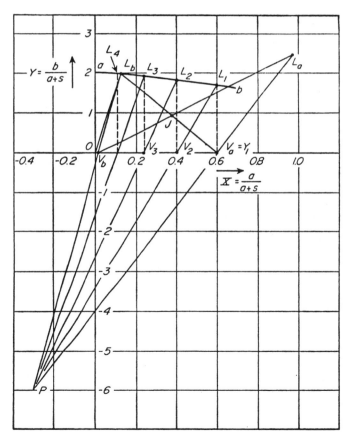

FIGURE 9.10   Ponchon-Savarit diagram for extractor.

SECTION 10
CRYSTALLIZATION

James R. Beckman, Ph.D.

Associate Professor and Associate Chair
Department of Chemical and Materials Engineering
Arizona State University
Tempe, AZ

10.1 SOLID-PHASE GENERATION OF AN ANHYDROUS SALT BY COOLING

A 65.2 wt % aqueous solution of potassium nitrate originally at 100°C (212°F) is gradually cooled to 10°C (50°F). What is the yield of KNO_3 solids as a function of temperature? How many pounds of KNO_3 solids are produced at 10°C if the original solution weighed 50,000 lb (22,680 kg)?

Calculation Procedure

1. Convert weight percent to mole percent.
In order to use Fig. 10.1 in the next step, the mole fraction of KNO_3 in the original solution must be determined. The calculations are as follows:

Compound	Pounds in original solution	÷	Molecular weight	=	Moles	Mole percent
KNO_3	0.652		101.1		0.00645	25.0
H_2O	0.348		18.0		0.01933	75.0
Total	1.000				0.02578	100.0%

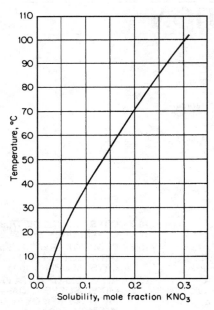

FIGURE 10.1   Solubility of KNO_3 in water versus temperature.

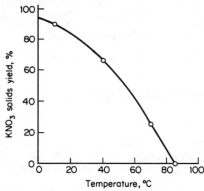

FIGURE 10.2   Yield of KNO_3 versus temperature (Procedure 10.1).

2. *Calculate yield of solids versus temperature.*
Figure 10.1 shows the composition of saturated KNO_3 solution as a function of temperature. The solids formed during cooling will be 100 percent KNO_3, because KNO_3 is anhydrous. The yield of solids is the ratio of KNO_3 solidified to the KNO_3 originally dissolved. As can be seen from Fig. 10.1, no solids are formed from a 25 mol % solution until the solution is cooled to 85°C. As cooling proceeds, solid KNO_3 continues to form while the (saturated) solution concentration continues to decline. At 70°C, for instance, the solution will contain 20 mol % KNO_3 (80 mol % H_2O). If 100 mol of the original solution is assumed, then originally there were 25 mol KNO_3 and 75 mol H_2O. This amount of water present does not change during cooling and solids formation. At 70°C there are therefore $[(0.20 \ KNO_3)/(0.80 \ H_2O)]$ (75 mol H_2O) = 18.8 mol KNO_3 dissolved, or $25 - 18.8 = 6.2$ mol KNO_3 solids formed. Therefore, the crystal yield at 70°C is $(6.2/25)(100 \ \text{percent}) =$ 24.8 percent.

   Similarly, at about 40°C, the solubility of KNO_3 is 10 percent, which leaves $[(0.10 \ KNO_3)/(0.90 \ H_2O)]$ (75 mol H_2O) = 8.3 mol KNO_3 in solution. Therefore, $25 - 8.3 = 16.7$ mol KNO_3 will have precipitated by the time the solution has cooled to that temperature. Consequently, the yield of solids at 40°C is $16.7/25 = 66.8$ percent. Finally, at 10°C, the KNO_3 solubility drops to 3 mol %, giving a yield of 91 percent. Figure 10.2 summarizes the yield of KNO_3 solids as a function of temperature.

3. *Calculate the weight of solids at 10°C.*
The weight of solids formed at 10°C is the solids yield (91 percent) multiplied by the weight of KNO_3 initially present in the 100°C mother liquor. The weight of KNO_3 initially in the mother liquor of a 50,000-lb solution is 50,000 lb × 0.652 = 32,600 lb. The weight of KNO_3 solids formed at 10°C is 32,600 lb × 0.91 = 29,670 lb (13,460 kg).

Related Calculations.   This method can be used to calculate the yield of any anhydrous salt from batch or steady-state cooling crystallizers. For hydrated salts, see Procedure 10.3.

10.2   SOLID-PHASE GENERATION OF AN ANHYDROUS SALT BY BOILING

A 70°C (158°F) aqueous solution initially containing 15 mol % KNO_3 is to be boiled so as to give a final yield of solid KNO_3 of 60 percent. How much of the initial water must be boiled off? What is the final liquid composition?

Calculation Procedure

1. *Find the final liquid composition.*
Use Fig. 10.1 to determine the solubility of KNO_3 in saturated water solution at 70°C. From the figure, the KNO_3 solubility is 20 mol %.

2. *Calculate the amount of water boiled off.*
Take a basis of 100 mol of initial solution. Then 15 mol KNO_3 and 85 mol H_2O were initially present. To give a solids yield of 60 percent, then, 0.60×15 mol = 9 mol KNO_3 must be precipitated from the solution, leaving $15 - 9 = 6$ mol KNO_3 in the solution. The solubility of KNO_3 at 70°C is 20 mol %, from step 1. Therefore, the amount of water still in solution is 6 mol × (0.80/0.20) = 24 mol H_2O, requiring that $85 - 24 = 61$ mol had to be boiled off. The percent water boiled off is 61/85 = 72 percent.

Related Calculations. This method can be used to determine the amounts of water to be boiled from boiling crystallizers that yield anhydrous salts. For hydrated salts, see Procedure 10.4.

10.3 SOLID-PHASE GENERATION OF A HYDRATED SALT BY COOLING

A 35 wt % aqueous $MgSO_4$ solution is originally present at 200°F(366 K). If the solution is cooled (with no evaporation) to 70°F (294 K), what solid-phase hydrate will form? If the crystallizer is operated at 10,000 lb/h (4540 kg/h) of feed, how many pounds of crystals will be produced per hour? What will be the solid-phase yield?

Calculation Procedure

1. *Determine the hydrate formation.*
As the phase diagram (Fig. 10.3) shows, a solution originally containing 35 wt % $MgSO_4$ will, when cooled to 70°C, form a saturated aqueous solution containing 27 wt % $MgSO_4$ (corresponding to point A) in equilibrium with $MgSO_4 \cdot 7H_2O$ hydrated solids (point B). No other hydrate can exist at equilibrium under these conditions. Now since the molecular weights of $MgSO_4$ and $MgSO_4 \cdot 7H_2O$ are 120 and 246, respectively, the solid-phase hydrate is (120/246)(100) = 48.8 wt % $MgSO_4$; the rest of the solid phase is H_2O in the crystal lattice structure.

2. *Calculate the crystal production rate and the solid-phase yield.*
Let L be the weight of liquid phase formed and S the weight of solid phase formed. Then, for 10,000 lb/h of feed solution, $L + S = 10,000$, and (by making a material balance for the $MgSO_4$) $0.35(10,000) = 0.27L + 0.488S$. Solving these two equations gives $L = 6330$ lb/h of liquid phase and $S = 3670$ lb/h (1665 kg/h) of $MgSO_4 \cdot 7H_2O$.

Now the solid-phase yield is based on $MgSO_4$, not on $MgSO_4 \cdot 7H_2O$. The 3670 lb/h of solid phase is 48.8 wt % $MgSO_4$, from step 1, so it contains 3670(0.488) = 1791 lb/h $MgSO_4$. Total $MgSO_4$ introduced into the system is 0.35 (10,000) = 3500 lb/h. Therefore, solid-phase yield is 1791/3500 = 51.2 percent.

As a matter of interest, the amount of H_2O removed from the system by solid (hydrate) formation is 3670(1.0 − 0.488) = 1879 lb/h.

Related Calculations. This method can be used to calculate the yield of any hydrated salt from a batch or a steady-state cooling crystallizer.

In step 2, L and S can instead be found by applying the inverse lever-arm rule to line segments $\overline{AB}$ and $\overline{AC}$ in Fig. 10.3. Thus, $S/(S + L) = S/10,000 = \overline{AB}/\overline{AC} = (0.35 - 0.27)/(0.488 - 0.27)$; therefore, $S = 3670$ lb/h.

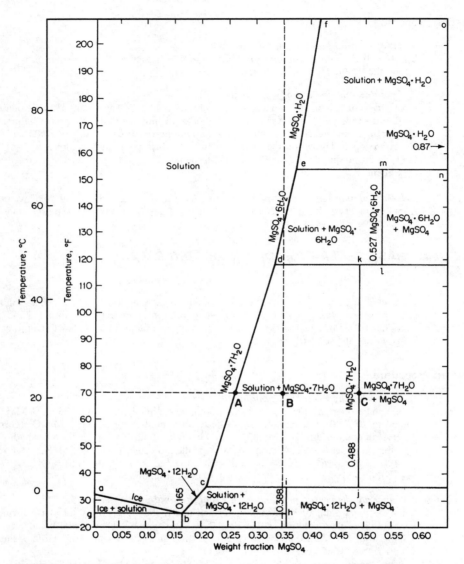

FIGURE 10.3   Phase diagram for $MgSO_4 \cdot H_2O$. (*From Perry—Chemical Engineers' Handbook, McGraw-Hill, 1963.*)

10.4 SOLID-PHASE GENERATION OF A HYDRATED SALT BY BOILING

Consider 40,000 lb/h (18,150 kg/h) of a 25 wt % $MgSO_4$ solution being fed at 200°F (366 K) to an evaporative crystallizer that boils off water at a rate of 15,000 lb/h (6800 kg/h). The crystallizer is operated at 130°F (327 K) under vacuum conditions. Determine the solid-phase composition, solid-phase production rate, and solid-phase yield. Also calculate the required energy addition rate for the process.

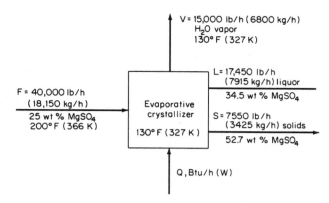

FIGURE 10.4   Flow diagram for evaporative crystallizer (Procedure 10.4).

Calculation Procedure

1. *Determine the hydrate formation (solids composition).*
Since 15,000 lb/h of water is removed, the product slurry will have an overall $MgSO_4$ composition of $0.25 \times 40,000$ lb/$(40,000 - 15,000$ lb$) = 40.0$ wt % $MgSO_4$. From Fig. 10.3, a system at 130°F and overall $MgSO_4$ composition of 40 wt % will yield $MgSO_4 \cdot 6H_2O$ solids in equilibrium with a 34.5 wt % $MgSO_4$ liquor. Since the molecular weights of $MgSO_4$ and $MgSO_4 \cdot 6H_2O$ are 120 and 228, respectively, the solid-phase hydrate is $(120/228)(100) = 52.7$ wt % $MgSO_4$, with 47.3 wt % water.

2. *Calculate the solids production rate.*
Let L be the weight of liquid phase formed and S the weight of solid phase formed. Then, for 40,000 lb/h of feed solution with 15,000 lb/h of water boil-off, $S + L = 40,000 - 15,000$, and (by making a material balance for the $MgSO_4$) $0.25(40,000) = 0.527S + 0.345L$. Solving these two equations gives $L = 17,450$ lb/h of liquid phase and $S = 7550$ lb/h (3425 kg/h) of $MgSO_4 \cdot 6H_2O$ solids.

3. *Calculate the solid-phase yield.*
The solid-phase yield is based on $MgSO_4$, not on $MgSO_4 \cdot 6H_2O$. From step 1, the 7550 lb/h of solid phase is 52.7 wt % $MgSO_4$, so it contains 7550 $(0.527) = 3979$ lb/h $MgSO_4$. Total $MgSO_4$ introduced into the system is $0.25(40,000) = 10,000$ lb/h. Therefore, solid-phase yield is $3979/10,000 = 39.8$ percent.

4. *Calculate the energy addition rate.*
Figure 10.4 shows the mass flow rates around the evaporative crystallizer, as well as an arrow symbolizing the energy addition. An energy balance around the crystallizer gives $Q = VH_V + LH_L + SH_S - FH_F$, where the Hs are the stream enthalpies. From Fig. 10.5, $H_L = -32$ Btu/lb, $H_S = -110$ Btu/lb (extrapolated to 52.7 percent), and $H_F = 52$ Btu/lb. The value for the water vapor, H_V, takes a little more work to get. The enthalpy basis of water used in Fig. 10.5 is 32°F liquid; this can be deduced from the fact that the figure shows an enthalpy value of 0 for pure water (that is, 0 wt % $MgSO_4$ solution) at 32°F. The basis of most steam tables is 32°F liquid water. From such a steam table an H_V value of about 1118 Btu/lb can be obtained for 130°F vapor water (the pressure correction is minor and can be neglected). Therefore, $Q = 15,000 \times 1118 + 17,450 \times (-32) + 7550 \times (-110) - 40,000 \times 52 = 13.3 \times 10^6$ Btu/h (3900 kW) energy addition to the crystallizer. Energy addition per pound of solids produced is $13.3 \times 10^6/7550 = 1760$ Btu (1860 kJ).

Related Calculations.   This method can be used to calculate the yield, boiling (if any), and energy addition to an evaporative or cooling crystallizer that produces any hydrated or anhydrous crystal solid.

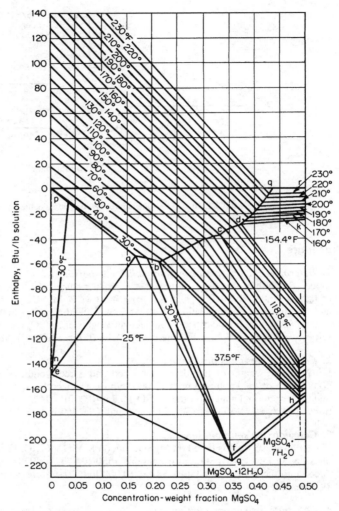

FIGURE 10.5   Enthalpy-concentration diagram for MgSO$_4$ · H$_2$O system. (Note: 1 Btu/lb = 2.326 kJ/kg.) (*From Perry—Chemical Engineers' Handbook, McGraw-Hill, 1963.*)

10.5   SEPARATION OF BENZENE AND NAPHTHALENE BY CRYSTALLIZATION

A 100,000 lb/h (4536 kg/h) 70°C (158°F) feed containing 80 wt % naphthalene is fed to a cooling crystallizer. At what temperature should the crystallizer operate for maximum naphthalene-only solids production? At this temperature, what is the solids yield of naphthalene? What is the total energy removed from the crystallizer? Naphthalene solids are removed from the mother liquor by centrifugation, leaving some of the solids liquor (10 wt % of the solids) adhering to the solids. After the solids are melted, what is the final purity of the naphthalene?

Calculation Procedure

1. Determine the appropriate operating temperature for the crystallizer.
Figure 10.6 shows the mutual solubility of benzene and naphthalene. Most of the naphthalene can be crystallized by cooling to (i.e., operating the crystallizer at) the eutectic temperature of −3.5°C (25.7°F), where the solubility of naphthalene in the liquor is minimized to 18.9 wt %. (If one attempted to operate below this temperature, the whole system would become solid.)

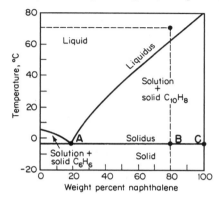

FIGURE 10.6 Phase diagram for the simple eutectic system naphthalene-benzene.

2. Calculate the solids yield.
The solids yield is the ratio of the naphthalene solids produced (corresponding to point C in Fig. 10.6) to the naphthalene in the feed liquid (point B). Point A corresponds to the naphthalene remaining in the mother liquor. Then, using the inverse lever-arm rule, we find the naphthalene solids rate S as follows: $S = 100,000(\overline{AB}/\overline{AC}) = 100,000 \times (0.8 - 0.189)/(1.0 - 0.189) = 75,300$ lb/h. (This leaves $100,000 - 75,300 = 24,700$ lb/h in the mother liquor.) The solids yield is $75,300/(100,000 \times 0.8) = 94.1$ percent. The flows are shown in Fig. 10.7.

3. Calculate the energy removal.
An energy balance around the crystallizer (see Fig. 10.5) gives $Q = LH_L + SH_S - FH_F$, where Q is the heat added (or the heat removed, if the solved value proves to be negative); L, S, and F are the flow rates for mother liquor, solid product, and feed, respectively; and H_L, H_S, and H_F are the enthalpies of those streams relative to some base temperature. Select a base temperature T_R of 70°C, so that $H_F = 0$.

For specifics of setting up an energy balance, see Procedure 2.7. From handbooks, the heat of fusion of naphthalene is found to be 64.1 Btu/lb, and over the temperature range considered here, the heat capacities of liquid benzene and naphthalene can be taken as 0.43 and 0.48 Btu/(lb)(°F), respectively.

Then, for the mother liquor (which consists of 18.9 wt % naphthalene and 81.1 wt % benzene), $H_L = (-3.5°C - 70°C)(1.8°F/°C)[0.48(0.189) + 0.43(0.811)] = -58.1$ Btu/lb. For the product naphthalene, which must cool from 70°C to −3.5°C and then solidify, $H_S = (-3.5°C - 70°C)(1.8°F/°C)(0.48) - 64.1 = -127.6$ Btu/lb.

Therefore, the heat added to the crystallizer is $Q = 24,700(-58.1) + 75,300(-127.6) - 100,000(0) = -11.0 \times 10^6$ Btu/h (3225 kW). Since the value for Q emerges negative, this is the energy *removed* from the crystallizer.

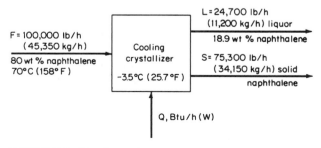

FIGURE 10.7 Flow diagram for cooling crystallizer (Procedure 10.5).

4. *Calculate the purity of the naphthalene obtained by melting the product crystals.*
The weight of mother liquor adhering to the solids is 10 percent of 75,300, or 7530 lb. The total amount of naphthalene present after melting is therefore $75,300 + 0.189(7530) = 76,720$ lb. The weight of benzene present (owing to the benzene content of the mother liquor) is $7530(1.0 - 0.189) = 6100$ lb. The product purity is therefore $76,720/(76,720 + 6100) = 92.6$ percent naphthalene.

Related Calculations.   This method can be used to separate organic mixtures having components of different freezing points, such as the xylenes. Organic separations by crystallization have found industrial importance in situations in which close boilers have widely separated freezing temperatures. Less energy is related to freezing as opposed to boiling processes because of the low ratio of heat of fusion to heat of vaporization.

10.6 ANALYSIS OF A KNOWN CRYSTAL SIZE DISTRIBUTION (CSD)

A slurry contains crystals whose size-distribution function is known to be $n = 2 \times 10^5 L \exp(-L/10)$, where n is the number of particles of any size L (in μm) per cubic centimeter of slurry. The crystals are spherical, with a density of 2.5 g/cc. Determine the total number of crystals. Determine the total area, volume, and mass of the solids per volume of slurry. Determine the number-weighted average, the length-weighted average, and the area-weighted average particle size of the solids. What is the coefficient of variation of the particles? Generate a plot of the cumulative weight fraction of particles that are undersize in terms of particle size L.

Calculation Procedure

1. *Calculate the total number of particles per volume of slurry.*
This step and the subsequent calculation steps require finding $\int_0^\infty n L^j \, dL$, that is (from the equation for n above), $2 \times 10^5 \int_0^\infty L^{j+1} \exp(-L/10) \, dL$, where L is as defined above and j varies according to the particular calculation step. From a table of integrals, the general integral is found to be $2 \times 10^5 [(j+1)!/(1/10)^{j+2}]$.

   For calculating the number of particles, $j = 0$, and the answer is the zeroth moment (designated M_0) of the distribution. Thus the total number N_T of particles is $2 \times 10^5[(0+1)!/(1/10)^{0+2}] = 2 \times 10^7$ particles per cubic centimeter of slurry.

2. *Calculate the first moment of the distribution.*
This quantity, M_1, which corresponds to the "total length" of the particles per cubic centimeter of slurry, is not of physical significance in itself, but it is used in calculating the averages in subsequent steps. It corresponds to the integral in step 1 when $j = 1$. Thus, $M_1 = 2 \times 10^5[(1+1)!/(1/10)^{1+2}] = 4 \times 10^8$ μm per cubic centimeter of slurry.

3. *Calculate the total area of the particles per volume of slurry.*
The total area $A_T = k_A M_2$, where k_A is a shape factor (see below) and M_2 is the second moment of the distribution, i.e., the value of the integral in step 1 when $j = 2$. Some shape factors are as follows:

Crystal shape	Value of k_A
Cube	6
Sphere	π
Octahedron	$2\sqrt{3}$

   In the present case, then, $A_T = \pi(2 \times 10^5)[(2+1)!/(1/10)^{2+2}] = 3.77 \times 10^{10}$ μm^2 (377 cm^2) per cubic centimeter of slurry.

4. *Calculate the total volume of crystals per volume of slurry.*

The volume of solids per volume of slurry $V_T = k_V \int_0^\infty nL^3 dL = k_V M_3$, where k_V is a so-called volume shape factor (see below) and M_3 is the third moment of the distribution, i.e., the value of the integral in step 1 when $j = 3$. Some volume shape factors are as follows:

Crystal shape	Value of k_V
Cube	1
Sphere	$\pi/6$
Octahedron	$\sqrt{2}/3$

In the present case, then, $V_T = (\pi/6)(2 \times 10^5)[(3 + 1)!/(1/10)^{3+2}] = 2.51 \times 10^{11} \mu m^3$ (0.251 cm^3) per cubic centimeter of slurry.

5. *Calculate the total mass of solids per volume of slurry.*

Total mass of solids $M_T = \rho_S V_T$, where ρ_S is the crystal density. Thus, $M_T = (2.5 g/cm^3)(0.251$ cm^3 per cubic centimeter of slurry) = 0.628 g per cubic centimeter of slurry.

6. *Calculate the average crystal size.*

The number-weighted average crystal size $\overline{L}_{1,0} = M_1/M_0$. Thus, $\overline{L}_{1,0} = (4 \times 10^8)/(2 \times 10^7) = 20 \mu m$. The length weighted average $\overline{L}_{2,1} = M_2/M_1$. Thus, $\overline{L}_{2,1} = 2 \times 10^5[(2 + 1)!/(1/10)^{2+2}]/(4 \times 10^8) = (12 \times 10^9)/(4 \times 10^8) = 30 \mu m$. And the area-weighted average $\overline{L}_{3,2} = M_3/M_2$. Thus, $\overline{L}_{3,2} = (2 \times 10^5)[(3 + 1)!/(1/10)^{3+2}]/(12 \times 10^9) = 40 \mu m$.

7. *Calculate the variance of the particle size distribution.*

The variance σ^2 of the particle size distribution equals $\int_0^\infty (\overline{L}_{1,0} - L)^2 nd\,L/M_0 = M_2/M_0 - (\overline{L}_{1,0})^2$. Thus, $\sigma^2 = (12 \times 10^9)/(2 \times 10^7) - 20^2 = 200 \mu m^2$.

8. *Calculate the coefficient of variation for the particle size distribution.*

The coefficient of variation c.v. equals $\sigma/\overline{L}_{1,0}$, where σ (the standard deviation) is the square root of the variance from step 7. Thus, c.v. $= 200^{1/2}/20 = 0.71$.

9. *Calculate and plot the cumulative weight fraction that is undersize.*

The weight fraction W undersize of a crystal size distribution is $W = \rho_s k_V \int_0^L nL^3 dL/M_T = \int_0^L nL^3 dL/M_3 = 1 - [(L/10)^4/24 + (L/10)^3/6 + (L/10)^2/2 + L/10 + 1] \exp(-L/10)$. A plot of this function (Fig. 10.8) has the characteristic S-shaped curvature.

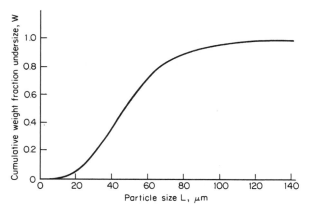

FIGURE 10.8 Cumulative weight fraction undersize versus particle size (Procedure 10.6).

Related Calculations.   This procedure can be used to calculate average sizes, moments, surface area, and mass of solids per volume of slurry for any known particle size distribution. The method can also be used for dry-solids distributions, say, from grinding operations. See Procedure 10.7 for an example of a situation in which the size distribution is based on an experimental sample rather than on a known size-distribution function.

10.7   CRYSTAL SIZE DISTRIBUTION OF A SLURRY SAMPLE

The first three columns of Table 10.1 show a sieve-screen analysis of a 100-cc (0.0001-m^3 or 0.0035-ft^3) slurry sample. The crystals are cubic and have a solids density ρ_s of 1.77 g/cc (110.5 lb/ft^3). Calculate the crystal size distribution n of the solids, the average crystal size, and the coefficient of variation of the crystal size distribution.

Calculation Procedure

1. Calculate the weight fraction retained on each screen.
The weight fraction ΔW_i retained on screen i equals the weight retained on that screen divided by the total solids weight, that is, 29.87 g. For instance, the weight fraction retained on screen 28 is $0.005/29.87 = 0.000167$. The weight fractions retained on all the screens are shown in the fourth column of Table 10.1.

2. Calculate the screen average sizes.
The screen average size L_i for a given screen reflects the average size of a crystal retained on that screen. Use the average of the size of the screen and the screen above it. For instance, the average crystal size of the solids on screen 28 is $(701 + 589)/2 = 645/\mu m$. The averages for all the screens appear in the fifth column of Table 10.1.

3. Calculate the size difference between screens.
The size difference between screens ΔL_i is the difference between the size of the screen in question and the size of the screen directly above it. For instance, the size difference ΔL_i for screen 60 is (295 μm) − (248 μm) = 47 μm. Size differences for all the screens appear in the sixth column of Table 10.1.

TABLE 10.1   Crystal Size Distribution of a Slurry Sample (Procedure 10.7)

Sieve screen analysis			Summary of crystal size distribution analysis			
Tyler mesh	Opening, μm	Weight retained, g	Weight fraction retained ΔW_i	Average screen size L_i, μm	ΔL_i, μm	n_i, no./ (cm^3) (μm)
(24) ficticious screen	(701)	—	—	—	—	—
28	589	0.005	0.000167	645	112	9.39×10^{-4}
32	495	0.016	0.00536	542	94	0.0605
35	417	0.096	0.00321	456	78	0.0733
42	351	0.315	0.0106	384	66	0.479
48	295	1.61	0.0539	323	56	4.827
60	248	3.42	0.1145	272	47	20.46
65	208	7.56	0.253	228	40	90.19
80	175	8.21	0.275	192	33	199.0
100	147	5.82	0.195	161	28	282.0
115	124	2.47	0.0827	136	23	241.5
150	104	0.32	0.0107	114	20	61.0
170	88	0.025	0.000837	96	16	9.99
200	74	0.0076	0.000254	81	14	5.77
		Total 29.87				

4. Calculate the third moment of the crystal size distribution.
The third moment M_3 of the crystal size distribution equals $M_T/\rho_s k_V$, where M_T is the total weight of the crystals and k_V is the volume shape factor; see step 4 of Procedure 10.6. In this case, $M_3 = 29.87/[1.77(1)(100 \text{ cc})] = 0.169 \text{ cm}^3$ solids per cubic centimeter of slurry.

5. Calculate the crystal size distribution.
The crystal size distribution for the ith screen n_i equals $10^{12} M_3 \Delta W_i / (L_i^3 \Delta L_i)$, in number of crystals per cubic centimeter per micron. For instance, for screen 60, $n_i = 10^{12}(0.169)(0.1145)/[272^3(47)] = 20.46$ crystals per cubic centimeter per micron. The size distributions for all other screens appear in the seventh column of Table 10.1.

6. Calculate the zeroth, first, and second moments of the crystal size distribution.
The zeroth moment M_0 is calculated as follows: $M_0 = \Sigma_i \, n_i \, \Delta L_i = 2.64 \times 10^4$ crystals per cubic centimeter. The first moment M_1 is calculated by $M_1 = \Sigma_i \, n_i \, L_i \Delta L_i = 4.64 \times 10^6 \, \mu m/cm^3$. The second moment M_2 is calculated by $M_2 = \Sigma_i \, n_i \, L_i^2 \Delta L_i = 8.62 \times 10^8 \, \mu m^2/cm^3$. [The third moment M_3 can be calculated by $M_3 = \Sigma_i \, n_i \, L_i^3 \Delta L_i = 0.169 \times 10^{12} \, \mu m^3/cm^3$ (0.169 cm^3/cm^3), which agrees with the calculation of the third moment from step 4.]

7. Calculate the average crystal size.
The number-weighted average crystal size is $\overline{L}_{1.0} = M_1/M_0 = (4.64 \times 10^6)/(2.64 \times 10^4) = 176 \, \mu m$
The length-weighted average crystal size is $\overline{L}_{2.1} = M_2/M_1 = (8.62 \times 10^8)/(4.64 \times 10^6) = 186 \, \mu m$.
The area-weighted average crystal size is $\overline{L}_{3.2} = M_3/M_2 = (0.169 \times 10^{12})/(8.63 \times 18^8) = 196 \, \mu m$.

8. Calculate the variance.
The variance of the crystal size distribution is $\sigma^2 = M_2/M_0 - (\overline{L}_{1.0})^2 = (8.62 \times 10^8)/(2.64 \times 10^4) - 176^2 = 1676 \, \mu m^2$.

9. Calculate the coefficient of variation.
The coefficient of variation c.v. $= \sigma/\overline{L}_{1.0} = 1676^{1/2}/176 = 0.23$.

Related Calculations. This procedure can be used to analyze either wet or dry solids particle size distributions. Particle size distributions from grinding or combustion and particles from crystallizers are described by the same mathematics. See Procedure 10.6 for an example of a situation in which the size distribution is based on a known size-distribution function rather than on an experimental sample.

10.8 PRODUCT CRYSTAL SIZE DISTRIBUTION FROM A SEEDED CRYSTALLIZER

A continuous crystallizer producing 25,000 lb/h (11,340 kg/h) of cubic solids is continuously seeded with 5000 lb/h (2270 kg/h) of crystals having a crystal size distribution as listed in Table 10.2. Predict the product crystal size distribution if nucleation is ignored. If the residence time of solids in the crystallizer is 2 h, calculate the average particle-diameter growth rate G.

Calculation Procedure

1. Calculate the crystal-mass-increase ratio.
The crystal-mass-increase ratio is the ratio of crystallizer output to seed input; in this case, 25,000/5000 = 5.0.

2. Calculate the increase ΔL in particle size.
The increase in weight of a crystal is related to the increase in particle diameter. For any given screen size, that increase ΔL is related to the initial weight ΔM_s and initial size L_s of seed particles

TABLE 10.2   Size Distribution of Seed Crystals (Procedure 10.8)

Tyler mesh	Weight fraction retained ΔW_i	Average size Li, μm (from Table 10.1)
(65)	—	—
80	0.117	192
100	0.262	161
115	0.314	136
150	0.274	114
170	0.032	96
200	0.001	81
	Total 1.000	

corresponding to that screen and to the product weight ΔM_p of particles corresponding to that screen, by McCabe's ΔL law:

$$\Delta M_p = \left(1 + \frac{\Delta L}{L_s}\right)^3 \Delta M_s$$

This equation can be solved for ΔL by trial and error. From step 1, and summing over all the screens, $\Sigma \Delta M_p / \Sigma \Delta M_s = 5.0$. The trial-and-error procedure consists of assuming a value for ΔL, calculating ΔM_p for each screen, summing the values of ΔM_p and ΔM_s, and repeating the procedure until the ratio of these sums is close to 5.0.

For a first guess, assume that $\Delta L = 100$ μm. Assuming that the total seed weight is 1.0 (in any units), this leads to the results shown in Table 10.3. Since $\Sigma \Delta M_p$ is found to be 5.26, the ratio $\Sigma \Delta M_p / \Sigma \Delta M_s$ emerges as 5.26/1.0 = 5.26, which is too high. A lower assumed value of ΔL is called for. At final convergence of the trial-and-error procedure, ΔL is found to be 96 μm, based on the results shown in the first five columns of Table 10.4.

This leads to the crystal size distribution shown in the last two columns of the table. The sixth column, weight fraction retained ΔW_i, is found (for each screen size) by dividing ΔM_p by $\Sigma \Delta M_p$. The screen size (seventh column) corresponding to each weight fraction consists of the original seed-crystal size L_s plus the increase ΔL.

3. Calculate the growth rate.
The average particle-diameter growth rate G can be found thus:

$$G = \frac{\Delta L}{\text{(elapsed time)}} = \frac{96 \,\mu\text{m}}{(2\,\text{h})(60\,\text{min/h})} = 0.8 \,\mu\text{m/min}$$

TABLE 10.3   Results from (Incorrect) Guess that $\Delta L = 100$ μm (Procedure 10.8). Basis: Total seed mass = 1.0

Tyler mesh	Seed mass ΔM_s (from Table 10.2)	Seed size L_s (from Table 10.2)	$(1 + \Delta L/L_s)^3$	Calculated product mass Δm_p
80	0.117	192	3.52	0.412
100	0.262	161	4.26	1.12
115	0.314	136	5.23	1.64
150	0.274	114	6.61	1.81
170	0.032	96	8.51	0.272
200	0.001	81	11.16	0.011
	$\Sigma \Delta M_s = 1.000$			$\Sigma \Delta M_p = 5.26$

$$\frac{\Sigma \Delta M_p}{\Sigma \Delta M_s} = \frac{5.26}{1.00} = 5.26 \quad \text{which is too high}$$

TABLE 10.4 Results from (Correct) Guess that $\Delta L = 96$ μm and Resulting Crystal Size Distribution (Procedure 10.8). Basis: Total seed mass = 1.0

Tyler mesh	Seed mass ΔM_s	Seed size L_s	$(1 + \Delta L/L_s)^3$	Calculated product mass ΔM_p	Calculated weight fraction retained ΔW_i	Product screen size $(L_s + \Delta L)$
80	0.117	192	3.37	0.395	0.079	288
100	0.262	161	4.07	1.066	0.214	257
115	0.314	136	4.96	1.56	0.312	232
150	0.274	114	6.25	1.71	0.342	210
170	0.032	96	8.00	0.256	0.051	192
200	0.001	81	10.4	0.0104	0.002	177
	$\Sigma \Delta M_s = 1.000$			$\Sigma \Delta M_p = 4.997$		

$$\frac{\Sigma \Delta M_p}{\Sigma \Delta M_s} = \frac{4.997}{1.000} = 4.997 \quad \text{which is close enough to 5.0}$$

Related Calculations. This method uses McCabe's ΔL law, which assumes total growth and no nucleation. For many industrial situations, these two assumptions seem reasonable. If significant nucleation is present, however, this method will overpredict product crystal size.

The presence of nucleation can be determined by product screening: If particles of size less than the seeds can be found, then nucleation is present. In such a case, prediction of product crystal size distribution requires a knowledge of nucleation kinetics; see Randolph and Larson [3] for the basic mathematics.

10.9 ANALYSIS OF DATA FROM A MIXED SUSPENSION–MIXED PRODUCT REMOVAL CRYSTALLIZER (MSMPR)

The first three columns of Table 10.5 show sieve data for a 100-cc slurry sample containing 21.0 g of solids taken from a 20,000-gal (75-m³) mixed suspension-mixed product removal crystallizer (MSMPR) producing cubic ammonium sulfate crystals. Solids density is 1.77 g/cm³, and the density of the clear liquor leaving the crystallizer is 1.18 g/cm³. The hot feed flows to the crystallizer at 374,000 lb/h (47 kg/s). Calculate the residence time τ, the crystal size distribution function n, the growth rate $\underline{G}$, the nucleation density n^0, the nucleation birth rate B^0, and the area-weighted average crystal size $L_{3,2}$ for the product crystals.

Calculation Procedure

1. Calculate the density of the crystallizer magma.
The slurry density in the crystallizer is the same as the density of the product stream. Select as a basis 100 cm³ slurry. The solids mass is 21.0 g; therefore, the solids volume is (21.0 g)/(1.77 g/cm³) = 11.9 cm³. The clear-liquor volume is $100 - 11.9 = 88.1$ cm³, and its mass is (88.1 cm³)(1.18 g/cm³) = 104 g. Therefore, the density of the slurry is (104 g + 21 g)/100 cm³ = 1.25 g/cm³ (78.0 lb/ft³).

2. Calculate the residence time in the crystallizer.
The residence time τ in the crystallizer is based on the outlet conditions (which are the same as in the crystallizer). Thus, τ = (volume of crystallizer)/(outlet volumetric flow rate) = (20,000 gal)/[(374,000 lb/h)/(78.0 lb/ft³)(0.1337 ft³/gal)] = 0.557 h = 33.4 min.

TABLE 10.5 Crystal Size Distribution from an MSMPR Crystallizer (Procedure 10.9)

Screen number	Tyler mesh	Weight fraction retained ΔW_i	Summary of crystal size distribution analysis				
			Screen size, μm	Average screen size L_i, μm	ΔL_i, μm	n_i	ln n_i
1	24	0.081	701	—	—	—	—
2	28	0.075	589	645	112	0.297	−1.21
3	32	0.120	495	542	94	0.954	−0.047
4	35	0.100	417	456	78	1.61	0.476
5	42	0.160	351	384	60	5.60	1.72
6	48	0.110	295	323	56	6.94	1.94
7	60	0.102	248	272	47	12.8	2.55
8	65	0.090	208	228	40	22.6	3.12
9	80	0.060	175	192	33	30.6	3.42
10	100	0.040	147	161	28	40.7	3.71
11	115	0.024	124	136	23	49.4	3.90
12	150	0.017	104	114	20	68.3	4.22
13	170	0.010	88	96	16	84.1	4.43
14	200	0.005	74	81	14	80.0	4.38
—	fines	0.006	—	—	—	—	—
		Total 1.000					

3. Calculate the third moment of the solids crystal size distribution.

The third moment M_3 of the crystal size distribution equals $M_T/\rho_s k_V$, where M_T is the weight of crystals, ρ_s is the solids density, and k_V is the volume shape factor; see step 4 of Procedure 10.6. Thus, $M_3 = 21.0$ g/[(1.77 g/cm^3)(1)(100 cm^3)] = 0.119 cm^3 solids per cubic centimeter of slurry.

4. Calculate the crystal size distribution function n.

The crystal size distribution for the ith sieve tray is $n_i = 10^{12} M_3 \Delta W_i/(L_i^3 \Delta L_i)$, where ΔW_i is the weight fraction retained on the ith screen, L_i is the average screen size of material retained on the ith screen (see Procedure 10.7, step 2), and ΔL_i is the difference in particle sizes on the ith screen (see Procedure 10.7, step 3). For instance, for the Tyler mesh 100 screen, $n_{10} = 10^{12}(0.119)(0.040)/(161^3 \times 28) = 40.7$ crystals per cubic centimeter per micron. Table 10.5 shows the results for each sieve screen.

5. Calculate the growth rate G.

The growth rate for an MSMPR can be calculated from the slope of an ln n versus L diagram (Fig. 10.9). Here the slope equals $-[1/(G\tau)] = (\ln n_2 - \ln n_1)/(L_2 - L_1) = (-0.6 - 5.4)/(600 - 0) = -0.010$ μm^{-1} or $G\tau = 100$ μm. Then the growth rate $G = 100$ μm/ 33.4 min = 3.0 μm/min.

6. Calculate the nucleation density n^0.

The nucleation density n^0 is the value of n at size $L = 0$. From Fig. 10.9, ln n^0 at size equal to zero is 5.4. So $n^0 = \exp(\ln n^0) = \exp 5.4 = 221$ particles per cubic centimeter per micron.

7. Calculate the nucleation birth rate B^0.

The nucleation birth rate is $B^0 = n^0 G = 221(3.0) = 663$ particles per cubic centimeter per minute.

8. Calculate area-weighted average size $\overline{L}_{3,2}$.

As shown in Procedures 10.6 and 10.7, the area-weighted average size equals M_3/M_2. However, for an MSMPR, the area-weighted average particle size also happens to equal $3G\tau$. Thus, $\overline{L}_{3,2} = 3(3$ μm/min)(33.4 min) = 300 μm.

Related Calculations. Use this procedure to calculate the crystal size distribution from both class I and class II MSMPR crystallizers. This procedure cannot be used to calculate growth rates and nucleation with crystallizers having either fines destruction or product classification.

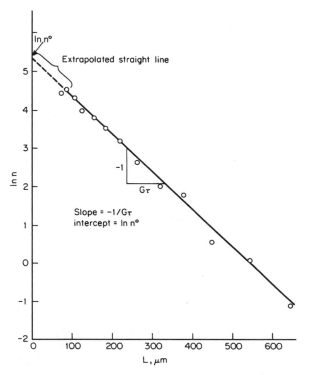

FIGURE 10.9   Ln n versus L for an MSMPR crystallizer (Procedure 10.9).

10.10   PRODUCT SCREENING EFFECTIVENESS

Figure 10.10 shows the sieve-screen analysis of a feed slurry, overflow slurry, and underflow slurry being separated by a 600-μm classifying screen. Calculate the overall effectiveness of the classifying screen.

Calculation Procedure

1. Calculate solids mass fractions in each stream.
On Fig. 10.10, draw a vertical line through the abscissa that corresponds to the screen size, that is, 600 μm. This line will intersect each of the three cumulative-weight-fraction curves. The ordinate corresponding to each intersection gives the mass fraction of total solids actually in that stream which would be in the overflow stream if the screen were instead perfectly effective. Thus the mass fraction in the feed x_F is found to be 0.28, the fraction in the overflow x_o is found to be 0.77, and the fraction in the underflow x_u is found to be 0.055.

2. Calculate solids-overflow to total-solids-feed ratio.
The ratio of total overflow-solids mass to total feed-solids mass q equals $(x_F - x_u)/(x_o - x_u)$; that is, $q = (0.28 - 0.055)/(0.77 - 0.055) = 0.315$.

3. Calculate the screen effectiveness based on oversize.
The screen effectiveness based on oversize material E_o equals $q(x_o/x_F)$. Thus, $E_o = 0.315$ (0.77/0.28) = 0.87.

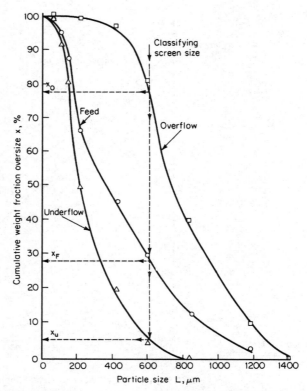

FIGURE 10.10   Cumulative weight fraction oversize versus particle size (Procedure 10.10).

4. Calculate the screen effectiveness based on undersize.
The screen effectiveness based on undersize material E_u equals $(1 - q)(1 - x_u)/(1 - x_F)$. Thus, $E_u = (1 - 0.315)(1 - 0.055)/(1 - 0.28) = 0.90$.

5. Calculate the overall screen effectiveness.
Overall screen effectiveness E is the product of E_o and E_u. Thus, $E = 0.90(0.87) = 0.78$.

Related Calculations.   This method can be used for determining separation effectiveness for classifying screens, elutriators, cyclones, or hydroclones in which a known feed of a known crystal size distribution is segregated into a fine and a coarse fraction. If a cut size cannot be predetermined, assume one at a time and complete the described effectiveness analysis. The assumed cut size that gives the largest effectiveness is the cut size that best describes the separation device.

*10.11   HEAT REMOVAL REQUIRED IN CRYSTALLATION

A 32.5% solution of $MgSO_4$ at 120°F (48.9°C) is cooled, without appreciable evaporation, to 70°F (21.1°C) in a Swenson-Walker crystallizer. How much heat must be removed from the solution per ton of crystals?

Calculation Procedure

1. *Use an enthalpy-concentration diagram to determine the heat absorbed.*
The initial solution is represented by the point on Fig. 10.11 at a concentration of 0.325 in the undersaturated-solution field on the 120°F (48.9°C) isotherm. The enthalpy coordinate of this point is −33.0 Btu/lb (−76.8 kJ/kg). The point for the final magma lies on the 70°F (21.1°C) isotherm in area *cihb* at concentration 0.325. The enthalpy coordinate of this point is −78.4. Per 100 lb (45.4 kg) of original solution the heat *absorbed* by the solution is:

$$100(−78.4 + 33.0) = −4540 \text{ Btu} (−4790 \text{ kJ})$$

This is a heat *evolution* of 4540 Btu (4790 kJ).

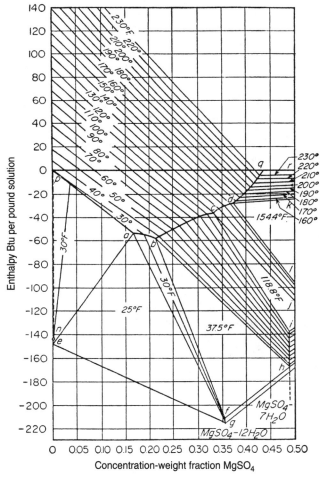

FIGURE 10.11 Enthalpy-concentration diagram, system MgSO$_4$–H$_2$O.
(By permission, from Chemical Engineers' Handbook, 3d ed., by J. H. Perry. Copyright, 1950. McGraw-Hill.)

2. *Find the heat evolved by the crystals.*
The split of the final magma between crystals and mother liquor may be found by the center-of-gravity principle applied to the 70°F(21.1°C) isotherm in Fig.10.11. The concentration of the mother liquor is 0.259, and that of the crystals is 0.488. Then, per 100 lb (45.5 kg) of magma, the crystals are

$$100 \, \frac{0.325 - 0.259}{0.488 - 0.259} = 28.8 \, \text{lb} \, (13.1 \, \text{kg})$$

The heat evolved per ton of crystals is $(4540/28.8)2000 = 315{,}000$ Btu (330,750 kJ).

In many industrial crystallization processes, the crystals and mother liquor are in contact long enough to reach equilibrium, and the mother liquor is saturated at the final temperature of the process. The yield of the process can then be calculated from the concentration of the original solution and the solubility at the final temperature. If appreciable evaporation occurs during the process, this must be known or estimated.

When the rate of crystal growth is slow, considerable time is required to reach equilibrium. This is especially true when the solution is viscous or where the crystals collect in the bottom of the crystallizer so there is little crystal surface exposed to the supersaturated solution. In such situations, the final mother liquor may retain appreciable supersaturation, and the actual yield will be less than that calculated from the solubility curve.

If the crystals are anhydrous, calculation of the yield is simple, as the solid phase contains no solvent. When the crop contains water of crystallization, account must be taken of the water accompanying the crystals, since this water is not available for retaining solute in solution. Solubility data are usually given either in parts by mass of anhydrous material per hundred parts by mass of total solvent or in mass percent anhydrous solute. These data ignore water of crystallization. The key to calculations of yields of hydrated solutes is to express all masses and concentrations in terms of hydrated salt and free water. Since it is this latter quantity that remains in the liquid phase during crystallization, concentrations or amounts based on free water may be subtracted to give a correct result.

This procedure is the work of Warren L. McCabe and Julian C. Smith in their book *Unit Operations of Chemical Engineering*, McGraw-Hill, 1956.

REFERENCES

1. Foust et al.—*Principles of Unit Operations,* Wiley.
2. Perry—*Chemical Engineers' Handbook,* McGraw-Hill.
3. Randolph and Larson—*Theory of Particulate Processes,* Academic Press.
4. Mullin—*Crystallization,* CRC Press.
5. Bamforth—*Industrial Crystallization,* Macmillan.
6. Institute of Chemical Engineers—*Industrial Crystallization,* Hodgson.
7. Felder and Rousseau—*Elementary Principles of Chemical Processes,* Wiley.

SECTION 11
ABSORPTION AND STRIPPING

K. J. McNulty, Sc.D.
Technical Director
Research and Development
Koch Engineering Co., Inc.
Wilmington, MA

11.1 HYDRAULIC DESIGN OF A PACKED TOWER

Gas and liquid are to be contacted countercurrently in a packed tower. The approach to flooding is not to exceed 80 percent as defined at constant liquid loading. What column diameter should be used for 2-in Koch Flexiring (FR) packing (i.e., 2-in slotted-ring random packing), and what diameter should be used for Koch Flexipac (FP) Type 2Y packing (i.e., structured packing having $1/2$-in crimp height, with flow channels inclined at 45° to the axis of flow)? The packings are to be of stainless steel. Calculate and compare the pressure drops per foot of packing for these two packings. The operating conditions are

Maximum liquid rate: 150,000 lb/h
Maximum gas rate: 75,000 lb/h
Liquid density: 62.4 lb/ft^3
Gas density: 0.25 lb/ft^3

Calculation Procedure

1. Calculate the tower diameter.
Various methods are available for the design and rating of packed towers. The method shown here is an extension of the CVCL model, which is more fundamentally sound than the generalized pressure

drop correlation (GPDC). The basic flooding equation is

$$C_{VF}^{1/2} + sC_{LF}^{1/2} = c \tag{11.1}$$

where C_{VF} is the vapor or gas capacity factor at flooding, C_{LF} is the liquid capacity factor at flooding, and s and c are constants for a particular packing. The capacity factors for the vapor and liquid are defined as

$$C_V = U_g[\rho_g/(\rho_l - \rho_g)]^{1/2} \tag{11.2}$$

$$C_L = U_L[\rho_l/(\rho_l - \rho_g)] \tag{11.3}$$

where U_g is the superficial velocity of the gas based on the empty-tower cross-sectional area, U_l is the superficial liquid velocity, ρ_g is the gas density, and ρ_l is the liquid density. In U.S. engineering units, C_V and U_g are in feet per second, and C_L and U_L are in gallons per minute per square foot. The superficial velocities are related to the mass flow rates as follows:

$$U_g = w_g/\rho_g A \quad \text{and} \quad U_l = w_i/\rho_l A \tag{11.4}$$

where w_g and w_l are, respectively, the mass flow rates of gas and liquid, and A is the tower cross-sectional area. Here, U_l has units of feet per second; this can be converted to the more commonly used U_L (with an uppercase subscript) having units of gallons per minute per square foot, by multiplying by 7.48 gal/ft^3 and 60 s/min.

For flooding that is defined at constant liquid loading L, the flooding capacity factor C_{LF} is the same as the design capacity factor C_L. The gas capacity factor at flood, C_{VF}, is equal to the design capacity factor divided by the fractional approach to flooding, f: $C_{VF} = C_V/f$. For this problem, f equals 0.80. Substituting the definitions and given values for this problem into Eq. (11.1), bearing in mind that for a circular cross section $A = \pi D^2/4$ and solving for tower diameter D gives

$$D = \frac{1}{c}\sqrt{\frac{4}{\pi}}\left\{\left[\frac{75{,}000}{(3600)(0.8)(0.25)}\sqrt{\frac{0.25}{(62.4 - 0.25)}}\right]^{1/2} + s\left[\frac{(150{,}000)(7.48)}{(62.4)(60)}\sqrt{\frac{62.4}{(62.4 - 0.25)}}\right]^{1/2}\right\} \tag{11.5}$$

where the quantity 3600 converts from hours to seconds, 60 converts from hours to minutes, and 7.48 converts from cubic feet to gallons.

It remains to determine values of s and c for the packings of interest. These values can be calculated from air-water pressure-drop data published by packing vendors. Table 11.1 provides values of s and c for several random and structured packings of stainless steel construction. With the appropriate constants from the table for 2-in FR and FP2Y substituted into Eq. (11.5), the diameter emerges as 4.96 ft for 2-in FR and 5.00 ft for FP2Y. Thus, for the conditions stated in this example, the capacities of these two packings are essentially identical. In subsequent calculations, a tower diameter of 5.0 ft will be used for both packings.

2. Calculate the loadings, the capacity factors, and the superficial F factor at the design conditions. For the design conditions given and a tower of 5 ft diameter, the loadings and capacity factors are calculated as follows from Eqs. (11.2), (11.3), and (11.4):

$$U_L = \frac{w_l}{\rho_l A} = \frac{(150{,}000)(4)}{\pi(62.4)(5)^2}\frac{(7.48)}{(60)} = 15.26 \text{ gal/(min)(ft}^2)$$

$$C_L = 15.26\sqrt{\frac{62.4}{62.4 - 0.25}} = 15.29 \text{ gal/(min)(ft}^2)$$

TABLE 11.1 Tower-Packing Constants for Hydraulic Design and Rating of Packed Towers

Generic shape	Typical brand	Nominal size	Void fraction, ϵ	K_1, (in $wc \cdot s^2$)/lb	s, [ft³/(gal/min)(s)]$^{1/2}$	c, (ft/s)$^{1/2}$
Random: slotted ring	Flexiring Packing	1 in	0.95	0.14	0.016	0.67
		1.5 in	0.96	0.11	0.015	0.71
		2 in	0.98	0.069	0.040	0.74
		3.5 in	0.98	0.044	0.034	0.78
Structured: corrugated sheet	Flexipac Packing	Type 1Y ¼ in, 45°	0.97	0.088	0.054	0.69
		Type 2Y ½ in, 45°	0.99	0.041	0.044	0.75
		Type 3Y 1 in, 45°	0.99	0.019	0.044	0.87
		Type 4Y 2 in, 45°	0.99	0.012	0.035	0.88

Note: Flexiring and Flexipac are trademarks of Koch Engineering Co., Inc.

$$U_g = \frac{w_g}{\rho_g A} = \frac{(75,000)(4)}{\pi(0.25)(5)^2}\frac{1}{3600} = 4.24 \text{ ft/s}$$

$$C_V = U_g\sqrt{\frac{\rho_g}{\rho_l - \rho_g}} = 4.24\sqrt{\frac{0.25}{62.4 - 0.25}} = 0.269 \text{ gal/(min)(ft}^2)$$

The superficial F factor F_s is calculated as follows:

$$F_S = U_g\sqrt{\rho_g} = 4.24(0.25)^{1/2} = 2.12 \text{ (ft/s)(lbm/ft}^3)^{1/2}$$

3. Construct the dry-pressure-drop line on log-log coordinates (optional).

For turbulent flow, the gas-phase pressure drop for frictional loss, contraction and expansion loss, and directional change loss are all proportional to the square of the superficial F factor. For the dry packing the pressure drop can be calculated from the equation

$$\Delta P/\Delta Z = K_1 F_s^2 \tag{11.6}$$

where ΔZ represents a unit height of packing (e.g., 1 ft) and K_1 is a constant whose value depends on the packing size and geometry. Values of K_1 can be determined from pressure-drop data published by packing vendors and are included in Table 11.1. For the two packings of this problem, the values of K_1 are 0.069 for 2-in FR and 0.041 for FP2Y. Using Eq. (11.6) with the design F factor calculated in step 2 gives the following. For 2-in FR,

$$\Delta P/\Delta Z = (0.069)(2.12)^2 = 0.310 \text{ in } wc/\text{ft}$$

For FP2Y, $\qquad\qquad \Delta P/\Delta Z = (0.041)(2.12)^2 = 0.184 \text{ in } wc/\text{ft}$

This calculation indicates that for the same nominal capacity, the random packing has a 68 percent higher pressure drop per foot than the structured packing.

To construct the dry-pressure-drop line, plot $\Delta P/\Delta Z$ vs. F_s on log-log coordinates. Select appropriate values of F_s (e.g., 1 and 3 for this example) and use Eq. (11.6) to calculate the corresponding pressure drop per foot (e.g., for 2-in FR, 0.069 in wc/ft at $F_s = 1$ and 0.621 in/ft at $F_s = 3$; and for FP2Y, 0.041 in/ft at $F_s = 1$ and 0.369 in/ft at $F_s = 3$). Plot the points on log-log coordinates and connect them with a straight line.

4. Construct the wet-pressure-drop line, on log-log coordinates, for pressure drops below column loading.

For most packings of commercial interest, the relationship between the wet and dry pressure drops can be given by

$$\frac{(\Delta P/\Delta Z)_W}{(\Delta P/\Delta Z)_D} = \frac{1}{\left(1 - \dfrac{0.02\, U_L^{0.8}}{\epsilon}\right)^{2.5}} \tag{11.7}$$

where the subscripts W and D refer to wet and dry, respectively, and ϵ is the void fraction of the dry packed bed. Values of ϵ for the various packings are given in Table 11.1. The liquid loading, U_L in Eq. (11.7), is in units of gallons per minute per square foot. For the random packing,

$$\frac{(\Delta P/\Delta Z)_W}{(\Delta P/\Delta Z)_D} = \frac{1}{\left(1 - \dfrac{0.02(15.26)^{0.8}}{0.98}\right)^{2.5}} = 1.64$$

For the structured packing, the numerical result is the same.

To construct the wet-pressure-drop line, select values of F_s that span the design value. For example, at $F_s = 1$, the pressure drop for random packing is $(1.64)(0.069)(1) = 0.113$ in wc/ft, and that

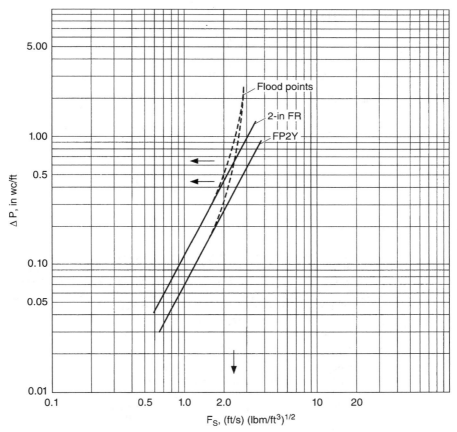

FIGURE 11.1 Pressure-drop curves for Procedure 11.1.

for structured packing is $(1.64)(0.041)(1) = 0.0672$ in wc/ft. At $F_s = 3$, the pressure drop for random packing is $(1.64)(0.069)(9) = 1.018$ in wc/ft, and that for structured packing is $(1.64)(0.041)(9) = 0.605$ in wc/ft. Plot these points on log-log coordinates and connect the points for each packing with a straight line. The solid lines of Fig. 11.1 are the wet-pressure-drop lines for the two packings.

5. Determine the pressure drop and superficial F factor at flooding.
For the air-water system at ambient conditions, flooding occurs at a pressure drop of about 2 in wc/ft. The pressure drop at flooding for other systems can be calculated by

$$\Delta P_F = (2 \text{ in } wc/\text{ft})[(\rho_l - \rho_g)/(62.4 - 0.075)] \qquad (11.8)$$

Employing Eq. (11.8) with the density values in this problem gives a pressure drop at flooding of 1.99 in wc/ft.

The superficial F factor at flooding F_{SF} is determined with the aid of the flooding equation, Eq. (11.1). Since the pressure drop curve is being constructed for a constant liquid loading, the value of C_{LF} in Eq. (11.1) is 15.29 gal/(min)(ft^2) as determined in step 2. With appropriate values for s and c, solve Eq. (11.1) for C_{VF} and determine F_{SF} from

$$F_{SF} = C_{VF}(\rho_l - \rho_g)^{1/2}$$

Thus, for 2-in FR,

$$C_{VF} = [0.74 - (0.040)(15.29)^{1/2}]^2 = 0.340 \text{ ft/s}$$

and accordingly $F_{SF} = 0.340(62.4 - 0.25)^{1/2} = 2.68 \text{ (ft/s)(lbm/ft}^3)^{1/2}.$

For FP2Y,

$$C_{VF} = [0.75 - (0.044)(15.29)^{1/2}]^2 = 0.334 \text{ ft/s}$$

and

$$F_{SF} = 0.334(62.4 - 0.25)^{1/2} = 2.63 \text{ (ft/s)(lbm/ft}^3)^{1/2}$$

Locate the flood points for the two packings on the pressure drop plot at the coordinates ΔP_F and F_{SF}. See Fig. 11.1.

6. *Draw the wet-pressure-drop curve for the loading region.*
The "loading region" covers a range of F factors from flooding downward to roughly half of the flooding F factor. This is the region in which the gas flow causes additional liquid holdup in the packing and produces a pressure drop higher than the one indicated by the straight lines of Fig. 11.1.

For each packing, sketch an empirical curve such that: (a) its upper end passes through the point F_{SF}, ΔP_F with a slope approaching infinity, and (b) its lower end becomes tangent to the straight, wet-pressure-drop line at an abscissa value of $F_{SF}/2$. These two pressure-drop curves for the loading regions are shown by the dashed lines of Fig. 11.1.

7. *Determine the pressure drops from the curves.*
The overall pressure-drop curve for each packing consists of the solid curve at low gas loadings and the dashed curve at loadings between the load point and the flood point. At the design F factor of 2.12 (ft/s)(lbm/ft^3)$^{1/2}$, the pressure drops determined from the curves are

For 2-in FR,   0.65 in *wc*/ft

For FP2Y,   0.44 in *wc*/ft

Thus, the pressure drop of the random packing is 48 percent higher than that of the structured packing at the design conditions.

Related Calculations.   It is usual practice in distillation operations to keep the liquid-to-vapor ratio constant as the throughput is varied. When this is the case, the percent of flood is usually defined at constant L/V rather than at constant L. The procedures for solving for the tower diameter are similar to those in step 1, except that the liquid-capacity factor at flooding is instead given by $C_{LF} = C_L/f$. For the present case, this would introduce a factor of 0.8 in the denominator of the second term in the brackets of Eq. (11.5).

11.2   *HYDRAULIC RATING OF A PACKED TOWER*

A 5.5-ft-diameter tower is to be used to countercurrently contact a vapor stream and a liquid stream. The mass flow rates are 150,000 lb/h for both. The liquid density is 50 lb/ft^3, and the gas density 1 lb/ft^3. Determine the approach to flooding at constant liquid loading L and at constant liquid-to-vapor ratio (L/V) for Flexipac type 2Y (FP2Y) structured packing.

Calculation Procedure

1. Calculate the liquid and vapor capacity factors, C_L and C_V.
For information about these factors, see Procedure 11.1. Use Eqs. (11.2), (11.3), and (11.4) from that problem to make the required calculation, bearing in mind that for a circular cross s ection, $A = \pi D^2/4$:

$$C_L = \frac{(150000)(4)}{(50)(\pi)(5.5)^2} \frac{7.48}{60} \sqrt{\frac{50}{50-1}} = 15.90 \text{ gal/(min)(ft}^2)$$

$$C_V = \frac{(150000)(4)}{(1)(\pi)(5.5)^2} \frac{1}{3600} \sqrt{\frac{1}{50-1}} = 0.250 \text{ ft/s}$$

2. Determine the flooding constants s and c.
From Table 11.1, the values of the flooding constants for FP2Y are

$$s = 0.044$$
$$c = 0.75$$

3. Determine the approach to flooding at constant L.
Use Eq. (11.1) with $C_{VF} = C_V/f$ and $C_{LF} = C_L$, where f is the fractional approach to flooding. Solve for f to give

$$f = \{(0.250)^{1/2}/[0.75 - (0.044)(15.90)^{1/2}]\}^2 = 0.757, \text{ or } 75.7 \text{ percent}$$

4. Determine the approach to flooding at constant L/V.
Use Eq. (11.1) with $C_{VF} = C_V/f$ and $C_{LF} = C_L/f$ where f is the fractional approach to flooding. Solve for f to give

$$f = \{[(0.250)^{1/2} + (0.044)(15.90)^{1/2}]/0.75\}^2 = 0.811, \text{ or } 81.1 \text{ percent}$$

Related Calculations.   To complete the hydraulic rating, the pressure drop at the design conditions is determined in the same way as for Procedure 11.1.

11.3   REQUIRED PACKING HEIGHT FOR ABSORPTION WITH STRAIGHT EQUILIBRIUM AND OPERATING LINES

An air stream containing 2% ammonia by volume (molecular weight = 28.96) is to be treated with water to remove the ammonia to a level of 53 ppm (the odor threshold concentration). The tower is to operate at 80 percent of flood defined at constant liquid to gas ratio (L/G). The liquid to vapor ratio is to be 25 percent greater than the minimum value. The absorption is to occur at 80°F and atmospheric pressure. What is the height of packing required for 2-in Flexiring random packing (2-in FR), and what height is required for Flexipac Type 2Y structured packing (FP2Y)? Determine the height using three methods: individual transfer units, overall transfer units, and theoretical stages.

Calculation Procedure

1. Construct the equilibrium line.
The equilibrium line relates the mole fraction of ammonia in the gas phase to that in the liquid phase when the two phases are at equilibrium. Equilibrium is assumed to exist between the two phases only at the gas-liquid interface. For dilute systems, Henry's law will apply. It applies for liquid mole fractions less than 0.01 in systems in general, and, as can be seen in Procedure 11.5, for the ammonia-water system it applies to liquid mole fractions as high as about 0.03. (Equilibrium data for this system are given in Perry's *Chemical Engineers' Handbook,* 4th ed., McGraw-Hill, New York, 1963, p. 14.4.)

From the data at low concentration, determine the Henry's law constant as a function of temperature for the temperature range of interest. To interpolate over a limited temperature range, fit the data to an equation of the form $\log_{10} \text{He} = a + (b/T)$. For the data cited.

$$\log_{10} \text{He} = 5.955 - (1778 \text{ K})/T \tag{11.9}$$

where T is absolute temperature in Kelvins and He is the Henry's law constant in atmospheres. Applying this equation at 80°F gives an He value of 1.06 atm.

Henry's law constant is defined as

$$\text{He} = Py_i/x_i \tag{11.10}$$

where P is the total pressure in atmospheres and y_i and x_i are, respectively, the gas-phase and liquid-phase mole fractions of ammonia in equilibrium with each other at the interface. The slope of the equilibrium line on the x-y operating diagram is

$$m = y_i/x_i = \text{He}/P \tag{11.11}$$

For operation at 1 atm, the equilibrium line is a straight line of slope 1.06 passing through the origin of the plot as shown in Fig. 11.2.

2. Locate the operating line of minimum slope.

The operating line gives the relationship between the bulk gas and liquid concentrations throughout the tower. A material balance around the tower is as follows:

$$L_M(x_1 - x_2) = G_M(y_1 - y_2)$$

or

$$L_M/G_M = (y_1 - y_2)/(x_1 - x_2) \tag{11.12}$$

where L_M and G_M are the liquid and gas molar fluxes (e.g., in pound-moles per square foot second), respectively, and the subscripts 1 and 2 refer to the bottom and top of the tower, respectively. For dilute systems, L_M and G_M can be taken as constants over the tower. From Eq. (11.12), the slope of the operating line is L_M/G_M, which for this example may be considered constant. This gives a straight operating line that can be constructed on the x-y diagram from a knowledge of the mole fractions at the top and bottom of the tower.

The operating line of minimum slope is the operating line that just touches the equilibrium line at one end (in this case, the bottom) of the tower. At the top of the tower, $y_2 = 53/10^6 = 5.3 \times 10^{-5}$ and $x_2 = 0$. At the bottom of the tower, $y_1 = 0.020$ and, by Eq. (11.11), $x_1 = 0.020/1.06 = 0.01887$. The slope of the operating line of minimum slope is $(L_M/G_M)_{\min} = (0.020 - 5.3 \times 10^{-5})/(0.01887 - 0) = 1.057$. Because the conditions at the top of the tower put the end of the operating line so close to the origin of the x-y diagram, the operating line of minimum slope is essentially coincident with the equilibrium line. In Fig. 11.2, the operating line of minimum slope would lie just above the equilibrium line but would be indistinguishable from it unless the plot in the region of the origin were greatly expanded.

3. Construct the actual operating line.

The design specifies operating at an L/G value of 1.25 times the minimum value determined in step 2. Accordingly, $L_M/G_M = (1.25)(1.057) = 1.32$. The values of y_1, y_2, and x_2 are given and remain the same as for step 2. The value of x_1 is determined from Eq. (11.12): $x_1 = (0.020 - 5.3 \times 10^{-5})/1.32 = 1.51 \times 10^{-5}$. Construct a straight line through the end points x_1, y_1 and x_2, y_2 to give the operating line as shown in Fig. 11.2.

4. Determine the liquid and gas loadings.

The liquid and gas loadings are defined by the liquid-to-gas ratio determined in step 3 and by the stipulation that the tower is to operate at an approach to flooding of 80 percent at constant L/G. For flooding defined at constant L/G, Eq. (11.1) becomes

$$(C_V/f)^{1/2} + s(C_L/f)^{1/2} = c \tag{11.13}$$

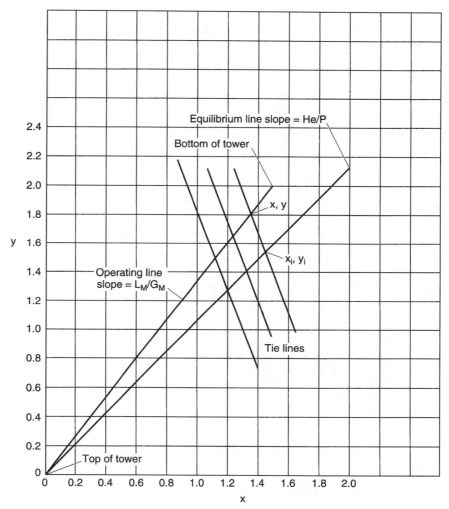

FIGURE 11.2 x-y operating diagram for Procedure 11.3.

From the definitions of C_V and C_L, their ratio can be related to the ratio of molar fluxes by

$$C_L/C_V = (L_M/G_M)(M_{wl}/M_{wg})(\rho_g/\rho_l)^{1/2} (7.48)(60) \ (\text{gal} \cdot \text{s})/(\text{min})(\text{ft}^3) \qquad (11.14)$$

In this equation, ρ_g is the gas density, ρ_l is the liquid density (62.3 lb/ft³), M_{wl} is the molecular weight of the liquid (18.02), M_{wg} is the molecular weight of the gas, and the numerical values 7.48 and 60 are the number of gallons per cubic foot and the number of seconds per minute, respectively. Use of the ideal-gas law yields a gas density of 0.0735 lb/ft³. Accordingly, Eq. (11.14) becomes $C_L/C_V = (1.32)(18.02/28.96)(0.0735/62.3)^{1/2} (448.8) = 12.7 \ (\text{gal} \cdot \text{s})/(\text{min}) \ (\text{ft}^3)$. Therefore, C_L in Eq. (11.13) can be replaced by 12.7 C_V. Upon rearrangement of Eq. (11.13) and substitution of the appropriate values of c and s from Table 11.1, we have

For 2-in FR,
$$C_V = \left(\frac{(0.74)(0.80)^{1/2}}{1 + 0.040\sqrt{12.7}} \right)^2 = 0.336 \text{ ft/s}$$

For FP2Y,
$$C_V = \left(\frac{(0.75)(0.80)^{1/2}}{1 + 0.044\sqrt{12.7}} \right)^2 = 0.336 \text{ ft/s}$$

$$C_L = 12.7\,C_V = 4.27 \text{ gal/(min)(ft}^2)$$

Thus, for this case, the two packings give the same loadings and would require the same tower diameter to operate at 80 percent of flood.

From the definition of the capacity factors and the F factor (see Procedure 11.1), alternative expressions for the gas and liquid loadings are

$$F_S = (0.336)(62.3 - 0.0735)^{1/2} = 2.65 \text{ (ft/s)(lb/ft}^3)^{1/2},$$

$$U_g = 2.65/(0.0735)^{1/2} = 9.77 \text{ ft/s},$$

and $$U_L = (4.27)[(62.3 - 0.0735)/62.3]^{1/2} = 4.27 \text{ gal/(min)(ft}^2)$$

5. Determine the HTUs for the packings at the design conditions.
For these packings, data available from the vendor give the values of H_g, the height of a transfer unit for the gas film, and H_l, the height of a transfer unit for the liquid film. At the design conditions, these values are

Packing	H_g, ft	H_l, ft
2-in FR	1.73	0.87
FP2Y	1.14	0.55

6. Determine the major resistance to mass transfer.
The major resistance to mass transfer can be in either the gas film or the liquid film. The film with the greater resistance will exhibit the greater driving force in consistent units of concentration. For dilute systems, the rate of transport from bulk gas to bulk liquid per unit lower volume $N_A\alpha$ is given, at a particular elevation in the tower, by

$$N_A\alpha = k_g\alpha(P/RT)(y - y_i) = k_l\alpha\rho_{Ml}(x_i - x) \qquad (11.15)$$

where $k_g\alpha$ and $k_l\alpha$ are the gas-side and liquid-side mass transfer coefficients, respectively, both in units of s^{-1}, P/RT and ρ_{Ml} are the gas and liquid molar densities, respectively, (lb · mol/ft^3), and $y - y_i$ and $x_i - x$ are the gas and liquid driving forces, respectively. The ratio of driving force R_{DF} in the gas to that in the liquid is

$$R_{DF} = (y - y_i)/(x_i - x)m = k_l\alpha\rho_{Ml}/k_g\alpha(P/RT)m \qquad (11.16)$$

where m is the slope of the equilibrium line.

This equation can be put into a more useful form. The stripping factor λ is defined as the ratio of the slope of the equilibrium line to that of the operating line:

$$\lambda = m/(L_M/G_M) \qquad (11.17)$$

For dilute systems, the height of a transfer unit for the gas resistance H_g and that for the liquid resistance H_l are related to their respective mass transfer coefficients by

$$H_g = U_g/k_g\alpha \quad \text{and} \quad H_l = U_l/k_l\alpha \qquad (11.18)$$

Noting that $L_M = U_l\rho_{Ml}$ and $G_M = U_g(P/RT)$, we can use Eqs. (11.17) and (11.18) to convert Eq. (11.16) to

$$R_{DF} = H_g/\lambda H_l \qquad (11.19)$$

If R_{DF} is greater than unity, the gas phase provides the major resistance to mass transfer; if R_{DF} is less than unity, the liquid phase provides the major resistance to mass transfer. From Eq. (11.17), $\lambda = 1.06/1.32 = 0.80$. For 2-in FR, $R_{DF} = 1.73/(0.80)(0.87) = 2.48$. For FP2Y, $R_{DF} = 1.14/(0.80)(0.55) = 2.59$. For both packings, the gas phase provides the major resistance to mass transfer. Therefore, greater precision will he obtained in the calculation of the packed height by using gas-phase transfer units.

7. Construct tie lines on the x-y operating diagram.

Tie lines are straight lines that connect corresponding points on the operating and equilibrium lines. The intersection of the tie line with the operating line gives the bulk concentration at a particular point in the tower; the intersection of the tie line with the equilibrium line gives the interfacial concentration at the same point in the tower. The slope of the tie line *TLS* is obtained from Eq. (11.15):

$$TLS = (y - y_i)/(x - x_i) = -k_l a \rho_{Ml}/k_g a(P/RT) = (H_g/H_l)(L_M/G_M) \qquad (11.20)$$

Various tie lines can be drawn between the operating and equilibrium line as shown in Fig. 11.2. (The diagram ignores the slight difference between the slopes for 2-in FR and for FP2Y.) These lines are useful in understanding the relationships between the various concentrations in the operating diagram. In the general case, the tie lines are essential in relating the bulk and interfacial concentrations so that the mass-transfer equations can be integrated. For complete gas-phase control, the tie line will be vertical; for complete liquid-phase control, the tie line will be horizontal.

8. Determine the equations for the height of packing.

Since the gas phase provides the major resistance to mass transfer, use the rate equation for transport across the gas film in the material balance on the gas phase in a differential height of the packed bed. Upon integration, this leads to the general equation for the height of packing:

$$Z = \int_{y_2}^{y_1} \frac{U_g}{k_g \alpha y_{BM}} \frac{y_{BM}\, dy}{(1 - y)(y - y_i)} \qquad (11.21)$$

where Z is the height of packing and y_{BM} is the log-mean average concentration of the nondiffusing gas (air in this case) between the bulk and the interface at a particular elevation in the tower. This quantity is considered in more detail in Procedure 11.5. For dilute systems such as the current example, y_{BM} and $(1 - y)$ are both approximately unity throughout the tower. With this simplification, Eq. (11.21) becomes

$$Z = \int_{y_2}^{y_1} \frac{U_g}{k_g a} \frac{dy}{(y - y_i)} = H_g \int_{y_2}^{y_1} \frac{dy}{(y - y_i)} = H_g N_g \qquad (11.22)$$

In this equation, it is assumed that H_g is constant over the height of the packed bed and can therefore be removed from the integral. The remaining integral function of y is the number of transfer units based on the gas film resistance N_g.

For dilute systems in which the equilibrium line and operating lines are straight, various analytical solutions of Eq. (1 1.22) are available. These may be listed as follows:

Individual Gas-Side Transfer Units

$$Z = H_g N_g$$

$$N_g = \frac{1 + (1/R_{DF})}{1 - \lambda} \ln\left[(1 - \lambda)\left(\frac{y_1 - mx_2}{y_2 - mx_2} \right) + \lambda \right]$$

$$R_{DF} = \frac{H_g}{\lambda H_l}$$

$$\lambda = \frac{m}{L_M/G_M} = \frac{He/P}{L_M/G_M}$$

(11.23)

When the absorption is completely gas-phase limited, R_{DF} approaches infinity and the $1/R_{DF}$ term of the above equation for N_g drops out.

Overall Gas-Side Transfer Units

The overall gas-side transfer units are calculated by assuming that all of the resistance to mass transfer is in the gas phase. The effect of liquid-phase resistance is taken into account by adjusting the height of the transfer unit from H_g to H_{og}. The equations are

$$Z = H_{og} N_{og}$$

$$H_{og} = H_g + \lambda H_l$$

$$N_{og} = \frac{1}{1 - \lambda} \ln \left[(1 - \lambda) \frac{y_1 - mx_2}{y_2 - mx_2} + \lambda \right] \tag{11.24}$$

$$\lambda = \frac{m}{L_M / G_M} = \frac{\text{He}/P}{L_M / G_M}$$

Overall Gas-Side Transfer Units Using the Log-Mean Driving Force

An alternative expression can be used to calculate N_{og} based on the log-mean concentration driving force across the tower. The equations are

$$N_{og} = \frac{y_1 - y_2}{(y - y^*)_{LM}} \tag{11.25}$$

$$(y - y^*)_{LM} = \frac{(y_1 - mx_1) - (y_2 - mx_2)}{\ln\left[(y_1 - mx_1)/(y_2 - mx_2)\right]}$$

where y is the bulk concentration and y^* is the interfacial concentration, assuming that all the resistance to mass transfer is in the gas phase, i.e., assuming that the tie lines of Fig. 11.2 are vertical. The liquid mole fraction at the bottom of the tower x_1 is determined by material balance around the tower.

Theoretical Stages

An analytical solution is also available for theoretical stages as opposed to transfer units, and the height equivalent to a theoretical plate (HETP). The number of theoretical stages or plates N_p can be determined by counting the steps between the operating line and the equilibrium line as is done with distillation problems (see Procedure 8.1 in Section 8, Distillation), but a more convenient analytical solution is as follows:

$$Z = (\text{HETP})N_p$$

$$\text{HETP} = H_{og} \frac{\ln \lambda}{\lambda - 1}$$

$$H_{og} = H_g + \lambda H_l \tag{11.26}$$

$$N_p = \frac{\ln\left[(1 - \lambda)\dfrac{(y_1 - mx_2)}{(y_2 - mx_2)} + \lambda\right]}{\ln\left(\dfrac{1}{\lambda}\right)}$$

TABLE 11.2 Values of Variables and Calculated Results for Packing Height

Parameter	Value for 2-in FR	Value for FP2Y
H_g	1.73 ft	1.14 ft
H_l	0.87 ft	0.55 ft
m	1.06	1.06
λ	0.80	0.80
R_{DF}	2.48	2.59
y_1	0.020	0.020
y_2	5.3×10^{-5}	5.3×10^{-5}
x_1	0.0151	0.0151
x_2	0	0
N_g	30.4	30.0
N_{og}	21.67	21.67
N_{og} (log mean)	21.67	21.67
N_p	19.42	19.42
H_{og}	2.43 ft	1.58 ft
HETP	2.71 ft	1.76 ft
Z	52.6 ft	34.2 ft

All of these forms are mathematically equivalent and give identical results for the height of the packing.

9. Calculate the height of packing.
Table 11.2 gives the values of the variables and the calculated results for Eqs. (11.23) through (11.26). The various sets of equations all yield the same height of packing. The height for the random packing is 54 percent greater than that of the structured packing. An appropriate safety factor should be added to the height of both packings for this and for subsequent examples.

11.4 PACKING HEIGHT FOR STRIPPING WITH STRAIGHT EQUILIBRIUM AND OPERATING LINES

A stream of groundwater flowing at a rate of 700 gal/min and containing 100 ppm of trichloroethylene (TCE) is to be stripped with air to reduce the TCE concentration to 5 ppb (drinking water quality). The tower is to be packed with 2-in polypropylene slotted rings and is to operate at 30 percent of flooding defined at constant liquid rate L, at a gas flow rate four times the minimum. Determine the tower diameter and height. Assume isothermal operation at 50°F and 1 atm. The density of the liquid stream is 62.4 lb/ft³; that of the gas stream is 0.0778 lb/ft³.

Calculation Procedure

1. Construct the equilibrium line.
For dilute solutions, Henry's law will apply. In general, it will be applicable for pressures under 2 atm and liquid mole fractions less than 0.01. For this example, the liquid mole fraction x is (100 lb TCE/10^6 lb soln) $(18/132) = 1.37 \times 10^{-5}$, where 18 and 132 are the molecular weights of water and trichloroethylene, respectively. Therefore, Henry's law will apply.

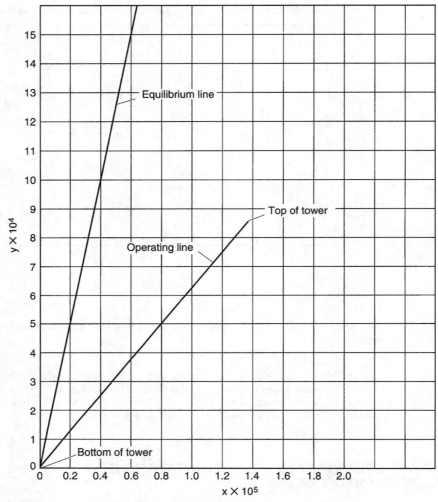

FIGURE 11.3 *x-y* operating diagram for Procedure 11.4.

The Henry's law constant can be obtained from the literature (e.g., C. Munz and P. V. Roberts, *Journal of the American Water Works Association.* 79(5), pp. 62–69, 1987). For a temperature of 50°F, a value of 248 atm is calculated.

For operation at 1 atm, accordingly, the equilibrium line is a straight line of slope 248 as shown in Fig. 11.3.

2. *Locate the operating line for minimum gas-phase molar flow rate,* G_M.

Because this system is dilute, the operating line will be straight. At the bottom of the tower, the concentration of TCE in the air coming into the tower may be assumed to be zero. The liquid mole fraction at 5 ppb, x_1 is (5 lb/10^9 lb)(18/132), or 6.85×10^{-10}. This places the coordinates for the bottom of the operating line very close to the origin of the *x-y* operating diagram. (For stripping, the operating line lies below the equilibrium line. For absorption, as seen in the preceding example, it lies above the equilibrium line.) The liquid mole fraction at the top of the tower x_2 is 1.37×10^{-5} as calculated in step 1.

The maximum slope of the operating line occurs when one end of it (in this case the top) intersects the equilibrium line. From Eq. (11.12), $y_2 = (248)(1.37 \times 10^{-5})$, or 0.00340. If we use these coordinates in Eq. (11.12), the maximum slope of the operating line is essentially the same as the slope of the equilibrium line, 248. Since the liquid-phase molar flow rate L_M is constant for this example, the maximum operating-line slope defines the minimum value of G_M.

3. *Locate the operating line.*
The actual gas loading is to be four times the minimum value. Thus, the slope of the operating line is one-fourth that of the operating line of maximum slope, or essentially one-fourth that of the equilibrium line. Thus, $L_M/G_M = 248/4 = 62$. By Eq. (11.12) with $y_1 = 0$, we calculate y_2 as (62) $(1.37 \times 10^{-5} - 6.85 \times 10^{-10})$, or 8.49×10^{-4}. Connect the coordinates at the bottom and the top of the tower (x_1, y_1 and x_2, y_2, respectively) with a straight line to give the operating line as shown in Fig. 11.3.

4. *Calculate the gas and liquid loadings and the tower diameter.*
Use Eq. (11.1) with $C_{VF} = C_V/f$ and $C_{LF} = C_L$ where f is the fractional approach to flooding, defined at constant liquid loading for this example. Express C_L in terms of C_V using Eq. (11.14). With $L_M/G_M = 62$, $\rho_l = 62.4$ lb/ft^3, and $\rho_g = 0.0778$ lb/ft^3, Eq. (11.14) gives $C_L/C_V = 611$ (gal · s)/(min)(ft^3).

Values of s and c for 2-in slotted rings in polypropylene may be obtained from the packing vendor or may be determined from air-water pressure-drop curves for the packing. The values are 0.042 and 0.72, respectively, in the units of Table 11.1. (Note that these values for polypropylene packing are close to the corresponding values for the same size of stainless steel packing in Table 11.1.) Substituting these values into Eq. (11.1) (as in step 4 of the previous example) gives

$$(C_V/0.3)^{1/2} + 0.042(611C_V)^{1/2} = 0.72$$

from which $C_V = 0.0632$ ft/s. And $C_L = 611\ C_V = 38.6$ gal/(min)(ft^2). From the definition of the capacity factors and F factor, other measures of the gas and liquid throughput may be given as $F_s = 0.499$ (ft/s) (lb/ft^3)$^{1/2}$, $U_g = 1.79$ ft/s, and $U_L = 38.6$ gal/(min)(ft^2).

Determine the tower diameter from the given flow rate and the liquid loading calculated above. The tower area is (700 gal/min)/[38.6 gal/(min)(ft^2)] = 18.1 ft^2. The tower diameter D is [(18.1) $(4)/\pi]^{1/2} = 4.8$ ft.

5. *Determine the HTU values for the liquid and gas resistances.*
Determining the values of H_l and H_g to use is usually the most uncertain part of absorption and stripping calculations. Values can often be obtained from the packing vendors. Generalized correlations are available in the literature, but these may not always give reliable results for aqueous systems. Published data on absorption or stripping of highly or sparingly soluble gases such as ammonia and carbon dioxide can be used, with appropriate adjustments. This method is illustrated in Procedure 11.6, where the following values are calculated for the conditions of this present example: $H_l = 3.48$ ft, and $H_g = 0.54$ ft.

6. *Determine the controlling resistance to mass transfer.*
The controlling resistance is determined from the ratio of the gas to the liquid driving force given by Eq. (11.19). From Eq. (11.17), $\lambda = 248/62 = 4.0$. With the HTUs from step 5, Eq. (11.19) gives $R_{DF} = 0.54/(4.0)(3.48) = 0.039$. Because R_{DF} is less than 1, the major resistance to mass transfer is on the liquid side of the interface. In this case, 96 percent of the resistance is in the liquid. Therefore, the best precision in the calculation of packing height will be obtained by using liquid film transfer units N_l or N_{ol}.

7. Calculate the number of transfer units.
For dilute systems in which the equilibrium and operating lines are both straight, the equations for the number of individual liquid-phase transfer units N_l and the number of overall liquid-phase transfer units N_{ol} are

$$N_l = \frac{1 + R_{DF}}{1 - (1/\lambda)} \ln \left\{ \left(1 - \frac{1}{\lambda}\right) \left[\frac{x_2 - (y_1/m)}{x_1 - (y_1/m)}\right] + \frac{1}{\lambda} \right\}$$

$$N_{ol} = \frac{1}{1 - (1/\lambda)} \ln \left\{ \left(1 - \frac{1}{\lambda}\right) \left[\frac{x_2 - (y_1/m)}{x_1 - (y_1/m)}\right] + \frac{1}{\lambda} \right\}$$

(11.27)

Substituting the appropriate values into these equations gives

$$N_l = \frac{1 + 0.039}{1 - (1/4.0)} \ln \left[\left(1 - \frac{1}{4.0}\right) \left(\frac{1.37 \times 10^{-5} - 0}{6.85 \times 10^{-10} - 0}\right) + \frac{1}{4.0} \right] = 13.32$$

$$N_{ol} = \frac{1}{1 - (1/4.0)} \ln \left[\left(1 - \frac{1}{4.0}\right) \left(\frac{1.37 \times 10^{-5} - 0}{6.85 \times 10^{-10} - 0}\right) + \frac{1}{4.0} \right] = 12.82$$

8. Calculate the height of packing required.
Based on individual liquid-transfer units, $Z = H_l N_l = (3.48 \text{ ft})(13.32) = 46.3$ ft. Based on overall liquid transfer units, $Z = H_{ol} N_{ol}$. The height of an overall transfer unit based on the liquid resistance is calculated as follows:

$$H_{ol} = H_l + H_g/\lambda$$

(11.28)

From this equation, $H_{ol} = 3.48 + (0.54/4.0) = 3.615$ ft, and $Z = (3.615 \text{ ft})(12.82) = 46.3$ ft.

Related Calculations. When the equilibrium and/or operating lines are curved, Eqs. (11.27) and (11.28) do not apply exactly. In this case, it is necessary to base the design on the individual number of transfer units for the liquid resistance and integrate graphically to determine N_l. This is illustrated in the following example.

11.5 PACKED HEIGHT FOR ABSORPTION WITH CURVED EQUILIBRIUM AND OPERATING LINES

An air stream containing 30% ammonia is to be contacted with a water stream containing 0.1% ammonia by weight (thus its mole fraction x_2 is 0.00106) in a tower packed with 2-in stainless steel Flexing (FR) random packing. The air leaving the tower is to contain 1% ammonia. The tower is to be designed to operate at 80% of flood defined at constant liquid-to-gas ratio L/G, and this ratio is to be twice the minimum value. Although significant heat effects would normally occur for this absorption, assume isothermal operation at 80°F and 1 atm total pressure.

Calculation Procedure

1. Construct the equilibrium line.
Equilibrium data for this system are given in *Perry's Chemical Engineers' Handbook*, 4th ed., p. 14.4, McGraw-Hill, New York (1963). From the data given, calculate values of y at temperatures of 10, 20, 30, and 40°C for a particular range of x values. For a particular x, fit the values of y to a regression equation of the form of Eq. (11.9) [i.e., $\log y = A + (B/T)$], and calculate the y value at 80°F (26.7°C). The equilibrium data interpolated in this way to 80°F are plotted to give the equilibrium line in Fig. 11.4. Henry's law is shown by the short dashed line in this figure. There is good

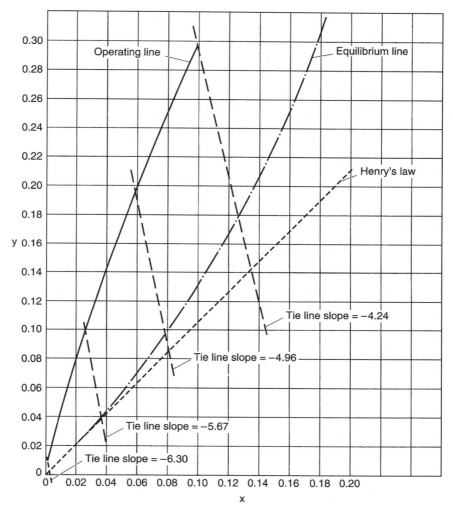

FIGURE 11.4   *x-y* operating diagram for Procedure 11.5.

agreement between the equilibrium curve and Henry's law below a liquid mole fraction of about 0.03 but increasing divergence as concentration increases above this value.

2. *Construct the limiting operating line.*

For concentrated solutions, significant changes will occur in the molar fluxes of gas and liquid. Therefore, base the material balance around the tower on the molar fluxes of the ammonia-free water and the ammonia-free air, L'_M and G'_M, respectively. Typical units are pound-moles of water (or air) per square foot second. The material balance then becomes

$$L'_M \left(\frac{x_1}{1 - x_1} - \frac{x_2}{1 - x_2} \right) = G'_M \left(\frac{y_1}{1 - y_1} - \frac{y_2}{1 - y_2} \right) \qquad (11.29)$$

The unknowns in this equation are L'_M, G'_M, and x_1. The limiting operating line is that which just touches the equilibrium line at the conditions corresponding to the bottom of the tower. From Fig. 11.4, x_1 is determined to be 0.177 from the equilibrium curve at $y_1 = 0.30$. Equation (11.29) is then solved for L_M'/G_M', and the values for the liquid and gas mole fractions are entered to give $(L_M'/G_M')_{min} = 1.955$.

It is not essential to actually construct the limiting operating line on the operating diagram. If desired, it can be constructed as shown below for the actual operating line, except that $(L_M'/G_M')_{min}$ is used rather than L_M'/G_M'.

3. Construct the actual operating line.

The actual L/G is specified to be two times the minimum L/G. Therefore, $L_M'/G_M' = 2(L_M'/G_M')_{min} = (2)(1.955) = 3.91$. Since the operating line is curved, corresponding values of x and y are needed at various heights between the top and bottom of the packing. These are determined by making a material balance around the top (or bottom) of the packing, where x and y are the mole fractions leaving the material balance control volume at some point in the bed. Equation (11.29) then becomes

$$\frac{y}{1-y} = \frac{L_M'}{G_M'}\left(\frac{x}{1-x} - \frac{x_2}{1-x_2}\right) + \frac{y_2}{1-y_2} = 3.91\left(\frac{x}{1-x} - 0.00106\right) + 0.0101 \quad (11.30)$$

Select various values of x between x_1 and x_2, substitute into Eq. (11.30), and solve for the corresponding values of y. Plot these values on the x-y operating diagram and draw a smooth curve through the points to give the operating line shown in Fig. 11.4. Application of Eq. (11.30) for $y = y_1$ gives $x_1 = 0.0975$ for the mole fraction of ammonia in the liquid leaving the tower. This defines the end of the operating line.

4. Determine the gas and liquid loadings at the bottom and top of the tower.

The maximum loadings will occur at the bottom of the tower, where the approach to flooding is specified to be 80 percent defined at constant L/G. In order to apply Eq. (11.1), the gas and liquid densities must be determined, and the ratio of C_L/C_V must be related to L_M'/G_M'. From the average molecular weight of the gas entering the tower (25.38) and the ideal-gas law, the gas density is 0.0644 lb/ft^3. From the concentration of ammonia in the liquid leaving the tower and from published data for specific gravity of aqueous ammonia solutions, the liquid density is 59.9 lb/ft^3.

The value of L_M/G_M at the bottom of the tower is given by

$$L_M/G_M = (L_M'/G_M')(1 - y_1)/(1 - x_1) = (3.91)(1 - 0.30)/(1 - 0.0975) = 3.03$$

The ratio of C_L to C_V is given by Eq. (11.14), with the average molecular weight of the liquid equal to 17.92 and the average molecular weight of the gas equal to 25.38. Substituting the appropriate values into Eq. (11.14) gives $C_L/C_V = 31.5$ (gal · s)/(min)(ft^3).

In Eq. (11.1), $C_{VF} = C_V/f$ and $C_{LF} = C_L/f$. The values of c and s for 2-in Flexiring packing are 0.74 and 0.040, respectively (see Table 11.1). Substituting these values into Eq. (11.1) gives $(C_V/0.80)^{1/2} + 0.040(31.5 \, C_V/0.80)^{1/2} = 0.74$. This gives $C_V = 0.292$ ft/s and $C_L = 31.5 \, C_V = 9.20$ gal/(min)(ft^2). The loadings at the bottom of the lower in other units are calculated from the defining equations as $F_S = 2.26$ (ft/s)(lb/ft^3)$^{1/2}$, $U_g = 8.90$ ft/s, $U_L = 9.20$ gal/(min)(ft^2), and $U_l = 0.0205$ ft/s.

Calculate the loadings at the top of the tower from the loadings at the bottom of the tower:

$$U_{gt} = U_{gb}\frac{1-y_1}{1-y_2} = 8.90\frac{(1-0.3)}{(1-0.01)} = 6.29 \text{ ft/s}$$

$$U_{lt} = U_{lb}\frac{\rho_{lb}}{M_{wlb}}\frac{(1-x_1)}{(1-x_2)}\frac{M_{wlt}}{\rho_{lt}} = 0.0205\left(\frac{59.9}{17.9}\right)\left(\frac{1-0.0975}{1-0.00106}\right)\left(\frac{18.02}{62.2}\right) = 0.01796 \text{ ft/s}$$

where the subscripts t and b refer to the top and bottom of the tower, respectively, and the 18.02 and 62.2 are the molecular weight and density of the liquid at the top of the tower. In terms of other units for loading, the F factor at the top is $(6.29)(0.0732)^{1/2} = 1.70$ (ft/s)(lb/ft^3)$^{1/2}$ and $U_L = (0.01796)(7.48)(60) = 8.06$ gal/(min)(ft^2). (Note that the superficial velocity in feet per second is equivalent to the volumetric flux in cubic feet per square foot-second.)

5. *Determine the individual HTUs for the liquid and gas resistances at the bottom and top of the tower.*

Vendor data for carbon dioxide desorption from water for 2-in FR packing, when corrected to the temperature and system physical properties for this example, give $H_{lb} = 0.83$ ft and $H_{lt} = 0.80$ ft.

Vendor data for ammonia absorption into water for 2-in FR packing, when corrected for temperature, give $H_{gb} = 1.42$ ft and $H_{gt} = 1.25$ ft.

The values of H_l at bottom and top and H_g at bottom and top are close enough to permit the use of average values. The average values are $H_l = 0.82$ ft and $H_g = 1.34$ ft.

6. *Construct the tie lines.*

The tie line slope (TLS) is established using procedures similar to those in Procedure 11.3. Equations (11.15) and (11.16) also apply to concentrated systems. For such systems, however, $k_g a$ and $k_l a$ are inversely proportional to y_{BM} and x_{BM}, respectively, which are the log mean mole fractions, between bulk and interface, of the nondiffusing component (air or water, respectively).

For concentrated systems, y_{BM} and x_{BM} can vary considerably across the tower. Therefore, in Eq. (11.21) the numerator and denominator are multiplied by y_{BM}. The first term under the integral is H_g for concentrated systems. This term is theoretically independent of concentration and pressure and is equal to the H_g for dilute systems, for which y_{BM} equals 1 as pointed out in Procedure 11.3. The variation of y_{BM} is included in the second term of Eq. (11.21), which integrates to N_g for concentrated systems.

When the loadings at the top and bottom of the tower are such that the value of H_g for dilute systems does not vary appreciably over the tower, as in this example, the first term under the integral of Eq. (11.21) can be considered to be constant at its average value and can be removed from the integral. The equations for a concentrated system are

$$Z = H_g N_g$$

$$H_g = \frac{U_g}{k_g a y_{BM}} \qquad H_l = \frac{U_l}{k_i a x_{BM}}$$

$$N_g = \int_{y_2}^{y_1} \frac{y_{BM} \, dy}{(1 - y)(y - y_i)} \tag{11.31}$$

$$\text{TLS} = -\frac{k_l a \rho_{Ml}}{k_g a (P/RT)} = -\frac{H_g}{H_l} \frac{L_M}{G_M} \frac{y_{BM}}{x_{BM}} = -\frac{H_g}{H_l} \frac{L'_M}{G'_M} \frac{(1 - y)}{(1 - x)} \frac{y_{BM}}{x_{BM}}$$

$$y_{BM} = \frac{(1 - y) - (1 - y_i)}{\ln\left(\dfrac{1 - y}{1 - y_i}\right)} \qquad x_{BM} = \frac{(1 - x) - (1 - x_i)}{\ln\left(\dfrac{1 - x}{1 - x_i}\right)}$$

In applying these equations, it is helpful to remember that the H_g for concentrated systems is the same as that for dilute systems at the same gas and liquid loadings. It is the $k_g a$ that varies with concentration and requires the introduction of y_{BM} to keep H_g constant.

Tie lines tie together corresponding concentrations on the operating line (bulk concentrations) and the equilibrium line (interfacial concentrations). The tie line slope is determined from the final equality for TLS in Eq. (11.31). The values of H_g and H_l are the average values determined in step 5 (1.34 ft and 0.82 ft, respectively). The value of L'_M/G'_M was determined in step 3 (3.91). Various tie lines are to be constructed over the entire concentration range covered by the operating line. A particular tie line will cross the operating line at x and y, the values of which are used in Eq. (11.31) to calculate y_{BM} and x_{BM}, and in the equation for TLS.

Unfortunately, the equations cannot be solved explicitly, because the equations for y_{BM} and x_{BM} contain the interfacial concentrations, which are unknown until the tie line slope is established.

A trial-and-error procedure is required to construct the tie lines. The procedure is illustrated here for the tie line that intersects the operating line at $y = 0.20$:

1. Locate y on the operating line: $y = 0.20$.
2. Determine x from the operating line or from the material balance, Eq. (11.29); $x = 0.059$.
3. Estimate the tie line slope assuming $y_{BM}/x_{BM} = 1$. Thus, $TLS_1 = -(1.34/0.82)(3.91)(1 - 0.20)/(1 - 0.059) = -5.43$.
4. Draw the estimated tie line on the operating diagram.
5. At the intersection of the estimated tie line and the equilibrium line, determine the estimated interfacial concentrations: $y_{i1} = 0.096$ and $x_{i1} = 0.078$.
6. Calculate the estimated values of y_{BM} and x_{BM}. Thus, $y_{BM1} = [(1 - 0.20) - (1 - 0.096)]/\ln[(1 - 0.20)/(1 - 0.096)] = 0.851$. Similarly, $x_{BM1} = 0.931$.
7. Calculate the second-iteration tie line slope. Thus, $TLS_2 = -(1.34/0.82)(3.91)[(1 - 0.20)/(1 - 0.059)][(0.851)/(0.931)] = -4.96$.
8. Repeat steps 4 through 7 until there is no further change in the tie line slope. For this case, the third iteration gives no further change and accordingly the slope is -4.96.
9. Draw the final tie line on the operating diagram. Slope $= -4.96$ passing through $y = 0.20$, $x = 0.059$. See Fig. 11.4.

This procedure is repeated at various intervals along the operating line. The more tie lines that are constructed, the more accurate the integration to determine N_g. Figure 11.4 shows 4 of the 20 tie lines constructed for the solution of this example.

7. *Determine the controlling resistance for mass transfer.*
For concentrated solutions, the ratio of the gas-phase driving force to that in the liquid phase in equivalent concentration units is

$$R_{DF} = (H_g/\lambda H_l)(y_{BM}/x_{BM}) \tag{11.32}$$

This differs from Eq. (11.19) for dilute solutions only by the ratio of y_{BM} to x_{BM}. Based on the average values, $H_g/H_l = 1.34/0.82 = 1.634$. Values of the other quantities in Eq. (11.32) vary over the tower. λ is the ratio of the slope of the equilibrium line to that of the operating line. Values of λ can be obtained directly from the operating diagram by determining the slope of the operating line at one end of a particular tie line and the slope of the equilibrium line at the other end of the same tie line. Values of x_{BM} and y_{BM} can be determined with the values of x and x_i and y and y_i at either end of the same tie line.

At the bottom of the tower, the slope of the operating line is 2.48 and the slope of the equilibrium line (determined at the point where it intersects a tie line of slope -4.24) is 1.98, y_{BM} is 0.76, and x_{BM} is 0.89. Accordingly, $R_{DFb} = (1.634)(2.48/1.98)(0.76/0.89) = 1.75$. Using the corresponding values at the top of the tower gives $R_{DFt} = 5.90$. As these values are both greater than 1, the gas phase provides the major resistance to mass transfer throughout the tower. Therefore, using N_g to calculate the number of transfer units in the tower will give the greater precision.

8. *Calculate the number of transfer units by graphic integration.*
From Eq. (11.31), N_g is equal to the area under a plot of $f(y)$ vs. y between the limits of y_2 and y_1 where $f(y)$ is $y_{BM}/[(1 - y)(y - y_i)]$. For each point of intersection between the operating line and a tie line, $f(y)$ is calculated and plotted against y. The plot is shown in Fig. 11.5. The area under the curve from $y_2 = 0.01$ to $y_1 = 0.30$ is 5.74.

9. *Calculate the height of packing.*
The height of packing is $Z = H_g N_g = (1.34 \text{ ft})(5.74) = 7.7$ ft.

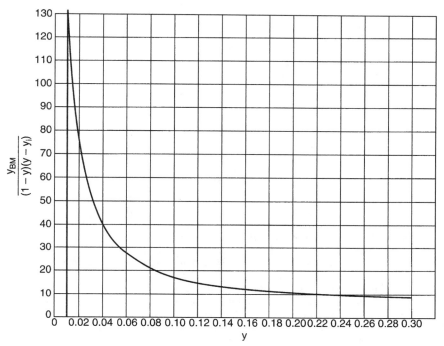

FIGURE 11.5   Graphic integration to determine number of transfer units for Procedure 11.5.

11.6 DETERMINING HTUs FOR AIR STRIPPING FROM DATA ON MODEL SYSTEMS

Determine the H_l and H_g to be used in Procedure 11.4 from data published for 2-in Pall rings on carbon dioxide stripping from 23°C water and on ammonia absorption into 20°C water. (See R. Billet and J. Mackowiak, *Chem. Eng. Technol.* 11:213–227, 1988.)

Calculation Procedure

1. Determine H_l for the model system at an equivalent liquid loading.
Below the load point, H_l is insensitive to the gas loading. Therefore, H_l can be correlated as a function of liquid loading only. For the packing of interest, the published data show a linear log-log relationship between $k_l\alpha$ and the volumetric liquid loading. This relationship can be described by the equation $k_l\alpha = 0.360\ U_l^{0.81}$, where $k_l\alpha$ is the mass transfer coefficient in s^{-1} and U_l is the superficial liquid velocity in feet per second (or equivalently, the superficial liquid flux in cubic feet per square foot-second). Since $H_l = U_l/k_l\alpha$, the equation for H_l becomes $H_l = 2.78\ U_l^{0.19}$, where H_l is in feet and U_l is in feet per second. For the liquid loading of Procedure 11.4, $U_l = 0.086$ ft/s ($U_L = 38.6$ gal/(min)(ft²) and $H_l = 2.78(0.086)^{0.19} = 1.74$ ft. This is the value of H_l for air stripping carbon dioxide from 23°C water at the given liquid loading.

2. Correct the H_l to the actual system physical properties.
The value of H_l is corrected from the model system to the actual system via the relationship that H_l is proportional to the square root of the Schmidt number (Sc = $\mu/\rho\mathcal{D}$, where μ is viscosity, ρ is density, and $\mathcal{D}$ is diffusion coefficient).

The quantities required for the Schmidt number are as follows. For the carbon dioxide-water system (CO_2-H_2O) at 23°C,

$$\mu = 0.936 \text{ cP} \quad \rho = 62.4 \text{ lb/ft}^3 \quad \mathcal{D} = 1.83 \times 10^{-5} \text{ cm}^2/\text{s}$$

For the trichloroethylene-water system $(TCE-H_2O)$ at 50°F,

$$\mu = 1.31 \text{ cP} \quad \rho = 62.4 \text{ lb/ft}^3 \quad \mathcal{D} = 6.4 \times 10^{-6} \text{ cm}^2/\text{s}$$

$$\frac{(H_l)_{TCE-H_2O}}{(H_l)_{CO_2-H_2O}} = \left[\frac{\left(\dfrac{\mu}{\rho\mathcal{D}}\right)_{TCE-H_2O}}{\left(\dfrac{\mu}{\rho\mathcal{D}}\right)_{CO_2-H_2O}}\right]^{1/2} = \left[\frac{(1.31)}{(0.936)}\frac{(1.83\times10^{-5})}{(6.4\times10^{-6})}\right]^{1/2} = 2.0$$

Therefore, the value of H_l for trichloroethylene-water is $(1.74 \text{ ft})(2.0) = 3.48$ ft.

3. Determine the value of H_g for the model system at equivalent loadings.

Determination of H_g is somewhat more complicated, because the cited reference gives H_{og} rather than H_g and these both depend on both the gas and liquid loadings. The data indicate that H_{og} is proportional to $U_l^{-0.48}$ and to $F_s^{0.38}$. For Procedure 11.4, the F factor is essentially 0.50 $(\text{ft/s})(\text{lb/ft}^3)^{1/2}$. For this F factor, the cited reference gives H_{og} as 0.80 ft. This value is for the ammonia-air-water system at 20°C and a liquid loading of 6.14 gal/(min)(ft²). Correcting this value to the proper liquid loading gives $H_{og} = (0.80 \text{ ft})(38.6/6.14)^{-0.48} = 0.33$ ft.

H_g and H_{og} are related by the equation $H_{og} = H_g + \lambda H_l$, where λ is the ratio of the slope of the equilibrium line to that of the operating line, as discussed in Procedure 11.3, step 6. For the ammonia-air-water system at 293 K, Eq. (11.9) in Procedure 11.3 gives He = 0.77 atm. From Eq. (11.11) of Procedure 11.3, the slope of the equilibrium line m is He/P, i.e., 0.77. From Procedure 11.4 $L_M/G_M = 62$. Accordingly, $\lambda = (0.77/62) = 0.0124$. The value of H_l determined in step 1 is 1.74 ft for the carbon dioxide-water system at 23°C. This value is corrected to the ammonia-water system at 20°C by using the square root of the Schmidt-number ratio. The values for the Schmidt number for the carbon dioxide-water system are given in step 2. For the ammonia-water system at 20°C, $\mu = 1.005$ cP, $\rho = 62.4$ lb/ft³, and $\mathcal{D} = 2.06 \times 10^{-5}$ cm²/s. The value of H_l for the ammonia-water system is $[1.74 \text{ ft}][(1.005/0.936)(1.83 \times 10^{-5}/2.06 \times 10^{-5})]^{1/2} = 1.70$ ft. Then, $H_g = H_{og} - \lambda H_l = 0.33 - (0.0124)(1.70) = 0.31$ ft.

4. Correct H_g to the actual system properties.

The H_g for the model system is corrected to that for the actual system using the relationship that H_g is proportional to the square root of the Schmidt number. The values are as follows. For ammonia-air at 20°C,

$$\mu = 0.018 \text{ cP}$$
$$\rho = 0.0752 \text{ lb/ft}^3$$
$$\mathcal{D} = 0.222 \text{ cm}^2/\text{s}$$

For trichloroethylene-air at 50°F,

$$\mu = 0.018 \text{ cP}$$
$$\rho = 0.0778 \text{ lb/ft}^3$$
$$\mathcal{D} = 0.070 \text{ cm}^2/\text{s}$$

The value of H_g for the actual system is, accordingly,

$$H_g = [0.31 \text{ ft}][(0.0752/0.778)(0.222/0.070)]^{1/2} = 0.54 \text{ ft}$$

11.7   PACKED HEIGHT FOR ABSORPTION WITH EFFECTIVELY INSTANTANEOUS IRREVERSIBLE CHEMICAL REACTION

A tower packed with Flexipac Type 2Y structured packing (FP2Y) is to be used to remove 99.5% of the ammonia from an air stream containing an ammonia mole fraction (y_1) of 0.005. The lower is to operate isothermally at 100°F and 1 atm. The ammonia is to be absorbed in an aqueous solution of nitric acid. The tower diameter has been selected to give an F factor of 2.0 $(ft/s)(lb/ft^3)^{1/2}$ and a liquid loading of 15 gal/(min)(ft^2). The density of the air stream is 0.0709 lb/ft^3, and that of the water is 62.0 lb/ft^3. What concentration of acid should be used to ensure that the maximum absorption rate is obtained throughout the entire tower? Compare the height of packing required for absorption with chemical reaction to that for physical absorption.

Calculation Procedure

1. Identify the reaction and estimate its rate and reversibility.
The absorption of ammonia with the liquid-phase reaction between dissolved ammonia and a strong acid can be represented by

$$NH_3(g) \leftrightarrow NH_3(aq)$$

$$NH_3(aq) + H^+ \leftrightarrow NH_4^+$$

In general, reactions that consist simply of the transfer of hydrogen ions can be considered to occur "instantaneously." Accordingly, the reaction rate in this example will be considered instantaneous.

All reactions are, to some extent, reversible. This is particularly true of ionic reactions that occur in aqueous solution, as is the case here. The designation of a reaction as "irreversible" (as in this example) is not to say that the reaction cannot proceed in the reverse direction. Rather, it signifies that under the conditions of the absorption, the equilibrium lies strongly in favor of the reaction products.

The criterion for a reaction being effectively irreversible with respect to the absorption of a gas is that the concentration of the unreacted gas in the solution [e.g., $NH_3(aq)$] is so small that its partial pressure at the interface is much less than the partial pressure of the absorbing gas [e.g., $NH_3(g)$] in the gas phase. When this criterion is satisfied, the interfacial mole fraction of the absorbing gas y_i can be considered to be zero. This obviates the need for an equilibrium line or an x-y operating diagram.

The second reaction shown above will be shifted in favor of the product by a large equilibrium constant and by a sufficient concentration of hydrogen ion in solution. For now, the reaction will be assumed to be effectively irreversible, although this assumption will be checked in step 5.

2. Determine the gas and liquid loadings.
From the loadings given for this example and the densities of air and water, calculate the molar fluxes. The superficial gas velocity is (see Procedure 11.1, step 2) $U_g = F_s / (\rho_g)^{1/2} = 2/(0.0709)^{1/2} = 7.51$ ft/s. The molar flux of gas is $G_M = U_g(P/RT) = (7.51)(1)/[(0.73)(560)] = 0.0184$(lb $\cdot$ mol)/(ft^2)(s). The molar flux of liquid is $L_M = U_l \rho_{Ml} = (15)(62.0)/[(7.48)(60)(18)] = 0.115$ (lb $\cdot$ mol)/(ft^2)(s). From these values, the liquid to gas ratio is $L_M/G_M = 0.115/0.0184 = 6.25$.

Note that in the calculation of the liquid molar flux, we assume that the acid concentration is so low that we can use the molecular weight of water as the liquid molecular weight. This assumption later proves to be valid.

3. Determine the HTUs for the packing at the design conditions.
When corrected to the design conditions of this example, vendor data on model systems for this packing give $H_g = 0.84$ ft; $H_l = 0.58$ ft.

4. *Calculate the required concentration of absorbent.*
Enough nitric acid must be used to maximize the rate of absorption throughout the tower. As the acid concentration is increased, the reaction plane moves closer and closer to the interface until, at a particular concentration, the reaction plane coincides with the interface. When this condition prevails, the concentration of unreacted $NH_3(aq)$ at the interface becomes zero, the absorption becomes completely limited by the gas film, and the rate of absorption is maximized because y_i is also zero. When this condition prevails throughout the tower, the maximum rate of absorption will be obtained.

The rate of absorption is equal to the rate of mass transfer through the gas film at the interface, which, at steady state, is equal to the rate of mass transfer through the liquid film. (The results of the film theory are used here because of their simplicity and ease of use in engineering calculations.) The rate of absorption is given by

$$\bar{R}a = k_g a(P/RT)(y - y_i) = k_i a \rho_{Ml} x_i [1 + (\mathcal{D}_B / z \mathcal{D}_A)(x_B / x_i)] \tag{11.33}$$

where $\mathcal{D}_B$ = diffusion coefficient of reactant (e.g., H^+) in solution
$\mathcal{D}_A$ = diffusion coefficient of dissolved gas (e.g., NH_3) in solution
ρ_{Ml} = molar density, mol/ft^3
z = number of moles of B (e.g., H^+) reacting with 1 mol of A (e.g., NH_3)
x_B = mole fraction of reactant (e.g.. H^+) in bulk liquid

Other mole fractions in Eq. (11.33) refer to the absorbing gas, as with previous notation. The term in parentheses on the right of Eq. (11.33) is the enhancement factor for chemical reaction in the liquid film. As x_B increases, x_i and y_i decrease until they become effectively zero.

From Eq. (11.10), $x_i = y_i P/\text{He}$. Substituting for x_i in Eq. (11.33) and solving for y_i gives

$$y_i = \frac{k_g a(P/RT)y - k_i a \rho_{Ml} \dfrac{\mathcal{D}_B x_B}{\mathcal{D}_A z}}{k_g a(P/RT) + k_i a \rho_{Ml}(P/\text{He})} \tag{11.34}$$

It is clear from Eq. (11.34) that y_i will be zero if the second term in the numerator is greater than the first. This leads to the desired criterion for x_B:

$$x_B \geq \frac{H_l}{H_g} \frac{G_M}{L_M} \frac{\mathcal{D}_A}{\mathcal{D}_B} zy$$

where Eq. (11.18), Procedure 11.3, has been used to convert from mass transfer coefficients to HTUs. The liquid-phase diffusion coefficients of NH_3 and H^+ (actually H_3O^+) are approximately equal, and the value of z is 1. The maximum value of y occurs at the bottom of the tower, where $y_1 = 0.005$. Therefore, the minimum value of x_B that will maximize the rate of absorption is $x_B = (0.58/0.84)(0.005/6.25) = 0.000552$.

This is the minimum concentration at the bottom of the tower. The concentration of nitric acid in the liquid feed to the top of the tower must be sufficient to react with all of the ammonia absorbed in addition to providing for x_{B1} to be 0.000552. By material balance, the mole fraction of H^+ in the feed to the tower is $x_{B2} = [(0.005 - 2.5 \times 10^{-5})/6.25] + 0.000552 = 0.00135$. This mole fraction is converted to concentration by multiplying by the liquid molar density [55.10 (g · mol)/L] to give $[B]_2 = 0.0743$ (g · mol)/L, which corresponds to a pH of 1.13. At the bottom of the tower, $[B]_1 = 0.0304$ (g · mol)/L, which gives a pH of 1.52.

5. Check the assumption of irreversibility.
For dissociation of the ammonium ion,

$$NH_4^+ \leftrightarrow H^+ + NH_3(aq)$$

$$K_a = \frac{[H^+][NH_3(aq)]}{[NH_4^+]} = 5.75 \times 10^{-10} (g \cdot mol)/L \quad \text{at } 25°C \tag{11.35}$$

$$[NH_3(aq)] = 5.75 \times 10^{-10} \frac{[NH_4^+]}{[H^+]}$$

The concentrations of NH_4^+ and H+ must be such that the concentration of ammonia calculated by Eq. (11.35) exerts a partial pressure which is negligible compared to the gas-phase partial pressure of ammonia. The minimum gas-phase partial pressure occurs at the top of the tower. The maximum concentration NH_4^+ and the minimum concentration of H^+ occur at the bottom of the tower. From the concentrations calculated in step 4, $[H^+] = 0.0304$ (g · mol)/L and $[NH_4^+] = 0.0742 - 0.0304 = 0.0438$ (g · mol)/L, and by Eq. (11.35), $[NH_3(aq)] = 8.28 \times 10^{-10}$ (g · mol)/L. Dividing by the liquid molar density gives the mole fraction $x = 8.28 \times 10^{-10}/55.10 = 1.50 \times 10^{-11}$. From Eq. (11.10), with He = 1.724 atm at 100°F, this mole fraction is equivalent to a partial pressure p of $(1.724)(1.50 \times 10^{-11})$ or 2.59×10^{-11} atm. This is clearly much less than the partial pressure at either the inlet (0.005 atm) or the outlet (2.5×10^{-5} atm) of the tower. Therefore, the reaction is effectively irreversible.

The above calculation assumes that the liquid makes only a single pass through the tower. In practice, a more effective way of treating the gas would be to recirculate the liquid with a small feed of concentrated acid to the recirculation loop and a small bleed of the recirculation loop contents. This would permit the NH_4^+ concentration to build up in the recirculation loop. However, even if this concentration built up to the solubility limit of ammonium nitrate ($NH_4 NO_3$), about 30 (g · mol)/L, the partial pressure exerted by the dissolved ammonia at the top of the tower would still be nearly two orders of magnitude below the partial pressure of ammonia in the gas at the top of the tower. Therefore, for all practical conditions, the reaction is effectively irreversible, given the acid concentrations calculated in step 4.

6. Calculate the number of gas-phase transfer units.
For dilute systems in which the interfacial mole fraction y_i is kept at zero by virtue of the chemical reaction, the expression for N_g from Eq. (11.22), Procedure 11.3, becomes

$$N_g = \int_{y_2}^{y_1} \frac{dy}{y} = \ln \frac{y_1}{y_2}$$

Substituting the appropriate values for y gives

$$N_g = \ln(0.005/2.5 \times 10^{-5}) = 5.30$$

7. Calculate the height of packing required.
The height of packing required is $Z = H_g N_g = (0.84 \text{ ft})(5.30) = 4.45$ ft.

8. Calculate the height of packing for physical absorption at the same conditions.
At 100°F, Eq. (11.9), Procedure 11.3, gives He = 1.72 atm. Equation (11.11) gives $m = 1.72$. By Eq. (11.17), $\lambda = 1.72/6.25 = 0.275$. Equation (11.19) gives $R_{DF} = 0.84/[(0.275)(0.58)] = 5.27$. And Eq. (11.23) for N_g gives

$$N_g = \frac{1 + (1/5.27)}{1 - 0.275} \ln\left[(1 - 0.275)\left(\frac{0.0050 - 0}{2.5 \times 10^{-5} - 0} \right) + 0.275 \right] = 8.17$$

The required height of packing is $Z = H_g N_g = (0.84 \text{ ft})(8.17) = 6.86$ ft. The same result is obtained with Eq. (11.24).

For this example, absorption with chemical reaction gives a 34 percent reduction in the height of the packing relative to absorption without chemical reaction.

*11.8   PRELIMINARY PROCESS DESIGN OF AN ABSORBER

Make a preliminary process design for an absorber to be used (in conjunction with a stripper, and auxiliary equipment) for recovering a crude ethane from a by product gas.

Data

1. Gas available, 250,000 SCF/h
2. Gas composition

H_2...70 mol %
CH_4.. 16 mol %
C_2H_6..12 mol %
C_3H_8..2 mol %

3. Absorption oil,
specific gravity 60/60°F (15.6/15.6°C) = 0.825 (40° A.P.I.)

Approximate molecular weight, 200

4. Ethane recovery desired, 95%
5. Cooling water available at 75°F (23.9°C)

Trial assumptions

1. Operating pressure, 400 psia.
2. Inlet gas and oil temperatures, 85°F (29.4°C)
3. Exit temperature of oil, 90°F (32.2°C)
4. Dissolved gas in oil returned from stripper

CH_4... 0.00 mol %
C_2H_4... 0.10 mol %
C_3H_8... 0.20 mol %

5. Absorption of hydrogen is negligible.
6. Design for 10 theoretical plates based on base conditions.
7. Equilibrium constant data published by Brown (*Natural Gasoline Supply Men's Association Technical Manual,* 4th ed., 1942.)

Calculation Procedure

1. *Determine the gas charged on a 1h basis.*

Gas charged, $\dfrac{250,000}{379}$ = 660 lb moles. Contains

Hydrogen	$660 \times 0.70 = 462$ lb moles
Methane	$660 \times 0.16 = 105.6$ lb moles
Ethane	$660 \times 0.12 = 79.2$ lb moles
Propane	$660 \times 0.02 = 13.2$ lb moles

2. *Find the limiting recovery for gas in equilibrium with oil.*
Limiting recovery for gas in equilibrium with oil

Ethane: $k = \dfrac{y}{x} = 1.46$

$$x = 0.001 \quad \text{(inlet oil)}$$
$$y^* = kx = 0.00146$$

Propane: $k = 0.47$

$$x = 0.002$$
$$y^* = 0.002 \times 0.47 = 0.0009$$

Methane: $k = 9.4$ but insufficient oil will be circulated for complete absorption

Assume: $100 \times \dfrac{1.46}{9.4} \times 1.2 = 19$ percent absorption

Residue gas (for equilibrium)

$$
\begin{aligned}
H_2 &= & 462 \text{ lb moles} \\
CH_4 &= (1 - 0.19) \times 105.6 = \ldots\ldots\ldots\ldots & \underline{85.5} \\
& & 547.5
\end{aligned}
$$

$$C_2H_6 = \frac{0.00146}{1 - 0.00146 - 0.0009} \times 547.5 = 0.80 \text{ lb mole}$$

$$C_3H_8 = \frac{0.0009}{1 - 0.00146 - 0.0009} \times 547.5 = 0.48 \text{ lb mole}$$

Fractional limiting recoveries

$$C_2H_6 = \frac{79.2 - 0.80}{79.2} = 0.990$$

$$C_3H_8 = \frac{13.2 - 0.48}{13.2} = 0.964$$

3. *Base the design on conditions at the gas inlet.*

Design based on conditions at gas inlet, absorption efficiency required for 95 percent absorption of ethane, $E = \dfrac{0.95}{0.99} = 0.96$.

From a plot of Kremser absorber characteristics, where kG = the product of the vaporization constant and mols of total gas per hr,

$$E = 0.96; n = 10; A = 1.18 = \frac{L}{kG}, \text{ or } L = 1.18 \, kG.$$

At assumed temperature of base = 90°F (32.2°C), $k = 1.51$

Total liquid rate, $L = 1.18 \times 1.51 \times 660 = 1178$ lb moles

Absorption factors for propane

$$k = 0.50$$

$$A = \frac{1178}{0.5 \times 660} = 3.55$$

E is nearly 1

Fraction propane absorbed = $1 \times 0.964 = 0.964$

Absorption factor for methane,

$$k = 9.5$$

$$A = \frac{1178}{9.5 \times 660} = 0.188 = \text{fraction absorbed}$$

4. *Find the gas quantities absorbed.*
Quantities absorbed

$$H_2 = \text{none}$$
$$CH_4 = 0.188 \times 105.6 = \quad 19.9 \text{ lb moles}$$
$$C_2H_6 = 0.95 \times 79.2 \quad = \quad 75.2$$
$$C_3H_8 = 0.964 \times 13.2 \quad = \quad \underline{12.7}$$
$$\text{Total} \dots\dots\dots\dots \quad = 107.8 \text{ lb moles}$$

5. *Analyze the inlet oil.*

$$\text{Inlet oil} = 1178 - 108 = 1070 \text{ lb moles}$$

$$1070 \times 200 = 214{,}000 \text{ lb } (97{,}156 \text{ kg})$$

$$\frac{214{,}000}{0.825 \times 8.33} = 31{,}100 \text{ gal } (520 \text{ gpm}) (32.8 \; L/s)$$

$$\frac{31{,}100}{250} = 124 \text{ gal/1000 SCF}$$

6. *Find the temperature at the base of the column.*
Temperature at base of column
  Heats of solution from Perry—*Chemical Engineering Handbook*

$$CH_4 = \text{small, neglect}$$
$$C_2H_6 = 75.2 \times 5500 = 420{,}000 \text{ Btu (rounded values)}$$
$$C_3H_8 = 12.7 \times 6500 = \underline{80{,}000} \text{ Btu}$$
$$\text{Total} \quad 500{,}000 \text{ Btu } (527{,}500 \text{ kJ})$$

Specific heat of oil from Perry = 0.467 Btu/(lb)(°F) (1.96 kJ/kg k)

$$\text{Temperature rise} = \frac{500{,}000}{0.467 \times 214{,}000} = 5°F(-15°C)$$

Temperature = 85 + 5 = 90°F (32.2°C).
This checks the temperature used.

7. *Base the design on conditions at the top of the column.*
Design based on conditions at top of column

$$\text{Liquid charged} = 1070 \text{ lb moles}$$
$$\text{Gas leaving} = 660 - 98 = 562 \text{ lb moles}$$

Equilibrium constant 85°F, (29.4°C), 400 psig for C_2H_6, $k = 1.46$.
Absorption factor for ethane

$$A = \frac{L}{kG} = \frac{1070}{1.46 \times 562} = 1.30$$

From a Kremser plot of absorber characteristics with 10 trays, $E = 0.97$.
   This indicates that 10 theoretical plates would actually give somewhat better than the absorption efficiency of 0.96 required.

8. *Find the number of plates required for this absorber.*
Corrected number of theoretical plates

$$A_E = \sqrt{A_B(A_T + 1) + 0.25} - 0.50$$
$$= \sqrt{1.18 \times 2.30 + 0.25} - 0.50 = 1.23$$

For absorption efficiency = 0.96 and A_E = 1.23, eight theoretical plates are indicated by a Kremser plot of absorber tray characteristics.

Related Calculations.   These calculations show that a circulation of 520 gpm (32.8 L/s) of lean oil and eight theoretical plates comprise a satisfactory process design. Calculations for more or less theoretical plates and different pressures are required before the best process design could be stated. Such factors are, however, seldom very critical, i.e., small changes do not affect costs greatly. For preliminary considerations, less detailed calculations may be employed. The best short cuts vary with circumstances and cannot be recommended in advance. The reader can readily see that many of the corrections made could have been omitted without undue loss of accuracy. Trial assumption 4 in procedure 11.8 would constitute the basis for stripper design.

It is usually necessary to take the heat of absorption into account to secure the temperature at the base before accurately computing the absorption factor. If the heat capacity of the solvent is large compared to that of the gas, the gas will leave the column nearly at the temperature of the inlet liquid. The heat of solution then will all go to heating the liquid, and the temperature rise of the liquid times its heat capacity will equal the heat of solution. If the heat capacity of the gas is comparable with that of the liquid, it will absorb some of the heat also.

The absorption efficiency is the fraction of the soluble gas component that is absorbed by a solvent initially free from dissolved gas. In case the inlet liquid contains some dissolved gas, it can then at best reduce the exit gas only to the equilibrium concentration. One may compute this concentration from the equilibrium involved. From this the maximum possible fractional absorption of the gas may be computed. *This fraction times the absorption efficiency E, is the net fraction absorbed.* Repeating this in specific units will give the number of plates required. Nomenclature:

Equilibrium concentration exit gas, kx_1
Minimum quantity of gas not absorbed, G_2kx_1
Quantity of gas in inlet stream, G_1y_1
Maximum possible fraction absorbed, $1 - \dfrac{G_2kx_1}{G_1y_1}$
Fraction actually absorbed, $E\left(1 - \dfrac{G_2kx_1}{G_1y_1}\right)$

This procedure is the work of Loyal Clark, Chemical Engineer.

*11.9 ABSORPTION TOWER FLOW AND ABSORPTION RATE

A mixture of ammonia and air at a pressure of 745 mmHg and a temperature of 40°C (104°F) contains 4.9 % NH_3 by volume. The gas is passed at a rate of 100 cfm (2.8 cm^3/min) through an absorption tower in which only ammonia is removed. The gases leave the tower at a pressure of 740 mmHg, a temperature of 20°C, and contain 0.13 % NH_3 by volume. Find the rate of flow of gas leaving the tower and the weight of ammonia absorbed in the tower per minute.

Calculation Procedure

1. *Determine the gas quantities at the top and the base of the tower.*
At top, 740 mm, 20°C, 0.13 % NH_3 by volume

At base, 745 mm, 40°C, 4.90 % NH_3 by volume 100 cfm (2.8 m^3/min) gas
100 % − 4.90 % = 95.1 % = 0.951.

$$\text{Air, cfm} = 100 \times 0.951 = 95.1 \text{ cfm } (2.7 \text{ m}^3/\text{min}).$$

2. *Compute the mole-volume of the ammonia gas.*

$$(mV_1) = (mV_0) \times \frac{T_1}{T_0} \times \frac{P_0}{P_1} = \text{mole-volume at other than standard condition}$$

$$(mV_1) \text{ for air at 745 mm and 40°C } = 359 \times \frac{(273 + 40)}{273} \times \frac{760}{745}$$
$$= 421 \text{ cu ft/mol (11.9 cu m/mol)}$$

$$\frac{95.1}{421.0} = 0.226 \text{ mol of air flowing per minute}$$

$$\frac{\text{Mole } NH_3}{\text{Mole air}} \text{ at entrance} = \frac{0.049}{0.951} = 0.0515$$

$$\frac{\text{Mole } NH_3}{\text{Mole air}} \text{ at exit} = \frac{0.0013}{0.9987} = 0.0013$$

Mole ammonia absorbed per mole air flowing
$$= 0.0515 - 0.0013 = 0.0502$$

3. *Find the rate of flow of the gas at the exit.*
Rate of flow of gas at exit in cfm

$$V = 0.226 \times 359 \times \frac{(273 + 20)}{273} \times \frac{760}{740} = 89.5 \text{ cfm } (2.5 \text{ m}^3/\text{min})$$

4. *Determine the weight of ammonia absorbed per minute.*

$$\text{Weight of ammonia absorbed per minute} = 0.226 \times 0.0502 \times 17$$
$$= 0.193 \text{ lb (0.087 kg) } NH_3 \text{ per minute}$$

where atomic weight $NH_3 = 17$

This procedure is the work of William S. La Londe, Jr., P.E.

SECTION 12
LIQUID AGITATION

David S. Dickey, Ph.D.
MixTech, Inc.
Dayton, OH

12.1 POWER REQUIRED TO ROTATE AN AGITATOR IMPELLER

For a pitched-blade turbine impeller that is 58 in (1.47 m) in diameter and has four 12-in-wide (0.305-m) blades mounted at a 45° angle, determine the power required to operate the impeller at 84 r/min (1.4 r/s) in a liquid with a specific gravity of 1.15 (1150 kg/m^3) and a viscosity of 12,000 cP (12 Pa · s). What size standard electric motor should be used to drive an agitator using this impeller?

Calculation Procedure

1. Determine the turbulent power number for impeller geometry.
Power number N_P is a dimensionless variable [5] which relates impeller power P to such operating variables as liquid density ρ, agitator rotational speed N, and impeller diameter D as follows:

$$N_P = \frac{P}{\rho N^3 D^5}$$

A conversion factor (see below) is used when working with English engineering units; no factor is necessary for SI metric units. For a given impeller geometry, the power number is a constant for conditions of turbulent agitation. Values of turbulent power numbers for some agitator impellers are shown in Fig. 12.1.

The pitched-blade impeller described in this example is similar to the four-blade impeller shown in the figure, except that the blade width-to-diameter ratio W/D is not exactly ⅕. To correct for the effect of a nonstandard W/D on a four-blade impeller, a factor of actual W/D to standard W/D raised

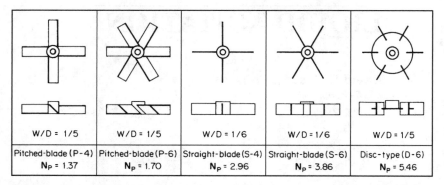

W/D = 1/5	W/D = 1/5	W/D = 1/6	W/D = 1/6	W/D = 1/5
Pitched-blade (P-4)	Pitched-blade (P-6)	Straight-blade (S-4)	Straight-blade (S-6)	Disc-type (D-6)
N_P = 1.37	N_P = 1.70	N_P = 2.96	N_P = 3.86	N_P = 5.46

FIGURE 12.1 Values of turbulent power number N_P for various impeller geometries. Note: W/D is actual blade-width-to-impeller-diameter ratio.

to the 1.25 power must be applied to the standard turbulent power number $N_P = 1.37$. Thus the turbulent power number for a 58-in-diameter impeller with a 12-in blade width is

$$N_P = 1.37[(12/58)/1/5]^{1.25} = 1.37(1.034)^{1.25} = 1.43$$

(The correction factor for nonstandard W/D on a six-blade impeller is the actual W/D to standard W/D raised to the 1.0 power, or simply the ratio actual to standard W/D.)

2. Determine the power number at process and operating conditions.
Turbulence in agitation can be quantified with respect to another dimensionless variable, the impeller Reynolds number. Although the Reynolds number used in agitation is analogous to that used in pipe flow, the definition of impeller Reynolds number and the values associated with turbulent and laminar conditions are different from those in pipe flow. Impeller Reynolds number N_{Re} is defined as

$$N_{Re} = \frac{D^2 N\rho}{\mu}$$

where μ is the liquid viscosity. In agitation, turbulent conditions exist for $N_{Re} > 20,000$ and laminar conditions exist for $N_{Re} < 10$.

Power number is a function of Reynolds number as well as impeller geometry. A correction factor based on N_{Re} accounts primarily for the effects of viscosity on power. The Reynolds number is computed from the definition found in the previous paragraph and the conditions given in the problem statement; a factor of 10.7 makes the value dimensionless when English engineering units are used:

$$N_{Re} = \frac{10.7D^2 N\rho}{\mu} = \frac{10.7(58)^2(84)(1.15)}{12,000} = 290$$

The viscosity power factor for $N_{Re} = 290$ is found to be 1.2 from Fig. 12.2. The power number for the impeller described in the example is the viscosity factor times the turbulent power number (from the previous step): $N_P = 1.2(1.43) = 1.72$.

3. Compute the shaft horsepower required to rotate the impeller.
Horsepower requirements can be determined by rearranging the definition of power number into $P = N_P\rho N^3 D^5$ and using the value of the power number determined in the previous step. The result must be divided by a factor of 1.524×10^{13} to convert units and give an answer in horsepower: $P = 1.72(1.15)(84)^3(58)^5/(1.524 \times 10^{13}) = 50.5$ hp (37.7 kW).

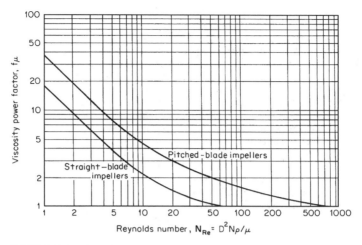

FIGURE 12.2   Viscosity power factor as a function of impeller Reynolds number.

4. *Select a standard motor horsepower.*

A typical turbine impeller-type agitator consists of a motor, a specially designed gear reducer, a shaft, and one or more impellers. Although losses through the gear reducer are typically only 3 to 8 percent, slight deviations in actual speed (which enters the power calculation cubed) and fluctuations in process conditions (density and viscosity) make motor loadings in excess of 85 percent of calculated impeller power unadvisable. Therefore, the calculated impeller power of 50.5 hp and a motor loading of 85 percent would indicate a minimum motor power: $P_{motor} = 50.5/0.85 = 59.4$. The next larger commercially available motor is 60 hp (or for metric sizes, 45 kW). Thus, a 60-hp motor should be used on an agitator designed to operate the 58-in-diameter impeller at 84 r/min as described in the example.

Related Calculations.   Impeller power requirements are relatively independent of mixing-tank diameter. However, the power numbers shown in Fig. 12.1 assume fully baffled conditions, which for a cylindrical tank would require four equally spaced (at 90°) vertical plate-type baffles. The baffles should extend the full height of the vertical wall (i.e., the straight side) of the tank and should be one-twelfth to one-tenth the tank diameter in width.

For relatively high viscosities, the liquid itself prevents uncontrolled swirling. Therefore, when the liquid viscosity is greater than 5000 cP (5 Pa · s), no baffles are required for most applications. For impellers located less than one impeller diameter from the bottom of the tank, an additional correction factor for the power number may be necessary [3].

The power number provides important design information about the correct motor size necessary to operate an impeller at a given speed. However, these calculations do not give any indication of whether or not the agitation produced is adequate for process requirements. The following example shows a method for determining the horsepower and speed required to achieve a given process result.

12.2   *DESIGNING AN AGITATOR TO BLEND TWO LIQUIDS*

A process requires the addition of a concentrated aqueous solution with a 1.4 specific gravity (1400 kg/m³) and a 15-cP (0.015 Pa · s) viscosity to a polymer solution with a 1.0 specific gravity (1000 kg/m³) and an 18,000-cP (18 Pa · s) viscosity. The two liquids are completely miscible and result in a final solution with a 1.1 specific gravity (1100 kg/m³) and a 15,000-cP (15 Pa · s) viscosity. The final batch volume will be 8840 gal (33.5 m³), and the mixing will take place in a 9.5-ft-diameter (2.9-m) flat-bottom tank. Design the agitation system.

Calculation Procedure

1. *Determine the required agitation intensity.*
Design of most agitation equipment is based on previous experience and a knowledge of the amount of liquid motion produced by a rotating impeller in a given situation. Although absolute rules for agitator design do not exist, good guidelines are available as a starting point for most process applications. The type of liquid motion used for most blending applications is a recirculating-flow pattern with good top-to-bottom motion. Axial-flow impellers, such as the pitched-blade turbines shown in Fig. 12.1, produce the desired liquid motion when operated in a baffled tank. Baffles are required to prevent excessive swirling of low-viscosity liquids (<5000 cP, or 5 Pa · s).

One measure of the amount of liquid motion in an agitated tank is velocity. However, by the very nature of mixing requirements, liquid velocities must be somewhat random in both direction and magnitude. Since actual velocity is difficult to measure and depends on location in the tank, an artificial, defined velocity called "bulk velocity" has been found to be a more practical measure of agitation intensity. "Bulk velocity" is defined as the impeller pumping capacity (volumetric flow rate) divided by the cross-sectional area of the tank. For consistency, the cross-sectional area is based on an "equivalent square batch tank diameter." A "square batch" is one in which the liquid level is equal to the tank diameter.

From previous design experience, the magnitude of bulk velocity can be used as a measure of agitation intensity for most problems involving liquid blending. Bulk velocities in the range from 0.1 to 1.0 ft/s (0.03 to 0.3 m/s) are typical of those found in agitated tanks. An agitator that produces a bulk velocity of 0.1 ft/s is normally the smallest agitator that will move liquid throughout the tank. An agitator capable of producing a bulk velocity of 1.0 ft/s is the largest practical size for most applications. Between these typical limits of bulk velocity, increments of 0.1 ft/s provide 10 levels[†] of agitation intensity that are associated with typical process results, as shown in Table 12.1.

In this example, the two fluids to be mixed have a specific gravity difference of 0.4 and a viscosity ratio of 1200. On the basis of the process capabilities associated with bulk velocities of 0.2 and 0.6 ft/s in Table 12.1, a bulk velocity of 0.4 ft/s (0.12 m/s) should be adequate for this example. Special circumstances, such as a reaction taking place or experience with a similar process, may influence the selection of a bulk velocity.

2. *Compute required impeller pumping capacity.*
To determine the required pumping capacity, the bulk velocity (0.4 ft/s) must be multiplied by the appropriate cross-sectional area. Since a "square batch" is assumed for the design basis of bulk velocity, an equivalent tank diameter T_{eq} is computed by rearranging the formula for the volume of a cylinder with the height equal to the diameter, that is,

$$\frac{\pi}{4} T_{eq}^3 = V$$

For the final batch volume of 8840 gal and the conversion of units,

$$T_{eq} = \left[8840 \text{ gal } (231 \text{ in}^3/\text{gal}) \frac{4}{\pi} \right]^{1/3} = 6.65(8840)^{1/3}$$

$$= 137.5 \text{ in } (3.49 \text{ m})$$

A 137.5-in-diameter tank has a cross-sectional area of $(\pi/4)(137.5 \text{ in})^2 = 14{,}849 \text{ in}^2$ or 103 ft^2, so the required impeller pumping capacity is bulk velocity times cross-sectional area: (0.4 ft/s)(103 ft^2) = 41.2 ft^3/s or 2472 ft^3/min (1.17 m^3/s). Geometry of the actual tank will be taken into consideration by location and number of impellers after the horsepower and speed of the agitator are determined.

[†]Called ChemScale levels by Chemineer-Kenics.

TABLE 12.1 Agitation Results Associated with Bulk Velocities

Bulk velocity, ft/s (m/s)	Description
0.1 (0.03) ↓ 0.2 (0.06)	Bulk velocities of 0.1 and 0.2 ft/s (0.03 and 0.06 m/s) are characteristic of applications requiring a minimum of liquid motion. Bulk velocity of 0.2 ft/s (0.06 m/s) will • Blend miscible liquids to uniformity if specific gravity differences are less than 0.1 • Blend miscible liquids to uniformity if the viscosity of the most viscous is less than 100 times that of any other • Establish liquid motion throughout the batch • Produce a flat but moving liquid surface
0.3 (0.09) ↓ 0.6 (0.18)	Bulk velocities between 0.3 and 0.6 ft/s (0.09 and 0.18 m/s) are characteristic of most agitation used in chemical processes. Bulk velocity of 0.6 ft/s (0.18 m/s) will • Blend miscible liquids to uniformity if the specific gravity differences are less than 0.6 • Blend miscible liquids to uniformity if the viscosity of the most viscous is less than 10,000 times that of any other • Suspend trace solids (<2%) with settling rates of 2 to 4 ft/min (0.01 to 0.02 m/s) • Produce surface rippling at low viscosities
0.7 (0.21) ↓ 1.0 (0.30)	Bulk velocities between 0.7 and 1.0 ft/s (0.21 and 0.30 m/s) are characteristic of applications requiring a high degree of agitation, such as critical reactors. Bulk velocity of 1.0 ft/s (0.30 m/s) will • Blend miscible liquids to uniformity if the specific gravity differences are less than 1.0 • Blend miscible liquids to uniformity if the viscosity of the most viscous is less than 100,000 times that of any other • Suspend trace solids (<2%) with settling rates of 4 to 6 ft/min (0.02 to 0.03 m/s) • Produce surging surface at low viscosities

Source: From Ref. 7.

3. *Select impeller diameter and determine required agitator speed.*
The pumping capacity Q for a pitched-blade impeller with four blades ($N_p = 1.37$) can be related to other mixing parameters by the correlation shown in Fig. 12.3. The correlation is between two dimensionless variables: pumping number (Q/ND^3) and Reynolds number ($D^2N\rho/\mu$). Since impeller diameter D and rotational speed N appear in both variables, an iterative solution may be required. A convenient approach to such a solution is as follows:

a. *Select an impeller diameter.* The impeller diameter must be some fraction of the tank diameter, typically between 0.2 and 0.6. For this calculation, an impeller-to-tank-diameter ratio (D/T) of 0.4 will be used. Based on the equivalent tank diameter (137.5 in), an impeller with a 0.4(137.5) = 55 in (1.4 m) diameter will be used. (Blade width will be 11 in, corresponding to a W/D of ⅕.)

b. *Compute initial estimate of impeller Reynolds number.* To compute impeller Reynolds number ($D^2N\rho/\mu$), an initial estimate of rotational speed must be made to begin the iterative solution. Let us assume 100 r/min. Using fluid properties for the final batch, 1.1 specific gravity, and 15,000 cP, the initial estimate of Reynolds number becomes $N_{Re} = 10.7(55)^2(100)(1.1)/15,000 = 237$. (The coefficient 10.7 is a conversion factor to make the value dimensionless.)

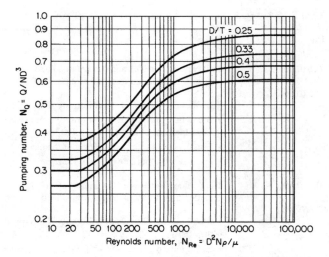

FIGURE 12.3   Pumping number as a function of impeller Reynolds number for pitched-blade impeller ($N_P = 1.37$). (*From Chemical Engineering, 1976.*)

   c. *Determine pumping number and compute speed.*   From the correlation for pumping number (Fig. 12.3), at a Reynolds number of 237 and a D/T of 0.4, the pumping number is $N_Q = 0.44$. By rearranging the definition of pumping number [$Q/(ND^3)$] and using the value obtained from the correlation (0.44), we can calculate a speed for the required pumping capacity of 2472 ft³/min and the impeller diameter of 55 in (4.583 ft):

$$N = \frac{Q}{N_Q D^3} = \frac{2472}{0.44(4.583)^3} = 58.4 \text{ r/min}$$

The estimated and calculated speeds do not match, and the pumping number is not constant for this Reynolds number range, so an iterative solution for the speed must continue. [In the turbulent range ($N_{Re} > 20{,}000$), where pumping number is constant, no iteration is required and the calculated speed is correct for the design.]

   d. *Perform an iterative calculation for agitator speed.*   Successive calculations of Reynolds number (based on the previously estimated speed), pumping number, and agitator speed, similar to steps 3b and 3c, will converge as follows:

Iteration	Reynolds number	Pumping number	Speed, r/min
2	139	0.38	67.6
3	160	0.40	64.2
4	152	0.39	65.8
5	156	0.395	65.0

Thus a speed of 65 r/min is necessary to provide the pumping capacity of 2472 ft³/min when using a 55-in-diameter impeller.

4. *Select standard speed and motor horsepower.*

Although design calculations have determined that an agitator speed of 65 r/min is required, only certain standard output speeds[†] are available with typical industrial gear reducers. The closest standard speed is 68 r/min. If 68 r/min is used instead of the calculated 65 r/min, the bulk velocity will increase to about 0.42 ft/s, a change imperceptible with respect to agitator performance.

The horsepower required to rotate a 55-in-diameter impeller (11-in blade width) at 68 r/min can be computed for the process fluid, using the technique described in Procedure 12.1. The turbulent power number is 1.37, from Fig. 12.1. The Reynolds number at 68 r/min becomes 161. The viscosity correction factor for this value is 1.35, from Fig. 12.2, which gives a power number N_P of $1.37(1.35) = 1.85$ for the design conditions. From the power number, impeller power can be computed: $P = 1.85(1.1)$ $(68)^3(55)^5/(1.524 \times 10^{13}) = 21.1$ hp. With an 85 percent loading for the motor, a minimum motor horsepower would be $21.1/0.85 = 24.8$ hp, so a 25-hp (18.5-kW) motor would be required. If the next larger standard motor is substantially larger than the minimum motor horsepower, the impeller diameter may be increased by an inch or two to fully utilize the available motor capacity.

5. *Specify the number and location of impellers.*

The calculations carried out in the previous steps show that a 25-hp agitator operating at 68 r/min will provide sufficient agitation to solve the problem by creating a bulk velocity of 0.4 ft/s. However, these calculations essentially ignore the fact that the process will be carried out in a $9\frac{1}{2}$-ft-diameter tank. The idea behind this final step in the design procedure is that 25 hp at 68 r/min will provide the desired agitation if the number and location of impellers is suitable for the batch height, as related by the ratio of liquid level to tank diameter Z/T.

According to Table 12.2, a 9½-ft-diameter tank holds 44.1 gal/in of liquid level. Therefore, 8840 gal will fill the tank to $8840/44.1 = 200$ in. The resulting liquid-level-to-tank-diameter ratio is $Z/T = 200/114 = 1.75$. The following guidelines for number and location of impellers should be applied:

Viscosity, cP (Pa · s)	Maximum level, Z/T	Number of impellers	Impeller clearance	
			Lower	Upper
<25,000 (<25)	1.4	1	$Z/3$	—
<25,000 (<25)	2.1	2	$T/3$	$(2/3)Z$
>25,000 (>25)	0.8	1	$Z/3$	—
>25,000 (>25)	1.6	2	$T/3$	$(2/3)Z$

Since the liquid viscosity is 15,000 cP (<25,000 cP) and the liquid level gives a Z/T of 1.75, two impellers should be used to provide liquid motion throughout the tank.

To properly load the 25-hp motor with a dual impeller system, each impeller should be sized for 25 hp/2 = 12.5 motor hp, or at 85 percent loading, $0.85(12.5 \text{ hp}) = 10.6$ impeller hp. By assuming the same viscosity correction factor (1.35) for the dual impeller size, an initial estimate can be made for power number, $N_P = 1.37(1.35) = 1.85$, and impeller diameter, $D = [1.524 \times 10^{13} P/(N_P \rho N^3)]^{1/5} =$ $\{1.524 \times 10^{13}(10.6)/[1.85(1.1)(68)^3]\}^{1/5} = 47.9$ in. Using this value to compute Reynolds number gives $N_{Re} = 10.7(47.9)^2(68)(1.1)/15,000 = 122$. For $N_{Re} = 122$, the viscosity correction factor is 1.47 from Fig. 12.2, or a power number, $N_P = 1.37(1.47) = 2.01$. Recomputing impeller diameter, that is, $D = \{1.524 \times 10^{13}(10.6)/[2.01(1.1)(68)^3]\}^{1/5} = 47.1$, shows that two 47.1-in-diameter (1.20-m) impellers (with 9.4-in blade width) are equivalent to one 55-in-diameter impeller. The lower impeller should be located $T/3 = 114/3 = 38$ in (0.965 m) off bottom and the upper impeller $(2/3)Z = (2/3)200 = 133$ in (3.39 m) off bottom. Had the tank been 11 or 12 ft in diameter, only one impeller would have been required. Liquid levels that result in $Z/T < 0.4$ are difficult to agitate.

[†]Standard speeds for common agitator drives are 230, 190, 155, 125, 100, 84, 68, 56, 45, and 37 r/min based on actual speeds of nominal 1800 and 1200 r/min motors and standard gear reductions, which are a geometric progression of the $\sqrt{1.5}$ for enclosed, helical, and spiral bevel gearing (American Gear Manufacturers' Association Standard 420.04, December 1975, p. 29).

TABLE 12.2 Capacity Data for Cylindrical Vessels

Vessel diameter		Volume of cylindrical vessel		Depth and volume of vessel head			
				Standard dished head		ASME flanged and dished head	
		Straight side, gal/in	Square-batch, gal	Depth, in	Volume, gal	Depth, in	Volume, gal
ft-in	in						
3ft	36	4.40	159	4.9	11	6.0	16
3 ft 6 in	42	5.99	252	5.7	18	7.2	25
4 ft	48	7.83	376	6.5	27	8.0	37
4 ft 6 in	54	9.91	535	7.3	38	9.0	53
5 ft	60	12.2	734	8.1	52	10	78
5 ft 6 in	66	14.8	977	8.9	70	11	104
6 ft	72	17.6	1269	9.7	90	12	135
6 ft 6 in	78	20.7	1631	11	114	14	170
7 ft	84	24.0	2041	11	142	15	212
7 ft 6 in	90	27.5	2478	12	174	15	261
8 ft	96	31.3	3007	13	212	16	314
8 ft 6 in	102	35.3	3607	14	254	18	375
9 ft	108	39.6	4287	15	301	19	446
9 ft 6 in	114	44.1	5035	15	353	20	524
10 ft	120	48.9	5873	16	414	21	612
10 ft 6 in	126	54	6799	17	480	22	705
11 ft	132	59	7817	18	560	23	806
11 ft 6 in	138	65	8932	20	665	24	926
12 ft	144	70	10148	20	735	25	995

Note: 1.0 ft = 0.3048 m; 1.0 in = 0.0254 m; 1.0 gal/in = 0.149 m^3/m; 1.0 gal = 3.785 × 10^{-3} m^3.

Related Calculations. Repeating the same design calculations but starting with a different impeller diameter will result in other horsepower-speed combinations capable of producing the same bulk velocity (0.4 ft/s). For instance, the following combinations also satisfy the design requirements:

Impeller diameter, in (m)	Motor horsepower	Speed, r/min
48 (1.219)	40	100
50 (1.270)	30	84
58 (1.321)	20	56
62 (1.412)	15	45

As is the case for many agitator design problems, there are several horsepower, speed, and impeller-diameter combinations that solve the problem by producing equivalent results. From the standpoint of energy conservation, large impellers usually require less horsepower to do a given job.

Agitation problems that require other process results, such as the suspension of solids or the dispersion of gas, use other design criteria [8, 9].

12.3 *TIME REQUIRED FOR UNIFORM BLENDING*

About 150 gal (0.57 m^3) of strong acid must be added to 10,000 gal (37.85 m^3) of slightly caustic waste held in a 12-ft-diameter (3.66-m) tank. The waste has a specific gravity ρ of 1.2 (1200 kg/m^3) and a viscosity μ of 500 cP (0.5 Pa · s). Determine the time required for neutralization if the tank is agitated by a 1-hp (0.75-kW) agitator operating at a rotational speed N of 68 r/min and having a pitched-blade impeller with diameter D of 30 in (0.762 m).

Calculation Procedure

1. *Compute the Reynolds number.*

Acid-base neutralizations typically are very fast reactions, so the time required for mixing is usually the limiting factor. Although both liquid motion and molecular diffusion are involved in liquid mixing, the liquid motion dominates the apparent rate of mixing. Impeller agitation creates both large-scale flow patterns and small-scale turbulence, which in combination give efficient and rapid mixing. The effect of turbulent flow patterns is to reduce the distances required for diffusion to almost the molecular scale.

One practical method for quantifying the complicated mixing process in an agitated tank is to measure the time required for a tracer material to blend to uniformity. Such measurements for blend time may use acid-base neutralization with a color-change indicator, a dye tracer, or an ionic salt with an electrode detector. Properly accounting for measurement accuracy, all these methods give essentially the same results for time required to go from extreme segregation to a high degree (>99 percent) of uniformity.

This measured blend lime t_b can be expressed as a dimensionless variable by forming a product $t_b N$ with agitator speed. This form of dimensionless blend time, multiplied by the impeller-to-tank-diameter ratio D/T to the 2.3 power, is shown as a function of impeller Reynolds number in Fig. 12.4.

The independent variable for the correlation is impeller Reynolds number ($D^2 N \rho / \mu$), which takes into account the effects of liquid properties on the blend time. To compute the value of the Reynolds number, a coefficient of 10.7 is necessary to put the given units in dimensionless form. Thus, $N_{Re} = 10.7(30)^2(68)(1.2)/500 = 1572$.

2. *Determine dimensionless blend time and D/T.*

A dimensionless blend time $t_b N (D/T)^{2.3} = 18$ is found for a Reynolds number of 1600, from Fig. 12.4. This form of dimensionless blend time takes into account the main geometric effects, as embodied in the impeller-to-tank-diameter ratio, $D/T = 30/144 = 0.208$.

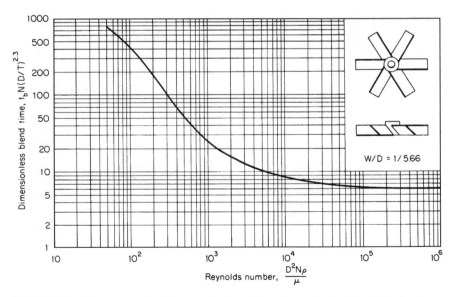

FIGURE 12.4 Dimensionless blend time as a function of Reynolds number for pitched-blade impellers.

3. *Compute blend time.*
Blend time can be computed directly from rearrangement of the definition and the values for dimensionless blend time, for D/T, and for agitator speed:

$$t_b = \frac{18}{N(D/T)^{2.3}} = \frac{18}{68(0.208)^{2.3}}$$
$$= 9.80 \text{ min (588 s)}$$

The computed blend time is about 10 min, a value that is accurate to ±10 percent for successive observations of the same process and affected slightly by the location of the acid addition. A reasonable design value for the blend time might be twice the calculated value, or 20 min (1200 s).

Related Calculations.   If the waste-neutralization process were continuous and the pH adjustment were relatively small (<3 pH units), a conservative design residence time in the tank might be 10 times the computed blend time, or 100 min.

12.4   *HEAT TRANSFER IN AN AGITATED TANK*

As part of the final blending operation in a continuous process, it is necessary to cool an oil product from 125°F (325 K) to 100°F (311 K) at the rate of 800 gal/h (0.84×10^{-3} m³/s).
The oil has the following physical properties at 100°F:

Viscosity $\mu = 1200$ cP (1.2 Pa · s)
Specific gravity $\rho = 0.89$ (890 kg/m³)
Heat capacity $C_p = 0.52$ Btu/(lb)(°F) [2175 J/(kg)(K)]
Thermal conductivity $k = 0.079$ Btu/(h)(ft)(°F) [0.137 W/(m)(K)]

The tank diameter T is 9 ft (2.74 m), and the vessel is designed to operate at a 5000-gal (18.9-m³) capacity. The tank bottom is a standard dished head and the straight side of the tank is fully jacketed. The jacket-side heat-transfer coefficient is estimated to be $h_o = 180$ Btu/(h)(ft²)(°F) [1021 W/(m²) (K)]. The wall thickness and its heat-transfer resistance are assumed to be negligible. If the agitator is 1.5 hp (1.1 kW) [1.15 impeller hp (0.858 kW)] operating at a speed N of 56 r/min, with a 38-in-diameter (D) (0.97-m) impeller, estimate the average cooling water temperature required to cool the oil. Also determine what effect increasing the agitator speed to 100 r/min (assuming appropriate changes were made in the agitator) would have on the temperature of the oil if all other conditions from the first part of the problem remained unchanged.

Calculation Procedure

1. *Compute the process-side heat-transfer coefficient.*
The correlations for inside (process-side) heat-transfer coefficient in an agitated tank are similar to those for heat transfer in pipe flow, except that the impeller Reynolds number and geometric factors associated with the tank and impeller are used and the coefficients and exponents are different. A typical correlation for the agitated heat-transfer Nusselt number ($N_{\text{Nu}} = h_i T/k$) of a jacketed tank is expressed as

$$N_{\text{Nu}} = 0.85 N_{\text{Re}}^{0.66} N_{\text{Pr}}^{0.33} \left(\frac{Z}{T}\right)^{-0.56} \left(\frac{D}{T}\right)^{0.13} \left(\frac{\mu}{\mu_w}\right)^{0.14}$$

All terms in the expression are dimensionless.

The correlation for heat transfer is evaluated with the respective dimensionless groups. With the units stated in the example, the impeller Reynolds number ($N_{Re} = D^2 N \rho / \mu$) requires a conversion factor of 10.7 to make it dimensionless; thus, $N_{Re} = 10.7(38)^2(56)(0.89)/1200 = 642$. The Prandtl number ($N_{Pr} = C_p \mu / k$) requires a conversion factor of 2.42; thus $N_{Pr} = 2.42(0.52)(1200)/0.079 = 19{,}115$. The liquid-level-to-tank-diameter ratio Z/T requires determination of the liquid level for a 5000-gal batch in the tank. A standard dished head holds 301 gal and is 15 in deep, from Table 12.2. The remaining $5000 - 301 = 4699$ gal fills the cylindrical part of the tank at a rate of 39.6 gal/in of height, or to a height of $4699/39.6 = 119$ in. The total liquid level is $Z = 119 + 15 = 134$ in and $Z/T = 134/108 = 1.24$. The impeller-to-tank-diameter ratio D/T is $38/108 = 0.35$. The viscosity ratio will be assumed to be unity ($\mu/\mu_w = 1$) because of lack of data and the very small exponent on the term.

Combining all these values according to the correlation gives a value for the Nusselt number of $N_{Nu} = 0.85(642)^{0.66}(19{,}115)^{0.33}(1.24)^{-0.56}(0.35)^{0.13}(1)^{0.14} = 1212$. The value for the inside heat-transfer coefficient h_i is obtained from the Nusselt number using conductivity and tank diameter (ft); thus, $h_i = N_{Nu} k / T = 1212(0.079)/9 = 10.6$ Btu/(h)(ft^2)(°F) [60.1 W/(m^2)(K)].

2. Compute overall heat-transfer coefficient.

The overall heat-transfer coefficient U_o is simply the series resistance to heat transfer for the inside and outside coefficients. The overall coefficient U_o is $(1/h_i + 1/h_o)^{-1} = (1/10.6 + 1/180)^{-1} = 10.0$ Btu/(h)(ft^2)(°F) [56.7 W/(m^2)(K)].

3. Determine total heat load for agitated heat transfer.

The process requires 25°F cooling of 8000 gal/h of oil. Since the oil has a specific gravity of 0.89, its density is $(0.89)(8.337$ lb water/gal$) = 7.42$ lb/gal. The product of volumetric flow, density, heat capacity, and temperature change equals the heat load for cooling the oil: (800 gal/h)(7.42 lb/gal) [0.52 Btu/(lb)(°F)](25°F) = 77,168 Btu/h (22.6 kW). In addition, the power input of the agitator (1.15 hp) also must be dissipated in the form of heat: (1.15 hp)[2545 Btu/(h)(hp)] = 2927 Btu/h. The total heat load q is $77{,}168 + 2927 = 80{,}095$ Btu/h (23.5 kW).

4. Compute required coolant temperature.

The coolant temperature can be determined from heat load and the heat-transfer coefficient because a sufficient temperature difference must exist to drive the heat-transfer rate, that is, $q = U_o A (T_i - T_o)$. The available heat-transfer area A is the jacketed vertical wall in contact with the liquid, since the bottom head is not jacketed; thus, $A = \pi D Z_{ss} = \pi(9$ ft$)(119/12$ ft$) = 280$ ft^2. The temperature difference is $(T_i - T_o) = q/(U_o A) = 80{,}095/[10.0(280)] = 28.6$°F. To provide a temperature difference of 28.6°F with respect to the process temperature, the average coolant temperature must be 71.4°F (295 K).

5. Determine the effect of increased agitator speed.

Two effects must be considered when the agitator speed is increased: (1) improved heat-transfer coefficient, and (2) increased power input. The agitator speed enters the heat-transfer correlation in the Reynolds number. For 100 r/min, the Reynolds number N_{Re} becomes $10.7(38)^2(100)(0.89)/1200 = 1146$, which increases the Nusselt number to 1777 and the inside heat-transfer coefficient h_i to $1777(0.079)/9 = 15.6$ Btu/(h)(ft^2)(°F) [88.5 W/(m^2)(K)]. The overall coefficient U_o becomes $(1/15.6 + 1/180)^{-1} = 14.4$ Btu/(h)(ft^2)(°F) [81.7 W/(m)^2(K)], or a 44 percent increase from the lower speed. The horsepower increase associated with increased speed is substantial, because power is roughly proportional to speed cubed (the effect of Reynolds number on power requirement is negligible between $N_{Re} = 600$ and 1200 for a pitched-blade impeller; see Fig. 12.2). The horsepower at 100 r/min is approximately $(1.15$ hp$)(100/56)^3 = 6.55$ hp, which results in a heat load of (6.55 hp)[2545 Btu/(h)(hp)] = 16,670 Btu/h. With the increased power input, the total heat load is $q = 77{,}168 + 16{,}670 = 93{,}838$ Btu/h (27.5 kW).

The combined effects of increased heat transfer and increased heat load can be seen by determining the resultant temperature difference, that is, $(T_i - T_o) = q/(U_o A) = 93{,}838/[14.4(280)] = 23.3$°F. For the same jacket temperature of 71.4°F, the process temperature would be reduced to $71.4 + 23.3 = 94.7$°F (307.8 K) by doubling the agitator speed.

The cost of improved heat transfer by increased speed must be measured against the increased cooling-water requirements and increased operating power and capital cost for the agitator. In general, these increased costs more than offset the benefits of improved heat transfer. Therefore, most agitators designed for heat transfer provide moderate blending (0.2 to 0.3 ft/s bulk velocity) for optimal operation.

Related Calculations.   See also Section 7 for situations involving both heat transfer and mixing.

12.5 SCALE-UP FOR AGITATED SOLIDS SUSPENSION

An agitator must be designed for a solids-suspension operation to be carried out in a 6000-gal (22.7-m³) tank that is 10 ft (3.05 m) in diameter and has a standard dished bottom. The material to be suspended is insoluble in the liquid and has a particle-size range from 30 to 200 μm with an actual specific gravity ρ of 3.8 (3800 kg/m³). The liquid is a mineral oil with a specific gravity of 0.89 (890 kg/m³) and a viscosity of 125 cP (0.125 Pa · s). The suspension is 30 wt % solids and must be sufficiently agitated to give particle uniformity of the large particles to at least three-fourths the liquid level.

Calculation Procedure

1. Compute suspension density.
To form 1 lb of 30 wt % suspension, 0.3 lb solids must be added to 0.7 lb liquid. The liquid has a density of 0.89(8.337 lb water/gal) = 7.42 lb/gal. Similarly, the solids must displace liquid at 3.8(8.337) = 31.68 lb/gal. Thus, 1 lb of the suspension will have a volume of 0.7/7.42 + 0.3/31.68 = 0.1038 gal, or a density of (1 lb)/0.1038 gal = 9.63 lb/gal, which is the same as a specific gravity of 9.63/8.337 = 1.16 (1160 kg/m³).

2. Determine batch height.
A 6000-gal batch in a 10-ft-diameter tank with a dished head will put 414 gal in the 16-in-deep dished head (see Table 12.2). The remaining 6000 − 414 = 5586 gal will fill the vertical-wall portion of the tank at a rate of 48.9 gal/in, for a total liquid depth of (5586 gal)/(48.9 gal/in) + 16 in = 130 in (3.3 m).

3. Use an experimental model to determine required agitation intensity.
Although physical-property data are available for solid particles, liquid, and suspension, the agitated-suspension characteristics of a relatively wide range of particle sizes in a slightly viscous liquid are almost impossible to predict without making experimental measurements in a small-scale agitated tank. The most direct approach to small-scale testing is to construct a geometrically similar model of the large-scale equipment. Assume that a 1-ft-diameter tank ($^1/_{10}$ scale) is available for such tests. By applying the scale factor ($^1/_{10}$) to the liquid level for the large tank, a (130 in)/10 = 13-in liquid level should be tested in the model.

The testing should determine the intensity of agitation necessary to obtain the desired level of suspension uniformity. Suppose that by visual observation and sample analysis, a pitched-blade impeller with a diameter D of 4 in (with four blades, each 0.8 in wide) operating at a speed N of 465 r/min was found to produce the level of agitation necessary for uniform suspension to three-fourths the total liquid level. With data about impeller diameter and agitator speed in a small tank, it should be possible to scale up performance to the large-scale tank.

4. Scale-up experimental results for solids suspension.
To maintain geometric similarity with scale-up, all length dimensions must remain in the same proportion between the small and large equipment. If the large tank diameter is 10 times as large as the small tank, then the impeller diameter should be 10(4 in) = 40 in for the large tank. Similarly, the blade width for the impeller should be 10(0.8 in) = 8 in for a four-blade impeller. Although geometric similarity is not strictly necessary for all agitation scale-up problems, satisfactory results are usually obtained and scale-up relationships are relatively simple.

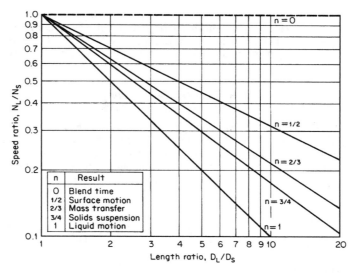

FIGURE 12.5 Scale-up rules for geometric similarity: speed ratio as a function of length ratio.

Since impeller diameter is established by geometric similarity, the agitator speed for the large-scale equipment must be determined to satisfy process requirements. Large-scale speed N_L can be computed by the following relationship:

$$N_L = N_S \left(\frac{D_S}{D_L} \right)^n$$

where N_S is the small-scale speed. The value of the exponent n depends on the type of process result that must be duplicated, since different process results scale up differently. Some typical exponents, their effect on speed, and their scale-up significance are shown in Fig. 12.5.

For equivalent solids suspension, an exponent of $^3/_4$ will be used to scale up from the small-scale solids suspension speed of 465 r/min:

$$N_L = 465 \left(\frac{4}{40} \right)^{3/4} = 82.7 \text{ r/min}$$

5. *Select standard speed and motor horsepower.*

Although scale-up calculations predict an 82.7 r/min operating speed for the large-scale agitator, only certain standard output speeds are commercially available. The nearest standard speed is 84 r/min. To determine power requirements, use the procedure outlined in Procedure 12.1. The turbulent-power number is first determined from Fig. 12.1: $N_P = 1.37$. The Reynolds number is then computed for the process conditions: $N_{Re} = 10.7(40)^2(84)(1.16)/125 = 13,345$, which is sufficiently turbulent that no correction factor from Fig. 12.2 need be applied to the power number. The calculated impeller power is $1.37(1.16)(84)^3(40)^5/(1.524 \times 10^{13}) = 6.33$ hp. Considering process variations, drive losses, and so forth, an 85 percent motor loading means that a minimum motor horsepower of (6.33 hp)/0.85 = 7.45 hp is required. The next larger standard motor is 7.5 hp (5.5 kW), which is used to drive a 40-in-diameter (1.02-m) impeller at 84 r/min to satisfy the solids-suspension requirements.

Related Calculations.   Scale-up problems for agitator design are not always obvious with respect to which scale-up exponent should be used. Problems involving chemical reactions, where kinetics and mixing interact, are often difficult to scale up accurately. Therefore, medium-sized pilot-plant reactors may be necessary to improve the understanding of how mixing influences performance.

12.6 *AGITATOR DESIGN FOR GAS DISPERSION*

Pilot-scale testing of an aerobic fermentation process has determined that maximum cell growth rate will consume 16.2 lb O_2 per hour per 1000 gal (5.4×10^{-4} kg O_2 per second per cubic meter) at a temperature of 120°F (322 K), providing the gas rate is sufficient to keep oxygen depletion of the air to less than 10 percent and the O_2 concentration in the broth is at least 2.4 mg/L (2.4×10^{-3} kg/m^3). The overall mass-transfer coefficient $k_L a(s^{-1})$ for the process is assumed to behave as in an ionic solution: $k_L a = (2.0 \times 10^{-3})(P/V)^{0.7} u_s^{0.2}$, where P/V is power per volume (in watts per cubic meter) and u_s is superficial gas velocity (in meters per second). Design an agitator to carry out the fermentation in a 10,000-gal (37.9-m^3) batch in a 12-ft-diameter (3.66-m) tank. The fermentation broth is initially waterlike and has a specific gravity of 1.0 (1000 kg/m^3) throughout the process.

Calculation Procedure

1. *Determine superficial gas velocity.*
The maximum oxygen uptake rate (16.2 lb/h per 1000 gal) will mean that 162 lb O_2 per hour will be required for the 10,000-gal batch. To keep oxygen depletion to less than 10 percent, 10 times the O_2 demand must flow through the tank, that is, 10(162) = 1620 lb O_2 per hour. Water-saturated air at 120°F contains 21.3 wt % O_2 and has a density of 0.065 lb/ft^3 at 1 atm (*Handbook of Tables for Applied Engineering Science*, CRC Press). To supply 1620 lb O_2 per hour, 1620/0.213 = 7606 lb air per hour must flow through the tank. This flow rate represents (7606 lb/h)/[(0.065 lb/ft^3)(60 min/h)] = 1950 ft^3/min (0.92 m^3/s) at 1 atm (101.3 kPa).
   To design the agitator, the gas density at the impeller location should be used for computing the superficial gas velocity. Because of liquid head, this necessitates a pressure correction. The total liquid level for 10,000 gal in a 12-ft-diameter tank is roughly 12 ft (see Table 12.2). If the impeller is located one-sixth of the liquid level off bottom, as it should be for gas dispersion, the additional static liquid head is 10 ft of water that must be added to the atmospheric pressure (34 ft of water). The pressure correction makes the actual volumetric flow rate of gas Q_A equal (1950 ft^3/min) (34 ft of water)/(34 + 10 ft of water) = 1507 ft^3/min. Flow rate divided by cross-sectional area of tank, that is, $\pi(12 \text{ ft})^2/4 = 113 \text{ ft}^2$, gives a superficial gas velocity u_s of (1507 ft^3/min)/113 ft^2 = 13.3 ft/min or 0.22 ft/s (0.067 m/s).

2. *Determine overall mass-transfer coefficient.*
As a design level for overall mass transfer, the coefficient should be based on the minimum concentration gradient and the maximum transfer rate. The minimum gradient will exist near the liquid surface, where the oxygen saturation concentration in the liquid is minimum because of the minimum total pressure and the low concentration of oxygen there. The volume (mole) percent oxygen in air with 10 percent depletion is (18.4 mol O_2 − 1.8 mol O_2)/(100 mol gas − 1.8 mol gas) = 0.169 mol O_2 per mol of gas, which gives an oxygen partial pressure of 0.169 atm. The Henry's law solubility constant H for O_2 at 120°F is 5.88×10^4 (*Handbook of Tables for Applied Engineering Science*, CRC Press) for partial pressure in atmospheres and oxygen concentration in mole fraction dissolved oxygen. Thus the mole fraction oxygen dissolved in the water is $0.169/5.88 \times 10^4 = 2.87 \times 10^{-6}$ mol O_2 per mole of liquid. This equals [2.87×10^{-6} mol O_2/1 mol H_2O][32 g O_2/1 mol O_2)/ (18 g H_2O/1 mol H_2O)][10^3 mg O_2/1 g O_2][10^3 g H_2O/1 L H_2O] = 5.10 mg O_2 per liter (5.10×10^{-3} kg O_2 per cubic meter).
   The maximum oxygen-transfer rate is 16.2 lb O_2 per hour per 1000 gal (5.4×10^{-4} kg O_2 per second per cubic meter) and the minimum concentration gradient between saturation (5.10×10^{-3} kg O_2 per

cubic meter) and bulk concentration (2.4×10^{-3} kg O_2 per cubic meter) is $(5.1 \times 10^{-3}) - (2.4 \times 10^{-3}) =$ 2.7×10^{-3} kg O_2 per cubic meter. The overall mass-transfer coefficient $k_L a$ is computed by dividing the transfer rate by the concentration gradient:

$$\frac{5.4 \times 10^{-4} \text{ kg/(s)(m}^3) \text{ O}_2}{2.7 \times 10^{-3} \text{ kg/m}^3 \text{ O}_2} = 0.2 \text{ s}^{-1}$$

3. Compute required power per volume for agitation.

Using the correlation for overall mass-transfer coefficient, that is, $k_L a = 2.0 \times 10^{-3} (P/V)^{0.7} u_s^{0.2}$, the design value for mass-transfer coefficient (0.2 s^{-1}), and the superficial gas velocity (0.067 m/s), the required agitation intensity can be computed: $P/V = [0.2 \text{ s}^{-1}/(0.067 \text{ m/s})^{0.2}(2.0 \times 10^{-3})]^{1/0.7} =$ 1558 W/m^3 (7.91 hp per 1000 gal).

4. Determine minimum impeller size to prevent flooding.

Flooding in an agitated gas dispersion occurs when the impeller power and pumping capacity are insufficient to control the gas flow rate. A flooding correlation for minimum power per volume and superficial gas velocity is shown in Fig. 12.6 for several impeller-to-tank-diameter ratios (D/T). For a superficial gas velocity of 0.22 ft/s and 7.91 hp per 1000 gal, the minimum D/T is less than 0.25. So any impeller larger than 0.25 $T = 0.25(144$ in$) = 36$ in should produce sufficient agitation to overcome flooding.

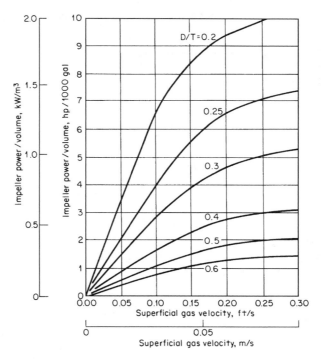

FIGURE 12.6 Minimum impeller-power requirement to overcome flooding as a function of superficial gas velocity and D/T. (*From Chemical Engineering, 1976.*)

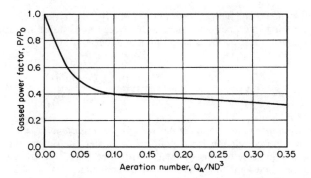

FIGURE 12.7 Gassed power factor as a function of aeration number. (*From Chemical Engineering, 1976.*)

5. Determine impeller size for required power input.
The impeller required for the process must draw 7.91 hp per 1000 gal or 79.1 hp for the 10,000-gal batch. This power level must be achieved at the gas flow rate required by the process, which is, $Q_A = 1507$ ft^3/min. The power required to operate an agitator impeller for gas dispersion can be much less than the power required for a liquid without gas. The ratio of power with gas to power without P/P_0 is shown in Fig. 12.7 and is a function of the dimensionless aeration number $N_{Ae} = Q_A/ND^3$.

For gas-dispersion applications, a radial discharge impeller, such as the straight-blade and disk style impellers in Fig. 12.1, should be used. A straight-blade impeller with a power number N_P of 3.86 will be used for this design. Assume that the operating speed will be 100 r/min.

Since impeller diameter is unknown, aeration number and, therefore, gassed power factor cannot be determined; however, an estimate for power factor P/P_0 of 0.4 is usually a good initial estimate. By rearranging the expression for power number, that is, $N_P = P/(\rho N^3 D^5)$, an expression for impeller diameter can be derived. The factor P/P_0 must be introduced for the effect of gas and a conversion for units makes

$$D = \left[\frac{1.524 \times 10^{13} P}{(P/P_0)N_P N^3} \right]^{1/5} = \left[\frac{1.524 \times 10^{13}(79.1)}{0.4(3.86)(100)^3} \right]^{1/5} = 60.0 \text{ in}$$

With a value for D of 60.0 in (5.0 ft), the aeration number can be computed:

$$N_{Ae} = \frac{1507 \text{ ft}^3/\text{min}}{(100 \text{ r/min})(5.0 \text{ ft})^3} = 0.121$$

The revised estimate for P/P_0 is 0.38 from Fig. 12.7, which gives an impeller diameter of 60.7 in. Therefore, a 60.7-in-diameter (1.54-m) straight-blade impeller with a 10.1-in (0.257-m) blade width operating at 100 r/min should satisfy the process requirements.

6. Select standard size motor.
A motor that is loaded to 85 percent by a 79.1-hp impeller will require a minimum size of (79.1 hp)/0.85 = 93.1 hp, which means a 100-hp (75-kW) motor. This motor and impeller assembly is correctly sized for conditions with the design gas flow. However, because of the gassed power factor, that is, $P/P_0 = 0.38$, should the gas supply be lost for any reason, the impeller power would increase to 78/0.38 = 205 hp and seriously overload the motor. To avoid this problem, some method (typically electrical control) prevents motor operation without the gas supply. When the gas supply is off, the control either stops the agitator motor or, in the case of a two-speed motor, goes to a lower speed.

Related Calculations.   The majority of gas-dispersion applications are sized on the basis of power per volume. In aerobic fermentation, levels of 5 to 12 hp per 1000 gal (1 to 2.4 kW/m^3) are typical, while for aerobic waste treatment, levels of 1 to 3 hp per 1000 gal (0.2 to 0.6 kW/m^3) are more common, primarily because of the concentrations and oxygen requirements of the microorganisms. For more on fermentation, see Section 17.

12.7 SHAFT DESIGN FOR TURBINE AGITATOR

A 15-hp (11.2-kW) agitator operating at 100 r/min has been selected for a process application. The vessel geometry requires two impellers, both of 36-in (0.91 m) diameter, the upper one located 80 in (2.03 m) below the agitator drive and the lower one 130 in (3.30 m) below the drive. Pitched-blade turbine impellers having four blades are to be used. The shaft is to be stainless steel, having an allowable shear stress of 6000 psi (41,370 kPa), an allowable tensile stress of 10,000 psi (68,950 kPa), a modulus of elasticity of 28,000,000 psi (193,000,000 kPa), and a density of 0.29 lb/in^3(8027 kg/m^3), which represents a weight of 0.228d^2 per linear inch, where d is shaft diameter in inches. The bearing span for support of the agitator shaft is to be 16 in. What shaft diameter is required for this application?

Calculation Procedure

1. Determine the hydraulic loads on the shaft.
To rotate the agitator, the shaft must transmit torque from the drive to the impellers. The actual torque required should be found by the sum of the horsepower required by each impeller. However, process conditions can change, so it is better to assume that the full motor power can be outputted from the drive. Maximum torque τ can be found by dividing motor horsepower P_{motor} by shaft speed N. A conversion factor of 63,025 makes the answer come out in English engineering units:

$$\tau = P_{motor}/N = (63,025)(15)/100 = 9454 \text{ in} \cdot \text{lb}$$

The output shaft must be large enough to transmit 9454 in · lb (1068 N · m) at the drive. Only half the torque need be transmitted in the shaft below the first impeller, because the lower impeller should require only half the total power.

If the hydraulic forces on individual impeller blades were always uniformly distributed, torsion considerations would constitute the only significant strength requirement for the shaft. However, real loads on impeller blades fluctuate, due to the shifting flow patterns that contribute to process mixing. For shaft design, using a factor of three-tenths (0.3) approximates the imbalanced force acting at the impeller diameter. This factor (0.3) is typical for pitched-blade turbines with four blades and may be different for other types of impellers. A higher factor should be used if the mixer is subjected to external loads, such as flow from a pump return.

The imbalanced forces result in a bending moment M on the shaft. Such moments must be summed for hydraulic loads at each shaft extension L_n where an impeller is located and must be adjusted for the impeller diameter D_n:

$$M = \sum_n 0.3(P_n/N)(L_n/D_n)$$

In the present example having two equal-size impellers, the motor power is split in half and the moments are calculated at each impeller location:

$$M = 0.3[(63,025)(15/2)/100][80/36] + 0.3[(63,025)(15/2)/100][130/36]$$
$$= 8272 \text{ in} \cdot \text{lb}$$

Thus, a maximum bending moment of 8272 in · lb (935 N · m) occurs just below the agitator drive, at the top of the shaft extension. The hydraulic loads on the shaft result in a torque and bending moment that the shaft must be strong enough to handle.

2. *Determine the minimum shaft diameter for strength.*
Since the torque and bending moment may act simultaneously on the shaft, these loads must be combined and resolved into shear and tensile stresses on the shaft. The minimum shaft diameter must be the larger of the shaft diameters required by either shear- or tensile-stress limits. The shaft diameter for shear stress d_s can be calculated as follows:

$$d_s = [16(\tau^2 + M^2)^{1/2}/\pi\sigma_s]^{1/3}$$

where σ_s is the allowable shear stress. As already noted, the σ_s value recommended for carbon steel and stainless steel typically used in agitator applications is 6000 psi (41,370 kPa). This stress value is low enough to prevent permanent distortions and to minimize the possibility of fatigue failures. The minimum shaft diameter for shear strength is, accordingly,

$$d_s = [16(9454^2 + 8272^2)^{1/2}/\pi(6000)]^{1/3} = 2.201 \text{ in}$$

The minimum shaft diameter for tensile strength d_t can be calculated using a similar expression:

$$d_t = [16[M + (\tau^2 + M^2)^{1/2}]\pi\sigma_t]^{1/3}$$

where σ_t is the allowable tensile stress. The minimum shaft diameter for tensile strength is, accordingly,

$$d_t = [16(8272 + (9454^2 + 8272^2)^{1/2}]/\pi(10,000)]^{1/3} = 2.197 \text{ in}$$

The minimum shaft diameter for shear and tensile limits is 2.201 in (0.056 m). The next larger standard size is 2.5 in (0.0635 m), which provides an adequate initial design.

3. *Calculate the natural frequency of the agitator shaft.*
Shaft strength is not the only limit to agitator shaft design—a long shaft may not be stiff enough to prevent uncontrolled vibrations. A given overhung shaft, of the sort typically used with top-entering agitators, will oscillate at a natural frequency, similar to the vibration of a tuning fork. If the operating speed of the agitator is too close to that frequency, destructive oscillations may occur. Most large agitator shafts are designed with the operating speed less than the first natural frequency, so that even as the agitator is started and stopped, excessive vibrations should not occur.

A typical formula for calculating the first natural frequency (critical speed) of an agitator shaft considers the shaft stiffness, the shaft length, the weights of impellers and shaft, and the rigidity of the shaft mounting:

$$N_c = 37.8d^2(E_y/\rho_m)^{1/2}/LW_e^{1/2}(L + L_b)^{1/2}$$

where N_c = critical speed, r/min
d = shaft diameter, in
L = shaft extension, in
W_e = equivalent weight (lb) of impellers and shaft at shaft extension
L_b = spacing (in) of bearings that support shaft
E_y = modulus of elasticity, lb/in^2
ρ_m = density, lb/in^3

E_y and ρ are two material properties that characterize the stiffness of the shaft. Substituting the modulus and density values given in the statement of the problem reduces the expression to

$$N_c = 371,400 d^2 / L W_e^{1/2} (L + L_b)^{1/2}$$

For a two-impeller situation, W_e can be calculated as follows:

$$W_e = W_l + W_u (L_u/L)^3 + w_s(L/4)$$

where W_l and W_u are, respectively, the weights of the lower and upper impellers, the upper impeller is located at shaft extension L_u, and w_s is the unit weight of the shaft as given in the statement of the problem. Data on impeller weight can be furnished by the mixer vendor, measured directly by the user, or estimated from dimensions provided by the vendor. For the present case, assume that the impeller hub for a 2.5-in shaft weighs 25 lb (11.4 kg) and a set of blades for a 36-in impeller weighs 34.5 lb (15.7 kg), so each impeller for this agitator weighs 59.5 lb (27.0 kg). Thus,

$$W_e = 59.5 + 59.5(80/130)^3 + [0.228(2.5)^2][130/4] = 120 \text{ lb}$$

and accordingly,

$$N_c = 371,400(2.5)^2 / [130(120)^{1/2}(130 + 16)^{1/2}] = 135 \text{ r/min}$$

This means that with the proposed 2.5-in shaft, an operating speed near 135 r/min must be avoided. In practice, the design limit for operating speed should be no higher than 65 percent of critical speed. This conservative margin is necessary because of many factors that might reduce the critical speed or increase loads on the shaft. For instance, the critical-speed calculation assumes that the agitator drive support is rigid, but in fact tank nozzles and support structures have some flexibility that reduces the natural frequency. Furthermore, dynamic loads on the impeller, such as those induced by operating near the liquid level, may make the effects of natural frequency more significant.

In the present case, the 100-r/min operating speed given in the statement of the problem is 74 percent of critical speed and therefore is too close to critical speed for safe operation.

4. Redesign the shaft to avoid critical-speed problems.
A larger shaft diameter should overcome critical-speed problems, provided the gear reducer will accept the larger shaft. The larger shaft diameter increases shaft stiffness and thus increases the natural frequency. However, a large shaft also means more weight in the impeller hubs and in the shaft.

For the present case, try a 3.0-in shaft. Assume that vendor information indicates a hub weight of 40 lb (18.2 kg) for such a shaft, thus increasing the impeller weights to 74.5 lb (33.8 kg). Accordingly, the new equivalent weight is

$$W_e = 74.5 + 74.5(80/130)^3 + [0.228(3.0)^2][130/4] = 159 \text{ lb}$$

And the new critical speed is

$$N_c = 371,400(3.0)^2 / [(130)(159)^{1/2}(130 + 16)^{1/2}] = 169 \text{ r/min}$$

The stated operating speed of 100 r/min is only 59 percent of this critical speed, so the 3.0-in shaft should operate safely.

5. Explore other alternatives for solving the critical-speed problem.
The seemingly obvious answer to a critical-speed problem is to reduce the operating speed of the agitator, and this option should always be checked out. However, it can introduce complications of its own. In particular, the lower speed will change the process performance of the agitator, and accordingly a larger impeller will be required for meeting performance requirements. Because larger impellers weigh more, the critical-speed determination must be made anew.

Apart from its effect on critical speed, such an impeller will affect horsepower requirements. Furthermore, the speed change may require a new gear reducer.

The critical-speed problem may in some cases be solved hand in hand with a common problem related to dynamic loads. One source of dynamic loads on an agitator shaft is the waves and vortices that occur when an impeller operates near the liquid surface, such as when a tank fills or empties. Adding stabilizer fins to the impeller blades will help reduce some of these loads. Such fins also permit the agitator to operate closer to critical speed, perhaps at 80 percent rather than 65 percent.

Suppose, for instance, that the 2.5-in shaft is used and stabilizer fins add 16 lb (7.3 kg) to the weight of the lower impeller. The equivalent weight increases accordingly:

$$W_e = (59.5 + 16) + 59.5(80/130)^3 + [0.228(2.5)^2][130/4] = 136 \text{ lb}$$

and the critical speed decreases:

$$N_c = 371,400(2.5)^2/[130(136)^{1/2}(130 + 16)^{1/2}] = 127 \text{ r/min}$$

The operating speed of 100 r/min is now 79 percent of critical, which is just within the aforementioned 80 percent ceiling.

Another alternative to avoid critical-speed problems is the use of a shorter shaft. Reducing the shaft length and impeller extensions by 10 in (0.25 m) reduces equivalent weight to

$$W_e = 59.5 + 59.5(70/120)^3 + [0.228(2.5)^2][120/4] = 114 \text{ lb}$$

At 114 lb (51.8 kg), the critical speed increases to

$$N_c = 371,400(2.5)^2/[120(114)^{1/2}(120 + 16)^{1/2}] = 155 \text{ r/min}$$

At 155 r/min, the operating speed is slightly less than 65 percent of critical speed. If the reduced shaft length can be achieved by reducing the mounting height, the impeller location and performance remain unchanged. Otherwise, review process conditions. Other design changes, such as a smaller shaft below the upper impeller, may reduce equivalent weight and increase critical speed.

The additional cost of the material for a larger-diameter shaft is rarely a major factor for carbon or stainless steel, but may be sizeable for special alloys. The additional cost of a larger mechanical seal, for pressurized applications, may add considerable cost regardless of the shaft material.

12.8 *VISCOSITY DETERMINATION FROM IMPELLER POWER*

A helical ribbon impeller, 45 in (1.14 m) diameter, is operated in a 47-in (1.19-m) diameter reactor; estimate the fluid viscosity from torque readings. The impeller is a single-turn helix with a 1 : 1 pitch, so the height of the impeller is the same as the diameter, 45 in (1.14 m). It is a double-flight helix, each blade of which is 4.5 in (0.114 m) wide.

The agitator on the reactor is instrumented with a tachometer and torque meter. In the early stages of a polymerization, the impeller is operated at 37 r/min and the torque reading is 460 in · lb (52 N · m). As the viscosity increases during polymerization, the agitator is slowed to 12 r/min. The reaction is stopped when the torque reaches 27,300 in · lb (3085 N · m). As a final check on the polymer, an additional torque reading of 20,600 in · lb (2328 N · m) is taken at 8 r/min. Assume that the polymer has a specific gravity of 0.92 throughout the polymerization. What is the apparent viscosity of the polymer at the early stage of the process and at the two final conditions?

Calculation Procedure

1. *Estimate viscous power number for the helix impeller.*
A helical-ribbon impeller, also called a helix impeller, is used primarily when high-viscosity fluids are being processed. Most of the power data on such impellers have been obtained in the laminar and transitional flow ranges. The effect on power of common geometry factors, i.e., impeller diameter D, tank diameter T, helix pitch P, impeller height H, and helix (blade) width W, can be incorporated into a correlation for a (dimensionless) viscous power number:

$$N_P^* = 96.9[D/(T - D)]^{0.5}[1/P][H/D][W/D]^{0.16}$$

For the impeller described in this problem the power number is

$$N_P^* = 96.9[45/(47 - 45)]^{0.5}[1/1][45/45][4.5/4.5]^{0.16} = 318$$

The viscous power number is *defined* in terms of power P, viscosity μ, shaft speed N, and impeller diameter D:

$$N_P^* = P/\mu N^2 D^3 = N_P N_{Re}$$

Thus, viscous power number N_P^* is related to turbulent power number N_P by the factor of the Reynolds number N_{Re}. The viscous power number is chosen as a correlating value because it has a constant value in the viscous, low–Reynolds number range, less than 60 ($N_{Re} < 60$). Using the viscous power number in the laminar range eliminates fluid density from the correlation, which is appropriate.

2. *Estimate viscosity at early stages of polymerization.*
The torque τ measurement, combined with shaft speed, can be converted to power:

$$P = \tau N/63,025 = (460)(37)/63,025 = 0.27 \text{ hp}$$

In this equation, 63,025 is a conversion factor for dimensional consistency.
At 0.27 hp (0.20 kW), a constant viscous power number can be used to predict an apparent viscosity, with the aid of a 6.11×10^{-15} units-conversion factor:

$$\mu_a = P/N_P^* N^2 D^3 = 0.27/(6.11 \times 10^{-15})(318)(37)^2(45)^3$$
$$= 1114 \text{ cP}$$

Now, a viscosity of 1114 cP is low for viscous (laminar) flow conditions to exist. So, it is prudent to check the magnitude of N_{Re}:

$$N_{Re} = D^2 N\rho/\mu = (10.7)(45)^2(37)(0.92)/1114 = 662$$

where 10.7 is for dimensional consistency when ρ is specific gravity.
While not fully turbulent, a Reynolds number of 662 is not laminar either. It is in the transitional range, so a correction factor from Fig. 12.8 must be applied to the viscous power number. This is done in step 3.
The term "apparent viscosity" refers to a viscosity that has been back-calculated from impeller torque or horsepower. A true-viscosity reading should be measured at a fixed and known shear rate. The effective shear rate developed by a mixing impeller is really a distribution of different shear rates. This distribution is probably most closely related to the shear between the helix blade and the tank wall, but other shear rates may affect power. If viscosity is shear-dependent, as often happens

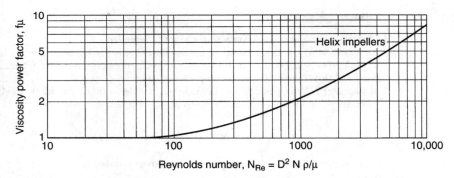

FIGURE 12.8 Viscosity power factor for viscous power number of a helix impeller as a function of impeller Reynolds number.

with high-viscosity polymers, the velocity distribution will affect the apparent viscosity. Different impeller types may give different apparent viscosities at the same shaft speed.

Different speeds will provide different shear rates and perhaps different apparent viscosities for many viscous fluids. Apparent viscosities measured by a mixing impeller are more useful for mixer design than those obtained with a viscometer, because the appropriate distribution of shear rates are included in the measurements.

3. Reestimate viscosity for intermediate Reynolds number.

In the intermediate–Reynolds number range, $60 < N_{Re} < 20,000$ (for a helix impeller), the viscous power number is not constant, nor is the turbulent power number. Figure 12.8, a graph of the viscosity power factor as a function of the Reynolds number, can be used to correct the viscous power number in the transitional range.

At a Reynolds number of 662, the viscosity power factor f_μ is approximately 1.8. Applying this factor to the power number used in the apparent-viscosity calculation of step 2 enables the estimation of another apparent viscosity:

$$\mu_a = 0.27/(6.11 \times 10^{-15})(1.8)(318)(37)^2(45)^3 = 619 \text{ cP}$$

Another estimate of Reynolds number based on a viscosity of 619 cP gives a value of 1191. At a Reynolds number of 1191, the power factor becomes 2.3, which leads to an apparent viscosity of 484. A few more iterations reach an estimated viscosity of 370 cP, based on a correction factor of 3.0 at a Reynolds number of 1994.

Several iterations are required, because power becomes less dependent on viscosity as conditions approach the turbulent range, typically $N_{Re} > 20,000$. In the turbulent range, power is independent of viscosity, and impeller power cannot be used to estimate viscosity.

4. Determine apparent viscosities at the end of the process.

As the polymerization proceeds, the viscosity increases. Higher viscosity means higher torque. Reducing the agitator speed is necessary to keep the torque and power within the capabilities of the agitator. At the end of the process, with the agitator turning at 12 r/min, the torque reaches 27,300 in · lb (3085 N · m) and the power required (see step 2) is

$$P = (27,300)(12)/63,025 = 5.20 \text{ hp}$$

At 5.20 hp (3.88 kW), the apparent viscosity is

$$\mu_a = 5.20/(6.11 \times 10^{-15})(318)(12)^2(45)^3 = 204,000 \text{ cP}$$

At 204,000 cP, the Reynolds number is only

$$N_{Re} = (10.7)(45)^2(12)(0.92)/204,000 = 1.2$$

which is well into the viscous (laminar) range. At these conditions, power and torque are proportional to viscosity at a set speed and the viscosity power factor is 1.

The same calculations at 8 r/min and 20,600 in · lb (2328 N · m) show the following:

$$P = (20,600)(8)/63,025 = 2.61 \, hp$$
$$\text{and} \quad \mu_a = 2.61/(6.11 \times 10^{-15})(318)(8)^2(45)^3 = 230,000 \, cP$$

Thus, the fluid appears to be more viscous at the lower speed. Another view of the fluid properties is that as agitator speed increases, the shear rate increases and the viscosity decreases. This fluid behavior is called "shear thinning" and is typical of many polymers.

Thus, a properly instrumented reactor may be used as a viscometer for high-viscosity fluids.

12.9 NONGEOMETRIC SCALE-UP FOR LIQUID AGITATION

A process involving a water-like liquid must be scaled up from an agitated 18-in (0.46-m) diameter, 15-gal (0.057-m³) pilot-scale reactor to a 120-in (3.05-m) diameter, 7000-gal (26.5-m³) large-scale reactor. The pilot-scale reactor has a 18-in (0.46-m) straight side and the large-scale reactor will have a 168-in (4.27-m) straight side. Both reactors have ASME dished heads on the top and bottom. Successful process performance was obtained in the pilot scale with two 6.0-in- (0.15-m-) diameter pitched-blade turbines operating at 350 rev/min (5.83 rev/s). It is proposed that the large-scale reactor use hydrofoil impellers instead of pitched-blade turbines, for improved liquid motion. Each pitched-blade turbine has a turbulent power number of 1.37 and each hydrofoil has a power number of 0.3. Past scale-up experiences with similar processes, but with geometrically similar tanks, were successful when impeller tip speed was held constant.

Calculation Procedure

1. *Calculate the liquid levels.*
Liquid level is usually measured from the center of the bottom of the vessel head. So, the head volume must be subtracted from the total volume to calculate the straight-side liquid level; then the head depth must be added to the straight-side liquid level.

In the pilot-scale reactor, an ASME head of 18-in (0.46-m) diameter has a volume of 2.04 gal (0.0077 m³) and a depth to the straight side of 3.16 in (0.080 m). Tables for head depths and volumes can be found in many engineering handbooks. The straight side holds (15 gal − 2.04 gal), or 12.96 gal (0.049 m³). Working from the formula for the volume of a cylinder, which is (base area) × (height), the height is calculated by dividing the volume by the cross-sectional area. The cylindrical volume is 12.96 gal (0.049 m³), which is (12.96 gal) (231 in³/gal) or 2994 in³. The cross-sectional area of the tank is $\pi T^2/4 = \pi 18^2/4 = 254.5$ in² (0.164 m²). Dividing cylinder volume by cross-sectional area, 2994 in³/254.5 in², gives 11.76 in (0.30 m) liquid level in the straight side of the tank. The total liquid level is the sum of the straight-side level and the head depth: 11.76 + 3.16 = 14.9 in (0.379 m). Because from 5% to 10% of the calculated tank volume is typically filled by tank internals, such as impellers, shaft, and baffles, the actual liquid level may be closer to 1.10(14.9) = 16.4 in (0.417 m).

Similar calculations can be done for the large-scale tank. A 120-in (3.05-m) diameter ASME head has a volume of 605 gal (2.29 m³) and a depth of 20.7 in (0.53 m). The 7000-gal volume has 7000 − 605 = 6395 gal (24.2 m³) in the straight side, which leaves a straight-side liquid level of (6395)(231)/(π 120²/4) = 1,477,245/11,310 = 130.6 in (3.32 m). The total liquid level is 130.6 + 20.7 = 151.3 in

(3.84 m). If the liquid level is adjusted for tank internals, the expected liquid level is 1.10(151.3) = 166 in (4.23 m). This liquid depth will easily fit in a tank with a 168-in (4.27 m) straight side.

Note that the ratio of liquid level to tank diameter on the pilot scale is 16.4/18.0 or 0.91, whereas the ratio for the large-scale vessel is 166/120 or 1.38. For geometric similarity, all such length ratios should instead be equal. Since the ratios are not equal in this case, it is necessary to use nongeometric scale-up.

2. *Consider the options and issues involved in direct scale-up for reactors that are not geometrically similar, and apply them to the present example.*
Although a direct scale-up based on equal tip speed is an option, many unanticipated changes may occur in the absence of geometric similarity. For instance, two hydrofoil impellers of 40-in (1.02-m) diameter could be chosen for the large-scale reactor. A simple tip-speed scale-up would require an adjustment to the pilot-scale speed by the inverse ratio of the impeller diameters: (350 rev/min) (6/40) = 52.5 rev/min (0.875 rev/s). As a basic check, impeller tip speed for the pilot-scale reactor can be calculated by multiplying the circumference at the blade tip by the rotational speed: π (6 in)(350 rev/min) = 6597 in/min or 550 ft/min (2.79 m/s). The same calculation for the large-scale reactor gives π (40 in) (52.5 rev/min) = 6597 in/min or 550 ft/min (2.79 m/s). These scale-up results have the same tip speed. Unfortunately, however, we cannot be sure that the arbitrarily chosen hydrofoil diameter is correct, nor whether the liquid-level ratio is important, nor whether other factors need to be considered.

One means for considering such other effects in scale-up is to calculate other basic values, such as power per volume and torque per volume, both of which can be used to describe agitation intensity. Procedure 12.1 goes through some details of calculating impeller power. For the situation under consideration here, since the fluid is described as waterlike, a specific gravity of 1.0 and a viscosity of 1.0 cp are appropriate. With these characteristics, a correction for Reynolds number effects is not needed, and a constant power number can be applied in the calculations. (A more comprehensive power calculation might include effects of such factors as viscosity, density, off-bottom clearance, and impeller spacing; these additional corrections should be made at any point in the calculations when they are significant. However, in this example, and in many real applications, such corrections are not essential for scale-up.)

Using the power number of 1.37 for the pitched-blade turbine, impeller power in the pilot-scale reactor can be calculated as shown in Procedure 12.1 by $1.37(1.0)(350)^3(6)^5/(1.524 \times 10^{13})$ = 0.030 hp (22.35 W). Since two impellers are used, double this value gives a fair representation of the total power: 2(0.030) = 0.060 hp (44.7 W). Power per volume in the pilot scale is 0.060 hp/15 gal = 0.004 hp/gal (787 W/m^3). Power per volume is often reported with respect to 1000 gal for reasonable size values: (1000)(0.060)/15 = 4.00 hp/1000 gal (787 W/m^3). Torque is simply power divided by speed; including a units-conversion factor (63,025), the impeller torque is 63,025(0.060 hp)/(350 rev/min) = 10.8 in-lb (1.22 N-m). On a thousand-gallon basis, the torque per volume is (1000)(10.8)/15 = 720 in-lb/1000 gal (21.5 N-m/m^3).

The same calculations for the large-scale reactor with two 40-in hydrofoil impellers start with the power calculation. The specified hydrofoil impeller has a power number of 0.30. (Because of individual geometries, this value may differ depending on the manufacture and impeller style.) For the proposed hydrofoil impellers, the power for two impellers is $(2)0.30(1.0)(52.5)^3(40)^5/(1.524 \times 10^{13})$ = 0.583 hp (435 W). The power per volume is 1000(0.583)/7000 = 0.08 hp/1000 gal (16.4 W/m^3). This scale-up results in only 0.08/4.0 = 0.02 or only 2% of the power per volume used in the pilot-scale reactor. The torque for the large scale is 63,025(0.583)/52.5 = 700 in-lb (79 N-m). Torque per volume is 1000(700)/7000 = 100 in-lb/1000 gal (2.8 N-m/m^3). In this case, large-scale torque per volume is 100/720 = 0.14 or 14% of the value in the pilot scale.

Power per volume is often associated with agitation for mass transfer, and torque per volume is often related to liquid velocities. If these values change drastically on scale-up—as is the case in the present example—the agitation intensity will also change significantly. Admittedly, the values will change even under geometric similarity; but torque per unit volume should not be less than one-third the original value, unless the pilot scale was tested with much more than the minimum level of agitation.

The drastic changes in power and torque per unit volume experienced in the present example signal the need to adopt step-by-step nongeometric scale-up. A systematic approach, as shown in the steps that follow, will identify problems and suggest possible remedies. The order of the scale-up steps is not critical, but the one used in this example works well. Not all of the steps are required in all scale-up problems and some steps can be used for other applications.

3. Conduct a geometric-similarity scale-up.
Perhaps paradoxically, a good starting point even for nongeometric scale-up is making a geometric-similarity one. For one thing, the rotational-speed changes are predictable. And, at least one key variable can be held constant. For geometric similarity, the ratios of all the length dimensions are the same from the small to the large scale. Thus a tank diameter scale-up from 18 in (0.46 m) to 120 in (3.05 m) represents a geometric ratio of 120/18 = 6.67. Applying this ratio to the impeller diameter gives a large-scale-impeller diameter of (6)(120/18) = 40 in (1.02 m). For tip speed (πDN) to remain constant, the rotational speed of the pilot scale must be adjusted in inverse proportion to the geometric ratio: (350)(6/40) = 52.5 rev/min (0.88 rev/s).

For all of the geometric ratios to be the same, the liquid level in the large-scale vessel has to be (16.4)(120/18) = 109 in (2.77 m). Based on the same large-scale head dimensions and volume calculations discussed in the first step of the example (and taking into account the extra 10% for internals), the geometric-similarity volume in a 120-in (3.05-m) diameter tank is calculated as 4444 gal (16.8 m^3).

To verify this scale-up step, calculate tip speed as π(40)(52.5)/12 = 550 ft/min (2.79 m/s). The power required for two impellers is (2)(1.37)(1.0)(52.5)^3(40)^5/(1.524 × 10^{13}) = 2.66 hp (1987 W). Power per volume is 1000(2.66)/4444 = 0.60 hp/1000 gal (118 W/m^3). Torque is 63,025 (2.66)/52.5 = 3193 in-lb (360 N-m) and torque per volume is 1000 (3193)/4444 = 720 in-lb/1000 gal (21.5 N-m/m^3).

A geometric-similarity scale-up with constant tip speed also keeps the torque/volume constant. Because liquid velocities (torque/volume) are an important measure of agitation intensify in liquid agitation problems, keeping torque/volume constant is more important than keeping power/volume constant. For this scale change, power per volume is 0.6/4.0 = 0.15 or 15% of the pilot scale. A power/volume scale-up for this scale change would be a very conservative design, but one that is potentially appropriate for certain applications.

4. Apply a volume adjustment to the results of step 3.
Since geometric similarity scale-up resulted in a volume of only 4444 gal (16.8 m^3) in a 120-in (3.05-m) diameter tank and the design specification called for 7000 gal (26.5 m^3), a higher liquid level is required. Based on the calculations in step 1, the open-tank liquid level would be 99.1 in (2.52 m). Allowing 10% for internals, the actual liquid level will be about 109 in (2.77 m).

For equal tip speed and impellers with geometric similarity, the rotational speed remains the same in spite of the added liquid level. Thus, the power and torque remain the same for the same impellers, speed, and fluid properties. However, the volume change affects power per volume: 1000(2.66 hp)/7000 gal = 0.38 hp/1000 gal (75.0 W/m^3). Similarly, the torque per volume is reduced to 1000(3193 in-lb)/7000 gal = 456 in-lb/1000 gal (13.6 N-m/m^3). Both changes represent a substantial reduction in agitation intensity from the pilot-scale results.

5. Adjust the step 4 results to gain constant torque/volume.
Because equal tip speed with geometric similarity gives equal torque/volume, those two different scale-up criteria are often confused with each other when evaluating previous scale-up experience. In the present example, it seems appropriate to supplement step 4's volume adjustment with an adjustment for equal torque per unit volume. Equal torque per unit volume should keep velocity magnitudes similar, thus giving the appearance of similar liquid agitation intensity. To maintain the pilot-vessel torque per volume of 720 in-lb/1000 gal (21.5 N-m/m^3) in the 7000-gal (26.5-m^3) vessel, the large-vessel torque must be 7000 (720)/1000 = 5040 in-lb (569 N-m). Combining the formulas for power and torque gives torque as $N_p \rho N^2 D^5$, which is divided by a conversion factor of 2.418 × 10^8. Rearranging this torque formula and solving for impeller speed gives an equal-torque-per-unit-volume speed of

$$N = \left[\frac{2.418 \times 10^8 (5040)}{2(1.37)(1.0)(40.0)^5} \right]^{1/2} = 65.9 \text{ rev/min } (1.10 \text{ rev/s})$$

At 65.9 rev/min (1.10 rev/s) the power is $(2)(1.37)(1.0)(65.9)^3(40.0)^5/1.524 \times 10^{13} = 5.27$ hp (3927 W) and power per volume is 0.75 hp/1000 gal (148 W/m³). However, tip speed now becomes $\pi(65.9)(40.0/12) = 690$ ft/min (3.51 m/s). Whether this increase in tip speed is an advantage or disadvantage depends on the process. Without further pilot-scale testing, which will be discussed later, insufficient information is available to choose the preferred criterion.

6. *Adjust the step 5 results to gain constant tip speed in addition to constant torque per unit volume.*
In the past two steps, the volume increase from 4444 gal (16.8 m³) to 7000 gal (26.5 m³) has been accomplished while holding either tip speed or torque per unit volume constant. However, with a change in the impeller diameter, both tip speed and torque per unit volume can be held constant simultaneously. Since turbulent conditions imply a constant power number, and since geometrically similar impellers imply the same power number, the mathematical expressions for tip speed and torque per volume can be simplified to expressions involving rotational speeds, impeller diameters, and liquid volumes. In what follows, the subscripts 1 and 2 represent the characteristics for the small and large volumes, respectively.

Constant Tip Speed

$$N_1 D_1 = N_2 D_2$$

Constant Torque per Volume

$$\frac{N_1^2 D_1^5}{V_1} = \frac{N_2^2 D_2^5}{V_2}$$

Eliminating Constant Tip Speed

$$N_1^2 D_1^2 \frac{D_1^3}{V_1} = N_2^2 D_2^2 \frac{D_2^3}{V_2}$$

$$\frac{D_1^3}{V_1} = \frac{D_2^3}{V_2}$$

Solving for Impeller Diameter

$$D_2 = D_1 \left(\frac{V_2}{V_1}\right)^{1/3}$$

$$D_2 = 40.0 \left(\frac{7000}{4444}\right)^{1/3}$$

$$= 46.5 \text{ in (1.18 m)}$$

Solving for Rotational Speed

$$N_2 = N_1 \left(\frac{D_1}{D_2}\right)$$

$$N_2 = 52.5 \left(\frac{40.0}{46.5}\right)$$

$$= 45.1 \text{ rev/min (0.75 rev/s)}$$

The power required to rotate two impellers of 46.5-in (1.18-m) diameter at 45.1 rev/min (0.75 rev/s) is $(2)(1.37)(1.0)(45.1)^3(46.5)^5/1.524 \times 10^{13} = 3.59$ hp (2689 W) or 1000 (3.59)/7000 = 0.51 hp/1000 gal (102 W/m³). Torque is 63,025(3.59)/45.1 = 5017 in-lb (567 N-m) and torque per volume is 1000 (5,017)/7000 = 717 in-lb/1000 gal (21.4 N-m/m³). Tip speed is $\pi(45.1)$ (46.5/12) = 550 ft/min (2.79 m/s). Thus, with a small adjustment in impeller diameter, both torque per volume and tip speed can be held constant. Since liquid level has changed, geometric similarity no longer applies and an impeller diameter adjustment may be an acceptable option.

7. *Investigate changing the number of impellers.*
When the volume changes from 4444 gal (16.8 m³) to 7000 gal (26.5 m³), the volume ratio is 7000/4444 = 1.56. Since the volume ratio is approximately 1.5, consider 1.5 times the number of impellers, or three impellers. More impellers in a tall tank are a good answer simply because the agitation effects are spread through the height of the vessel. The result with three impellers looks much like the geometric similarity scale-up with an additional half tank on the top.

For three impellers with a constant-tip-speed adjustment from the geometric similarity scale-up, the power is $3(1.37)(1.0)(52.5)^3(40.0)^5/1.524 \times 10^{13} = 4.00$ hp (2980 W) or 1000(4.00)/7000 = 0.57 hp/1000 gal (112 W/m³). Torque is 63,025(4.00)/52.5 = 4800 in-lb (542 N-m) and torque per volume is 1000(4800)/7000 = 685 in-lb/1000 gal (20.5 N-m/m³). The addition of a third impeller for the volume adjustment leaves tip speed constant and both power per volume and torque per volume almost unchanged.

This solution appears best for the initial changes using the same style impellers. However, we can still investigate the effect of changing impeller type.

8. *Investigate changing impeller type while keeping tip speed constant.*
A change in impeller type requires a new power number for the calculations. The typical hydrofoil impeller in this problem has a power number of 0.3, compared with the 1.37 power number for the pitched-blade turbines considered in the six previous steps. Following the success of the pitched-blade design with three impellers, consider replacing these impellers with hydrofoil impellers. To keep tip speed constant, the impeller diameters and rotational speed can remain unchanged from those of the pitched-blade-turbine case. Thus, power is $3(0.3)(1.0)(52.5)^3(40.0)^5/1.524 \times 10^{13} = 0.875$ hp (653 W) and power per volume is 1000(0.875)/7000 = 0.13 hp/1000 gal (24.6 W/m³). Torque is 63,025(0.875)/52.5 = 1050 in-lb (119 N-m) and torque per volume is 1000(1050)/7000 = 150 in-lb/1000 gal (4.5 N-m/m³). Although hydrofoil impellers are more efficient than pitched-blade turbines for most blending problems, in this instance they are providing only one-fifth of the original torque per unit volume, which is not going to provide as much agitation intensity as was found in the pilot scale.

9. *Change impeller type with equal tip speed and adjusted torque per volume.*
Calculations similar to those of step 6 can be used to keep both tip speed and torque per volume constant when the impeller type is changed. Because volume does not change, that factor is removed from the calculations, but the change in impeller type means that power number must be included. In addition, a hydrofoil impeller is more efficient at creating liquid motion than a pitched-blade turbine; therefore, the torque required by the hydrofoil impeller will be assumed to be half that of the pitched-blade turbine. The subscript 1 will be used for the pitched-blade turbine and the subscript 2 will be used for the hydrofoil impeller.

Constant Tip Speed

$$N_1 D_1 = N_2 D_2$$

Half the Pitched-Blade Torque

$$\frac{N_{P_1} N_1^2 D_1^5}{2} = N_{P_2} N_2^2 D_2^5$$

The relative amounts of torque required by different impellers are not fixed values; they must be evaluated for individual situations. This complication demonstrates that equal values for properties or characteristics do not need to be applied in all situations.

Eliminating Constant Tip Speed

$$N_1^2 D_1^2 \frac{N_{P_1} D_1^3}{2} = N_2^2 D_2^2 N_{P_2} D_2^3$$

$$\frac{N_{P_1} D_1^3}{2} = N_{P_2} D_2^3$$

Solving for Impeller Diameter

$$D_2 = D_1 \left(\frac{N_{P_1}}{2 N_{P_2}} \right)^{1/3}$$

$$D_2 = 40.0 \left(\frac{1.37}{2(0.3)} \right)^{1/3}$$

$$= 52.7 \text{ in } (1.34 \text{ m})$$

The power number for individual impellers should be used in this calculation. If different style or size impellers were used at different locations, separate calculations for each impeller would be required.

Solving for Rotational Speed

$$N_2 = N_1 \left(\frac{D_1}{D_2} \right)$$

$$N_2 = 52.5 \left(\frac{40.0}{52.7} \right)$$

$$= 39.9 \text{ rev/min } (0.66 \text{ rev/s})$$

The power for three hydrofoil impellers of 52.7-in (1.34-m) diameter operating at 39.9 rev/min (0.66 rev/s) is $(3)(0.3)\ (1.0)\ (39.9)^3\ (52.7)^5/1.524 \times 10^{13} = 1.52$ hp (1131 W). Power per unit volume is $1000(1.52)/7000 = 0.22$ hp/1000 gal (42.7 W/m^3). Torque is $63,025\ (1.52)/39.9 = 2399$ in-lb (271 N-m) and torque per volume is $1000\ (2,399)/7000 = 343$ in-lb/1000 gal (10.2 N-m/m^3). Tip speed is $\pi (39.9)(52.7/12) = 550$ ft/min (2.79 m/s).

These results are consistent with a scale-up from the pilot-scale operation. Although power per volume is reduced and impeller to tank diameter ratio is increased, torque per volume is about half and tip speed is the same. With any realistic scale-up, some factors unavoidably must change, while the important ones are held constant. In a different situation and process, a different variable, such as power per volume, might be held constant. Other variables, such as blend time, could be calculated, at each step in the scale-up.

An important and practical consideration in any nongeometric-similarity scale-up is the idea of making only one change at a time, usually starting with a geometric similarity scale-up to the large-size tank diameter. Then at each step, while adjusting conditions on the large scale, relevant values should be calculated to be sure that unwanted or unacceptable changes do not occur. This process may lead to trial changes with unacceptable results. Such changes must be reexamined for possible alternatives. The solution to this example demonstrates several methods for making design change; in different situations, steps may be eliminated or added as necessary. The steps demonstrated in this problem can be used for a variety of other situations, such as changing impeller size or type in an existing application or changing operating liquid level.

No specific scale-up method can be applied to all problems. Each situation is different. Nongeo-metric scale-up can be done and is necessary in many situations. However, certain applications, such as solids suspension, which is geometry dependent, should follow geometric similarity scale-up.

Related Calculations.   When considering whether tip speed or torque per volume was the preferred method of scale-up for this example, pilot-scale results were not available to suggest possible differences. Some simple comparisons, using different size impellers in the pilot scale, may give better insight into the process behavior. For instance, instead of making all of the tests with two 6-in (0.15-m) diameter impellers, additional tests could have been made with two 7-in (0.18-m) diameter impellers.

With a change in impeller diameter, speed adjustment based on equal tip speed, equal torque, or equal power all give different rotational speeds for comparison. Examining a change from the 6-in (0.15-m) diameter impeller operating at 350 rev/min (5.83 rev/s) to a 7-in (0.18-m) diameter impeller shows these differences.

Equal Tip Speed

$$N_1 D_1 = N_2 D_2$$

$$N_2 = N_1 \left(\frac{D_1}{D_2} \right)$$

$$N_2 = 350 \left(\frac{6}{7} \right)$$

$$= 300 \text{ rev/min } (5.0 \text{ rev/s})$$

Equal Power

$$N_1^3 D_1^5 = N_2^3 D_2^5$$

$$N_2 = N_1 \left(\frac{D_1}{D_2} \right)^{5/3}$$

$$N_2 = 350 \left(\frac{6}{7} \right)^{5/3}$$

$$= 271 \text{ rev/min } (4.5 \text{ rev/s})$$

Equal Torque

$$N_1^2 D_1^5 = N_1^2 D_1^5$$

$$N_1 = N_2 \left(\frac{D_1}{D_2} \right)^{5/2}$$

$$N_1 = 350 \left(\frac{6}{7} \right)^{5/2}$$

$$= 238 \text{ rev/min } (4.0 \text{ rev/s})$$

All these calculations assume that all of the fluid properties are the same and the impellers are geometrically similar in turbulent conditions so that the power numbers remain constant. For conditions with variable values, more extensive calculations are necessary. Other changes may occur owing to different impeller to tank diameter ratios and the resulting change in flow patterns. Small changes to impeller diameters are strongly suggested.

*12.10   TEMPERATURE ANALYSIS OF HEATED AGITATED VESSEL

An agitated vessel, Fig. 12.9, is charged with 5200 lb (2364 kg) of nitrobenzene at 75°F (23.9°C). An internal coil with 40 ft² of heat-transfer surface contains steam condensing at 250°F (121.1°C). The average specific heat of the nitrobenzene is 0.61 Btu/(lb)(°F) (2.6 kJ/kg °C). If U = 165 Btu/(ft²)(h)(°F)(936.9 W/m² °C) (a) how long will it take to heat the liquid to 200°F (93.3°C)? (b) what will the liquid temperature be 1 h after the steam is turned on? (c) If the steam is turned off at the end of 1 h and 20 gal/min of cooling water admitted at 60°F (15.6°C) to the coil, how long will it take to cool the nitrobenzene from 228 (108.9°C) to 90°F (32.2°C)? Assume U = 150.

Calculation Procedure

1. Use a heating-time equation to determine the heat-up time.

$$\theta_T = \frac{mc_p}{UA} \ln \frac{t_h - t_a}{t_h - t_b}$$

where:

$$c_p = 0.61 \quad m = 5200 \text{ lb (2364 kg)} \quad t_h = 250°F (121.1°C)$$

$$t_a = 75°F (23.9°C) \quad t_b = 200°F (93.3°C) \quad U = 165 \quad A = 40$$

$$\theta_T = \frac{0.61 \times 5200}{165 \times 40} \ln \frac{250 - 75}{250 - 200} = 0.602 \text{ h}$$

2. Find the liquid temperature.
The equation above can be used to find the liquid temperature at any time.
Since $\theta_T = 1$,

$$1 = \frac{0.61 \times 5200}{165 \times 40} \ln \frac{250 - 75}{250 - t_b}$$

$$\ln \frac{250 - 75}{250 - t_b} = \frac{40 \times 165}{0.61 \times 5200} = 2.08$$

$$\frac{250 - 75}{250 - t_b} = 8.015$$

From this, $t_b = 229°F$ (109.4°C).

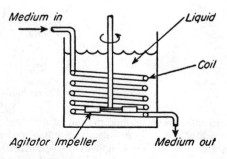

FIGURE 12.9   Agitator vessel having unsteady-state heat transfer.

3. Determine the time for the agitator to cool.
The following equation applies

$$\ln \frac{t_{ha} - t_{ca}}{t_{hb} - t_{ca}} = \frac{w_c c_{pc}}{m c_p} \left(1 - \frac{1}{K} \right) \theta_T$$

This equation may be solved for the time θ_T or the final liquid temperature t_{hb} if the other quantities are known.

The new quantities needed are

$$c_{pc} = 1.0 \quad t_{ca} = 60°F\ (15.6°C) \quad t_a = 229°F\ (109.4°C) \quad t_b = 90°F\ (32.2°C)$$

$$w = 20 \times 60 \times 8.33 = 10,000\ \text{lb/h}\ (4545\ \text{kg/h})$$

$$K = e^{(150 \times 40)/10,000} = e^{0.60} = 1.822$$

Substituting the values in the equation above

$$\ln \frac{229 - 60}{90 - 60} = \frac{10,000}{0.61 \times 5200} \left(1 - \frac{1}{1.822} \right) \theta_T$$

From this, $\theta_T = 1.21$ h.

Related Calculations.   The two equations given here can be used for a variety of agitator designs. This procedure is the work of Warren L. McCabe and Julian C. Smith, as given in their book *Unit Operations of Chemical Engineering*, McGraw-Hill, 1956.

REFERENCES

1. Uhl and Gray, eds.—*Mixing, Theory and Practice*, vol. 1, Academic Press, 1966; vol. 3, Academic Press, 1986.

2. Nagata—*Mixing, Principles and Applications*, Wiley.

3. Bates, Fondy, and Corpstein—*Ind. Eng. Chem. Process Des. Dev.* 2:310, 1963.

4. Gates, Henley, and Fenic—*Chem. Eng.* Dec. 8, 1975, p. 110.

5. Dickey and Fenic—*Chem. Eng.* Jan. 5, 1976, p. 139.

6. Dickey and Hicks—*Chem. Eng.* Feb. 2, 1976, p. 93.

7. Hicks, Morton, and Fenic—*Chem. Eng.* Apr. 26, 1976, p. 102.

8. Gates, Morton, and Fondy—*Chem. Eng.* May 24, 1976, p. 144.

9. Hicks and Gates—*Chem. Eng.* July 19, 1976, p. 141.

10. Hill and Kime—*Chem. Eng.* Aug. 2, 1976, p. 89.

11. Ramsey and Zoller—*Chem. Eng.* Aug. 30, 1976, p. 101.

12. Meyer and Kime—*Chem. Eng.* Sept. 27, 1976, p. 109.

13. Rautzen, Corpstein, and Dickey—*Chem. Eng.* Oct. 25, 1976, p. 119.

14. Hicks and Dickey—*Chem. Eng.* Nov. 8, 1976, p. 127.

15. Gates, Hicks, and Dickey—*Chem. Eng.* Dec. 6, 1976, p. 165.

16. van't Riet—*Ind. Eng. Chem. Process Des. Dev.* 18:357, 1979.

17. Aerstin and Street—*Applied Chemical Process Design*, Plenum.

18. Holland and Chapman—*Liquid Mixing and Processing in Stirred Tanks*, Reinhold.

19. Brodkey, ed.—*Turbulence in Mixing Operations*, Academic Press.

20. Millich and Carraher, eds.—*Interfacial Synthesis*, vol. 1: *Fundamentals*, Marcel Dekker.

21. Oldshue—*Chem. Eng.* June 13, 1983, p. 82.

22. Coble and Dickey—*AIChE Equipment Testing Procedure, Mixing Equipment (Impeller Type)*, AIChE, 1987.

23. Tatterson—*Fluid Mixing and Gas Dispersion in Agitated Tanks*, McGraw-Hill.

SECTION 13
SIZE REDUCTION[†]

Ross W. Smith, Ph.D.
Department of Chemical and Metallurgical Engineering
University of Nevada at Reno
Reno, NV

13.1 SIZE DISTRIBUTIONS OF CRUSHED OR GROUND MATERIAL

The data in the first three columns of Table 13.1 show the size distributions of a quartzgold ore obtained by sieving the material. What is the cumulative weight percent passing each sieve size? Also, construct a log-log plot of the cumulative percent versus particle size, and express the resulting relationship as an equation.

Calculation Procedure

1. Compute the cumulative weight percent passing each sieve size.
Add up the weights of samples retained on each screen (the weights are shown in the third column of Table 13.1). Express each as a percent of the total, obtaining the percents shown in the fourth column. Sum the cumulative amounts retained and subtract from 100, obtaining the cumulative percents passing. These are shown in the fifth column.

2. Express the relationship between cumulative percent passing and particle size as an equation.
Plot cumulative weight percent passing (y) as a function of particle size (x) in microns (as denoted by size of sieve opening) on log-log paper, as in Fig. 13.1.

[†]Procedure 13.10 is adapted from an article in *Chemical Engineering* magazine.

TABLE 13.1   Input Data and Calculated Results on Ore Size Distribution (Procedure 13.1)

Tyler mesh size	Sieve opening, μm	Weight of retained sample, g	Percent retained (by calculation)	Cumulative percent passing (by calculation)
4	4950	0	0	100.0
6	3350	7.20	3.0	97.0
8	2360	27.84	11.6	85.4
10	1700	36.96	15.4	70.0
14	1180	43.68	18.2	51.8
20	850	34.08	14.2	37.6
28	600	27.60	11.5	26.1
35	425	16.08	6.7	19.4
48	300	15.36	6.4	13.0
65	212	8.64	3.6	9.4
100	150	6.48	2.7	6.7
150	106	4.56	1.9	4.8
200	75	3.84	1.6	3.2
270	53	2.16	0.9	2.3
Pan	—	5.52	2.3	—

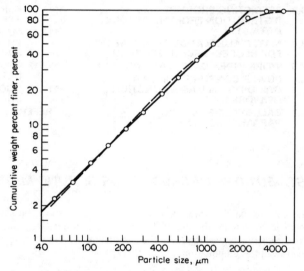

FIGURE 13.1   Cumulative weight percent of particles finer than a given size as a function of the particle size (Procedure 13.1).

The straight-line portion of the plot can be expressed as $y = cx^b$. This can be recast into a form that is especially useful for other size-reduction calculations (e.g., see Procedure 13.2), namely, the Gaudin-Schumann equation, that is, $y = 100(x/k)^a$, where a (the distribution modulus) is a constant for a particular size distribution, and k (the size modulus) is the 100 percent size, in microns, of the extrapolated straight-line portion of the plot. By applying least-squares curve fitting to the log-log plot, the values of a and k can be obtained, yielding $y = 100(x/2251)^{1.003}$.

Related Calculations.   Sometimes the Rosin-Rammler equation is used to represent the size distribution graphically. The dashed line on Fig. 13.1 corresponds to the Rosin-Rammler equation in the form $y = 100 - 100 \exp[-(x/A)^b]$, where, in this case, $A = 1558$ and $b = 1.135$.

TABLE 13.2   Ball-Mill Test Data on Cement Rock (Procedure 13.2)

Tyler mesh size	Sieve opening, μm	Cumulative mass fraction finer than mesh size after noted grinding time			
		1 min	2 min	4 min	6 min
8	2360	1.000	1.000	1.000	1.000
10	1700	0.450	0.695	0.910	0.966
48	300	0.080	0.175	0.337	0.523
100	150	0.046	0.100	0.196	0.300
200	75	0.030	0.062	0.126	0.195
400	38	0.018	0.046	0.076	0.110

Feed size: −8, +10 mesh (8 × 10 mesh)

13.2 BREAKAGE AND SELECTION FUNCTIONS FROM BATCH BALL-MILL TESTS

Determine the breakage characteristics of a cement rock by using selection and breakage functions, calculating these on the basis of the ball-mill test data in Table 13.2. (A "selection function" is a parameter that represents the resistance of some size fraction to being produced during breakage. The "breakage functions" are related quantities that determine the breakage-product size distribution for material broken in this size fraction.)

Calculation Procedure

1. Calculate the selection function that pertains to the size fraction of the feed material.
This selection function S_1 is defined by the equation

$$M_1(t) = M_1(0)\exp(-S_1 t)$$

where $M_1(t)$ is the mass fraction of feed remaining after time t. Referring to Table 13.2, note that for this 8 × 10 mesh feedstock, $M_1(t)$ equals 1 minus the cumulative mass fraction finer than no. 10 mesh at time t. Using the data in Table 13.2, we can therefore determine S_1 by plotting log mass fraction of feed versus grinding time, determining the slope of the resulting straight line by least-squares curve fitting and multiplying the slope by −2.303. The resulting value is 0.577 min^{-1}.

2. Determine the production rate constant.
In batch ball milling, the changes in particulate distribution with time can be characterized [1] by

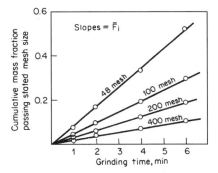

FIGURE 13.2 Cumulative mass fraction of ground material as a function of grinding time (Procedure 13.2).

$$\frac{dY_i(t)}{dt} = \overline{F}_i$$

where t is time, $Y_i(t)$ is cumulative mass fraction finer than size x_i at time t for short grinding times, and $\overline{F}_i$ is cumulative zero-order production rate constant for size x_i and the production of fine sizes much smaller than feed size. If, then, a linear plot is made of cumulative mass fraction versus (short) grinding time, the $\overline{F}_i$ constants can be determined for each fine size from the slopes of the curves generated. Figure 13.2 shows such a plot for the present data. From the slopes, Table 13.3 can be assembled.

TABLE 13.3 Calculated Production Rate Constants (Procedure 13.2)

Particle size x		
Mesh	Microns	Production rate constant $\bar{F}_i$
48	300	0.0869
100	150	0.0500
200	75	0.0325
400	38	0.0183

3. Determine the breakage functions that pertain to the size fraction of the feed material.
Cumulative breakage functions can be calculated [1] from the relation $B_{i,j} = \bar{F}_i/S_j$. Therefore, in the case of the feed-size selection function S_1, the equation is $B_{i,1} = \bar{F}_i/S_1$, and the breakage functions are as follows:

$$B_{48\text{mesh}, 1} = \frac{\bar{F}_{48\text{mesh}}}{S_1} = \frac{0.0869}{0.577} = 0.151$$

$$B_{100\text{mesh}, 1} = \frac{0.0500}{0.577} = 0.087$$

$$B_{200\text{mesh}, 1} = 0.056 \quad \text{and} \quad B_{400\text{mesh}, 1} = 0.032$$

4. Determine the selection and breakage functions for other size fractions of the same material.
Selection functions for other size intervals may be calculated via the relation

$$S_j = S_1\left[\frac{(X_j X_{j+1})^{1/2}}{(X_1 X_2)^{1/2}}\right]^a$$

where X_1 and X_2 are the sieve-opening sizes that define the size fraction of the feed material, X_j and X_{j+1} are the sieve openings that define the size fraction whose selection function is now to be calculated, and a is the slope of the log-log plot of the zero-order production rate constants $\bar{F}_i$ against particle size x_i, in microns, from Table 13.3. (That plot is not shown here.) The slope, determined by least-squares curve fitting, is 0.741. This slope, a, happens to be the same as the distribution modulus in the Gaudin-Schumann equation (see Procedure 13.1).

For instance, the selection function for the size fraction, -10, $+14$ mesh (corresponding to sieve openings of 1700 and 1180 μm, respectively) can be calculated thus:

$$S_2 = 0.577\left\{\frac{[1700(1180)]^{1/2}}{[2360(1700)]^{1/2}}\right\}^{0.741} = 0.446$$

Similar calculations can be made to find selection functions for other size intervals. Then, cumulative breakage functions can be calculated by the relationship noted in step 3, namely, $B_{i,j} = \bar{F}_i/S_j$.

13.3 PREDICTING PRODUCT SIZE DISTRIBUTION FROM FEEDSTOCK DATA

Given the feed size distribution, the breakage functions, and the selection functions (probabilities of breakage) for a feedstock to a grinding operation, as shown in Table 13.4, predict the size distribution for the product from the operation.

TABLE 13.4 Data on Feedstock to Grinding Operation (Procedure 13.3)

Size range	Size interval Tyler Mesh	Micron size	Feed size distribution F, percent	Breakage matrix B						Selection matrix S
1	−6, +8	−3350, +2360	24	0.18	0	0	0	0	0	1.00
2	−8, +10	−2360, +1700	16	0.22	0.18	0	0	0	0	0.81
3	−10, +14	−1700, +1180	12	0.16	0.22	0.18	0	0	0	0.60
4	−14, +20	−1180, +850	10	0.12	0.16	0.22	0.18	0	0	0.47
5	−20, +28	−850, +600	7	0.10	0.12	0.16	0.22	0.18	0	0.35
6	−28, +35	−600, +425	5	0.08	0.10	0.12	0.16	0.22	0.18	0.24
	−35	−425	26							

TABLE 13.5 Estimating the Percentage of Broken and Unbroken Particles (Procedure 13.3)

Size range	Feed size distribution F, percent	Selection functions S	Particles broken $(S)(F)$, percent	Particles not broken $(F) − (S)(F)$, percent
1	24	1.00	24.00	0
2	16	0.81	12.96	3.04
3	12	0.60	7.20	4.80
4	10	0.47	4.70	5.30
5	7	0.35	2.45	4.55
6	5	0.24	1.20	3.80
−35 mesh	26			

Calculation Procedure

1. *Predict the weight percentages of broken and unbroken particles within each size range.*
In each size range, multiply the feed size percentage F by the corresponding selection function S; this product gives an estimate as to the percentage of feed particles within the range that become broken. Then subtract the product from the feed size percentage; this difference is an estimate as to the particles that remain unbroken. These operations are shown in Table 13.5. In matrix notation (for consistency with subsequent steps), the S's can be considered to form an $n \times n$ diagonal matrix, and the F's an $n \times 1$ matrix.

2. *Predict the size distribution of the product of the breakage of broken particles.*
This product does not exist by itself as a separate entity, of course, because the particles that become broken remain mixed with those which stay unbroken. Even so, the distribution can be calculated by postmultiplying the $n \times n$ lower triangular matrix of breakage functions B in Table 13.4 by the percentage of particles broken, the $n \times 1$ matrix $(S)(F)$, as calculated in step 1. This postmultiplication works out as follows:

Size range	Amount of breakage product in size range
1	0.18(24.00) = 4.32
2	0.22(24.00) + 0.18(12.96) = 7.61
3	0.16(24.00) + 0.22(12.96) + 0.18(7.20) = 7.99
4	0.12(24.00) + 0.16(12.96) + 0.22(7.20) + 0.18(4.70) = 7.38
5	0.10(24.00) + 0.12(12.96) + 0.16(7.20) + 0.22(4.70) + 0.18(2.45) = 6.59
6	0.08(24.00) + 0.10(12.96) + 0.12(7.20) + 0.16(4.70) + 0.22(2.45) + 0.18(1.20) = 5.59

3. *Within each size range, sum up the percent of particles that remained unbroken throughout the operation (from step 1) and percent of particles in that range which resulted from breakage (from step 2).*

These sums work out as follows. The −35 mesh percentage is found by difference (i.e., it is the residual):

Size interval	Total-product size distribution, percent
1	0 + 4.32 = 4.32
2	3.04 + 7.61 = 10.65
3	4.80 + 7.99 = 12.79
4	5.30 + 7.38 = 12.68
5	4.55 + 6.58 = 11.13
6	3.80 + 5.59 = 9.39
−35 mesh	39.04

Related Calculations. If the grinding system includes a classification step that recycles the larger particles (e.g., those in size ranges 1 and 2) to the mill instead of allowing them to leave with the product, the operation is known as "closed-circuit grinding." In such a case, the preceding sequence can be expanded into an iterative procedure. In essence, steps 1 through 3 are applied anew to the material in size ranges 1 and 2, yielding a "final product" size distribution for this second round of breakage. Steps 1 through 3 are then applied a third time to the material in size ranges 1 and 2 from the second round of breakage; the procedure is repeated until virtually no material remains in those two size ranges.

13.4 MATERIAL-BALANCE CALCULATIONS FOR CLOSED-CIRCUIT GRINDING

In the grinding operation of Fig. 13.3, a ball mill is in closed circuit with a hydrocyclone classifier. The mass flow rates of the classifier feed, oversize, and undersize are denoted by the symbols A, O, and U, respectively. The fineness of classifier feed a, of oversize o, and of undersize u, all expressed as percentages passing a 200 mesh sieve, are 58.5, 48.2, and 96.0 percent, respectively. Undersize is produced at a rate of 20.3 tons/h. What is the percent circulating load? What are the flow rates of classifier feed and oversize?

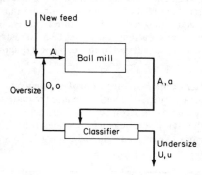

FIGURE 13.3 Ball-mill-classifier circuit (Procedure 13.4).

Calculation Procedure

1. *Find the percent circulating load.*
The percent circulating load L is defined by the relationship $L = 100O/U$. By material-balance algebra, then, $L = 100(u - a)/(a - o)$, which is $100(96.0 - 58.5)/(58.5 - 48.2) = 364$ percent.

2. *Find the flow rate of oversize.*
From step 1, $O = LU/100$, which is $364(20.3)/100 = 73.9$ tons/h.

3. *Find the flow rate of classifier feed.*
By material-balance algebra, $A = O + U$, which is $73.9 + 20.3 = 94.2$ tons/h.

13.5 WORK-INDEX CALCULATIONS

A limestone ore is to be ground in a conventional ball mill to minus 200 mesh. The Bond grindability of the ore for a mesh-of-grind of 200 mesh (75 μm) is determined by laboratory test to be 2.73 g per revolution. The 80 percent passing size of the feed to the test is 1970 μm; the 80 percent passing size of the product from it is 44 μm. Calculate the work index W_i for the material at this mesh-of-grind. (The "work index," defined as the energy needed to reduce ore from infinite size to the state where 80 percent will pass a 100 mesh screen, is a parameter that is useful in calculating size-reduction power requirements; see Procedure 13.6.)

Calculation Procedure

1. *Employ an empirical formula that yields the work index directly.*
The formula [3] is

$$W_i = \frac{44.5}{P_1^{0.23} G^{0.82} \left(\dfrac{10}{P^{0.5}} - \dfrac{10}{F^{0.5}} \right)}$$

where W_i is the work index, in kilowatt hours per ton, G is grindability, in grams per revolution, P is the 80 percent passing size of the product of the grindability test, in microns, F is the 80 percent passing size of the feed to the grindability test, in microns, and P_1 is the size of the mesh-of-grind of the grindability test, in microns. Thus,

$$W_i = \frac{44.5}{75^{0.23} 2.73^{0.82} \left(\dfrac{10}{44^{0.5}} - \dfrac{10}{1970^{0.5}} \right)}$$

$$= 5.64 \text{ kWh/ton}$$

Related Calculations. Although the Bond grindability test is run dry, the work index calculated above is for wet grinding. For dry grinding, the work index must be multiplied by a factor of ⁴/₃.

A crushability work index can also be empirically calculated. A Bond crushability test (based on striking the specimen with weights) indicates the crushing strength C per unit thickness of material, in foot pounds per inch. This is related to work index W_i, in kilowatthours per ton, by the formula $W_i = 2.59C/S$, where S is the specific gravity.

13.6 POWER CONSUMPTION IN A GRINDING MILL AS A FUNCTION OF WORK INDEX

The Bond work index for a mesh-of-grind of 200 mesh for a rock consisting mainly of quartz is 17.5 kWh/ton. How much power is needed to reduce the material in a wetgrinding ball mill from an 80 percent passing size of 1100 μm to an 80 percent passing size of 80 μm?

Calculation Procedure

1. *Employ a formula based on the Bond third theory of comminution.*
The formula is

$$W = 10W_i(P^{-0.5} - F^{-0.5})$$

where W is power required, in kilowatthours per ton, W_i is work index, in kilowatthours per ton, P is 80 percent passing size of product, in microns, and F is 80 percent passing size of feed, in microns.

Thus,

$$W = 10(17.5)(80^{-0.5} - 1100^{-0.5})$$
$$= 14.3 \text{ kWh/ton}$$

Related Calculations.   The Charles-Holmes equation [4], that is,

$$W = 100^r W_i (P^{-r} - F^{-r})$$

is in principle more accurate than the Bond third theory formula illustrated above. However, it requires determining the parameter r by running grindability tests on the rocks in question at two or more meshes-of-grind. If a single Bond grindability test is run at a mesh-of-grind close to that of the maximum product size from the proposed grinding operation, the Bond third theory equation should be almost as accurate.

13.7   BALL-MILL OPERATING PARAMETERS

A ball mill having an inside diameter of 12 ft and an inside length of 14 ft is to be used to grind a copper ore. Measurement shows that the distance between the top of the mill and the leveled surface of the ball charge is 6.35 ft. What is the weight of balls in the mill? What is the critical speed of the mill (the speed at which the centrifugal force on a ball in contact with the mill wall at the top of its path equals the force due to gravity)? At what percent of critical speed should the mill operate?

Calculation Procedure

1. Calculate the volume percent of the mill occupied by the balls.
The volume percent can be calculated from the relationship $V_p = 113 - 126H/D$, where V_p is percent of mill volume occupied by grinding media, H is distance from top to leveled surface, and D is mill inside diameter. Thus, $V_p = 113 - 126(6.35)/12 = 46.3$ percent.

2. Calculate the weight of the balls in the mill.
It can be assumed that loose balls weigh 290 lb/ft^3. (Rods would weigh 390 lb/ft^3, and silica pebbles would weigh 100 lb/ft^3.) Then, weight of balls equals (290)(volume of mill occupied by balls) = $290\pi(D/2)^2 LV_p/100 = 290(3.14)(12/2)^2(14)(46.3/100) = 213,000$ lb.

3. Calculate the proper mill speed.
This can be estimated from the equation $N_o = 57 - 40 \log D$, where N_o is proper speed, in r/min, and D is mill inside diameter, in feet. Thus, $N_o = 57 - 40 \log 12 = 57 - 40(1.079) = 13.8$ r/min. [The relation $N_o = 57 - 40 \log D$ is only approximate. In actual practice, it will be found that short mills ($L < 2D$) often tend to run at slightly higher speeds, and long mills often tend to operate at slightly lower speeds.]

4. Calculate the critical mill speed.
This can be estimated from the equation $N_c = 76.6/D^{1/2}$, where N_c is critical speed, in r/min, and D is mill inside diameter, in ft. Thus, $N_c = 76.6/12^{1/2} = 22.1$ r/min.

5. Calculate the percent of critical speed at which the mill should be operated.
This follows directly from steps 3 and 4. Thus, percent of critical speed equals $100(13.8)/22.1 = 62.4$ percent.

13.8   MAXIMUM SIZE OF GRINDING MEDIA

A taconite ore is to be ground wet in a ball mill. The mill has an internal diameter of 13 ft (3.96 m) and is run at 68 percent of critical speed. The work index of the ore is 12.2 kWh/ton, and its specific gravity is 3.3. The 80 percent passing size of the ore is 5600 μm. What is the maximum size of grinding media (maximum diameter of balls) to be used for the operation?

Calculation Procedure

1. *Employ an empirical formula that yields the maximum size directly.*
The maximum size grinding media for a ball mill (whether for initial startup or for makeup) may be calculated from the formula

$$M = \left(\frac{F}{K}\right)^{1/2}\left(\frac{SW_i}{100C_s D^{1/2}}\right)^{1/3}$$

where M is maximum size of balls, in inches, F is 80 percent passing size of feed to the mill, in microns, S is specific gravity of the ore, W_i is work index of the ore, in kilowatt hours per ton, D is inside diameter of the ball mill, in feet, C_s is fraction of critical speed of the mill, and K is a constant (350 for wet grinding or 330 for dry grinding).

Thus,

$$M = \left(\frac{5600}{350}\right)^{1/2}\left[\frac{3.3(12.2)}{100(0.68)(13)^{1/2}}\right]^{1/3}$$

$$= 2.19 \text{ in or about } 2\tfrac{1}{4} \text{ in } (0.057 \text{ m})$$

Related Calculations.   The maximum-diameter rod to be fed to a rod mill can be calculated from the empirical equation

$$R = \frac{F^{0.75}}{160[W_i S/(C_s D)^{1/2}]^{1/2}}$$

where R is maximum diameter of rod, in inches, and the other variables are as in the example.

13.9 *POWER DRAWN BY A GRINDING MILL*

A ball mill with an inside diameter of 12 ft (3.66 m) is charged with 129 tons (117,000 kg) of balls that have a maximum diameter of 3 in (0.076 m) and occupy 46.3 percent of the mill volume. The mill is operated wet at 62.4 percent of critical speed. What is the horsepower needed to drive the mill at this percentage of critical speed?

Calculation Procedure

1. *Employ an empirical formula that yields the horsepower directly.*
The formula [5] is

$$hp = 1.341\ W_b\{D^{0.4}C_s(0.0616 - 0.000575V_p) - 0.1(2)^{[(C_s - 60)/10]-1}\}$$

where hp is the required horsepower, W_b is the weight of the ball charge, in tons, D is the inside mill diameter, in feet, C_s is the percentage of critical speed at which the mill is operated, and V_p is the percentage of mill volume occupied by balls. Thus,

$$hp = 1.341(129)\{12^{0.4}(62.4)[0.0616 - 0.000575(46.3)] - 0.1(2)^{[(62.4 - 60)/10]-1}\}$$

$$= 1.341(129)(5.84) = 1010 \text{ hp}\,(753 \text{ kW})$$

Related Calculations. For large-diameter mills using makeup balls of relatively small maximum size, it is often necessary to subtract a "slump correction" [4] from the braced term of the empirical relation above. The formula for this correction is $[12D - 60B(D - 8)]/240B$, where D is inside mill diameter, in feet, and B is the largest size of makeup ball, in inches. For the present example, the slump correction is $[12(12) - 60(3)(12 - 8)]/240(3) = -0.8$. Thus the required horsepower becomes $1.341(129)[5.84 - (-0.8)] = 1150$ hp (858 kW).

13.10 *WATER REQUIREMENTS FOR CLOSED-CIRCUIT MILL SYSTEM*

A closed-circuit grinding system employs a high-efficiency air classifier wherein the classifier feed (i.e., the mill discharge) is exposed to outside air rather than recirculated air, thus reducing the product-cooling load. The system is shown in Fig. 13.4.

Fresh-feed rate N and mill production rate P are each 200,000 lb/h (90,900 kg/h). The circulating load L is 150 percent (i.e., 1.5). The flow rate of 80°F (300 K) ambient air to the classifier A is 221,000 lb/h (100,500 kg/h). The fresh feed enters the mill at 160°F (344 K). The flow rate of 80°F (300 K) sweep air S to the mill is 53,000 lb/h (24,100 kg/h). Mill power input is 4000 hp.

How much 70°F(294 K) cooling water must be sprayed into the mill to keep the product temperature from exceeding 150°F(339 K)?

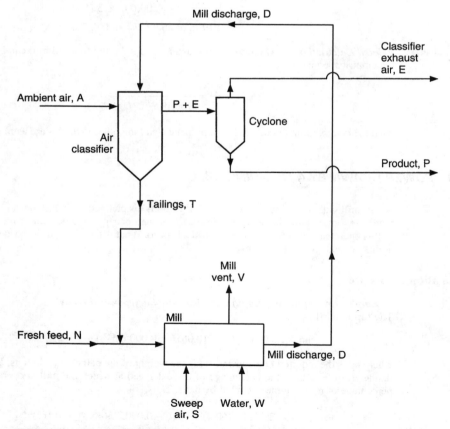

FIGURE 13.4 Closed-circuit grinding system with high-efficiency air classifier (Procedure 13.10).

Use 0.25 Btu/(lb)(°F) as the specific heat of the air, and 0.19 Btu/(lb)(°F) as the specific heat of the fresh feed, mill discharge D, tailings T, and product. Assume that the fractional heat losses in the mill and classifier circuits are 20 percent and 12 percent, respectively. Assume that the tailings are 15 degrees hotter than the product (i.e., that they are 165°F), that the classifier exhaust E is 2 degrees cooler than the product (i.e., that the exhaust is 148°F), and that the mill vent is 20 degrees cooler than the mill discharge. Assume that the water achieves its cooling via vaporization in the mill, and that amount of water vapor leaving with the mill vent is negligibly small.

Calculation Procedure

1. Determine the mill-discharge flow rate D.
Use the equation $D = P(1 + L)$ where, as noted in the statement of the problem, L is the circulating load. Thus, $D = 200,000(1 + 1.5) = 500,000$ lb/h.

2. Calculate the tailings flow rate T.
The equation is $T = PL$. Thus, $T = 200,000(1.5) = 300,000$ lb/h.

3. Determine the enthalpies of the fresh feed N, product P, tailings T, ambient air to classifier A, and classifier exhaust E.
For all of these streams, use the general formula $H = cm(t - t_o)$, where H is enthalpy in Btu's per hour, c is specific heat as given in the statement of the problem, m is mass flow rate, and t and t_o are, respectively, the stream temperature and reference temperature in degrees Fahrenheit. For arithmetical simplicity, use 0°F as the reference temperature throughout. Then

$$H_N = 0.19(200,000)(160 - 0) = 6.1 \times 10^6 \text{ Btu/h}$$
$$H_P = 0.19(200,000)(150 - 0) = 5.7 \times 10^6 \text{ Btu/h}$$
$$H_T = 0.19(300,000)(165 - 0) = 9.4 \times 10^6 \text{ Btu/h}$$
$$H_A = 0.25(221,000)(80 - 0) = 4.4 \times 10^6 \text{ Btu/h}$$
$$H_E = 0.25(221,000)(148 - 0) = 8.2 \times 10^6 \text{ Btu/h}$$

4. Estimate the heat loss H_K from the classifier circuit.
Use the equation $H_K = [p_K/(1 - p_K)](H_E - H_A)$, where p_K is the fractional heat loss (12 percent, or 0.12). Thus,

$$H_K = [0.12/(1 - 0.12)][(8.2 - 4.4) \times 10^6] = 0.5 \times 10^6 \text{ But/h}$$

5. Estimate the heat loss H_M from the mill circuit.
The equation is $H_M = (p_M)(\text{power input to mill})$, where p_M is the fractional heat loss (20 percent, or 0.2). Thus,

$$H_M = (0.2)(4000 \text{ hp})[2545 \text{ Btu/(h)(hp)}] = 2.0 \times 10^6 \text{ Btu/h}$$

6. Determine H_D, the enthalpy of the mill-discharge stream.
This is done by making an energy balance around the classifier and solving it for H_D. The energy balance is $H_D + H_A = H_P + H_T + H_E + H_K$. Accordingly, $H_D = (5.7 + 9.4 + 8.2 + 0.5 - 4.4) \times 10^6 = 19.4 \times 10^6$ Btu/h.

7. Calculate the mill-discharge temperature t_D.
Solve the general enthalpy equation (step 3) for t. Thus, $t_D = H_D/cm = (19.4 \times 10^6)/(0.19)(0.5 \times 10^6) = 204°F$, with m being the mill-discharge rate determined in step 1.

8. *Estimate H_V, the mill-vent enthalpy.*
The mill-vent rate equals the sweep-air rate, 53,000 lb/h. From the statement of the problem, the mill-vent temperature is $204 - 20$, i.e., $184°F$. Accordingly, $H_V = 0.25(53,000)(184 - 0) = 2.4 \times 10^6$ Btu/h.

9. *Determine H_W, the enthalpy of the water sprayed into the mill.*
The water must remove the heat introduced via the fresh feed, the tailings, the sweep air, and the mill power input H_I, less the heat removed via the mill discharge and the mill vent and less the mill heat losses. Thus, $H_W = H_N + H_T + H_S + H_I - H_D - H_V - H_M$. Now, $H_I = (4000 \text{ hp})[2545 \text{ Btu/(h)(hp)}] = 10.2 \times 10^6$ Btu/h, and $H_S = (0.25)(53,000)(80 - 0) = 1.1 \times 10^6$ Btu/h. Accordingly. $H_W = (6.1 + 9.4 + 1.1 + 10.2 - 19.4 - 2.4 - 2.0) \times 10^6 = 3.0 \times 10^6$ Btu/h.

10. *Calculate the required water rate W.*
From enthalpy tables, determine the enthalpy difference between the water entering the mill and the vapor leaving. For the purpose of this example, assume that the difference is 1100 Btu/lb. Then the amount of water needed to satisfy the enthalpy requirement calculated in step 9 is

$$(3.0 \times 10^6 \text{ Btu/h})/(1100 \text{ lb/h}) = 2700 \text{ lb/h}(1225 \text{ kg/h})$$

This is significantly lower cooling duty than would be the case with a conventional closed-circuit grinding system.

Related Calculations.   Three related problems are (1) How much will the product temperature change during the hottest part of the year if the water flow rate is substantially raised at the same time? (2) Determine the required water rate assuming that the mill vent V is sent to the classifier instead of being discharged to the atmosphere; (3) Assume that the water rate determined in Procedure (step 2) is at a maximum; how much will the product temperature increase during the hottest period of the year? All three of these call for trial-and-error solution.
   Note: This example is adapted from an article by Ivan Klumpar of Badger Engineers, Inc., in *Chemical Engineering,* March 1992.

*13.11 MAKING A PRELIMINARY CHOICE OF SIZE-REDUCTION MACHINERY

Make a preliminary choice of primary and secondary crushers for a hard chemical having a 40-in (101.6-cm) feed and a 4-in (10.2-cm) secondary feed. Determine the hph (kWh) input per ton of material crushed if 70 percent of the output of the secondary crusher is to pass through a No.200 sieve.

Calculation Procedure

1. *Choose the type of primary and secondary crushers from available data.*
Use the data in Table 13.6 to make a preliminary selection of the type of crusher to use, choosing from Table 13.7. Table 13.7 shows that with a 40-in (01.6-cm) primary feed, a jaw crusher would probably be suitable, following the guidelines in the last column of Table 13.6.
   For the secondary crusher, use the same procedure, entering with a 4-in (10.1-cm) feed. Table 13.6 guidelines indicate that a toothed-shredder would probably be suitable. With both types of crushers you would consult with the manufacturer to see if your choice was suitable for the chemical material being crushed.

2. Estimate the power required by the secondary crusher.
Use Fig. 13.5 to estimate the hph (kWh) per ton of this material. For a 70 percent sieve pass-through, 15 hph (11.2 kW) per ton of chemical material crushed.

TABLE 13.6 Guide to Selection of Crushing and Grinding Equipment[†]

| Size-reduction operation | Hardness of material | Size[‡] | | | | Reduction ratio[¶] | Types of equipment |
| | | Range of feeds, in[§] | | Range of products, in[§] | | | |
		Max.	Min.	Max.	Min.		
Crushing:							
Primary	Hard	60	12	20	4	3 to 1	A to D
		20	4	5	1	4 to 1	
Secondary . . .	Hard	5	1	1	0.2	5 to 1	A to F
		1.5	0.25	0.185	0.033	7 to 1	
				(4)	(20)		
	Soft	20	4	2	0.4	10 to 1	C to G
Grinding:							
Pulverizing:							
Coarse	Hard	0.185	0.033	0.023	0.003	10 to 1	D to I
		(4)	(20)	(28)	(200)		
Fine	Hard	0.046	0.0058	0.003	0.00039	15 to 1	H to K
		(14)	(100)	(200)	(1250)		
Disintegration:							
Coarse	Soft	0.5	0.065	0.023	0.003	20 to 1	F, I
Fine	Soft	0.156	0.0195	0.003	0.00039	50 to 1	I to K
		(5)	(32)	(200)	(1250)		

[†]Perry—*Chemical Engineers' Handbook,* McGraw-Hill, 1984.
[‡]85 percent by weight smaller than the size given.
[§]Sieve number in parentheses.
[¶]Higher reduction ratios for closed-circuit operations.
Note: To convert inches to centimeters, multiply by 2.54.

Related Calculations.   Use this general procedure to choose–in a preliminary manner–chemical crushing equipment. The chosen equipment must be carefully studied before a final choice is made. Consulting with several manufacturers can help you reach your final decision on the equipment to use.

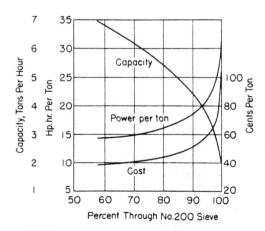

FIGURE 13.5 Variation in capacity, power, and cost of grinding relative to fineness of product. *(Perry—Chemical Engineers' Handbook, McGraw-Hill, 1984.)*

TABLE 13.7 Types of Size-Reduction Equipment[†]

A. Jaw crushers
   1. Blake
   2. Overhead eccentric
   3. Dodge
B. Gyratory crushers
   1. Primary
   2. Secondary
   3. Cone
C. Heavy-duty impact mills
   1. Rotor breakers
   2. Hammer mills
   3. Cage impactors
D. Roll crushers
   1. Smooth rolls (double)
   2. Toothed rolls (single and double)
E. Dry pans and chaser mills
F. Shredders
   1. Toothed shredders
   2. Cage disintegrators
   3. Disk mills
G. Rotary cutters and dicers
H. Media mills
   1. Ball, pebble, rod, and compartment mills:
      a. Batch
      b. Continuous
   2. Autogenous tumbling mills
   3. Stirred ball and sand mills
   4. Vibratory mills
I. Medium peripheral-speed mills
   1. Ring-roll and bowl mills
   2. Roll mills, cereal type
   3. Roll mills, paint and rubber types
   4. Buhrstones
J. High-peripheral-speed mills
   1. Fine-grinding hammer mills
   2. Pin mills
   3. Colloid mills
   4. Wood-pulp beaters
K. Fluid-energy superfine mills
   1. Centrifugal jet
   2. Opposed jet
   3. Jet with anvil

[†]Perry—*Chemical Engineers'* Handbook, McGraw-Hill, 1984.

*13.12 CRUSHER POWER INPUT DETERMINATION

A chemical process requires the crushing of 240 tons/h (217.7 t/h) of quartz. The quartz feed used is such that 80 percent passes a 3-in (7.6-cm) screen and 80 percent of the product must pass ¼-in (0.64-cm) screen. Determine the power input to the crusher.

Calculation Procedure

1. Compute the crusher capacity in tons/min.
Use the relation $t_m = t_h/60$, where t_m = crusher capacity, tons/min; t_h = crusher capacity, tons/h. Substituting yields $t_m = 240/60 = 4$ tons/min (3.63 t/min).

2. Determine the material work index.
The work index for any material that will be crushed is the total energy, kWh/ton, needed to reduce the feed to a size so that 80 percent of the product will pass through a 100-μm screen. Standard references such as Perry's—*Chemical Engineers' Handbook* list work indexes for various materials. For quartz having a specific gravity of 2.65, Perry gives the work index $W_i = 13.57$ kWh/ton (14.96 kWh/t).

3. Compute the raw-material and product mesh sizes.
Use the relation $d_r = s/12$, where d_r = mesh size, ft, for feed; s = mesh opening measure used, in. For the product, $d_p = s/12$, where the symbols are the same as before except that the mesh opening is that used for the product. Substituting gives $d_r = 3/12, = 0.25$; $d_p = 0.25/12 = 0.0208$ (0.0064 m).

4. Compute the required power input to the crusher.
Use the relation $hp = 1.46 t_m W_i (1/d_p^{0.5} - 1/d_r^{0.5})$, where the symbols are as given earlier. Substituting gives $hp = 1.46(4)(13.57)(1/0.208^{0.5} - 1/0.25^{0.5}) = 391$ hp (291.6 kW). A 400-hp (298.3 kW) motor would be used to drive this crusher.

Related Calculations.   Use this general procedure, known as the Bond crushing law and work index, to determine the power input required for commercially available grinders and crushers of all types. The result obtained is valid for all usual preliminary calculations.

*13.13   COOLING-WATER FLOW RATE FOR CHEMICAL-PLANT MIXERS

A kneader used in a chemical plant requires 300-hp (223.7-kW) input per 1000 gal (3785.0 L) of material kneaded. If this kneader handles 3000 lb (1360.8 kg) of a chemical having a density of 65 lb/ft^3 (1041.2 kg/m^3), determine the quantity of cooling water required in gal/min and gal/h if the maximum allowable temperature rise of the water during passage through the kneader is 25°F (13.9°C).

Calculation Procedure

1. Convert the kneader load to gallons.
Since the power-input requirements of chemical mixers are normally stated in hp/gal, the kneader load must be converted to gal. Use the relation, load, gal = load weight, lb (7.48 gal/ft^3 water)/load density, lb/ft^3. For this kneader, load, gal = 3000(7.48)/65 = 345 gal (1306.0 L).

2. Compute the required power input.
Use this relation: power input hp = hp input per 1000 gal (load, gal)/1000. For this kneader, power input $hp = 300(345)/1000 = 103.5$ hp (77.2 kW).

3. Compute the heat that must be removed.
Since 1 hp = 2545 Btu/h (745.9 W), the heat that must be removed = (103.5 hp)(2545) = 263,407.5 Btu/h (77,145.5 W).

4. Compute the cooling-water flow rate.
With an allowable temperature rise of 25°F (–3.9°C), and a specific heat of 1 Btu/(h · °F) (0.293 W), the cooling-water flow rate required = (263,407.5 Btu/h)/[(25°F)(1.0)(8.33 lb/gal of water)(60 min/h)] = 21.1 gal/min, or 21.2(60 min/h) = 1265 gal/h (4788 L/h).

Related Calculations.   Use the general procedure given here for any of the usual chemical mixers, such as paddles, turbines, propellers, disks, cones, change cans, dispersers, tumbling mixers, mixing rolls, masticators, pug mills, and mixer-extruders. Consult Perry's—*Chemical Engineers' Handbook* for suitable power-input data for mixers of various types.

REFERENCES

1. Herbst and Fuerstenau—*Trans. SME/AIME 241*:538, 1968.
2. Lynch—*Mineral Crushing and Grinding Circuits,* Elsevier.
3. Bond—*Crushing and Grinding Calculations,* Allis-Chalmers.
4. Smith and Lee—*Trans. SME/AIME 241*:91, 1968.
5. Smith—*Mining Engineering,* April 1961.
6. Klumpar Ivan—*Chemical Engineering,* March 1992.

SECTION 14
FILTRATION[†]

Frank M. Tiller, Ph.D.
M.D. Anderson Professor
Department of Chemical Engineering
University of Houston
Houston, TX

Wenfang Leu, Ph.D.
Research Scientist
Department of Chemical Engineering
University of Houston
Houston, TX

14.1 BASIS FOR FILTRATION CALCULATIONS

Mass Balance

An overall filtration material balance based on a unit area is

Mass of slurry = mass of cake + mass of filtrate

or

$$\frac{w_c}{s} = \frac{w_c}{s_c} + \rho v$$

where w_c is total mass of dry-cake solids per unit area, v is filtrate volume per unit area, s and s_c are, respectively, the mass fraction and average mass fraction of solids in the slurry and cake, and ρ is

---

[†]Procedure 14.7 is from *Chemical Engineering* magazine, September 1998, pp. 159ff.

the density of filtrate. Solving for w_c yields

$$w_c = \frac{\rho s}{1 - s/s_c} v = cv \tag{14.1}$$

where c is concentration expressed by mass of dry cake per unit volume of filtrate.

The value of c can be obtained from Eq. (14.1) (that is, $c = w_c/v$) if it is possible to obtain the mass of solids in the cake. However, draining the slurry can often lead to difficulties in accurate determination of the cake mass. An alternative approach is to consider the cake thickness.

The cake thickness L can be related to the cake mass w_c by

$$w_c = \rho_s(1 - \epsilon_{av})L$$

where ρ_s is the true density of the solid and ϵ_{av} is the average porosity of the cake. Combining the two w_c equations produces

$$L = \frac{c}{\rho_s(1 - \epsilon_{av})} v = c_L v \tag{14.2}$$

where c_L is the ratio of cake thickness L to the per-unit area filtrate volume v. The thickness L is in fact the primary parameter related to filter design. Spacing of leaves, frame thickness, and minimum cake thickness for removal from vacuum drum filters all depend on a knowledge of L.

Rate Equations

In filtration theory, Darcy's law is used in the form

$$\frac{dp_L}{dw} = \frac{-dp_s}{dw} = \mu\alpha q$$

or

$$\frac{dp_L}{dx} = \frac{-dp_s}{dx} = \mu\rho_s(1 - \epsilon)\alpha q$$

where p_L is hydraulic pressure, p_s is solid compressive or effective pressure, w is mass of cake per unit area in the cake-thickness range from 0 to x, μ is viscosity of the filtrate, α is local specific filtration resistance, and q is superficial velocity of liquid. Solid compressive pressure is defined by $p_s = p - p_L$, where p is filtration pressure.

With respect to the cake cross section, p_s is zero at the cake-slurry interface and reaches its maximum at the cake-medium interface. Conversely, p_L has its maximum value (equal to p) at the cake-slurry interface, whereas at the cake-medium interface it consists solely of p_1, the pressure required to overcome the resistance R_m of the medium.

Integration of the preceding dp_L/dw expression with the assumption that q is constant throughout the cake gives

$$\mu q w_c = \mu cqv = \frac{\Delta p_c}{\alpha_{av}} \tag{14.3}$$

where $\Delta p_c = p - p_1 = p - \mu q R_m$. Substitution and rearrangement give

$$\frac{dv}{dt} = q = \frac{p}{\mu(\alpha_{av} w_c + R_m)} \tag{14.4}$$

The latter equation can be used to calculate v or L as a function of time t, once the filtration mode is specified, e.g., constant-pressure filtration, with p constant; constant-rate filtration, with q constant; or centrifugal-pump filtration, with q as a function of p.

14.2 CONSTANT-PRESSURE FILTRATION

The first two columns of Table 14.1 show laboratory data[†] on filtering calcium silicate with an average particle size of 6.5 μm in a 0.04287-m² (0.460-ft²) plate-and-frame filter press operating at a pressure p of 68.9 kPa (10 lbf/in²) and a slurry-solid mass fraction s of 0.00495. The cake had an average moisture content corresponding to a cake mass fraction of solids S_c of 0.2937. Viscosity of the water μ was 0.001 Pa · s (1.0 cP). The densities of liquid and solid ρ and ρ_s, respectively, were 1000 and 1950 kg/m³. Calculate the average specific and medium resistances, and set up an equation relating cake thickness L to filtration time t.

Calculation Procedure

1. Calculate v, the filtrate volume per unit filtration area, and plot v versus t and log v versus log t. Values of v based on a 0.04287-m² area are shown in col. 3 of Table 14.1. The plots are presented in Figs. 14.1 and 14.2. (These figures also include 16 data points that are omitted from the table for simplicity.) The slope of the logarithmic plot can be taken as essentially 0.5 when t exceeds 120 s. Beyond 120 s, the curve of $p/(\rho q_{av})$ versus w_c in step 5 below should be straight enough to give an adequate value of α_{av} at the full applied pressure.

TABLE 14.1 Data and Calculated Values for Constant-Pressure Filtration (Procedure 14.2)

Elapsed time t, s (1)	Filtrate volume V, m³ (2)	Volume per unit area v, m³/m² (3)	Interpolated v, m³/m² (4)	$w_c = cv$, kg/m² (5)	$q = \frac{\Delta v}{\Delta t}$, m/s (6)	$\frac{p}{\mu q_{av}} = \frac{pt}{\mu v}$, 1/m (7)	$\frac{p}{\mu q} = \frac{p\Delta t}{\mu \Delta v}$, 1/m (8)
0	0	0		0		—	
			0.012	0.06	2.56×10^{-3}		2.69×10^{10}
9	10×10^{-4}	0.023		0.12		2.70×10^{10}	
			0.035	0.18	2.40		2.87
19	20	0.047		0.24		2.79	
			0.059	0.30	1.84		3.74
31.5	30	0.070		0.35		3.10	
			0.082	0.41	1.28		5.38
49.5	40	0.093		0.47		3.67	
			0.105	0.53	1.17		5.89
70	50	0.117		0.59		4.12	
			0.129	0.65	1.00		6.89
93	60	0.140		0.71		4.58	
			0.152	0.77	8.52×10^{-4}		8.09
120	70	0.163		0.82		5.07	
			0.175	0.88	7.50		9.19
152	80	0.187		0.94		5.60	
			0.199	1.00	6.57		1.05×10^{11}
187	90	0.210		1.06		6.14	
			0.222	1.12	5.75		1.20
227	100	0.233		1.17		6.71	
			0.245	1.23	5.58		1.23
270	110	0.257		1.30		7.24	

[†]M. Hosseini, M.Sc. thesis, University of Manchester, 1977.

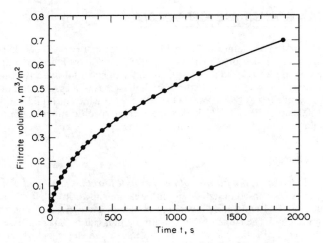

FIGURE 14.1 Filtrate volume versus elapsed time (Procedure 14.2).

2. Calculate w_c, the total mass of dry-cake solids per unit area.
Use Eq. (14.1). Thus,

$$w_c \frac{1000(0.00495)}{1 - 0.00495/0.2937} v = 5.04v$$

The values of w_c thus calculated are shown in col. 5 of Table 14.1.

3. Calculate $p/(\mu q_{av})$.
The average rate is simply $q_{av} = v/t$. Then, $p/(\mu q_{av}) = (6.89 \times 10^7)t/v$. Values are listed in col. 7 of Table 14.1.

4. Calculate $p/(\mu q)$.
The instantaneous rate $q = dv/dt$ must be obtained. Inasmuch as v versus t data pertaining to constant-pressure filtration are parabolic, use the following property of parabolas to obtain the slope:

$$\left(\frac{dv}{dt} \right)_{(v_1 + v_2)/2} = \frac{\Delta v}{\Delta t}$$

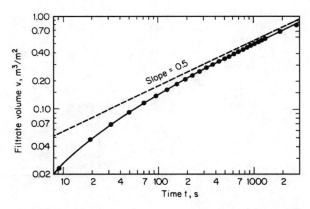

FIGURE 14.2 Filtrate volume versus elapsed time (logarithmic scales) (Procedure 14.2).

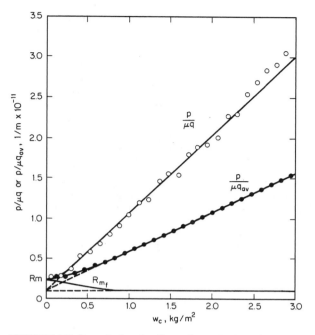

FIGURE 14.3 Determination of medium resistance R_m (Procedure 14.2).

This equation states that the angle of the tangent taken at the midpoint $(v + \Delta v/2)$ of a volume interval (not midpoint with respect to time) equals the angle of the secant. The rule is valid regardless of the size of Δv and is generally best applied to smoothed data.

Between 9 and 19 s, $\Delta v/\Delta t = (0.047 - 0.023)/10 = 0.00240$ m/s. This value corresponds to dv/dt at $v = (0.047 + 0.023)/2 = 0.035$ m. Since any size interval can be used, we find that between 93 and 270 s, $\Delta v/\Delta t = (0.257 - 0.140)/177 = 0.00066$ m/s. The value of v to be used is $(0.257 + 0.140)/2 = 0.199$ m.

In Table 14.1, values of $(v_1 + v_2)/2$ and the corresponding derivatives are shown in cols. 4 and 6, respectively. The $p/(\mu q)$ values in col. 8 are obtained from the expression $6.89 \times 10^7/q$.

5. *Plot $p/(\mu q)$ and $p/(\mu q_{av})$ versus w_c.*

These plots are shown in Fig. 14.3. They enable the calculation of the medium resistance R_m and the specific filtration resistance α_{av} because Eq. (14.4) can be rearranged into the form $p/(\mu q) = \alpha_{av}w_c + R_m$, and there also can be written a similar expression (valid only when α_{av} and R_m are constant): $p/(\mu q_{av}) = (\alpha_{av}/2)w_c + R_m$. Both equations have the same y intercept, namely, R_m. The slope of the first is twice that of the second. It is advisable to plot both lines in order to reach a compromise on the slopes and intercept.

It is important to take enough v and t data to generate the initial curved portions of the plots. The specific filtration resistance is smaller at the start of filtration, and the slopes of both plots have their minimum value when t and w_c equal zero. If the first four points taken during the first 50 s were omitted, only the straight-line portions of the plots would be present; if these were extrapolated along the dotted lines to $w_c = 0$, the resulting value of the y intercept would be false.

6. *Determine the medium resistance R_m.*

R_m is the (true) y intercept in Fig. 14.3, approximately 0.24×10^{11} m^{-1}. (However, use this result with caution, because it is hard to establish operating conditions at $t = 0$.) It frequently happens that the intercept of the $p/(\mu q)$ line is negative. Such a result generally implies (1) migration of fine particles, with subsequent blinding of medium or cake, or (2) sedimentation on a horizontal filter surface facing up.

7. Calculate α_{av}.
In this step, do *not* rely on finding a single, constant slope of either of the plots in Fig. 14.3. Such a method would be correct only if the entire plot (including its initial portion) were straight and the false medium resistance were the true medium resistance. Instead, use the first equation in step 5, rearranged into the form $\alpha_{av} = [p/(\mu q) - R_m]/w_c$. The resulting value of α_{av} will vary, with the difference from value to value being greatest when w_c is small. As w_c increases, the value of α_{av} does approach the slope of the $p/(\mu q)$ plot in its straight-line portion, namely, 0.97×10^{11}.

Thus when $w_c = 0.88$ kg/m^2 and $p/(\mu q) = 0.92 \times 10^{11}$ m^{-1}, $\alpha_{av} = [(0.92 \times 10^{11}) - (0.24 \times 10^{11})]/0.88 = 0.77 \times 10^{11}$ m/kg (1.15×10^{11} ft/lb). And when $w_c = 3.0$ kg/m^2, $\alpha_{av} = [(3.03 \times 10^{11}) - (0.24 \times 10^{11})]/3.0 = 0.93 \times 10^{11}$ m/kg.

The pressure drop across the cake is given by the equation $\Delta p_c = p - \mu q\, R_m$. Thus, at these two points, it equals 50.9 and 63.4 kPa (7.38 and 9.20 lbf/in^2), respectively.

8. Calculate the average ϵ_{av}.
Average porosity can be calculated from the equation

$$\epsilon_{av} = \frac{1}{1 + (\rho/\rho_s)s_c/(1 - s_c)} = \frac{1}{1 + (1000/1950)0.2937/(1 - 0.2937)} = 0.824$$

9. Obtain equations for v versus t and L versus t.
In Fig. 14.3, the $p/(\mu q_{av})$ plot is an excellent straight line for w_c values greater than about 0.8 kg/m^2, i.e., for filtration times greater than about 120 s. Therefore, data *in this range* can be accurately represented by an equation based on the assumption that α_{av} and R_m are constants and are the slope and intercept of the straight-line portion of the plot, respectively. Thus, $\alpha_{av} = 0.97 \times 10^{11}$ m/kg, and R_m, found by extending the straight line leftward until it intercepts the vertical axis, is 0.10×10^{11} m^{-1}. Then, substituting $w_c = cv$ and integrating Eq. (14.4) yields the parabola

$$v^2 + \frac{2R_m}{c\alpha_{av}}v = \frac{2p}{\mu c\alpha_{av}}t$$

Noting from step 2 that $c = 5.04$ kg/m^3 and substituting and rearranging, we obtain the relationship $t = 3548v^2 + 145v$. And relating L to v via Eq. (14.2), where $c_L = 5.04/1950(1 - 0.824) = 0.0147$, we obtain $t = (1.64 \times 10^7)L^2 + (9.9 \times 10^3)L$. In these equations, v is in cubic meters per square meter and L is in meters.

As for filtration times of less than 120 s—for instance, with continuous drum or disk filters, where filtration time would normally be less than 60 s—the initial α_{av} of 0.11×10^{11} m/kg and the true R_m of 0.24×10^{11} m^{-1} will yield a reasonable representation of the data. Thus the equations become $t = 402v^2 + 348v$ and $t = (1.86 \times 10^6)L^2 + (2.4 \times 10^4)L$.

Related Calculations. In constant-rate (as opposed to constant-pressure) filtration, $v = qt$ and $w_c = cqt$. Then, from Eq. (14.3), the average specific filtration resistance $\alpha_{av} = (p - p_1)/(\mu cq^2t)$, where P_1 is the pressure at the interface of the filter medium and the cake. Constant-rate filtrations are usually operated at above 10 lbf/in^2 (69 kPa), and this equation is accurate enough for most purposes. At higher and higher pressures, it becomes more and more acceptable to neglect p_1, which leads to the pressure-time relationship $p = \alpha_{av}\mu cq^2t$.

Constant-rate filtration is employed sometimes when an improperly used centrifugal pump may break down the slurry particles. In fact, however, centrifugal pumps are most often chosen for filtration operations. The following example shows the relevant calculations.

14.3 CENTRIFUGAL-PUMP FILTRATION

A 2% (by weight) aqueous slurry containing solids with an average density of 202.6 lb/ft³ (3244 kg/m³) is to be filtered in a 500-ft² (46.45-m²) filter using a centrifugal pump having the following performance characteristics:

Pressure:										
lbf/in²	15	20	25	30	35	40	45	50	55	60
kPa	103	138	172	207	241	276	310	345	379	414
Flow rate:										
gal/min	500	482	457	420	375	308	232	155	78	0
m³/s	0.0315	0.0304	0.0288	0.0265	0.0237	0.0194	0.0146	0.0098	0.0049	0

The pump has a throttling valve that relates pressure drop to flow rate Q as follows: Δp (throttling) = $15(Q/500)^2$. The temperature varies from 20 to 27°C (68 to 80.6°F). A series of constant-pressure tests yielded the data on α_{av} and $(1 - \epsilon_{av})$ shown in Fig. 14.4, α_{av} being the average specific filtration resistance and ϵ_{av} the average porosity of the cake. Find cake thickness as a function of time.

Calculation Procedure

1. Select the approach to be used.
If both pressure and filtration rate vary, as in this case, it is necessary to impose the pump characteristics on the filtration equations. No simple formulas can be obtained to relate p to t; instead, a relatively easy numerical integration is used.

Equation (14.3) can be rearranged into the form

$$v = \frac{\Delta p_c}{\mu c q \alpha_{av}}$$

the terms being defined as at the beginning of this section. Now q is a function of p, and α_{av} is a function of Δp_c. Once v has been obtained as a function of p and q, t can be obtained by integration:

$$t = \int_0^v \frac{dv}{q}$$

If data from a series of constant-pressure tests yield values of α_{av} and the average solid fraction s_c, the first equation in this paragraph can be used to find the volume-versus-rate relationship, which can then be used to find the time by integration.

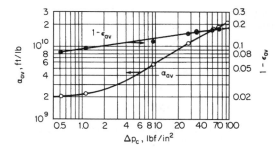

FIGURE 14.4   Average specific filtration resistance and average porosity (Procedure 14.3). (Note: 1 lbf/in² = 6.895 kPa; 1 ft/lb = 6.72 m/kg.)

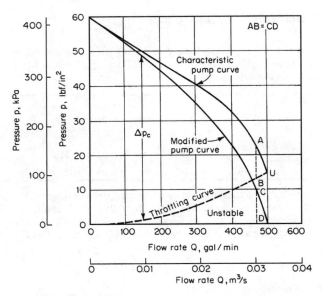

FIGURE 14.5 Centrifugal-pump curves (Procedure 14.3).

2. Construct a modified pump curve.
The characteristic pump curve is plotted as the upper line in Fig. 14.5. Below point U, the pump is unstable and must be throttled so that the pressure does not fall to too low a value. Inasmuch as the throttling pressure is not available to the filter, it must be subtracted from the characteristic curve to give the modified pump curve. From the equation in the statement of the problem, plot the throttling curve. Then, assuming a negligible filter-medium resistance, the pressure drop Δp_c across the cake equals the difference between the characteristic and throttling curves. Plot this difference, labeling it the "modified pump curve."

3. Calculate c, the concentration of cake, i.e., the mass of dry cake per unit volume of filtrate.
As indicated in Eq. (14.1), $c = \rho s/(1 - s/s_c)$, or in the present case, $c = 62.3(0.02)/(1 - 0.02/s_c)$, where 62.3 is the density of water in lbm/ft^3. Now, $s_c = \rho_s(1 - \epsilon_{av})/[\rho_s(1 - \epsilon_{av}) + \rho\epsilon_{av}]$, or in the present case, since ρ_s is given as 202.6 lbm/ft^3, $s_c = 1/[1 + 0.308\epsilon_{av}/(1 - \epsilon_{av})]$.
    Select various values for Δp_c; from Fig. 14.4, determine ϵ_{av}. From the equations in the preceding paragraph, calculate s_c and c. The results are shown in the fourth, fifth, and sixth columns of Table 14.2.

4. Calculate C_L, the ratio of cake thickness L to the filtrate volume per unit filter area v.
The values of c_L can be calculated from the equation

$$c_L = \frac{c}{\rho_s(1 - \epsilon_{av})}$$

Note that both c and c_L thus vary with Δp_c. The calculated values of c_L appear in Table 14.2.

5. Calculate q, the superficial velocity (velocity based on unit filter area) corresponding to the values of Δp_c selected in step 3.
This operation employs Fig. 14.3. The modified pump curve relates Δp_c to Q, the flow rate in gallons per minute. For a given Δp_c, find Q and multiply it by (1 min/60 s) × (1 ft^3/7.481 gal)(1/500 ft^2 of filter area) to find q. The calculated values of q appear in Table 14.2.

TABLE 14.2   Data and Calculated Results for Centrifugal-Pump Filtration (Procedure 14.3)

Δp_c, lbf/in²	Q, gal/min	q, ft/s	ϵ_{av}	s_c	c, lbm/ft³	c_L	$\alpha_{av} \times 10^{-9}$, ft/lbm	v, ft³/ft²	L, in	$1/q$, s/ft
0.5	499	0.00224	0.915	0.232	1.364	0.0792	2.1	0.54	0.51	446
1	498	0.00222	0.905	0.254	1.352	0.0702	2.2	1.04	0.88	450
5	487	0.00217	0.880	0.307	1.332	0.0548	3.8	3.14	2.06	461
10	469	0.00209	0.869	0.329	1.327	0.0500	5.7	4.36	2.62	478
15	446	0.00199	0.860	0.346	1.322	0.0466	7.2	5.46	3.05	503
20	418	0.00186	0.853	0.359	1.319	0.0443	9.0	6.24	3.32	538
25	384	0.00171	0.849	0.366	1.318	0.0431	10.0	7.65	3.96	585
30	345	0.00154	0.846	0.371	1.317	0.0422	12.0	8.50	4.30	649

Note: 1 lbf/in² = 6.895 kPa; 1 gal/min = 6.3×10^{-5} m³/s; 1 ft/s = 0.3048 m/s; 1 lbm/ft³ = 16.02 kg/m³; 1 ft/lbm = 0.67 m/kg; 1 ft³/ft² = 0.3048 m³/m²; 1 in = 0.0254 m; 1 s/ft = 3.28 s/m.

6. Calculate v, the volume of filtrate per unit of filter area.
Use the first equation in step 1. Thus,

$$v = \frac{\Delta p_c}{\mu c q \alpha_{av}}$$

$$= \frac{(144 \text{ in}^2/\text{ft}^2)[32.17 \text{ lbm . ft}/(\text{lbf} \cdot \text{s}^2)](\Delta p_c \text{ lbf/in}^2)}{[0.000672 \text{ lbm}/(\text{ft . s})](c \text{ lbm/ft}^3)(q \text{ ft/s})(\alpha_{av} \text{ ft/lbm})}$$

$$= 6.893 \times 10^6 \frac{\Delta p_c}{c q \alpha_{av}}$$

Use Fig. 14.4 to obtain α_{av} for each value of Δp_c. These values of α_{av} are shown in Table 14.2, and so are the calculated values of v.

7. Calculate L, the cake thickness.
The equation is $L = c_L v$. The values of L thus calculated are shown in Table 14.2, in inches.

8. Plot L, Δp_c, and l/q against v.
The values are taken from Table 14.2. The resulting smoothed plots appear in Fig. 14.6.

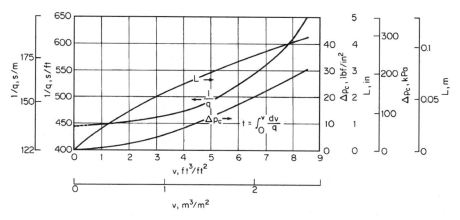

FIGURE 14.6   Cake thickness, pressure drop, and reciprocal of superficial velocity versus volume of filtrate: centrifugal-pump filtration (Procedure 14.3).

9. Find filtration time t.
The time is found by taking the area under the curve of $1/q$ versus v. The result of this integration is shown in the third and fourth columns of Table 14.3. Values of Δp_c and L are repeated in the table for convenience.

TABLE 14.3   Determination of Filtration Time for Centrifugal-Pump Filtration (Procedure 14.3)

v, ft³/ft²	Incremental area, s	t, s	t, min	Δp_c, lbf/in²	L, in
1	446	446	7.4	0.9	0.80
2	451	897	15.0	2.6	1.47
3	457	1354	22.6	5.0	2.00
4	466	1820	30.3	8.3	2.47
5	482	2302	38.4	12.6	2.90
6	507	2809	46.8	17.2	3.32
7	540	3349	55.8	22.3	3.70
8	582	3931	65.5	27.5	4.10

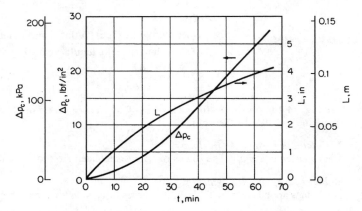

FIGURE 14.7   Cake thickness and pressure drop as a function of time; centrifugal-pump filtration (Procedure 14.3).

10. Show cake thickness as a function of time.
This relationship is shown graphically in Fig. 14.7. The figure also shows Δp_c as a function of time.

Related Calculations.   Filtrate volume as a function of time can be calculated by dividing each value of L in Fig. 14.7 by c_L and multiplying by the filter area.
   Note that this pump proves to be somewhat large for the filter. If the cake thickness is restricted to 2.5 in (0.064 m), the rate would never drop below 95 percent of the initial pump rate.

14.4   FILTER-CAKE WASHING

A filter cake was washed at a rate of 0.2 gal/(ft²)(min) [0.0081 m³/(m²)(min)] with pure water to remove the soluble salts present in the voids. The cake had a thickness of 2.0 in (0.051 m) and the following compositions:

	Mass fraction	Density, lbm/ft^3 (kg/m^3)
Inert solids	0.4789	88.6 (1420)
Water	0.4641	62.4 (1000)
Soluble salts	0.0570	85.1 (1363)

Data for instantaneous wash concentration versus time were taken as shown in Table 14.4. At the end of the washing period the cake was analyzed and found to have mass fraction 0.24% of salts on a moisture-free basis. How much water must be used if it is permissible to leave mass fraction 0.67% of soluble material on a moisture-free basis?

Calculation Procedure

1. *Convert mass fractions into volume fractions and find average porosity.*
The volume fraction x_i of component i (inert solid, water, or soluble salts) can be calculated by

$$ x_i = \frac{y_i/\rho_i}{\sum (y_i/\rho_i)} $$

where y_i, and ρ_i are the mass fraction and density of component i, respectively. Then, the initial composition of cake is 40.00% inert solid, 55.04% water, and 4.96% soluble salts by volume. The average porosity ϵ_{av} of the cake is thus also known to be $0.5504 + 0.0496 = 0.6$.
  The volume fraction of soluble material on a moisture-free basis ψ_v can be obtained from

$$ \psi_v = \frac{\psi_m/\rho_{salt}}{\psi_m/\rho_{salt} + (1 - \psi_m)/\rho_s} $$

where ψ_m is the mass fraction of soluble material on a moisture-free basis and the subscript s refers to the inert solids. Thus the volume fractions of salts on a moisture-free basis are 0.25% at the end of 120 min and 0.70% permissible.

TABLE 14.4   Data on Filter-Cake Washing (Procedure 14.4)

Time t, min	Volume of wash v_w, gal/ft^2	Wash concentration C_w, lbm solute/gal
0	0	0.740
1	0.2	0.739
2	0.4	0.740
3	0.6	0.687
4	0.8	0.480
5	1.0	0.266
6	1.2	0.144
8	1.6	0.0575
10	2.0	0.0313
15	3.0	0.0158
20	4.0	0.0115
40	8.0	0.00624
60	12.0	0.00312
90	18.0	0.00119
120	24.0	0.00070

Note: 1.0 gal/ft^2 = 0.041 m^3/m^2; 1.0 lbm solute/gal = 120.02 kg solute/m^3.

2. Calculate the average density of the cake, mass of dry inert solid, initial and final mass of soluble salts, and void volume.

Average cake density = $\sum$(density of component i)(volume of fraction of component i)
$$= 88.6(0.4000) + 62.4(0.5504) + 85.1(0.0496)$$
$$= 74.0 \text{ lbm/ft}^3 \text{ (1186 kg/m}^3\text{)}$$

Mass of dry inert solid per unit area of filtration = (inert solid density)(cake thickness)
$$\times \text{(volume fraction of dry inert solid)}$$
$$= 88.6(2/12)(0.4)$$
$$= 5.91 \text{ lbm/ft}^2 \text{ (28.8 kg/m}^2\text{)}$$

Initial mass of solute per unit area of filtration = (solute density)(cake thickness)
$$\times \text{(volume fraction of solute)}$$
$$= 85.1(2/12)(0.0496)$$
$$= 0.703 \text{ lbm/ft}^2 \text{ (3.43 kg/m}^2\text{)}$$

Final mass of solute per unit area of filtration = (mass of dry inert solid)$[\psi_m/(1 - \psi_m)]$
$$= 5.91(0.0024)/0.9976$$
$$= 0.0142 \text{ lbm/ft}^2 \text{ (0.0692 kg/m}^2\text{)}$$

Void volume per unit area of filtration = (ϵ_{av}) (cake thickness)
$$= 0.6(2/12) = 0.1 \text{ ft}^3/\text{ft}^2$$
$$= 0.748 \text{ gal/ft}^2 \text{ (0.0305 m}^3/\text{m}^2\text{)}$$

3. Calculate the average wash concentration $C_{w,av}$.
The instantaneous wash concentration C_w, shown as the third column in Table 14.4, is plotted against volume of wash (the second column in Table 14.4) in Fig. 14.8. Now, by a material balance for the solute in the cake,

$$\epsilon_{av} L(C_0 - C_{av}) = \int_0^{v_w} C_w dv_w = C_{w,av} v_w$$

where L is cake thickness (and therefore $\epsilon_{av} L$ is void volume per unit area of filtration, calculated in step 2), C_0 is initial concentration of solute in the cake, and v_w is volume of wash per unit area of filtration. Therefore, the total (cumulative) amount of solute removed from the cake may be determined by integrating the instantaneous-concentration-versus-volume data as shown in the first three columns of Table 14.5. Dividing the cumulative amount of solute removed by the cumulative volume of wash used yields the average wash concentration, shown as the fourth column in Table 14.5 and plotted in Fig. 14.8.

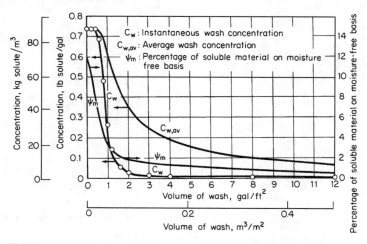

FIGURE 14.8   Wash concentration and percent of soluble material versus volume of wash (Procedure 14.4).

TABLE 14.5 Results of Cake Washing (Procedure 14.4)

Range of wash volume, gal/ft²	Solute removed, lbm/ft²	Conditions at the end of the interval of wash volume				
		Cumulative solute removed, lbm/ft²	Average wash concentration, lbm/gal	Mass of solute remaining, lbm/ft²	Present solute on moisture-free basis	Average cake concentration, lbm/gal
0	0	0	—	0.8438	12.49	1.128
0.0–0.1	0.0740	0.0740	0.740	0.7698	11.52	1.029
0.1–0.2	0.0740	0.1480	0.740	0.6958	10.53	0.930
0.2–0.3	0.0740	0.2220	0.740	0.6218	9.52	0.831
0.3–0.4	0.0740	0.2960	0.740	0.5478	8.48	0.732
0.4–0.5	0.0730	0.3690	0.738	0.4748	7.44	0.635
0.5–0.6	0.0705	0.4395	0.733	0.4043	6.40	0.541
0.6–0.7	0.0650	0.5045	0.721	0.3393	5.43	0.454
0.7–0.8	0.0545	0.5590	0.699	0.2848	4.60	0.381
0.8–0.9	0.0420	0.6010	0.669	0.2428	3.95	0.325
0.9–1.0	0.0315	0.6325	0.633	0.2113	3.45	0.282
1.0–1.1	0.0226	0.6551	0.596	0.1887	3.09	0.252
1.1–1.2	0.0163	0.6714	0.560	0.1724	2.83	0.230
1.2–1.3	0.0126	0.6840	0.526	0.1598	2.63	0.214
1.3–1.4	0.0100	0.6940	0.496	0.1498	2.47	0.200
1.4–1.5	0.0079	0.7019	0.468	0.1419	2.34	0.190
1.5–1.6	0.0063	0.7082	0.443	0.1356	2.24	0.181
1.6–1.7	0.0050	0.7132	0.420	0.1306	2.16	0.175
1.7–1.8	0.0042	0.7174	0.399	0.1264	2.09	0.169
1.8–1.9	0.0036	0.7210	0.379	0.1228	2.04	0.164
1.9–2.0	0.0032	0.7242	0.362	0.1196	1.98	0.160
2.0–3.0	0.02190	0.7461	0.249	0.0977	1.63	0.131
3.0–4.0	0.01350	0.7596	0.190	0.0842	1.40	0.113
4.0–5.0	0.01035	0.7700	0.154	0.0738	1.23	0.099
5.0–6.0	0.00888	0.7788	0.130	0.0650	1.09	0.087
6.0–7.0	0.00762	0.7864	0.112	0.0574	0.96	0.077
7.0–8.0	0.00662	0.7931	0.099	0.0507	0.85	0.068
8.0–9.0	0.00575	0.7988	0.089	0.0450	0.76	0.060
9.0–10.0	0.00490	0.8037	0.080	0.0401	0.67	0.054
10.0–11.0	0.00425	0.8080	0.073	0.0358	0.60	0.048
11.0–12.0	0.00350	0.8115	0.068	0.0323	0.54	0.043
12.0–13.0	0.00302	0.8145	0.063	0.0293	0.49	0.039
13.0–14.0	0.00258	0.8171	0.058	0.0267	0.45	0.036
14.0–15.0	0.00222	0.8193	0.055	0.0245	0.41	0.033
15.0–16.0	0.00182	0.8211	0.051	0.0227	0.38	0.030
16.0–17.0	0.00152	0.8226	0.048	0.0212	0.36	0.028
17.0–18.0	0.00132	0.8240	0.046	0.0198	0.33	0.026
18.0–19.0	0.00118	0.8252	0.043	0.0186	0.31	0.025
19.0–20.0	0.00102	0.8262	0.041	0.0176	0.30	0.024
20.0–21.0	0.00100	0.8272	0.039	0.0166	0.28	0.022
21.0–22.0	0.00092	0.8281	0.038	0.0157	0.26	0.021
22.0–23.0	0.00080	0.8289	0.036	0.0149	0.25	0.020
23.0–24.0	0.00075	0.8296	0.035	0.0142	0.24	0.019

Note: 1.0 gal/ft² = 0.041 m³/m²; 1.0 lbm/ft² = 4.88 kg/m²; 1.0 lbm/gal = 119.8 kg/m³.

4. Calculate the mass of solute remaining in the cake at each moment.
Since 24 gal/ft^2 (0.98 m^3/m^2) of wash was used in the run, the total amount of solute removed was 0.8296 lb/ft^2, as shown in Table 14.5. The amount remaining in the cake is 0.0142 lb/ft^2, from step 2. Thus the total amount present in the system initially must have been 0.8296 + 0.0142 = 0.8438 lb/ft^2 (4.11 kg/m^2). (Since step 2 shows that only 0.703 lb/ft^2 was present in the *cake* initially, the rest must have been in the feed lines to the filter press; it is important to keep this complication in mind when dealing with problems of this kind.)

Then the initial cake concentration C_0 may be calculated from the true initial amount of solute divided by void volume, that is 0.8438/0.1 = 8.438 lbm/ft^3 (135 kg/m^3), and the mass of solute remaining in the cake may be calculated by subtracting the cumulative amount of solute removed from the true initial mass of solute, 0.8438 lbm/ft^2, as shown in the fifth column of Table 14.5. Since the mass of solid remaining equals 5.91 $\psi_m/(1 - \psi_m)$, the values in that column can be employed to calculate ψ_m; the results (on a percentage basis) appear as the sixth column in the table.

Thus, if it is permissible to retain 0.67 percent of soluble material, on a moisture-free basis, the table shows that slightly over 10 gal of wash water must be used per square foot (slightly over 0.41 m^3/m^2).

Related Calculations.   The average concentration of solute in the cake consists of mass of solute remaining divided by void volume. These values appear as the final column in Table 14.5.

14.5 *ROTARY-VACUUM-DRUM FILTERS*

The salient features of rotary-drum filtration are illustrated in Fig. 14.9, where a cylindrical drum having a permeable surface is revolving counterclockwise partially submerged in a slurry. A pressure differential is usually maintained between the outer and inner surfaces by means of a vacuum pump. However, the drum might be enclosed and operated under pressure. In addition to the vacuum or pressure, each point on the periphery of the drum is subjected to a hydrostatic head of slurry.

Continuous multicompartment drum filters, as illustrated in Fig. 14.9, are normally used on materials that are relatively concentrated and easy to filter. Rates of cake buildup are in the range of 0.05 in/min (0.0013 m/min) to 0.05 in/s (0.0013 m/s). Submergence normally runs from 25 to 75 percent (40 percent being quite common), with rotation speeds from 0.1 to 3 r/min. With these conditions, filtration times could range from 5 s to 7.5 min, the great majority of industrial filtration falling within these limits.

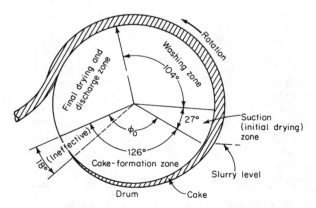

FIGURE 14.9   Rotary-vacuum-drum filter (Procedure 14.6).

Drum diameters typically are 6 to 12 ft (1.83 to 3.66 m), although larger values may be encountered. With 40 percent submergence and a 12-ft diameter, the hydrostatic head ranges up to 5.7 ft (1.74 m), which is a significant fraction of the driving force in vacuum filtration. Since the slurry will have a density greater than water, the effective head may be as high as 7 ft (2.13 m).

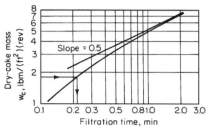

FIGURE 14.10 Dry-cake mass as a function of filtration time (Procedure 14.6). [Note: 1 lbm/(ft²)(r) = 4.88 kg/(m²)(r).]

The drum of radius r rotates at an angular velocity of ω rad/s (N r/s). The portion of the drum submerged in the slurry is subtended by an arc ϕ_0. The remaining part of the drum is utilized for washing, drying, and discharge. Filtration through a given portion of the drum is assumed to begin at the instant that portion enters the slurry; in practice, however, there is a time lag in establishing the full vacuum because of the need to maintain a vacuum seal as each compartment enters the slurry.

Dry-cake mass per area is shown in Fig. 14.10 as a function of elapsed time during a given revolution of the drum.

Cake Washing

Experimental wash curves represented as fraction of solute remaining versus the wash ratio j (ratio of wash to void volume of cake) can be plotted semilogarithmically as in Fig. 14.11 (the solid line). No experimental point will fall on the left of the maximum theoretical curve (the dotted line), which represents perfect displacement.

Cake wash time is the most difficult variable to correlate. Filtration theory suggests three possible correlations: (1) wash time versus $w_c v_w$; where w_c and v_w are total mass of dry solids and volume of wash, each per area of filtration; (2) wash time versus jw_c; and (3) wash lime per form time versus wash volume per form volume. Fortunately, the easiest correlation (no. 1) usually gives satisfactory results. This curve starts as a straight line, but often falls off as the volume of wash water increases, as in Fig. 14.12.

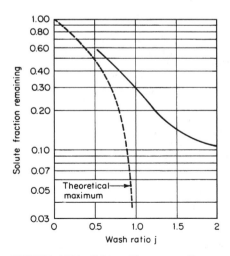

FIGURE 14.11 Cake-washing curves (Procedure 14.6).

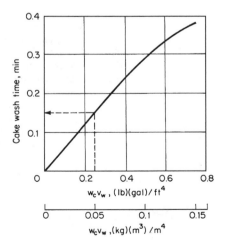

FIGURE 14.12 Cake-wash time correlation with mass of dry solids w_c and volume of wash v_w per unit of area (Procedure 14.6).

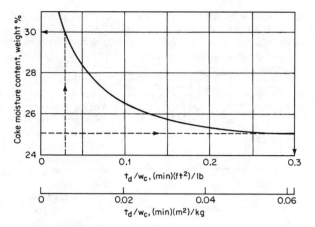

FIGURE 14.13 Correlating factor for cake moisture content and air rate (Procedure 14.6).

Cake Moisture Content, Air Rate

Experience has shown that the following factor is useful in correlating cake-moisture-content data [1]:

$$\text{Correlating factor} = \frac{\text{ft}^3/\text{min}}{\text{ft}^2} \frac{\Delta p_c}{w_c} \frac{t_d}{\mu}$$

where $\text{ft}^3/(\text{min})(\text{ft}^2)$ is air rate through the filter cake, t_d is drying time, Δp_c is pressure drop across the cake, and μ is liquid viscosity.

Figure 14.13 shows the general shape of the curve. The correlating factor chosen for design should be somewhere past the knee of the curve. Values to the left approach an unstable operating range, wherein a small change in operating conditions can result in a relatively drastic change in cake moisture content.

If runs are made at constant temperature and vacuum, the pressure drop and viscosity terms can be dropped from the expression. Often air-rate data are not available, but correlations can be obtained without air rates, particularly if the cakes are relatively non porous. The correlating factor is then reduced to the simplified term t_d/w_c, which involves only drying time and cake weight per unit area per revolution. A substantial degree of data scatter is normally encountered in the moisture-content correlation. Any point selected on the correlation will represent an average operating condition. To ensure that cake moisture content will not exceed a particular value, the correlating factor at the desired minimum should be multiplied by 1.2 before calculating the required drying time.

Air rate through the cake—and thus, vacuum-pump capacity—can be determined from measurements of flow rate as a function of time. Integration of these data over the times involved in the first and second stages of drying in continuous filters yields vacuum-pump-capacity data.

14.6 DESIGN OF A ROTARY-VACUUM-DRUM FILTER

A drum filter as illustrated in Fig. 14.9 is to be used for filtering, washing, and drying a cake having the properties given by Figs. 14.10 through 14.13. Air rate through the cake is determined from measurements of flow rate as a function of time with a rotameter as follows:

	Time, min								
	0.05	0.1	0.2	0.3	0.4	0.6	0.8	1.0	1.5
ft³/(min)(ft²)	2.5	4.2	5.9	6.8	7.45	8.2	8.6	8.75	9.2
m³/(min)(m²)	0.762	1.28	1.80	2.07	2.27	2.50	2.62	2.67	2.80

These data are also plotted in Fig. 14.14.

The following conditions and specifications are assumed: slurry contains 40% solids by weight; solute in the liquid is 2%; final cake moisture is 25%; wash ratio (wash volume per void volume) is 1.5; cake mass w_c (in lbm/ft²) is $7.2L$, where L is cake thickness in inches; maximum submergence is 35 percent or 126°; effective submergence is 30 percent or 108°; maximum washing arc is 29 percent or 104°; suction (initial drying) arc is 7.5 percent or 27°; discharge and resubmergence arc is 25 percent or 90°; and minimum cake thickness is $\frac{1}{8}$ in (0.0032 m). Determine the relevant design parameters for the filter.

Calculation Procedure

1. Calculate the cake mass, find the filtration time for a thickness of 0.25 in (0.0064 m), and determine the minimum cycle needed for cake formation.
The cake mass is given by $w_c = 7.2(0.25) = 1.8$ lbm/ft². From Fig. 14.10, filtration time is found to be 0.22 min, and so the filtration rate is $(1.8/0.22)(60) = 491$ lbm/h per square foot of drum surface. With an effective submergence of 30 percent, the minimum cycle based on cake formation is $0.22/0.3 = 0.73$ min/r, which corresponds to 1.37 r/min.

2. Check to see if initial drying or washing can be done within the time available during the minimum cycle from step 1.
Minimum suction time elapses during passage through 27° (7.5 percent) of the perimeter. Therefore, drying time is $0.075(0.73) = 0.06$ min, and the correlating factor is $t_d/w_c = 0.06/1.8 = 0.033$. Based on Fig. 14.13, the dewatered but unwashed (D/u) cake will have a moisture content of 30%. Then with a wash ratio of 1.5, liquid in D/u cake equals $(30/70)(1.8) = 0.77$ lbm/(ft²)(r), and quantity of wash equals $1.5(0.77) = 1.155$ lbm/(ft²)(r), or, at 8.33 lbm/gal, 0.14 gal/(ft²)(r). To calculate wash time, $w_c v_w = 1.8(0.14) = 0.25$. From Fig. 14.12, the required wash time is 0.15 min. This corresponds to an arc of $0.15/0.73 = 0.21$; i.e., to 21 percent of the circumference. Since up to 29 percent of the circumference can be used, washing offers no problems.

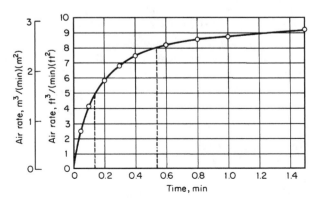

FIGURE 14.14 Air rate as a function of time (Procedure 14.6).

3. Check the drying time and determine the cycle time.

For a final moisture content of 25%, the simplified correlating factor t_d/w_c from Fig. 14.13 (taking into account the 1.2 factor mentioned above) is approximately 1.2(0.25) = 0.3 (the 25% moisture content is employed to enter the graph along the ordinate; the 0.25 is read from the graph along its abscissa). With $w_c = 1.8$, $t_d = 0.54$ min and 0.54/0.73 = 0.739, which takes up nearly three-quarters of the circumference for drying, so a lower speed must be used.

As a first estimate, note that since 25 percent of the arc is needed for discharge and resubmergence, the maximum arc for washing plus final drying is given by 75 − (cake-formation arc) − (suction arc) = 75 − 30 − 7.5 = 37.5 percent. Using the originally calculated washing plus drying times of 0.54 + 0.15 = 0.69 min, then 0.69/0.375 = 1.84 min/r and the washing arc is 0.15/1.84, which is 8.15 percent or 29°.

4. Repeat the calculations of step 2.

Minor adjustments can be made by recalculating each quantity with each change in conditions. Thus initial drying time equals 1.84(0.075) = 0.14 min, so t_d/w_c = 0.14/1.8 = 0.08. From Fig. 14.13, D/u moisture is 27%, and accordingly, the liquor in the D/u cake is (27/73)(1.8) = 0.67 lbm/(ft^2)(r). The quantity of wash becomes 1.5(0.67) = 1.0 lbm/(ft^2)(r), or 0.12 gal/(ft^2)(r). Then $w_c v_w$ = 1.8(0.12) = 0.22, and from Fig. 14.12, the wash time becomes 0.14 min.

5. Summarize the filtration cycle.

The cycle time is now (0.14 + 0.54)/0.375 = 1.81 min/r, equivalent to 0.55 r/min. The required effective submergence is (0.22/1.81)(100) = 12.2 percent. This is much less than the 30 percent available. The filter valve bridge must delay the start of vacuum or the slurry level can be reduced. If the level is reduced, additional initial drying time will be available, thereby reducing the angle required for washing. The design cycle is as follows:

Operation	Minutes
Form	0.22
Initial dry	0.14
Wash	0.14
Final dry	0.54
Discharge and resubmergence	0.77
Total time	1.81

6. Calculate cake thickness and filtration rate.

Cake thickness is given by L (in inches) = w_c/7.2 = 1.8/7.2 = 0.25 in. Taking into account the effective submergence of 12.2 percent, we calculate the filtration rate as 491 (0.122) = 59.9 lbm/(h)(ft^2) [0.081 kg/(s)(m^2)]. Experience suggests applying a scale-up factor of 0.8: 0.8(59.9) = 47.9 lbm/(h)(ft^2) [0.065 kg/(s)(m^2)]. This is not intended as a safety factor to allow for increased production; instead, it corrects for deviation owing to the size of the test equipment, to media blinding, and to similar factors.

7. Calculate the efficiency of solute recovery.

Assume that Fig. 14.11 applies. With $j = 1.5$, the fraction remaining is 0.145. To be on the safe side, use a value of 0.2. The following calculations are needed:

Solute in feed = (60/40)(0.02) = 0.03 lb solute per pound of feed

Solute in D/u cake = (27/73)(0.02) = 0.0074 lb solute per pound of cake

Solute in washed cake = 0.0074(0.2) = 0.0015 lb solute per pound of washed cake

The fractional recovery, then, equals (0.030 − 0.0015)/0.03 = 0.95. Using 0.145 instead of 0.2 would have yielded a figure of 0.964.

8. Calculate the air rate.

The air rate can be calculated based on the data previously presented and shown in Fig. 14.14. During the 0.14 min of initial drying, the average rate is found to be 2.95 (ft^3/min)/(ft^2)(r). The average rate during the final 0.54 min of drying is 5.85 (fl^3/min)/(ft^2)(r). The total air rate is given by 0.14(2.95) + 0.54(5.85) = 3.57 (ft^3/min)/(ft^2)(r). Since there are 1.81 min/r, the air rate is 3.57/1.81 = 1.97 (ft^3/min)/ft^2 [0.01 (m^3/s)/m^2].

14.7 PRESSURE DROP AND SPECIFIC DEPOSIT FOR A GRANULAR-BED FILTER

For filtering dust streams, a fixed-granular-bed filter with a length, L, of 1 m is filled with glass spheres of 5-mm (0.005-m) diameter, d_m. The dust to be captured is limestone with a 0.1-mm mean diameter, d_d. The limestone density, ρ_d, is 2700 kg/m^3, and the gas velocity, u, is 0.5 m/s. The initial porosity, ϵ_0 of the filter is 0.4, and the filtration process stops when the porosity has dropped to a level, ϵ, of 0.3. The gas has a viscosity, μ_g, of 17.2 × 10^{-6} N-s/m^2 and a density, ρ_g, of 1.293 kg/m^3. Determine the pressure drop through the filter, ΔP, and the mass of deposited dust per unit volume (also known as the specific deposit), m.

Calculation Procedure

1. Calculate the direct interception number, N_I, as well as $(1 - \epsilon_0)/(1 - \epsilon)$.
N_I is defined as d_d/d_m. In this case, it equals 0.1 mm/5 mm, or 0.02. And $(1 - \epsilon_0)/(1 - \epsilon) = (1 - 0.4)/(1 - 0.3) = 0.857$.

2. Find the parameter f.
Use Fig. 14.15. From an abscissa of 0.857 and the line corresponding to N_I of 0.02, read along the ordinate a value of 8 for f.

   Caution: Do not use Fig. 14.15 or Fig. 14.16 if the gas velocity is greater than 1 m/s, because the dust particles may tend to "bounce" off the surface of the granular medium.

3. Calculate the Reynolds particle number, Re_m.
The relationship is as follows:

$$Re_m = (u\rho_g d_m)/\mu_g$$

So,

$$Re_m = (0.5)(1.293)(0.005)/(17.2 \times 10^{-6}) = 188$$

4. Find the parameter F.
Use Fig. 14.16. From an abscissa of 188/8, or 23.5, and the line corresponding to an ϵ of 0.3, read along the ordinate a value of 190 for F.

5. Calculate the pressure drop through the filter.
Use the relationship $\Delta P = L\rho_g u^2 f F/d_m$. Thus,

$$\Delta P = (1)(1.293)(0.5)3(8)(190)/0.005 = 96,200 \, \text{Pa}$$

6. Calculate the specific deposit.
Use the relationship

$$m = \rho_d(\epsilon_0 - \epsilon) = 2700(0.4 - 0.3) = 270 \, \text{kg/m}^3$$

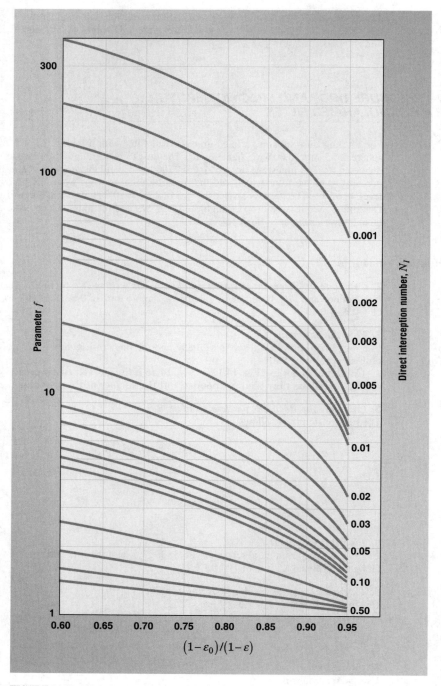

FIGURE 14.15   Chart for determining parameter f for use in ΔP calculation.

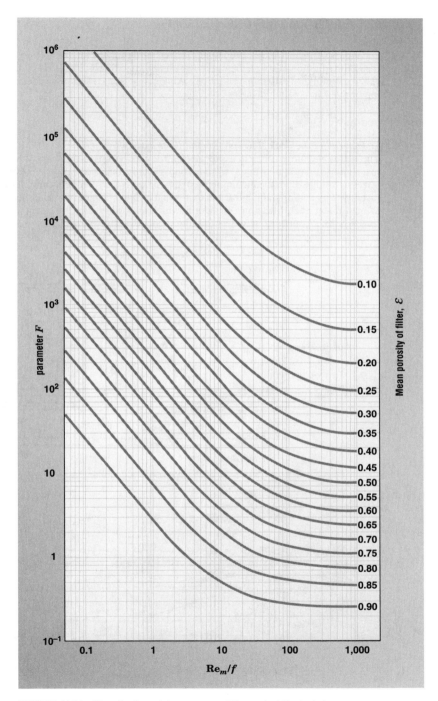

FIGURE 14.16 Chart for determining parameter F for use in ΔP calculation.

*14.8   SIZING OF A TRAVELING-BRIDGE FILTER

Secondary effluent from a municipal wastewater treatment facility is to receive tertiary treatment, including filtration, through the use of traveling bridge filters. The average daily flow rate is 4.0 Mgd (2778 gal/min) (15,140 m^3/d) and the peaking factor is 2.5. Determine the size and number of traveling bridge filters required.

Calculation Procedure

1. *Determine the peak flow rate for the filter system.*
The traveling bridge filter is a proprietary form of a rapid sand filter. This type of filter is used mainly for filtration of effluent from secondary and advanced wastewater treatment facilities. In the traveling bridge filter, the incoming wastewater floods the filter bed, flows through the filter medium (usually sand and/or anthracite), and exits to an effluent channel via an underdrain and effluent ports located under each filtration cell. During the backwash cycle, the carriage and the attached hood (see Fig. 14.17) move slowly over the filter bed, consecutively isolating and backwashing each cell.

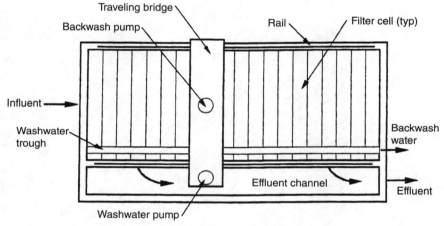

Traveling Bridge Filter Plan

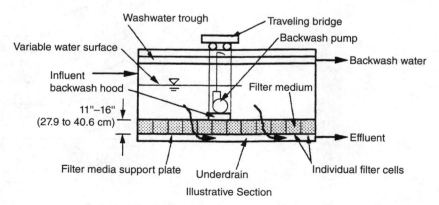

Illustrative Section

FIGURE 14.17   Traveling bridge filter.

The washwater pump, located in the effluent channel, draws filtered wastewater from the effluent chamber and pumps it through the effluent port of each cell, forcing water to flow up through the cell thereby backwashing the filter medium of the cell. The backwash pump located above the hood draws water with suspended matter collected under the hood and transfers it to the backwash water trough. During the backwash cycle, wastewater is filtered continuously through the cells not being backwashed.

Filtration in a traveling bridge filter is accomplished at a hydraulic loading typically in the range of 1.5 to 3.0 gal/min per square foot of filter surface (1.02 to 2.04 L/s · m²) at average daily flow. The maximum hydraulic loading used for design is typically 4.0 gal/min/ft² (2.72 L/s · m²) at peak flow. The peak hydraulic loading is used to size the traveling bridge filter.

The peak flow for this treatment facility is calculated as follows:

$$\text{Peak flow} = (\text{peaking factor})(\text{average daily flow}) = (2.5)(4.0 \text{ Mgd})$$

$$= 10 \text{ Mdg} = 6944 \text{ gal/min} (438.2 \text{ L/s})$$

2. Find the required filter surface area.
Filter surface area required is calculated using

$$\text{Filter surface area required (ft}^2) = \frac{\text{Peak flow (gal/min)}}{\text{Hydraulic loading (gal/min} \cdot \text{ft}^2)}$$

Using values from above

$$\text{Filter surface area required (ft}^2) = \frac{6944 \text{ gal/min}}{4.0 \text{ gal/min} \cdot \text{ft}^2} = 1736 \text{ ft}^2 (161.3 \text{ m}^2)$$

3. Determine the number of filters required.
Standard filter widths available from various manufacturers are 8, 12, and 16 ft (2.44, 3.66, and 4.88 m). Using a width of 12 ft (3.66 m) and length of 50 ft (15.2 m) per filter, the area of each filter is:

$$\text{Area of each filter} = (12)(50) = 600 \text{ ft}^2 (55.7 \text{ m}^2)$$

The number of filters required is 1736/600 per filter = 2.89. Use three filters for a total filter area of 1800 ft² (167.2 m²).

It must be kept in mind that most state and local regulations stipulate that rapid sand filters shall be designed to provide a total filtration capacity for the maximum anticipated flow with at least one of the filters out of service. Therefore, four traveling bridge filters should be provided, each with filtration area dimensions of 12 ft wide × 50 ft long (3.66 × 15.2 m).

The media depth for traveling bridge filters ranges from 11 to 16 in (27.9 to 40.6 cm). Dual media may be used with 8 in (20.3 cm) of sand underlying 8 in (20.3 cm) of anthracite.

4. Find the hydraulic loading under various service conditions.
The hydraulic loadings with all filters in operation is:

$$\text{Average flow} = \frac{2778 \text{ gal/min}}{4(600 \text{ ft}^2)} = 1.16 \text{ gal/min} \cdot \text{ft}^2 (0.79 \text{ L/s} \cdot \text{m}^2)$$

$$\text{Peak flow} = \frac{6944 \text{ gal/min}}{4(600 \text{ ft}^2)} = 2.89 \text{ gal/min} \cdot \text{ft}^2 (1.96 \text{ L/s} \cdot \text{m}^2)$$

Hydraulic loading with one filter out of service (three active filters) is:

$$\text{Average flow} = \frac{2778 \text{ gal/min}}{3(600 \text{ ft}^2)} = 1.54 \text{ gal/min} \cdot \text{ft}^2 \text{ (1.05 L/s} \cdot \text{m}^2)$$

$$\text{Peak flow} = \frac{6944 \text{ gal/min}}{3(600 \text{ ft}^2)} = 3.86 \text{ gal /min} \cdot \text{ft}^2 \text{ (2.62 L/s} \cdot \text{m}^2)$$

These hydraulic loadings are acceptable and may be used in specifying the traveling bridge filter.

The amount of backwash water produced depends upon the quantity and quality of influent to the filter. The backwash pumps are usually sized to deliver approximately 25 gal/min (1.58 L/s) during the backwash cycle. Backwash water is generally returned to the head of the treatment facility for reprocessing.

This procedure is the work of Kevin D. Wills, M.S.E., P.E., who, at the time of its preparation, was consulting engineer, Stanley Consultants, Inc.

*14.9 SIZING A POLYMER DILUTION/FEED SYSTEM

Equalized secondary effluent of 1.0 Mgd (3785 m^3/d) from a municipal wastewater treatment facility is to undergo coagulation and flocculation in a direct filtration process. The coagulant used will be an emulsion polymer with 30 percent active ingredient. Size the polymer dilution/feed system including the quantity of dilution water required, and the amount of neat (as supplied) polymer required.

Calculation Procedure

1. *Determine the daily polymer requirements.*
Depending on the quality of settled secondary effluent, organic polymer addition is often used to enhance the performance of tertiary effluent filters in a direct filtration process. See *Design of a Rapid Mix Basin and Flocculation Basin.* Because the chemistry of the wastewater has a significant effect on the performance of a polymer, the selection of a type of polymer for use as a filter aid generally requires experimental testing. Common test procedures for polymers involve adding an initial polymer dosage to the wastewater (usually one part per million, ppm) of a given polymer and observing the effects. Depending on the effects observed, the polymer dosage should be increased or decreased by 0.5 ppm increments to obtain an operating range. A polymer dosage of 2 ppm (two parts polymer per 1×10^6 parts wastewater) will be used here.

In general, the neat polymer is supplied with approximately 25 to 35 percent active polymer, the rest being oil and water. As stated above, a 30 percent active polymer will be used for this example. The neat polymer is first diluted to an extremely low concentration using dilution water, which consists of either potable water or treated effluent from the wastewater facility. The diluted polymer solution usually ranges from 0.005 to 0.5 percent solution. The diluted solution is injected into either a rapid mix basin or directly into a pipe. A 0.5 percent solution will be used here.

The gallons per day (gal/day) (L/d) of active polymer required is calculated using the following:

$$\text{Active polymer (gal/day)} = \text{(wastewater flow, Mgd)}$$
$$\text{(active polymer dosage, ppm)}$$

Using the values outlined above

$$\text{Active polymer} = (1.0 \text{ Mgd})(2 \text{ ppm}) = 2 \text{ gal/day active polymer (pure polymer)}$$
$$= 0.083 \text{ gal/h (0.31 L/h)}$$

2. Find the quantity of dilution water required.
The quantity of dilution water required is calculated using the following:

$$\text{Dilution water (gal/h)} = \frac{\text{active polymer, gal/h}}{\% \text{ solution used (as a decimal)}}$$

Therefore, using the values obtained above

$$\text{Dilution water} = \frac{0.083 \text{ gal/h}}{0.005} = 16.6 \text{ gal/h (62.8 L/h)}$$

3. Find the quantity of neat polymer required.
The quantity of neat polymer required is calculated as follows:

$$\text{Neat polymer (gal/h)} = \frac{\text{active polymer, gal/h}}{\% \text{ active polymer in emulsion as supplied}}$$

Using the values obtained above

$$\text{Neat polymer} = \frac{0.083 \text{ gal/h}}{0.30} = 0.277 \text{ gal/h (1.05 L/h)}$$

This quantity of neat polymer represents the amount of polymer used in its "as supplied" form. Therefore, if polymer is supplied in a 55 gal (208.2 L) drum, the time required to use one drum of polymer (assuming polymer is used 24 h/d, 7 d/week) is:

$$\text{Time required to use one drum of polymer} = \frac{55 \text{ gal}}{0.277 \text{ gal}} \cong 200 \text{ h} = 8 \text{ days}$$

This procedure is the work of Kevin D. Wills, M.S.E., P.E., who, at the time of its preparation, was consulting engineer, Stanley Consultants, Inc.

*14.10   DESIGN OF A PLASTIC MEDIA TRICKLING FILTER

A municipal wastewater with a flow of 1.0 Mgd (694 gal/min) (3785 m^3/d) and a BOD$_5$ of 240 mg/L is to be treated in a single-stage plastic media trickling filter without recycle. The effluent wastewater is to have a BOD$_5$ of 20 mg/L. Determine the diameter of the filter, the hydraulic loading, the organic loading, the dosing rate, and the required rotational speed of the distributor arm. Assume a filter depth of 25 ft (7.6 m). Also assume that a treatability constant ($k_{20/20}$) of 0.075 (gal/min)$^{0.5}$/ft^2 was obtained in a 20-ft (6.1-m) high test filter at 20°C (68°F). The wastewater temperature is 30°C (86°F).

Calculation Procedure

1. Adjust the treatability constant for wastewater temperature and depth.
Due to the predictable properties of plastic media, empirical relationships are available to predict performance of trickling filters packed with plastic media. However, the treatability constant must first be adjusted for both the temperature of the wastewater and the depth of the actual filter.

(a) *Adjustment for temperature.* The treatability constant is first adjusted from the given standard at 20°C (68°F) to the actual wastewater temperature of 30°C (86°F) using the following equation:

$$k_{30/20} = k_{20/20}\theta^{T-20}$$

where $k_{30/20}$ = treatability constant at 30°C (86°F) and 20 ft (6.1 m) filter depth
$k_{20/20}$ = treatability constant at 20°C (68°F) and 20 ft (6.1 m) filter depth
θ = temperature activity coefficient (assume 1.035)
T = wastewater temperature

Using above values

$$k_{30/20} = [0.075 \,(\text{gal/min})^{0.5}/\text{ft}^2](1.035)^{30-20} = 0.106 \,(\text{gal/min})^{0.5}/\text{ft}^2$$

(b) *Adjustment for depth.* The treatability constant is then adjusted from the standard depth of 20 ft (6.1 m) to the actual filter depth of 25 ft (7.6 m) using the following equation:

$$k_{30/25} = k_{30/20}(D_1/D_2)^x$$

where $k_{30/25}$ = treatability constant at 30°C (86°F) and 25 ft (7.6 m) filter depth
$k_{30/20}$ = treatability constant at 30°C (86°F) and 20 ft (6.1 m) filter depth
D_1 = depth of reference filter (20 ft) (6.1 m)
D_2 = depth of actual filter (25 ft) (7.6 m)
x = empirical constant (0.3 for plastic medium filters)

Using above values

$$k_{30/25} = (0.106 \,(\text{gal/min})^{0.5}/\text{ft}^2)(20/25)^{0.3}$$
$$= 0.099 \,(\text{gal/min})^{0.5}/\text{ft}^2 \,[0.099\,(\text{L/s})^{0.5}/\text{m}^2]$$

2. Size the plastic media trickling filter.
The empirical formula used for sizing plastic media trickling filters is:

$$\frac{S_e}{S_i} = \exp[-kD(Q_v)^{-n}]$$

where S_e = BOD$_5$ of settled effluent from trickling filter (mg/L)
S_i = BOD$_5$ of influent wastewater to trickling filter (mg/L)
k_{20} = treatability constant adjusted for wastewater temperature and filter depth = $(k_{30/25})$
D = depth of filter (ft)
Q_v = volumetric flow rate applied per unit of filter area (gal/min · ft^2) (L/s · m^2) = Q/A
Q = flow rate applied to filter without recirculation (gal/min) (L/s)
A = area of filter (ft^2) (m^2)
n = empirical constant (usually 0.5)

Rearranging and solving for the trickling filter area (A)

$$A = Q\left[\frac{-\ln(S_e/S_i)}{(k_{30/25})D}\right]^{1/n}$$

Using values from above, the area and diameter of the trickling filter are:

$$A = 694 \text{ gal/min} \left[\frac{-\ln(20/240)}{(0.099)25 \text{ ft}} \right]^{1/0.5} = 699.6 \text{ ft}^2 \Rightarrow \text{diameter} = 29.9 \text{ ft} (9.1 \text{ m})$$

3. Calculate the hydraulic and organic loading on the filter.
The hydraulic loading (Q/A) is then calculated:

$$\text{Hydraulic loading} = 694 \text{ gal/min}/699.6 \text{ ft}^2 = 0.99 \text{ gal/min} \cdot \text{ft}^2 (0.672 \text{ L/s} \cdot \text{m}^2)$$

For plastic media trickling filters, the hydraulic loading ranges from 0.2 to 1.20 gal/min $\cdot$ ft^2 (0.14 to 0.82 L/s $\cdot$ m^2).

The organic loading to the trickling filter is calculated by dividing the BOD$_5$ load to the filter by the filter volume as follows:

$$\text{Organic loading} = \frac{(1.0 \text{ Mgd})(240 \text{ mg/L})(8.34 \text{ lb} \cdot \text{L/mg} \cdot \text{Mgal})}{(699.6 \text{ ft}^2)(25 \text{ ft})(10^3 \text{ ft}^3/1000 \text{ ft}^3)}$$

$$= 144 \frac{\text{lb}}{10^3 \text{ ft}^3 \cdot \text{d}} (557 \text{ kg/m}^2 \cdot \text{d})$$

For plastic media trickling filters, the organic loading ranges from 30 to 200 lb/10^3 ft^3 $\cdot$ d (146.6 to 977.4 kg/m^2 $\cdot$ d).

4. Determine the required dosing rate for the filter.
To optimize the treatment performance of a trickling filter, there should be a continual and uniform growth of biomass and sloughing of excess biomass. To achieve uniform growth and sloughing, higher periodic dosing rates are required. The required dosing rate in inches per pass of distributor arm may be approximated using the following:

$$\text{Dosing rate} = (\text{organic loading, lb/10}^3 \text{ ft}^3 \cdot \text{d})(0.12)$$

Using the organic loading calculated above, the dosing rate is:

$$\text{Dosing rate} = (114 \text{ lb/10}^3 \text{ ft}^3 \cdot \text{d})(0.12) = 13.7 \text{ in/pass} (34.8 \text{ cm/pass})$$

Typical dosing rates for trickling filters are listed in Table 14.6. To achieve the typical dosing rates, the speed of the rotary distributor can be controlled by (1) reversing the location of some of

TABLE 14.6 Typical Dosing Rates for Trickling Filters[†]

Organic loading lb BOD$_5$/10^3 ft^3 $\cdot$ d (kg/m^2 $\cdot$ d)	Dosing rate, in/pass (cm/pass)
<25 (122.2)	3 (7.6)
50 (244.3)	6 (15.2)
75 (366.5)	9 (22.9)
100 (488.7)	12 (30.5)
150 (733.0)	18 (45.7)
200 (977.4)	24 (60.9)

[†]Metcalf and Eddy, *Wastewater Engineering: Treatment, Disposal, and Reuse*, 3rd ed.

the existing orifices to the front of the distributor arm, (2) adding reversed deflectors to the existing orifice discharges, and (3) by operating the rotary distributor with a variable-speed drive.

5. *Determine the required rotational speed of the distributor.*
The rotational speed of the distributor is a function of the instantaneous dosing rate and may be determined using the following:

$$n = \frac{1.6(Q_T)}{(A)(DR)}$$

where n = rotational speed of distributor (rpm)
Q_T = total applied hydraulic loading rate (gal/min · ft^2) (L/s · m^2) = $Q + Q_R$
Q = influent wastewater hydraulic loading rate (gal/min · ft^2) (L/s · m^2)
Q_R = recycle flow hydraulic loading rate (gal/min · ft^2) (L/s · m^2)
  Note: recycle is assumed to be zero in this example.
A = number of arms in rotary distributor assembly
DR = dosing rate (in/pass of distributor arm)

Assuming two distributor arms (two or four arms are standard), and using values from above, the required rotational speed is:

$$n = \frac{1.6(0.99 \text{ gal/min} \cdot \text{ ft}^2)}{(2)(13.7) \text{ in/pass}} = 0.058 \text{ rpm}$$

This equates to one revolution every 17.2 min.

This procedure is the work of Kevin D. Wills, M.S.E., P.E., who, at the time of its preparation, was consulting engineer, Stanley Consultants, Inc.

REFERENCES

1. Nelson and Dahlstrom—*Chem. Eng. Prog. 53*:320, 1957.

2. Tiller—*Chem. Eng. Prog. 51*:282, 1955.

3. Tiller and Crump—*Chem. Eng. Prog. 73*:65, Oct. 1977.

4. Tiller, Crump, and Ville—*Proceedings of the Second World Filtration Congress (London)*, Sept. 1979.

5. Macías-Machín, A., and Santana, D. F., "Calculate Delta P of a Fixed-Bed Filter," *Chemical Engineering,* September 1998, pp. 159–162.

SECTION 15
AIR POLLUTION CONTROL[†]

Louis Theodore, Eng.Sc.D.

Professor
Department of Chemical Engineering
Manhattan College
Bronx, NY

15.1 EFFICIENCY OF PARTICULATE-SETTLING CHAMBER

A particulate-settling chamber is installed in a small heating plant that uses a traveling grate stoker. Determine the overall collection efficiency of the chamber, given the following operating conditions, chamber dimensions, and particle-size distribution:

Chamber width: 10.8 ft (3.29 m)
Chamber height: 2.46 ft (0.75 m)
Chamber length: 15.0 ft (4.57 m)
Volumetric flow rate of contaminated air stream: 70.6 std ft³/s (2.00 m³/s)
Flue-gas temperature: 446°F
Viscosity of air stream at 446°F: 1.75×10^{-5} lb/(ft)(s) [2.60×10^{-5} (N)(s)/m²]
Flue-gas pressure: 1 atm (101.3 kPa)
Particle concentration: 0.23 grains/std ft³ (8.13 grains/m³)
Particle specific gravity: 2.65
Standard operating conditions: 32°F, 1 atm (273 K, 101.3 kPa)

[†]Procedures 15.10 and 15.11 are adapted from *Chemical Engineering* magazine.

Particle-size distribution on the inlet dust:

Particle size range, μm	Average particle diameter, μm	grains/std ft³	grains/m³	Weight %
0–20	10	0.0062	0.219	2.7
20–30	25	0.0159	0.562	6.9
30–40	35	0.0216	0.763	9.4
40–50	45	0.0242	0.855	10.5
50–60	55	0.0242	0.855	10.5
60–70	65	0.0218	0.770	9.5
70–80	75	0.0161	0.569	7.0
80–94	87	0.0218	0.770	9.5
94+	94+	0.0782	2.763	34.0

Assume that the actual terminal settling velocity is one-half of the velocity given by Stokes' law.

Calculation Procedure

1. Express the collection efficiency E in terms of the particle diameter d_p by employing the terminal settling velocity for Stokes' law.
Since the actual terminal settling velocity is assumed to be one-half of the Stokes' law velocity,

$$v_t = gd_p^2\rho_p/36\mu$$

and

$$E = v_t BL/q = g\rho_p BLd_p^2/36\mu q_a$$

where v_t = terminal velocity, ft/s
   g = gravitational constant, 32.2 (lbm)(ft)/(s²)(lbf)
   d_p = particle diameter, ft
   ρ_p = particle density, lb/ft³
   μ = air viscosity, lb/(ft)(s)
   B = chamber width, ft
   L = chamber length, ft
   q = volumetric flow rate for the gas, actual ft³/s

2. Determine the particle density.
As the specific gravity is 2.65, the density is (2.65)(62.4 lb/ft³) = 165.4 lb/ft³ (2648 kg/m³).

3. Determine the actual volumetric flow rate.
Use Charles' law to convert from q_s, the flow rate at the standard-conditions temperature T_s, to q_a, the flow rate at the actual temperature T_a:

$$q_a = a_s(T_a/T_s) = 70.6[(446 + 460)/(32 + 460)] = 130 \text{ actual ft}^3/s$$

4. Express the collection efficiency in terms of d_p in micrometers.
The efficiency equation set out in step 1 is for d_p in feet. To adapt it for d_p in micrometers, note that there are 304,800 μm in a foot and accordingly divide the equation by the conversion factor $(304,800)^2$. Thus,

$$E = g\rho_p BLd_p^2/36\mu q_a$$
$$= (32.2)(165.4)(10.8)(15)d_p^2/(36)(1.75 \times 10^{-5})(130)(304,800)^2$$
$$= 1.14 \times 10^{-4}d_p^2$$

5. *Calculate the collection efficiency at each average particle size given in the statement of the problem.*

Applying the equation from step 4 and multiplying the answer by 100 (to convert the efficiency from a decimal fraction to a percent) gives the following results:

Average particle diameter, μm	Efficiency, %
10	1.1
25	7.1
35	14
45	23
55	34
65	48
75	64
87	86
94	100

6. *Calculate the overall collection efficiency.*

This is the sum of each of the efficiencies from the previous step multiplied by the weight fraction of the corresponding particle size in the mixture; in other words,

$$E = \Sigma w_i E_i$$

Thus, from the weight fractions given in the statement of the problem,

$$E = (0.027)(1.1) + (0.069)(7.1) + (0.094)(14) + (0.105)(23) + (0.105)(34)$$
$$+ (0.095)(48) + (0.070)(64) + (0.095)(86) + (0.34)(100) = 59.0\%$$

Related Calculations. Instead of following steps 5 and 6 as stated, one can calculate the efficiency at a variety of arbitrary particle diameters, graph the results, and then read the efficiencies for the actual particle diameters from the graph.

15.2 *EFFICIENCY OF CYCLONE SEPARATOR*

A cyclone separator 2 ft (0.62 m) in diameter, with an inlet width of 0.5 ft (0.15 m) and rated at providing 4.5 effective turns is being considered for removing particulates from offgases from a gravel dryer. Gases to the cyclone have a loading of 0.5 grains/ft^3 (17.7 grains/m^3), with an average particle diameter of 7.5 μm. Specific gravity of the particles is 2.75. Operating temperature is 70°F (294 K) at which the air viscosity is 1.21×10^{-5} lb/(ft)(s) [1.80×10^{-5}(N)(s)/m^2]. Inlet velocity to the cyclone is 50 ft/s (15.2 m/s). The local air-pollution authority requires that the maximum total loading of the cyclone effluent be 0.1 grains/ft^3 (3.53 grains/m^3). Can this cyclone meet that criterion?

Calculation Procedure

1. *Calculate the particle density.*

As the specific gravity of the particles is 2.75, the particle density ρ_p is (2.75)(62.4) or 171.6 lb/ft^3 (2747 kg/m^3).

2. *Calculate the cut diameter.*
The cut diameter d_{pc} for a given cyclone and given gas to be treated is the diameter of the particle that would be collected at 50 percent efficiency by the cyclone. It can be found from the equation

$$d_{pc} = [9\mu B_c / 2\pi n_t\, v_i(\rho_p - \rho)]^{0.5}$$

where μ = air viscosity, lb/(ft)(s)
   B_c = cyclone inlet width, ft
   n_t = number of effective turns provided by cyclone
   v_i = inlet gas velocity, ft/s
   ρ_p = particle density, lb/ft^3
   ρ = gas density, lb/ft^3

In this example, the gas density can be assumed negligible in comparison with the particle density, so use ρ_p instead of the density difference.
   Thus,

$$d_{pc} = [9(1.21 \times 10^{-5})(0.5)/2\pi(4.5)(50)(171.6)]^{0.5} = 1.5 \times 10^{-5}\,\text{ft}$$
$$= 4.57\mu\text{m}$$

3. *Calculate the ratio of average particle diameter to the cut diameter.*
Thus, $d_p/d_{pc} = 7.5/4.57 = 1.64$.

4. *Determine the collection efficiency using Lapple's curve.*
Refer to Fig. 15.1. For the particle-size ratio of 1.64 on the abscissa, the curve yields an efficiency of 0.72, or 72 percent, on the ordinate.

5. *Determine the collection efficiency required by the air-pollution-control authority.*
This efficiency is, simply, (inlet loading − outlet loading)/(inlet loading). In the present example, it equals $(0.5 - 0.1)/(0.5)$, i.e., 0.8, or 80 percent.

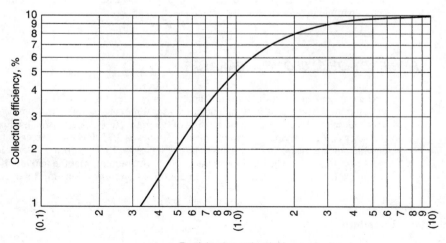

FIGURE 15.1   Collection efficiency as a function of particle size ratio.

6. *Does the cyclone meet the requirements of the pollution-control authority?*
As an 80 percent efficiency is required whereas the cyclone achieves only a 72 percent efficiency, the cyclone does not meet the requirements.

Related Calculations.   If the cyclone is of conventional design and its inlet width is not given, it is relatively safe to assume that the width is one-quarter the cyclone diameter.
Instead of using Lapple's curve, one can solve for cyclone efficiency by using the equation

$$E = 1.0/[1.0 + (d_{pc}/d_p)^2]$$

where d_{pc} is the cut diameter and d_p the average particle diameter.

15.3 *SIZING AN ELECTROSTATIC PRECIPITATOR*

A duct-type electrostatic precipitator is to be used to clean 100,000 actual ft³/min (47.2 actual m³/s) of an industrial gas stream containing particulates. The proposed design of the precipitator consists of three bus sections (fields) arranged in series, each having the same amount of collection surface. The inlet loading has been measured as 17.78 grains/ft³ (628 grains/m³), and a maximum outlet loading of 0.08 grains/ft³ (2.8 grains/m³) (both volumes corrected to dry standard conditions and 50 percent excess air) is allowed by the local air-pollution regulations. The drift velocity for the particulates has been experimentally determined in a similar installation, with the following results:

First section (inlet):   0.37 ft/s (0.11 m/s)

Second section (middle):   0.35 ft/s (0.107 m/s)

Third section (outlet):   0.33 ft/s (0.10 m/s)

Calculate the total collecting surface required. And find the total mass flow rate of particulates captured in each section.

Calculation Procedure

1. *Calculate the required total collection efficiency E based on the given inlet and outlet loading.*
The equation is

$$E = 1 - \text{(outlet loading)/(inlet loading)}$$

Thus,

$$E = 1 - 0.08/17.78 = 0.9955, \text{ or } 99.55\%$$

2. *Calculate the average drift velocity w.*
Thus, $w = (0.37 + 0.35 + 0.33)/3 = 0.35$ ft/s (0.107 m/s).

3. *Calculate the total surface area required.*
Use the Deutsch-Anderson equation:

$$E = 1 - \exp(-wA/q)$$

where E = collection efficiency
w = average drift velocity
A = required surface area
q = gas flow rate

Rearrange the equation as follows:

$$A = \ln(1 - E)/(w/q)$$

For consistency between w and q, convert q from ft³/min to ft³/s:

$$100,000/60 = 1666.7 \text{ ft}^3/\text{s}$$

Then

$$A = -\ln(1 - 0.9955)/(0.35/1666.7) = 25{,}732 \text{ ft}^2 \ (2393 \text{ m}^2)$$

4. Calculate the collection efficiencies of each section.
Use the Deutsch-Anderson equation (from step 3) directly. For the first section, $E_1 = 1 - \exp$ $[-(25{,}732)(0.37)/(3)(1666.7)] = 0.851$. Similarly, E_2 for the second section is found to be 0.835, and E_3 for the third section 0.817.

5. Calculate the mass flow rate $\dot{m}$ of particulates captured by each section.
For the first section, the equation is

$$\dot{m} = (E_1)(\text{inlet loading})(q)$$

Thus,

$$\dot{m} = (0.851)(17.78)(100{,}000) = 1.513 \times 10^6 \text{ grains/min} = 216.1 \text{ lb/min } (1.635 \text{ kg/s})$$

For the second section, the equation is

$$\dot{m} = (1 - E_1)(E_2)(\text{inlet loading})(q)$$

Thus,

$$\dot{m} = (1 - 0.851)(0.835)(17.78)(100{,}000) = 2.212 \times 10^5 \text{ grains/min}$$
$$= 31.6 \text{ lb/min } (0.239 \text{ kg/s})$$

And for the third section, the equation is

$$\dot{m} = (1 - E_1)(1 - E_2)(E_3)(\text{inlet loading})(q)$$

which yields 5.10 lb/min (0.039 kg/s).
   The total mass captured is the sum of the amounts captured in each section, i.e., 252.8 lb/min (1.91 kg/s). It is not surprising that a full 85 percent of the mass is captured in the first section.

15.4   EFFICIENCY OF A VENTURI SCRUBBING OPERATION

A gas stream laden with fly ash is to be cleaned by a venturi scrubber using a liquid-to-gas ratio (q_L/q_G) of 8.5 gal per 1000 actual cubic feet. The fly ash has a particle density of 43.7 lb/ft³ (700 kg/m³). The collection-efficiency k factor equals 200 ft³/gal. The throat velocity is 272 ft/s (82.9 m/s), and the gas viscosity is 1.5×10^{-5} lb/(ft)(s)[2.23×10^{-5} (N)(s)/m²]. The particle-size distribution is as follows:

d_{pi}, μm	Percent by weight, w_i, %
<0.10	0.01
0.1–0.5	0.21
0.6–1.0	0.78
1.1–5.0	13.0
6.0–10.0	16.0
11.0–15.0	12.0
16.0–20.0	8.0
>20.0	50.0

Calculation Procedure

1. Determine the mean droplet diameter d_o.
Use the Nukiyama-Tanasawa equation:

$$d_o = (16{,}400/v) + 1.45(q_L/q_G)^{1.5}$$

where d_o is the mean droplet diameter in micrometers and v is the throat velocity in feet per second. Thus,

$$d_o = (16{,}400/272) + 1.45(8.5)^{1.5} = 96.23\ \mu m = 3.16 \times 10^{-4}\ ft\ (9.63 \times 10^{-5} m)$$

2. Express the inertial impaction number ψ_1 in terms of particle diameter d_p.
The relationship is

$$\psi_1 = d_p^2 \rho_p v / 9\mu d_o$$

where ρ_p is particle density in lb/ft³ and μ is gas viscosity in lb/(ft)(s). Thus,

$$\psi_1 = d_p^2 (43.7)(272)/(9)(1.5 \times 10^{-5})(3.16 \times 10^{-4})$$
$$= 2.78 \times 10^{11} d_p^2 \quad \text{with } d_p \text{ in feet}$$

To make the expression suitable for the steps that follow, express it with d_p in micrometers, by dividing by the feet-to-micrometers conversion factor squared (because d_p is squared):

$$2.78 \times 10^{11} d_p^2 / (3.048 \times 10^5)^2 = 3.002 d_p^2 \quad \text{with } d_p \text{ in micrometers}$$

3. Express the individual collection efficiency in terms of d_p.
Use the Johnstone equation:

$$E_i = 1 - \exp(-kq_L \psi_1^{0.5}/q_G)$$
$$= 1 - \exp(-2.94 d_{pi})$$

4. *Calculate the overall collection efficiency.*
This is done by (a) determining the midpoint size of each of the particle-size ranges in the statement of the problem, (b) using the expression from step 3 to calculate the collection efficiency corresponding to that midpoint size, (c) multiplying each collection efficiency by the percent representation w_i for that size range in the statement of the problem, and (d) summing the results. These steps are embodied in the following table:

d_{pi}, μm	E_i	w_i, %	$E_i w_i$,%
0.05	0.1367	0.01	0.001367
0.30	0.586	0.21	0.123
0.80	0.905	0.78	0.706
3.0	0.9998	13.0	12.998
8.0	0.9999	16.0	16.0
13.0	0.9999	12.0	12.0
18.0	0.9999	8.0	8.0
20.0	0.9999	50.0	50.0

Then $E = \Sigma E_i w_i = 99.83\%$.

Related Calculations. In situations where the particles are so small that their size approaches the length of the mean free path of the fluid molecules, the fluid can no longer be regarded as a continuum; that is, the particles can "fall between" the molecules. That problem can be offset by applying a factor, the Cunningham correction factor, to the calculation of the inertial-impact-number expression in step 2.

15.5 SELECTING A FILTER BAG SYSTEM

It is proposed to install a pulse-jet fabric-filter system to remove particulates from an air stream. Select the most appropriate bag from the four proposed below. The volumetric flow rate of the air stream is 10,000 std ft³/min (4.72 m³/s) (standard conditions being 60°F and 1 atm), the operating temperature is 250°F (394 K), the concentration of pollutants is 4 grains/ft³ (141 grains/m³), the average air-to-cloth ratio is (2.5 ft³/min)/ft², and the required collection efficiency is 99 percent.
Information on the four proposed bags is as follows:

Bag designation	A	B	C	D
Tensile strength	Excellent	Above average	Fair	Excellent
Recommended maximum temperature, °F	260	275	260	220
Cost per bag	$26.00	$38.00	$10.00	$20.00
Standard size	8 in by 16 ft	10 in by 16 ft	1 ft by 16 ft	1 ft by 20 ft

Note: No bag has an advantage from the standpoint of durability.

Calculation Procedure

1. *Eliminate from consideration any bags that are patently unsatisfactory.*
Bag D is eliminated because its recommended maximum temperature is below the operating temperature for this application. Bag C is also eliminated, because a pulse-jet fabric-filter system requires that the tensile strength of the bag be at least above average.

2. *Convert the given flow rate to actual cubic feet per minute.*
The flow rate as stated corresponds to flow at 60°F, whereas the actual flow q will be at 250°F. Accordingly,

$$q = (10,000)(250 + 460)/(60 + 460) = 13,654 \text{ actual ft}^3/\text{min } (6.44 \text{ actual m}^3/\text{s})$$

3. *Establish the filtering velocity v_f.*
The air-to-cloth ratio is $(2.5 \text{ ft}^3/\text{min})/\text{ft}^2$. Dimensional simplification converts this to 2.5 ft/min (0.0127 m/s).

4. *Calculate the total filtering area required.*
This equals the actual volumetric flow rate (from step 2) divided by the filtering velocity (from step 3). Thus,

$$\text{Total filtering area} = 13,654/2.5 = 5461.6 \text{ ft}^2 (507.9 \text{ m}^2)$$

5. *Calculate the filtering area available per bag.*
Consider the operating bag to be in the form of a cylinder, whose wall constitutes the filtering area. This is accordingly calculated from the formula $A = \pi Dh$, where A is area, D is bag diameter, and h is bag height. The two bags still under consideration are Bag A and Bag B. The area of each is as follows:

For Bag A, $A = \pi(8/12)(16) = 33.5 \text{ ft}^2 (3.12 \text{ m}^2)$

For Bag B, $A = \pi(10/12)(16) = 41.9 \text{ ft}^2 (3.90 \text{ m}^2)$

6. *Determine the number of bags required.*
The total filtering area required is 5461.6 ft². Accordingly, if Bag A is selected, the number of bags needed is 5461.6/33.5, i.e., 163 bags. If instead Bag B is selected, the number needed is 5461.6/41.9, i.e., 130 bags.

7. *Determine the total bag cost.*
If Bag A is used, the total cost is (163)($26.00), i.e., $4238. If instead Bag B is used, the total cost is (130)($38.00), i.e., $4940.

8. *Select the most appropriate bag.*
Since the total cost for Bag A is less than that for Bag B, select Bag A.

15.6   *SIZING A CONDENSER FOR ODOR-CARRYING STEAM*

The discharge gases from a meat-rendering plant consist mainly of atmospheric steam, plus a small fraction of noncondensable odor-carrying gases. The stream is to pass through a condenser to remove the steam before the noncondensable gases go to an incinerator or adsorber. Estimate the size of a condenser to treat 60,000 lb/h (7.55 kg/s) of this discharge stream. Assume that the overall heat-transfer coefficient is 135 Btu/(h) (ft²)(°F) [765 W/(m²)(K)], that the enthalpy of vaporization for the steam is 1000 Btu/lb, and that the cooling water enters the condenser at 80°F (300 K) and leaves at 115°F (319 K).

Calculation Procedure

1. *Determine the heat load Q for the condenser.*
This equals the flow rate times the enthalpy of vaporization:

$$Q = (60,000)(1000) = 6.0 \times 10^7 \text{ Btu/h } (17.6 \times 10^6 \text{ W})$$

2. *Estimate the log-mean temperature-difference driving force.*
The formula is LMTD = $(t_g - t_l)/\ln(t_g - t_l)$, where LMTD is log-mean temperature difference, t_g is the maximum temperature difference between steam and cooling water, and t_l, is the minimum temperature difference between them. The maximum difference is $(212 - 80)$, or 132; the minimum is $(212 - 115)$, or 97. Accordingly, LMTD = $(132 - 97)/\ln(132/97) = 113.6°F$.

3. *Calculate the required area of the condenser.*
The formula is $A = Q/U$ (LMTD), where A is area, Q is heat load, and U is overall heat-transfer coefficient. Thus, $A = (6.0 \times 10^7)/(135)(113.6)$, i.e., 3912 ft² (363.8 m²).

Related Calculations. A comprehensive design procedure for condensers, including several examples of its application, was originally developed by the author and can be found in several of the author's *Theodore Tutorials*. Refer also to Section 7 of this handbook, which deals with heat transfer.
    Note: This material is original to the author. Some of it has been published elsewhere without authorization.

15.7 AMOUNT OF ADSORBENT FOR A VOC ADSORBER

Determine the required height of adsorbent in an adsorption column that treats a degreaser-ventilation stream contaminated with trichloroethylene (TCE). Design and operating data are as follows:

Volumetric flow rate of contaminated air: 10,000 std ft³/min
    (4.72 m³/s), standard conditions being 60°F and 1 atm

Operating temperature: 70°F (294 K)

Operating pressure: 20 psia (138 kPa)

Adsorbent: activated carbon

Bulk density ρ_B of activated carbon: 36 lb/ft³ (576 kg/m³)

Working capacity of activated carbon: 28 lb TCE per 100 lb carbon

Inlet concentration of TCE: 2000 ppm (by volume)

Molecular weight of TCE: 131.5

The adsorption column is a vertical cylinder with an inside diameter of 6 ft (1.8 m) and a height of 15 ft (4.57 m). It operates on the following cycle: 4 h in the adsorption mode, 2 h for heating and desorbing, 1 h for cooling, 1 h for standby. An identical column treats the contaminated gas while the first one is not in the adsorption mode. The system is required to recover 99.5 percent of the TCE by weight.

Calculation Procedure

1. *Determine the actual volumetric flow rate of the contaminated gas stream.*
The flow rate as stated corresponds to flow at 60°F and 1 atm, whereas the actual flow q is at 70°F and 20 psia. Accordingly,

$$q = 10,000[(70 + 460)/(60 + 460)][14.7/20]$$
$$= 7491 \text{ actual ft}^3/\text{min, or } 4.5 \times 10^5 \text{ actual ft}^3/\text{h (3.54 actual m}^3/\text{s)}$$

2. Calculate the volumetric flow rate of TCE.
This flow rate q_{TCE} equals the inlet concentration y_{TCE} of TCE in the gas times the gas flow rate. Thus,

$$q_{TCE} = (y_{TCE})(q) = (2000 \times 10^{-6})(4.5 \times 10^5)$$
$$= 900 \text{ actual ft}^3/\text{h } (0.007 \text{ m}^3/\text{s})$$

3. Convert the volumetric flow rate of TCE into mass flow rate.
For the conversion, rearrange the ideal-gas law, bearing in mnd that mass equals the number of moles times the molecular weight. Thus $\dot{m}$, the mass flow rate in pounds per hour, equals $q_{TCE}(PM/RT)$, where P is the pressure, M the molecular weight, T the absolute temperature, and R the gas constant. Accordingly,

$$\dot{m} = (900)(131.5)/(10.73)(70 + 460) = 416.2 \text{ lb/h } (0.052 \text{ kg/s})$$

4. Determine the mass of TCE to be adsorbed during the 4-h period.
This equals, simply, the required degree of adsorption times the amount of TCE that will pass through the system in 4 h:

$$\text{TCE adsorbed} = (416.2 \text{ lb/h})(4 \text{ h})(0.995) = 1656.6 \text{ lb}$$

5. Calculate the volume of activated carbon required.
To obtain this volume v_{AC}, divide the amount of TCE to be adsorbed by the adsorption capacity of the carbon, and convert from mass to volume by taking into account the bulk density of the carbon:

$$v_{AC} = (\text{TCE to be adsorbed})/(28 \text{ lb TCE}/100 \text{ lb carbon})(\text{bulk density})$$
$$= (1656.6)/(28/100)(36) = 164 \text{ ft}^3 (4.64 \text{ m}^3)$$

6. Find the required height of the carbon in the adsorber.
This height z equals the volume of carbon divided by the cross-sectional area of the column:

$$z = 164/[\pi(D^2/4)] = 164/[\pi(6^2/4)] = 5.8 \text{ ft } (1.77 \text{ m})$$

Note: This material is original to the author. Some of it has been published elsewhere without authorization.

15.8 PERFORMANCE OF AN AFTERBURNER

It is proposed to use a natural-gas-fired, direct-flame afterburner to incinerate toluene in the effluent gases from a lithography plant. The afterburner system is as shown in Fig. 15.2. The flow rate of the 300°F (422 K) effluent is 7000 std ft^3/min (3.30 m^3/s), standard conditions being 60°F and 1 atm; its toluene content is 30 lb/h (0.0038 kg/s). After passing through the afterburner preheater, the gas enters the afterburner at 738°F (665 K). The afterburner is essentially a horizontal cylinder, 4.2 ft (1.28 m) in diameter and 14 ft (4.27 m) long; it can be assumed to incur heat losses at 10 percent in excess of the calculated heat load. Gases leaving the afterburner are at 1400°F (1033 K).

When reviewing plans for such an installation, the local air-pollution–control agency knows from experience that in order to meet emission standards, the afterburner must operate at 1300 to 1500°F, that the residence time in the vessel must be 0.3 to 0.5 s, and that the velocity within it must be 20 to 40 ft/s. Can this afterburner satisfy those three criteria?

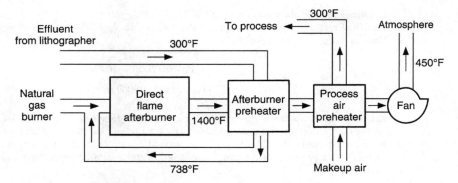

FIGURE 15.2   Natural-gas-fired afterburner (Procedure 15.8).

Use the following data:

Gross heating value of natural gas: 1059 Btu/std ft^3

Volume of combustion products produced per standard cubic foot of natural gas burned: 11.5 std ft^3 (11.5 m^3 per standard cubic meter)

Average available heat from natural gas between 738 and 1400°F: 600 Btu/std ft^3 (22,400 kJ/m^3)

Molecular weight of toluene: 92

Average heat capacity C_{p1} of effluent gases between 0 and 738°F: 7.12 Btu/(lb · mol)(°F)[29.8 kJ/(kg · mol)(K)]

Average heat capacity C_{p2} of effluent gases between 0 and 1400°F: 7.38 Btu/(lb · mol)(°F)[30.9 kJ/(kg · mol)(K)]

Volume of air required to combust natural gas: 10.33 std ft^3 air/std ft^3 natural gas (10.33 m^3/m^3 natural gas)

Calculation Procedure

1. Convert the gas flow rate to the molar basis.
Since 1 lb-mol of gas at 32°F and 1 atm occupies 359 ft^3, at 60°F it will occupy (359)[(460 + 60)/(460 + 32)], i.e., 379 ft^3. Accordingly, the molar gas flow rate $\dot{n}$ is (7000 actual ft^3/min)/(379 actual ft^3/lb · mol), i.e., 18.47 lb · mol/min (0.139 kg · mol/s).

2. Calculate the total heat load (heating rate) required to raise the gas stream from 738 to 1400°F.
Since the heat capacity data are given with 0°F as a basis, this heat load Q must be based on first cooling the gases from 738 to 0°F, then heating them from 0 to 1400°F Thus,

$$Q = \dot{n}[C_{p2}(1400 - 0) - C_{p1}(738 - 0)]$$
$$= 18.47[(7.38)(1400) - (7.12)(738)] = 93{,}790 \text{ Btu/min (1648 kW)}$$

3. Determine the actual required heat load, taking into account the 10 percent heat loss.
Thus, actual heat load = 1.1 Q = (1.1)(93,790) = 103,169 Btu/min (1813 kW).

4. Find the rate of natural gas needed to satisfy this heat load.
This is, simply, the heat load divided by the available heat. Thus,

$$\text{Rate of natural gas} = 103{,}169/600 = 171.9 \text{ std ft}^3/\text{min (0.081 m}^3/\text{s)}$$

5. Determine the volumetric flow rate of the combustion products of the natural gas.
This flow rate q_1 equals the natural gas rate times the volume of combustion products produced per standard cubic foot of natural gas. Thus,

$$q_1 = (171.9)(11.5) = 1976 \text{ std ft}^3/\text{min } (0.932 \text{ m}^3/\text{s})$$

Note that the 11.5 figure already takes into account the air needed to burn the natural gas.

6. Calculate the total volumetric flow rate through the afterburner.
This flow rate q_T equals that of the effluent from the lithography plant plus that of the combustion products from step 5. Thus,

$$q_T = 7000 + 1976 = 8976 \text{ std ft}^3/\text{min } (4.233 \text{ m}^3/\text{s})$$

For the subsequent steps, this figure must be converted to actual cubic feet per minute. Since the afterburner operates at 1400°F, this equals $(8976)[(1400 + 460)/(60 + 460)]$, or 32,106 actual ft^3/min (15.14 m^3/s).

7. Determine the cross-sectional area of the afterburner.
As the cross section is circular, this area S equals $\pi D^2/4$, where D is the diameter of the afterburner. Thus, $S = \pi (4.2)^2/4 = 13.85$ ft^2 (4.22 m^2).

8. Calculate the velocity through the afterburner.
This residence time t equals the length of the vessel divided by the velocity. Thus, $t = 14/38.6 = 0.363$ s.

9. Find the residence time within the afterburner.
This residence time t equals the length of the vessel divided by the velocity. Thus, $t = 14/38.6 = 0.363$ s.

10. Ascertain whether the afterburner meets the agency's criteria.
The afterburner operates at 1400°F (1033 K), so it meets the temperature criterion. The residence time of 0.363 s meets the residence-time criterion. And the gases pass through the vessel at 38.6 ft/s (11.8 m/s), which satisfies the velocity criterion. Thus, all three criteria are satisfied.

Related Calculations.   The determination of the natural gas rate is discussed in more detail in the Wiley-Interscience text *Introduction to Hazardous Waste Incineration* by Theodore and Reynolds.
    Note: This material is original to the author. Some of it has been published elsewhere without authorization.

15.9   *PRELIMINARY SIZING OF AN ABSORBER FOR GAS CLEANUP*

Describe how one can make a rough estimate of the required diameter and height for a randomly packed absorption tower to achieve a given degree of gas cleanup without detailed information on the properties of the dirty gas, knowing only that the absorbent is water (or has properties similar to those of water) and that the pollutant has a strong affinity with the absorbent.

Calculation Procedure

1. Estimate a diameter for the tower.
Use the rule of thumb that superficial gas velocity through the tower (i.e., velocity assuming that the tower is empty) should be about 3 to 6 ft/s (1 to 2 m/s). If we assume, for instance, a velocity of

4 ft/s, then the tower cross section S equals the actual volumetric flow rate of the dirty gas divided by 4. Then the diameter D can be found from the formula $D = (4S/\pi)^{0.5}$.

To illustrate, assume that the volumetric flow rate is 60 actual ft^3/s (1.7 m^3/s). Then a suitable cross section would be about 60/4, i.e., 15 ft^2 (1.4 m^2), and a suitable diameter about $[(4)(15)/\pi]^{0.5}$, i.e., about 4.4 ft (1.3 m). Given the imprecision of the superficial-gas–velocity guideline, it is appropriate to round this figure off to 4 ft.

2. Choose a packing size for the tower.
If D is about 3 ft (1 m), use a packing whose diameter is 1 in. If D is under 3 ft, use smaller packing; if D is greater than 3 ft, use larger.

Continuing the illustration from step 1, since the tower is to be about 4 ft, use packing larger than 1 in, for instance, 1.5 in.

3. Estimate a height for the tower.
The height of an absorption tower equals the product of H_{OG}, the height of a gas transfer unit, and N_{OG}, the number of transfer units needed. It is also prudent to multiply this product by a safety factor of 1.25 to 1.50.

Since equilibrium data are not available, assume that the slope of the equilibrium curve (see Section 11 for discussion of this curve) approaches zero. This is not an unreasonable assumption for most solvents that preferentially absorb (or react with) the pollutant. For that condition, the value of N_{OG} approaches $\ln(y_1/y_2)$, where y_1 and y_2 represent the inlet and outlet concentrations, respectively. Accordingly, if for instance the required degree of gas cleanup is 99 percent, then $N_{OG} = \ln[1/(1 - 0.99)] = 4.61$.

As for H_{OG}, since the solvent is either water or similar to it, use the values that are normally encountered for aqueous systems:

Packing diameter, in	H_{OG} for plastic packing, ft	H_{OG} for ceramic packing, ft
1.0	1.0	2.0
1.5	1.25	2.5
2.0	1.5	3.0
3.0	2.25	4.5
3.5	2.75	5.5

Continue the illustration from step 2. Assume that the required cleanup is in fact 90 percent, that 1.5-in ceramic packing is to be used, and that a safety factor of 1.4 is to be used in the height calculation. Then the estimated height equals (safety factor)$(N_{OG})(H_{OG})$, i.e., $(1.4)(4.61)(2.5)$, or 16 ft (4.9 m).

Related Calculations. Apart from the rule of thumb concerning superficial velocity (see step 1), be aware of similar guidelines that pertain to mass flow rate through the absorption tower. For plastic packing, the liquid and gas flow rates are both typically around 1500 to 2000 lb/h per square foot of tower cross-sectional area; for ceramic packing, the corresponding range is about 500 to 1000 lb/h.

As a rough estimate of pressure drop for the gas flow through the packing, it is about 0.15 to 0.4 in of water per foot of packing.

For more detail on absorption, see Section 11.

Note: This material is original to the author. Some of it has been published elsewhere without authorization.

15.10 ESTIMATING HAZARD DISTANCES FROM ACCIDENTAL RELEASES

Hydrogen sulfide is accidentally released at the rate of 2 kg/s from a 15-m vent at night, from a plant located in an urban area. The wind speed is 3 m/s. Estimate the hazard distance for this release if the hazard level of concern for hydrogen sulfide is 0.042 mg/L (42 mg/m^3).

Calculation Procedure

1. Determine the hazard distance, HD, without taking vent height or wind speed into account.
Use the formula: $HD = 16,800(Q_1/C_1)^{0.73}$, where Q_1 is the release rate in kilograms per second, C_1 is the ground-level concentration of concern in milligrams per cubic meter, and HD is in meters. Thus, $HD = 16,800(2/42)^{0.73} = 1820$ m.

2. Determine the correction factor to take into account the elevation of the vent.
The height correction factor is $1.12 - 0.014H$, where H is the height in meters, assuming that H lies between 10 and 20 m. In this case, the correction factor is $1.12 - 0.014(15) = 0.91$.

3. Determine the correction factor for wind speed.
The result in step 1 assumes a (nighttime) wind speed of 1.5 m/s. For correcting other wind speeds, the correction factor equals $1.32U^{-0.68}$, where U is the wind speed. In this case, the correction factor is $1.32(3)^{-0.68} = 0.62$.

4. Determine the hazard distance corrected for height and wind speed.
The result is, from steps 1, 2, and 3, as follows:

$$1,820(0.91)(0.62) = 1027 \text{ m}$$

Related Calculations.   For daytime releases from elevated vents, the equation for step 1 is $HD = 2431(Q_1/C_1)^{0.65}$ and the height correction factor (for heights of 10 to 20 m) is $1.09 - 0.0096H$; and for wind speeds other than 5 m/s (not other than 1.5 m/s), the correction factor is $2.85(U) - 0.65$. For more information, see *Chemical Engineering,* August 1998, pp. 121ff. For information on calculating ground-level concentrations of unburned gas released from flares, see Procedure 15.11.

15.11 *GROUND-LEVEL CONCENTRATION OF UNBURNED FLAMMABLE GAS*

Estimate the maximum ground-level concentration, C, if a flammable gas is accidentally released unburned from a flare, if the release rate to the atmosphere, Q, is 200,000 lb/h (25,200 g/s), the exit velocity, V_{ex}, is 275 ft/s (83.8 m/s), and flare tip diameter, d, is 1.5 ft (0.46 m). The flare stack height, H, is 200 ft (61 m). Assume that the wind speed, U, is 10 ft/s (3.1 m/s). The molecular weight, MW, of the gas is 54.

Calculation Procedure

1. Calculate the momentum plume rise (the plume rise due to gas exit velocity), ΔH.
Use the formula, $\Delta H = 3d\, V_{ex}/U$. Thus,

$$\Delta H = 3(0.46)(83.8)/3.1 = 37.3 \text{ m}$$

2. Calculate the effective flare stack height, H'.
Use the formula

$$H' = H + \Delta H = 61 + 37.3 = 98.3 \text{ m}$$

3. Estimate the maximum ground-level concentration in grams per cubic meter.
Use the approximation formula $C = 0.23\, Q/U\, (H')^2$, where C is in grams per cubic meter, Q is in grams per second, U is in meters per second, and H' is in meters. Thus,

$$C = 0.23(25,200)/(3.1)(98.3)^2 = 19 \text{ g/m}^3, \text{ or } 190 \text{ mg/m}^3$$

4. Convert the result to parts per million.
Use the formula

$$\text{parts per million} = [\text{milligrams per cubic meter}][24.45/(\text{molecular weight})]$$

Thus,

$$\text{parts per million} = 190(24.45/54) = 86 \text{ ppm}$$

Related Calculations. For more information, see *Chemical Engineering,* December 1998, pp. 133ff. For a method to calculate ground-level hazard distances for accidental releases of hazardous pollutants, see Procedure 15.10.

*15.12   ESTIMATING SIZE AND COST OF VENTURI SCRUBBERS

Determine the size and cost of a venturi scrubber to handle 100,000 actual ft³/min (47.2 m³/s) of gas entering the venturi in an air-pollution control system. The scrubber, Fig. 15.3, is to remove all particles larger than 0.6 μm.

Calculation Procedure

1. Determine the scrubber base cost.
The cost of a venturi scrubber depends on the volumetric flow, operating pressure, and materials of construction. Figure 15.4 gives flange-to-flange costs—covering the venturi, elbow, separator, pumps, and controls—for base systems [i.e., constructed of 0.125-in (0.32-cm) carbon steel] for different volumetric flow rates. Do not extend the curve or equation beyond 200,000 actual ft³/min (94.4 m³/s).

Whether a thickness other than 0.125 in (0.32 cm) is required can be determined from Fig. 15.5 the thickness being a function of design operating pressure and shell diameter. The curves in Fig. 15.5 include a safety factor of 2, but no allowance is made for corrosion and erosion. Figure 15.5 is used after the base cost is determined from Fig. 15.4.

Entering Fig. 15.4 at 100,000 ft³/min (47.2 m³/s) at the bottom and projecting vertically upward to the curve, we find a base cost of $39,400. From the equation, the exact cost would be $39,417.

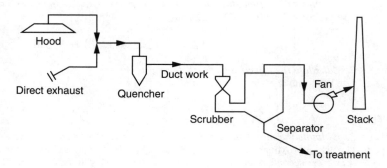

FIGURE 15.3   Scrubber-separator in an air-pollution control system. (*Chemical Engineering.*)

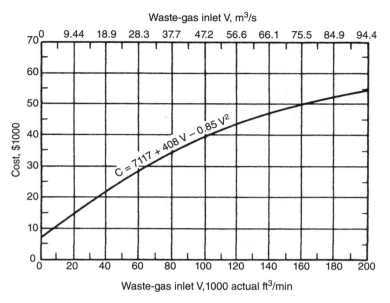

FIGURE 15.4   Base scrubber cost, flange-to-flange construction of 0.125-in (0.32-cm) carbon steel. (*Chemical Engineering magazine and Fuller Company.*)

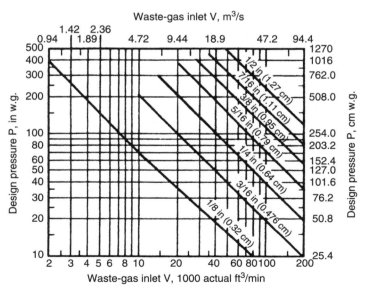

FIGURE 15.5   Flow rate and design pressure dictate scrubber metal thickness. (*Chemical Engineering magazine and Fuller Company.*)

2. Find the pressure drop in the scrubber.

For an efficiency that removes all particles larger than 0.6 μm, Fig. 15.6 shows that a 35-in water gage (w.g.) (88.9-cm w.g.) pressure drop is needed. Note that as the particle size removed decreases, the pressure drop required for particle removal increases.

3. Compute the metal thickness and scrubber cost.

At the 35-in w.g. (88.9-cm w.g.) pressure drop, Fig. 15.5 shows that a scrubber metal thickness of 0.25 in (0.64 cm) is required. (Always round up to the next standard metal thickness when you use this procedure.) For 0.25-in (0.64-cm) carbon steel, Fig. 15.7 gives a cost adjustment factor of 1.6. Hence, the scrubber cost will be $1.6 \times \$39,400 = \$63,040$.

If the scrubber metal is to be 304 or 316 stainless steel, multiply the above cost estimate, $63,040, by 2.3 or 3.2, respectively. And if the scrubber is to be equipped with a fiberglass lining to reduce wear of the metal, multiply the Fig. 15.4 estimate of $39,400 by 0.15 to obtain $5910, and add this to $63,040 to arrive at an estimate of $68,950.

4. Determine the cost of a rubber-lined scrubber.

If the scrubber is to be lined with 0.1857-in (0.476-cm) rubber, or any other thickness rubber, determine the internal surface area of the scrubber from Fig. 15.8. For 100,000 actual ft^3/min (47.2 m^3/s), the scrubber internal area is 1500 ft^2 (139.4 m^2). The unit cost for lining a scrubber with this thickness rubber is $4.69 per ft^2 ($50.48 per m^2). Hence, the total cost = (1500 ft^2)($4.69/$ft^2$) = $7035. Adding this to $63,040 gives an estimate of $70,075.

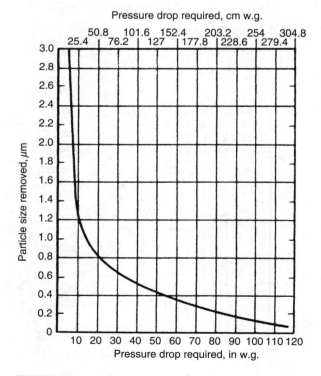

FIGURE 15.6 Correlation gives efficiency performance of venturi scrubbers. (*Chemical Engineering magazine and AIChE.*)

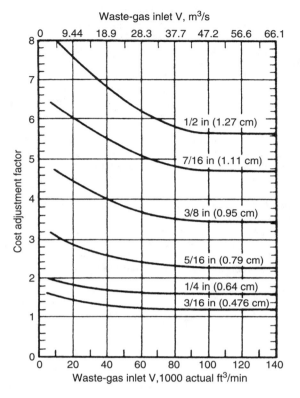

FIGURE 15.7 Factors adjust scrubber cost for flow rate and metal thickness. (*Chemical Engineering magazine and Fuller Company.*)

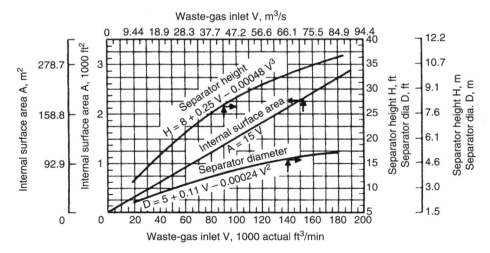

FIGURE 15.8 Scrubber internal surface area and separator dimensions. (*Chemical Engineering magazine and Fuller Company.*)

5. *Determine the number and diameter of the scrubber trays.*
If the separator is to be equipped for gas cooling, the number of trays needed must be determined, based on an average of 5 lb (2.3 kg) of water removed per square foot of tray area, with an outlet gas temperature about 40°F (22.2°C) higher than the inlet water temperature. [This is valid for typical scrubber outlet-gas temperatures of 200°F (93.3°C) or less, cooling-water temperature of about 70°F (21.1°C), and superficial gas velocities of 600 ft/min (3.05 m/s).] The total water to be removed is determined from the difference between the absolute humidities of the inlet and outlet gas streams.

Find the diameter of each tray from Fig. 15.8 for a flow of 100,000 ft³/min as 13.5 ft (4.1 m). Figure 15.9 gives a cost of $14,000 per tray. This includes the cost of the tray plus the cost of additional separator height to contain the tray.

If the separator requires six trays to achieve the dehumidification required, the $70,075 estimate will be increased by $84,000 (= 6 × $14,000) to $154,075, say $154,000. (If the chart in Fig. 15.4 were used instead of the equation, the total estimate would be $153,000 after rounding.) To update the estimate to current costs, use a suitable cost index, as detailed below.

Related Calculations.   Venturi scrubbers are highly efficient in removing submicron dust particles from gas streams.

Basically, the gas stream accelerates in the converging section of the venturi to maximum velocity in the throat, where it is sprayed by a scrubbing liquor. The faster velocity of the gas stream atomizes the liquor and promotes collisions between the particles and the droplets. Agglomeration in the diverging section produces droplets, with entrapped particles, of a size easily removed by mechanical means.

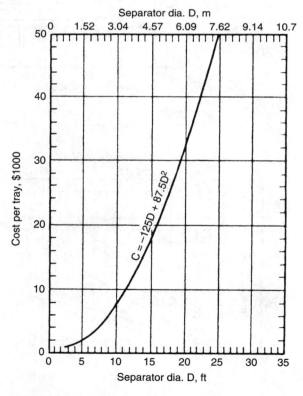

FIGURE 15.9   Cost per tray of separator internal cooler. (*Chemical Engineering magazine and Fuller Company.*)

Collection efficiency depends on the venturi pressure drop, which is a function of gas-stream throat velocity and scrubbing-liquor flow rate. The smaller the particles, the higher the pressure drop required. Venturis are normally operated at pressure drops between 6 in (15.2 cm) and 80 in (203.2 cm) water gage, depending on the characteristics of the dust, and at liquor flow rates of 3 to 20 gal/min (0.19 to 1.26 L/s) per 1000 actual ft³/min (0.47 m³/s). Collection efficiencies range from 99+ percent for 1-μm and larger particles to between 90 and 99 percent for those less than 1 μm.

Precise pressure drops can be obtained from Fig. 15.6, which represents 100 percent removal of particle sizes indicated for a particular pressure drop. For instance, to remove all particles 0.4 μm (10^{-6} m) and larger requires a pressure drop of 55 in (139.7 cm) w.g. (Also $P_d = 15.4d^{-139}$, with d the diameter in micrometers, also gives the pressure drop.)

A separator—normally a cylindrical tank having a low tangential inlet and a centered top gas outlet—located immediately downstream of the scrubber removes the agglomerated liquor drops from the gas stream by a cyclonic motion that forces them to impinge on the tank wall. Slurry settles into a bottom cone, from which most of it is sent to the water treatment facility, with the cleaner liquid above the sediment being recycled to the venturi (Fig. 15.3).

In hot processes, a considerable amount of water is vaporized in the scrubber and upstream equipment (particularly the quencher). Unless this vapor is removed, it must be handled by the fan (commonly a radial-tip fan), which therefore must be of higher horsepower and so is more costly to operate.

A gas cooler can be incorporated into the separator to cool and dehumidify the gas stream. Such a cooler can be one of several types, including one in which the gas stream passes through spray banks of cooling water and then impinges on baffles, and another in which the stream rises through perforated holes or bubble caps in trays flooded with cooling water.

*15.13 SIZING VERTICAL LIQUID-VAPOR SEPARATORS

Find the diameter needed for a vertical vessel to separate a liquid having a density $d = 58.0$ lb/ft³ (928.6 kg/m³) from 2000 mol/h of vapor having a molecular weight of 25.0 at an operating temperature of 300°F (148.7°C) and 250 lb/in² (gage) (1723.5 kPa). The compressibility factor $Z = 1.0$ for the vapor.

Calculation Procedure

1. Find the vapor volumetric flow rate V.
Using the gas law $PV = nRTZ$, let $V = $ ft³/s and $n = 2000/3600$ mol/s. Solving yields $V = nRTZ/P = (2000/3600)(10.73)(760)(1.0)/264.7 = 17.1$ ft³/s (48,393 m³/s). In this equation, $R = $ the gas constant; $T = $ absolute temperature, °R $= 460 + $ operating temperature, °F; $P = $ absolute pressure of the vapor, lb/in² (abs).

2. Determine the density of the vapor d_v.
Use the relation $d_v = $ (mol/h) (molecular weight)/volumetric flow rate, lb/h. Or, $d_v = (2000)(25)/(17.1)(3600$ s/h$) = 0.812$ lb/ft³ (13.0 kg/m³).

3. Compute the terminal vapor velocity v_t.
Use the relation $v_t = K'[(d - d_v)/d_v]^{0.5}$, where K' is a constant which ranges between 0.1 and 0.35, with 0.227 being the value for many satisfactory designs and recommended except when special considerations are warranted. Substituting, we find $v_t = 0.227[(58.0 - 0.812)/0.812]^{0.5} = 1.91$ ft/s (0.58 m/s).

4. Find the allowable vapor velocity v_a.
Use the relation $v_a = 0.15v_r$, where the constant 0.15 is based on an allowable vapor velocity of 15 percent of v_t to ensure good liquid disentrainment during the normal flow surges. For usual

designs, researchers have determined that v_a should be 15 percent of v_r. By substituting, $v_a = 0.15(1.91) = 0.286$ ft/s (0.087 m/s).

5. Determine the separator cross-sectional area and diameter.

The separator cross-sectional area $A = V/v_a = 17.1/0.286 = 59.8$ ft^2 (5.6 m^2). Then the separator diameter $D = [(4)(59.8)/\pi]^{0.5} = 8.7$ ft (2.65 m). A diameter of 9 ft (2.74 m) would be chosen.

Related Calculations.   Vertical liquid-vapor separators are used primarily to disengage a liquid from a vapor when the volume of the first is small compared with that of the second. The separation is accomplished by providing an environment (i.e., a vessel) in which the liquid particles are directed by the force of gravity rather than the force of the flowing vapor.

Devices have been developed to agglomerate liquid particles in a vapor stream and enhance disentrainment. Some act as baffles, causing multiple changes in the direction of vapor flow. Inertia keeps the liquid particles from changing direction, and they impinge on the baffles. As the particles coalesce on the baffles, they agglomerate into droplets, which fall because of gravity.

Other devices, such as packing and grids, provide a large surface area for liquid coalescence and agglomeration. One such device that has gained wide acceptance—because it is highly efficient and relatively inexpensive and causes negligible pressure drop—is the mist elimination pad. Usually a mesh formed by knitting metal wire, it comes in a variety of standard thicknesses and densities. For general process-separator and compressor-suction knockout-pot services, a stainless-steel pad of 4-in (10.2-cm) thickness and nominal 9-lb/ft^3 (144.1-kg/m^3) density is the most economical.

A separator equipped with a mist eliminator can be considerably smaller in diameter than one not having it. Indeed, design practice permits ignoring the 15 percent safety factor and letting the allowable vapor velocity be equal to the terminal vapor velocity (that is, $v_a = v_t$).

In the separator above, therefore, the required cross-sectional area of the vertical separator A now becomes $A = 18.1/1.91 = 9.0$ ft^2 (2.74 m^2). And the separator diameter D becomes $D = [(4 \times 9.0)/\pi]^{0.5} = 3.4$ ft (1.04 m). A diameter of 3 ft 6 in (1.07 m) would now be chosen.

The height of the liquid level in a vertical separator, Fig. 15.10, depends primarily on the residence time dictated by process considerations. Suppose that the residence time for the above separator is chosen as 5 min. For the 3.5-ft (1.07-m) diameter separator, the cross-sectional area is $A = (\pi/4)(3.5)^2 = 9.62$ ft^2 (0.89 m^2).

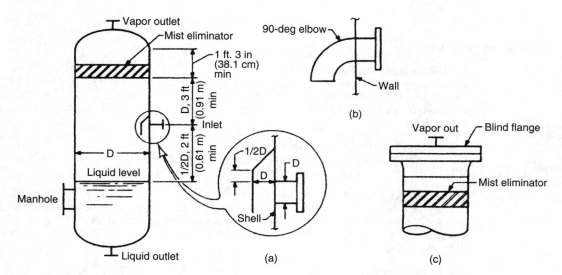

FIGURE 15.10   (a) Inlet configuration, diameter $\geq$ 30 in (9.1 cm). (b) Inlet configuration, diameter < 30 in (9.1 cm). (c) Top head configuration of pipe separators, diameter < 30 in (9.1 cm). (*Chemical Engineering.*)

Assuming that liquid is entering the separator at a rate of 2000 gal/h (2.1 L/s), or 4.64 ft^3/min (0.00219 m^3/s), we find the liquid level for a 5-min residence time is L = (ft^3/min) (residence time, min)/A = 4.46(5)/9.62 = 2.32 ft (0.71 m). Choose a sump height of 2.5 ft (0.76 m). A vertical separator is usually specified when a short liquid holdup time is permitted.

The following procedure is standard in the process design of vertical liquid-vapor separators. A standardized design procedure and vessel configuration saves much engineering time. A separator is usually relatively inexpensive, and the application of a rigorous, sophisticated procedure to achieve an optimum design is seldom warranted. Only in special cases, such as when a separator is built of extra thick laminated shells, does it become economical to attain an optimum design, because the saving in fabrication cost can be significant.

The standard procedure stipulates the following:

1. The allowable vapor velocity v_a in a separator shall be equal to the terminal velocity v_t, calculated by rounding up the vessel diameter to the nearest 6 in (15.2 cm), when a mist eliminator is used. However, v_a shall be no greater than 15 percent of v_t when the separator is not equipped with a mist eliminator.

2. The disengaging space, the distance between any inlet and the bottom of the mist elimination pad (see Fig. 15.10), shall be equal to the diameter of the separator. However, when the diameter of the separator is less than 3 ft 0 in (0.91 m), the height of the disengaging space shall be a minimum of 3 ft (0.91 m).

3. The distance between the inlet nozzle and the maximum liquid level shall be equal to one-half the vessel diameter, or a minimum of 2 ft (0.61 m).

4. The dimension between the top tangent line of the separator and the bottom of the mist elimination pad shall be a minimum of 1 ft 3 in (38.1 cm) (Fig. 15.10).

5. Vessel diameters 3 ft 0 in (0.91 m) and larger shall be specified in increments of 6 in (15.2 cm). Diameters of shell plate vessels shall be specified as inside diameters. Vessel lengths shall be specified in 3-in (7.6-cm) increments.

6. Separators of 30-in (76.2-cm) diameter and smaller shall be specified as fabricated from pipe. Diameter dimensions shall represent pipe outside diameters. Top heads shall be specified as full-diameter flanges, with blind flange covers (Fig. 15.10). Bottom heads shall be standard heads or pipe caps.

7. Inlets shall have an internal arrangement to divert flow downward. Vessels 3 ft 0 in (0.91 m) and larger shall have a hood, attached to the shell, covering the inlet nozzle (Fig. 15.10).

8. Outlets shall have antivortex baffles.

9. Mist elimination pads shall be specified as 4-in (10.2-cm) thick, nominal 9-lb/ft^3 (144.1-kg/m^3) density, and stainless steel. Spiral-wound pads are not acceptable.

The method given here is valid for vertical separators used in process, chemical, petrochemical, power, marine, and a variety of other plants. This procedure is the work of Arthur Gerunda, Vice President of Commercial Development, The Heyward-Robinson Co., as reported in *Chemical Engineering* magazine.

*15.14 SIZING A HORIZONTAL LIQUID-VAPOR SEPARATOR

Design a horizontal vessel to separate 7000 gal/h (7.36 L/s) of liquid having a density of 60 lb/ft^3 (960.6 kg/m^3) from 1000 mol/h of vapor having a molecular weight of 28 if the holding time for the liquid is to be 8 min when the operating temperature is 100°F (37.8°C), the operating pressure is 300 lb/in^2 (gage) (2068.2 kPa), and Z = 1.0.

Calculation Procedure

1. Find the volumetric flow rate V.
Using the gas law as in the previous calculation procedure with $n = 1000/3600$ mol/s and $A = 1.0$, we get $V = (1000/3600)(10.73)(560)(1.0) = 5.3$ ft^3/s (15,004 m^3/s).

2. Compute the density of the vapor d_v.
As in the previous calculation procedure, $d_v = $ (mol/h) (molecular weight)/(V) (3600) $= 1.47$ lb/ft^3 (23.5 kg/m^3).

3. Determine the terminal vapor velocity v_t.
Use the relation $v_t = 0.227[(d_l - d_v)/d_v]^{0.5}$, where $d_l = $ density of liquid. Or $v_t = 0.227[(60 - 1.47)/1.47]^{0.5} = 1.43$ ft/s (0.44 m/s). The constant 0.227 is obtained in the same way as described in the previous calculation procedure.

4. Decide what sets the size of the separator.
Either the rate of liquid separation from the vapor or the liquid holding time will set the size of the separator. By liquid separation, $D = [V/3(\pi/4)(0.15)(V_t)]^{0.5}$ where $D = $ separator diameter, ft. Or, $D = [5.3/(4\pi/4)(0.15 \times 1.43)]^{0.5} = 2.8$ ft (0.85 m).

By holding time, $D = [t_h V_i/3(\pi/4) f]^{0.333}$, where $t_h = $ holding time, min; $V_i = $ volumetric flow rate of the liquid, ft^3/min; $f = $ fraction of the separator area occupied by the liquid. Assuming an L/D ratio, i.e., separator length/diameter, of 4, and with 7000 gal/h $= 15.6$ ft^3/min (0.01 m^3/s), make a first approximation with an assumed liquid-space area f of 0.70: $D = [8(15.6)/4(\pi/4)(0.70)]^{0.333} = 3.84$ ft (1.17 m).

Since the larger diameter is set by the holding time, this is the determining factor in the sizing of the separator.

Next, examine a 4-ft (1.22-m) diameter vessel with $f = 0.70$. With $L/D = 4$, $L = 4D = 4(4) = 16$ ft (4.9 m).

With the area of the vapor space $= 30$ percent (i.e., $1.00 - 0.70$), the fractional height of the vapor space $f_{hv} = 0.342$, from the geometry of the tank. The height of the vapor space then is 0.342 (46 in) $= 15.7$ in (39.9 cm), where the 46 in (116.8 cm) is the approximate actual internal diameter of the vessel. Make the vapor-space height 18 in (45.7 cm). This gives a vapor-space area fraction f_{av} of 0.36 and a liquid-space area fraction f_{al} of 0.64.

The holding time t_h now is $t_h = [(\pi/4)(4^2)(0.64)(16)]/15.6 = 8.25$ min. Hence, the final separator size is as follows: diameter $= 4$ ft (1.22 m), length $= 16$ ft (4.88 m), and liquid height $= 2.5$ ft (0.76 m). As with a vertical separator, when a horizontal separator is equipped with a mist elimination pad, the allowable vapor velocity v_a can be taken to be the same as the terminal velocity v_t in the vessel diameter calculations. Figure 15.11 shows some typical arrangements of mist eliminators in horizontal liquid-vapor separators.

Related Calculations.   The chief concern in designing a horizontal liquid separator is to have the vapor velocity sufficiently low to give the liquid particles just enough time to settle out before the vapor leaves the vessel. Figure 15.12 shows the approximate traverse of a liquid particle for which the minimum time has been allowed for its disentrainment from the vapor. Indicated in the cross section are the fraction of area f_{av} and height f_{hv} taken up by the vapor space.

As with a vertical separator, empirical findings have shown that for safe design the allowable vapor velocity v_a in a horizontal separator should be no greater than 15 percent of the calculated terminal velocity v_t. Another restriction found necessary is that f_{av} be no less than 15 percent of the cross-sectional area.

For horizontal separators L/D ratios are dictated by economics and plot restrictions. As a general guide, the following provides economic designs:

Operating pressure		L/D ratio
lb/in^2 (gage)	kPa	
0–250	0–1723.5	3.0
251–500	1730.4–3447.0	4.0
501 and higher	3423.9 and higher	5.0

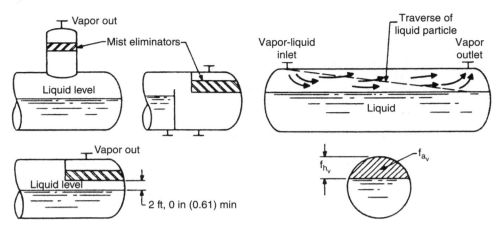

FIGURE 15.11 Mist eliminators in horizontal separators. (*Chemical Engineering.*)

FIGURE 15.12 Traverse of liquid particle. (*Chemical Engineering.*)

For a first trial size, set the liquid level at the centerline of the separator, so that $f_h = f_a = 0.5$.
Now, $D = [V/(3)(\pi/4)(0.15 \times v_t)]^{0.5}$, or $D = [V/(0.35 \times v_t)]^{0.5}$.

These equations provide a good starting point for trial-and-error calculations to determine the size of a liquid-vapor horizontal separator operating at less than 251 lb/in² (gage) (1730.4 kPa). Note that the terminal vapor velocity v_t has, in effect, been replaced by the allowable vapor velocity v_a, because v_t is multiplied by the safety design factor, 0.15.

The following specifications are generally standard in the design of horizontal separators: (1) The maximum liquid level shall provide a minimum vapor space height of 15 in (38.1 cm) but not be below the centerline of the separator. (2) The volume of dished heads is not taken into account in vessel sizing calculations. (3) Inlet and outlet nozzles shall be located as closely as practical to the vessel tangent lines. (4) Liquid outlets shall have antivortex baffles.

When the size of a horizontal separator is set by the holdup time for the liquid, the diameter of the vessel must be determined by trial-and-error calculations. If f_{al} = the fraction of area occupied by the liquid, the holdup time t_h is given by $t_h = [(\pi/4)D^2 f_{al} L] V_r$. Here, L = vessel length and V_t = volumetric flow rate of the liquid. When the operating pressure is below 251 lb/in² (gage) (1730.4 kPa), $L/D = 3.0$. Solving for D yields $D = [t_h V_t/3(\pi/4)f_{al}]^{0.333}$.

This procedure is the work of Arthur Gerunda, Vice President, Commercial Development, Heyward-Robinson Co., as reported in *Chemical Engineering* magazine.

*15.15 EFFECTIVE STACK HEIGHT FOR DISPOSING CHEMICAL-PLANT GASES AND VAPORS

Acetylene (molecular weight 26) is being emitted from a process at 100 lb/min (0.76 kg/s). What stack height is needed to achieve an allowable downwind concentration of 40 percent of the lower explosive limit of 2.5?

Calculation Procedure

1. *Convert the flow rate to moles.*
To convert from lb/min to mol/h, use the expression $M = 60L/m$, where M = flow rate, mol/h; L = flow rate, lb/min; m = molecular weight of the flowing gas or vapor. Substituting gives $M = 60 (100)/26 = 230$ mol/h.

2. Determine the allowable concentration of the vapor.

Use the relation $A = E_L L_e$, where A = allowable concentration downwind of the stack, in percent. Or, $A = 2.5(0.40) = 1.00$ vol%.

3. Find the required stack height.

Use Fig. 15.13 to determine the stack height, entering at the top of the chart at the lower explosive limit of 1 vol% and projecting vertically downward to the contaminant flow rate of 230 mol/h. The intersection is just below the stack height of 25 ft (7.6 m). Use a height of 25 ft (7.6 m) because this height is accepted in industry as the minimum allowable. This is the effective height for a 1-mi/h (1.6-km/h) wind.

4. Find the distance downwind from the stack.

The distance downwind from the stack where the maximum concentration will be found varies with the turbulence conditions in the area. Turbulence parameters as given by Bosanquet-Pearson are as follows:

	p	q	p/q
Low turbulence	0.02	0.04	0.50
Average turbulence	0.50	0.08	0.63
Moderate turbulence	0.10	0.16	0.63

Note: p and q are the vertical and horizontal dimensionless diffusion coefficients, respectively.

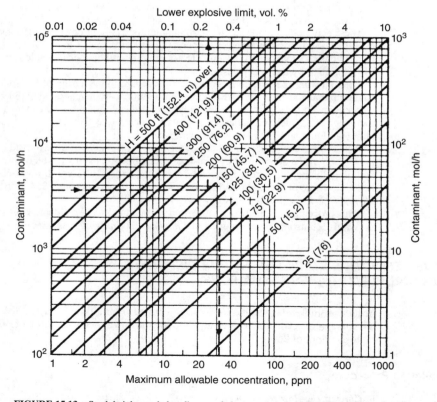

FIGURE 15.13   Stack height needed to dispose of plant gases and vapors. (*Chemical Engineering.*)

The distance downwind from the stack for maximum concentration of the effluent is given by $d = H/2p$, where d = distance, ft (m); H = stack height, ft (m); p = vertical diffusion coefficient. For low turbulence, $d = 25/2(0.02) = 625$ ft (190.6 m). For moderate turbulence, $d = 25/2(0.1) = 125$ ft (38.1 m).

Related Calculations. When designing emission control systems, be sure to consult the local air-pollution control ordinance (if any) for the criteria which must be met. If an odorous pollutant is being emitted, the design basis will be the concentration below the odor threshold (Table 15.1 at ground level outside the plant.

Do not design for emission directly to the atmosphere in areas where atmospheric temperature inversions are known to occur. Use, instead, a closed system to rid the plant of the gas or vapor. Extreme care in design is required to avoid the possibility of a legal nuisance suit. Be sure that all applicable ordinances are reviewed before any final design work is begun.

Figure 15.13 is based on the well-known Bosanquet-Pearson formula. Two solutions plotted on Fig. 15.13 are for a contaminant flow rate of 3800 mol/h and a lower explosive limit of 0.25 vol%, requiring a stack height of 150 ft (45.7 m); and a contaminant flow rate of 20 mol/h and a maximum allowable concentration of 31 ppm, requiring a stack height of 75 ft (22.9 m).

As a general guide for stack design, the following relations are given by the above formula:

1. Concentration of an effluent downwind from a source is directly proportional to the discharge quantity.

2. It is impossible to alter materially the downwind ground-level concentration of a contaminant by diluting the effluent.

3. Concentrations downwind of a stack are inversely proportional to wind speed; doubling the wind speed cuts pollutant concentration by half.

4. Pollutant concentration is inversely proportional to the square of the stack height. Doubling the stack height reduces the maximum ground-level concentration to one-fourth the previous level.

5. Location of the maximum ground-level concentration depends on atmospheric stability. When atmospheric conditions are unstable—i.e., wind speeds are low or there is an inversion—the maximum concentration occurs close to the stack. As the wind speed increases and the inversion disappears, the maximum ground-level concentrations move farther away from the stack.

6. Figure 15.13 is based on a wind speed of 1 mi/h (1.6 km/h) and $p/q = 1$. Further, a smooth, level terrain was ensured for the equation from which the chart is plotted.

7. The effective height to which the plume from a stack rises before it begins to turn downward is the actual height of the stack plus the plume rise created by the sum of the exit velocity and the difference in density above the plume. Use the relation $H_v = 4.77(Q_1 V/1.5)^{0.5}$, where H_v = plume rise due to exit velocity, ft; V = stack exit velocity, ft/s; $Q_1 = Q_v T_1/530$, in which Q_v = stack exit volume, ft^3/s; $T_1 = 18.3$ (molecular weight of contaminant).

This procedure can be used for a variety of gases and vapors, including acetylene, ammonia, amylene, benzene, butane, carbon monoxide, ethylene, hydrogen sulfide, methanol, propane, and sulfur dioxide. Pertinent data for these effluents are given in Table 15.1.

This calculation procedure is the work of John D. Constance, Consultant, as reported in *Chemical Engineering* magazine.

Where plant gases or vapors pollute the local environment, expensive pollution-abatement equipment may be required. On the west coast of the United States certain chemical and refining plants are gaining emissions credits for their stacks by eliminating pollution elsewhere.

TABLE 15.1   Air-Pollution Control Criteria

Substance	Maximum ground concentration, ppm[†]	Lower explosive limit, ppm	Odor threshold, ppm
Acetylene	—	2.5	—
Ammonia	100	15.5	53
Amylene	—	1.7	2.3
Benzene	50	1.4	1.5
Butane	—	1.9	5,000
Carbon monoxide	100	12.5	Odorless
Ethylene	—	2.8	—
Hydrogen sulfide	30	4.3	0.1
Methanol	200	6.7	410
Propane	—	2.1	20,000
Sulfur dioxide	10	—	3.0

[†]8-h exposure.

REFERENCES

1. Theodore and Feldman—*Theodore Tutorial: Air Pollution Control Equipment for Particulates,* Research-Cottrell.
2. Theodore, Reynolds, and Richman—*Theodore Tutorial: Air Pollution Control Equipment for Gaseous Pollutants,* Research-Cottrell.
3. Theodore and McGuinn—*U.S. EPA Instructional Problem Workbook: Air Pollution Control Equipment.*
4. *Air Pollution Control Equipment: Selection, Design, Operation and Maintenance,* ETS International (Roanoke, VA).
5. Theodore, Reynolds, and Taylor—*Accident and Emergency Management,* Wiley-Interscience.
6. Theodore and McGuinn—*Pollution Prevention,* Van Nostrand Reinhold.
7. Theodore, personal notes.
8. Kumar, A., "Estimating Hazard Distances from Accidental Releases," *Chemical Engineering,* August 1998, pp. 121–128.
9. Kumar, A., "Design and Operate Flares Safely," *Chemical Engineering,* December 1998, pp. 133–138.

SECTION 16
WATER POLLUTION CONTROL[†]

[†]The material in this section is adapted from Metcalf and Eddy, *Wastewater Engineering*, 4th edition, McGraw-Hill.

16.1 ANALYSIS OF SOLIDS DATA

The following test results were obtained for a wastewater sample taken at the headworks to a wastewater-treatment plant:

Tare mass of evaporating dish = 53.5433 g
Mass of evaporating dish plus residue after evaporation at 105°C = 53.5794 g
Mass of evaporating dish plus residue after ignition at 550°C = 53.5625 g
Tare mass of Whatman GF/C filter after drying at 105°C = 1.5433 g
Mass of Whatman GF/C filter and residue after drying at 105°C = 1.5554 g
Mass of Whatman GF/C filter and residue after ignition at 550°C = 1.5476 g

All of the tests were performed using a sample size of 50 mL. Determine the concentration of total solids, total volatile solids, suspended solids, volatile suspended solids, total dissolved solids, and total volatile dissolved solids. The samples used in the solids analyses were all evaporated, dried, or ignited to constant weight.

Calculation Procedure

1. Determine total solids.

$$TS = \frac{\left(\begin{array}{c}\text{mass of evaporating}\\\text{dish plus residue, g}\end{array}\right) - \left(\begin{array}{c}\text{mass of evaporating}\\\text{dish, g}\end{array}\right)}{\text{sample size, L}}$$

$$= \frac{[(53.5794 - 53.5433)\ \text{g}](10^3\ \text{mg/g})}{0.050\ \text{L}} = 722\ \text{mg/L}$$

2. Determine total volatile solids.

$$TVS = \frac{\left(\begin{array}{c}\text{mass of evaporating}\\\text{dish plus residue, g}\end{array}\right) - \left(\begin{array}{c}\text{mass of evaporating dish}\\\text{plus residue after ignition, g}\end{array}\right)}{\text{sample size, L}}$$

$$= \frac{[(53.5794 - 53.5625)\ \text{g}](10^3\ \text{mg/g})}{0.050\ \text{L}} = 388\ \text{mg/L}$$

3. Determine the total suspended solids.

$$TSS = \frac{\left(\begin{array}{c}\text{residue on filter}\\\text{after drying, g}\end{array}\right) - \left(\begin{array}{c}\text{tare mass of filter}\\\text{after drying, g}\end{array}\right)}{\text{sample size, L}}$$

$$= \frac{[(1.5554 - 1.5433)\ \text{g}](10^3\ \text{mg/g})}{0.050\ \text{L}} = 242\ \text{mg/L}$$

4. Determine the volatile suspended solids.

$$VSS = \frac{\left(\begin{array}{c}\text{residue on filter}\\\text{after drying, g}\end{array}\right) - \left(\begin{array}{c}\text{residue on filter}\\\text{after ignition, g}\end{array}\right)}{\text{sample size, L}}$$

$$= \frac{[(1.5554 - 1.5476)\ \text{g}](10^3\ \text{mg/g})}{0.050\ \text{L}} = 156\ \text{mg/L}$$

5. *Determine the total dissolved solids.*

$$TDS = TS - TSS = 722 - 242 = 480 \text{ mg/L}$$

6. *Determine the volatile dissolved solids.*

$$VDS = TVS - VSS = 338 - 156 = 182 \text{ mg/L}$$

16.2 CONVERSION OF GAS CONCENTRATION UNITS

The off gas from a wastewater force main (i.e., pressure sewer) was found to contain 9 ppm_v (by volume) of hydrogen sulfide (H_2S). Determine the concentration in $\mu g/m^3$ and in mg/L at standard conditions (0°C, 101.325 kPa). The molecular weight of H_2S is 34.08.

Calculation Procedure

1. *Compute the concentration in $\mu g/L$.*
The following relationship, based on the ideal gas law, is used to convert between gas concentrations expressed in ppm_v and $\mu g/m^3$:

$$\mu g/m^3 = \frac{(\text{concentration, } ppm_v)(\text{mw, g/mol of gas})(10^6 \mu g/g)}{(22.414 \times 10^{-3} \text{ m}^3/\text{mol of gas})}$$

Now, 9 ppm_v = 9 $m^3/10^6 \text{ m}^3$, so

$$\mu g/m^3 = \left(\frac{9 \text{ m}^3}{10^6 \text{ m}^3}\right)\left(\frac{(34.08 \text{ g/mol H}_2\text{S})}{(22.4 \times 10^{-3} \text{ m}^3/\text{mol of H}_2\text{S})}\right)\left(\frac{10^6 \mu g}{g}\right) = 13{,}693 \ \mu g/m^3$$

2. *Compute the concentration in mg/L.*
The concentration in mg/L is

$$\left(\frac{13{,}693 \ \mu g}{m^3}\right)\left(\frac{1 \text{ mg}}{10^3 \ \mu g}\right)\left(\frac{1 \text{ m}^3}{10^3 \text{L}}\right) = 0.0137 \text{ mg/L}$$

Related Calculations. If gas measurements, expressed in $\mu g/L$, are made at other than standard conditions, the concentration must be corrected to standard conditions, using the ideal gas law, before converting to ppm.

16.3 SATURATION CONCENTRATION OF OXYGEN IN WATER

What is the saturation of oxygen in water in contact with dry air at 1 atm and 20°C?

Calculation Procedure

1. *Establish the partial pressure of oxygen in air, p_g.*
Dry air contains about 21% oxygen by volume. Therefore, $p_g = 0.21$ mol O_2/mol air.

TABLE 16.1   Henry's Law Constants at 20°C, Unitless Henry's Law Constants at 20°C, and Temperature-Dependent Coefficients

Parameter	Henry's constant, atm	Henry's constant, unitless	Temperature coefficients A	B
Air	66,400	49.68	557.60	6.724
Ammonia	0.75	5.61×10^{-4}	1887.12	6.315
Carbon dioxide	1420	1.06	1012.40	6.606
Carbon monoxide	53,600	40.11	554.52	6.621
Chlorine	579	0.43	875.69	5.75
Chlorine dioxide	1500	1.12	1041.77	6.73
Hydrogen	68,300	51.10	187.04	5.473
Hydrogen sulfide	483	0.36	884.94	5.703
Methane	37,600	28.13	675.74	6.880
Nitrogen	80,400	60.16	537.62	6.739
Oxygen	41,100	30.75	595.27	6.644
Ozone	5300	3.97	1268.24	8.05
Sulfur dioxide	36	2.69×10^{-2}	1207.85	5.68

Source: Adapted in part from Montgomery (1985), Cornwell (1990), and Hand et al. (1998).

2. Find the mole fraction of oxygen in the water, x_g.
From the Table 16.1, at 20°C, Henry's constant is

$$H = 4.11 \times 10^4 \frac{\text{atm (mol gas/mol air)}}{\text{(mol gas/mol water)}}$$

The value of x_g is

$$x_g = \frac{P_T}{H} p_g$$

where P_T equals the total pressure. So,

$$x_g = \frac{1.0 \text{ atm}}{4.11 \times 10^4 \dfrac{\text{atm (mol gas/mol air)}}{\text{(mol gas/mol water)}}} (0.21 \text{ mol gas/mol air})$$

$$= 5.11 \times 10^{-6} \text{ mol gas/mol water}$$

3. Find the number of moles of oxygen per liter, n_g.
One liter of water contains 1000 g/(18 g/mol) = 55.6 mol, thus

$$\frac{n_g}{n_g + 55.6} = 5.11 \times 10^{-6}$$

Because the number of moles of dissolved gas in a liter of water is much less than the number of moles of water,

$$n_g + 55.6 \approx 55.6$$

and so

$$n_g \approx (55.6)\, 5.11 \times 10^{-6}$$

$$\approx 2.84 \times 10^{-4} \text{mol } O_2/L$$

4. *Determine the saturation concentration of oxygen.*

$$C_s \approx \frac{(2.84 \times 10^{-4} \text{ mol } O_2/L)(32 \text{ g/mol } O_2)}{(1 \text{ g}/10^3 \text{ mg})}$$

$$\approx 9.09 \text{ mg/L}$$

16.4 DETERMINATION OF BIOCHEMICAL OXYGEN DEMAND (BOD) FROM LABORATORY DATA

In a seeded 5-day BOD test conducted on a wastewater sample, 15 mL of the waste sample was added directly into a 300-mL BOD incubation bottle. The initial dissolved oxygen level, D_1, of the diluted sample was 8.8 mg/L, and the final level, D_2, after 5 days was 1.9 mg/L. The corresponding initial and final dissolved oxygen levels of the seeded dilution water, B_1 and B_2, were 9.1 and 7.9, respectively. What is the 5-day BOD (BOD_5) of the wastewater sample?

Calculation Procedure

Use the equation

$$BOD_5, \text{mg/L} = \frac{(D_1 - D_2) - (B_1 - B_2)f}{P}$$

where f is the fraction of seeded dilution waster volume in the sample to the volume or seeded dilution water in the seed control, and P is the wastewater sample volume divided by the combined volume. Then

$$f = [(300 - 15)/300] = 0.95$$
$$P = 15/300 = 0.05$$
$$BOD_5, \text{mg/L} = \frac{(8.8 - 1.9) - (9.1 - 7.9)0.95}{0.05} = 115.2 \text{ mg/L}$$

16.5 CALCULATION OF THEORETICAL OXYGEN DEMAND (ThOD)

Determine the ThOD for glycine $[CH_2(NH_2)COOH]$ using the following assumptions:

1. In the first step, the organic carbon and nitrogen are converted to carbon dioxide (CO_2) and ammonia (NH_3), respectively.
2. In the second and third steps, the ammonia is oxidized sequentially to nitrite and nitrate.
3. The ThOD is the sum of the oxygen required for all three steps.

Calculation Procedure

1. *Write a balanced reaction for the carbonaceous oxygen demand.*

$$CH_2(NH_2)COOH + \frac{3}{2}O_2 \rightarrow NH_3 + 2CO_2 + H_2O$$

2. *Write balanced reactions for the nitrogenous oxygen demand.*

$$NH_3 + \frac{3}{2}O_2 \rightarrow HNO_2 + H_2O$$

$$HNO_2 + \frac{1}{2}O_2 \rightarrow HNO_3$$

$$\overline{NH_3 + 2O_2 \rightarrow HNO_3 + H_2O}$$

3. *Determine the ThOD.*

$$ThOD = (3/2 + 4/2) \text{ mol } O_2/\text{mol glycine}$$
$$= 7/2 \text{ mol } O_2/\text{mol glycine} \times 32 \text{ g/mol } O_2$$
$$= 112 \text{ g } O_2/\text{mol glycine}$$

16.6   DEVELOPMENT OF BOD SUSTAINED MASS-LOADING VALUES

Develop a sustained BOD peak mass-loading curve for a treatment plant with a design flow rate of 1 m^3/s (22.8 Mgal/d). Assume that the long-term daily average BOD concentration is 200 g/m^3.

Calculation Procedure

1. *Compute the daily mass-loading value for BOD.*

$$\text{Daily BOD mass loading, kg/d} = \frac{(200 \text{ g/m}^3)(1 \text{ m}^3/\text{s})(86{,}400 \text{ s/d})}{(10^3 \text{ g/kg})} = 17{,}280 \text{ kg/d}$$

2. *Construct a computation table.*
Set up a computation table for the development of the necessary information for the peak sustained BOD mass-loading curve (see following table).

3. *Get peaking factors and mass loading rates.*
Obtain peaking factors for the sustained peak BOD loading rate from Fig. 16.1a, and determine the sustained mass-loading rates for various time periods [see table, columns (1), (2), and (3)].

4. *Develop data for the sustained mass-loading curve and prepare a plot of the resulting data.*
See Fig. 16.2.

Length of sustained peak, d (1)	Peaking factor[†] (2)	Peak BOD mass loading, kg/d (3)	Total mass loading, kg[‡] (4)
1	2.4	41,472	41,472
2	2.1	36,288	72,576
3	1.9	32,832	98,496
4	1.8	31,104	124,416
5	1.7	29,376	146,880
10	1.4	24,192	241,920
15	1.3	22,464	336,960
20	1.25	21,600	432,000
30	1.21	19,872	596,160
365	1.0	17,280	

[†]From Fig. 16.la.
[‡]Col. 1 × Col. 3 = Col. 4.

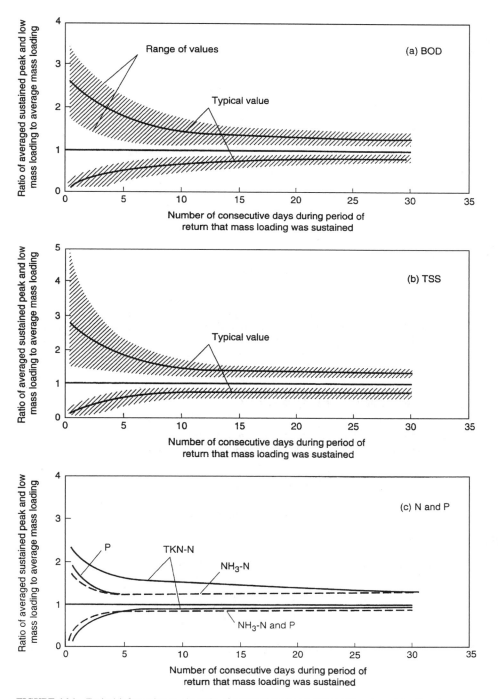

FIGURE 16.1 Typical information on the ratio of averaged peak and low-constituent mass loadings to average mass loadings for (a) BOD, (b) TSS, and (c) nitrogen and phosphorus.

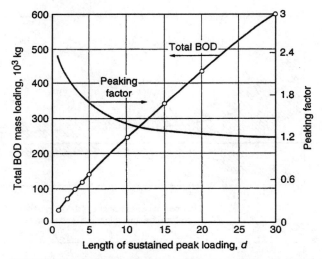

FIGURE 16.2 Peaking factor and total BOD versus length of sustained peak loading.

The interpretation of the curve plotted for this example is as follows. If the sustained peak loading period were to last for 10 days, the total amount of BOD that would be received at a treatment facility during the 10-day period would be 241,695 kg. The corresponding amounts for sustained peak periods of 1 and 2 days would be 41,401 and 72,451 kg, respectively. Computations for an example of this type can be facilitated by using a personal computer spreadsheet program.

16.7 HEADLOSS BUILDUP IN COARSE SCREENS

Determine the buildup of headloss, h, through a bar screen when 50 percent of the flow area is blocked off by the accumulation of coarse solids. Assume the following conditions apply:

> approach velocity, $v = 0.6$ m/s
> velocity through clean bar screen, $V = 0.9$ m/s
> open area for flow through clean bar screen $= 0.19$ m^2
> headloss coefficient for a clean bar screen, $C = 0.7$

Calculation Procedure

1. Compute the clean water headloss through bar screen.
Use the equation

$$h_L = \frac{1}{C}\left(\frac{V^2 - v^2}{2g}\right)$$

where g is the acceleration due to gravity, to get

$$h_L = \frac{1}{0.7}\left[\frac{(0.9 \text{ m/s})^2 - (0.6 \text{ m/s})^2}{2(9.81 \text{ m/s}^2)}\right] = 0.033 \text{ m}$$

2. Estimate the headloss through the clogged bar screen.
Reducing the screen area by 50 percent results in a doubling of the velocity.

The velocity through the clogged bar screen is

$$V_c = 0.9 \text{ m/s} \times 2 = 1.8 \text{ m/s}$$

Assuming the flow coefficient for the clogged bar screen is approximately 0.6, the estimated headloss is

$$h_L = \frac{1}{0.6}\left[\frac{(1.8 \text{ m/s})^2 - (0.6 \text{ m/s})^2}{2(9.81 \text{ m/s})^2}\right] = 0.24 \text{ m}$$

Related Calculations.   Where mechanically cleaned coarse screens are used, the cleaning mechanism typically is actuated by the buildup of headloss. Headloss is determined by measuring the water level before and after the screen. In some cases, the screen is cleaned at predetermined time intervals, as well as at a maximum head differential.

16.8 DETERMINATION OF FLOW RATE EQUALIZATION VOLUME REQUIREMENTS AND EFFECTS ON BOD MASS LOADING

For the flow rate and BOD concentration data given in the following table, determine (1) the in-line storage volume required to equalize the flow rate and (2) the effect of flow equalization on the BOD mass-loading rate.

Time period	Given data — Average flow rate during time period, m³/s	Given data — Average BOD concentration during time period, mg/L	Derived data — Cumulative volume of flow at end of time period, m³	Derived data — BOD mass loading during time period, kg/h
M–1	0.275	150	990	149
1–2	0.220	115	1782	91
2–3	0.165	75	2376	45
3–4	0.130	50	2844	23
4–5	0.105	45	3222	17
5–6	0.100	60	3582	22
6–7	0.120	90	4014	39
7–8	0.205	130	4752	96
8–9	0.355	175	6030	223
9–10	0.410	200	7506	295
10–11	0.425	215	9036	329
11–N	0.430	220	10,584	341
N–1	0.425	220	12,114	337
1–2	0.405	210	13,572	306
2–3	0.385	200	14,958	277
3–4	0.350	190	16,218	239
4–5	0.325	180	17,388	211
5–6	0.325	170	18,558	199
6–7	0.330	175	19,746	208
7–8	0.365	210	21,060	276
8–9	0.400	280	22,500	403
9–10	0.400	305	23,940	439
10–11	0.380	245	25,308	335
11–M	0.345	180	26,550	224
Average	0.307			213

Note: m³/s × 35.3147 = ft³/s.
m³ × 35.3147 = ft³.
mg/L = g/m³.

Calculation Procedure

1. Determine the volume of the basin required for the flow equalization.
The first step is to develop a cumulative volume curve of the wastewater flow rate expressed in cubic meters. The cumulative volume curve is obtained by converting the average flow rate (q_i) during each hourly period to cubic meters, using the expression

$$\text{volume, m}^3 = (q_i, \text{m}^3/\text{s})(3600 \text{ s/h})(1.0 \text{ h})$$

and then cumulatively summing the hourly values to obtain the cumulative flow volume. For example, for the first two time periods shown in the data table, the corresponding hourly volumes are as follows:

$$V_{M-1} = (0.275 \text{ m}^3/\text{s})(3600 \text{ s/h})(1.0 \text{ h}) = 990 \text{ m}^3$$
$$V_{1-2} = (0.220 \text{ m}^3/\text{s})(3600 \text{ s/h})(1.0 \text{ h}) = 792 \text{ m}^3$$

The cumulative flow, expressed in m³, at the end of each time period is determined as follows:

$$V_1 = 990 \text{ m}^3 \text{ (at the end of the first time period M–1)}$$
$$V_2 = 990 + 792 = 1782 \text{ m}^3 \text{ (at the end of the second time period 1–2)}$$

The cumulative flows for all the hourly time periods are computed in a similar manner (see derived data in data table).
The second step is to prepare a plot of the cumulative flow volume, as shown in the following diagram. As will be noted, the slope of the line drawn from the origin to the endpoint of the inflow mass diagram represents the average flow rate for the day, which in this case is equal to 0.307 m³/s.

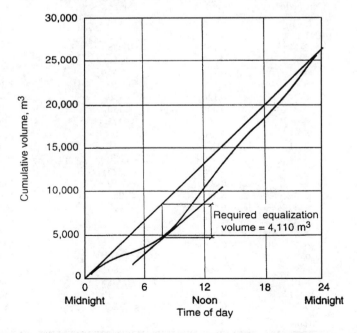

The third step is to determine the required storage volume. The required storage volume is determined by drawing a line parallel to the average flow rate tangent to the low point of the inflow mass diagram. The required volume is represented by the vertical distance from the point of tangency to the straight line representing the average flow rate. Thus, the required volume is

$$V = 4110 \text{ m}^3 (145,100 \text{ ft}^3)$$

2. Determine the effect of the equalization basin on the BOD mass-loading rate.

Although there are alternative computation methods, perhaps the simplest way is to perform the necessary computations starting with the time period when the equalization basin is empty. Because the equalization basin is empty at about 8:30 A.M., the necessary computations will be performed starting with the 8–9 time period.

The first step is to compute the liquid volume in the equalization basin at the end of each time period. The volume required is obtained by subtracting the equalized hourly flow rate expressed as a volume from the inflow flow rate also expressed as a volume. The volume corresponding to the equalized flow rate for a period of 1 h is 0.307 m³/s × 3600 s/h = 1106 m³. Using this value, the volume in storage is computed using the following expression:

$$V_{sc} = V_{sp} + V_{ic} - V_{oc}$$

where V_{sc} = volume in the equalization basin at the end of current time period
V_{sp} = volume in the equalization basin at the end of previous time period
V_{ic} = volume of inflow during the current time period
V_{oc} = volume of outflow during the current time period

Thus, using the values in the original data table, the volume in the equalization basin for the time period 8–9 is as follows:

$$V_{sc} = 0 + 1278 \text{ m}^3 - 1106 \text{ m}^3 = 172 \text{ m}^3$$

For time period 9–10:

$$V_{sc} = 172 \text{ m}^3 + 1476 \text{ m}^3 - 1106 \text{ m}^3 = 542 \text{ m}^3$$

The volume in storage at the end of each time period has been computed in a similar way (see the following computation table).

Time period	Volume of flow during time period, m³	Volume in storage at end of time period, m³	Average BOD concentration during time period, mg/L	Equalized BOD concentration during time period, mg/L	Equalized BOD mass loading during time period, kg/h
8–9	1278	172	175	175	193
9–10	1476	542	200	197	218
10–11	1530	966	215	210	232
11–N	1548	1408	220	216	239
N–1	1530	1832	220	218	241
1–2	1458	2184	210	214	237
2–3	1386	2464	200	209	231
3–4	1260	2618	190	203	224
4–5	1170	2680	180	196	217
5–6	1170	2746	170	188	208
6–7	1188	2828	175	184	203
7–8	1314	3036	210	192	212
8–9	1440	3370	280	220	243

(Continued)

Time period	Volume of flow during time period, m³	Volume in storage at end of time period, m³	Average BOD concentration during time period, mg/L	Equalized BOD concentration during time period, mg/L	Equalized BOD mass loading during time period, kg/h
9–10	1440	3704	305	245	271
10–11	1368	3966	245	245	271
11–M	1242	4102	180	230	254
M–1	990	3986	150	214	237
1–2	792	3972	115	196	217
2–3	594	3160	75	179	198
3–4	468	2522	50	162	179
4–5	378	1794	45	147	162
5–6	360	1048	60	132	146
6–7	432	374	90	119	132
7–8	738	0	130	126	139
Average					213

Note: m³ × 35.3147 = ft³.
kg × 2.2046 = lb.
g/m³ = mg/L.

The second step is to compute the average concentration leaving the storage basin. Using the following expression, which is based on the assumption that the contents of the equalization basin are mixed completely, the average concentration leaving the storage basin is

$$X_{oc} \frac{(V_{ic})(X_{ic}) + (V_{sp})(X_{sp})}{V_{ic} + V_{sp}}$$

where X_{oc} = average concentration of BOD in the outflow from the storage basin during the current time period, g/m³ (mg/L)
V_{ic} = volume of wastewater inflow during the current period, m³
X_{ic} = average concentration of BOD in the inflow wastewater volume, g/m³
V_{sp} = volume of wastewater on storage basin at the end of the previous time period, m³
X_{sp} = concentration of BOD in wastewater in storage basin at the end of the previous time period, g/m³

Using the data given in column 2 of the previous computation table, the effluent concentration is computed as follows:

$$X_{oc} = \frac{(1278 \text{ m}^3)(175 \text{ g/m}^3) + (0)(0)}{1278 \text{ m}^3} = 175 \text{ g/m}^3 \quad \text{(for the time period } 8-9)$$

$$X_{oc} = \frac{(1476 \text{ m}^3)(200 \text{ g/m}^3) + (172 \text{ m}^3)(175 \text{ g/m}^3)}{(1476 + 172) \text{ m}^3} = 197 \text{ g/m}^3 \quad \text{(for the time period } 9-10)$$

All the concentration values computed in a similar manner are reported in the previous computation table.

The third step is to compute the hourly mass-loading rate using the following expression:

$$\text{mass-loading rate, kg/h} = \frac{(X_{oc}, \text{g/m}^3)(q_i, \text{m}^3/\text{s})(3600 \text{ s/h})}{(10^3 \text{ g/kg})}$$

For example, for the time period 8–9, the mass-loading rate is

$$\frac{(175 \text{ g/m})(0.307 \text{ m}^3/\text{s})(3600 \text{ s/h})}{(10^3 \text{ g/kg})} = 193 \text{ kg/h}$$

All hourly values are summarized in the computation table. The corresponding values without flow equalization are reported in the original data table.

The effect of flow equalization can be shown best graphically by plotting the hourly unequalized and equalized BOD mass loading (see the following plot). The following flow rate ratios, derived from the data presented in the table given in the problem statement and the computation table prepared in this step, are also helpful in assessing the benefits derived from flow equalization:

Ratio	BOD mass loading	
	Unequalized	Equalized
$\dfrac{\text{Peak}}{\text{Average}}$	$\dfrac{439}{213} = 2.06$	$\dfrac{271}{213} = 1.27$
$\dfrac{\text{Minimum}}{\text{Average}}$	$\dfrac{17}{213} = 0.08$	$\dfrac{132}{213} = 0.62$
$\dfrac{\text{Peak}}{\text{Minimum}}$	$\dfrac{439}{17} = 25.82$	$\dfrac{271}{132} = 2.05$

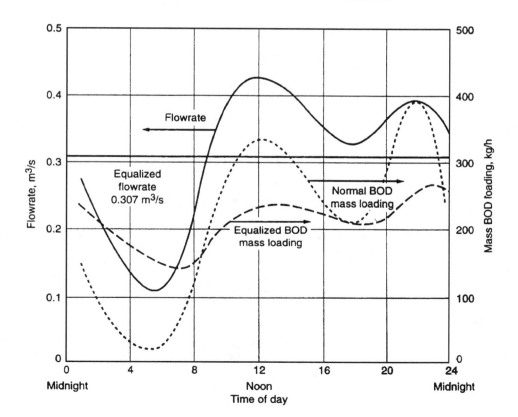

Related Calculations.   Where in-line equalization basins are used, additional damping of the BOD mass-loading rate can be obtained by increasing the volume of the basins. Although the flow to a treatment plant was equalized in this example, flow equalization would be used, more realistically, in locations with high infiltration or inflow or peak stormwater flows.

16.9   POWER REQUIREMENTS AND PADDLE AREA FOR A WASTEWATER FLOCCULATOR

Determine the theoretical power requirement, P, and paddle area, A, theoretically required to achieve an average velocity gradient, G, of 50/s in a 3000 m^3 tank. The water temperature is 15°C, the dynamic viscosity, μ, is 1.139×10^{-3} N · s/m^2, the density is 999.1 kg/m^3, the coefficient of drag, C_D, for rectangular paddles is 1.8, the paddle-tip velocity, v, is 0.6 m/s, and the relative velocity, v_p, of the paddles with respect to the fluid is 0.75v.

Calculation Procedure

1. *Determine the theoretical power requirement.*
Use the equation

$$
\begin{aligned}
P &= G^2 \mu V \\
&= (50/s)^2 (1.139 \times 10^{-3} \text{ N · s/m}^2)(3000 \text{ m}^3) \\
&= 8543 \text{ kg · m}^2/\text{s}^3 = 8543 \text{ W} \\
&= 8.543 \text{ kW}
\end{aligned}
$$

2. *Determine the required paddle area.*
Use the equation

$$
\begin{aligned}
A &= \frac{2P}{C_D \, \rho v_P^3} \\
&= \frac{2(8543 \text{ kg/m}^2 \cdot \text{s}^3)}{1.8 \,(999.1 \text{ kg/m}^3)(0.75 \times 0.6 \text{ m/s})^3} \\
&= 104.3 \text{ m}^2
\end{aligned}
$$

16.10   SIZING AN ACTIVATED-SLUDGE SETTLING TANK

The settling curve shown in the following diagram was obtained for an activated sludge with an initial solids concentration, C_0, of 3000 mg/L. The initial height of the interface in the settling column was at 0.75 m (2.5 ft). Determine the area required to yield a thickened-solids concentration, C_u, of 12,000 mg/L with a total flow of 3800 m^3/d (1 Mgal/d). Determine also the solids loading (kg/m^2 · d) and the overflow rate (m^3/m^2 · d).

Calculation Procedure

1. *Determine the area required for thickening.*
Determine the value of H_u, the depth at which the solids are at the desired thickened-solids concentration:

$$
H_u = \frac{C_0 H_0}{C_u}
$$

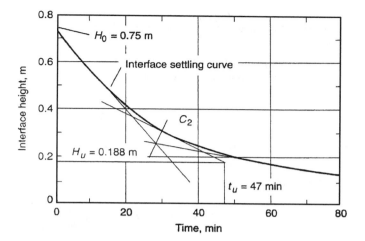

where H_0 is the initial height of the interface, so

$$H_u = \frac{(3000 \text{ mg/L})(0.75 \text{ m})}{(12,000 \text{ mg/L})} = 0.188 \text{ m}$$

On the settling curve, a horizontal line is constructed at $H_u = 0.188$ m. A tangent is constructed to the settling curve at C_2, the midpoint of the region between hindered and compression settling. Bisecting the angle formed where the two tangents meet determines point C_2. The intersection of the tangent at C_2 and the line $H_u = 0.188$ m determines t_u. Thus $t_u = 47$ min, and the required area is

$$A = \frac{Qt_u}{H_0} = \left[\frac{(3800 \text{ m}^3/\text{d})}{(24 \text{ h/d})(60 \text{ min/h})} \right] \left(\frac{47 \text{ min}}{0.75 \text{ m}} \right) = 165 \text{ m}^2$$

2. *Determine the area required for clarification.*
First determine the interface subsidence velocity, v. The subsidence velocity is determined by computing the slope of the tangent drawn from the initial portion of the interface settling curve. The computed velocity represents the unhindered settling rate of the sludge.

$$v = \left(\frac{0.75 \text{ m} - 0.3 \text{ m}}{29.5 \text{ m}} \right) \left(\frac{60 \text{ min}}{\text{h}} \right) = 0.92 \text{ m/h}$$

Then determine the clarification rate. Because the clarification rate is proportional to the liquid volume above the critical sludge zone, it may be computed as follows:

$$Q = 3800 \text{ m}^3/\text{d} \left(\frac{0.75 \text{ m} - 0.188 \text{ m}}{0.75 \text{ m}} \right) = 2847 \text{ m}^3/\text{d}$$

The required area is obtained by dividing the clarification rate by the settling velocity:

$$A = \frac{Q_c}{v} = \frac{(2847 \text{ m}^3/\text{d})}{(24 \text{ h/d})(0.91 \text{ m/h})} = 129 \text{ m}^2$$

3. *Determine the controlling area.*
The controlling area is the thickening area (165 m^2) because it exceeds the area required for clarification (129 m^2).

4. *Determine the solids loading.*
The solids loading is computed as follows:

$$\text{solids, kg/d} = \frac{(3800 \text{ m}^3/\text{d})(3000 \text{ g/m}^3)}{(10^3 \text{ g/kg})} = 11{,}400 \text{ kg/d}$$

$$\text{solids loading} = \frac{(11{,}400 \text{ kg/d})}{165 \text{ m}^2} = 69.1 \text{ kg/m}^2 \cdot \text{d}$$

5. *Determine the hydraulic loading rate.*

$$\text{Hydraulic loading rate} = \frac{(3800 \text{ m}^3/\text{d})}{165 \text{ m}^2} = 23.0 \text{ m}^3/\text{m}^2 \cdot \text{d}$$

16.11 *STRIPPING OF BENZENE IN THE ACTIVATED-SLUDGE PROCESS*

Determine the amount of benzene that can be stripped in a complete-mix activated-sludge reactor equipped with a diffused-air aeration system. Assume the following conditions apply:

> wastewater flow rate, $Q = 4000$ m^3/d
> aeration tank volume, $V = 1000$ m^3
> depth of aeration tank = 6 m
> air flow rate = 50 m^3/min at standard conditions
> oxygen mass-transfer rate = 6.2/h
> influent concentration of benzene = 100 μg/m^3
> H, Henry's constant = 5.49×10^{-3} m$^3 \cdot$ atm/mol (see Table 16.2)
> n, coefficient for mass-transfer equation in step 3 = 1.0
> temperature = 20°C
> oxygen diffusivity, $D_{O_2} = 2.11 \times 10^{-5}$ cm^2/s
> benzene diffusivity, $D_{voc} = 0.96 \times 10^{-5}$ cm^2/s

Calculation Procedure

1. *Determine the quantity of air referenced to the middepth of the aeration tank.*
This represents the depth for an average bubble size:

$$Q_g = (50 \text{ m}^3/\text{min}) \times \frac{10.33}{10.33 + 3} = 38.7 \text{ m}^3/\text{min}$$

2. *Determine the air/liquid ratio Q_g/Q.*

$$Q = \frac{(4000 \text{ m}^3/\text{d})}{(1440 \text{ min/d})} = 2.78 \text{ m}^3/\text{min}$$

$$\frac{Q_g}{Q} = \frac{(38.7 \text{ m}^3/\text{min})}{(2.78 \text{ m}^3/\text{min})} = 13.9$$

TABLE 16.2 Physical Properties of Selected Volatile and Semivolatile Organic Compounds[†]

Compounds	mw	mp, °C	bp, °C	vp, mmHg	vd	sg	Sol, mg/L	C, g/m³	H, m³ · atm/mol	log K_{ow}
Benzene	78.11	5.5	80.1	76	2.77	0.8786	1780	319	5.49×10^{-3}	2.1206
Chlorobenzene	112.56	−45	132	8.8	3.88	1.1066	500	54	3.70×10^{-3}	2.18–3.79
o-Dichlorobenzene	147.01	18	180.5	1.60	5.07	1.036	150	N/A	1.7×10^{-3}	3.3997
Ethylbenzene	106.17	−94.97	136.2	7	3.66	0.867	152	40	8.43×10^{-3}	3.13
1,2-Dibromoethane	187.87	9.8	131.3	10.25	0.105	2.18	2699	93.61	6.29×10^{-4}	N/A
1,1-Dichloroethane	98.96	−97.4	57.3	297	3.42	1.176	7840	160.93	5.1×10^{-3}	N/A
1,2-Dichloroethane	98.96	−35.4	83.5	61	3.4	1.25	8690	350	1.14×10^{-3}	1.4502
1,1,2,2-Tetrachloroethane	167.85	−36	146.2	14.74	5.79	1.595	2800	13.10	4.2×10^{-4}	2.389
1,1,1-Trichloroethane	133.41	−32	74	100	4.63	1.35	4400	715.9	3.6×10^{-3}	2.17
1,1,2-Trichloroethane	133.4	−36.5	133.8	19	N/A	N/A	4400	13.89	7.69×10^{-4}	N/A
Chloroethene	62.5	−153	−13.9	2548	2.15	0.912	6000	8521	6.4×10^{-2}	N/A
1,1-Dichloroethene	96.94	−122.1	31.9	500	3.3	1.21	5000	2640	1.51×10^{-1}	N/A
c-1,2-Dichloroethene	96.95	−80.5	60.3	200	3.34	1.284	800	104.39	4.08×10^{-3}	N/A
t-1,2-Dichloroethene	96.95	−50	48	269	3.34	1.26	6300	1428	4.05×10^{-3}	N/A
Tetrachloroethene	165.83	−22.5	121	15.6	N/A	1.63	160	126	2.85×10^{-2}	2.5289
Trichloroethene	131.5	−87	86.7	60	4.54	1.46	1100	415	1.17×10^{-2}	2.4200
Bromodichloromethane	163.8	−57.1	90	N/A	N/A	1.971	N/A	N/A	2.12×10^{-3}	N/A
Chlorodibromomethane	208.29	<−20	120	50	N/A	2.451	N/A	N/A	8.4×10^{-4}	N/A
Dichloromethane	84.93	−97	39.8	349	2.93	1.327	20,000	1702	3.04×10^{-3}	N/A
Tetrachloromethane	153.82	−23	76.7	90	5.3	1.59	800	754	2.86×10^{-2}	2.7300
Tribromomethane	252.77	8.3	149	5.6	8.7	2.89	3130	7.62	5.84×10^{-4}	N/A
Trichloromethane	119.38	−64	62	160	4.12	1.49	7840	1027	3.10×10^{-3}	1.8998
1,2-Dichloropropane	112.99	−100.5	96.4	41.2	3.5	1.156	2600	25.49	2.75×10^{-3}	N/A
2,3-Dichloropropene	110.98	−81.7	94	135	3.8	1.211	insol.	110	N/A	N/A
t-1,3-Dichloropropene	110.97	N/A	112	99.6	N/A	1.224	515	110	N/A	N/A
Toluene	92.1	−95.1	110.8	22	3.14	0.867	515	110	6.44×10^{-3}	2.2095

Source: [†]Data were adapted from Lang et al. (1987).
All values are reported at 20°C.

Note: mw = molecular weight, mp = melting point, bp = boiling point, vd = vapor density, sg = specific gravity, Sol = solubility, C_s = saturation concentration in air, H = Henry's law constant, log K_{ow} = logarithm of the octanol–water partition coefficient, N/A = not available.

3. *Estimate the mass-transfer coefficient for benzene.*
Use the equation

$$K_L a_{\text{VOC}} = K_L a_{O_2} \left(\frac{D_{\text{VOC}}}{D_{O_2}} \right)^n$$

where $K_L a_{\text{VOC}}$ is the system mass-transfer coefficient and $K_L a_{O_2}$ is the system oxygen mass-transfer coefficient. Then

$$K_L a_{\text{VOC}} = (6.2/\text{h}) \left[\frac{(0.96 \times 10^{-5} \text{ cm}^2/\text{S})}{(2.11 \times 10^{-5} \text{ cm}^2/\text{S})} \right]^{1.0}$$

$$= 2.82/\text{h} = 0.047/\text{min}$$

4. *Determine the dimensionless value of Henry's constant.*
Use the equation

$$H_u = \frac{H}{RT}$$

to get

$$H_u = \frac{(0.00549 \text{ m}^3 \cdot \text{atm/mol})}{(0.000082057 \text{ atm} \cdot \text{m}^3/\text{mol} \cdot \text{K})[(273.15 + 20)\text{K}]} = 0.228$$

5. *Determine the saturation parameter ϕ.*
Use the equation

$$\phi = \frac{(K_L a)_{\text{VOC}} V}{H_u Q_g}$$

to get

$$\phi = \frac{(0.047/\text{min} \times 1000 \text{ m}^3)}{(0.228 \times 38.7 \text{ m}^3/\text{min})} = 5.33$$

6. *Determine the fraction of benzene removed from the liquid phase.*
Use the equation

$$1 - \frac{C_e}{C_i} = 1 - \left[1 + \frac{Q_g}{Q}(H_u)(1 - e^{-\phi}) \right]^{-1}$$

where C_e is the benzene concentration in the effluent and C_i is its concentration in the influent. Then

$$1 - \frac{C_e}{C_i} = 1 - [1 + 13.9(0.228)(1 - e^{-5.33})]^{-1}$$

$$= 1 - 0.32 = 0.68$$

16.12   ESTIMATION OF SLUDGE VOLUME FROM CHEMICAL
PRECIPITATION OF UNTREATED WASTEWATER

Estimate the mass and volume of sludge produced from untreated wastewater without and with the use of ferric chloride for the enhanced removal of total suspended solids (TSS). Also estimate the amount of lime required for the specified ferric chloride dose. Assume that 60 percent of the TSS is removed in the primary settling tank without the addition of chemicals, and that the addition of ferric chloride results in an increased removal of TSS to 85 percent. Also, assume that the following data apply to this situation:

Wastewater flow rate, m^3/d	1000
Wastewater TSS, mg/L	220
Wastewater alkalinity as $CaCO_3$, mg/L	136
Ferric chloride ($FeCl_3$) added, $kg/1000\ m^3$	40
Raw sludge properties	
Specific gravity	1.03
Moisture content, %	94
Chemical sludge properties	
Specific gravity	1.05
Moisture content, %	92.5

Calculation Procedure

1. Compute the mass of TSS removed without and with chemicals.
Determine the mass of TSS removed without chemicals:

$$M_{TSS} = \frac{0.6(220\ g/m^3)(1000\ m^3/d)}{(10^3\ g/kg)} = 132.0\ kg/d$$

Determine the mass of TSS removed with chemicals:

$$M_{TSS} = \frac{0.85(220\ g/m^3)(1000\ m^3/d)}{(10^3\ g/kg)} = 187.0\ kg/d$$

2. Determine the mass of ferric hydroxide $[Fe(OH)_3]$ produced from the addition of 40 $kg/1000\ m^3$ of ferrous sulfate ($FeCl_3$).
The relevant reaction is

$$
\begin{array}{ccccccccc}
2\times 162.2 & & 3\times 100\ (\text{as } CaCO_3) & & 2\times 106.9 & & & & \\
2FeCl_3 & + & 3Ca(HCO_3)_2 & \leftrightarrow & 2Fe(OH)_3 & + & 3CaCl_2 & + & 6CO_2 \\
\text{Ferric} & & \text{Calcium} & & \text{Ferric} & & \text{Calcium} & & \text{Carbon} \\
\text{chloride} & & \text{bicarbonate} & & \text{hydroxide} & & \text{sulfate} & & \text{dioxide} \\
\text{(soluble)} & & \text{(soluble)} & & \text{(insoluble)} & & \text{(soluble)} & & \text{(soluble)}
\end{array}
$$

where 162.2, 100, and 106.9 are molecular weights.
So,

$$Fe(OH)_3\ formed = 40 \times \left(\frac{2 \times 106.9}{2 \times 162.2} \right) = 26.4\ kg/1000\ m^3$$

3. *Determine the mass of lime required to convert the ferric chloride to ferric hydroxide Fe(OH)$_3$.*
The relevant equation is

$$\underset{\substack{\text{Ferric}\\\text{chloride}\\\text{(soluble)}}}{\underset{2\text{FeCl}_3}{2 \times 162.2}} + \underset{\substack{\text{Calcium}\\\text{hydroxide}\\\text{(slightly soluble)}}}{\underset{3\text{Ca(OH)}_2}{3 \times 56(\text{as CaO})}} \leftrightarrow \underset{\substack{\text{Ferric}\\\text{hydroxide}\\\text{(insoluble)}}}{\underset{2\text{Fe(OH)}_3}{2 \times 106.9}} + \underset{\substack{\text{Calcium}\\\text{chloride}\\\text{(soluble)}}}{\underset{3\text{CaCl}_2}{3 \times 111}}$$

So,

$$\text{lime required} = 40 \times \left(\frac{3 \times 56}{2 \times 162.2} \right) = 20.7 \text{ kg/1000 m}^3$$

Because there is sufficient natural alkalinity no lime addition will be required.

4. *Determine the total amount of sludge on a dry basis resulting from chemical precipitation.*

$$\text{Total dry solids} = 187 + 26.4 = 213.4 \text{ kg/1000 m}^3$$

5. *Determine the total volume of sludge resulting from chemical precipitation.*
Assume that the sludge has a specific gravity of 1.05 and a moisture content of 92.5 percent. Then

$$V_s = \frac{(213.4 \text{ kg/d})}{(1.05)(1000 \text{ kg/m}^3)(0.075)} = 2.71 \text{ m}^3\text{/d}$$

6. *Determine the total volume of sludge without chemical precipitation.*
Assume that the sludge has a specific gravity of 1.03 and a moisture content of 94 percent. Then

$$V_s = \frac{(132 \text{ kg/d})}{(1.03)(1000 \text{ kg/m}^3)(0.06)} = 2.1 \text{ m}^3\text{/d}$$

7. *Prepare a summary table of sludge masses and volumes without and with chemical precipitation.*

	Sludge	
Treatment	Mass, kg/d	Volume, m^3/d
Without chemical precipitation	132.0	2.13
With chemical precipitation	213.4	2.71

Related Calculations.   The magnitude of the sludge disposal problem when chemicals are used is evident from a review of the data presented in the summary table given in step 7. Even larger volumes of sludge are produced when lime is used as the primary precipitant.

16.13   DETERMINATION OF ALUM DOSAGE FOR PHOSPHORUS REMOVAL

Determine the amount of liquid alum required to precipitate phosphorus in a wastewater that contains 8 mg P/L. Also determine the required alum storage capacity if a 30-day supply is to be stored at the treatment facility. Based on laboratory testing, 1.5 mol of Al will be required per mol of P. The flow rate is 12,000 m^3/d. The following data are for the liquid alum supply:

> Formula for liquid alum $Al_2(SO_4)_3 \cdot 18H_2O$
> Alum strength = 48 percent
> Density of liquid alum solution = 1.2 kg/L

Calculation Procedure

1. Determine the weight of aluminum (Al) available per liter of liquid alum.
The weight of alum per liter is

$$alum/L = (0.48)(1.2 \text{ kg/L}) = 0.576 \text{ kg/L}$$

The weight of aluminum per liter is calculated as follows:

> molecular weight of alum = 666.5
>
> aluminum/L = $(0.58 \text{ kg/L})(2 \times 26.98/666.5) = 0.0466$ kg/L

2. Determine the weight of Al required per unit weight of P.
The relevant equation is

$$Al^3 + H_nPO_4^{3-n} \leftrightarrow AlPO_4 + nH^+$$

So,

> theoretical dosage = 1.0 mol Al per 1.0 mol P
>
> aluminum required = 1.0 kg $\times$ (mw Al/mw P)
>
> = 1.0 kg $\times$ (26.98/30.97) = 0.87 kg Al/kg P

3. Determine the amount of alum solution required per kg P.

$$\text{Alum dose} = 1.5 \times \left(\frac{0.87 \text{ kg Al}}{1.0 \text{ kg P}} \right) \left(\frac{\text{L alum solution}}{0.0466 \text{ kg}} \right)$$

$$= 28.0 \text{ L alum solution/kg P}$$

4. Determine the amount of alum solution required per day.

$$\text{Alum} = \frac{(12{,}000 \text{ m}^3/\text{d})(8 \text{ g P/m}^3)(28.0 \text{ L alum/kg P})}{(10^3 \text{ g/kg})}$$

$$= 2688 \text{ L alum solution/d}$$

5. Determine the required alum solution storage capacity based on average flow.

$$\text{Storage capacity} = (2688 \text{ L alum solution/d})(30 \text{ d})$$

$$= 80{,}640 \text{ L} = 80.6 \text{ m}^3$$

16.14 ESTIMATION OF SLUDGE VOLUME FROM THE CHEMICAL PRECIPITATION OF PHOSPHORUS WITH LIME IN A PRIMARY SEDIMENTATION TANK

Estimate the mass and volume of sludge produced in a primary sedimentation tank from the precipitation of phosphorus with lime. Assume that 60 percent of the total suspended solids (TSS) is removed without the addition of lime and that the addition of 400 mg/L of $Ca(OH)_2$ results in an increased removal of TSS to 85 percent. Assume the following data apply:

Wastewater flow rate, m^3/d	1000
Wastewater TSS, mg/L	220
Wastewater volatile TSS, mg/L	150
Wastewater PO_4^{3-} as P, mg/L	10
Wastewater total hardness as $CaCO_3$, mg/L	241.3
Wastewater Ca^{2+}, mg/L	80
Wastewater Mg^{2+}, mg/L	10
Effluent PO_4^{3-} as P, mg/L	0.5
Effluent Ca^{2+}, mg/L	60
Effluent Mg^{2+}, mg/L	0
Chemical sludge properties	
Specific gravity	1.07
Moisture content, %	92.5

Calculation Procedure

1. Compute the mass and volume of solids removed without chemicals.
Assume that the sludge contains 94 percent moisture and has a specific gravity of 1.03.
    Determine the mass of TSS removed:

$$M_{TSS} = \frac{0.6(220 \text{ g/m}^3)(1000 \text{ m}^3/\text{d})}{(10^3 \text{ g/kg})} = 132 \text{ kg/d}$$

Determine the volume of sludge produced:

$$V_s = \frac{(132 \text{ kg/d})}{(1.03)(1000 \text{ kg/m}^3)(0.06)} = 2.14 \text{ m}^3/\text{d}$$

2. Using the equations summarized in the table, determine the mass of $Ca_5(PO_4)_3OH$, $Mg(OH)_2$, and $CaCO_3$ produced from the addition of 400 mg/L of lime.
First determine the mass of $Ca_5(PO_4)_3OH$ formed.

    I. Determine the moles of P removed:

$$\text{moles P removed} = \frac{(10 - 0.5) \text{ mg/L}}{(30.97 \text{ g/mol})(10^3 \text{ mg/g})}$$
$$= 0.307 \times 10^{-3} \text{ mol/L}$$

Reaction	Chemical species in sludge
Lime, $CaCO_3$	
1. $10Ca^{2+} + 6PO_4^{3-} + 2OH^- \leftrightarrow Ca_{10}(PO_4)_6(OH)_2$	$Ca_{10}(PO_4)_6(OH)_2$
2. $Mg^{2+} + 2OH^- \leftrightarrow Mg(OH)_2$	$Mg(OH)_2$
3. $Ca^{2+} + CO_3^{2-} \leftrightarrow CaCO_3$	$CaCO_3$
Alum, Al (III)	
1. $Al^{3-} + PO_4^{3-} \leftrightarrow AlPO_4$	$AlPO_4$
2. $2Al^{3-} + 3OH^- \leftrightarrow Al(OH)_3$	$Al(OH)_3$
Iron, Fe (III)	
1. $Fe^{3+} + PO_4^{3-} \leftrightarrow FePO_4$	$FePO_4$
2. $Fe^{3+} + 3OH^- \leftrightarrow Fe(OH)_3$	$Fe(OH)_3$

II. Determine the moles of $Ca_5(PO_4)_3OH$ formed:

$$\text{moles } Ca_5(PO_4)_3OH \text{ formed} = 1/3 \times 0.307 \times 10^{-3} \text{mol/L}$$
$$= 0.102 \times 10^{-3} \text{ mol/L}$$

III. Determine the mass of $Ca_5(PO_4)_3OH$ formed:

$$\text{mass } Ca_5(PO_4)_3OH = 0.102 \times 10^{-3} \text{ mol/L} \times 502 \text{ g/mol} \times 10^3 \text{ mg/g}$$
$$= 51.3 \text{ mg/L}$$

Then determine the mass of $Mg(OH)_2$ formed.

I. Determine the moles of Mg^{2+} removed:

$$\text{moles } Mg^{2+} \text{ removed} = \frac{(10 \text{ mg/L})}{(24.31 \text{ g/mol})(10^3 \text{ mg/g})}$$
$$= 0.411 \times 10^{-3} \text{ mol/L}$$

II. Determine the mass of $Mg(OH)_2$ formed:

$$\text{moles } Mg(OH)_2 = 0.411 \times 10^{-3} \text{ mol/L} \times 58.3 \text{ g/mol} \times 10^3 \text{ mg/g}$$
$$= 24.0 \text{ mg/L}$$

Then determine the mass of $CaCO_3$ formed.

I. Determine the mass of Ca^{2+} in $Ca_5(PO_4)_3(OH)$:

$$\text{mass } Ca^{2+} \text{ in } Ca_5(PO_4)_3(OH) = 5(40 \text{ g/mol})(0.102 \times 10^{-3} \text{ mol/L})(10^3 \text{ mg/g})$$
$$= 20.4 \text{ mg/L}$$

II. Determine the mass of Ca^{2+} added in the original dosage:

$$\text{mass } Ca^{2+} \text{ in } Ca(OH)_2 = \frac{(40 \text{ g/mol})(400 \text{ mg/L})}{(74 \text{ g/mol})}$$

$$= 216.2 \text{ mg/L}$$

III. Determine the mass of Ca present as $CaCO_3$:

$$Ca^{2+} \text{ in } CaCO_3 = \text{Ca in } Ca(HO)_2 + (Ca^{2+} \text{ in influent wastewater})$$
$$- (Ca^{2+} \text{ in } Ca_5(PO_4)_3OH) - (Ca^{2+} \text{ in effluent wastewater})$$
$$= 216.2 + 80 - 20.4 - 60$$
$$= 215.8 \text{ mg/L}$$

IV. Determine the mass of $CaCO_3$:

$$\text{mass } CaCO_3 = \frac{(100 \text{ g/mol})(215.8 \text{ mg/L})}{(40 \text{ g/mol})}$$

$$= 540 \text{ mg/L}$$

3. *Determine the total mass of solids removed as a result of the lime dosage.*
The mass of TSS in wastewater is

$$M_{TSS} = \frac{0.85(220 \text{ g/m}^3)(1000 \text{ m}^3/\text{d})}{(10^3 \text{ g/kg})} = 187 \text{ kg/d}$$

The masses of the chemical solids are

$$M_{Ca5(PO4)3OH} = \frac{(51.2 \text{ g/m}^3)(1000 \text{ m}^3/\text{d})}{(10^3 \text{ g/kg})} = 51.2 \text{ kg/d}$$

$$M_{Mg(OH)2} = \frac{(24 \text{ g/m}^3)(1000 \text{ m}^3/\text{d})}{(10^3 \text{ g/kg})} = 24.0 \text{ kg/d}$$

$$M_{CaCO3} = \frac{(540 \text{ g/m}^3)(1000 \text{ m}^3/\text{d})}{(10^3 \text{ g/kg})} = 540 \text{ kg/d}$$

Hence the total mass of solids removed is

$$M_T = (187 + 51.3 + 24 + 540) \text{ kg/d}$$
$$= 802.3 \text{ kg/d}$$

4. *Determine the total volume of sludge resulting from chemical precipitation.*
Assume that the sludge has a specific gravity of 1.07 and a moisture content of 92.5 percent. Then

$$V_s = \frac{(802.3 \text{ kg/d})}{(1.07)(1000 \text{ kg/m}^3)(0.075)} = 10.0 \text{ m}^3/\text{d}$$

5. *Prepare a summary table of sludge masses and volumes without and with chemical precipitation.*

	Sludge	
Treatment	Mass, kg/d	Volume, m^3/d
Without chemical precipitation	132.0	2.14
With chemical precipitation	802.3	10.0

16.15   DETERMINATION OF REACTION POTENTIAL

Determine whether hydrogen sulfide (H_2S) can be oxidized with hydrogen peroxide (H_2O_2). The pertinent half reactions from Table 16.3 are as follows:

$$H_2S \leftrightarrow S + 2H^+ + 2e^- \quad E° = -0.14$$

$$H_2O_2 + 2H^+ + 2e^- \leftrightarrow 2H_2O \quad E° = +1.776$$

TABLE 16.3   Selected Standard Electrode Potentials for Oxidation-Reduction Half Reactions

Half reaction	Oxidation potential,[†] V
$Li^+ + e^- \rightarrow Li$	−3.03
$K^+ + e^- \rightarrow K$	−2.92
$Ba^{2+} + 2e^- \rightarrow Ba$	−2.90
$Ca^{2+} + 2e^- \rightarrow Ca$	−2.87
$Na^+ + e^- \rightarrow Na$	−2.71
$Mg(OH)_2 + 2e^- \rightarrow Mg + 2OH^-$	−2.69
$Mg^{2+} + 2e^- \rightarrow Mg$	−2.37
$Al^{3+} + 3e^- \rightarrow Al$	−1.66
$MnO_4^- + 8H^+ + 5e^- \rightarrow Mn^{2+} + 4H_2O$	−1.51
$Mn^{2+} + 2e^- \rightarrow Mn$	−1.18
$2H_2O + 2e^- \rightarrow H_2 + 2OH^-$	−0.828
$Zn^{2+} + 2e^- \rightarrow Zn$	−0.763
$Fe^{2+} + 2e^- \rightarrow Fe$	−0.440
$Cd^{2+} + 2e^- \rightarrow Cd$	−0.40
$Ni^{2+} + 2e^- \rightarrow Ni$	−0.250
$S + 2H^+ + 2e^- \rightarrow H_2S$	−0.14
$Pb^{2+} + 2e^- \rightarrow Pb$	−0.126
$2H^+ + 2e^- \rightarrow H_2$	0.000
$Cu^{2+} + e^- \rightarrow Cu^+$	+0.15
$N_2 + 4H^+ + 3e^- \rightarrow NH_4^+$	+0.27
$Cu^{2+} + 2e^- \rightarrow Cu$	+0.34
$I_2 + 2e^- \rightarrow 2I^-$	+0.54
$O_2 + 2H^+ + 2e^- \rightarrow H_2O_2$	+0.68
$Fe^{3+} + e^- \rightarrow Fe^{2+}$	+0.771
$Ag^+ + e^- \rightarrow Ag$	+0.799
$ClO^- + H_2O + 2e^- \rightarrow Cl^- + 2OH^-$	+0.90
$Br_2(aq) + 2e^- \rightarrow 2Br^-$	+ 1.09
$O_2 + 4H^+ + 4e^- \rightarrow 2H_2O$	+1.229
$Cl_2(g) + 2e^- \rightarrow 2Cl^-$	+ 1.360
$H_2O_2 + 2H^+ + 2e^- \rightarrow 2H_2O$	+ 1.776
$O_3 + 2H^+ + 2e^- \leftrightarrow O_2 + H_2O$	+2.07
$F_2 + 2H^+ + 2e^- \rightarrow 2HF$	+2.87

Source: Adapted in part from Bard (1996) and Benefield et al. (1982).
[†]Reported values will vary depending on source.

Calculation Procedure

1. Determine the overall reaction by adding the two half reactions.

$$H_2S \leftrightarrow S + 2H^+ + 2e^-$$

$$\underline{H_2O_2 + 2H^+ + 2e^- \leftrightarrow 2H_2O}$$

$$H_2S + H_2O_2 \leftrightarrow S + 2H_2O$$

2. Determine the $E^\circ_{reaction}$ for the overall reaction.

$$E^\circ_{reaction} = E^\circ_{H_2O_2{}^{2+}, H_2O} - E^\circ_{s^{2-}, s}$$

$$= (1.78) - (-0.14) = +1.92 \text{ V}$$

Because the $E^\circ_{reaction}$ for the reaction is positive, the reaction will proceed as written.

16.16   *OBSERVED BIOMASS YIELD AND OXYGEN CONSUMPTION*

The aerobic complete-mix biological treatment process without recycle, as shown in the drawing, receives wastewater with a biodegradable soluble COD (bsCOD) concentration of 500 g/m³. The flow rate is 1000 m³/d and the reactor effluent bsCOD and suspended volatile solids (VSS) concentrations are 10 and 200 g/m³, respectively. Based on these data:

What is the observed yield in g VSS/g COD removed?
What is the amount of oxygen used in g O_2/g COD removed and in g/d?

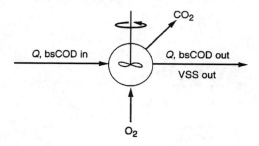

Calculation Procedure

1. Determine the observed yield.
Assume the following general reaction is applicable:

$$\text{organic matter} + O_2 + \text{nutrients} \rightarrow C_5H_7NO_2 + CO_2 + H_2O$$
$$\phantom{\text{organic matter}}\text{500g COD/m}^3 \phantom{+ O_2 + \text{nutrients} \rightarrow} \text{200 g VSS/m}^3$$

The g VSS/d produced is

$$\text{g VSS/d} = 200 \text{ g/m}^3 (1000 \text{ m}^3/d) = 200,000 \text{ g VSS/d}$$

The g bsCOD/d removed is

$$\text{g COD/d} = (500 - 10) \text{ g COD/m}^3 (1000 \text{ m}^3/d)$$
$$= 490,000 \text{ g COD/d}$$

The observed yield is then

$$Y_{obs} = \frac{(200,000 \text{ g VSS/d})}{(490,000 \text{ g COD/d})} = 0.41 \text{ g VSS/g COD removed}$$

2. Determine the amount of oxygen used per g bsCOD removed.
Prepare a steady-state COD mass balance around the reactor:

$$\text{accumulation} = \text{inflow} - \text{outflow} + \text{conversion}$$
$$0 = COD_{in} - COD_{out} - \text{oxygen used (expressed as COD)}$$

Now,

$$\text{oxygen used} = COD_{in} - COD_{out}$$

and

$$COD_{in} = 500 \text{ g COD/m}^3 \ (1000 \text{ m}^3\text{/d}) = 500,000 \text{ g COD/d}$$
$$COD_{out} = bsCOD_{out} + \text{biomass } COD_{out}$$

where
$$bsCOD_{out} = 10 \text{ g/m}^3 \ (1000 \text{ m}^3\text{/d}) = 10,000 \text{ g COD/d}$$
$$\text{biomass } COD_{out} = 200,000 \text{ g VSS/d } (1.42 \text{ g COD/g VSS})$$
$$= 284,000 \text{ g COD/d}$$

So,

$$\text{total } COD_{out} = 10,000 \text{ g/d} + 284,000 \text{ g/d} = 294,000 \text{ g COD/d}$$

The oxygen used is

$$\text{oxygen used} = 500,000 \text{ g COD/d} - 294,000 \text{ g COD/d}$$
$$= 206,000 \text{ g COD/d} = 206,000 \text{ g O}_2\text{/d}$$

The amount of oxygen used per unit COD removed is then

$$\text{oxygen/COD} = (206,000 \text{ g/d})/(490,000 \text{ g/d}) = 0.42 \text{ g O}_2\text{/g COD}$$

Related Calculations.   The general COD balance that accounts for cell production and COD oxidation is

$$\text{g COD cells} + \text{g COD oxidized} = \text{g COD removed}$$
$$(0.41 \text{ g VSS/g COD})(1.42 \text{ g O}_2\text{/g VSS}) + 0.42 \text{ g O}_2\text{/g COD} = 1.0 \text{ g O}_2\text{/g COD}$$

16.17 DETERMINATION OF BIOMASS AND SOLIDS YIELDS

For an industrial wastewater activated-sludge process, the amount of bsCOD in the influent wastewater is 300 g/m³ and the influent nonbiodegradable suspended volatile solids (nbVSS) concentration is 50 g/m³. The influent flow rate is 1000 m³/d, the biomass concentration is 2000 g/m³, the reactor bsCOD concentration is 15 g/m³, and the reactor volume is 105 m³. If the cell debris fraction, f_d, is 0.10, determine the net biomass yield, the observed solids yield, and the biomass fraction in the mixed-liquor volatile suspended solids (MLVSS). Use the kinetic coefficients given in the table that follows.

Coefficient	Unit	Value[†]	
		Range	Typical
k	g bs COD/g VSS · d	2–10	5
K_s	mg/L BOD	25–100	60
	mg/L bsCOD	10–60	40
Y	mg VSS/mg BOD	0.4–0.8	0.6
	mg VSS/mg bsCOD	0.3–0.6	0.4
k_d	g VSS/g VSS · d	0.06–0.15	0.10

[†]Values reported are for 20°C.

Calculation Procedure

1. Determine the net biomass yield.
Use the equation

$$Y_{bio} = -r_g/r_{su},$$

where r_g and r_{su} are rates of biomass growth and substrate utilizatization.

First, solve for r_{su} using the information given in the table and the following equation, in which X is biomass concentration, S is growth-limiting substrate concentration, and K_s is the substrate concentration at one-half the maximum specific substrate-utilization rate:

$$r_{su} = -\frac{k\,X\,S}{K_S + S}$$

$$= -\frac{(5/d)(2000 \text{ g/m}^3)(15 \text{ g bsCOD/m}^3)}{(40 + 15) \text{ g/m}^3}$$

$$= -2727 \text{ g bsCOD/m}^3 \cdot \text{d}$$

Then determine the net biomass production rate r_g using the equation

$$r_g = -Y\,r_{su} - k_d X$$

where k_d is endogenous decay coefficient. This gives

$$r_g = -(0.40 \text{ g VSS/g bsCOD})(-2727 \text{ g bsCOD/m}^3 \cdot \text{d})$$
$$- (0.10 \text{ g VSS/g VSS} \cdot \text{d})(2000 \text{ g VSS/m}^3)$$
$$= 891 \text{ g VSS/m}^3 \cdot \text{d}$$

Now calculate the net biomass yield:

$$Y_{bio} = -r_g/r_{su} = (891 \text{ g VSS/m}^3 \cdot \text{d})/(2727 \text{ g bsCOD/m}^3 \cdot \text{d})$$
$$= 0.33 \text{ g VSS/g bsCOD}$$

2. Determine VSS production rate.
Use the equation

$$r_{XT, \text{VSS}} = -Y r_{su} - k_d X + f_d(k_d)X + QX_{o,i}/V$$

where f_d is the fraction of biomass that remains as cell debris, Q is the influent flow rate, $X_{o,i}$ is the influent nbVSS concentration, and V is the reactor volume. Then,

$$\begin{aligned}
r_{XT, \text{VSS}} &= 891 \text{ g VSS/m}^3 \cdot \text{d} \\
&+ (0.10 \text{ g VSS/g VSS})(0.10 \text{ g VSS/g VSS} \cdot \text{d})(2000 \text{ g VSS/m}^3) \\
&+ (1000 \text{ m}^3/\text{d})(50 \text{ g VSS/m}^3)/105 \text{ m}^3 \\
&= (891 + 20 + 476) \text{ g VSS/m}^3 \cdot \text{d} \\
&= 1387 \text{ g VSS/m}^3 \cdot \text{d}
\end{aligned}$$

3. Calculate the observed solids yield.
Use the equation

$$Y_{\text{obs}} = -r_{XT, \text{VSS}}/r_{su}$$

Then

$$\begin{aligned}
Y_{\text{obs}} &= -(1387 \text{ g VSS/m}^3 \cdot \text{d})/(-2727 \text{ g bsCOD/m}^3 \cdot \text{d}) \\
&= 0.51 \text{ g VSS/g bsCOD}
\end{aligned}$$

4. Calculate the active biomass fraction in the MLVSS.
Use the equation

$$F_{X, \text{act}} = (-Y r_{su} - k_d X)/r_{XT, \text{VSS}}$$

to get

$$\begin{aligned}
F_{X, \text{act}} &= (891 \text{ g VSS/m}^3 \cdot \text{d})/(1387 \text{ g VSS/m}^3 \cdot \text{d}) \\
&= 0.64
\end{aligned}$$

Thus, accounting for the nbVSS in the wastewater influent and cell debris produced, the MLVSS contains 64 percent active biomass.

16.18 PREDICTING METHANE PRODUCTION FROM AN ANAEROBIC REACTOR

An anaerobic reactor, operated at 35°C, processes a wastewater stream with a flow of 3000 m^3/d and a bsCOD concentration of 5000 g/m^3. At 95 percent bsCOD removal and a net biomass synthesis yield of 0.04 g VSS/g COD used, what is the amount of methane produced in m^3/d?

Calculation Procedure

1. *Prepare a steady-state mass balance for COD to determine the amount of the influent COD converted to methane.*
The required steady-state mass balance is

$$
0 \ = \ \begin{matrix} \text{Influent} \\ \text{COD} \\ \ \end{matrix} \ - \ \begin{matrix} \text{portion of} \\ \text{influent COD} \\ \text{in effluent} \end{matrix} \ - \ \begin{matrix} \text{influent COD} \\ \text{converted to} \\ \text{cell tissue} \end{matrix} \ - \ \begin{matrix} \text{influent COD} \\ \text{converted to} \\ \text{methane} \end{matrix}
$$

$$
COD_{in} \ = \ COD_{eff} \ + \ COD_{VSS} \ + \ COD_{methane}
$$

Determine the values of the individual mass balance terms:

$$
COD_{in} = (5000 \ g/m^3)(3000 \ m^3/d) = 15{,}000{,}000 \ g/d
$$

$$
COD_{eff} = (1 - 0.95)(5000 \ g/m^3)(3000 \ m^3/d) = 750{,}000 \ g/d
$$

$$
COD_{VSS} = (1.42 \ g \ COD/g \ VSS)(0.04 \ g \ VSS/g \ COD)(0.95)(15{,}000{,}000 \ g/d)
$$

$$
= 809{,}400 \ g/d
$$

Solve for the COD converted to methane:

$$
COD_{methane} = 15{,}000{,}000 - 750{,}000 - 809{,}400 = 13{,}440{,}600 \ g/d
$$

2. *Determine the amount of methane produced at 35°C.*
Determine the volume of gas occupied by 1 mol of gas at 35°C:

$$
V = \frac{nRT}{P}
$$

$$
= \frac{(1 \ mol)(0.082057 \ atm \cdot L/mol \cdot K)[(273.15 + 35)K]}{1.0 \ atm}
$$

$$
= 25.29 \ L
$$

The CH_4 equivalent of COD converted under anaerobic conditions is $(25.29 \ L/mol)/(64 \ g \ COD/mol \ CH_4) = 0.40 \ L \ CH_4/g \ COD$.
   The amount of methane produced is then

$$
CH_4 \ production = (13{,}440{,}600 \ g \ COD/d)(0.40 \ L \ CH_4/g \ COD)(1 \ m^3/10^3 \ L)
$$

$$
= 5376 \ m^3/d
$$

At 65 percent methane

$$
total \ gas \ flow = (5376 \ m^3/d)/0.65
$$

$$
= 8271 \ m^3/d
$$

It is important to determine the volume occupied by the gas at the actual operating temperature.

16.19 *WASTEWATER CHARACTERIZATION EVALUATION*

Given the following wastewater characterization results, determine concentrations for the following:

1. bCOD (biodegradable COD)
2. nbpCOD (nonbiodegradable particulate COD)

3. sbCOD (slowly biodegradable COD)

4. nbVSS (nonbiodegradable volatile suspended solids)

5. iTSS (inert total suspended solids)

6. nbpON (nonbiodegradable particulate organic nitrogen)

7. total degradable TKN (Kjeldahl nitrogen)

The influent wastewater characteristics are as follows:

Constituent	Concentration, mg/L
BOD	195
sBOD	94
COD	465
sCOD (soluble COD)	170
rbCOD (readily biodegradable COD)	80
TSS	220
VSS	200
TKN	40
NH_4-N	26
Alkalinity	200 (as $CaCO_3$)

The activated-sludge effluent data are as follows:

Constituent	Concentration, mg/L
sCODe	30
sON	1.2

Calculation Procedure

1. *Determine biodegradable (bCOD).*
Use the equation

$$bCOD = {\sim}1.6(BOD)$$

to get

$$bCOD = 1.6(195 \text{ mg/L}) = 312 \text{ mg/L}$$

2. *Determine the nbpCOD.*
First determine the nbCOD using the equation

$$nbCOD = COD - bCOD$$

So,

$$nbCOD = (465 - 312) \text{ mg/L} = 153 \text{ mg/L}$$

Then determine the nbpCOD using the equation

$$nbpCOD = nbCOD - sCODe$$

to get

$$nbpCOD = (153 - 30)\ mg/L = 123\ mg/L$$

3. *Determine the sbCOD.*
Use the equation

$$sbCOD = bCOD - rbCOD$$

to get

$$sbCOD = (312 - 80)\ mg/L = 232\ mg/L$$

4. *Determine the nbVSS.*
First determine the bpCOD/pCOD ratio using the equation

$$\frac{bpCOD}{pCOD} = \frac{(bCOD/BOD)(BOD - sBOD)}{COD - sCOD}$$

where bpCOD is biodegradable particulate COD. So,

$$\frac{bpCOD}{pCOD} = \frac{1.6(195 - 94)\ mg/L}{(465 - 170)\ mg/L} = 0.55$$

Then determine the nbVSS using

$$nbVSS = \left[1 - \left(\frac{bpCOD}{pCOD}\right)\right]VSS$$

which gives

$$nbVSS = (1 - 0.55)(200\ mg/L) = 90\ mg/L$$

5. *Determine the inert TSS.*

$$iTSS = TSS - VSS = (220 - 200)\ mg/L = 20\ mg/L$$

6. *Determine the nbpON.*
First determine the organic N content of VSS using the equation

$$f_N = \frac{(TKN - sON - NH_4\text{-}N)}{VSS}$$

So,

$$f_N = \frac{(40 - 1.2 - 26)\ mg/L}{200\ mg/L} = 0.064$$

Then determine the nbpON using

$$nbpON = f_N(nbVSS)$$

which gives

$$nbpON = 0.064(90\ mg/L) = 5.8\ mg/L$$

7. *Determine total degradable TKN.*

$$bTKN = TKN - nbpON - sON$$
$$= (40 - 5.8 - 1.2) \, mg/L$$
$$= 33 \, mg/L$$

16.20   *TRICKLING FILTER SIZING USING NRC EQUATIONS*

A municipal wastewater having a BOD of 250 g/m³ is to be treated by a two-stage trickling filter. The desired effluent quality is 25 g/m³ of BOD. If both of the filter depths, D_e, are to be 1.83 m and the recirculation ratio, R, is 2:1, find the required filter diameters. Assume the following design assumptions apply:

Flow rate = 7570 m³/d
Wastewater temperature = 20°C
BOD removal in primary sedimentation = 35 percent
$E_1 = E_2$

Calculation Procedure

1. *Compute E_1 and E_2.*

$$\text{Overall efficiency} = \left\{ \frac{[(200 - 25) \, g/m^3]}{(200 \, g/m^3)} \right\} (100) = 87.5 \text{ percent}$$

$$E_1 + E_2(1 - E_1) = 0.875$$
$$E_1 = E_2 = 0.646$$

2. *Compute the recirculation factor F.*
Use the equation

$$F = \frac{1 + R}{(1 + R/10)^2} = \frac{1 + 2}{(1.2)^2} = 2.08$$

3. *Compute the BOD loading for the first filter, W_1.*

$$\text{Primary effluent BOD} = (1.0 - 0.35)(250 \, g/m^3) = 163 \, g/m^3$$
$$W_1 = (163 \, g/m^3)(7570 \, m^3/d)(1 \, kg/10^3 \, g) = 1234 \, kg \, BOD/d$$

4. *Compute the volume for the first stage, V_1.*
Use the equation

$$E_1 = \frac{100}{1 + 0.4432 \sqrt{\dfrac{W_1}{V_1 F}}}$$

So,

$$64.6 = \cfrac{100}{1 + 0.4432\sqrt{\cfrac{1234}{V_1(2.08)}}}$$

and

$$V_1 = 388 \text{ m}^3$$

5. Compute the diameter of the first filter, D_1, from the cross-sectional area, A_1.

$$A_1 = \frac{V_1}{D_e} = \frac{388 \text{ m}^3}{1.83 \text{ m}} = 212 \text{ m}^2 = \frac{\pi}{4}D_1^2$$

$$D_1 = 16.4 \text{ m}$$

6. Compute the BOD loading for the second-stage filter, W_2.

$$W_2 = (1 - E_1)W_1 = (1 - 0.646)(1234 \text{ kg BOD/d}) = 437 \text{ kg BOD/d}$$

7. Compute V_2, the volume of the second-stage filter.
Use the equation

$$E_2 = \cfrac{100}{1 + \cfrac{0.4432}{1 - E_1}\sqrt{\cfrac{W_2}{V_2 F}}}$$

So,

$$64.6 = \cfrac{100}{1 + \cfrac{0.4432}{1 - 0.646}\sqrt{\cfrac{437}{V_2(2.08)}}}$$

and

$$V_2 = 1096 \text{ m}^3$$

8. Compute the diameter of the second filter, D_2.

$$A_2 = \frac{V_2}{D_e} = \frac{1096 \text{ m}^3}{1.83 \text{ m}} = 599 \text{ m}^2$$

$$D_2 = 27.6 \text{ m}$$

9. Compute the BOD loading to each filter.
For the first-stage filter:

$$\text{BOD loading} = \frac{(1234 \text{ kg/d})}{388 \text{ m}^3} = 3.18 \text{ kg/m}^3 \cdot \text{d}$$

For the second-stage filter:

$$\text{BOD loading} = \frac{(437 \text{ kg/d})}{1096 \text{ m}^3} = 0.40 \text{ kg/m}^3 \cdot \text{d}$$

10. *Compute the hydraulic loading to each filter.*
For the first-stage filter:

$$\text{hydraulic loading} = \frac{(1 + 2)(7570 \text{ m}^3/\text{d})}{(1440 \text{ min/d})(212 \text{ m}^2)}$$

$$= 0.0744 \text{ m}^3/\text{m}^2 \cdot \text{min}$$

For the second-stage filter:

$$\text{hydraulic loading} = \frac{(1 + 2)(7570 \text{ m}^3/\text{d})}{(1440 \text{ min/d})(599 \text{ m}^2)}$$

$$= 0.0263 \text{ m}^3/\text{m}^2 \cdot \text{min}$$

Related Calculations.   To accommodate standard rotary distributor mechanisms, the diameters of the two filters should be rounded to the nearest 1.5 m (5 ft). To reduce construction costs, the two trickling filters are often made the same size. Where two filters of equal diameter are used, the removal efficiencies will be unequal. In many cases, the hydraulic loading rate will be limited by state standards.

16.21   *DETERMINATION OF FILTER-MEDIUM SIZES*

A dual-medium filter bed composed of sand and anthracite is to be used for the filtration of settled secondary effluent. If the effective size of the sand in the dual-medium filter is to be 0.55 mm, determine the effective size of the anthracite to avoid significant intermixing.

Calculation Procedure

1. *Summarize the properties of the filter media.*
For sand,

$$\text{effective size} = 0.55 \text{ mm}$$

$$\text{specific gravity} = 2.65 \text{ (see table)}$$

For anthracite,

$$\text{effective size} = \text{to be determined, mm}$$

$$\text{specific gravity} = 1.7 \text{ (see table)}$$

Filter material	Specific gravity	Porosity, α	Sphericity[†]
Anthracite		0.56–0.60	0.40–0.60
Sand	1.4–1.75	0.40–0.46	0.75–0.85
Garnet	2.55–2.65	0.42–0.55	0.60–0.80
Ilmenite	3.8–4.3	0.40–0.55	
Fuzzy filter medium	4.5	0.87–0.89	

Source: Adapted in part from Cleasby and Logsdon (1999).
[†]Sphericity is defined as the ratio of the surface area of an equal volume sphere to the surface area of the filter medium particle.

2. *Compute the effective size of the anthracite.*
Use the equation

$$d_1 = d_2 \left(\frac{\rho_2 - \rho_w}{\rho_1 - \rho_w} \right)^{0.667}$$

where ρ is density. The result is identical if the densities are replaced by the numerical values for the specific gravities. Thus,

$$d_1 = 0.55 \text{ mm} \left(\frac{2.65 - 1}{1.7 - 1} \right)^{0.667}$$

$$= 0.97 \text{ mm}$$

Related Calculations.   Another approach that can be used to assess whether intermixing will occur is to compare the fluidized bulk densities of the two adjacent layers (e.g., upper 450 mm of sand and lower 100 mm of anthracite).

16.22   DETERMINATION OF MEMBRANE AREA REQUIRED FOR DEMINERALIZATION

A brackish water having a TDS concentration of 3000 g/m^3 is to be desalinized using a thin-film composite membrane having a flux rate coefficient, k_w, of 1.5×10^{-6} s/m and a mass-transfer rate coefficient, k_i, of 1.8×10^{-6} m/s. The product water is to have a TDS of no more than 200 g/m^3. The flow rate, Q_p, is to be 0.010 m^3/s. The net operating pressure ($\Delta P_a - \Delta \Pi$) will be 2500 kPa. Assume the recovery rate, r, will be 90 percent. Estimate the rejection rate and the concentration of the concentrate stream.

Calculation Procedure

1. *Set out the basis for solving the problem.*
The problem involves determination of the membrane area required to produce 0.010 m^3/s of water and the TDS concentration of the permeate. If the permeate TDS concentration is well below 200 g/m^3, blending of feed and permeate will reduce the required membrane area.

2. *Estimate membrane area A.*
Use the equation

$$F_w = k_w (\Delta P_a - \Delta \Pi)$$

where F_w is the flux of water to get

$$F_w = (1.5 \times 10^{-6} \text{ s/m})(2500 \text{ kg/m}^2) = 3.75 \times 10^{-3} \text{ kg/m}^2 \cdot \text{s}$$

Since $Q_p = F_w \times A$, then

$$A = \frac{(0.010 \text{ m}^3/\text{s})(10^3 \text{ kg/m}^3)}{(3.75 \times 10^{-3} \text{ kg/m}^2 \cdot \text{s})} = 2667 \text{ m}^2$$

3. *Estimate permeate TDS concentration, C_p.*
Use the equation

$$F_i = k_i \Delta C_i = \frac{Q_p C_P}{A}$$

where ΔC_i is the solution concentration gradient

$$Q_p C_p = k_i \left(\left[\frac{C_f + C_c}{2} \right] - C_p \right) A$$

Assume $C_c \approx C_f$ where C_f is solute concentration in feed, and solve for C_p:

$$C_p = \frac{k_i A C_f}{Q_p + k_i A}$$

Assume $Q_p = rQ_f$. Then

$$C_p = \frac{(1.8 \times 10^{-6} \text{ m/s})(2667 \text{ m}^2)(3.0 \text{ kg/m}^3)}{(0.01)(0.9) + (1.8 \times 10^{-6} \text{ m/s})(2667 \text{ m}^3)} = 0.152 \text{ kg/m}^3$$

The permeate solute concentration is lower than necessary. It may be possible to reduce the area by blending.

4. Estimate the rejection rate R.
Use the equation

$$R, \% = \frac{C_f - C_p}{C_f} \times 100$$

Then

$$R = \frac{(3.0 \text{ kg/m}^3 - 0.152 \text{ kg/m}^3)}{(3.0 \text{ kg/m}^3)} \times 100 = 95\%$$

5. Estimate the concentrate stream TDS.
Use the equation

$$C_c = \frac{Q_f C_f - Q_p C_p}{Q_c}$$

Then

$$C_c = \frac{(0.1 \text{ L})(3.0 \text{ kg/m}^3) - (0.9 \text{ L})(0.152 \text{ kg/m}^3)}{0.1 \text{ L}} = 31.4 \text{ kg/m}^3$$

16.23 AREA AND POWER REQUIREMENTS FOR ELECTRODIALYSIS

Determine the area and power required to demineralize 4000 m^3/d of treated wastewater to be used for industrial cooling water using an electrodialysis unit composed of 240 cells. Assume the following data apply:

> TDS concentration = 2500 mg/L
> cation and anion concentration = 0.010 g-eq/L
> efficiency of salt removal η = 50 percent
> current efficiency E_c = 90 percent
> CD/N ratio = 500 mA/cm^2

where CD is current density and N is the normality of the solution,

> resistance R = 5.0 Ω

Calculation Procedure

1. *Calculate the current.*
Use the equation

$$I = \frac{FQN\eta}{\eta E_c}$$

where F is Faraday's constant and Q is flow rate. Then

$$Q = (4000 \text{ m}^3/\text{d})(10^3 \text{ L/m}^3)/(86,400 \text{ s/d}) = 46.3 \text{ L/s}$$

$$I = \frac{(96,485 \text{ A} \cdot \text{s/g-eq})(46.3 \text{ L/s})(0.010 \text{ g-eq/L})(0.50)}{240 \times 0.90}$$

$$= 103.4 \text{ A}$$

2. *Determine the power required.*
Use the equation

$$P = R(I)^2$$

to get

$$P = (5.0 \text{ }\Omega)(103.4 \text{ A})^2 = 53,477 \text{ W} = 53.5 \text{ kW}$$

3. *Determine the required surface area.*
First determine the current density:

$$CD = (500)(\text{normality}) = 500 \text{ mA/cm}^2 \times 0.010 = 50 \text{ mA/cm}^2$$

The required area is

$$\text{area} = \frac{(103.4 \text{ A})(10^3 \text{ mA/A})}{(50 \text{ mA/cm}^2)} = 2068 \text{ cm}^2$$

Then determine area of membrane assuming a square configuration will be used:

$$\text{area per membrane} = \sqrt{2068 \text{ cm}^2} \approx 45 \text{ cm}^2$$

Comment.   The actual performance will have to be determined from pilot tests.

16.24   *ANALYSIS OF ACTIVATED-CARBON ADSORPTION DATA*

Determine the Freundlich and Langmuir isotherm coefficients for the following adsorption test data on granular activated carbon (GAC). The liquid volume used in the batch adsorption tests was 1 L. The initial concentration of the adsorbate in solution was 3.37 mg/L. Equilibrium was obtained after 7 days.

Mass of GAC, m, g	Equilibrium concentration of adsorbate in solution, C_e, mg/L
0.0	3.37
0.001	3.27
0.010	2.77
0.100	1.86
0.500	1.33

The Freundlich isotherm is defined as follows:

$$\frac{x}{m} = K_f C_e^{1/n}$$

where x/m = mass of adsorbate adsorbed per unit mass of adsorbent, mg adsorbate/g activated carbon

K_f = Freundlich capacity factor, (mg absorbate/g activated carbon)(L water/mg adsorbate)$^{1/n}$

C_e = equilibrium concentration of adsorbate in solution after adsorption, mg/L

$1/n$ = Freundlich intensity parameter

The constants in the Freundlich isotherm can be determined by plotting log (x/m) versus log C_e and making use of the equation

$$\log\left(\frac{x}{m}\right) = \log K_f + \frac{1}{n}\log C_e$$

The Langmuir adsorption isotherm is defined as

$$\frac{x}{m} = \frac{abC_e}{1 + bC_e}$$

where x/m = mass of adsorbate adsorbed per unit mass of adsorbent, mg adsorbate/g activated carbon

a, b = empirical constants

C_e = equilibrium concentration of adsorbate in solution after adsorption, mg/L

The constants in the Langmuir isotherm can be determined by plotting $C_e/(x/m)$ versus C_e and making use of the equation

$$\frac{C_e}{(x/m)} = \frac{1}{ab} + \frac{1}{a}C_e$$

Calculation Procedure

1. Derive the values needed to plot the Freundlich and Langmuir adsorption isotherms using the batch adsorption test data.

Adsorbate concentration, mg/L					
C_o	C_e	$C_o - C_e$	m, g	x/m,[†] mg/g	$C_e/(x/m)$
3.37	3.37	0.00	0.000	—	—
3.37	3.27	0.10	0.001	100	0.0327
3.37	2.77	0.60	0.010	60	0.0462
3.37	1.86	1.51	0.100	15.1	0.1232
3.37	1.33	2.04	0.500	4.08	0.3260

[†] $q_e = \frac{x}{m} = \frac{(C_o - C_e)V}{m}$

2. *Plot the Freundlich and Langmuir adsorption isotherms.*
Use the data developed in step 1 (see following figures).

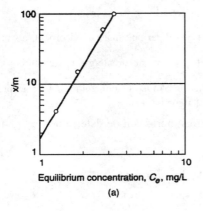

Equilibrium concentration, C_e, mg/L

(a)

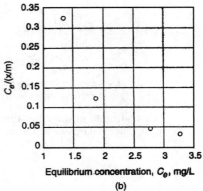

Equilibrium concentration, C_e, mg/L

(b)

3. *Determine the adsorption isotherm coefficients.*
For the Freundlich coefficients, when x/m versus C_e is plotted on log–log paper, the intercept when $C_e = 1.0$ is the value of (x/m) and the slope of the line is equal to $1/n$. Thus, $x/m = 1.55$, and $K_f = 1.55$. When $x/m = 1.0$, $C_e = 0.9$ and $1/n = 0.26$. Thus,

$$\frac{x}{m} = 1.5\,C_e^{0.26}$$

Because the plot for the Langmuir isotherm is curvilinear, use of the Langmuir adsorption isotherm is inappropriate.

16.25 ESTIMATE THE REQUIRED OZONE DOSE FOR A TYPICAL SECONDARY EFFLUENT

Estimate the ozone dose needed to disinfect a filtered secondary effluent to an MPN (most probable number) value of 240/100 mL using the following disinfection data obtained from pilot-scale installation. Assume the starting coliform concentration will be 1×10^6/100 mL and that the ozone transfer efficiency is 80 percent.

Test number	Initial coliform count N_o, MPN/100 mL	Ozone transferred, mg/L	Final coliform count N, MPN/100 mL	$-\log(N/N_o)$
1	95,000	3.1	1500	1.80
2	470,000	4.0	1200	2.59
3	3,500,000	4.5	730	3.68
4	820,000	5.0	77	4.03
5	9,200,000	6.5	92	5.00

The relevant equations are

$$N/N_o = 1 \quad \text{for } U < q \tag{16.1}$$

$$N/N_o = [(U)/q]^{-n} \quad \text{for } U > q \tag{16.2}$$

where N = number of organisms remaining after disinfection
N_o = number of organisms present before disinfection
U = utilized (or transferred) ozone dose, mg/L
n = slope of dose response curve
q = value of x intercept when $N/N_o = 1$ or log $N/N_o = 0$ (assumed to be equal to the initial ozone demand)

The required ozone dosage must be increased to account for the transfer of the applied ozone to the liquid. The required dosage can be computed with the following expression:

$$D = U\left(\frac{100}{TE}\right) \tag{16.3}$$

where D = total required ozone dosage, mg/L
U = utilized (or transferred) ozone dose, mg/L
TE = ozone transfer efficiency, %

Typical ozone transfer efficiencies vary from about 80 to 90 percent.

Calculation Procedure

1. Determine the coefficients in Eq. (16.1) using the pilot-plant data.
Linearize Eq. (16.1) and plot the log inactivation data versus the ozone dose on log–log paper to determine the constants in

$$N/N_o = [(U)/q]^{-n}$$
$$\log(N/N_o) = -n \log(U/q)$$

The required log–log plot is as follows:

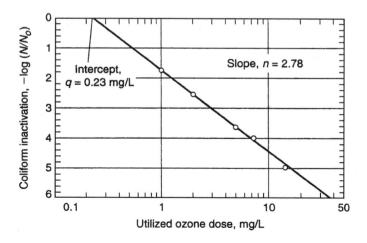

The required coefficients are

$$q = 0.23 \text{ mg/L}$$
$$n = 2.78$$

2. *Determine the ozone dose required to achieve an effluent coliform concentration of 240 MPN/100 mL.*
Rearrange Eq. (16.2) to solve for U:

$$U = q(N/N_o)^{-1/n}$$

Then

$$U = (0.23 \text{ mg/L})(240/10^6)^{-1/2.78} = 4.61 \text{ mg/L}$$

3. *Determine the dose that must he applied.*
Use Eq. (16.3), and assume a transfer efficiency of 80 percent. Then

$$D = U\left(\frac{100}{TE}\right) = (4.61 \text{ mg/L})\left(\frac{100}{80}\right) = 5.76 \text{ mg/L}$$

16.26   *DESIGN OF A UV DISINFECTION SYSTEM*

Design a UV disinfection system that will deliver a minimum design dose of 100 mJ/cm^2. Assume for the purpose of this example that the following data apply:

Wastewater characteristics:

- Minimum design flow = 6000 m^3/d = 4167 L/min (diurnal low flow).
- Maximum design flow = 21,000 m^3/d = 14,584 L/min (peak hour flow with recycle streams).
- Minimum transmittance = 55 percent.

System characteristics:

- Lamp configuration is horizontal.
- Validated system performance = 100 mJ/cm^2 within range of 20–43 L/min · lamp.
- System headloss coefficient = 1.8 (manufacturer-specific).
- Lamp/sleeve diameter = 23 mm.
- Cross-sectional area of quartz sleeve = 4.15 × 10^{-4} m^2.
- Lamp spacing = 75 mm (center to center).
- One standby UV bank will be required per channel.

Calculation Procedure

1. *Determine the number of UV channels.*
The manufacturer has provided validated information for a three-bank system at a flow range of 20 to 80 L/min · lamp. The system is capable of applying a dose of 100 mJ/cm^2 within the range of 20 to 43 L/min · lamp.

   Therefore, the system has an approximate 2:1 flow variation capacity for the design dose under consideration.

Use two channels. From 4167 to 8000 L/min use one channel. From 8000 to 14,584 L/min split the flow between two channels such that each channel is loaded between 4000 and 7300 L/min.

2. *Determine the number of lamps required per channel.*
At 8000 L/min, the total number of required lamps is

$$\text{lamps required at 8000 L/min} = \frac{(8000 \text{ L/min})}{(43 \text{ L/lamp} \cdot \text{min})} = 186 \text{ lamps}$$

3. *Determine the minimum number of lamps per bank.*

$$\frac{186 \text{ lamps}}{3 \text{ banks}} = 62 \frac{\text{lamps}}{\text{bank}}$$

4. *Configure the UV disinfection system.*
Typically, 2, 4, 8, or 16 lamps per module are available. For an 8-lamp module, eight modules are required per bank for a total of 64 lamps per bank.

5. *Check that the design falls within the manufacturer recommended range.*
At low flow,

$$\frac{(4167 \text{ L/min})}{192 \text{ lamps}} = 21.7 \text{ L/lamp} \cdot \text{min}$$

At high flow,

$$\frac{(14,584 \text{ L/min})}{384 \text{ lamps}} = 38.0 \text{ L/lamp} \cdot \text{min}$$

Both of these hydraulic loading rates fall within the acceptable range for the UV disinfection system provided by the manufacturer.

6. *Check whether the headloss for the selected configuration is acceptable.*
First determine the channel cross-sectional area:

$$\text{cross-sectional area of channel} = (8 \times 0.075 \text{ m}) \times (8 \times 0.075 \text{ m})$$
$$= 0.36 \text{ m}^2$$

Then determine the net channel cross-sectional area by subtracting the cross-sectional area of the quartz sleeves (4.15×10^{-4} m^2/lamp):

$$A_{\text{channel}} = 0.36 \text{ m}^2 - [(8 \times 8) \text{ lamps/bank}] \times (4.15 \times 10^{-4} \text{ m}^2/\text{lamp})$$
$$= 0.33 \text{ m}^2$$

Now determine the maximum velocity in each channel:

$$v_{\text{channel}} = \frac{(14,584 \text{ L/min})}{(2 \text{ channel})(0.33 \text{ m}^2)(10^3 \text{ L/m}^3)(60 \text{ s/min})} = 0.37 \text{ m/s}$$

The headloss per UV channel is then

$$h_{channel} = 1.8\frac{v^2}{2g}$$

$$= 1.8\frac{(0.37 \text{ m/s})^2(10^3 \text{ mm/m})}{2(9.81 \text{ m/s}^2)}(4 \text{ banks}) = 50 \text{ mm}$$

Note that four banks were used to determine system headloss. Use of four banks includes a redundant bank of lamps in each channel. Because the clear spacing between quartz sleeves is 52 mm (75 mm − 23 mm), the headloss cannot exceed 26 mm total (one-half the clear spacing between the quartz sleeves) without exposing the uppermost row of lamps to the air. To allow for the calculated 50 mm of total headloss, each UV channel will require a stepped channel bottom. A 24-mm step between the second and third bank of lamps is required to allow for the expected headloss and to allow the third and fourth banks of lamps to be set lower.

7. *Summarize the system configuration.*
The minimum required system utilizes two channels, each channel containing four banks of lamps in series, three operational banks and one redundant bank. Each bank contains eight modules, each of which contains eight lamps.

16.27 *ESTIMATION OF BLOWDOWN WATER COMPOSITION*

Reclaimed water with the chemical characteristics given in the table is being considered for use as makeup water for a cooling tower. Calculate the composition of the blowdown flow if five cycles of concentration are to be used. Assume that the temperature of the hot water entering the cooling tower is 50°C (120°F) and the solubility of $CaSO_4$ is about 2200 mg/L as $CaCO_3$ at this temperature.

Constituent	Concentration, mg/L
Total hardness (as $CaCO_3$)	118
$\quad$ Ca^{2+} (as $CaCO_3$)	85
$\quad$ Mg^{2+} (as $CaCO_3$)	33
Total alkalinity (as $CaCO_3$)	90
$\quad$ SO_4^{2-}	20
$\quad$ Cl^-	19
$\quad$ SiO_2	2

Note: Molecular weights—$CaCO_3 = 100$, $CaSO_4 = 136$, $H_2SO_4 = 98$, and $SO_4 = 96$.

Calculation Procedure

1. *Determine the total hardness in the circulating water.*
When the total alkalinity is less than the total hardness, Ca and Mg are also present in forms other than carbonate hardness.

Setting the cycles of concentration equal to 5, we obtain the total hardness in circulating water of

$$C_b = (\text{cycles of concentration})(C_m)$$

where C_b and C_m are salt concentration in blowdown and makeup water, respectively. Thus,

$$C_b = 5 \times 118 = 590 \text{ mg/L } CaCO_3$$

2. Determine the total amount of H_2SO_4 that must be added to convert the $CaCO_3$ to $CaSO_4$.
To convert from $CaCO_3$ to $CaSO_4$, sulfuric acid is injected into the circulating water and the following reaction occurs:

$$CaCO_3 \; + \; H_2SO_4 \; \rightarrow \; CaSO_4 \; + \; H_2O \; + \; CO_2$$
$$\quad 100 \qquad\quad 98 \qquad\qquad 136$$

where 100, 98, and 136 are the molecular weights.

The alkalinity in the circulating water, if not converted into sulfates, is $5 \times 90 = 450$ mg/L as $CaCO_3$. If 10 percent of the alkalinity is left unconverted to avoid corrosion, the amount of alkalinity remaining is $0.1 \times 450 = 45$ mg/L as $CaCO_3$. The amount of alkalinity that must be converted is $450 - (0.1 \times 450) = 405$ mg/L as $CaCO_3$.

The amount of sulfate that must be added for the conversion is

$$SO_4^{2-} = (405 \text{ mg/L})\left(\frac{96}{100}\right) = 389 \text{ mg/L}$$

Converting to mg/L $CaSO_4$ yields

$$CaSO_4 = (389 \text{ mg/L})\left(\frac{136}{96}\right) = 551 \text{ mg/L}$$

3. Determine the required sulfuric acid concentration in the circulating water.

$$H_2SO_4 = (389 \text{ mg/L } SO_4)\left(\frac{98}{96}\right) = 397 \text{ mg/L}$$

4. Determine the sulfate concentration in the circulating water contributed by the makeup water.
Sulfate from makeup water is $5 \times 20 = 100$ mg/L as SO_4. If combined with Ca^{2+}, the concentration is

$$CaSO_4 = (100 \text{ mg/L})\left(\frac{136}{96}\right) = 142 \text{ mg/L}$$

5. Determine the amount of additional $CaSO_4$ formation that is permissible.
The solubility of $CaSO_4$ at 50°C (120°F) is about 2200 mg/L. In the circulating water, 142 mg/L $CaSO_4$ was originally present after five cycles of concentration and 551 mg/L was formed by the addition of sulfuric acid. Therefore, an additional 1507 mg/L of $CaSO_4$ formation is theoretically permissible before the solubility limit is exceeded. The cycles of concentration could have been carried out much higher before $CaSO_4$ would precipitate in the system.

6. Determine the concentrations of Cl and SiO_2 in the circulating water.
For chloride,

$$Cl^- = 5 \times 19 = 95 \text{ mg/L}$$

For silica,

$$SiO_2 = 5 \times 2 = 10 \text{ mg/L}$$

7. *Summarize the composition of the blowdown flow after five cycles of concentration.*

Parameter	Concentration, mg/L	
	Initial	Final
Total hardness (as $CaCO_3$)	118	590
Total alkalinity (as $CaCO_3$)	90	45
SO_4^{2-}	20	489
Cl^-	19	95
SiO_2	2	10

16.28 DESIGN A GRAVITY THICKENER FOR COMBINED PRIMARY AND WASTE-ACTIVATED SLUDGE

Design a gravity thickener for a wastewater-treatment plan having primary and waste-activated sludge with the following characteristics:

Type of sludge	Specific gravity	Solids, %	Flow rate, m³/d
Average design conditions:			
Primary sludge	1.03	3.3	400
Waste activated	1.005	0.2	2250
Peak design conditions:			
Primary sludge	1.03	3.4	420
Waste activated	1.005	0.23	2500

Calculation Procedure

1. *Compute the dry solids at peak design conditions.*
For the primary sludge,

$$\text{kg/dry solids} = (420 \text{ m}^3/\text{d})(1.03)(0.034 \text{ g/g})(10^3 \text{ kg/m}^3)$$
$$= 14,708 \text{ kg/d}$$

For the waste-activated sludge,

$$\text{kg/dry solids} = (2500 \text{ m}^3/\text{d})(1.005)(0.0023 \text{ g/g})(10^3 \text{ kg/m}^3)$$
$$= 5779 \text{ kg/d}$$

Therefore,

$$\text{combined sludge mass} = 14,708 + 5779 = 20,487 \text{ kg/d}$$
$$\text{combined sludge flow rate} = 2500 + 420 = 2920 \text{ m}^3/\text{d}$$

2. *Compute solids concentration of the combined sludge.*
Assume the specific gravity of the combined sludge is 1.02. Then

$$\% \text{ solids} = \frac{(20,487 \text{ kg/d})}{(2926 \text{ m}^3/\text{d})(1.02)(10^3 \text{ kg/m}^3)} \times 100\% = 0.69\%$$

TABLE 16.4 Typical Concentrations of Unthickened and Thickened Sludges and Solids Loadings for Gravity Thickeners

Type of sludge or biosolids	Solids concentration, %		Solids loading	
	Unthickened	Thickened	lb/ft$^2 \cdot$ d	kg/m$^2 \cdot$ d
Separate:				
Primary sludge	2–6	5–10	20–30	100–150
Trickling-filter humus sludge	1–4	3–6	8–10	40–50
Rotating biological contactor	1–3.5	2–5	7–10	35–50
Air-activated sludge	0.5–1.5	2–3	4–8	20–40
High-purity oxygen-activated sludge	0.5–1.5	2–3	4–8	20–40
Extended aeration-activated sludge	0.2–1.0	2–3	5–8	25–40
Anaerobically digested primary sludge from primary digester	8	12	25	120
Combined:				
Primary and trickling-filter humus sludge	2–6	5–9	12–20	60–100
Primary and rotating biological contactor	2–6	5–8	10–18	50–90
Primary and waste-activated sludge	0.5–1.5	4–6	5–14	25–70
	2.5–4.0	4–7	8–16	40–80
Waste-activated sludge and trickling-filter humus sludge	0.5–2.5	2–4	4–8	20–40
Chemical (tertiary) sludge:				
High lime	3–4.5	12–15	24–61	120–300
Low lime	3–4.5	10–12	10–30	50–150
Iron	0.5–1.5	3–4	2–10	10–50

Source: Adapted from WEF (1996).

3. Compute surface area based on solids loading rate.

Because the solids concentration is between 0.5% and 1.5%, select a solids loading rate of 50 kg/ m$^2 \cdot$ d from Table 16.4. Then

$$\text{area} = \frac{(20,487 \text{ kg/d})}{50 \text{ kg/m}^2 \cdot \text{d}} = 409.7 \text{ m}^2$$

4. Compute hydraulic loading rate.

$$\text{Hydraulic loading} = \frac{(2920 \text{ m}^3/\text{d})}{409.7 \text{ m}^2} = 7.13 \text{ m}^3/\text{m}^2 \cdot \text{d}$$

5. Compute diameter of thickener.

Assume two thickeners. Then

$$\text{diameter} = \sqrt{\frac{4 \times 409.7 \text{ m}^2}{2 \times \pi}} = 16.15 \text{ m}$$

Related Calculations. The hydraulic loading rate of 7.13 m^3/m$^2 \cdot$ d at peak design flow is at the lower end of the recommended rate. To prevent septicity and odors, dilution water should be provided. Calculation of the dilution water requirements for average design flow is a homework problem. The thickener size of 16.15 m is within the maximum size of 20 m customarily recommended by thickener equipment manufacturers for use in municipal wastewater treatment. In actual design, round the thickener diameter to the nearest 0.5 m, or, in this case, 16 m.

16.29 BELT-FILTER PRESS DESIGN

A wastewater-treatment plant produces 72,000 L/d of thickened biosolids containing 3 percent solids. A belt-filter press installation is to be designed based on a normal operation of 8 h/d and 5 d/week, a belt-filter press loading rate of 275 kg/m · h, and the following data:

Total solids in dewatered sludge = 25 percent.
Total suspended solids concentration in filtrate = 900 mg/L = 0.09 percent.
Washwater flow rate = 90 L/min per m of belt width.
Sepcific gravities of sludge feed, dewatered cake, and filtrate are 1.02, 1.07, and 1.01, respectively.

Compute the number and size of belt-filter presses required and the expected solids capture, in percent. Determine the daily hours of operation required if a sustained 3-day peak solids load occurs.

Calculation Procedure

1. *Compute average weekly sludge production rate.*

$$\text{Wet biosolids} = (72,000 \text{ L/d})(7 \text{ d/week})(10^3 \text{ g/L})(1 \text{ kg}/10^3 \text{ g})(1.02)$$
$$= 514,080 \text{ kg/week}$$
$$\text{dry solids} = 514,080 \times 0.03 = 15,422 \text{ kg/week}$$

2. *Compute daily and hourly dry solids-processing requirements.*

$$\text{Daily rate} = 15,442 \text{ kg/week}/5 \text{ operating d/week}$$
$$= 3084 \text{ kg/d}$$
$$\text{hourly rate} = 3084/8 = 385.5 \text{ kg/h (per 8 h operating d)}$$

3. *Compute belt-filter press size.*

$$\text{Belt width} = \frac{(385.5 \text{ kg/h})}{(275 \text{ kg/m} \cdot \text{h})} = 1.40 \text{ m}$$

Use one 1.5-m belt-filter press and provide one of identical size for standby.

4. *Compute filtrate flow rate by developing solids balance and flow balance equations.*
First, develop the daily solids balance equation:

$$\text{solids in sludge feed} = \text{solids in sludge cake} + \text{solids in filtrate}$$
$$3084 = (S \text{ L/d})(1.07)(0.25) + (F \text{ L/d})(1.01) \times (0.0009)$$
$$3084 = 0.2675 \, S + 0.00091 \, F$$

where S = sludge cake flow rate, L/d
F = filtrate flow rate, L/d

Then develop the flow rate equation:

$$\text{sludge flow rate} + \text{washwater flow rate} = \text{filtrate flow rate} + \text{cake flow rate}$$
$$\text{daily sludge flow rate} = (72,000 \text{ L/d})(7/5) = 100,800 \text{ L/d}$$
$$\text{washwater flow rate} = (90 \text{ L/min} \cdot \text{m})(1.5 \text{ m})(60 \text{ min/h})(8 \text{ h/d})$$
$$= 64,800 \text{ L/d}$$

So

$$100,800 + 64,800 = 165,600 = F + S$$

Now solve the mass balance and flow rate equations simultaneously to get

$$F = 154,600 \text{ L/d}$$

5. Determine solids capture.

$$\text{Solids capture} = \frac{\text{solids in feed} - \text{solids in filtrate}}{\text{solids in feed}} \times 100\%$$

$$= \frac{(3084 \text{ kg/d}) - [(154,600 \text{ L/d})(1.01)(0.0009)(10^3 \text{ g/L})(1 \text{ kg}/10^3 \text{ g})]}{(3084 \text{ kg/d})} \times 100\%$$

$$= 95.4\%$$

6. Determine operating requirements for sustained peak biosolids load.
First determine the peak 3-day load. From Fig. 16.3, the ratio of peak to average mass loading for three consecutive days is 2. The peak load is $72,000(2) = 144,000$ L/d.

Then determine the daily operating time requirements, neglecting sludge in storage:

$$\text{dry solids/d} = 144,000/\text{d} \ (1.02)(0.03)$$

$$= 4406 \text{ kg/d}$$

$$\text{operating time} = \frac{(4406 \text{ kg/d})}{(275 \text{ kg/m} \cdot \text{h})(1.5 \text{ m})} = 10.7 \text{ h}$$

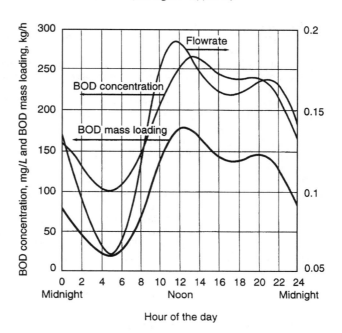

FIGURE 16.3 Illustration of diurnal wastewater flow, BOD, and mass-loading variability.

The operating time can be accomplished by running the standby belt-filter press in addition to the duty press or by operating the duty press for an extended shift.

Comment. The value of sludge storage is important in dewatering applications because of the ability to schedule operations to suit labor availability most efficiently. Scheduling sludge dewatering operations during the day shift is also desirable if sludge has to be hauled off-site.

*16.30   DESIGN OF AN AEROBIC DIGESTER

An aerobic digester is to be designed to treat the waste sludge produced by an activated-sludge wastewater treatment facility. The input waste sludge will be 12,823 gal/d (48.5 L/d) (input 5 d/ week only) of thickened waste-activated sludge at 5.0 percent solids. Assume the following apply:

1. The minimum liquid temperature in the winter is 15°C (59°F), and the maximum liquid temperature in the summer is 30°C (86°F).
2. The system must achieve a 40 percent volatile suspended solids (VSS) reduction in the winter.
3. Sludge concentration in the digester is 70 percent of the incoming thickened-sludge concentration.
4. The volatile fraction of digester suspended solids is 0.8.

Calculation Procedure

1. *Find the daily volume of sludge for disposal.*
Factors that must be considered in designing aerobic digesters include temperature, solids reduction, tank volume (hydraulic retention time), oxygen requirements, and energy requirements for mixing.

Because the majority of aerobic digesters are open tanks, digester liquid temperatures are dependent on weather conditions and can fluctuate extensively. As with all biological systems, lower temperatures retard the process, whereas higher temperatures accelerate it. The design of the aerobic digester should provide the necessary degree of sludge stabilization at the lowest liquid operating temperature and should supply the maximum oxygen requirements at the maximum liquid operating temperature.

A major objective of aerobic digestion is to reduce the mass of the solids for disposal. This reduction is assumed to take place only with the biodegradable content (VSS) of the sludge, although there may be some destruction of the inorganics as well. Typical reduction in VSS ranges from 40 to 50 percent. Solids destruction is primarily a direct function of both basin liquid temperature and sludge age, as indicated in Fig. 16.4. The plot relates VSS reduction to degree-days (temperature × sludge age).

To ensure proper operation, the contents of the aerobic digester should be well mixed. In general, because of the large amount of air that must be supplied to meet the oxygen requirement, adequate mixing is usually achieved. However, mixing power requirements should always be checked.

The aerobic digester will operate 7 days per week, unlike the thickening facilities which operate intermittently due to larger operator attention requirements. The thickened sludge is input to the digester at 12,823 gal/d (48.5 L/d), 5 days per week. However, the volume of the sludge to be disposed of daily by the digester will be lower due to its operation 7 days per week (the "bugs" do not take the weekends off). Therefore the volume of sludge to be disposed of daily (Q_i) is:

$$Q_i = (12,823 \text{ gal/d})(5/7) = 9159 \text{ gal/d} = 1224 \text{ ft}^3/\text{d} \ (34.6 \text{ m}^3/\text{d})$$

2. *Determine the required VSS reduction.*
The sludge age required for winter conditions is obtained from Fig. 16.4 using the minimum winter temperature and required VSS reduction.

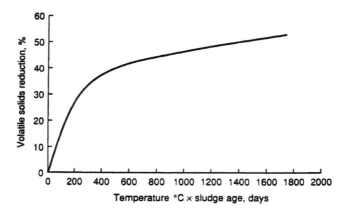

FIGURE 16.4 VSS reduction in aerobic digester vs. liquid temperature ×
sludge age. (*Metcalf and Eddy—Wastewater Engineering: Treatment, Disposal,
and Reuse, 3rd ed., McGraw-Hill.*)

To achieve a 40 percent VSS reduction in the winter, the degree-days required from Fig. 16.4 is
475°C · d. Therefore, the required sludge age is 475°C · d/15°C = 31.7 days. During the summer,
when the liquid temperature is 30°C, the degree-days required is (30°C)(31.7 days) = 951°C · d.
From Fig. 16.4, the VSS reduction will be 46 percent.

The total mass of solids processed by the digester will be 3933 lb/d (1785.6 kg/d) which is the
total mass of solids wasted from the treatment facility. The total mass of VSS input to the digester is:

$$(0.8)(3933 \text{ lb/d}) = 3146 \text{ lb/d} (1428.3 \text{ kg/d})$$

Therefore, during the winter:

- VSS reduction = (3146 lb/d)(0.40) = 1258 lb VSS reduced/d (571.1 kg/d).
- Digested (stabilized) sludge leaving the digester = 3933 lb/d − 1258 lb/d = 2675 lb/d (1214.5 kg/d).

3. *Compute the volume of digested sludge.*
The volume of digested sludge is:

$$V = \frac{W_s}{(\rho)(\text{s.g.})(\% \text{ solids})}$$

where V = sludge volume (ft³) (m³)
  W_s = weight of sludge (lb) (kg)
  ρ = density of water (62.4 lb/ft³) (994.6 kg/m³)
  s.g. = specific gravity of digested sludge (assume s.g. = 1.03)
% solids = percent solids expressed as a decimal (incoming sludge: 5.0%)

Therefore, the volume of the digested sludge is:

$$V = \frac{2675 \text{ lb/d}}{(62.4 \text{ lb/ft}^3)(1.03)(0.05)} = 832 \text{ ft}^3/\text{d} = 6223 \text{ gal/d} (23.6 \text{ L/d})$$

During the summer:

- VSS reduction = (3146 lb/d)(0.46) = 1447 lb VSS reduced/d (656.9 kg/d).
- Digested (stabilized) sludge leaving the digester = 3933 lb/d − 1447 lb/d = 2486 lb/d (1128.6 kg/d).
- Volume of digested sludge:

$$V = \frac{2486 \text{ lb/d}}{(62.4 \text{ lb/ft}^3)(1.03)(0.05)} = 774 \text{ ft}^3/\text{d} = 5790 \text{ gal/d (21.9 L/d)}$$

4. *Find the oxygen and air requirements.*

The oxygen required to destroy the VSS is approximately 2.3 lb O_2/lb VSS (kg/kg) destroyed. Therefore, the oxygen requirements for winter conditions are:

$$(1258 \text{ lb VSS/d})(2.3 \text{ lb } O_2/\text{lb VSS}) = 2893 \text{ lb } O_2/\text{d (1313.4 kg/d)}$$

The volume of air required at standard conditions (14.7 lb/in² and 68°F) (96.5 kPa and 20°C) assuming air contains 23.2 % oxygen by weight and the density of air is 0.075 lb/ft³ is:

$$\text{volume of air} = \frac{2893 \text{ lb } O_2/\text{d}}{(0.075 \text{ lb/ft}^3)(0.232)} = 166,264 \text{ ft}^3/\text{d (4705.3 m}^3/\text{d)}$$

For summer conditions:

- Oxygen required = (1447 lb/d)(2.3 lb O_2/d) = 3328 lb O_2/d (1510.9 kg/d).
- Volume of air = 3328 lb O_2/d/(0.075 lb/ft³)(0.232) = 191,264 ft³/d (5412.8 m³/d).

Note that the oxygen transfer efficiency of the digester system must be taken into account to get the actual volume of air required. Assuming diffused aeration with an oxygen transfer efficiency of 10 percent, the actual air requirements at standard conditions are:

- Winter: volume of air = 166,264 ft³/d/(0.1)(1400 min/d) = 1155 ft³/min (32.7 m³/min)
- Summer: volume of air = 191,264 ft³/d/(0.1)(1440 min/d) = 1328 ft³/min (37.6 m³/min)

To summarize winter and summer conditions:

Parameter	Winter	Summer
Total solids in, lb/d (kg/d)	3933 (1785.6)	3933 (1785.6)
VSS in, lb/d (kg/d)	3146 (1428.3)	3146 (1428.3)
VSS reduction, (%)	40	46
VSS reduction, lb/d (kg/d)	1258 (571.1)	1447 (656.9)
Digested sludge out, gal/d (L/d)	6223 (23.6)	5790 (21.9)
Digested sludge out, lb/d (kg/d)	2675 (1214.5)	2486 (1128.6)
Air requirements @ S.C., ft³/min (m³/min)	1155 (32.7)	1328 (37.6)

5. *Determine the aerobic digester volume.*

From the above analysis it is clear that the aerobic digester volume will be calculated using values obtained under the winter conditions analysis, while the aeration equipment will be sized using the 1328 ft³/min (37.6 m³/min) air requirement obtained under the summer conditions analysis.

The volume of the aerobic digester is computed using the following equation, assuming the digester is loaded with waste-activated sludge only:

$$V = \frac{Q_i X_i}{X(K_d P_v + 1/\theta_c)}$$

where V = volume of aerobic digester, ft^3 (m^3)

$\quad Q_i$ = influent average flow rate to the digester, ft^3/d (m^3/d)

$\quad X_i$ = influent suspended solids, mg/L (50,000 mg/L for 5.0% solids)

$\quad X$ = digester total suspended solids, mg/L

$\quad K_d$ = reaction rate constant, d^{-1}. May range from 0.05 d^{-1} at 15°C (59°F) to 0.14 d^{-1} at 25°C (77°F) (assume 0.06 d^{-1} at 15°C)

$\quad P_v$ = volatile fraction of digester suspended solids (expressed as a decimal) = 0.8 (80%) as stated in the initial assumptions

$\quad \theta_c$ = solids retention time (sludge age), d

Using values obtained above with winter conditions governing, the aerobic digester volume is:

$$V = \frac{(1224 \text{ ft}^3/\text{d})(50,000 \text{ mg/L})}{(50,000 \text{ mg/L})(0.7)[(0.06 \text{ d}^{-1})(0.8) + 1/31.7 \text{ d}]} = 21,982 \text{ ft}^3 \ (622.2 \text{ m}^3)$$

The air requirement per 1000 ft^3 (2.8 m^3) of digester volume with summer conditions governing is:

$$\text{volume of air} = \frac{1328 \text{ ft}^3/\text{min}}{21.982 \ 10^3 \text{ft}^3} = 60.41 \text{ ft}^3/\text{min}/10^3 \text{ft}^3 \ (0.97 \text{ m}^3/\text{min}/\text{Mm}^3)$$

The mixing requirements for diffused aeration range from 20 to 40 ft^3/min/10^{3}ft^3 (0.32 to 0.64 m^3/min/Mm^3). Therefore, adequate mixing will prevail.

This procedure is the work of Kevin D. Wills, M.S.E., P.E., who, at the time of its preparation, was consulting engineer, Stanley Consultants, Inc.

*16.31 DESIGN OF AN AERATED GRIT CHAMBER

Domestic wastewater enters a wastewater treatment facility with an average daily flow rate of 4.0 Mgal/d (15,140 L/d). Assuming a peaking factor of 2.5, size an aerated grit chamber for this facility including chamber volume, chamber dimensions, air requirement, and grit quantity.

Calculation Procedure

1. Determine the aerated grit chamber volume.

Grit removal in a wastewater treatment facility prevents unnecessary abrasion and wear of mechanical equipment such as pumps and scrappers, and grit deposition in pipelines and channels. Grit chambers are designed to remove grit (generally characterized as nonputrescible solids) consisting of sand, gravel, or other heavy solid materials that have settling velocities greater than those of the organic putrescible solids in the wastewater.

In aerated grit chamber systems, air introduced along one side near the bottom causes a spiral roll velocity pattern perpendicular to the flow through the tank. The heavier particles with their correspondingly higher settling velocities drop to the bottom, while the rolling action suspends the lighter organic particles, which are carried out of the tank. The rolling action induced by the air diffusers is independent of the flow through the tank. Then non–flow–dependent rolling action allows the aerated grit chamber to operate effectively over a wide range of flows. The heavier particles that settle on the bottom of the tank are moved by the spiral flow of the water across the tank bottom and into a grit hopper. Screw augers or air lift pumps are generally utilized to remove the grit from the hopper.

The velocity of roll governs the size of the particles of a given specific gravity that will be removed. If the velocity is too great, grit will be carried out of the chamber. If the velocity is too small, organic material will be removed with the grit. The quantity of air is easily adjusted by throttling the air discharge or using adjustable speed drives on the blowers. With proper adjustment, almost 100 percent grit removal will be obtained, and the grit will be well washed. Grit that is not well washed will contain organic matter and become a nuisance through odor emission and the attraction of insects.

Wastewater will move through the aerated grit chamber in a spiral path as illustrated in Fig. 16.5. The rolling action will make two to three passes across the bottom of the tank at maximum flow and more at lesser flows. Wastewater is introduced in the direction of the roll.

At peak flow rate, the detention time in the aerated grit chamber should range from 2 to 5 min. A detention time of 3 min will be used for this example. Because it is necessary to drain the chamber periodically for routine maintenance, two redundant chambers will be required. Therefore, the volume of each chamber is:

$$V(\text{ft}^3) = \frac{(\text{peak flow rate, gal/d})(\text{detention time, min})}{(7.48 \text{ gal/ft}^3)(24 \text{ h/d})(60 \text{ min/h})}$$

Using values from above, the chamber volume is:

$$V(\text{ft}^3) = \frac{(2.5)(4 \times 10^6 \text{ gal/d})(3 \text{ min})}{(7.48 \text{ gal/ft}^3)(24 \text{ h/d})(60 \text{ min/h})} = 2785 \text{ ft}^3 \ (78.8 \text{ m}^3)$$

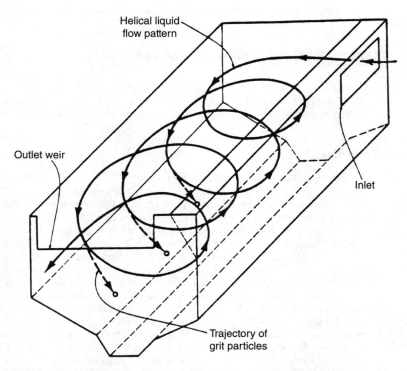

FIGURE 16.5   Aerated grit chamber. (*Metcalf and Eddy—Wastewater Engineering: Treatment, Disposal, and Reuse, 3rd ed., McGraw-Hill.*)

2. Determine the dimensions of the grit chamber.

Width-depth ratio for aerated grit chambers range from 1:1 to 5:1. Depths range from 7 to 16 ft (2.1 to 4.87 m). Using a width-depth ratio of 1.2:1 and a depth of 8 ft (2.43 m), the dimensions of the aerated grit chamber are:

$$\text{Width} = (1.2)(8 \text{ ft}) = 9.6 \text{ ft} (2.92 \text{ m})$$

$$\text{Length} = (\text{volume})/[(\text{width})(\text{depth})] = \frac{2785 \text{ ft}^3}{(8 \text{ ft})(9.6 \text{ ft})} = 36.3 \text{ ft} (11.1 \text{ m})$$

Length-width ratios range from 3:1 to 5:1. As a check, length to width ratio for the aerated grit chamber sized above is: 36.3 ft/9.6 ft = 3.78:1 which is acceptable.

3. Determine the air supply required.

The air supply requirement for an aerated grit chamber ranges from 2.0 to 5.0 ft^3/min/ft of chamber length (0.185 to 0.46 m^3/min · m). Using 5.0 ft^3/min/ft (0.46 m^3/min · m) for design, the amount of air required is:

$$\text{air required (ft}^3/\text{min)} = (5.0 \text{ ft}^3/\text{min/ft})(36.3 \text{ ft}) = 182 \text{ ft}^3/\text{min} (5.2 \text{ m}^3/\text{min})$$

4. Estimate the quantity of grit expected.

Grit quantities must be estimated to allow sizing of grit handling equipment such as grit conveyors and grit dewatering equipment. Grit quantities from an aerated grit chamber vary from 0.5 to 27 ft^3/ Mgal (3.74 to 201.9 m^3/L) of flow. Assume a value of 20 ft^3/Mgal (149.5 m^3/L). Therefore, the average quantity of grit expected is:

$$\text{volume of grit (ft}^3/\text{d)} = (20 \text{ ft}^3/\text{Mgal})(4.0 \text{ Mgal/d}) = 80 \text{ ft}^3/\text{d} (2.26 \text{ m}^3/\text{d})$$

Some advantages and disadvantages of the aerated grit chamber are listed below:

Advantages	Disadvantages
The same efficiency of grit removal is possible over a wide flow range.	Power consumption is higher than other grit removal processes.
Head loss through the grit chamber is minimal.	Additional labor is required for maintenance and control of the aeration system.
By controlling the rate of aeration, a grit of relatively low putrescible organic content can be removed.	Significant quantities of potentially harmful volatile organics and odors may be released from wastewaters containing these constituents.
Preaeration may alleviate septic conditions in the incoming wastewater to improve performance of downstream treatment units.	Foaming problems may be created if influent wastewater has surfactants present.
Aerated grit chambers can also be used for chemical addition, mixing, preaeration, and flocculation ahead of primary treatment.	

This procedure is the work of Kevin D. Wills, M.S.E., P.E., who, at the time of its preparation, was consulting engineer, Stanley Consultants, Inc.

*16.32   DESIGN OF A RAPID-MIX BASIN AND FLOCCULATION BASIN

One Mgal/d (3785 m³/d) of equalized secondary effluent from a municipal wastewater treatment facility is to receive tertiary treatment through a direct filtration process which includes rapid-mix with a polymer coagulant, flocculation, and filtration. Size the rapid-mix and flocculation basins necessary for direct filtration and determine the horsepower of the required rapid mixers and flocculators.

Calculation Procedure

1. Determine the required volume of the rapid-mix basin.

A process flow diagram for direct filtration of a secondary effluent is presented in Fig. 16.6. This form of tertiary wastewater treatment is used following secondary treatment when an essentially "virus-free" effluent is desired for wastewater reclamation and reuse.

The rapid-mix basin is a continuous mixing process in which the principal objective is to maintain the contents of the tank in a completely mixed state. Although there are numerous ways to accomplish continuous mixing, mechanical mixing will be used here. In mechanical mixing, turbulence is induced through the input of energy by means of rotating impellers such as turbines, paddles, and propellers.

The hydraulic retention time of typical rapid-mix operations in wastewater treatment range from 5 to 20 s. A value of 15 s will be used here. The required volume of the rapid-mix basin is calculated as follows:

$$\text{volume }(V) = (\text{hydraulic retention time})(\text{wastewater flow})$$

$$V = \frac{(15 \text{ s})(1 \times 10^6 \text{ gal/d})}{86,400 \text{ s/d}} = 174 \text{ gal} \cong 24 \text{ ft}^3 \ (0.68 \text{ m}^3)$$

2. Compute the power required for mixing.

The power input per volume of liquid is generally used as a rough measure of mixing effectiveness, based on the reasoning that more input power creates greater turbulence, and greater turbulence leads to better mixing. The following equation is used to calculate the required power for mixing:

$$G = \sqrt{\frac{P}{\mu V}}$$

where G = mean velocity gradient (s⁻¹)
  P = power requirement (ft · lb/s) (kW)
  μ = dynamic viscosity (lb · s/ft²) (Pa · s)
  V = volume of mixing tank (ft³) (m³)

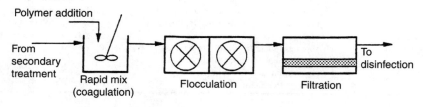

FIGURE 16.6   Process flow for direct filtration.

G is a measure of the mean velocity gradient in the fluid. G values for rapid mixing operations in wastewater treatment range from 250 to 1500 s^{-1}. A value of 1000 s^{-1} will be used here. For water at 60°F (15.5°C), dynamic viscosity is 2.36×10^{-5} lb · s/ft^2 (1.13×10^{-3} Pa · s). Therefore, the required power for mixing is computed as follows:

$$P = G^2\mu V = (1000 \text{ s}^{-1})^2(2.36 \times 10^{-5} \text{ lb · s/ft}^2)(24 \text{ ft}^3) = 566 \text{ ft · lb/s}$$
$$= 1.03 \text{ hp (0.77 kW)}$$

Use the next largest motor size available = 1.5 hp (1.12 kW). Therefore, a 1.5 hp (1.12 kW) mixer should be used.

3. Determine the required volume and power input for flocculation.

The purpose of flocculation is to form aggregates, or flocs, from finely divided matter. The larger flocs allow a greater solids removal in the subsequent filtration process. In the direct filtration process, the wastewater is completely mixed with a polymer coagulant in the rapid-mix basin. Following rapid mix, the flocculation tanks gently agitate the wastewater so that large flocs form with the help of the polymer coagulant. As in the rapid-mix basins, mechanical flocculators will be utilized.

For flocculation in a direct filtration process, the hydraulic retention time will range from 2 to 10 min. A retention time of 8 min will be used here. Therefore, the required volume of the flocculation basin is:

$$V = \frac{(8 \text{ min})(1 \times 10^6 \text{ gal/d})}{1440 \text{ min/d}} = 5556 \text{ gal} \cong 743 \text{ ft}^3 \text{ (21 m}^3)$$

G values for flocculation in a direct filtration process range from 20 to 100 s^{-1}. A value of 80 s^{-1} will be used here. Therefore, the power required for flocculation is:

$$P = G^2\mu V = (80 \text{ s}^{-1})^2(2.36 \times 10^{-5})(743 \text{ ft}^3)$$
$$= 112 \text{ ft · lb/s} = 0.2 \text{ hp (0.15 kW)}$$

Use the next largest motor size available = 0.5 hp (0.37 kW). Therefore, a 0.5 hp (0.37 kW) flocculator should be used.

It is common practice to taper the energy input to flocculation basins so that flocs initially formed will not be broken as they leave the flocculation facilities. In the above example, this may be accomplished by providing a second flocculation basin in series with the first. The power input to the second basin is calculated using a lower G value (such as 50 s^{-1}) and hence provides a gentler agitation.

Related Calculations.   If the flows to the rapid-mix and flocculation basin vary significantly, or turn down capability is desired, a variable speed drive should be provided for each mixer and flocculator. The variable speed drive should be controlled via an output signal from a flow meter immediately upstream of each respective basin.

It should be noted that the above analysis provides only approximate values for mixer and flocculator sizes. Mixing is in general a "black art," and a mixing manufacturer is usually consulted regarding the best type and size of mixer or flocculator for a particular application.

This procedure is the work of Kevin D. Wills, M.S.E., P.E., who, at the time of its preparation, was consulting engineer, Stanley Consultants, Inc.

*16.33 DESIGN OF AN ANAEROBIC DIGESTOR

A high-rate anaerobic digestor is to be designed to treat a mixture of primary and waste-activated sludge produced by a wastewater-treatment facility. The input sludge to the digester is 60,000 gal/d (227.1 m³/d) of primary and waste-activated sludge with an average loading of 25,000 lb/d (11,350 kg/d) of ultimate BOD (BOD_L). Assume the yield coefficient (Y) is 0.06 lb VSS/lb BOD_L (kg/kg), and the endogenous coefficient (k_d) is 0.03 d^{-1} at 35°C (95°F). Also assume that the efficiency of waste utilization in the digester is 60 percent. Compute the digester volume required, the volume of methane gas produced, the total volume of digester gas produced, and the percent stabilization of the sludge.

Calculation Procedure

1. Determine the required digester volume and loading.

Anaerobic digestion is one of the oldest processes used for the stabilization of sludge. It involves the decomposition of organic and inorganic matter in the absence of molecular oxygen.

The major applications of this process are in the stabilization of concentrated sludges produced from the treatment of wastewater.

In the anaerobic digestion process, the organic material is converted biologically, under anaerobic conditions, to a variety of end products including methane (CH_4) and carbon dioxide (CO_2). The process is carried out in an airtight reactor. Sludge, introduced continuously or intermittently, is retained in the reactor for varying periods of time. The stabilized sludge, withdrawn continuously or intermittently from the reactor, is reduced in organic and pathogen content and is nonputrescible.

In the high-rate digestion process, as shown in Fig. 16.7 the contents of the digester are heated and completely mixed. For a complete-mix flow through digester, the mean cell residence time (θ_c) is the same as the hydraulic retention time (θ).

In the United States, the use and disposal of sewage sludge is regulated under 40 CFR Part 503 promulgated February 1993. The new regulation replaces 40 CFR Part 257—the original regulation governing the use and disposal of sewage sludge, in effect since 1979. The new regulations state that "for anaerobic digestion, the values for the mean-cell-residence time and temperature shall be between 15 days at 35°C (95°F) to 55°C (131°F) and 60 days at 20°C (68°F)."

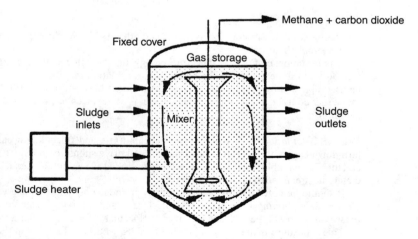

FIGURE 16.7 High-rate single-stage complete-mix anaerobic digester. (*Adapted from Metcalf and Eddy—Wastewater Engineering: Treatment, Disposal, and Reuse, 3rd ed., McGraw-Hill.*)

Therefore, for an operating temperature of 35°C (95°F), a mean cell residence time of 15 days will be used. The influent sludge flow rate (Q) is 60,000 gal/d = 8021 ft^3/d (226.9 m^3/d). The digester volume V required is computed using:

$$V = \theta_c Q \Rightarrow V = (15 \text{ days})(8021 \text{ ft}^3/\text{d}) = 120,315 \text{ ft}^3 \ (3404.9 \text{ m}^3)$$

The BOD entering the digester is 25,000 lb/d (11,350 kg/d). Therefore, the volumetric loading to the digester is:

$$\text{volumetric loading} = \frac{25,000 \text{ lb BOD/d}}{120,315 \text{ ft}^3} = 0.21 \text{ lb/ft}^3 \cdot \text{d} \ (3.37 \text{ kg/m}^3 \cdot \text{d})$$

For high-rate digesters, loadings range from 0.10 to 0.35 lb/ft^3 · d (1.6 to 5.61 kg/m^3 · d).
Assuming 60 percent waste utilization, the BOD_L exiting the digester is:

$$(25,000 \text{ lb/d})(1 - 0.6) = 10,000 \text{ lb/d} \ (4540 \text{ kg/d})$$

2. Compute the daily quantity of volatile solids produced.
The quantity of volatile solids produced each day is computed using:

$$P_x = \frac{Y[(\text{BOD}_{in}, \text{lb/d}) - (\text{BOD}_{out}, \text{lb/d})]}{1 + k_d \theta_c}$$

where P_x = volatile solids produced, lb/d (kg/d)
    Y = yield coefficient (lb VSS/lb BOD_L)
    k_d = endogenous coefficient (d^{-1})
    θ_c = mean cell residence time (d)

Using values obtained above, the volatile solids produced each day are:

$$P_x = \frac{(0.06 \text{ lb VSS/lb BOD}_L)[25,000 \text{ lb/d} - (10,000 \text{ lb/d})]}{1 + (0.03 \text{ d}^{-1})(15 \text{ d})}$$

$$= 621 \text{ lb/d} \ (281.9 \text{ kg/d})$$

3. Determine the volume of methane produced.
The volume of methane gas produced at standard conditions (32°F and 1 atm) (0°C and 101.3 kPa) is calculated using:

$$V_{CH_4} = 5.62 \text{ ft}^3/\text{lb}[(\text{BOD}_{in}, \text{lb/d}) - (\text{BOD}_{out}, \text{lb/d}) - 1.42 P_x]$$

where V_{CH_4} = volume of methane gas produced at standard conditions (ft^3/d) (m^3/d)

Using values obtained above:

$$V_{CH_4} = 5.62 \text{ ft}^3/\text{lb}[(25,000 \text{ lb/d}) - (10,000 \text{ lb/d}) - 1.42(621 \text{ lb/d})]$$

$$= 79,344 \text{ ft}^3/\text{d} \ (2245 \text{ m}^3/\text{d})$$

Since digester gas is approximately 2/3 methane, the volume of digester gas produced is:

$$(79,344 \text{ ft}^3/\text{d})/0.67 = 118,424 \text{ ft}^3/\text{d} \ (3351.4 \text{ m}^3/\text{d})$$

4. Calculate the percent stabilization.
Percent stabilization is calculated using:

$$\% \text{ stabilization} = \frac{[(BOD_{in}, \text{lb/d}) - (BOD_{out}, \text{lb/d}) - 1.42\, P_x]}{BOD_{in}, \text{lb/d}} \times 100$$

Using values obtained above:

$$\% \text{ stabilization} = \frac{[(25,000 \text{ lb/d}) - (10,000 \text{ lb/d}) - 1.42(621 \text{ lb/d})]}{25,000 \text{ lb/d}}$$

$$\times\, 100 = 56.5 \text{ percent of molecular oxygen}$$

Related Calculations.   This disadvantages and advantages of the anaerobic treatment of sludge, as compared to aerobic treatment, are related to the slow growth rate of the methanogenic (methane-producing) bacteria. Slow growth rates require a relatively long retention time in the digester for adequate waste stabilization to occur. With methanogenic bacteria, most of the organic portion of the sludge is converted to methane gas, which is combustible and therefore a useful end product. If sufficient quantities of methane gas are produced, the methane gas can be used to operate duel-fuel engines to produce electricity and to provide building heat.

This procedure is the work of Kevin D. Wills, M.S.E., P.E., who, at the time of its preparation, was consulting engineer, Stanley Consultants, Inc.

*16.34   DESIGN OF A CHLORINATION SYSTEM FOR WASTEWATER DISINFECTION

Chlorine is to be used for disinfection of a municipal wastewater. Estimate the chlorine residual that must be maintained to achieve a coliform count equal to or less than 200/100 mL in an effluent from an activated-sludge facility assuming that the effluent requiring disinfection contains a coliform count of 10^7/100 mL. The average wastewater flow requiring disinfection is 0.5 Mgal/d (1892.5 m^3/d) with a peaking factor of 2.8. Using the estimated residual, determine the capacity of the chlorinator. Per regulations, the chlorine contact time must not be less than 15 min at peak flow.

Calculation Procedure

1. Find the required residual for the allowed residence time.
The reduction of coliform organisms in treated effluent is defined by the following equation:

$$\frac{N_t}{N_0} = (1 + 0.23 C_t\, t)^{-3}$$

where N_t = number of coliform organisms at time t
N_0 = number of coliform organisms at time t_0
C_t = total chlorine residual at time t (mg/L)
t = residence time (min)

Using values for coliform count from above:

$$\frac{2 \times 10^2}{1 \times 10^7} = (1 + 0.23C_t t)^{-3}$$
$$\Rightarrow 2.0 \times 10^{-5} = (1 + 0.23C_t t)^{-3}$$
$$\Rightarrow 5.0 \times 10^4 = (1 + 0.23 C_t t)^3$$
$$\Rightarrow 1 + 0.23C_t t = 36.84$$
$$\Rightarrow C_t t = 155.8 \text{ mg} \cdot \text{min/L}$$

For a residence time of 15 min, the required residual is:

$$C_t = (155.8 \text{ mg} \cdot \text{min/L})/15 \text{ min} = 10.38 \text{ mg/L}$$

The chlorination system should be designed to provide chlorine residuals over a range of operating conditions and should include an adequate margin of safety. Therefore, the dosage required at peak flow will be set at 15 mg/L.

2. Determine the required capacity of the chlorinator.
The capacity of the chlorinator at peak flow with a dosage of 15 mg/L is calculated using:

$$Cl_2 \text{ (lb/d)} = (\text{dosage, mg/L})(\text{avg flow, Mgal/d})(P.F.)(8.34)$$

where Cl_2 = pounds of chlorine required per day (kg/d)
    dosage = dosage used to obtain coliform reduction
avg. flow = average flow
      P.F. = peaking factor for average flow
      8.34 = 8.34 lb · L/Mgal · mg

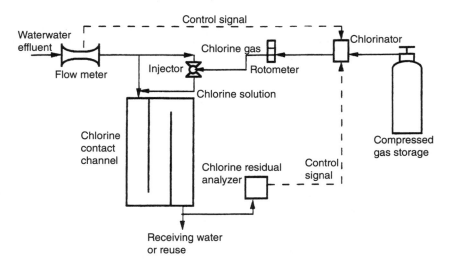

FIGURE 16.8 Compound-loop chlorination system flow diagram.

Using values from above:

$$Cl_2 \ (lb/d) = (15 \ mg/L)(0.5 \ Mgal/d)(2.8)(8.34) = 175 \ lb/d \ (79.5 \ kg/d)$$

The next largest standard chlorinator size is 200 lb/d (90.8 kg/d). Therefore, two 200 lb/d (90.8 kg/d) chlorinators will be used with one serving as a spare. Although the peak capacity will not be required during most of the day, it must be available to meet chlorine requirements at peak flow.

3. *Compute the daily consumption of chlorine.*
The average daily consumption of chlorine assuming an average dosage of 15 mg/L is:

$$Cl_2 \ (lb/d) = (15 \ mg/L)(0.5 \ Mgal/d)(8.34) = 62.5 \ lb/d \ (28.4 \ kg/d)$$

A typical chlorination flow diagram is shown in Fig. 16.8. This is a compound-loop system which means the chlorine dosage is controlled through signals received from both effluent flow rate and chlorine residual.

This procedure is the work of Kevin D. Wills, M.S.E., P.E., who, at the time of its preparation, was consulting engineer, Stanley Consultants, Inc.

REFERENCES

1. Bard—*Chemical Equilibrium*, Harper and Row, 1966.
2. Benfield et al.—*Process Chemistry for Water and Wastewater Treatment*, Prentice-Hall, 1982.
3. Cleasby and Logsdon—"Granular Bed and Precoat Filtration," in Letterman (ed.), *Water Quality and Treatment: A Handbook of Community Water Supplies*, 5th ed, McGraw-Hill, 1999.
4. Cornwell—"Air Stripping and Aeration," in Pontius (ed.) *Water Quality and Treatment: A Handbook of Community Water Supplies.* 4th ed, McGraw-Hill, 1990.
5. Hand et al.—"Air Stripping and Aeration," in Letterman (ed.), *Water Quality and Treatment: A Handbook of Community Water Supplies*, 5th ed, McGraw-Hill, 1999.
6. Lang et al.—*Trace Organic Constituents in Landfill Gas*, Dept. of Civil Engineering, Univ. of California-Davis, 1987.
7. Montgomery—*Water Treatment Principles and Design*, Wiley, 1985.
8. WEF—*Operation of Wastewater Treatment Plants*, 5th ed, Manual of Practice No. 11, Vol. 3, Chaps. 27–33, Water Environment Federation, 1996.

SECTION 17
BIOTECHNOLOGY

Robert L. Dream

Director, Pharmaceuticals & Biotechnology, Europe
Lockwood Greene
Somerset, NJ

17.1 DETERMINING THE LABORATORY-REACTOR SIZE NEEDED FOR SEEDING A BIOLOGICAL REACTION

Assuming a minimum 12% inoculum volume, what size of laboratory vessel would be required to initiate the seeding of a 20,000-L full-scale cell-culture bioreactor?

Calculation Procedure

1. Determine the size of reactor that would be required to seed the 20,000-L bioreactor.
Since the seed volume must represent 12% of the vessel before reaction starts, the bioreactor being specified in this step would have to have a size 12% that of the 20,000-L bioreactor, or (0.12)(20,000), or 2400 L.

2. Determine the size of bioreactor needed to seed the 2400-L bioreactor of step 1.
Applying the same logic as in step 1, we see that the bioreactor being sought in this second step must be sized at 12% of 2400 L, or (0.12)(2400), or 288 L.

3. Repeat step 2 successively until a bioreactor of reasonable laboratory volume is reached.
Twelve percent of 288 L is 34.6 L; then, 12% of 34.6 is 4.15 L; and 12% of 4.15 L is 500 mL. Thus, a 4.15-L laboratory vessel can be used if available. Otherwise, use a 500-mL vessel. The contents of the 500-mL vessel provide seeding for the 4.15-L vessel; the contents of the latter vessel then seed the 34.6-L bioreactor; the contents of this latter then seed the 288-L bioreactor; and so on.

17.2   ESTIMATING THE HOLDING TIME NEEDED FOR STERILIZING A BIOREACTOR

A bioreactor containing 20 m³ of medium at room temperature (25°C) is ready for sterilization by direct injection of saturated clean steam. The typical bacterial count of the medium is about $10 \times 10^{12}/m^3$, which needs to be reduced to a level so that the chance for contaminant surviving the sterilization is 1 in 1000, i.e., a level of 1×10^{-3} for the vessel.

The steam (345 kPa, absolute pressure) will be injected with a flow rate of 2500 kg/h, which will be stopped when the medium temperature reaches 122°C. During the holding time, the heat loss through the vessel is assumed negligible. After an appropriate holding time, the bioreactor will be cooled by passing 200 m³/h of 20°C water through the bioreactor jacket until the medium reaches 30°C. The jacket has a heat-transfer area of 10 m² and for this operation the average overall heat-transfer coefficient U for cooling is 2000 kJ/(h · m² · K). The heat-resistant bacterial spores in the medium can be characterized by an Arrhenius coefficient (k_{d0}) of 5.7×10^{39} h⁻¹ and an activation energy (E_d) of 2.834×10^5 kJ/kmol [Refs. 9, 10]. The thermal death constant (k_d) at 122°C is 197.6/h. The heat capacity and density of the medium are 4.187 kJ/kg · K and 1000 kg/m³, respectively. Estimate the required holding time.

Calculation Procedure

1. Determine the del factor that pertains to this sterilization operation.
The del factor, ∇, is the natural logarithm of the ratio of n_0, the original number of bacteria present, to n, the number allowed to remain. Thus,

$$\nabla = \ln(n_0/n) = \ln\{[(10 \times 10^{12}\,m^{-3})(20\ m^3)]/[1 \times 10^{-3}]\} = 39.8 \tag{17.1}$$

2. Determine the time t needed to heat the system from its initial temperature T_0 of 25°C to a temperature T of 122°C.

$$T = T_0 + \frac{H_{ms}t}{c(M + m_s t)} \tag{17.2}$$

where c is heat capacity, H_{ms} is the enthalpy of the saturated steam relative to that of the raw mixture, and M and m are the masses of medium and steam, respectively.

From the steam table, the enthalpies of saturated steam at 345 kPa and water at 25°C are 2731 and 105 kJ/kg, respectively. Therefore the enthalpy of saturated steam at 345 kPa relative to raw medium temperature 25°C is

$$H = 2731 - 105 = 2626\ kJ/kg \tag{17.3}$$

which can be used to calculate the time required to heat the medium from 25 to 122.1°C:

$$T = T_0 + \frac{(2626\ kJ/kg)(2500\ kg/h)t}{(4.187\ k \cdot JK)[(20\ m^3)(1000\ kg/m^3) + (2500\ kg/h)t]} \tag{17.4}$$

$$= T_0 + \frac{78.4\,t}{1 + 0.125t} \tag{17.5}$$

Trial-and-error solution of Eq. (17.5) when $T = 122°C$ (395 K) and $T_0 = 25°C$ (298 K) yields a t value of 1.46 h.

3. *Determine the del factor for the heating step, ∇_{heat}*
The thermal death of microorganisms typically follows a first-order process,

$$\frac{dn_v}{dt} = -kn_v = -k_0 e^{[-E/RT(t)]} n_v \tag{17.6}$$

or

$$\nabla = \ln\frac{n_0}{n} = \int_0^t k_d \, dt = k_{d0} \int_0^t e^{(-E_d/RT)} \, dt \tag{17.7}$$

where k_{d0} and E_d are Arrhenius coefficient and activation energy, respectively, T is absolute temperature, and t is elapsed time. Then,

$$\nabla_{heat} = 5.7 \times 10^{39} \int_0^{1.46} \exp\left[\frac{-2.834 \times 10^5}{8.318} \left(298 + \frac{78.4t}{1 + 0.125t} \right)^{-1} \right] dt \tag{17.8}$$

$$= 14.8 \tag{17.9}$$

4. *Determine the time needed to cool the mixture to 30°C (303 K) and the del factor for that cooling step, ∇_{cool}.*
During the cooling process, the change of temperature can be approximated by

$$T = T_{co} + (T_o - T_{co}) \exp\left[\left(1 - e^{(-uA/m_cC)} \right) m_c t/M \right] \tag{17.10}$$

$$= 293 + 102 e^{(-0.674)t} \tag{17.11}$$

where T_{co} is the temperature of the cooling medium, U is the overall heat-transfer coefficient [2000 kJ/ (h · m² · K)], A is the heat-transfer area (10 m²), and m_c and c refer to the mass and heat capacity of the 20°C coolant water. Then, solving for t when the final temperature is 303 K yields a cooling time of 3.45 h. Accordingly,

$$\nabla_{cool} = 5.7 \times 10^{39} \int_0^{3.45} \exp\left[-\frac{2.834 \times 10^5}{8.318[293 + 102 \exp(-0.674t)]} \right] dt \tag{17.12}$$

$$= 13.9$$

5. *Find the del factor for the holding time, ∇_{hold}, and, accordingly, the required holding time, t_{hold}.*
The del factor for the holding time (∇_{hold}) is

$$\nabla_{hold} = \nabla_{total} - \nabla_{heat} - \nabla_{cool} \tag{17.13}$$

$$= 39.8 - 14.8 - 13.9 \tag{17.14}$$

$$= 11.1 \tag{17.15}$$

The thermal death constant (k_d) is 197.6 h⁻¹ (at 122°C). Therefore

$$t_{hold} = \frac{\nabla_{hold}}{k_d} = \frac{11.1}{197.6} \tag{17.16}$$

$$= 0.056 \, \text{h} = 3.37 \, \text{min} \tag{17.17}$$

Related Calculations.   The value of k_d, the specific death rate, depends not only on the type of species but also on the physiological form of the cells. The temperature dependence of k_d can be assumed to follow the Arrhenius equation,

$$k_d = k_{d0}e^{[-E_d/RT(t)]} \tag{17.18}$$

where E_d is activation energy, which can be obtained from the slope of the k_d versus $1/T$ plot.

17.3 DETERMINING THE REQUIRED SIZE FOR A STERILIZER

A continuous sterilizer with a direct steam injector and a flash cooler will be used to sterilize a medium continuously. The medium will flow through the sterilizer at a rate of 2 m³/h. The typical bacterial count, n_{v0}, for the medium is 5×10^{12} m³; this must be lowered to such a level, n_v, that only one organism can survive during a month of continuous operation. The heat-resistant bacterial spores in the medium can be characterized by an Arrhenius coefficient, k_{d0}, of 5.7×10^{39} h⁻¹ [Refs. 9, 10]. The activation energy, E_d, is 2.834×10^5 kJ/kmol. The sterilizer will be fabricated with a pipe having an inner diameter of 0.102 m. Steam at 600 kPa gage pressure is available to bring the sterilizer to an operating temperature of 125°C. The physical properties of the medium at that temperature are heat capacity, c_p, of 4.187 kJ/(kg) · (K), density, ρ of 1000 kg/m³, and viscosity, μ, of 4 kg/(m) · (h). How long should the sterilizer pipe be, if ideal plug flow is assumed?

Calculation Procedure

1. Determine the relevant del factor (see Procedure 17.2).
Use Eq. (17.1), in which k_d is the death rate:

$$\nabla = \ln\frac{n_{v0}}{n_v} = \int_0^t k_d dt = k_{d0}\int_0^t e^{(-E_d/RT)}dt \tag{17.19}$$

or,

$$\nabla = \ln\frac{n_{v0}}{n_v} = \ln\left[\frac{(5 \times 10^{12}\,\text{m}^{-3})(2\,\text{m}^3/\text{h})(24\,\text{h/d})(30\,\text{days})}{1}\right]$$
$$= 36.51 \tag{17.20}$$

2. Determine the required holding time, τ_{hold}.
Since the temperature of the holding section is constant, Eq. (17.20) simplifies to

$$\nabla_{\text{heat}} = k_d\tau_{\text{hold}} \tag{17.21}$$

From the given data, k_d can be calculated by using the equation

$$k_d = k_{d0}e^{(-E_d/RT)} \tag{17.22}$$

to yield

$$k_d = 378.6\,\text{h}^{-1} \tag{17.23}$$

Therefore,

$$\tau_{hold} = \frac{\nabla}{k_d} = \frac{36.51}{378.6} = 0.10 \text{ h} \tag{17.24}$$

3. Determine the velocity of the medium through the sterilizer.
The velocity of the medium, u, is the volumetric flow rate divided by the cross-sectional area of the sterilizer:

$$u = [2 \text{ m}^3/\text{h}]/[(\pi/4)(0.102 \text{ m})^2] = 245 \text{ m/h}$$

4. Determine the required length of the sterilizer.
The length, L, equals the required holding time multiplied by the velocity of the medium:

$$L = u\tau_{hold} = (0.10 \text{ h})(245 \text{ m/h}) = 24.5 \text{ m}$$

17.4 FINDING THE MASS TRANSFER COEFFICIENT FOR DISSOLVING OXYGEN IN WATER

Calculate the mass transfer coefficient, k_L, for dissolution of oxygen from the air into 25°C water at 1 atm in a mixing vessel equipped with a flat-blade disk turbine and sparger. At those conditions, the diffusivity of oxygen in water, D_{AB}, is 2.5×10^{-9} m²/s, the viscosity of water is 8.904×10^{-4} kg/(m-s), and the density of water is 997.08 kg/m³. Use Calderbank and Moo-Young's correlations (see Ref. 14).

Calculation Procedure

1. Determine the density of air, ρ_{air}, at 25°C.
Use the ideal gas law:

$$\rho_{air} = (pressure)(molecular\ weight)/(gas\ constant)(temperature)$$
$$\rho_{air} = (1.01325 \times 10^5 \text{ Pa})(29 \text{ kg/kg-mol})/(8.314 \times 10^3)(298 \text{ K})$$
$$= 1.186 \text{ kg/m}^3$$

2. Calculate the relevant Schmidt number, N_{Sc}.
The relevant Schmidt number is the ratio of liquid viscosity to the product of liquid density and the diffusivity:

$$N_{Sc} = (8.904 \times 10^{-4})/(997.08)(2.5 \times 10^{-9}) = 357.2$$

3. Determine k_L for small bubbles.
Use the correlation,

$$k_L = 0.31 N_{Sc}^{-2/3} \left(\frac{\Delta\rho\mu_c g}{\rho_c^2} \right)^{1/3} \tag{17.25}$$

in which the density difference is between that of the water and the air, the subscript c refers to the liquid, and g is the acceleration of gravity. Then,

$$k_L = 0.31(357.2)^{-2/3}\left[\frac{(997.08 - 1.186)8.904 \times 10^{-4}(9.81)}{(997.08)^2}\right]^{1/3} \quad (17.26)$$

$$= 1.27 \times 10^{-4}\,\text{m/s} \quad (17.27)$$

4. *Determine k_L for large bubbles.*
For large bubbles, substituting in the equation

$$k_L = 0.42N_{Sc}^{-1/2}\left(\frac{\Delta\rho\mu_c g}{\rho_c^2}\right)^{1/3} \quad (17.28)$$

gives

$$k_L = 0.42(357.2)^{-1/2}\left[\frac{(997.08 - 1.186)8.904 \times 10^{-4}(9.81)}{(997.08)^2}\right]^{1/3} \quad (17.29)$$

$$= 4.58 \times 10^{-4}\,\text{m/s} \quad (17.30)$$

Comment.   Note that for small bubbles and large bubbles alike, the mass transfer coefficients are independent of both the mixer power consumption and the gas flow rate.

17.5   MAKING ESTIMATES FOR DIFFUSIVITY OF OXYGEN IN WATER, AND ASSESSING THE RESULTS

Estimate the diffusivity for oxygen in water, D_{AB} (with oxygen as component A and water as component B), at 25°C, using both the Wilke-Chang and the Othmer-Thakar correlations (see Refs. 7 and 5), and compare the findings with the experimental value of 2.5×10^{-9} m²/s. (which, itself, has a possible error of ±20%). Then, convert the experimental value to one corresponding to a temperature of 40°C. The molecular volume of oxygen, V_{bA}, is 0.0256 m³/kmol; the association factor, ξ, for water is 2.26; the viscosity of water at 25°C is 8.904×10^{-4} kg/(m-s); the viscosity of water at 40°C is 6.529×10^{-4} kg/(m-s).

Calculation Procedure

1. *Estimate D_{AB} via the Wilke-Chang correlation.*
The Wilke-Chang correlation is as follows:

$$D_{AB}^{\circ} = \frac{1.173 \times 10^{-16}(\xi M_B)^{0.5}T}{\mu V_{bA}^{0.6}} \quad (17.31)$$

with M_B representing molecular weight of water.

Then,

$$D_{AB}^\circ = \frac{1.173 \times 10^{-16}[2.26(18)]^{0.5}298}{8.904 \times 10^{-4}(0.0256)^{0.6}} \tag{17.32}$$

$$= 2.25 \times 10^{-9}\,\text{m/s} \tag{17.33}$$

2. Estimate D_{AB} via the Othmer-Thakar correlation.
The Othmer-Thakar correlation is as follows:

$$D_{AB}^\circ = \frac{1.112 \times 10^{-13}}{\mu^{1.1}V_{bA}^{0.6}} \tag{17.34}$$

Then,

$$D_{AB}^\circ = \frac{1.112 \times 10^{-13}}{(8.904 \times 10^{-4})^{1.1}(0.0256)^{0.6}} \tag{17.35}$$

$$= 2.27 \times 10^{-9}\,\text{m}^2/\text{s} \tag{17.36}$$

3. Assess the errors between these estimated values and the experimental one.
If we define the error between these predictions and the experimental value as

$$\%\,\text{error} = \frac{(D_{AB}^\circ)_{\text{predicted}} - (D_{AB}^\circ)_{\text{experimental}}}{(D_{AB}^\circ)_{\text{experimental}}} \times 100 \tag{17.37}$$

the resulting errors are −10% and −9.2%, respectively. Since the experimental result itself comes with an accuracy of ±20%, both estimates are satisfactory.

4. Convert the estimated value to one that corresponds to 40°C.
The Wilke-Chang correlation (see step 1) suggests that $(D_{AB}^\circ\mu/T)$ is constant for a given liquid system. We may use that assumption to estimate the diffusivity at 40°C:

$$D_{AB}^\circ \text{ at } 40°C = (D_{AB}^\circ \text{ at } 25°C)(\text{ratio of viscosities})(\text{ratio of absolute temperatures})$$
$$= [2.5 \times 10^{-9}][(8.904 \times 10^{-4})/(6.529 \times 10^{-4})][313/298]$$
$$= 3.58 \times 10^{-9}\,\text{m}^2/\text{s}$$

Alternatively, the Othmer-Thakar correlation (see step 2) suggests that $D_{AB}^\circ\mu^{1.1}$ is constant. Under this assumption,

$$D_{AB}^\circ \text{ at } 40°C = (D_{AB}^\circ \text{ at } 25°C)(\text{ratio of viscosities})^{1.1}$$
$$= [2.5 \times 10^{-9}][(8.904 \times 10^{-4})/(6.529 \times 10^{-4})]^{1.1}$$
$$= 3.52 \times 10^{-9}\,\text{m}^2/\text{s}$$

17.6 CALCULATING PARAMETERS FOR SPARGED BIOCHEMICAL VESSELS

A dished head tank of diameter $D_T = 1.22$ m is filled with water to an operating level equal to the tank diameter. The tank is equipped with four equally spaced baffles whose width is one-tenth of the tank diameter. The tank is agitated with a 0.36-m-diameter, flat, six-blade disk turbine. The impeller rotational speed is 2.8 rev/s. The sparging air enters through an open-ended tube situated below the impeller, and its volumetric flow, Q, is 0.00416 m³/s at 25°C. Calculate the following: the impeller power requirement, P_m; gas holdup (the volume fraction of gas phase in the dispersion), H; and Sauter mean diameter of the dispersed bubbles. The viscosity of the water, μ, is 8.904 × 10⁻⁴ kg/(m-s), the density, ρ, is 997.08 kg/m³, and, therefore, the kinematic viscosity, ν, is 8.93 × 10⁻⁷ m²/s. The interfacial tension for the air–water interface, σ, is 0.07197 kg/s². Assume that the air bubbles are in the range of 2–5 mm diameter.

Calculation Procedure

1. Calculate the impeller Reynolds number.
The formula for impeller Reynolds number is

$$N_{Re} = \frac{\rho N D_I^2}{\mu}$$

where N is impeller rotational speed and D_i is impeller diameter (for more information on impeller Reynolds number, see Section 12).
   Therefore,

$$N_{Re} = (997.08)(2.8)(0.36)^2 / 8.904 \times 10^{-4} = 406{,}357$$

2. Calculate P_{mo}, the impeller power that would be required if there were no gas sparging.
Rearrange the formula for turbulent power number, N_P (see Section 12), so as to solve for impeller power:

$$P_{mo} = (N_P)\rho N^3 D^5$$

Since the impeller Reynolds number is over 10,000, the impeller power number is constant at 6. Therefore, $P_{mo} = (6)(997.08)(2.8)^3(0.36)^5 = 794$ W.

3. Calculate P_m, the power required in the gas-sparged system.
P_m is calculated from the following equation:

$$\log_{10} \frac{P_m}{P_{mo}} = 192 \left(\frac{D_I}{D_T}\right)^{4.38} \left(\frac{D_I^2 N}{\nu}\right)^{0.115} \left(\frac{D_I N^2}{g}\right)^{1.96\left(\frac{D_I}{D_T}\right)} \left(\frac{Q}{N D_I^3}\right)$$

where g is the acceleration due to gravity. Therefore,

$$\log_{10} \frac{P_m}{794} = 192 \left(\frac{0.36}{1.22}\right)^{4.38} \left[\frac{(0.36^2)(2.8)}{8.93 \times 10^{-7}}\right]^{0.115} \left[\frac{(0.36)(2.8^2)}{9.81}\right]^{1.96\left(\frac{0.36}{1.22}\right)} \left[\frac{0.00416}{(2.8)(0.36^3)}\right]$$

Therefore, $P_m = 687$ W.

4. Calculate v, the volume of the gas–liquid system.
This is simply the volume of the fluid system:

$$v = \pi(1.22)^2(1.22)/4 = 1.43 \text{ m}^3$$

5. Calculate V_s, the superficial velocity of the sparged gas.
Use the formula, $V_s = 4Q/\pi(D_T)^2$.
Thus,

$$V_s = 4(0.00416)/3.1416(1.22)^2 = 0.00356 \text{ m/s}$$

6. Calculate H, the gas holdup.
Use the equation

$$H = \left(\frac{V_s H}{V_t}\right)^{1/2} + 2.16 \times 10^{-4}\left[\frac{(P_m/v)^{0.4}\rho_c^{0.2}}{\sigma^{0.6}}\right]\left(\frac{V_s}{V_t}\right)^{1/2}$$

where V_t, the terminal gas-bubble velocity during free rise is 0.265 m/s when the bubble size is in the range of 2–5 mm diameter. For high superficial gas velocities ($V_s > 0.02$ m/s), replace P_m and V_t with effective power input P_e and $(V_t + V_s)$, respectively (see Ref. 8).
Then,

$$H = \left(\frac{0.00356H}{0.265}\right)^{1/2} + 2.16 \times 10^{-4}\left[\frac{(687/1.43)^{0.4}(997.08)^{0.2}}{0.07197^{0.6}}\right]\left(\frac{0.00356}{0.265}\right)^{1/2}$$
$$= 0.023$$

7. Calculate the Sauter mean diameter, D_{32}.
The Sauter mean diameter is the diameter of a hypothetical droplet in which the ratio of droplet volume to droplet surface equals that of the entire dispersion. Use the formula

$$D_{32} = 4.15\left[\frac{\sigma^{0.6}}{(P_m/v)^{0.4}\rho_c^{0.2}}\right]H^{0.5} + 9.0 \times 10^{-4}$$

This gives

$$D_{32} = 4.15\left[\frac{0.07197^{0.6}}{(687/1.43)^{0.4}997.08^{0.2}}\right](0.023)^{0.5} + 9.0 \times 10^{-4}$$
$$= 0.00366 \text{ m}$$
$$= 3.66 \text{ mm}$$

Related Calculations. For calculation of the liquid-phase mass transfer coefficient, k_L, the following formula can be used [Ref. 14], where N_{Sc} is the Schmidt number and subscript c refers to the liquid phase:
For bubbles less than 2.5 mm in diameter,

$$k_L = 0.31N_{Sc}^{-2/3}\left(\frac{\Delta\rho\mu_c g}{\rho_c^2}\right)^{1/3}$$

For bubbles larger than 2.5 mm in diameter,

$$k_L = 0.42N_{Sc}^{-1/2}\left(\frac{\Delta\rho\mu_c g}{\rho_c^2}\right)^{1/3}$$

17.7 SIZING A FILTER FOR MICROFILTRATION OF A PROTEIN SOLUTION

We wish to concentrate and achieve a solvent switch for a solution by batch crossflow microfiltration. The flux, j_v, for the ceramic microfiltration membrane is 10 gal/(h-ft^2). The initial solution volume is 1800 gal; the final volume is 360 gal. The amount of protein present is 18.0 kg, and the molecular weight is 1213.43 g/mol. The pressure drop is 30 psi (essentially 2 atm) and the operating temperature is 277 K. Calculate the area, A, required to complete the filtration in 2 h.

Calculation Procedure

1. Set out the equation that governs the time needed for microfiltration.
The equation is as follows, with L_p representing solvent permeability and n_1 the number of moles of solute present:

$$t = \left(\frac{1}{AL_p \Delta p} \right) \left[(V_0 - V) + \left(\frac{RTn_1}{\Delta p} \right) \ln \left(\frac{V_0 - \dfrac{RTn_1}{\Delta p}}{V - \dfrac{RTn_1}{\Delta p}} \right) \right] \qquad (17.38)$$

2. Simplify the equation in step 1 by inserting known numerical values.
The term,

$$\frac{RTn_1}{\Delta P} = \frac{[0.082 \text{ L-atm/(gmol-K)}](277 \text{ K}) \left(\dfrac{18 \times 10^3 \text{ g}}{1213.43 \text{ g/gmol}} \right)}{2 \text{ atm}} = 168.33 \text{ L} \qquad (17.39)$$

And from Eq. (17.38), the term,

$$\Pi = \left[(V_0 - V) + \left(\frac{RTn_1}{\Delta p} \right) \ln \left(\frac{V_0 - \dfrac{RTn_1}{\Delta p}}{V - \dfrac{RTn_1}{\Delta p}} \right) \right] \qquad (17.40)$$

can be calculated as

$$\Pi = \left[(1800 \text{ gal} \times 3.78 \text{ L/gal} - 360 \text{ gal} \times 3.78 \text{ L/gal}) \right.$$

$$\left. + 168.33 \text{ L} \ln \left(\frac{1800 \text{ gal} \times 3.78 \text{ L/gal} - 168.33 \text{ L}}{360 \text{ gal} \times 3.78 \text{ L/gal} - 168.33 \text{ L}} \right) \right] \qquad (17.41)$$

$$= 1{,}516.41 \text{ gal} \qquad (17.42)$$

Therefore, Eq. (17.38) reduces to

$$t = \left(\frac{1}{AL_p \Delta p} \right) (1516.41 \text{ gal}) \qquad (17.43)$$

3. Set out, then rearrange the equation for flow rate, dV/dt, through the filter.

$$\frac{dV}{dt} = -Aj_v = -AL_p \Delta p \left(1 - \frac{RTc_1}{\Delta p} \right)$$ (17.44)

But,

$$n_1 = c_1 V$$ (17.45)

or

$$c_1 = \frac{n_1}{V} = \frac{\dfrac{18 \times 10^3\,g}{1213.43\,\text{g/gmol}}}{1800\,\text{gal} \times 3.78\,\text{L/gal}} = 0.002\,\text{gmol/L}$$ (17.46)

Substituting gives

$$\frac{RTc_1}{\Delta P} = \frac{[0.082\,\text{L-atm/(gmol-K)}](277°K)(0.002\,\text{gmol/L})}{2\,\text{atm}} = 0.02$$ (17.47)

4. Rearrange Eq. (17.38) and solve it for the required area.

$$A = \frac{\Pi}{L_p \Delta pt}$$ (17.48)

Therefore from Eqs. (17.44) and (17.47) we have

$$j_v = L_p \Delta p(0.98)$$ (17.49)

Then Eq. (17.48) reduces to

$$A = \frac{\Pi(0.98)}{j_v t}$$ (17.50)

$$= \frac{1516.49\,\text{gal}(0.98)}{(2\,h) \times [10\,\text{gal/(h-ft}^2)]}$$ (17.51)

And therefore,

$$A = 74.31\,\text{ft}^2$$

is the area required to cross flow 1800 gal to 360 gal in 2 h.

Related Calculations.   The same calculational approach can be used for nanofiltration.

In either case, if the membrane completely rejects the solute and the concentration polarization is negligible, then the concentration across the boundary layer is constant and small. Equation (17.38) consequently reduces to

$$t = \left(\frac{1}{AL_p \Delta p} \right)(V_0 - V)$$ (17.52)

or

$$t = \left(\frac{1}{AL_p \Delta p} \right)(1800 \text{ gal} - 360 \text{ gal}) \tag{17.53}$$

or

$$t = \left(\frac{1}{AL_p \Delta p} \right)(1440 \text{ gal}) \tag{17.54}$$

And because $RTc_l/\Delta P$ becomes small and negligible, and because $n_1 = c_1 V$, then

$$\frac{dV}{dt} = -Aj_v = -AL_p \Delta p \left(1 - \frac{RTc_l}{\Delta p} \right) \tag{17.55}$$

reduces to

$$j_v \cong L_p \Delta p \tag{17.56}$$

Then modify Eq. (17.54) to

$$A = \frac{1440 \text{ gal}}{L_p \Delta pt} \tag{17.57}$$

or

$$A = \frac{1440 \text{ gal}}{j_v t} \tag{17.58}$$

or

$$A = \frac{(1440 \text{ gal})}{(2 \text{ h}) \times [10 \text{ gal/(h-ft}^2)]} \tag{17.59}$$

And therefore,

$$A = 72 \text{ ft}^2$$

is the area required to cross flow 1800 gal to 360 gal in 2 h, when the concentration polarization is negligible.

The percent additional membrane area required because of concentration polarization is

$$\Delta A\% = \frac{74.31 \text{ ft}^2 - 72 \text{ ft}^2}{74.31 \text{ ft}^2} \frac{1}{100} = 3.12\%$$

17.8 ESTIMATING THE CAPACITY OF A WESTFALIA SEPARATOR TO REMOVE BACTERIA FROM BROTH

Estimate the throughput capacity of a Westfalia clarifier Type CSA 8 to separate *Escherichia Coli* bacteria from a fermentation broth. The specifications of the separator are as follows:

Bowl speed: $n = 9200$ L/min, that is, $\omega = 963$ s^{-1}
Number of disks: $z = 72$
Disk angle: $\varphi = 55°$
External radius of disk: $r_1 = 0.081$ m
Internal radius of disk: $r_2 = 0.036$ m

The characteristics of the *E. coli* are as follows:

Smallest *E. coli* bacteria to be separated: $d_{limit} = 0.8$ μm $= 0.8 \times 10^{-6}$ m
Solids density: $\rho_1 = 1.05$ g/cm^3 $= 1050$ kg/m^3
Density of nutrient: $\rho_2 = 1.02$ g/cm^3 $= 1020$ kg/m^3
Density difference: $\Delta\rho = 0.03$ g/cm^3 $= 30$ kg/m^3
Dynamic viscosity: $\eta = 1.02 \times 10^{-3}$ kg/m-s
Acceleration due to gravity: $g = 9.81$ m/s^2

Calculation Procedure

1. *Apply the following equation to arrive at the capacity Q.*

$$Q = \left(\frac{d_{limit}^2 \Delta\rho}{18\eta}\right) g \left(\frac{2\pi}{3g}\right)(\omega^2 z \tan\varphi)\left(r_1^3 - r_2^3\right) \tag{17.60}$$

Substituting the data in Eq. (17.60) we have

$$Q = \left[\frac{(0.8 \times 10^{-6}\,\text{m})^2 (30\,\text{kg/m}^3)}{18(1.02 \times 10^{-3}\,\text{kg/m-s})}\right] g \left(\frac{2\pi}{3g}\right)$$

$$\times (963\,\text{s}^{-1})(72)(\tan 55°)[(0.081\,\text{m})^3 - (0.036\,\text{m})^3] \tag{17.61}$$

$$= 1.01 \times 10^{-4}\,\text{m}^3/\text{s} = 365\,\text{L/h} \tag{17.62}$$

Related Calculations.   In biotechnology separations such as this, data such as those for the density difference, viscosity, and cell size distribution are typically not known with certainty, so the calculation here can be regarded as providing only an approximate value; for more precision, experimental means must be used. *E. coli* suspensions with cell sizes between 0.8 and 1.8 μm have in practice been separated, with efficiencies of over 98%, at rates of 200 to 400 L/h. This suggests that the approximation made here is a good one.

Equation (17.60) combines the formula for Stokes settling velocity, v_s, and that for the equivalent clarification area, Σ_T, also known as the comparison coefficient: $Q = (v_s)(\Sigma_T)$. The v_s term is calculated from the properties of the process material:

$$v_s = \frac{d_{limit}^2 \Delta\rho}{18\eta} g$$

The Σ_T term, by contrast, is made up of design data for the disk-type bowl:

$$\Sigma_T = \frac{2\pi}{3g}\omega^2 z \tan\varphi\left(r_1^3 - r_2^3\right)$$

It corresponds to the amount of surface that would be required in a sedimentation vessel to achieve the same results as in a centrifugal separator (such as the Westfalia). It can be used for comparing the capacities of separators of different size of design. Specifically, the capacities Q_1 and Q_2 of two separators are related as follows:

$$\frac{Q_1}{\Sigma_{T1}} = \frac{Q_2}{\Sigma_{T2}}$$

or

$$Q_1 = \frac{Q_2}{\Sigma_{T2}} \cdot \Sigma_{T1}$$

*17.9   FRACTIONATING COLUMN AND CONDENSER ANALYSIS FOR BIOTECHNOLOGY

A solution containing 35% by weight of ethanol (and 65% water) is supplied to a fractionating column. The feed is a saturated liquid. The top product contains 80% by weight of ethanol and the bottom product contains 5% by weight of ethanol. The condenser operates as a total condenser and the reflux enters the top plate at 120°F.

a. If 2 mol of liquid at 120°F are returned for each mole of product removed, determine the slope of the operating line to be used in a McCabe-Thiele diagram.
b. Compare the amount of heat that must be supplied in the reboiler per pound of feed when the reflux is returned at 120°F with that required when the reflux is returned as a saturated liquid.

Calculation Procedure

1. Draw a schematic flow diagram of the column and condenser.
Figure 17.1 shows the flow diagram of the column and condenser with the symbols for the equations used in the calculations.

2. Write the enthalpy balance equation for the process.

$$R_E = \frac{L_E}{P} = \text{external reflux ratio}$$

$$V_1 = L_E + P$$

$$\qquad = \text{material balance top of column and reflux splitter}$$

$$V_1 = R_E P + P = P(R_E + 1)$$

$$L_E + V_2 = L_1 + V_1 = \text{material balance top plate}$$

$$L_E H_{LE} + V_2 H_{V_2} = L_1 H_{L_1} + V_1 H_{V_1} = \text{enthalpy balance}$$

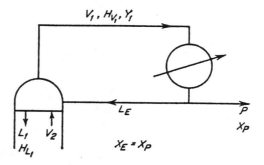

FIGURE 17.1   Schematic diagram of the column and condenser.

3. Find the external and internal flux ratios.

Assuming that the enthalpies of vapors V_1 and V_2 are equal

$$L_E H_{LE} + V_2 H_{V_1} = L_1 H_{L_1} + V_1 H_{V_I}$$

$$H_{V_1}(V_2 - V_1) = L_1 H_{L_1} - L_E H_{LE}$$

and therefore

$$(L_1 - L_E)(H_{V_1}) = L_1 H_{L_1} - L_E H_{LE}$$

$$L_1 H_{V_1} - L_1 H_{L_1} = L_E H_{V_1} - L_E H_{LE}$$

$$L_1 = L_E \frac{H_{V_1} - H_{LE}}{H_{V_1} - H_{L_1}}$$

The external reflux ratio is defined as $\frac{R_E}{R_E+1} = \frac{L_E}{V_1}$, and the internal reflux ratio is defined as $\frac{R_I}{R_I+1} = \frac{L_1}{V_2}$.

$$R_E V_1 = R_E L_E + L_E$$

$$V_1 = \frac{R_E L_E + L_E}{R_E}$$

and

$$\frac{R_I}{R_I + 1} = \frac{L_1}{V_2} = \frac{L_1}{L_1 + V_1 - L_E}$$

$$R_I L_1 + R_I V_1 - R_I L_E = R_I L_1 + L_1$$

$$V_1 = \frac{R_I L_E + L_1}{R_I}$$

and therefore,

$$\frac{R_E L_E + L_E}{R_E} = \frac{R_I L_E + L_1}{R_I}$$

$$R_I L_E = R_E L_I$$

$$R_I = \frac{R_E L_E}{L_E} \frac{H_{V_1} - H_{LE}}{H_{V_1} - H_{L_1}} = R_E \frac{H_{V_1} - H_{LE}}{H_{V_1} - H_{L_1}}$$

4. Determine the mass fractions for the vapor and liquid and slope of the operating line.
For a saturated vapor 0.8 mass fraction ethanol,

$$H_{V_1} = 595 \text{ Btu/lb}$$

For a saturated liquid 0.8 mass fraction ethanol

$$H_{LE} = 103 \text{ Btu/lb}$$

$$H_{LE} = 58 \text{ Btu/lb at } 120°F$$

Composition of liquid in equilibrium with 0.8 mass fraction vapor is 0.62 and

$$H_{L_1} = 115 \text{ Btu/lb}$$

$$R_E = 2\frac{595 - 58}{595 - 115} = 2\frac{537}{480} = 2.236$$

The slope of the operating line is 0.69.

5. Find the amounts of the top and bottom products.
    For 1 lb of feed the amounts of top and bottom product are obtained by means of a material balance.

$$0.35 = 0.8V + 0.05B = 0.75V + (0.05V + 0.05B)$$
$$1.0 = V + B \text{ and } 0.05 = (0.05V + 0.05B)$$

Therefore, $0.35 = 0.75V + 0.05$

and $0.75V = 0.30$, whence $V = 0.40$ lb and $B = 0.60$ lb.

For every pound of feed, 0.40 lb of top product are withdrawn and 0.80 lb of reflux are returned to the top plate. When the reflux is fed to the top plate as a saturated liquid, only the latent heat of vaporization which is removed in the condenser must be supplied in the reboiler or

$$0.80 \times (595 - 103) = 393.6 \text{ Btu/lb}$$

When the reflux is returned at 120°F, but with a composition of 0.8 mass fraction ethanol, its enthalpy is 58 Btu/lb. Therefore, $0.80 \times (595 - 58) = 429.6$ Btu/lb of feed must be supplied. When the "cold" liquid stream enters the top plate, it removes latent heat from vapors on the top plate and internal or operating reflux is greater than external reflux. Data on enthalpies of ethanol–water mixtures are found in engineering handbooks.
    This procedure is the work of William S. La Londe, Jr., P.E.

REFERENCES

1. Midler, M., Jr., and R. K. Finn—"A Model System for Evaluating Shear in the Design of Stirred Fermentors," *Biotechnology and Bioengineering 8* (1966): 71–84.

2. Croughan, M. S., J.-F. Hamel, and D. I. C. Wang—"Hydrodynamic Effects of Animal Cells Grown in Micro-Carrier Cultures," *Biotechnology and Bioengineering 29* (1987): 130–141.

3. Charm, S. E., and B. L. Wong—"Enzyme Inactivation with Shearing," *Biotechnology and Bioengineering 12* (1970): 1103–1109.

4. Hooker, B. S., J. M. Lee, and G. An—"The Response of Plant Tissue Culture to a High Shear Environment," *Enzyme and Microbial Technology 11* (1989): 984–490.

5. Othmer, D. F., and M. S. Thakar—"Correlating Diffusion Coefficients in Liquids," *Industrial and Engineering Chemistry 45* (1953): 589–593.

6. Rushton, J. H., E. W. Costich, and H. J. Everett—"Power Characteristics of Mixing Impellers, Part II," *Chemical Engineering Progress 46* (1950): 467–476.

7. Wilke, C. R., and P. Chang—"Correlation of Diffusion Coefficients in Dilute Solutions," *American Institute of Chemical Engineers Journal (AICHE J.) 1* (1955): 264–270.

8. Miller, D. N.—"Scale-Up of Agitated Vessels Gas-Liquid Mass Transfer," *American Institute of Chemical Engineers Journal (AICHE J.) 20* (1974): 445–453.

9. Deindoerfer, F. H., and A. E. Humphrey—"Analytical Method for Calculating Heat Sterilization Time," *Appl. Micro. 7* (1959a): 256–264.

10. Deindoerfer, F. H., and A. E. Humphrey—"Analytical Method for Calculating Heat Sterilization Time," *Appl. Micro. 7* (1959b): 264–270.

11. Calderbank, P. H.—"Physical Rate Processes in Industrial Fermentation, Part 1: Interfacial Area in Gas–Liquid Contacting with Mechanical Agitation," *Transactions of the Institution of Chemical Engineers 36* (1958): 443–459.

12. Calderbank, P. H.—"Physical Rate Processes in Industrial Fermentation, Part II: Mass Transfer Coefficients in Gas–Liquid Contacting with and without Mechanical Agitation," *Transactions of the Institution of Chemical Engineers 37* (1959): 173–185.

13. Calderbank, P. H., and S. J. R. Jones—"Physical Rate Processes in Industrial Fermentation—Part III: Mass Transfer Fluids to Solid Particles Suspended in Mixing Vessels," *Transactions of the Institution of Chemical Engineers 39* (1961): 363–368.

14. Calderbank, P. H., and M. B. Moo-Young—"The Continuous Phase Heat and Mass-Transfer Properties of Dispersions," *Chemical Engineering Science 16* (1961): 39–54.

15. Chandrasekharan, K., and P. H. Calderbank—"Further Observations on the Scale-up of Aerated Mixing Vessels," *Chemical Engineering Science 36* (1981): 819–823.

16. Aiba, S., A. E. Humphrey, and N. F. Millis—*Biochemical Engineering,* 2nd ed., pp. 242–246 and p. 263. Tokyo, Japan: University of Tokyo Press, 1973.

SECTION 18
COST ENGINEERING

William Vatavuk

Vatavuk Engineering
Durham, NC

Before examining the problems for this chapter, we need to present some basic cost engineering concepts. First, there are two general cost categories: total capital investment (TCI) and total annual cost (TAC). Also referred to as the capital cost or first cost, the TCI of a plant, process, operation, or similar activity includes all costs required to purchase the necessary equipment; the costs of the materials and labor needed to install that equipment (the direct installation costs); the costs for site preparation and buildings; and the costs for engineering, contractor fees, contingencies, and other indirect installation costs. The TCI also includes costs for land, working capital, and off-site facilities.

The equations that follow show how the various elements of the total capital investment are related.

$$TCI = \text{total depreciable investment} + \text{total nondepreciable investment} \tag{18.1}$$

$$\text{Total depreciable investment} = \text{total direct cost} + \text{total indirect cost}$$
$$+ \text{off-sites facilities cost} \tag{18.2}$$

where total direct cost = costs of equipment, direct installation, site preparation and buildings

$$\text{Total nondepreciable investment} = \text{costs of land and working capital} \tag{18.3}$$

The total direct cost (TDC) includes both the direct installation costs and the costs of site preparation and buildings. Further, the sum of the total direct cost and total indirect cost (or direct installation costs) is termed the "battery limits" cost. Finally, the battery limits cost plus the cost of off-site facilities (e.g., a railroad spur) comprise the total depreciable investment. Put simply, this is the portion of the TCI for which the firm is permitted to take a depreciation deduction on its corporate income tax return. The other portion of the *TCI,* namely, land and working capital, may not be depreciated. Hence, this portion is called the total nondepreciable investment.

The *TAC* consists of direct annual costs (*DAC*) and indirect annual costs (*IAC*). Also termed the "operating and maintenance" (O&M) cost, the *DAC* are those costs that tend to be proportional or partially proportional to the production rate, stream flow rate, or some other measure of activity. These include costs for raw materials, utilities (e.g., electricity, steam, or natural gas), labor, maintenance, royalties, and waste treatment and disposal. Conversely, the *IAC* are those costs whose values are generally independent of the activity level and that, in fact, would be incurred even if the process were not operating. The *IAC* include administrative and laboratory charges, property taxes, and insurance. (With some cost calculation methods, the indirect annual cost also includes the capital recovery cost [*CRC*]; however, this is not true in general. See the subsequent discussion on the equivalent-uniform-annual-revenue method.) Those interested in a more detailed discussion of cost engineering terminology should refer to *Basic Cost Engineering* [Ref. 1], *Estimating Costs of Air Pollution Control* [Ref. 2], or other introductory works.

If a process is to be profitable, the total annual cost must be offset by the revenue (*R*) the process generates. Moreover, if the revenue exceeds the *TAC,* income taxes will be owed. Finally, if income taxes are a consideration, so will be the depreciation of the process equipment. (Depreciation, an income tax book entry, should not be confused with the capital recovery cost; see later discussion on the *EUAR* method.) Space does not permit a thorough discussion of either income taxes or depreciation, as the rules governing the calculation of both are quite complex. For our purposes, the following equation, which encompasses all of these variables, will suffice:

$$NCF_k = (1 - t)(R_k - TAC_k) + t\, Dep_k \tag{18.4}$$

where t = combined state and federal marginal income tax rate
  Dep_k = depreciation deduction taken in year k of the project
  NCF_k = net cash flow in year k of the project
  TAC_k = total annual cost in year k
  R_k = revenue in year k

As Eq. (18.4) suggests, the revenue, *TAC,* and depreciation might be different for each year in the project's life. For instance, during the first years of a project, the depreciation allowance could be higher than in later years, especially if certain depreciation computation methods (such as double-declining balance) are used. Similarly, the revenue in the earlier years could be lower as the "bugs" are removed from the process and the product market share is achieved. In any event, the net cash flow (large/small, positive/negative) is the quantity that ties all of these variables together.

Typically, firms have a limited amount of funds to invest in new projects. How do they decide which of several competing projects to finance? These decisions are made via one or more "measures of merit." The three most commonly used measures are: (1) net present worth (*NPW*), (2) internal rate of return (*IRR*), and (3) equivalent uniform annual revenue (*EUAR*). Inextricably tied to each of these measures is the annual interest rate, i, or "opportunity cost." A succinct definition of the opportunity cost is: "A dollar today is worth more than the prospect of a dollar tomorrow."

In the present worth (or discounted cash flow) method, each annual net cash flow for a project is "discounted" to the beginning of the project (year zero), in order to express all cash flows on the same basis. This discounting is done via the following equation:

$$Discounted\ NCF_k = NCF_k / (1 + i)_k \tag{18.5}$$

where $1/(1 + i)_k$ = the discount factor
  i = annual after-tax discount (interest) rate

The sum of these discounted *NCF*s over the life of the project is the project's net present worth. The project with the highest positive net present worth would be the one selected. The discount rate used in Eq. (18.5) is the firm's marginal acceptable rate of return (*MARR*) for anticipated projects or its "hurdle rate." Note that the discount rate is expressed on an after-tax basis to be consistent with the *NCF*. Also, when comparing projects via the present worth method, all projects must have equal lives. Finally, by convention, *TCI* is incurred when the project comes on-line (in year zero). Hence, the *NCF* for year zero is equal to *minus TCI,* because, for that year, the revenue is zero and there are no other costs.

The internal rate of return method is a special case of the present worth method. With the *IRR,* the net present worth of each project first is arbitrarily set equal to zero, with the discount rate kept as an independent variable. Then, each *NPW* equation is solved (via an iterative procedure) for the unique discount rate/internal rate of return that yields an *NPW* of zero. The project with the highest positive *IRR* would be the one selected.

The third method, *EUAR,* is easy to use and to understand. It also allows the comparison of projects with unequal lives. However, it is cumbersome to use when the *NCF*s vary from year to year. But in special cases where the annual cash flow is constant, use of the *EUAR* method is recommended.

The *EUAR* is calculated according to the following equation:

$$EUAR = NCF - CRC \tag{18.6}$$

where $EUAR$ = equivalent uniform annual revenue
NCF = net cash flow
CRC = capital recovery cost

In this equation, note that the *CRC* is given a negative sign, because it is a negative cash flow item. The *CRC,* in turn, is the product of the *TCI* and the capital recovery factor, or

$$CRC = TCI[i(1 + i)^n/(1 + i)^n - 1)] \tag{18.7}$$

where i = discount (hurdle) rate
n = project life (years)

The term in braces is the capital recovery factor (*CRF*).

In essence, the *CRC* converts the *TCI* (a one-time cost) to a series of equal annual payments over the life of the project.

18.1 PLANT CAPITAL COST ESTIMATION VIA SCALING FACTOR

Given that the total capital investment (*TCI*) of a 50,000-ton/year polypropylene unit is $60,000,000 (in 2002 dollars), find the *TCI* required for a 75,000-ton/year polypropylene unit via the scaling factor method.

Calculation Procedure

1. Apply the appropriate power-function formula.
In the scaling factor method, the *TCI* is estimated via the following formula (a power function):

$$TCI_2 = TCI_1(C_2/C_1)^E \tag{18.8}$$

where TCI_1 and TCI_2 = total capital investment of existing and planned unit, respectively, in dollars
TCI_1 = $60,000,000
C_1, C_2 = capacity of existing and planned unit, respectively, in tons/year
C_1 = 50,000 and C_2 = 75,000
E = scaling exponent = 0.70 [Ref. 3]

Thus

$$TCI_2 = \$60,000(75,000/50,000)^{0.70} = \$80,000,000 \text{ (rounded)}$$

Related Calculations. The scaling factor method is an appropriate procedure for estimating the *TCI* only under the following conditions:

1. The existing and planned units are identical (or nearly so), in terms of processing steps, end products, major equipment items used, and other respects.
2. The desired estimate falls within the category of "order-of-magnitude/screening/scoping" cost estimates (i.e., those estimates with a presumed accuracy less precise than ± 30%).
3. The capacity of the planned unit falls within the capacity range for which the scaling exponent is valid. Rarely is the power function relationship between *TCI* and capacity a smooth curve over the entire capacity range. Typically, the scaling exponent increases in value with increasing capacity. However, as the scaling exponent approaches unity, it becomes less costly to build two units, each with half the capacity of the large plant, than to construct a single, large-capacity plant.
4. The costs of both the existing and planned units are expressed in dollars of the same period. In this example, the *TCI*s are in 2002 dollars. If the costs are not of the same vintage, the cost of the existing plant (which is likely older) will have to be adjusted to the same year dollars as that of the planned unit. However, unless the cost vintages are much different (e.g., 5 years or more), adjustments for escalation would be "fine tuning," compared to the relative inaccuracy of these scaling factor estimates. (See Procedure 18.2 for more on cost escalation.)

18.2 CAPITAL AND ANNUAL COST ESCALATION

The *TCI* of a 2500-ton/d oxygen plant constructed in 1998 was \$78,700,000. The direct annual ("O&M") costs for this facility (also in 1998 dollars) were as follows:

Natural gas:	\$3,450,000
Electricity:	1,650,000
Labor (w/overhead)	1,640,000
Maintenance:	2,360,000
Total O&M cost:	\$9,100,000

Find the total capital investment (capital) and O&M costs in 2001 dollars.

Calculation Procedure

1. *Scale up the capital cost, using an index.*
To adjust (escalate) these costs from 1998 to 2001, published generic escalation indices must be used, as no single index has been formulated specifically for oxygen plants. One index will be used to escalate the capital cost, while others will be employed to adjust the various O&M costs. To escalate the capital cost, the Chemical Engineering Plant Cost Index (CEPCI) will be used. Designed specifically for chemical process industry facilities, the CEPCI is updated and published monthly in *Chemical Engineering* magazine [Ref. 4].

The escalated capital cost is calculated via the following formula:

$$TCI \text{ in 2001 dollars} = (TCI \text{ in 1998 dollars})[(CEPCI \text{ for 2001}) / (CEPCI \text{ for 1998})]$$

where CEPCI is the annual index for the year in question and the ratio of the CEPCIs is the escalation factor.

Values of the CEPCI are

$$CEPCI\ (1998)\ =\ 398.5$$
$$CEPCI\ (2001)\ =\ 394.3$$

Substituting these index values and the given *TCI* into this formula, we obtain

$$TCI \text{ in 2001 dollars} = 78,700,000(394.3/389.5) = \$79,669,859, \text{ or } \$79,700,000 \text{ (rounded)}$$

2. Scale up the individual O&M costs using indexes.

Each of the O&M costs is escalated via the same formula, but by using a different index. The index to use for each cost is a Producer Price Index (PPI), compiled and published by the U.S. Department of Labor's Bureau of Labor Statistics [Ref. 5]. The PPI index numbers, names, and average 1998 and 2001 values used are as follows:

O&M cost	PPI number	PPI name	1998 avg.	2001 avg.	Escalation factor
Natural gas	wpu05310105	Fuels & related products/natural gas	106.3	217.8	2.0489
Electricity	wpu0543	Fuels & related products/indus. electric power	130.0	141.1	1.0854
Labor	eeu32280006	Chemicals & allied products/hourly earnings	17.09	18.61	1.0889

Because the maintenance cost typically is calculated as a percentage of the *TCI*, the CEPCI can also be used for its escalation:

$$\text{Maintenance in 2001 dollars} = (\text{maintenance in 1998 dollars})(394.3/389.5)$$

$$= \$2,389,083$$

Applying the escalation factors tabulated here to the 1998 costs, we obtain the following escalated O&M costs (in 2001 dollars):

Natural gas:	$7,068,705
Electricity:	1,790,910
Labor:	1,785,796

The total O&M cost in 2001 dollars is the sum of the individual escalated costs, or $13,034,500 (rounded).

Related Calculations

1. Although it is easy to escalate capital and annual costs, escalation over periods longer than 5–10 years is not recommended, as it tends to introduce additional inaccuracy to the estimates. For longer periods, rather than escalate the costs via an index, the cost estimator should obtain current cost quotes from equipment vendors and other sources (such as utilities).

2. Selecting the "right" escalation index is often difficult, as different published indices can appear to be applicable. For instance, when escalating the capital cost of a refinery hydrosulfurization unit, should one use the CEPCI, the Nelson-Farrar Refinery Index, or a combination of the two indexes? The decision often rests as much on engineering judgment as it does on hard data.

3. In this problem, note that escalating the capital cost from 1998 to 2001 dollars only adds another $1 million (about a 1% increase), whereas the escalated O&M cost is over 40% higher than the 1998 cost. Most of this increase is due to the doubling in natural gas prices over this 3-year period. By contrast, the electricity and labor costs only increased by about 9%. Nevertheless, as this problem shows, it would be most inadvisable to use a capital cost index to escalate O&M costs, and vice versa.

18.3   UNIT OPERATION TOTAL CAPITAL INVESTMENT

A packed gas-absorber column is installed in a sulfuric acid plant to control sulfur dioxide (SO_2) emissions. The absorber unit consists of the column, packing, ductwork, fan, and solvent feed pump. Pertinent input data are as follows:

- Stream specifications:

  Inlet waste gas flow rate, actual cubic feet per minute (acfm): 12,500

  Inlet waste gas temperature, °F: 100

  Inlet waste gas pressure, atmospheres: 1.00

  Inlet waste gas SO_2 concentration, mole fraction: 0.001871

  SO_2 removal efficiency, fractional: 0.990

  Solvent: aqueous ammonia (NH_4OH)

  Inlet SO_2 concentration in solvent: 0

- Packing:

  Packing type: 2-in ceramic Raschig rings, costing $20/ft^3 in 1991

  Surface-area-to-volume ratio, ft^2/ft^3: 28

- Column:

  Material of construction: fiber-reinforced plastic (FRP)

  Gas flow rate, lb-mol/h: 1835

  Solvent flow rate, lb-mol/h: 31.86

  Outlet solvent pollutant concentration, mole fraction: 0.0964

  Outlet gas pollutant concentration, mole fraction: 1.87447×10^{-5}

  Slope of equilibrium line: 0.001037396

  Absorption factor: 16.77

  Column cross-sectional area, A_c, ft^2: 30.93

  Column diameter, D_c, ft^2: 6.275

  Number of transfer units: 4.831

  Height of a transfer unit, ft: 2.377

  Packing depth, ft: 11.483

  Column total height, ft: 25.29

  Column surface area, ft^2: 560.4

  Packing volume, ft^3: 355.2

- Ductwork:

  Material of construction: FRP

  Gas velocity, ft/min: 2000

Diameter, in: 38.20

Length, in feet of equivalent straight duct: 150

- Fan, motor, and motor starter:

Fan design: backward-curved centrifugal

Material of construction: FRP

Wheel diameter, in: 45

- Pump, motor, and motor starter:

Pump design: single-stage centrifugal

Material of construction: cast iron/bronze fitted with stainless steel mechanical seal

Power, hp: 0.5

Find the total capital investment (TCI) of the gas absorber unit in 2001 dollars.

Calculation Procedure

1. *Estimate the cost of the column and packing.*

The solution to the problem begins with calculating the cost of each of the major equipment items (MEIs) comprising the gas absorber unit. The MEIs consist of the column and packing, ductwork, fan, and pump. We start with the cost of the column and packing. The sizing and costing procedure for the column and packing as presented here closely follows that in the U.S. Environmental Protection Agency (EPA) report, *EPA Air Pollution Control Cost Manual* [Ref. 6].

The cost of the column is a function of its surface area, which is calculated from the column diameter (D_c) and height (H_c) via the following formula:

$$S(\text{ft}^2) = \pi D_c (H_c + D_c/2)$$
$$= \pi(6.275)(25.29 + 6.275/2)$$
$$= 560.4\,\text{ft}^2 \tag{18.9}$$

Based on FRP fabrication, the column equipment cost (C_c) in 1991 dollars is

$$C_c = 161S = \$90,224$$

The packing cost, C_p, is a function of the packing volume, V_p, which is in turn a product of the packing depth, H_p, and column cross-sectional area, A_c:

$$V_p(\text{ft}^3) = H_p A_c = (11.483)(30.93) = 355.2\ \text{ft}^3 \tag{18.10}$$

The cost of this packing (Raschig rings) is $20/ft³ in 1991 dollars. The packing cost is the product of the unit cost and the packing volume, or $7104.

The column-plus-packing equipment cost is, therefore, $97,328 in 1991 dollars.

However, this cost in 1991 dollars has to be escalated to 2001 dollars. As shown in Procedure 18.2, the escalated cost is the product of the base cost and the escalation factor. The costs of gas absorbers can be escalated via one of the Vatavuk Air Pollution Control Cost Indexes (VAPCCIs). Updated quarterly since 1994 and published in *Chemical Engineering*, the VAPCCIs have been developed for gas absorbers and eight other types of air pollution control devices. The annual indices for gas absorbers for the years 1994 and 2001 are 100.8 and 114.4, respectively. In addition, the EPA study documenting the development of the VAPCCIs showed that gas absorber prices increased by 5.06% from 1991 to 1994 [Ref. 7]. Combining this increase with the VAPCCI data, we get

$$\text{Escalation factor} = 1.0506(114.4/100.8) = 1.192$$

The escalated cost of the column and packing is the product of this factor and the cost in 1991 dollars, or (1.192)(97,378), or

$$\text{Escalated cost} = \$116,075 \text{ in 2001 dollars}$$

2. Estimate the cost of the ductwork.
In calculating the ductwork cost, we use the equivalent straight duct run (see the list earlier in this section) that is long enough to cover the cost of the fittings (elbows, tees, etc.). Based on this equivalent length, the duct diameter (D_d), and a duct cost equation obtained from an EPA report [Ref. 8], the following relationship can be used:

$$\text{Unit duct (\$/ft, in 1993 dollars)} = 11.8e^{(0.0542D_d)} \tag{18.11}$$

where e is the base of natural logarithms. Substituting the duct diameter into this equation, we get

$$\text{Unit duct cost} = 11.8e^{[(0.05242)(38.20)]}$$
$$= \$93.55/\text{ft}$$

Lastly, multiply this unit cost by the equivalent straight duct length:

$$\text{Duct cost in 1993 dollars} = (93.55)(150) = \$14,033$$

Escalate this cost to 2001 dollars via the Producer Price Index for plastic pipe (pcu3084#l). The annual index values for 1993 and 2001 are 98.2 and 101.0, respectively. Then

$$\text{Escalated duct cost} = \$14,033(101.0/98.2) = \$14,433 \text{ in 2001 dollars}$$

3. Estimate the cost of the fan package.
The costs of a fan, motor, and motor starter (fan package) in 1988 dollars are combined in the following equation from Ref. 9:

$$\text{Fan package cost} = 42.3D_f^{1.20} \tag{18.12}$$

where D_f is the fan wheel diameter in inches (45 in). Substituting the wheel diameter gives

$$\text{Fan package cost in 1988 dollars} = 42.3(45)^{1.20} = \$4076$$

To escalate this cost to 2001 dollars, use the Producer Price Index for centrifugal fans and blowers (pcu3564#3). Index values for 1988 and 2001 are 106.8 and 151.1, respectively. Therefore,

$$\text{Fan Package cost in 2001 dollars} = \$4076(151.1/106.8) = \$5767$$

4. *Estimate the cost of the pump package.*

The cost of a pump package (pump, motor, and motor starter) in 1988 dollars is provided by the following relationship:

$$\text{Pump package cost} = 538(HP)^{0.438} \qquad (18.3)$$

where HP = pump horsepower = 0.5
Thus,

$$\text{Pump package cost in 1988 dollars} = 538(0.5)^{0.438} = \$397$$

To escalate this cost to 2001 dollars, use the Producer Price Index for industrial pumps (pcu3561#1). Annual values for 1988 and 2001 are 112.3 and 171.1, in turn.
Thus, the

$$\text{Pump package cost in 2001 dollars} = \$397(171.1/112.3) = \$605$$

5. *Estimate the total equipment cost of the absorber unit in 2001 dollars.*

The total gas absorber unit equipment cost (TEC) is the sum of the individual equipment costs:

$$TEC = \$116,075 + 14,433 + 5767 + 605$$
$$= \$136,880$$

6. *Estimate the purchased equipment cost.*

The total equipment cost (TEC) is the foundation of the total capital investment, which is computed from this equipment cost by adding fixed percentages or "factors" of it. Therefore, this category of total capital investment is known as a "factored estimate" or "study estimate." Factored or study estimates represent a step above scaling factor estimates (see Procedure 18.1), in that they require more effort to make but are more accurate (nominally ±30%). Unlike scaling factor estimates, for which the only required process input is the plant capacity, study estimates require a number of process-specific inputs, including a process flow sheet, preliminary energy and material balances, specifications for the major equipment items, preliminary ducting and piping layouts, and similar data.

As already indicated, the gas absorber sizing and equipment cost-estimating procedure used here closely follows that in the *EPA Air Pollution Control Cost Manual* [Ref. 6]. Accordingly, this manual's methodology for estimating the capital cost will also be followed. In that methodology, the purchased equipment cost (PEC) first is factored from the TEC. Then, the various direct and indirect installation costs are factored from the PEC. Finally, the TCI is calculated by adding the PEC to the direct and indirect installation costs. The same reasoning can be applied to TCI estimates that do not involve air-pollution-control facilities.

According to the *EPA Manual* procedure, the PEC is made up of three elements in addition to the total equipment cost: (1) freight cost, (2) sales taxes, and (3) instrumentation cost. The percentages the *EPA Manual* assigns to these costs are 5%, 3%, and 10% of the total equipment cost, respectively. For this problem, the costs are, accordingly, as follows:

Total equipment:	$136,880
Freight:	6844
Sales tax:	4106
Instrumentation:	13,688
PEC:	$161,518

7. *Estimate the direct and indirect installation costs.*
As previously explained, the various installation costs are factored from the *PEC*. The *EPA Manual* factors, which apply to a gas absorber installed under "typical conditions," along with the costs factored from the *PEC* here, are as follows:

Direct installation cost	Cost factor ($\times$ *PEC*)	Calculated cost
Foundations and supports	0.12	$ 19,382
Handling and erection	0.40	64,607
Electrical	0.01	1615
Piping	0.30	48,455
Insulation	0.01	1615
Painting	0.01	1615
Subtotal, direct installation	0.85	$137,289

Indirect installation cost	Cost factor $\times$ (*PEC*)	Calculated cost
Engineering	0.10	$16,152
Construction and field expense	0.10	16,152
Contractor fees	0.10	16,152
Start-up	0.01	1615
Performance test	0.01	1615
Contingencies	0.03	4846
Subtotal, indirect installation	0.35	$56,532

8. *Estimate the total capital investment.*
The total capital investment is the sum of the *PEC* and the direct and indirect installation costs: $161,518 + 137,289 + 56,532 = $355,339, or $355,300 (rounded).

18.4 PROCESS ANNUAL COSTS

A 70,000 ton/year, $22,000,000 plant was started up in 2001. Since then, it has been operating at capacity for an average of 7200 h/year. It is staffed by 25 full-time production workers (@$32/h), two maintenance workers (@$35/h), and four shift supervisors (@$39/h). All labor rates include payroll and plant overhead. The process consumes 530 kW of electricity (@$0.045/kWh), 50,000 gal/h of water (@$0.15/thousand gal), and 8.75 ton/h of feedstock (@$0.78/lb). Maintenance materials are estimated as equivalent to maintenance labor. Royalties amount to $4.20 per ton of product. The property taxes, insurance, and administrative charges total 5.2% of the total capital investment. Find the direct and indirect annual costs for the plant.

Calculation Procedure

1. *Determine the direct annual costs.*
The direct annual costs equal the sum of the costs of production and maintenance labor, maintenance materials, supervisory labor, electricity, water, feedstock, and royalties. Thus:

Production labor: (25 workers/h)(7200 h/year)($32/h) = $5,760,000

Maintenance labor: (2 workers/h)(7200 h/year)($35/h) = $504,000

Maintenance materials = maintenance labor = $504,000

Supervisory labor: (4 supervisors/h)(7200 h/year)($39/h) = $1,123,200

Electricity: (530 kW)(7200 h/year)($0.045/kWh) = $171,720

Water: (50 thousand gal/h)(7200 h/year)($0.15/thousand gal) = $54,000

Feedstock: (8.75 ton/h)(7200 h/year)($0.078/lb)(2000 lb/ton) = $9,828,000

Royalties: (70,000 ton/year)($4.20/ton) = $294,000

The total direct annual cost is the sum of these, or $18,238,920.

2. *Determine the indirect annual costs.*
The indirect annual costs, consisting of property taxes, insurance, and administrative charges are given in the problem statement as 5.2% of the total capital investment. Thus,

$$\text{Indirect annual costs} = (0.052)(\$22,000,000) = \$1,144,000$$

3. *Determine the total annual costs.*
From steps 1 and 2, the sum is $ 18,238,920 + 1,144,000, or $19,382,920.

18.5   NET PRESENT WORTH (NPW) ANALYSIS

A firm is evaluating two competing projects. The first is a new ("grass roots") inorganic chemicals plant, while the second is the expansion of a textile fibers facility. The process engineers have estimated the projected annual revenue, total capital investment, and total annual cost (without capital recovery) for each project, as follows:

Cost category	Cost (2002 dollars)	
	Inorganic chemicals plant	Textile fibers plant expansion
Revenue(projected)	$33,700,000	$30,900,000
Total capital investment	52,500,000	57,300,000
Total annual cost	25,100,000	21,500,000

Both the new and expanded plants would have an estimated life of 20 years. The firm's hurdle rate (marginal acceptable rate of return) is 12.5% (before tax), and its marginal state and federal corporate income tax rate is 52%. Assuming straight-line depreciation (with zero salvage) and that 100% of the investment is depreciable, calculate the net present worth of each project.

Calculation Procedure

1. *Find the annual depreciation for each project.*
Based on straight-line depreciation, the annual depreciations for the two projects would be simply their total capital investments divided by 20 years. Thus,

Annual depreciation for inorganic chemical plant: $52,500,000/20 = $2,625,000.

Annual depreciation for textile fibers plant: $57,300,000/20 = $2,865,000.

2. *Calculate the after-tax discount or hurdle rate.*
The after-tax discount rate, which applies to each of the two proposals, equals the before-tax rate multiplied by 1 minus the tax rate. Thus, the after-tax hurdle rate, $i*$, equals $12.5\% \, (1 - 0.52) = 6.0\%$.

3. *Calculate the annual undiscounted net cash flows (NCFs) for the inorganic chemicals plant.*
For the inorganic chemicals plant, for the beginning of the project (year zero), the net cash flow will be negative, as the only cash flow item will be the total capital investment. That is, NCF (year zero) = −$52,500,000. However, for each of the following years (year 1 through year 20, the end of the project) the NCF will be calculated by Eq. (18.4) from the introduction to this chapter:

$$NCF \text{ (year } 1, 2, \ldots, 20) = (R - TAC)(1 - t) + tDep$$

$$= (33,700,000 - 25,100,000)(1 - 0.52) + 0.52(2,625,000)$$

$$= \$5,493,000$$

4. *Calculate the net present worth (NPW) for the inorganic chemicals plant.*
Step 3 determined undiscounted $NCFs$. To determine the net present worth, NPW, it is necessary to determine the discounted NCF for each year in the project life and then their sum. The discounted NCF for year k is the product of the discount factor $[1/(1 + i*)k]$ and the undiscounted NCF. The discounted $NCFs$ for the various years, and their sum the NPW, are as follows:

Year	Discount factor (rounded)	NCF (discounted)
0	—	−52,500,000
1	0.9434	5,182,075
2	0.8900	4,888,750
3	0.8396	4,612,029
4	0.7921	4,350,970
5	0.7473	4,104,689
6	0.7050	3,872,348
7	0.6651	3,653,159
8	0.6274	3,446,376
9	0.5919	3,251,298
10	0.5584	3,067,263
11	0.5268	2,893,644
12	0.4970	2,729,853
13	0.4688	2,575,333
14	0.4423	2,429,559
15	0.4173	2,292,037
16	0.3936	2,162,299
17	0.3714	2,039,905
18	0.3503	1,924,438
19	0.3305	1,815,508
20	0.3118	1,712,743
	NPW:	$10,504,276

5. *Calculate the annual undiscounted net cash flows (NCFs) for the textile fibers plant.*
Follow the procedure of step 3, but apply it to the data for the textile fibers plant. Thus,

$$NCF \text{ for year zero} = -TCI = -\$57,300,000$$

$$\text{Undiscounted } NCF \text{ for years } 1,2,\ldots,20 = (30,900,000 - 21,500,000)(1 - 0.52)$$
$$+ 0.52(2,865,000) = \$6,001,800$$

6. Calculate the net present worth (NPW) for the textile fibers plant.
Follow the procedure of step 4, but apply it to the data for the textile fibers plant. The discounted NCFs for the various years for the textile fibers plant, and their sum the NPW, are as follows:

Year	Discount factor (rounded)	NCF (discounted)
0	—	−52,300,000
1	0.9434	5,662,075
2	0.8900	5,341,581
3	0.8396	5,039,227
4	0.7921	4,753,988
5	0.7473	4,484,894
6	0.7050	4,231,032
7	0.6651	3,991,540
8	0.6274	3,765,604
9	0.5919	3,552,456
10	0.5584	3,351,374
11	0.5268	3,161,673
12	0.4970	2,982,711
13	0.4688	2,813,878
14	0.4423	2,654,602
15	0.4173	2,504,341
16	0.3936	2,362,586
17	0.3714	2,228,855
18	0.3503	2,102,693
19	0.3305	1,983,673
20	0.3118	1,871,390
	NPW:	$10,504,276

7. Select the more attractive project.
Because the projected net present worth of the textile fibers plant expansion project is higher than that of the inorganic chemicals plant, the firm's funds would be better spent on it.

Related Calculations.   Under the conditions stated here, the preferred project would be the textile fibers plant expansion. However, this might not have been the case if we had assumed a different hurdle rate, plant life, revenues, or total annual costs.

Another key assumption is that the revenues and total annual costs are constant over the life of the project. That is, they are expressed in constant (2002) dollars. Neither the costs nor the revenues have been adjusted for inflation.

By accounting convention, depreciation is never adjusted for inflation, even if the other costs and revenues are so adjusted. Depreciation is assumed to be constant in this constant-dollar analysis as well. Technically, this assumption is incorrect, as the depreciation cash flows should have been adjusted. However, to minimize the example's complexity, this adjustment was not been made. Another simplification is the assumption of straight-line depreciation. Current U.S. tax laws are more generous, in that they allow for larger depreciation deductions in the earlier years of the project than later. In such cases, the depreciation deduction would vary from year to year.

18.6   *INTERNAL RATE OF RETURN (IRR) ANALYSIS*

Given the data for the inorganic chemicals plant and the textile fibers plant expansion in the problem statement for Procedure 18.5, use the internal rate of return method to determine which project the firm should fund.

Calculation Procedure

1. Take into account how this approach differs from that for NPW analysis.
The project lives, *TCIs*, *TACs*, depreciation, undiscounted net cash flows, plant lives, and tax rate are the same as those given for Procedure 18.5, on *NPW* analysis. However, in this present example, the hurdle (discount) rate is not an input. In Procedure 18.5, the hurdle rate was an input and, based on this rate, the net present worth was calculated for each project. In this example, by contrast, the net present worth is arbitrarily set to zero and the unique discount rate that produces a *NPW* of zero is solved for. This discount rate is the internal rate of return (*IRR*). The project with the higher internal rate of return is selected as the one to be funded.

Because the *NPW* is the sum of a geometric series, the *IRR* cannot be solved for algebraically. Instead, it is necessary to use an iterative procedure: Guess a value for the *IRR*, determine the sum of the discounted cash flows that result from your guessed *IRR*, and repeat the procedure until it leads to a sum that is virtually zero.

2. Find, by iteration, the IRR for the inorganic chemicals plant.
Iterations not shown here lead ultimately to an *IRR* of 8.3643% after-tax, which implies 17.4255% pre-tax (see Procedure 18.5, step 2). The employment of 8.3643 as after-tax hurdle rate to find the discounted net case flows for the inorganic chemicals plant leads to the following results:

Year	Discount factor (rounded)	NCF (discounted)
0	—	-$52,500,000
1	0.9228	5,069,014
2	0.8516	4,677,755
3	0.7859	4,316,695
4	0.7252	3,983,504
5	0.6692	3,676,031
6	0.6176	3,392,291
7	0.5699	3,130,452
8	0.5259	2,888,823
9	0.4853	2,665,845
10	0.4479	2,460,078
11	0.4133	2,270,193
12	0.3814	2,094,965
13	0.3520	1,933,262
14	0.3248	1,784,040
15	0.2997	1,646,336
16	0.2766	1,519,261
17	0.2552	1,401,994
18	0.2355	1,293,779
19	0.2174	1,193,917
20	0.2006	1,101,763
NPW:		-$1

3. Find, by iteration, the IRR for the textile fibers plant expansion.
Repeat step 2 with respect to the expansion of the textile fibers plant. This iteration leads to an after-tax *IRR* of 8.3792%, or 17.4568% pre-tax. The computation and summing of the discounted *NCF*s with a discount rate of 8.3792% leads to the following results:

Year	Discount factor (rounded)	*NCF* (discounted)
0	—	−$57,300,000
1	0.9227	5,537,776
2	0.8513	5,109,628
3	0.7855	4,714,581
4	0.7248	4,350,077
5	0.6688	4,013,755
6	0.6171	3,703,435
7	0.5693	3,417,107
8	0.5253	3,152,916
9	0.4847	2,909,151
10	0.4472	2,684,233
11	0.4127	2,476,704
12	0.3808	2,285,220
13	0.3513	2,108,540
14	0.3242	1,945,520
15	0.2991	1,795,104
16	0.2760	1,656,317
17	0.2546	1,528,260
18	0.2349	1,410,104
19	0.2168	1,301,083
20	0.2000	1,200,491
	NPW:	$2

4. *Select the more attractive project.*
Based on these results, we conclude that the textile fibers plant expansion would be the preferred project, as its after-tax *IRR* (about 8.38%) is slightly higher than the after-tax *IRR* for the inorganic chemicals plant (about 8.36%).

Related Calculations.   The difference between the two internal rates of returns is so small that, on a purely economic basis, the projects are virtually indistinguishable. By contrast, the difference in the projects' net present worths (see Procedure 18.5) is large enough to make the textile fibers expansion the clear choice for funding. As with the net present worth method, the internal rate of return procedure cannot be used unless the lifetimes of the competing projects are equal.

18.7 EQUIVALENT UNIFORM ANNUAL REVENUE (EUAR) ANALYSIS

Given the data for the inorganic chemicals plant and the textile fibers plant expansion provided in the problem statement of Procedure 18.5, employ equivalent uniform annual revenue (*EUAR*) analysis to determine which project the firm should fund.

Calculation Procedure

1. *Calculate the capital recovery factor.*
In this method, each of the undiscounted net cash flows (*NCF*s) for years 1 to 20 is algebraically added to the capital recovery cost (*CRC*) to obtain the equivalent uniform annual revenue (*EUAR*). The *CRC* is the product of the total capital investment, *TCI*, and the capital recovery factor, *CRF*,

which is defined and calculated by Eq. (18.7) of the introduction. The project with the higher *EUAR* will be the one to fund. The pertinent input data are as follows:

	Revenue/cost (2002 dollars)	
Revenue/cost category	Inorganic chemicals plant	Textile fibers plant expansion
Revenue (projected)	$33,700,000	$30,900,000
Total capital investment	52,500,000	57,300,000
Total annual cost	25,100,000	21,500,000
Net cash flow (undiscounted)	5,493,000	6,001,800

From Eq. (18.7) of the introduction, using the after-tax hurdle rate as calculated in Procedure 18.5, step 2, we obtain

$$CRF = i\,(1 + i)^n / (1 + i)^n - 1 = 0.06(1.06)^{20}/1.06^{20} - 1 = 0.08718$$

2. Calculate the capital recovery cost for the inorganic chemicals plant.

$$CRC = (TCI)(CRF) = (52,500,000)(0.08718) = \$4,576,950$$

3. Calculate the equivalent uniform annual revenue for the inorganic chemicals plant.

Equivalent uniform annual revenue = $NCF - CRC = 5,493,000 - 4,576,950 = \$916,050$

4. Calculate the capital recovery cost for the textile fibers plant expansion.

$$CRC = (TCI)(CRF) = (57,300,000)(0.08718) = \$4,995,414$$

5. Calculate the equivalent uniform annual revenue for the inorganic chemicals plant.

Equivalent uniform annual revenue = $NCF - CRC = 6,001,800 - 4,995,414 = \$1,006,386$

6. Determine the more attractive process.
Based on the *EUAR* measure, the textile fibers plant expansion would be the clear choice to fund, because its *EUAR* of $1,006,386 is larger than the $916,050 *EUAR* of the inorganic chemicals plant.

Related Calculations.   The *EUAR* for the textile fibers plant expansion is 9.86% higher than that of the inorganic chemicals plant. The textile fibers plant *NPW* is also 9.86% higher than the inorganic chemicals plant *NPW* (see Procedure 18.5). This is not a coincidence. In fact, the *EUAR* and *NPW* methods will always yield the same results and can be shown to be mathematically equivalent.

Even so, the *EUAR* method is best suited to those situations where (as in this case) the undiscounted net cash flows are constant. However, when they are not constant, each *NCF* must be discounted back to year zero, summed, and annualized by multiplying it by the *CRF*. Finally, this annualized *NCF* must be added to the capital recovery cost. By the time the analyst has done all of this, he or she could just as well have calculated the net present worths or internal rates of return of the competing projects.

18.8   *EVALUATING CAPITAL INVESTMENT ALTERNATIVES*

A portland cement plant ball mill emits particulate matter (PM) emissions that must be controlled to meet state air pollution regulations. Three PM control devices, each of which can control these emissions to the same level, are being evaluated: (1) a high-energy wet scrubber (scrubber), (2) an electrostatic precipitator (ESP), and (3) a fabric filter (baghouse). Unlike the wet scrubber, the ESP and the baghouse each recover salable cement dust, and, accordingly, revenue can be attributed to those two options. Two scenarios are visualized, with after-tax hurdle rates of 6% and 18%, respectively. The economic specifications for the three devices are as follows:

Parameter (all revenues and costs in 2002 dollars)	Control device		
	Scrubber	ESP	Baghouse
Life (years)	10	20	15
Marginal tax rate	0.52	0.52	0.52
Hurdle rate (after tax)			
— Scenario "A"	0.06	0.06	0,06
— Scenario "B"	0.18	0.18	0.18
Revenue	0	$290,000	$290,000
Total capital investment	$5,300,000	$9,750,000	$7,870,000
Total annual cost	2,770,000	1,840,000	2,345,000
Salvage value	0	0	0
Depreciation (straight line)	530,000	487,500	524,667

For each of the two scenarios, determine the most economical control device to control the dust from the ball mill.

Calculation Procedure

1. Select the most appropriate evaluation method.
Note that the control devices have different economic lives. Thus, neither the net present worth nor the internal rate of return method can be used, as both require that all options have the same economic life. However, the equivalent uniform annual revenue method can be used, as this restriction does not apply to it.

2. Determine the EUAR for each option, assuming a 6% hurdle rate.
Following the procedure of Procedure 18.7, calculate for each option the undiscounted net cash flow, the capital recovery factor, the capital recovery cost, and the EUAR. The results are as follows:

Cost/cost factor (2002 dollars)	Control device		
	Scrubber	ESP	Baghouse
NCF	−$1,054,000	−$490,500	−$713,573
CRF (rounded)	0.1359	0.0872	0.1030
CRC	720,100	850,049	810,317
EUAR	−$1,774,100	−$1,340,540	−$1,523,890

Under this scenario, the most economical control device is the ESP, as its EUAR is the largest (i.e., least negative) of the three. (Put another way, it has the lowest equivalent uniform annual cost.)

3. Determine the EUAR for each option, assuming an 18% hurdle rate.
Repeat step 2, but with the 18% hurdle rate. In this case, the results are as follows:

Cost/cosr factor (2002 dollars)	Control device		
	Scrubber	ESP	Baghouse
NCF	−$1,054,000	−$490,500	−$713,573
CRF (rounded)	0.2225	0.1868	0.1964
CRC	1,179,328	1,821,495	1,545,690
EUAR	−$2,233,328	−$2,311,995	−$2,259,263

Under this scenario, the most economical control device is the wet scrubber, as it has the least negative *EUAR*.

Related Calculations.   Clearly, the *EUAR* and the selection of the most economical control device depend on the hurdle rate, even when the other inputs are held constant. Moreover, a given firm's hurdle rate can vary according to general economic conditions, the expected risk associated with a project, and other factors. Thus, the control device selection in this hypothetical situation or in any similar, real-world situation could also be affected by these factors.

In this problem, the analysis is done on an after-tax basis. However, the analysis could just as well be performed on a before-tax basis, using a pre-tax hurdle rate. If so, the depreciation term would be zero and the net cash flow equation would simplify to

$$NCF = \text{revenue} - TAC$$

and the *EUAR* expression would reduce to

$$EUAR = \text{revenue} - TAC - CRC$$

*18.9   COST ESTIMATION OF CHEMICAL-PLANT HEAT EXCHANGERS AND STORAGE TANKS VIA CORRELATIONS

Using correlations, estimate the cost of a fixed-head, carbon-steel heat exchanger rated for 150 lb/in² (gage) (1034 kPa) having a total heat-transfer area of 1500 ft² (139.4 m²). Using the same approach, estimate the cost of a cone-roof storage tank made of carbon steel having a total capacity of 677,000 gal (2,562,445 L). Show how to update the costs from the base year (cost index = 200.8) to a year in which the cost index is 265.

Calculation Procedure

1. Compute the base cost of the head exchanger.
Using Table 18.1, substitute the area A in the relation $C_B - \exp[8.551 - 0.30863 \ln A + 0.06811 (\ln A)^2]$. Or, $C_B = \exp[8.551 - 0.30863 \ln 1500 + 0.06811 (\ln 1500)^2] = \$20,670$.

TABLE 18.1   Correlations for Costs of Heat Exchangers[†]

USCS units	SI units
Base cost for carbon-steel, floating-head, 100 lb/in² (gage) exchanger:	Base cost for carbon-steel, floating-head, 700-kN/m² exchanger:
$C_B = \exp[8.551 - 0.30863 \ln A + 0.06811 (\ln A)^2]$	$C_B = \exp[8.202 + 0.01506 \ln A + 0.06811 (\ln A)^2]$
Exchanger-type cost factor:	Exchanger-type cost factor:
Fixed-head: $F_D = \exp(-1.1156 + 0.0906 \ln A)$ Kettle reboiler: $F_D = 1.35$ U-tube: $F_D = \exp(-0.9816 + 0.0830 \ln A)$	Fixed-head: $F_D = \exp(-0.9003 + 0.0906 \ln A)$ Kettle reboiler: $F_D = 1.35$ U-tube: $F_D = \exp(-0.7844 + 0.0830 \ln A)$
Design-pressure cost factor:	Design-pressure cost factor:
100–300 lb/in² (gage): $F_P = 0.7771 + 0.04981 \ln A$ 300–600 lb/in² (gage): $F_P = 1.0305 + 0.07140 \ln A$ 600–900 lb/in² (gage): $F_P = 1.1400 + 0.12088 \ln A$	700–2100 kN/m²: $F_P = 0.8955 + 0.04981 \ln A$ 2100–4200 kN/m²: $F_P = 1.2002 + 0.07140 \ln A$ 4200–6200 kN/m²: $F_P = 1.4272 + 0.12088 \ln A$
A in ft²; lower limit: 150 ft², upper limit 12,000 ft²	A in m²; lower limit: 14 m², upper limit 1100 m²

[†]*Chemical Engineering.*

TABLE 18.2   Material-of-Construction Cost Factors for Heat Exchangers[†]

Material	USCS units, A in ft^2 $F_M = g_1 + g_2 \ln A$		SI units, A in m^2 $F_M = g_1 + g_2 \ln A$	
	g_1	g_2	g_1	g_2
Stainless steel 316	0.8608	0.23296	1.4144	0.23296
Stainless steel 304	0.8193	0.15984	1.1991	0.15984
Stainless steel 347	0.6116	0.22186	1.1388	0.22186
Nickel 200	1.5092	0.60859	2.9553	0.60859
Monel 400	1.2989	0.43377	2.3296	0.43377
Inconel 600	1.2040	0.50764	2.4103	0.50764
Incoloy 825	1.1854	0.49706	2.3665	0.49706
Titanium	1.5420	0.42913	2.5617	0.42913
Hastelloy	0.1549	1.51774	3.7614	1.51774

[†]*Chemical Engineering.*

2. Determine the exchanger-type cost factor for the heat exchanger.

Again, by using Table 18.1 for a fixed-head exchanger, $F_D = \exp(-1.1156 + 0.090606 \ln A)$, where F_D = exchanger-type cost factor. Substituting yields $F_D = \exp(-1.1156 + 0.090606 \ln 1500) = 0.6357$.

3. Find the design-pressure cost factor for the exchanger.

From Table 18.1, the design-pressure cost factor for a pressure in the 100 to 300-lb/in^2 (gage) range (700–2100-kPa range) is $F_P = 0.7771 + 0.04981 \ln A$. Substituting, we find $F_p = 0.7771 + 0.04981 \ln 1500 = 1.1414$.

4. Find the materials-of-construction cost factor.

The materials-of-construction cost factor, F_M, for carbon steel is unity, or 1.0. Factors for other materials of construction are shown in Table 18.2.

5. Compute the heat-exchanger cost.

Use the relation $C_E = C_B F_D F_p F_M$, where C_E = exchanger cost. Or, $C_E = (\$20,670)(0.6357)(1.1414)(1.0) = \$15,000$.

6. Update the heat-exchanger cost.

The base-year cost index—for 1976, the year on which the above costs are based—is 200.8. For the year in which the cost estimate is being made, the cost index is 265 (obtained from any of the standard, widely accepted cost indices). Updating the heat-exchanger cost reveals $C_{EU} = \$15,000(265/200.8) = \$19,796$. In this relation, the updated cost is $C_{EU} = C_E$ (current-year equipment cost index/base-year cost index).

7. Compute the storage-tank cost.

Using Table 18.3, apply the relation $C_B = \exp[11.362 - 0.6104 \ln V + 0.045355 (\ln V)^2]$, where C_B = base cost of field-erected tank in carbon steel; V = tank volume, gal. Substituting gives $C_B = \exp[11.362 - 0.6104 \ln 677.000 + 0.045355 (\ln 677,000)^2] = \$84,300$. Updating the cost, as before, we find $C_{BU} = \$84,300(265/200.8) = \$111,252$. Table 18.4 shows materials of construction cost factors for storage tanks.

Related Calculations.   The approach given here correlates the cost of shell-and-tube heat exchangers and heat-transfer area. This contrasts with cost estimation procedures that take into account shell diameter, number, and length of the tubes, types of heads, and other construction details. The accuracy of the simple correlation of cost versus area is sufficient for preliminary cost estimates.

TABLE 18.3   Correlations for Costs of Storage Tanks[†]

USCS units	SI units
Base cost for carbon-steel, shop-fabricated tanks:	Base cost for carbon-steel, shop-fabricated tanks:
$C_B = \exp [2.331 + 1.3673 \ln V - 0.063088 (\ln V)^2]$	$C_B = \exp [7.994 + 0.6637 \ln V - 0.063088 (\ln V)^2]$
V in gallons; lower limit: 1300 gal, upper limit: 21,000 gal	V in m^3; lower limit 5 m^3, upper limit: 80 m^3
Base cost for carbon-steel, field-erected tanks:	Base cost for carbon-steel, field-erected tanks:
$C_B = \exp [11.362 - 0.6104 \ln V + 0.045355 (\ln V)^2]$	$C_B = \exp [9.369 - 0.1045 \ln V + 0.045355 (\ln V)^2]$
V in gallons; lower limit: 21,000 gal, upper limit: 11,000,000 gal	V in m^3; lower limit: 80 m^3, upper limit: 45,000 m^3

[†]*Chemical Engineering.*

Correlations for base cost are given in both USCS and SI units in the accompanying tables. The base-cost basis for the equipment is given in each table. While heat-exchanger costs are based on area, storage-tank costs are based on the total tank volume. The tank volume is calculated (for the base cost) from residence time, a fixed overcapacity factor of 20%, and volumetric flow rate.

Omitted from the cost estimation procedure given here are the number and sizes of nozzles and manholes and other design details. These details cause variations in cost that are usually within the accuracy of preliminary estimates.

Data on the cost of shell-and-tube heat exchangers in a wide range of heat-transfer areas and design pressures were used in developing the correlations for 10 different materials of construction and three design types. PDQ$, Inc., supplied these and the cost data for cylindrical carbon-steel tanks having cone roofs and flat bottoms in a wide range of volumes. The cost of field-erected tanks includes the cost of platforms and ladders, but not of foundations and other installation materials (piping, electric instrumentation, etc.). The cost of the shop-fabricated tanks does not include any of the installation materials.

This procedure is the work of Armando B. Corripio, Louisiana State University, and Katherine S. Chrien and Lawrence B. Evans, both of the Massachusetts Institute of Technology, as reported in *Chemical Engineering* magazine.

TABLE 18.4   Material-of-Construction Cost Factors for Storage Tanks[†]

Material of construction	Cost factor F_M
Stainless steel 316	2.7
Stainless steel 304	2.4
Stainless steel 347	3.0
Nickel	3.5
Monel	3.3
Inconel	3.8
Zirconium	11.0
Titanium	11.0
Brick-and-rubber- or brick-and-polyester-lined steel	2.75
Rubber- or lead-lined steel	1.9
Polyester, fiberglass-reinforced	0.32
Aluminum	2.7
Copper	2.3
Concrete	0.55

[†]*Chemical Engineering,*

*18.10 ESTIMATING CHEMICAL-PLANT CENTRIFUGAL-PUMP AND ELECTRIC-MOTOR COST BY USING CORRELATIONS

Determine the cost of a ductile-steel pump to deliver 1430 gal/min (90.2 L/s) at a differential head of 77 ft · lbf/lb (230.2 J/kg). A horizontally split case one-stage pump running at 3550 r/min is specified. The specific gravity of the fluid being pumped is 0.952.

Calculation Procedure

1. Determine the size parameter S.
The size parameter is defined as $S = QH^{0.5}$, where Q = design capacity of the pump, gal/min (m³/s), and H is the required head for the pump, ft·lb/lb or J/kg. Substituting, we get $S = 1430(77)^{0.5} = 12,550$, closely.

2. Find the pump base cost C_B.
Use the base cost relation from Table 18.5, or $C_B = \exp[8.3949 - 0.6019 \ln S + 0.0519 (\ln S)^2]$. So $C_B = \exp[8.3949 - 0.6019 \ln 12,550 + 0.0519 (\ln 12,550)^2] = \$1536$.

3. Compute the pump design-type factor F_T.
Table 18.5 shows that the design-type factor for a one-stage, 3550-r/min HSC pump is found from $F_T = \exp[b_1 + b_2 \ln S + f_3 (\ln S)^2]$. Substituting the values given in the table, we see $F_T = \exp[0.0632 + 0.2744 \ln 12,550 - 0.0253 (\ln 12,550)^2] = 1.491$.

4. Find the materials-of-construction factor F_M.
From Table 18.6 for ductile iron, $F_M = 1.15$.

5. Compute the pump cost C_P with base plate and coupling.
Use the relation $C_P = C_B F_T F_M$, where the symbols are as given above. Thus, $C_p = (\$1536)(1.491)(1.15) = \$2630$.

TABLE 18.5   Correlations for Costs of Centrifugal Pumps[†]

USCS units				SI units			
Base cost for one-stage, 3550 r/min, VSC cast-iron pump:				Base cost for one-stage, 3550 r/min, VSC cast-iron pump:			
$C_B = \exp[8.3949 - 0.6019 \ln S + 0.0519(\ln S)^2]$				$C_B = \exp[7.2234 + 0.3451 \ln S + 0.0519(\ln S)^2]$			
Here, $S = Q\sqrt{H}$, Q in gal/min and H in ft·lbf/lb (ft of head).				Here, $S = Q\sqrt{H}$, with Q in m³/s, and H in J/kg or m²/s².			
Cost factor for pump type:				Cost factor for pump type:			
$F_T = \exp[b_1 + b_2 \ln S + b_3 (\ln S)^2]$				$F_T = \exp[b_1 + b_2 \ln S + b_3 (\ln S)^2]$			
Type	b_1	b_2	b_3	Type	b_1	b_2	b_3
One-stage, 1750-r/min, VSC	5.1029	−1.2217	0.0771	One-stage, 1750-r/min, VSC	0.3740	0.1851	0.0771
One-stage, 3550-r/min, HSC	0.0632	0.2744	−0.0253	One-stage, 3550-r/min, HSC	0.4612	−0.1872	−0.0253
One-stage, 1750-r/min, HSC	2.0290	−0.2371	0.0102	One-stage, 1750-r/min, HSC	0.7147	−0.0510	0.0102
Two-stage, 3550-r/min, HSC	13.7321	−2.8304	0.1542	Two-stage, 3550-r/min, HSC	0.7445	−0.0167	0.1542
Multistage, 3550-r/min, HSC	9.8849	−1.6164	0.0834	Multistage, 3550-r/min, HSC	2.0798	−0.0946	0.0834

[†]*Chemical Engineering* and Richardson Engineering Services, Inc.

TABLE 18.6   Cost Factors for Material of Construction[†]

Material	Cost factor F_M
Cast steel	1.35
304 or 316 fittings	1.15
Stainless steel, 304 or 316	2.00
Cast Gould's alloy no.20	2.00
Nickel	3.50
Monel	3.30
ISO B	4.95
ISO C	4.60
Titanium	9.70
Hastelloy C	2.95
Ductile iron	1.15
Bronze	1.90

[†]*Chemical Engineering.*
Source: Monsanto Co.'s FLOWTRAN pump-costing subprogram.

6. Determine the required horsepower for the motor.
Use the relation $P_B = pQH/33,000N_p$, where P_B = bhp input to pump; p = fluid density, lb/gal; N_P = pump efficiency, percent; other symbols as given earlier. In this method of cost estimating, $N_Q = -0.316 + 0.24015 \ (\ln Q) - 0.01199 \ (\ln Q)^2$.
    Find the fluid density from p = specific gravity (8.33 lb/gal) = 0.952(8.33) = 7.93 lb/gal (0.94 kg/L). The pump efficiency, from the previous relation, is $N_p = -0.316 + 0.24015 \ (\ln 1430) - 0.01199 \ (\ln 1430)^2 = 0.796$.
    Substituting in the power relation yields $P_B = 7.93 \ (1430)(77)/33,000(0.796) = 33.2$ hp (24.8 kW). A 40-hp (29.8·kW) motor is require for this pump.

7. Compute the cost of the electric motor.
Use the appropriate correlation from Table 18.7. Assume a 3600-r/min totally enclosed fan-cooled motor is needed. Then the motor cost $C_M = \exp \ [3.8544 + 0.8331 \ (\ln P_B) + 0.02399 \ (\ln P_B)^2] = \exp \ [3.8544 + 0.8331 \ (\ln 40) + 0.02399 \ (\ln 40)^2] = \$1410$.

8. Determine the total cost of the pump and motor.
Find the sum of $C_P + C_M$. Or, $C_P + C_M = \$2630 + \$1410 = \$4040$.

9. Compute the pump power consumption.
Use the relation $U_M + 0.80 = 0.0319 \ (\ln P_B) - 0.00182 \ (\ln P_B)^2$ to find the efficiency of the motor. By substituting, $N_M = 0.80 + 0.0319 \ (\ln 33.2) - 0.00182 \ln \ (33.2)^2 = 0.889$. Then the pump power consumption $P_C = P_B/N_M = 33.2/0.889 = 37.3$ hp (27.8 kW).

Related Calculations.   This procedure can be used for centrifugal pumps and electric motors in a variety of industries and applications provided the pump and motor are of the type listed in the tables. Typical industries and applications include chemical, petroleum, petrochemical, power, marine, air-conditioning, heating, and food processing.
    Data on the cost of centrifugal pumps and electric motors were taken from Vol. 4 of the data book by Richardson Engineering Services (Solana Beach, CA), *Process Plant Construction Estimating Standards.* The material-of-construction cost factors for pumps were taken from Monsanto Co.'s FLOWTRAN pump-costing subprograms.
    Although the cost of a pump includes the cost of the driver coupling, cost correlations for belt-, chain-, and variable-speed drive couplings were obtained from the U.S. Bureau of Mines equipment-costing program. These correlations were escalated from their original data of 1967 to the first quarter of 1979 by using the chemical engineering pumps and compressors index ratio

TABLE 18.7 Correlation for Cost of Electric Motors[†]

Cost of 60-Hz standard-voltage motor and insulation, discounted
$C_M = exp\,[a_1 + a_2\,ln\,P + a_3\,(ln\,P)^2]$
P is the nominal size in horsepower

	Coefficients				
	a_1	a_2	a_3	hp limits	kW limits
Open, drip-proof:					
3600 r/min	4.8314	0.09666	0.10960	1–7.5	0.75–5.6
	4.1514	0.53470	0.05252	7.5–250	5.6–186.5
	4.2432	1.03251	−0.03595	250–700	186.5–522.2
1800 r/min	4.7075	−0.01511	0.22888	1–7.5	0.75–5.6
	4.5212	0.47242	0.04820	7.5–250	5.6–186.5
	7.4044	−0.06464	0.05448	250–600	186.5–447.6
1200 r/min	4.9298	0.30118	0.12630	1–7.5	0.75–5.6
	5.0999	0.35861	0.06052	7.5–250	5.6–186.5
	4.6163	0.88531	−0.02188	250–500	186.5–373.0
Totally enclosed, fan-cooled:					
3600 r/min	5.1058	0.03316	0.15374	1–7.5	0.75–5.6
	3.8544	0.83311	0.02399	7.5–250	5.6–186.5
	5.3182	1.08470	−0.05695	250–400	186.5–298.4
1800 r/min	4.9687	−0.00930	0.22616	7.5–250	5.6–186.5
	4.5347	0.57065	0.04609		
1200 r/min	5.1532	0.28931	0.14357	1–7.5	0.75–5.6
	5.3858	0.31004	0.07406	7.5–350	5.6–261.1
Explosion-proof:					
3600 r/min	5.3934	−0.00333	0.15475	1–7.5	0.75–5.6
	4.4442	0.60820	0.05202	7.5–200	5.6–149.2
1800 r/min	5.2851	0.00048	0.19949	1–7.5	0.75–5.6
	4.8178	0.51086	0.05293	7.5–250	5.6–186.5
1200 r/min	5.4166	0.31216	0.10573	1–7.5	0.75–5.6
	5.5655	0.31284	0.07212	7.5–200	5.6–149.2

[†]*Chemical Engineering.*

of 270/11.2 = 2.43. All other cost data were for the first quarter of 1979, when the pumps and compressors index was 270 and the electrical equipment index was 175.5. To update the costs to the year in which an estimate is being made, simply apply the current index, as detailed in the preceding procedure.

Table 18.8 gives the flow, head, and power limits for the centrifugal pumps considered in this procedure. Table 18.9 shows the correlations for the cost of drive coupling for the pumps.

TABLE 18.8 Flow, Head, and Power Limits for Centrifugal Pumps[†]

	Flow, gal/min (m³/s)		Head, ft·lbf/lb (J/kg)		Motor hp, upper limit	Motor kW
	Lower limit	Upper limit	Lower limit	Upper limit		
One-stage, 3550-r/min, VSC	50 (0.00315)	900 (0.568)	50 (150)	400 (1200)	75	55.95
One-stage, 1750-r/min, VSC	50 (0.00315)	3500 (0.2208)	50 (150)	200 (600)	200	149.2
One-stage, 3550-r/min, HSC	100 (0.00631)	1500 (0.0946)	100 (300)	450 (1350)	150	111.9
One-stage, 1750-r/min, HSC	250 (0.01577)	5000 (0.3155)	50 (150)	500 (1500)	250	186.5
Two-stage, 3550-r/min, HSC	50 (0.00315)	1100 (0.0694)	300 (900)	1100 (3300)	250	186.5
Multistage. 3550-r/min, HSC	100 (0.00631)	1500 (0.0946)	650 (2000)	3200 (9600)	1450	1081.7

[†]*Chemical Engineering.*

TABLE 18.9 Correlations for Cost of Drive Coupling[†]

Cost of belt-drive coupling:

$$C_c = \exp(3.689 + 0.8917\ln P)$$

Cost of chain-drive coupling:

$$C_c = \exp(5.329 + 0.5048\ln P)$$

Cost of variable-speed-drive coupling:

$$C_c = 1/[1.562 \times 10^{-4} + (7.877 \times 10^{-4}/P)]$$

Upper limit = 75 hp; S = nominal motor size in hp

[†]*Chemical Engineering* and U. S. Bureau of Mines.

This procedure is the work of Armando B. Corripio, Louisiana State University; Katherine S. Chrien of J.S. Dweck, Consultant, Inc.; and Lawrence B. Evans, Massachusetts Institute of Technology, as reported in *Chemical Engineering* magazine.

REFERENCES

1. Humphries, K. K., and S. Katell—*Basic Cost Engineering.* Marcel Dekker, New York, 1981, pp. 17–33.
2. Vatavuk, W. M.—*Estimating Costs of Air Pollution Control.* CRC Press/Lewis Publishers, Boca Raton, FL, 1990, pp. 17–39.
3. Dysert, L.—"Sharpen Your Capital-Cost-Estimation Skills," *Chemical Engineering,* October 2001, pp. 70–81.
4. Vatavuk, W. M.—"Updating the *CE* Plant Cost Index," *Chemical Engineering*, January 2002, pp. 62–70.
5. Bureau of Labor Statistics—U.S. Department of Labor, Washington, DC (http://data.bls.gov/cgi-bin/srgate).
6. *EPA Air Pollution Control Cost Manual* (6th ed.)—Section 5.2, Chapter 1: "Wet Scrubbers for Acid Gas." U.S. Environmental Protection Agency, Research Triangle Park, NC, January 2002 (EPA 452/B-02-001).
7. *Escalation Indexes for Air Pollution Control Costs*—U.S. Environmental Protection Agency, Research Triangle Park, NC, October 1995 (EPA-452/R-95-006).
8. *Op. Cit., EPA Air Pollution Control Cost Manual* (6th ed.)—Section 2, Chapter 1: "Hoods, Ductwork, and Stacks."
9. *Op. Cit.—Estimating Costs of Air Pollution Control*, p. 70.
10. *Ibid.*, p. 137.

SECTION 19

ENERGY CONSERVATION AND POLLUTION CONTROL IN THE CHEMICAL PROCESSING INDUSTRIES

*19.1 FLASH-STEAM HEAT RECOVERY FOR COGENERATION IN CHEMICAL PROCESSING PLANTS

Fifty steam traps of various sizes in a chemical plant discharge a total of 95,000 lb/h (11.96 kg/s) of condensate from equipment operating at 150 lb/in^2 (gage) (1034.3 kPa) to a flash tank maintaining a pressure of 5 lb/in^2 (gage) (34.5 kPa) at a temperature close to the steam temperature. The remaining condensate is discharged. Determine the quantity, available heat, and temperature of the flash steam formed. What quantity of water would be heated by this steam in a hot-water heater having an overall efficiency of 85 percent if the temperature is raised from 40°F (4.4°C) to 140°F (55.6°C)? Determine the effect on flash steam and condensate outlet temperature for the flash tank if the terminal temperature difference (flash-down) is 25°F (13.9°C). What would the effect on flash steam be if the condensate in the steam traps is subcooled 13°F (7.2°C)?

Calculation Procedure

1. Sketch the complete condensate and flash-steam recovery system.
Refer to Fig. 19.1 for a typical installation.

2. Determine the percent of flash steam formed.
In Table 19.1, locate an initial steam pressure of 150 lb/in^2 (1034.3 kPa). Cross to the right to the 5-lb/in^2 (gage) (34.5-kPa) flash tank pressure column and read 14.8 percent of the condensate forms flash steam.

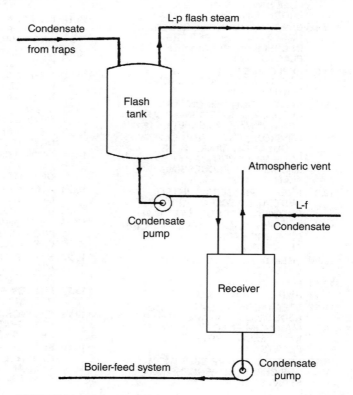

FIGURE 19.1 Complete condensate and flash-steam recovery system. (*Chemical Engineering.*)

3. Compute the quantity of flash steam formed.

This equals the percent of flash steam formed multiplied by the condensate discharge from the steam traps, or $(0.148)(95,000) = 14,060$ lb/h (1.8 kg/s).

4. Compute the available heat in the flash steam formed.

This equals the latent heat of evaporation for a flash tank pressure of 5 lb/in^2 (gage) (34.5 kPa) multiplied by the quantity of flash steam formed. From Table 19.1, the latent head of evaporation is 960 Btu/lb (2232.9 kJ/kg) at 5 lb/in^2 (gage) (34.5 kPa). Hence, the available heat is $(960)(14.060) = 13,500,000$ Btu/h (3955.5 kW).

5. Determine the flash-steam temperature.

This equals the saturated water temperature corresponding to the saturated flash tank pressure of 5 lb/in^2 (gage) (34.5 kPa). From Table 19.1 this value is shown as 228°F (108.9°C).

6. Compute the quantity of water heated in the hot-water heater.

If water were heated with the energy from the flash steam (assuming all of the flash steam could be used), this quantity would be equivalent to the (flash-steam available heat)(efficiency of the hot-water heater)(temperature increase) or $(13,500,000)(0.85)(100) = 114,750$ lb/h (14.4 kg/s). (*Note:* Additional heat is available in the condensed flash steam.)

7. Compute the effect on flash steam and remaining condensate temperature for the flash tank with a terminal temperature difference (flash down) of 25°F (13.9°C).

The temperature of the remaining condensate at the flash tank outlet equals the saturated water temperature plus flashdown, or $228 + 25 = 253$°F (122.8°C). (*Note:* The flashdown represents a loss to the system, and may be necessary due to sizing considerations.) The flashdown process will continue across the flashtank outlet, and until the temperature of the remaining condensate is 228°F (108.9°C), the quantity of flash steam formed as a consequence of flashdown will be reduced.

TABLE 19.1   Percent Flash Steam Formed

Initial steam pressure		Sat. temp		Flash-tank pressure, lb/in^2 (gage) (kPa)						
lb/in^2 (gage) (kPa)		°F	°C	0 (0)	5 (34.5)	10 (68.9)	50 (344.5)	100 (689)	125 (861.3)	150 (1033)
125	(861.1)	353	577.8	14.8	13.4	12.2	6.3	1.7	0	0
150	(1034.3)	366	601.2	16.8	14.8	13.7	7.8	2.3	1.6	0
175	(1206.5)	377	621.0	17.4	16.0	15.0	9.0	4.6	3.0	1.5
200	(1378.8)	388	640.8	18.7	17.5	16.2	10.4	6.0	4.4	2.8
				Total heat of flash steam, Btu/lb (kPa)						
				1500	1156	1160	1179	1189	1193	1195
				(2674.9)	(2693.9)	(2702.8)	(2747.1)	(2770.4)	(2779.7)	(2784.4)
				Latent heat of evaporation, Btu/lb (kJ/kg)						
				970	960	952	912	881	868	857
				(2260.1)	(2232.9)	(2218.2)	(2125)	(2052.7)	(2022.4)	(1996.8)
				Heat of liquid, Btu/lb (kJ/kg)						
				180	196	208	267	309	324	338
				(419.4)	(456.7)	(484.6)	(622.1)	(719.9)	(754.9)	(787.5)
				Saturated water temperature, °F (°C)						
				212	228	240	298	338	353	366
				(100)	(108.9)	(115.5)	(147.8)	(170)	(178.3)	(185.6)
				Volume of flash steam, ft^3/lb (m^3/kg)						
				26.8	20.0	16.3	6.6	3.9	3.2	2.7
				(1.67)	(1.25)	(1.02)	(0.41)	(0.24)	(0.20)	(0.17)

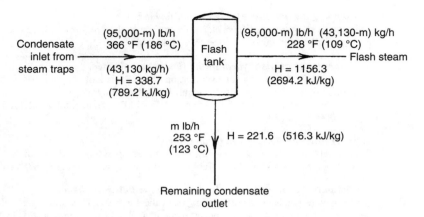

FIGURE 19.2   Flash-tank heat balance. (*Chemical Engineering.*)

Refer Fig. 19.2 for a schematic of the flash tank energy balance. Note that the values shown for enthalpy (H) are determined from steam-table data. Hence, since energy input equals energy output, $95,000(338.7) = (m)(221.6) + (95,000 - m)(1156.3)$. Solving for m, the quantity of remaining condensate is 83,100 lb/h (10.5 kg/s). Therefore, the quantity of flash steam is 95,000 – 83,100 or 11,900 lb/h (1.5 kg/s). And the flashdown reduces the flash steam quantity by the following: (14,060 – 11,900)/14,060 = 0.1536 or about 15.4 percent.

8.   *Compute the effect on flash-steam quantity if the condensate in the steam traps is subcooled by 13°F (7.2°C).*
From Table 19.1, read a saturated water temperature at 150 lb/in^2 (gage) as 366°F (185.6°C). Therefore, subcooling by 13°F (7.2°C) will reduce the condensate temperature to 366 – 13 = 353°F (178.3°C). Cross to 353°F (178.3°C) in the 5-lb/in^2 (gage) (34.5 kPa) flash-tank-pressure saturation-temperature column and read 13.4 percent of the condensate forms flash steam. The quantity of flash steam may then be computed as (95,000)(0.134) = 12,730 lb/h (1.6 kg/s). Hence, the subcooling reduces the flash-steam quantity by 14,060 – 12,730/14,060 = 0.095 or 9.5 percent. Note that interpolation may be used for intermediate temperature values.

Related Calculations.   This general procedure can be used for analyzing steam flow in flash tanks used in commercial, industrial, marine, and similar applications. With the great emphasis on conserving energy to reduce fuel costs, flash tanks are receiving greater attention than ever before. Since flash steam contains valuable heat, every effort possible is being made to recover this heat, consistent with the investment required for the recovery.

The method given here is the work of T. R. MacMillan, as reported in *Chemical Engineering* magazine.

*19.2   ENERGY CONSERVATION AND COST REDUCTION DESIGN FOR FLASH-STEAM USAGES IN CHEMICAL PROCESSING PLANTS

A plant has the steam layout shown in Fig. 19.3. Determine the dollar value of the flashed steam and what can be done about reducing the energy loss, if any. In this plant, steam from the boiler is condensed in the heat exchanger at 100 lb/in^2 (gage) (689 kPa) and 338°F (170°C). Process water is heated from 50°F (10°C) to 150°F (65.6°C), with heat transferred at the rate of 1-million Btu/h

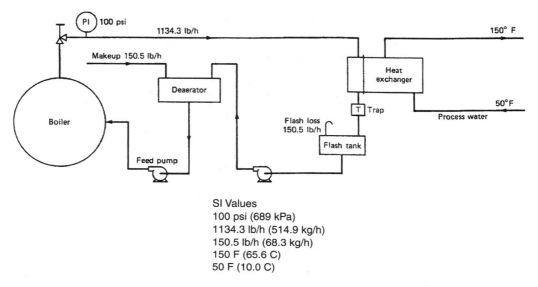

SI Values
100 psi (689 kPa)
1134.3 lb/h (514.9 kg/h)
150.5 lb/h (68.3 kg/h)
150 F (65.6 C)
50 F (10.0 C)

FIGURE 19.3 With steam valued at 58 per million Btu, venting flash steam results in an annual loss of almost 512,000. (*Chemical Engineering.*)

(293 kW). Condensate drains through a trap to a flash tank, where the flash steam is vented to the atmosphere. Analyze the benefits of reducing the supply steam pressure, and the financial benefits of recovering the flashed steam.

Calculation Procedure

1. Determine the amount of heat lost in the flashed steam.
From the steam tables, each pound of condensate at 100 lb/in^2 (gage) (69 kPa) contains 309.0 Btu (718.7 kJ/kg). At atmospheric pressure, each pound of condensate holds 180.2 Btu (419.2 kJ/kg) as sensible heat. The surplus, 309.0 − 180.2 = 128.8 Btu/lb (299.6 kJ/kg) flashes off 128.8 Btu/(970.6 Btu/lb) = 0.1327 lb of steam per pound of condensate (0.06 kg/kg), or 13.27 percent. In this relation the value 970.6 is the latent heat of the condensate at 14.7 lb/in^2 (gage) (101.3 kPa). Since the flash steam carries its total heat with it, 13.27 percent of the 1150.8 Btu (1214.1 J) is vented per pound of condensate, or 152.7 Btu/lb (355.2 kJ/kg).

Because 1-million Btu/h (1055 kJ) is transferred in the heat exchanger, and the latent heat of 100-lb/in^2 (gage) steam is 881.6 Btu/lb (2050.6 kJ/kg), the steam flow from the boiler is (1,000,000 Btu/h)/(881.6 Btu/lb) = 1134.3 lb/h (514.9 kg/h). The heat vented from the flash tank is (152.7 Btu/lb) (1134.3 lb/h) = 173.207 Btu/h (50.7 kW). Makeup water at 50°F (10°C) brings in 18 Btu (18.9 J) with each 13.27 percent of 1134.3 lb (514.9 kg), or (18 Btu/lb)(150.5 lb/h) = 2709.0 Btu/h (851.8 W). Thus, the heat loss is 173,207 Btu/h − 2709 Btu/h = 170,498 Btu/h (49.9 kW). This is more than 17 percent of the useful heat transferred to the exchanger, i.e., 170,498/1,000,000 = 0.170498.

2. Compute the annual dollar cost of the lost heat.
To determine the dollar cost multiply the cost of steam, per million Btu, by the hour loss, Btu, and the number of operating hours per year. Assuming continuous 24-h operation of this plant, the annual dollar cost with steam priced at $8/million Btu is: (170,498 Btu/h)(8760 h/year)($8/1,000,000 Btu steam cost) = $11,949 per year. This is nearly $1000 per month lost from the flash steam.

3. *Analyze an alternative plant layout to reduce the cost of the lost heat.*
To reduce the annual loss, an arrangement such as that in Fig. 19.4 is often proposed. A. pressure-reducing valve (PT) in the diagram, in the steam-supply line lowers the boiler steam pressure to, say, 10 lb/in^2 (gage) (68.9 kPa) instead of the 100 lb/in^2 (gage) (689 kPa) used in step 1. The latent heat at this pressure increases to 952.9 Btu/lb (2216.5 kJ/kg), as shown in the steam tables.

With 1,000,000 Btu/h (1055 kJ) transferred in the heat exchanger, as earlier, the steam flow rate will be (1,000,000)/(952.9 Btu/lb) = 1049.4 lb/h (476.4 kg/h). At 10 lb/in^2 (gage) (68.9 kPa) the sensible heat is 207.9 Btu/lb (483.6 kJ/kg) from the steam tables. At 0 lb/in^2, i.e., atmospheric pressure to which the flash tank exhausts, the sensible heat is 180.2 Btu/lb (419.2 kJ/kg). Then, the flash-steam percentage will be (207.9 Btu/lb − 180.2 Btu/lb)/(970.6 Btu/lb) = 0.02854, or 2.854.

The rate of flow of flash steam will be 0.02854 × 1049.4 = 29.95 lb/h (13.6 kg/h). The heat loss of this flash steam will be 29.95 lb/h × 1150.8 Btu/lb = 34,466 Btu/h (10.1 kW). Makeup water containing 18 Btu/lb (41.9 kJ/kg) enters at a rate of 29.95 lb/h (13.6 kg/h) providing 18 × 29.95 = 539.1 Btu/h (158 W). Then, the net heat loss = 34,466 − 539 = 33,927 Btu/h (9.94 kW), or 3.39 percent of the heat transferred in the exchanger.

Such a reduction in heat loss, amounting to about 136,571 Btu/h (40 kW), an annual saving of about $9570 (using the same steam cost as earlier), represents a substantial saving.

But at least one additional factor must be considered before installing the pressure-reducing valve. The heat exchanger that was large enough when supplied with 100-lb/in^2 (gage) (689-kPa) 338°F (170°C) steam would be too small when supplied with 10-lb/in^2 (gage) (68.9-kPa) 240°F (115.6°C) steam.

Temperatures at the exchanger for these two cases (assuming for simplicity that arithmetic mean temperature differences are sufficiently accurate) are shown in Fig. 19.5. The Δt across the system has dropped from 238°F (114°C) to 140°F (60°C). If the U value remains the same, the surface area of the exchanger would have to be increased by 238°F/140°F = 1.7 times.

If U = 150 Btu/(ft^2 · °F · h), from $Q = UA \Delta t$, A = 1,000,000/(150)(238) = 28 ft^2 (2.6 m^2). Hence, at the lower steam pressure, the exchanger surface area would have to be increased by

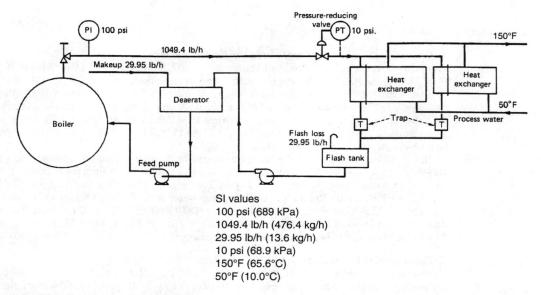

SI values
100 psi (689 kPa)
1049.4 lb/h (476.4 kg/h)
29.95 lb/h (13.6 kg/h)
10 psi (68.9 kPa)
150°F (65.6°C)
50°F (10.0°C)

FIGURE 19.4   Lowering the steam pressure by using a pressure-reducing valve cuts operating cost but increases capital cost. (*Chemical Engineering.*)

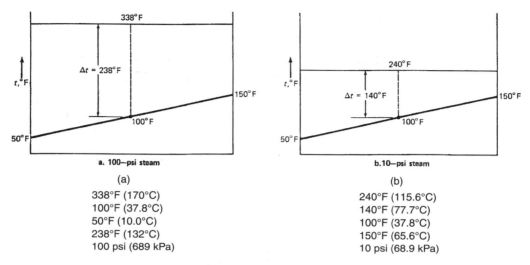

a. 100-psi steam   b.10-psi steam

(a)   (b)

338°F (170°C)	240°F (115.6°C)
100°F (37.8°C)	140°F (77.7°C)
50°F (10.0°C)	100°F (37.8°C)
238°F (132°C)	150°F (65.6°C)
100 psi (689 kPa)	10 psi (68.9 kPa)

FIGURE 19.5   Temperatures around the heat exchanger are indicated for before and after the reduction in steam pressure. (*Chemical Engineering.*)

$0.7 \times 28 = 19.6$ ft^2 (1.82 m^2). Such an exchanger would cost about \$550, \$1200 when installed. This would be in addition to the cost of about \$600 for buying and installing the pressure-reducing valve. If such alterations must be made to produce the annual saving of \$9570, other alternatives should also be considered.

4. *Determine if recovering the flash steam is economically worthwhile.*

The minimum heat-transfer area is needed if as much as possible of the heat flow is to be from the 100-lb/in^2 (gage) (689-kPa) 338°F(170°C) steam. Suppose, however, that the high-pressure condensate were not discharged to a flash venting tank but to a flash-steam recovery vessel. (For maximum economy, let the recovery system operate at atmospheric pressure.) If the flash steam were passed to a supplementary condenser fitted at the inlet side of the main exchanger (thus serving as a preheater), the atmospheric-pressure condensate would flow to the return pump without further heat loss, and the installation would be as shown in Fig. 19.6, with the temperature diagram shown in Fig. 19.7.

In Fig. 19.6, the latent heat of the 100-lb/in^2 (689-kPa) steam is 881.6 Btu/lb (2051 kJ/kg); flash steam at 0 lb/in^2 (gage) (0 kPa) has a latent heat of 970.6 Btu/lb (2258 kJ/kg). Each pound (0.45 kg) of condensate at 100 lb/in^2 (689 kPa) contains 309.0 Btu (718.7 kJ/kg); at atmospheric pressure the sensible heat of the condensate is 180.2 Btu/lb (419.1 kJ/kg). The surplus heat, $309.0 - 180.2 = 128.8$ Btu/lb (299.6 kJ/kg) flashes off $128.8/970.6 = 0.1327$ lb of steam per pound of condensate (0.06 kg/kg), or 13.2 percent of the condensate. The latent heat transferred $= 881.6 + 128.8 = 1010.4$ Btu (1065.9 J). The total steam flow with 1,000,000 Btu/h (293 kW) heat transfer is $1,000,000/1010.4 = 989.7$ lb (449.3 kg). The proportion of latent heat in the flash steam is $128.8/1010.4 = 0.1275$, or 12.75 percent. Temperature rise in the preheater $= (0.1275 \times 100°F) = 12.75°F$ (22.95°C). From this, the other temperatures in Fig. 19.7 can be derived.

Assuming again that the U value of 150 is maintained, the preheater surface area $= (127,000$ Btu/h)/[150 Btu/(ft$^2 \cdot$ °F $\cdot$ h)](155.6°F) $= 5.5$ ft^2 (0.51 m^2). Hence, the exchanger is slightly oversized.

In retrofitting, the 28-ft^2 (2.6-m^2) exchanger would allow the steam pressure to be less than 100 lb/in^2 (689 kPa). In a new installation, an exchanger having a surface area of $(0.1275 \times 1,000,000)/(1,000,000 \times 28) = 25$ ft^2 (2.32 m^2) could be installed.

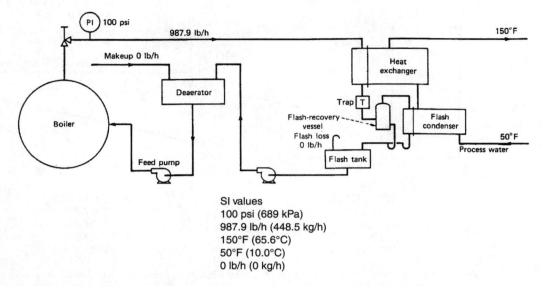

FIGURE 19.6 Recovering the flash steam eliminates the venting loss and reduces the additional capital investment. (*Chemical Engineering.*)

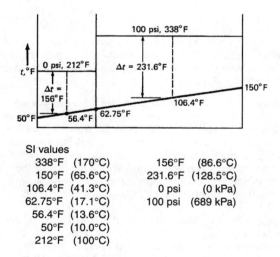

FIGURE 19.7 Temperatures of exchanger and preheater with flash-steam recovery. (*Chemical Engineering.*)

The cost of the installed preheater would be about $800, to which must be added the cost of the flash-recovery vessel at about $600. The following summarizes the choices:

Installation	Extra capital cost, $	Operating cost, $/year
Existing	0	11,964
Add pressure-reducing valve and extra 20 ft² (1.85 m²) of heat exchanger	1800	2394
Add flash-recovery vessel and extra 5.5 ft² (0.5 m²) of exchanger	1400	0

In an actual installation, some of the simplifications made here would be reassessed. However, it will generally remain the case that, with the recovery of flash steam, heat exchangers can operate at the greater efficiency afforded by higher-pressure, higher-temperature steam. Flash steam, recovered at the lowest practical pressure, can be used in a preheater, or in a separate, unconnected load.

Related Calculations.   With greater emphasis on lowering environmental air pollution, reducing flash-steam costs takes on more importance in plant analyses. During initial design studies, a choice must often be made between high- and low-pressure operating steam at both the boiler and at any heat exchangers in the system. For new installations, the economics of generating moderately high-pressure, rather than low-pressure, steam are usually fairly obvious.

Low-pressure boilers are physically larger and more costly than boilers producing the same quantity of steam at higher pressure. Also, steam produced at low pressure is often wetter than high-pressure steam, leading to lower heat-transfer rates in exchangers, even if water hammer is avoided.

The choice between high- and low-pressure steam may not always be so clear. Using high-pressure steam to get the benefits of lower capital costs of smaller heat-transfer areas can mean higher operating costs. Steam losses from flash tank vents are greater with high-pressure steam. The value of the steam lost can quickly exceed the capital-cost savings. Such considerations often lead to choosing lower-operating steam pressures.

Since flash-steam losses represent fuel used to generate this steam, every effort possible should be made to control flash steam. When flash losses are reduced, fuel consumption is cut. With lower fuel consumption, there is less atmospheric pollution. Reduced pollution lowers the cost of handling flash, sulfur compounds, and other boiler effluents. So there are many good reasons for limiting flash-steam losses in any plant.

The procedure presented here can be used for any plant using steam for processes, heating, power generation, or other heat-transfer purposes. Such plants include chemical, food, textile, manufacturing, marine, cogeneration, and central-station. All can benefit from reducing flash-steam losses, as described above.

This procedure is the work of Albert Armer, Technical Adviser and Sales Specialist, Spirax Sarco, Inc., as reported in *Chemical Engineering* magazine. SI values were added by the handbook editor to the calculations and illustrations.

*19.3   HEAT RECOVERY ENERGY AND FUEL SAVINGS IN CHEMICAL PROCESSING PLANTS

Determine the primary-fuel saving which can be produced by heat recovery if 150 M Btu/h (158.3 MJ/h) in the form of 650-lb/in^2 (gage) (4481.1-kPa) steam superheated to 750°F (198.9°C) is recovered. The projected average primary-fuel cost (such as coal, gas, oil, etc.) over a 12-year evaluation period for this proposed heat-recovery scheme is $0.75 per 10^6 Btu ($0.71 per million joules) lower heating value (LHV). Expected thermal efficiency of a conventional power boiler to produce steam at the equivalent pressure and temperature is 86 percent, based on the LHV of the fuel.

Calculation Procedure

1. Determine the value of the heat recovered during 1 year.
Entering Fig. 19.8 at the bottom at 1 year and project vertically to the curve marked $0.75 per 10^6 LHV. From the intersection with the curve, project to the left to read the value of the heat recovered as $5400 per year per M Btu/h recovered ($5094 per MJ).

2. Find the total value of the recovered heat.
The total value of the recovered heat = (hourly value of the heat recovered, $/10^6 Btu)(heat recovered, 10^6 Btu/h)(life of scheme, years). For this scheme, total value of recovered heat = ($5400) (150 × 10^6 Btu/h)(12 years) = $9,720,000.

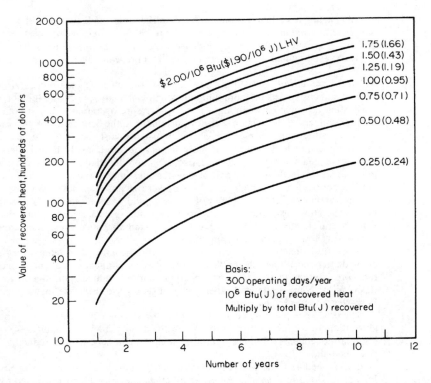

FIGURE 19.8   Chart yields value of 1 million Btu of recovered heat. This value is based on the projected average costs for primary fuel. (*Chemical Engineering.*)

3. *Compute the total value of the recovered heat, taking the boiler efficiency into consideration.*
Since the power boiler has an efficiency of 86 percent, the equivalent cost of the primary fuel would be $0.75/0.86 – $0.872 per 10^6 Btu ($0823 per million joules). The total value of the recovered heat if bought as primary fuel would be $9,720,000($0.872/$0.75) = $11,301,119. This is nearly $1 million a year for the 12-year evaluation period—a significant amount of money in almost any business. Thus, for a plant producing 1000 tons/day (900 t/day) of a product, the heat recovery noted above will reduce the cost of the product by about $3.14 per ton, based on 258 working days per year.

Related Calculations.   This general procedure can be used for any engineered installation where heat is available for recovery, such as power-generating plants, chemical-process plants, petroleum refineries, marine steam-propulsion plants, nuclear generating facilities, air-conditioning and refrigeration plants, building heating systems, etc. Further, the procedure can be used for these and any other heat-recovery projects where the cost of the primary fuel can be determined. Offsetting the value of any heat saving will be the cost of the equipment needed to effect this saving. Typical equipment used for heat savings include waste-heat boilers, insulation, heat pipes, incinerators, etc.

With the almost certain continuing rise in fuel costs, designers are seeking new and proven ways to recover heat. Ways which are both popular and effective include the following:

1. Converting recovered heat to high-pressure steam in the 600- to 1500-lb/in^2 (gage) (4137- to 10,343-kPa) range where the economic value of the steam is significantly higher than at lower pressures.

2. Superheating steam using elevated-temperature streams to both recover heat and add to the economic value of the steam.

3. Using waste heat to raise the temperature of incoming streams of water, air, raw materials, etc.

4. Recovering heat from circulating streams of liquids which might otherwise be wasted.

In evaluating any heat-recovery system, the following facts should be included in the calculation of the potential savings:

1. The economic value of the recovered heat should exceed the value of the primary energy required to produce the equivalent heat at the same temperature and/or pressure level. An efficiency factor must be applied to the primary fuel in determining its value compared to that obtained from heat recovery. This was done in the above calculation.

2. An economic evaluation of a heat-recovery system must be based on a projection of fuel costs over the average life of the heat-recovery equipment.

3. Environmental pollution restrictions must be kept in mind at all time because they may force the use of a more costly fuel.

4. Many elevated-temperature process streams require cooling over a long temperature range. In such instances, the economic analysis should credit the heat-recovery installation with the savings that result from eliminating non–heat-recovery equipment that normally would have been provided. Also, if the heat-recovery equipment permits faster cooling of a stream and this time saving has an economic value, this value must be included in the study.

5. Where heat-recovery equipment reduces primary-fuel consumption, it is possible that plant operations can be continued with the use of such equipment whereas without the equipment the continued operation of a plant might not be possible.

The above calculations and comments on heat recovery are the work of J. P. Fanaritis and H. J. Streich, both of Struthers Wells Corp., as reported in *Chemical Engineering* magazine.

Where the primary-fuel cost exceeds or is different from the values plotted in Fig. 19.8, use the value of $1.00 per 10^6 Btu (J) (LHV) and multiply the result by the ratio of (actual cost, dollars per 10^6 Btu/$1)($/J/$1). Thus, if the actual cost is $3 per 10^6 Btu (J), solve for $1 per 10^6 Btu (J) and multiply the result by 3. And if the actual cost were $0.80, the result would be multiplied by 0.8.

*19.4 CAPITAL COST OF COGENERATION HEAT-RECOVERY BOILERS IN CHEMICAL PROCESSING PLANTS

Use the Foster-Pegg method to estimate the cost of the gas-turbine heat-recovery boiler system shown in Fig. 19.9 based on these data: The boiler is sized for a Canadian Westinghouse 251 gas turbine; the boiler is supplementary fired and has a single gas path; natural gas is the fuel for both the gas turbine and the boiler; superheated steam generated in the boiler at 1200 lb/in^2 (gage) (8268 kPa) and 950°F (510°C) is supplied to an adjacent chemical process facility; 230-lb/in^2 (gage) (1585-kPa) saturated steam is generated for reducing NO$_x$ in the gas turbine; steam is also generated at 25 lb/in^2 (gage) (172 kPa) saturated for deaeration of boiler feedwater; a low-temperature economizer preheats underaerated feedwater obtained from the process plant before it enters the deaerator. Estimate boiler costs for two gas-side pressure drops: 14.4 in (36.6 cm) and 10 in (25.4 cm), and without, and with, a gas bypass stack. Table 19.2 gives other application data. *Note:* Since cogeneration will account for a large portion of future chemical-plant power generation, this procedure is important from an environmental standpoint. Many of the new chemical-plant cogeneration facilities planned today consist of gas turbines with heat-recovery boilers, as does the plant analyzed in this procedure.

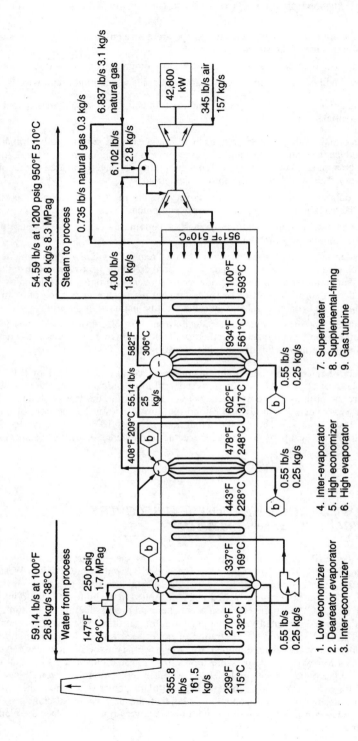

FIGURE 19.9 Gas-turbine and heat-recovery boiler system. (*Chemical Engineering.*)

1. Low economizer
2. Deareator evaporator
3. Inter-economizer
4. Inter-evaporator
5. High economizer
6. High evaporator
7. Superheater
8. Supplemental-firing
9. Gas turbine

19.12

Calculation Procedure

1. Determine the average LMTD of the boiler.

The average log mean temperature difference (LMTD) of a boiler is indicative of the relative heat-transfer area, as developed by R. W. Foster-Pegg, and reported in *Chemical Engineering* magazine. Thus, $LMTD_{avg} = Q_t/C_t$, where Q_t = total heat exchange rate of the boiler, Btu/s (W); C_t = conductance, Btu/s $\cdot$ °F (W). Substituting, using data from Table 19.2, $LMTD_{avg} = 81,837/1027 = 79.7$°F (26.5°C).

2. Compute the gas pressure drop through the boiler.

The gas pressure drop, ΔP inH_2O (cmH_2O) $= 5C_t/G$, where G = gas flow rate, lb/s (kg/s). Substituting, $\Delta P = 5(1027/355.8)$ with a gas flow of 355.8 lb/s (161.5 kg/s), as given in Fig. 19.9; then $\Delta P = 14.4$ inH_2O (36.6 cmH_2O). With a stack and inlet pressure drop of 3 inH_2O (7.6 cmH_2O) and a supplementary-firing pressure drop of 3 inH_2O (7.6 cmH_2O) given by the manufacturer, or determined from previous experience with similar designs, the total pressure drop $= 14.4 + 3.0 + 3.0 = 20.4$ inH_2O (51.8 cmH_2O).

3. Compute the system costs.

The conductance cost component, $Cost_{ts}$, is given by $Cost_{ts}$, in thousands of $ $ $ = 565 [(C_{sh}^{0.8} + C_1^{0.8} + \ldots + (C_n^{0.8}) + 2(C_n^{0.8})]$, where C = conductance, Btu/s $\cdot$ F(W), and the subscripts represent the boiler elements listed in Table 19.2. Substituting, $Cost_{ts} = 5.65(404.37) = \$2,285,000$ in base-year dollars. To update to present-day dollars, use the ratio of the base-year *Chemical Engineering* plant cost index (310) to the current year's cost index thus: Current cost = (today's plant cost index/310)(cost computed above).

The steam-flow cost component, $Cost_w$, in thousands of $ $ $ = 4.97(W_1 + W_2 + \ldots + W_n)$, where $Cost_w$ = cost of feedwater, $; W = feedwater flow rate, lb/s (kg/s); the subscripts 1, 2, and n denote different steam outputs. Substituting, $Cost_w = 4.97(59.14) = \$294,000$ in base-year dollars, with a total feedwater flow of 59.14 lb/s (26.9 kg/s).

The cost for gas flow includes connecting ducts, casing, stack, etc. It is proportional to the sum of the separate gas flows, each raised to the power of 1.2. Or, cost of gas flow, $Cost_g$, in thousands of $ $ $ = 0.236 (G_1^{1.2} + G_2^{1.2} + \ldots + G_n^{1.2})$. Substituting, $Cost_g = 0.236(355.8)^{1.2} = \$272,000$ with a gas flow of 355.8 lb/s (161.5 kg/s) and no bypass stack.

The cost of a supplementary-firing system for the heat-recovery boiler in base-year dollars is additional to the boiler cost. Typical fuels for supplementary firing are natural gas or No. 2 fuel oil, or both. The supplementary-firing system cost, $Cost_f$, in thousands of $ $ $ = B/1390 + 30N + 20$, where B = boiler firing capacity in Btu (kJ) high heating value; N = number of fuels burned. For this installation with *one* fuel, $Cost_f = 16,980/1390 + 30 + 20 = \$62,000$, rounded off. In this equation the 16,980 Btu/s (17,914 kJ/s) is the high heating value of the fuel and $N = 1$ since only *one* fuel is used.

TABLE 19.2   Data for Heat-Recovery Boiler[†]

	LMTD, °F	Q, Btu/s	C, Btu/s $\cdot$ °F	$C^{0.8}$ Btu/s $\cdot$ °F
Superheater	237	16,098	67.92	29.22
High evaporator	116	32,310	278.53	90.34
High economizer	40	11,583	290.3	93.39
Inter-evaporator	50.5	3277	64.89	28.17
Inter-economizer	37	9697	169.82	60.81
Deaerator evaporator	46	6130	134.43	50.44
Low economizer	131	2742	20.93	11.39
Additional for superheater material				29.22
Additional for low-economizer material				11.39
Total	81,837		1027	404.37

[†]See procedure for SI values in this table.
Source: Chemical Engineering.

The total boiler cost (with base gas ΔP and no gas bypass stack) = total material cost + erection cost, or $2,285,000 + 294,000 + 272,000 + 62,000 = $2,913,000 for the materials. A *budget estimate* for the cost of erection = 25 percent of the total material cost, or $0.25 \times \$2,913,000 = \$728,250$. Thus, the budget estimate for the erected cost = $2,913,000 + $728,250 = $3,641,250.

The estimated cost of the entire system—which includes the peripheral equipment, connections, startup, engineering services, and related erection—can be approximated at 100 percent of the cost of the major equipment delivered to the site, but not erected. Thus, the total cost of the boiler ready for operation is approximately twice the cost of the major equipment material, or 2(boiler material cost) = 2($2,913,000) = $5,826,000.

4. Determine the costs with the reduced pressure drop.

The second part of this analysis reduces the gas pressure drop through the boiler to 10 inH_2O (25.4 cmH_2O). This reduction will increase the capital cost of the plant because much of the equipment will be larger.

Proceeding as earlier, the total pressure drop, $\Delta P = 10 + 3 + 3 = 16$ inH_2O (40.6 cmH_2O). The pressure drop for normal solidity (i.e., normal tube and fin spacing in the boiler) is $\Delta P_1 = 14.4$ inH_2O (36.6 cmH_2O). For a different pressure drop, ΔP_2, the surface cost, C_s ($), is at ΔP_2, $C_s = [1.67(\Delta P_1/\Delta P_2)^{0.28} - 0.67]$ (C_s at P_1). Substituting, $C_s = 1.67(14.4/10)^{0.28} - 0.67 = 1.18 \times$ base cost from above. Hence, the surface cost for a pressure drop of 10 inH_2O (25.4 cmH_2O) = $1.18 \times$ ($2,285,000) = $2,696,300.

The total material cost will then be $2,696,300 + $272,000 + $62,000, using the data from above, or $3,324,300. Budget estimate for erection, as before = 1.25($3,324,300) = $4,155,375. And the estimated system cost, ready to operate = 2($3,324,300) = $6,648,600.

Adding for a gas bypass stack, the gas-flow component is the same as before, $272,000. Then the budget estimate of the installed cost of the gas bypass stack = 1.25($272,000) = $340,000. And the total cost of the boiler ready for operation at a gas-pressure drop of 10 inH_2O (25.4 cmH_2O) with a gas bypass stack = 2(53,324,000 + $272,000) = $7,192,600.

Related Calculations. To convert the costs found in this procedure to current-day costs, assume that the *Chemical Engineering* plant cost index today is 435, compared to the base-year index of 310. Then, today's cost, $ = (today's cost index/base-year cost index)(base-year plant or equipment cost, $). Thus, for the first installation, today's cost = (435/310)($5,826,000) = $8,175,194. And for the second installation, today's cost = (435/310)($7,192,600) = $10,092,842.

Boilers for recovering exhaust heat from gas turbines are very different from conventional boilers, and their cost is determined by different parameters. Because engineers are becoming more involved with cogeneration, the differences are important to them when making design and cost estimates and decisions.

In a conventional boiler, combustion air is controlled at about 110 percent of the stoichiometric requirement, and combustion is completed at about 3000°F (1649°C). The maximum temperature of the water (i.e., steam) is 1000°F (538°C), and the temperature difference between the gas and water is about 2000°F (1093°C). The temperature drop of the gas to the stack is about 2500°F (1371°C), and the gas/water ratio is consistent at about 1.1.

By contrast, the exhaust from a gas turbine is at a temperature of about 1000°F (538°C), and the difference between the gas and water temperatures averages 100°F (56°C). The temperature drop of the gas to the stack is a few hundred degrees, and the gas/water ratio ranges between 5 and 10. Because the airflow to a heat-recovery boiler is fixed by the gas turbine, the air varies from 400 percent of the stoichiometric requirement of the fuel to the turbine (unfired boiler) to 200 percent if the boiler is supplementary fired.

In heat-recovery boilers, the tubes are finned on the outside to increase heat capture. Fins in conventional boilers would cause excessive heat flux and overheating of the tubes. Although the lower gas temperatures in heat-recovery boilers allow gas enclosures to be uncooled internally insulated walls, the enclosures in conventional boilers are water-cooled and refractory-lined.

Because the exhaust from a gas turbine is free of particles and contaminants, gas velocities past tubes can be high, and fin and tube spacings can be close, without erosion or deposition. Because the products of combustion in a conventional boiler may contain sticky residues, carbon, and ash particles, tube spacing must be wider and gas velocities lower. Because of its configuration and absence of refractories, the heat-recovery boiler used with gas turbines can be shop-fabricated to a greater extent than conventional boilers.

These differences between conventional and heat-recovery boilers result in different cost relationships. With both operating on similar clean fuels, a heat-recovery boiler will cost more per pound of steam and less per square foot (m^2) of surface area than a conventional boiler. The cost of a heat-recovery boiler can be estimated as the sum of three major parameters, plus other optional parameters. Major parameters are: (1) the capacity to transfer heat ("conductance"), (2) steam flow rate, and (3) gas flow rate. Optional parameters are related to the optional components of supplementary firing and a gas bypass stack. The optional parameters will vary by the size of the installation, its use, and expected life.

This procedure is the work of R. W. Foster-Pegg, Consultant, as reported in *Chemical Engineering* magazine. Note that the costs computed by the given equations are in base-year dollars, which for this procedure were 1985 dollars. Therefore, they must be updated to current costs using the Chemical Engineering plant cost index.

*19.5 EXPLOSIVE-VENT SIZING FOR CHEMICAL PROCESSING PLANTS

Choose the size of explosion vents to relieve safely the maximum allowable overpressure of 0.75 lb/in² (5.2 kPa) in the building shown in Fig. 19.10 for an ethane/air explosion. Specify how the vents will be distributed in the structure.

Calculation Procedure

1. Determine the total internal surface area of Part A of the building.
Using normal length and width area formulas for Part A, we have: Building floor area = 100×25 = 2500 ft² (232.3 m²); front wall area = 12×100 = 1200 ft² − 12×20 = 960 ft² (89.2 m²); rear wall area = 12×100 = 1200 ft² (111.5 m²); end wall area = $2 \times 25 \times 12 + 2 \times 25 \times 3/2$ = 675 ft² (62.7 m²); roof area = $2 \times 3 \times 100$ = 600 ft² (55.7 m²). Thus, the total internal surface area of Part A of the building is $2500 + 960 + 1220 + 600$ = 5935 ft² (551.4 m²).

2. Determine the total internal surface area of Part B of the building.
Using area formulas, as before: Floor area = 50×20 = 1000 ft² (92.9 m²); side wall area = $2 \times 50 \times 12$ = 1200 ft² (111.5 m²); front wall area = 20×12 = 240 ft² (22.3 m²); roof area = 50×20 = 1000 ft² (92.9 m²); total internal surface area of Part B is $1000 + 1200 + 240 + 1000$ = 3440 ft² (319.6 m²).

3. Compute the vent area required.
Using the relation $A_v = CA_s/(P_{red})^{0.5}$, where A_v = required vent area, m²; C = deflagration characteristic of the material in the building, $(kPa)^{0.5}$, from Table 19.3. A_s = internal surface area of the structure to be protected, m². For this industrial structure, $A_v = 0147(551.4 + 319.6)/(5.17)^{0.5}$ = 180.1-m² (1939-ft²) total vent area.

The required vent area should be divided proportionally between Part A and Part B of the building, or Part A vent area = $180.1(551.4/871.0)$ = 114 m² (11227 ft²); Part B vent area = $180.1(319.6/871.0)$ = 66.1 m² (712 ft²).

The required vent area should be distributed equally over the external wall and roof areas in each portion of the building. Before making a final choice of the vent areas to be used, the designer should consult local and national fire codes. Such codes may require different vent areas, depending on a variety of factors such as structure location, allowable overpressure, and gas mixture.

Related Calculations.   This procedure is the work of Tom Swift, a consultant reported in *Chemical Engineering*. In his explanation of his procedure he points out that the word *explosion* is an imprecise term. The method outlined above is intended for those explosions known as deflagrations—exothermic

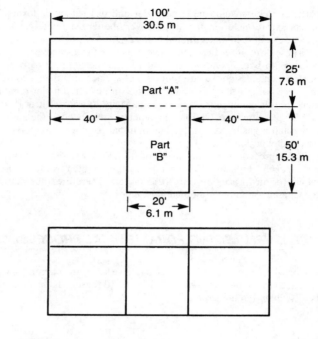

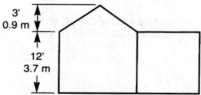

FIGURE 19.10   Typical industrial building for which explosion vents are sized.

TABLE 19.3   Parameters for Vent Area Equation[†]

Material	$\dfrac{S_u \rho_u}{G'}$	$\dfrac{P_{max}}{P_0}$
Methane	1.1×10^{-3}	8.33
Ethane	1.2×10^{-3}	9.36
Propane	1.2×10^{-3}	9.50
Pentane	1.3×10^{-3}	9.42
Ethylene	1.9×10^{-3}	9.39

Material	C, $(kPa)^{1/2}$
Methane	0.41
Ethane	0.47
Propane	0.48
Pentane	0.51
Ethylene	0.75
ST 1 dusts	0.26
ST 2 dusts	0.30

[†]Article cited in *Related Calculations*.

reactions that propagate from burning gases to unreacted materials by conduction, convection, and radiation. The great majority of structural explosions at chemical plants are deflagrations.

The equation used in this procedure is especially applicable to "low-strength" structures widely used to house chemical processes and other manufacturing operations. This equation is useful for both gas and dust deflagrations. It applies to the entire subsonic venting range. Nomenclature for Table 19.3 is given as follows:

A_s	Internal surface area of structure to be protected, m^2
A_v	Vent area, m^2
B	Dimensionless constant
C	Deflagration characteristic, $(kPa)^{1/2}$
C_D	Discharge coefficient
G'	Maximum subsonic mass flux through vent, $kg/m^2 \cdot s$
P_f	Overpressure, kPa
P_{max}	Maximum deflagration pressure in a sealed spherical vessel, kPa
P_0	Initial (ambient) pressure, kPa
P_{red}	Maximum reduced explosion pressure that a structure can withstand, kPa
S_u	Laminar burning velocity, m/s
γ_b	Ratio of specific heats of the combustion gases
ρ_u	Density of the unburnt gases, kg/m^3
λ	Turbulence enhancement factor

With increased interest in the environment by regulatory authorities, greater attention is being paid to proper control and management of industrial overpressures. Explosion vents that are properly sized will protect both the occupants of the building and surrounding structures. Therefore, careful choice of explosion vents is a prime requirement of sensible environmental protection.

*19.6 VENTILATION DESIGN FOR CHEMICAL PROCESSING PLANT ENVIRONMENTAL SAFETY

Determine the ventilation requirements to maintain interior environmental safety of a pump and compressor room in an oil refinery in a cool-temperate climate. Floor area of the pump and compressor room is 2000 ft² (185.8 m²) and room height is 15 ft (4.6 m); gross volume = 30,000 ft³ (849 m³). The room houses two pumps—one of 150 hp (111.8 kW) with a pumping temperature of 350°F (177°C), and one of 75 hp (55.9 kW) with a pumping temperature of 150°F (66°C). Also housed in the room is a 1000-hp (745.6-kW) compressor and a 50-hp (37.3-kW) compressor.

Calculation Procedure

1. Determine the hp-deg for the pumps.

The hp-deg = pump horsepower × pumping temperature. For these pumps, the total hp-deg = (150 × 350) + (75 × 150) = 63,750 hp-deg (19.789 kW-deg). Enter Fig. 19.11 on the left axis at 63,750 and project to the diagonal line representing the ventilation requirements for pump rooms in cool-temperate climates. Then extend a line vertically downward to the bottom axis to read the air requirement as 7200 ft³/min (203.8 m³/min).

The compressors require a total of 1050 hp (782.9 kW). Enter Fig. 19.11 on the right-hand axis at 1050 and project horizontally to cool-temperate climates for compressor and machinery rooms. From the intersection with this diagonal project vertically to the top axis to read 2200 ft³/min (62.3 m³/min) as the ventilation requirement.

Since the ventilation requirements of pumps and compressors are additive, the total ventilation-air requirement for this room is 7200 + 2200 = 9400 ft³/min (266 m³/min).

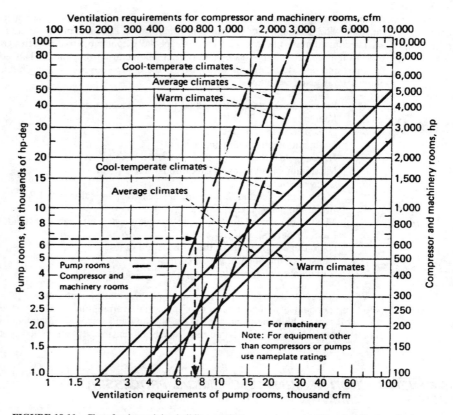

FIGURE 19.11   Chart for determining building ventilation requirements. (*Chemical Engineering.*)

2. Check to see if the computed ventilation flow meets the air-change requirements.
Use the relation $N = 60F/V$, where N = number of air changes per hour; F = ventilating-air flow rate, ft^3/min (m^3/min); V = room volume, ft^3 (m^3). Using the data for this room, $N = 60(9400)/30,000 = 18.8$ air changes per hour.

Figure 19.11 is based on a minimum of 10 air changes per hour for summer and 3 air changes per hour for winter. Since the 18.8 air changes per hour computed exceeds the minimum of 10 changes per hour on which the chart is based, the computed air flow is acceptable.

In preparing the chart in Fig.19.11 the climate lines are based on ASHRAE degree-day listings, namely: *cool, temperate climates*, 5000 degree-days and up; *average climates*, 2000 to 5000 degree-days; *warm climates*, 2000 degree-days maximum.

3. Select the total exhaust-fan capacity.
An exhaust fan or fans must remove the minimum computed ventilation flow, or 9400 ft^3/min (266 m^3/min) for this room. To allow for possible errors in room size, machinery rating, or temperature, choose an exhaust fan 10 percent larger than the computed ventilation flow. For this room the exhaust fan would therefore have a capacity of $1.1 \times 9400 = 10,340$ ft^3/min (292.6 m^3/min). A fan rated at 10,500 or 11,000 ft^3/min (297.2 or 311.3 m^3/min), depending on the ratings available from the supplier, would be chosen.

Related Calculations.   Ventilation is environmentally important and must accomplish two goals:
(1) Removal of excess heat generated by machinery or derived from hot piping and other objects;
(2) removal of objectionable, toxic, or flammable gases from process pumps, compressors, and
piping.

The usual specifications for achieving these goals commonly call for an arbitrary number of
hourly air changes for a building or room. However, these specifications vary widely in the number
of air changes required, and use inconsistent design methods for ventilation. The method given in
this procedure will achieve proper results, based on actual applications.

Because of health and explosion hazards, workers exposed to toxic or hazardous vapors and gases
should be protected against dangerous levels [threshold limit values (TLV)] and explosion hazards
[lower explosive limit (LEL)] by diluting workspace air with outside air at adequate ventilation rates.
If a workspace is protected by adequate ventilation rates for health (i.e., below TLV) purposes, the
explosion hazard (LEL) will not exist. The reason for this is that the health air changes far exceed
those required for explosion prevention.

To render a workspace safe in terms of TLV, the number of ft^3/min (m^3/min) of dilution air, A_d
required can be found from: $A_d = [1540 \times S \times T/(M \times \text{TLV})]\,K$, or in SI, $A_{dm} = \text{m}^3/\text{min} = 0.0283A_d$,
where S = gas or vapor expelled over an 8-h period, lb (kg); M = molecular weight of vapor or gas;
TLV = threshold limit value, ppm; T = room temperature, absolute °R (K); K = air-mixing factor
for nonideal conditions, which can vary from 3 to 10, depending on actual space conditions and the
efficiency of the ventilation-air distribution system.

If the space temperature is assumed to be 100°F (37.8°C) (good average summer conditions), the
above equation becomes $A_d = [862,400 \times S/(M \times \text{TLV})]\,K$.

For every pound (kg) of gas or vapor expelled of an 8-h period, when $S = 1$, the second equation
becomes $A_d = [862,400/(M \times \text{TLV})]K$. For values of S less or greater than unity, simple multiplica-
tion can be used.

It is only for ideal mixing that $K = 1$. Hence, K must be adjusted upward, depending on ventilation
efficiency, operation, and the particular system application.

If mixing is perfect and continuous, then each air change reduces the contaminant concentration
to about 35 percent of that before the air change. Perfect mixing is seldom attainable, however, so a
room mixing factor, K, ranging from 3 to 10 is recommended in actual practice.

The practical mixing factor for a particular workspace is at best an estimate. Therefore, some
flexibility should be built into the ventilation system in anticipation of actual operations. For small
enclosures, such as ovens and fumigation booths, K-values range from 3 to 5. If you are not familiar
with efficient mixing within enclosures, use a K-factor equal to 10. Then your results will be on
the safe side. Figure 19.11 is based on a K-value equal to 8 to 10. Table 19.4 gives K-factors for
ventilation-air distribution systems as indicated.

In some installations, heat generated by rotating equipment (pumps, compressors, blowers)
process piping, and other equipment can be calculated, and the outside-air requirements for dilution
ventilation determined. In most cases, however, the calculation is either too cumbersome and time
consuming or impossible.

Figure 19.11 was developed from actual practice in the chemical-plant and oil-refinery
businesses. The chart is based on a closed processing system. Hence, air quantities found from the
chart are not recommended if (1) the system is not closed or (2) if abnormal operating conditions

TABLE 19.4   *K*-values for Various Ventilation-Air Distribution Systems[†]

K-values	Distribution system
1.2–1.5	Perforated ceiling
1.5–2.0	Air diffusers
2.0–3.0	Duct headers along ceiling with branch headers pointing downward
3.0 and up	Window fans, wall fans, and the like

[†]*Chemical Engineering.*

prevail that permit the escape of excessive amounts of toxic and explosive materials into the work-place atmosphere.

For these situations, special ventilation measures, such as local exhaust through hoods, are required. Vent the exhaust to pollution-control equipment or, where permitted, directly outdoors.

Figure 19.11 and the procedure for determining dilution-air ventilation requirements were developed from actual tests of workspace atmospheres within processing buildings. Design and operating show that by supplying outside air into a building near the floor, and exhausting it high (through the roof or upper outside walls), safe and comfortable conditions can be attained. Use of chevron-type storm-proof louvers permits outside air to enter low in the room.

The chevron feature causes the air to sweep the floor, picking up heat, and diluting gases and vapors on the way up to the exhaust fan (Fig. 19.12).

In the system shown in Fig. 19.12 there are a number of features worth noting. With low-level distribution and adequate high exhaust, only the internal plant heat load (piping, equipment) is of importance in maintaining desirable workspace conditions. Wall and transmission heat loads are swept out of the building and do not reach the work areas. Even the temperature rise caused by the plant load occurs above the work level. Hence, low-level distribution of the supply air maintains the work area close to supply-air temperatures.

For any installation, it is good practice to check the ratio of hp-deg/ft^2 (kW-deg/m^2) of floor area. When this ratio exceeds 100, consider installing a totally enclosed ventilation system for cooling. This should be complete with ventilating fans taking outside air, preferably from a high stack, and discharging through ductwork into a sheet metal motor housing.

The result is the greater use of outside air for cooling through a confined system at a lower ventilation rate. Ventilation air flow needs may be obtained from the equipment manufacturer or directly from the chart (Fig. 19.11). The remainder of the building may be ventilated as usual, based either on the absence or equipment or on any equipment outside the ventilation enclosure. When designing the duct system, take care to prevent moisture entrainment with the incoming airstream.

This procedure is the work of John A. Constance, P.E., consultant, as reported in *Chemical Engineering.*

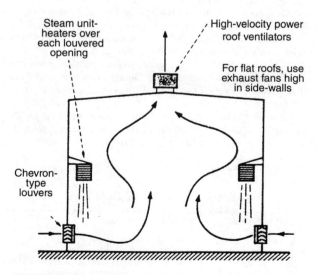

FIGURE 19.12   Ventilation system for effective removal of plant heat loads. (*Chemical Engineering.*)

*19.7 ENVIRONMENTAL AND SAFETY-REGULATION RANKING OF EQUIPMENT CRITICALITY IN CHEMICAL PROCESSING PLANTS

Rank the criticality of a chemical-plant boiler-feed pump operating at 250°F (121°C) and 100 lb/in^2 (68.9 kPa) if its mean time between failures (MTBF) is 10 months, and vibration is an important element in its safe operation. Use the National Fire Protection Association (NFPA) ratings of process chemicals for health, fire, and reactivity hazards. Show how the criticality of the unit is developed.

Calculation Procedure

1. Determine the hazard criticality rating (HCR) of the equipment.
Process industries of various types—chemical, petroleum, food, etc.—are giving much attention to complying with new process safety regulations. These efforts center on reducing hazards to people and the environment by ensuring the mechanical and electrical integrity of equipment.

To start a program, the first step is to evaluate the most critical equipment in a plant or factory. To do so, the equipment is first ranked on the basis of some criteria, such as the relative importance of each piece of equipment to the process or plant output.

The hazard criticality rating (HCR) can be determined from a listing such as that in Table 19.5. This tabulation contains the analysis guidelines for assessing the process chemical hazard (PCH) and the other hazards (O). The rankings for such a table of hazards should be based on the findings of an experienced team thoroughly familiar with the process being evaluated. A good choice for such a task is the plant's process hazard analysis (PHA) group. Since a team's familiarity with a process is highest at the end of a PHA study, the best time for ranking the criticality of equipment is toward the end of such safety evaluations.

From Table 19.5, the NFPA rating, N, of process chemicals for health, fire, and reactivity, is $N = 2$, because this is the highest of such ratings for health. The fire and reactivity ratings are 0, 0, respectively, for a boiler-feed pump because there are no fire or reactivity exposures.

The risk reduction factor (RF), from Table 19.5, is RF = 0, since there is the potential for serious burns from the hot water handled by the boiler-feed pump. Then, the process chemical hazard, PCH = $N - $ RF $= 2 - 0 = 2$.

The rating of other hazards, O, Table 19.5, is O = 1, because of the high temperature of the water. Thus, the hazard criticality rating, HCR = 2, found from the higher numerical value of PCH and O.

2. Determine the process criticality rating, PCR, of the equipment.
From Table 19.6, prepared by the PHA group using the results of its study of the equipment in the plant, PCR = 3. The reason for this is that the boiler-feed pump is critical for plant operation because its failure will result in reduced capacity.

3. Find the process and hazard criticality ranking, PHCR.
The alphanumeric PHC value is represented first by the alphabetic character for the category. For example, Category A is the most critical, while Category D is the least critical to plant operation. The first numeric portion represents the hazard criticality rating, HCR, while the second numeric part represents the process criticality rating, PCR. These categories and ratings are a result of the work of the PHA group.

From Table 19.7, the process and hazard criticality ranking, PHCR = B23. This is based on the PCR = 3 and HCR = 2, found earlier.

4. Generate a criticality list by ranking equipment using its alphanumeric PHCR values.
Each piece of equipment is categorized, in terms of its importance to the process, as: highest priority, Category A; high priority, Category B; medium priority, Category C; low priority, Category D.

Since the boiler-feed pump is critical to the operation of this process, it is a Category B, i.e., high priority item in the process.

TABLE 19.5 The Hazard Critically Rating (HCR) Is Determined in Three Steps[†]

Hazard criticality rating

1. Assess the process chemical hazard (PCH) by:
   - Determining the NFPA ratings (N) of process chemicals for:
     Health, fire, reactivity hazards
   - Selecting the highest value of N
   - Evaluating the potential for an emissions release (0–4):
     High (RF = 0): possible serious health, safety, or environmental effects
     Low (RF = 1): minimal effects
     None (RF = 4): no effects
   - *Then*, $PCH = N - RF$. (Round off negative values to zero.)
2. Rate other hazards (O) with an arbitrary number (0–4) if they are:
   - Deadly (4), if:
     Temperatures > 1000°F
     Pressures are extreme
     Potential for release of regulated chemicals is high
     Release causes possible serious health safety or environmental effects
     Plant requires steam turbine trip mechanisms, fired-equipment shutdown systems, or
       toxic- or combustible-gas detectors[†]
     Failure of pollution control system results in environmental damage[†]
   - Extremely dangerous (3), if:
     Equipment rotates at > 5000 rpm
     Temperatures > 500°F
     Plant requires process venting devices
     Potential for release of regulated chemicals is low
     Failure of pollution control system may result in environmental damage[†]
   - Hazardous (2), if:
     Temperatures > 300°F
     Extended failure of pollution control system may cause damage[†]
     Equipment rotates at > 3600 rpm
     Temperatures > 140°F or pressures 20 psig
   - Not hazardous (0), if:
     No hazards exist
3. Select the higher value of PCH and O as the hazard criticality rating

Equipment with spares drop one category rating. A spare is an inline unit that can be immediately serviced or be substituted by an alternative process option during the repair period.
[†]*Chemical Engineering.*

TABLE 19.6 The Process Criticality Rating (PCR)

	Process criticality rating
Essential (4)	The equipment is essential if failure will result in shutdown of the unit, unacceptable product quality, or severely reduced process yield.
Critical (3)	The equipment is critical if failure will result in greatly reduced capacity, poor product quality, or moderately reduced process yield.
Helpful (2)	The equipment is helpful if failure will result in slightly reduced capacity, product quality, or process yield.
Not critical (1)	The equipment is not critical if failure will have little or no process consequences.

Chemical Engineering.

5. Determine the criticality and repetitive equipment, CRE, value for this equipment.
This pump has an MTBF of 10 months. Therefore, Table 19.8, CRE = bl. Note that the CRE value will vary with the PCHR and MTBF values for the equipment.

TABLE 19.7 The Process and Hazard Criticality Rating

Process criticality rating	PHC rankings				
	Hazard criticality rating				
	4	3	2	1	0
4	A44	A34	A24	A14	A04
3	A43	B33	B23	B13	B02
2	A42	A32	C22	C12	C02
1	A41	B31	C21	CD11	D01

Note: The alphanumeric PHC value is represented first by the alphabetic character for the category (e.g., Category A is the most critical while D is the least critical). The first numeric portion represents the hazard criticality rating, and the second numeric part the process criticality rating.
Source: Chemical Engineering.

TABLE 19.8 The Criticality and Repetitive Equipment Values

PHCR	CRE values			
	Mean time between failures, months			
	0–6	6–12	12–24	>24
A	a1	a2	a3	a4
B	a2	b1	b2	b3
C	a3	b2	c1	c2
D	a4	b3	c2	d1

Source: Chemical Engineering.

6. Determine equipment inspection frequency to ensure human and environmental safety.
From Table 19.9, this boiler feed pump requires vibration monitoring every 90 days. With such monitoring it is unlikely that an excessive number of failures might occur to this equipment.

7. Summarize criticality findings in spreadsheet form.
When preparing for a PHCR evaluation, a spreadsheet, as shown in Table 19.10, listing critical equipment, should be prepared. Then, as the various rankings are determined, they can be entered in the spreadsheet where they are available for easy reference.

Enter the PCH, Other, HCR, PCR, and PHCR values in the spreadsheet, as shown. These data are now available for reference by anyone needing the information.

Related Calculations. The procedure presented here can be applied to all types of equipment used in a facility: fixed, rotating, and instrumentation. Once all the equipment is ranked by criticality, priority lists can be generated. These lists can then be used to ensure the mechanical integrity of

TABLE 19.9 Predictive Maintenance Frequencies for Rotating Equipment Based on Their CRE Values

CRE	Maintenance cycles frequency, days			
	7	30	90	360
a1, a2	VM	LT		
a3, a4		VM	LT	
b1, b3			VM	
c1, d1				VM

LT = lubrication sampling and testing, VM = vibration monitoring.
Chemical Engineering.

TABLE 19.10 Typical Spreadsheet for Ranking Equipment Criticality

Equipment number	Equipment description	Spreadsheet for calculating equipment PHCRS								
		NFPA rating								
		H	F	R	RF	PCH	Other	HCR	PCR	PHCF
TKO	Tank	4	4	0	0	4	0	4	4	A44
TKO	Tank	4	4	0	1	3	3	3	4	A34
PUIBFW	Pump	2	0	0	0	2	1	2	3	B23

Source: Chemical Engineering.

critical equipment by prioritizing predictive and preventive maintenance programs, inventories of critical spare parts, and maintenance work orders in case of plant upsets.

In any plant, the hazards posed by different operating units are first ranked and prioritized based on a PHA. These rankings are then used to determine the order in which the hazards need to be addressed. When the PHAs approach completion, team members evaluate the equipment in each operating unit using the PHCR system.

The procedure presented here can be used in any plant concerned with human and environmental safety. Today, this represents every plant, whether conventional or automated. Industries in which this procedure finds active use include chemical, petroleum, textile, food, power, automobile, aircraft, military, and general manufacturing.

This procedure is the work of V. Anthony Ciliberti, Maintenance Engineer, The Lubrizol Corp., as reported in *Chemical Engineering* magazine.

*19.8 ENERGY PROCESS-CONTROL SYSTEM SELECTION FOR CHEMICAL PROCESSING

A continuous industrial process contains four process centers, each of which has two variables that must be controlled. If a fast process-reaction rate is required with only small to moderate dead time, select a suitable mode of control. The system contains more than two resistance-capacity pairs. What type of transmission system would be suitable for this process?

Calculation Procedure

1. Compute the number of process capacities.
The number of process capacities = (number of process centers)(number of variables per center), or, for this system, $4 \times 2 = 8$ process capacities. This is defined as a *multiple* number of process capacities because the number controlled is greater than unity.

2. Analyze the process-time lags.
A small to moderate dead time is allowed in this process-control system. With such a dead-time allowance and with two or more resistance-capacity pairs in the system, a mode of control that provides for any number of process-time lags is desirable.

3. Select a suitable mode of control.
Table 19.11 summarizes the forms of control suited to processes having various characteristics. This table is a *general guide*—it provides, at best, an *approximate* aid in selecting control modes. Hence it is suitable for tentative selection of the mode of control. Final selection must be based on actual experience with similar systems.

Inspection of Table 19.11 shows that for a multiple number of processes with small to moderate dead time and any number of resistance-capacity pairs, a proportional plus reset mode of control

TABLE 19.11 Process Characteristics versus Mode of Control

Number of process capacities	Process reaction rate	Process time lags		Load changes		Suitable mode of control
		Resistance capacity (RC)	Dead time (transportation)	Size	Speed	
Single	Slow	Moderate to large	Small	Any	Any	Two-position; two-position with differential gap
				Moderate	Slow	Multiposition; proportional input
Single (self-regulating)	Fast	Small	Small	Any	Slow	Floating modes: single speed, multispeed
					Moderate	Proportional-speed floating
Multiple	Slow to moderate	Moderate	Small	Small	Moderate	Proportional position
Multiple	Moderate	Any	Small	Small	Any	Proportional plus rate
Multiple	Any	Any	Small to moderate	Large	Slow to moderate	Proportional plus reset
Multiple	Any	Any	Small	Large	Fast	Proportional plus reset plus rate
Any	Faster than that of the control system	Small or nearly zero	Small to moderate	Any	Any	Wideband proportional plus fast reset

Source: Considine, *Process Instruments and Controls Handbook*, McGraw-Hill.

is probably suitable. Further, this mode of control provides for any (i.e., fast or slow) reaction rate. Since a fast reaction rate is desired, the proportional plus reset method of control is suitable because it can handle any process-reaction rate.

4. *Select the type of transmission system to use.*
Four types of transmission systems are used for process control today: pneumatic, electric, electronic, and hydraulic. The first three types are by far the most common.

Pneumatic transmission systems use air at 3 to 20 lb/in² (gage) (20.7 to 137.9 kPa) to convey the control signal through small-bore metal tubing at distances ranging to several thousand feet. The air used in pneumatic systems must be clean and dry. To prevent a process from getting out of control, a constant supply of air is required. Pneumatic controllers, receivers, and valve positioners usually have small air-space volumes of 5 to 10 in³ (81.0 to 163.9 cm³). Air motors of the diaphragm or piston type have relatively large volumes: 100 to 5000 in² (1639 to 81,935 cm³).

Pneumatic control systems are generally considered to be spark-free. Hence, they find wide use in hazardous process areas. Also, control air is readily available, and it can be "dumped" to the atmosphere safely. The response time of pneumatic control systems may be slower than that of electric or hydraulic systems.

Electric and electronic control systems are fast-response with the signal conveyed by a wire from the sensing point to the controller. In hazardous atmospheres the wire must be protected against abrasion and breakage.

Hydraulic control systems are also rapid-response. These systems are capable of high power actuation. Slower-acting hydraulic systems use fluid pressures in the 50 to 100 lb/in² (344.8 to 689.5 kPa) range; fast-acting systems use fluid pressures to 5000 lb/in² (34,475 kPa).

Dirt and fluid flammability are two factors that may be disadvantages in certain hydraulic-control-system applications. However, new manufacturing techniques and nonflammable fluids are overcoming these disadvantages.

Since a fast response is desired in this process-control system, electric, electronic, or hydraulic transmission of the signals would be considered first. With long distances between the sensing points [say 1000 ft (305 m) or more], an electric or electronic system would probably be best.

Next, determine whether the systems being considered can provide the mode of control (step 2) required. If a system cannot provide the necessary mode of control, eliminate the system from consideration.

Before a final choice of a system is made, other factors must be considered. Thus, the relative cost of each type of system must be determined. Should an electric system prove too costly, the slightly slower response time of the pneumatic system might be accepted to reduce the initial investment.

Other factors influencing the choice of the type of a control system include type of controls, if any, currently used in the installation, skill and experience of the operating and maintenance personnel, type of atmosphere in which, and type of process for which, the controls will be used. Any of these factors may alter the initial choice.

Related Calculations. Use this general method to make a preliminary choice of controls for continuous processes, intermittent processes, air-conditioning systems, combustion-control systems, etc. Before making a final choice of any control system, be certain to weigh the cost, safety, operating, and maintenance factors listed above. Last, the system chosen *must be* able to provide the mode of control required.

19.9 PROCESS-ENERGY TEMPERATURE CONTROL SYSTEM SELECTION

A chemical-plant water-storage tank (Fig. 19.13) contains 500 lb (226.8 kg) of water at 150°F (65.6°C) when full. Water is supplied to the tank at 50°F (10°C) and is withdrawn at the rate of 25 lb/min (0.19 kg/s). Determine the process-time constant and the zero-frequency process gain if the thermal-sensing pipe contains 15 lb (6.8 kg) of water between the tank and thermal bulb and the

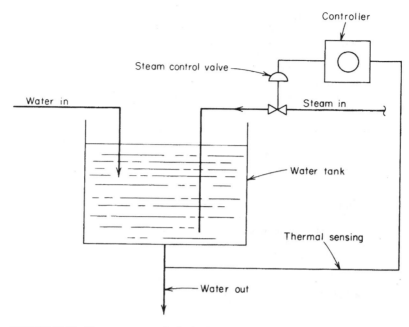

FIGURE 19.13 Temperature control of a simple process.

maximum steam flow to the tank is 8 lb/min (0.060 kg/s). The steam flow to the tank is controlled by a standard linear regulating valve whose flow range is 0 to 10 lb/min (0 to 0.076 kg/s) when the valve operator pressure changes from 5 to 30 lb/in² (34.5 to 206.9 kPa).

Calculation Procedure

1. Compute the distance-velocity lag.

The time in minutes needed for the thermal element to detect a change in temperature in the storage tank is the *distance-velocity lag*, which is also called the *transportation lag*, or *dead time*. For this process, the distance-velocity lag d is the ratio of the quantity of water in the pipe between the tank and the thermal bulb—that is, 15 gal (57.01 L)—and the rate of flow of water out of the tank—that is, 25 lb/min (0.114 kg/s)—or $d = 15/25 - 0.667$ min.

2. Compute the energy input to the tank.

This is a *transient-control process,* i.e., the conditions in the process are undergoing constant change instead of remaining fixed, as in *steady-state conditions.* For transient-process conditions the heat balance is $H_{in} = h_{out} + H_{stop}$, where H_{in} = heat input, Btu/min; H_{out} = heat output, Btu/min; H_{stor} = heat stored, Btu/min.

The heat input to this process is the enthalpy of vaporization h_{fg} Btu/(lb · min) of the steam supplied to the process. Since the regulating valve is linear, its sensitivity s is (flow-rate change, lb/min)/(pressure change, lb/in²). Or, by using the known valve characteristics, $s = (10 - 0)/(30 - 5) = 0.4$ (lb/min)/(lb/in²) [0.00044 kg/(kPa · s)].

With a change in steam pressure of p lb/in² (p' kPa) in the valve operator, the change in the rate of energy supply to the process is $H_{in} = 0.4$ (lb/min)/(lb/in²) $\times p \times h_{fg}$. Taking h_{fg} as 938 Btu/lb (2181 kJ/kg) gives $h_{in} = 375p$ Btu/min (6.6p' kW).

3. Compute the energy output from the system.

The energy output H_{out} = lb/min of liquid outflow × liquid specific heat, Btu/(lb · °F) × $(T_a - 150°F)$, where T_a = tank temperature, °F, at any time. When the system is in a state of equilibrium, the temperature of the liquid in the tank is the same as that leaving the tank or, in this instance, 150°F (65.6°C). But when steam is supplied to the tank under equilibrium conditions, the liquid tempera-ture will rise to $150 + T_r$, where T_r = temperature rise, °F (T_r, °C), produced by introducing steam into the water. Thus, the above equation becomes H_{out} = 25 lb/min × 1.0 Btu/(lb · °F) × T_r = $25T_r$ Btu/min ($0.44T_r$ kW).

4. Compute the energy stored in the system.

With the rapid mixing of the steam and water, H_{stor} = liquid storage, lb × liquid specific heat, Btu/(lb · °F) × $T_r q$ = 500 × 1.0 × $t_r q$, where q = derivative of the tank outlet temperature with respect to time.

5. Determine the time constant and process gain.

Write the process heat balance, substituting the computed values in $H_{in} = H_{out} + H_{stor}$, or $375p = 25T_r + 500T_r q$. Solving gives $T_r/p = 375/(25 + 500q) = 15/(1 + 20q)$.

The denominator of this linear first-order differential equation gives the process-system time con-stant of 20 min in the expression $1 + 20q$. Likewise, the numerator gives the zero-frequency process gain of 15°F/(lb/in^2) (1.2°C/kPa).

Related Calculations.   This general procedure is valid for any liquid using any gaseous heating medium for temperature control with a single linear lag. Likewise, this general procedure is also valid for temperature control with a double linear lag and pressure control with a single linear lag.

*19.10   ENERGY PROCESS-CONTROL VALVE SELECTION

Select a steam-control valve for a chemical-plant heat exchanger requiring a flow of 1500 lb/h (0.19 kg/s) of saturated steam at 80 lb/in^2 (gage) (551.6 kPa) at full load and 300 lb/h (0.038 kg/s) at 40 lb/in^2 (gage) (275.8 kPa) at minimum load. Steam at 100 lb/in^2 (gage) (689.5 kPa) is available for heating.

Calculation Procedure

1. Compute the valve-flow coefficient.

The valve-flow coefficient C_v is a function of the maximum steam flow rate through the valve and the pressure drop that occurs at this flow rate. In choosing a control valve for a process-control system, the usual procedure is to assume a maximum flow rate for the valve based on a considered judgment of the overload the system may carry. Usual overloads to not exceed 25 percent of the maximum rated capacity of the system. Using this overload range as a guide, assume that the valve must handle a 20 percent overload, or 0.20 (1500) = 300 lb/h (0.038 kg/s). Hence, the rated capacity of this valve should be 1500 + 300 = 1800 lb/h (0.23 kg/s).

The pressure drop across a steam-control valve is a function of the valve design, size, and flow rate. The most accurate pressure-drop estimate usually available is that given in the valve manu-facturer's engineering data for a specific valve size, type, and steam-flow rate. Without such data, assume a pressure drop of 5 to 15 percent across the valve as a first approximation. This means that the pressure loss across this valve, assuming a 10 percent drop at the maximum steam-flow rate, would be 0.10 × 80 = 8 lb/in^2 (gage) (55.2 kPa).

With these data available, compute the valve-flow coefficient from $C_v = WK/3(\Delta p\ P_2)^{0.5}$, where W = steam flow rate, lb/h; $K = 1 + (0.007 × °F$ superheat of the steam); p = pressure drop across the valve at the maximum steam-flow rate, lb/in^2; P_2 = control-valve outlet pressure at maximum steam flow rate, lb/in^2 (abs). Since the steam is saturated, it is not superheated and $K = 1$. Then $C_v = 1500/3(8 \times 94.7)^{0.5} = 18.1$.

2. Compute the low-load steam flow rate.

Use the relation $W = 3(C_v \, \Delta p \, P_2)^{0.5}/K$, where all the symbols are as before. Thus, with a 40-lb/in^2 (gage) (275.8-kPa) low-load heater inlet pressure, the valve pressure drop is $80 - 40 = 40$ lb/in^2 (gage) (275.8 kPa). The flow rate through the valve is then $W = 3(18.1 \times 40 \times 54.7)^{0.5}/1 = 598$ lb/h (0.75 kg/s).

Since the heater requires 300 lb/h (0.038 kg/s) of steam at the minimum load, the valve is suitable. Had the flow rate of the valve been insufficient for the minimum flow rate, a different pressure drop, i.e., a larger valve, would have to be assumed and the calculation repeated until a flow rate of at least 300 lb/h (0.038 kg/s) was obtained.

Related Calculations.   The flow coefficient C_v of the usual 1-in (2.5-cm) diameter double-seated control valve is 10. For any other size valve, the approximate C_v valve can be found from the product $10 \times d^2$, where d = nominal body diameter of the control valve. Thus, for a 2-in (5.1-cm) diameter valve, $C_v = 10 \times 2^2 = 40$. By using this relation and solving for d, the nominal diameter of the valve analyzed in steps 1 and 2 is $d = (d_v/10)^{0.5} = (18.1/10)^{0.5} = 1.35$ in (3.4 cm); use a 1.5-in (3.8-cm) valve because the next smaller standard control valve size, 1.25 in (3.2 cm), is too small. Standard double-seated control-valve sizes are ¾, 1, 1¼, 1½, 2, 2½, 3, 4, 6, 8, 10, and 12 in (1.9, 2.5, 3.2, 3.8, 5.1, 6.4, 7.6, 10.2, 15.2, 20.3, 25.4, 30.5 cm). Figure 19.14 shows typical flow-lift characteristics of popular types of control valves.

To size control valves for liquids, use a similar procedure and the relation $C_v = V(G/\Delta p)$, where V = flow rate through the valve, gal/min; Δp = pressure drop across the valve at maximum flow rate, lb/in^2; G = specific gravity of the liquid. When a liquid has a specific gravity of 100 SSU or less, the effect of viscosity on the control action is negligible.

To size control valves for gases, use the relation $C = Q(GT_a)^{0.5}/1360(\Delta p \, P_2)^{0.5}$, where Q = gas flow rate, ft^3/h at 14.7 lb/in^2 (abs) (101.4 kPa) and 60°F (15.6°C); T_a = temperature of the flowing gas, °F abs = 460 + °F; other symbols as before. When the valve outlet pressure P_2 is less than $0.5P_1$, where P_1 = valve inlet pressure, use the value of $P_1/2$ in place of $(\Delta p \, P_2)^{0.5}$ in the denominator of the above relation.

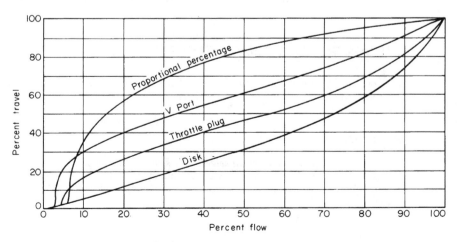

FIGURE 19.14   Flow-lift characteristics of control valves. (*Taylor Instrument Process Control Division of Sybron Corporation.*)

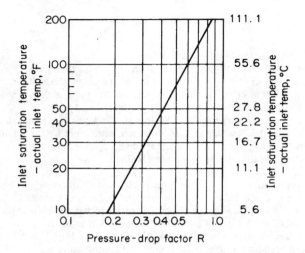

FIGURE 19.15 Pressure-drop correction factor for water in the liquid state. (*International Engineering Associates.*)

To size control valves for vapors other than steam, use the relation $C_v = W(v_2/\Delta p)^{0.5}/63.4$, where W = vapor flow rate, lb/h; v_2 = specific volume of the vapor at the outlet pressure P_2, ft^3/lb; other symbols as before. When P_2 is less than $0.5P_1$, use the value of $P_1/2$ in place of Δp and use the corresponding value of v_2 at $P_1/2$.

When the control valve handles a flashing mixture of water and steam, compute C_v by using the relation for liquids given above after determining which pressure drop to use in the equation. Use the *actual* pressure drop or the *allowable* pressure drop, whichever is smaller. Find the allowable pressure drop by taking the product of the supply pressure and the correction factor R, where R is obtained from Fig. 19.15. For a further discussion of control-valve sizing, see Considine—*Process Instruments and Controls Handbook*, McGraw-Hill, and G. F. Brockett and C. F. King—"Sizing Control Valves Handling Flashing Liquids," Texas A & M Symposium.

*19.11 FLOW CONTROL-VALVE SIZE AND CAPACITY SELECTION

Show how to select control-valve size and capacity for valves handling liquids, gases, steam and vapors. Modify the selection for various viscosities.

Calculation Procedure

1. Show the equations to use to select valve size and capacity.
Equations 19.1 and 19.2, below, are for determining valve size and valve capacity for various flowing conditions. In each equation the valve-flow coefficient, C_v, is an integral part of the calculation. Thus,

To Determine valve size

    Data given: all flowing conditions.
    Solve for: C_v and then select valve size for type of valve under consideration from manufacturers' tables of C_v rating versus valve size.

To Determine valve capacity

Data given: C_v and flowing conditions.

Solve for: capacity V, in U.S. gpm of liquid

capacity Q, in ft³/h of gas at 14.7 psia and 60°F

capacity W, in lb/h of steam

Liquids. The basic equations for valve size and valve capacity for liquids are:

$$C_v = V\sqrt{\frac{G}{(\Delta P)}} \tag{19.1}$$

$$V = C_v\sqrt{\frac{(\Delta P)}{G}} \tag{19.2}$$

where V = flow, gpm (U.S.)

ΔP = pressure drop at maximum flow, psi

G = specific gravity (water 1.0)

C_v = valve-flow coefficient

When flowing temperature is above 200°F, use specific gravity and quantity at flowing condition. When the viscosity exceeds 100 SSU or 20 cs, check the viscosity correction.

2. Show the basic equations for valve size and capacity for gases and vapors.

$$C_v = \frac{Q\sqrt{GTa}}{1360\sqrt{(\Delta P)P_2}} \tag{19.3}$$

$$Q = \frac{1360C_v\sqrt{(\Delta P)P_2}}{\sqrt{GTa}} \tag{19.4}$$

where Q = quantity, ft³/h at 14.7 psia and 60°F

ΔP = pressure drop at maximum flow, psi $(P_1 - P_2)$

P_1 = inlet pressure at maximum flow, psia

P_2 = outlet pressure at maximum flow, psia

G = specific gravity (air = 1.0)

T_a = flowing temperature absolute (460 + °F)

C_v = valve flow coefficient

When P_2 is less than ½ P_1, use the value of $P_1/2$ in place of $\sqrt{(\Delta P)P_2}$.

Steam. The basic equations for valve size and valve capacity for steam are:

$$C_v = \frac{WK}{3\sqrt{(\Delta P)P_2}} \tag{19.5}$$

$$W = \frac{3C_v\sqrt{(\Delta P)P_2}}{K} \tag{19.6}$$

where W = lb/h of steam

ΔP = pressure drop at maximum flow, psi

P_1 = inlet pressure at maximum flow, psia

P_2 = outlet pressure at maximum flow, psia

$K = 1 + (0.0007 \times$ °F superheat$)$

C_v = valve-flow coefficient

When P_2 is less than $\frac{1}{2}P_1$, use the value of $P_1/2$ in place of $\sqrt{(\Delta P)P_2}$.

The steam formula has been set up using $1/0.00225P_2$ in plane of the specific volume, to eliminate the need for steam tables.

The flow of compressible fluids through a restriction reaches a saturation velocity when the differential pressure is increased to approximately 50 percent of the inlet pressure. This critical pressure ratio varies with the composition of the fluid. The average value of one-half the absolute inlet pressure is well within the tolerance established by the formulas.

Vapors other than steam. The fundamental equations (weight basis) for valve size and valve capacity for vapors other than steam are:

$$C_v = \frac{W}{63.4}\sqrt{\frac{v_2}{(\Delta P)}} \tag{19.7}$$

$$W = 63.4C_v\sqrt{\frac{(\Delta P)}{v_2}} \tag{19.8}$$

where W = lb/h of vapor
ΔP = pressure drop at maximum flow, psi $(P_1 - P_2)$
v_2 = specific volume (ft^3/lb at outlet pressure P_2)
P_1 = inlet pressure at maximum flow, psia
P_2 = outlet pressure at maximum flow, psia
C_v = valve-flow coefficient

When P_2 is less than $\frac{1}{2}P_1$, use the value of $P_1/2$ in place of (ΔP) and use v_2 corresponding to $P_1/2$.

3. *Apply viscosity corrections for liquids.*

Solve for C, assuming no viscosity effect. Next, solve for factor R from Eq. (a) or (b) in Fig. 19.16. Then read the correction factor at the intercept of factor R in Fig. 19.16. Lastly, multiply the C value, as determined earlier, by the correction factor from Fig. 19.16. Use the corrected C value to select the valve size from valve-size tables.

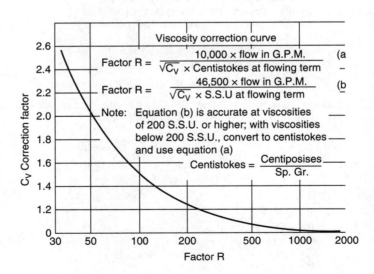

FIGURE. 19.16   Viscosity correction factors.

TABLE 19.12 Examples of Valve Size and C

Double-seated valve size, in	C_v rating full open	Double-seated valve size, in	C_v rating full open
¾	5.6	3	90
1	10	4	160
1¼	16	6	360
1½	23	8	640
2	40	10	1000
2½	62	12	1440

4. Estimate C values for various control-valve sizes.
For estimating purposes, the conventional 1-in (2.54-cm) double-seated control valve has a C of 10. All other sizes can be approximated by multiplying 10 times the square of the nominal body size, giving the ratings shown in Table 19.12.

The ratings given in the table are representative, but when sizing a particular make of valve, the figures published by the manufacturer should be used. Angle valves, butterfly valves, Saunders-type valves, single-seated valves, and special reduced port valves have C_v ratings which cannot be estimated and must be obtained from the manufacturer's test data.

This procedure is from Considine—*Process Instruments and Controls Handbook*, McGraw-Hill, 1957, in a subsection authored by Glenn F. Brockett, Chief Sales Engineer, Fisher Governor Company.

*19.12   STEAM-CONTROL AND PRESSURE-REDUCING VALVE SIZING FOR MAXIMUM ENERGY SAVINGS IN CHEMICAL PROCESSING PLANTS

Dry saturated steam at 30 lb/in² (abs) (206.9 kPa) will flow at the rate of 1000 lb/h (0.13 kg/s) through a single-seat pressure-reducing throttling valve. The desired exit pressure is 20 lb/in² (abs) (137.9 kPa) at the valve outlet. Select a valve of suitable size.

Calculation Procedure

1. Determine the critical pressure for the valve.
Critical pressure exists in a valve and piping system when the pressure at the valve outlet is 58 percent, or less, of the absolute inlet pressure for saturated steam (55 percent for hydrocarbon vapors and superheated steam). Thus, for this system the critical outlet pressure is $P_c = 0.58P_i$, where $P_c =$ critical pressure for the system, lb/in² (abs) (kPa); $P_i =$ inlet pressure, lb/in² (abs) (kPa). Or, $P_c = 0.58(30) = 17.4$ lb/in² (abs) (119.9 kPa).

Since the outlet pressure, 20 lb/in² (abs) (137.9 kPa), is greater than the critical pressure, the flow through the valve is noncritical.

2. Find the density of the outlet steam.
Assume adiabatic expansion of the steam from 30 lb/in² (abs) (206.9 kPa) to 20 lb/in² (abs) (137.9 kPa). (This is a valid assumption for a throttling process such as that which takes place in a pressure-reducing valve.) Using the steam tables, we find the density of the steam at the outlet pressure of 20 lb/in² (abs) (137.9 kPa) is 0.05 lb/ft³ (0.8 kg/m³).

3. Compute the valve-flow coefficient c_v.
Use the relation $C_v = W/63.5\sqrt{(P_i - P_2)\rho}$, where $C_v =$ valve-flow coefficient, dimensionless; $W =$ steam (or vapor or gas) flow rate, lb/h (kg/s); $P_i =$ valve inlet pressure, lb/in² (abs); $p =$ density

of the vapor or gas flowing through the valve, lb/ft^3 (kg/m^3). For this valve, $C_v = 1000/63.5$ $\sqrt{(30 - 20)0.05} = 22.3$, say 22.0 because C_v valves are usually stated in even numbers for larger-size valves.

4. *Select the control valve to use.*
At normal operating conditions, most engineers recommend that the flow through the valve not exceed 80 percent of the maximum flow possible. Thus the valve selected should have a C_v equal to or greater than the computed $C_v/0.80$. Thus, for this valve, choose a unit having a C_v equal to or greater than $22/0.80 = 27.5$. From Table 19.13 choose a 2-in (5.08-cm) single-seat valve having a C_v of 36.

The operating C_v of any valve is $C_{vo} = C_{vf}/C_{vs}$, where $C_{vf} = C_v$ value computed by the formula in step 3 and $C_{vs} = C_v$ of actual valve selected. Or, for this valve, $C_{vo} = 22/36 = 0.61$.

To avoid wire drawing, which occurs when the valve plug operates too close to the valve seat, C_{vo} values of less than 0.10 should not be used. Since $C_{vo} = 0.61$ for this valve, wire drawing will not occur.

Related Calculations. To speed up the determination of C_v, Fig. 19.17 can be used instead of the formula in step 3. This is a performance-tested chart valid for steam-control valves for blast-heating coils, tank heaters, pressure-reducing stations, and any other installations—stationary, mobile, or marine—where steam flow and pressure are to be regulated. The approach can also be used for valves handling gases other than steam.

The valve coefficient C_v is conventional; it equals the gallons per minute (liters per second) of clear cold water at 60°F (15.6°C) that will pass through the flow restriction (valve or orifice) while undergoing a pressure drop of 1 lb/in^2 (7.0 kPa). The C_v value is the same for liquids, gases, and steam. Tables listing C_v values versus valve size and type are published by the various valve manufacturers. General C_v values not limited to any manufacturer are given in Table 19.13 for a variety of valve types and sizes.

Note in Fig. 19.17 the relations for the density of the steam of various valve outlet pressures. In these relations P_c = critical pressure, lb/in^2 (abs) (kPa), as defined earlier. The solution given in Fig. 19.17 is for a flow rate of 200 lb/h (0.025 kg/s) of steam having a density of 0.08 lb/ft^3 (1.25 kg/m^3) at the valve outlet with a 10-lb/in^2 (abs) (68.9-kPa) pressure drop through the valve, giving a C_v of 3.8.

This procedure is the work or John D. Constance, P.E., as reported in *Chemical Engineering* magazine.

TABLE 19.13 Flow Coefficients for Steam-Control Valves

Size		Straight-through throttling		Straight-through on-off	Straight-through regulators	
in	cm	Single seat	Double seat	Single seat	Single seat	Double seat
⅛	0.32	0.23				
¼	0.64	0.78				
⅜	0.95	1.7				
½	1.27	3.2				
¾	1.91	5.4	7.2	7	3.6	4.3
1	2.54	9	12	12	6	7.2
1¼	3.18	14	18	18	9	10.8
1½	3.81	21	28	27	14	16.8
2	5.08	36	48	42	24	28.8
2½	6.35	54	72	65	36	432
3	7.62	75	100	93	50	60
4	10.2	124	165	170	83	99
6	15.2	270	360	380	180	216
8	20.3	480	640	660	320	384
10	25.4	750	1000	1100	500	600
12	30.5	1080	1440	1550	720	864

Source: Chemical Engineering.

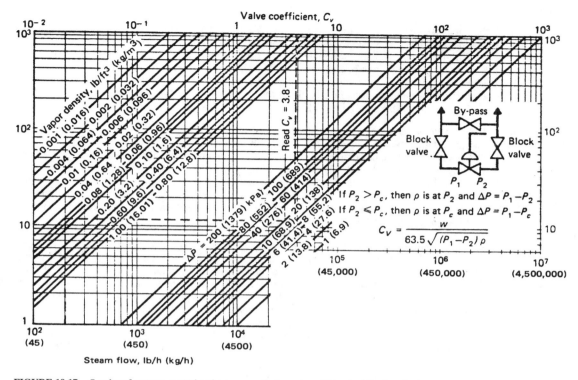

FIGURE 19.17 C_v valves for steam-control and pressure-reducing valves. (*Chemical Engineering.*)

19.13 CONTROL-VALVE SELECTION FOR CHEMICAL PROCESSING PLANTS

Select a steam control valve for a heat exchanger requiring a flow of 1500 lb/h (0.19 kg/s) of saturated steam at 80 psig (551.6 kPag) at full load and 300 lb/h (0.038 kg/s) at 40 psig (275.8 kPag) at minimum load. Steam at 100 psig (689.5 kPag) is available for heating.

Calculation Procedure

1. Compute the valve flow coefficient.
The valve flow coefficient C_v is a function of the maximum steam flow rate through the valve and the pressure drop that occurs at this flow rate. When choosing a control valve for a process control system, the usual procedure is to assume a maximum flow rate for the valve based on a considered judgment of the overload the system may carry. Usual overloads do not exceed 25 percent of the maximum rated capacity of the system. Using this overload range as a guide, assume that the valve must handle a 20 percent overload, or 0.20(1500) = 300 lb/h (0.038 kg/s). Hence, the rated capacity of this valve should be 1500 + 300 = 1800 lb/h (0.23 kg/s).

The pressure drop across a steam control valve is a function of the valve design, size, and flow rate. The most accurate pressure-drop estimate that is usually available is that given in the valve manufacturer's engineering data for a specific valve size, type, and steam flow rate. Without such data, assume a pressure drop of 5 to 15 percent across the valve as a first approximation. This means that the pressure loss across this valve, assuming a 10 percent drop at the maximum steam flow rate, would be $0.10 \times 80 = 8$ psig (55.2 kPag).

With these data available, compute the valve flow coefficient from $C_v = W K / 3 (\Delta p \, P_2)^{0.5}$, where W is steam flow rate, in lb/h, K equals $1 + (0.0007 \times °\text{F}$ superheat of the steam$)$, p is pressure drop across the valve at the maximum steam flow rate, in lb/in², and P_2 is control-valve outlet pressure at maximum steam flow rate, in psia. Since the steam is saturated, it is not superheated, and $K = 1$. Then, $C_v = 1500 / 3 (8 \times 94.7)^{0.5} = 18.1$.

2. Compute the low-load steam flow rate.
Use the relation $W = 3 (C_v \Delta p \, P_2)^{0.5} / K$, where all the symbols are as before. Thus, with a 40-psig (275.5-kPag) low-load heater inlet pressure, the valve pressure drop is $80 - 40 = 40$ psig (275.8 kPa). The flow rate through the valve is then $W = 3(18.1 \times 40 \times 54.7)^{0.5} / 1 = 598$ lb/h (0.75 kg/s).

Since the heater requires 300 lb/h (0.038 kg/s) of steam at the minimum load, the valve is suitable. Had the flow rate of the valve been insufficient for the minimum flow rate, a different pressure drop, i.e., a larger valve, would have to be assumed and the calculation repeated until a flow rate of at least 300 lb/h (0.038 kg/s) was obtained.

Related Calculations. The flow coefficient C_v of the usual 1-in-diameter (2.5-cm) double-seated control valve is 10. For any other size valve, the approximate C_v valve can be found from the product $10 \times d^2$, where d is nominal body diameter of the control valve, in inches. Thus, for a 2-in-diameter (5.1-cm) valve, $C_v = 10 \times 2^2 = 40$. Using this relation and solving for d, the nominal diameter of the valve analyzed in steps 1 and 2 is $d = (C_v / 10)^{0.5} = (18.1/10)^{0.5} = 1.35$ in (3.4 cm); use a 1.5-in (3.8-cm) valve because the next smaller standard control valve size, 1.25 in (3.2 cm), is too small. Standard double-seated control-valve sizes are: ¾, 1, 1¼, 1½, 2, 2½, 3, 4, 6, 8, 10, and 12 in. Figure 19.18 shows typical flow-lift characteristics of popular types of control valves.

To size control valves for liquids, use a similar procedure and the relation $C_v = V (G / \Delta p)$, where V is flow rate through the valve, in gal/min, Δp is pressure drop across the valve at maximum flow rate, in lb/in², and G is specific gravity of the liquid. When a liquid has a specific gravity of 100 SSU or less, the effect of viscosity on the control action is negligible.

To size control valves for gases, use the relation $C_v = Q (G T_a)^{0.5} / [1360 (\Delta p \, P_2)^{0.5}]$, where Q is gas flow rate, in ft³/h at 14.7 psia (101.4 kPa) and 60°F (15.6°C), T_a is temperature of the flowing gas, in °F abs $= 460 + F$; other symbols as before. When the valve outlet pressure P_2 is less than $0.5P_1$,

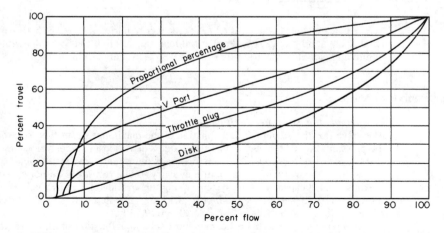

FIGURE 19.18 Flow-lift characteristics for control valves. (*Taylor Instrument Process Control Division of Sybron Corporation.*)

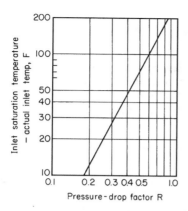

FIGURE 19.19 Pressure-drop correction factor for water in the liquid state. (*International Engineering Associates.*)

where P_1 is valve inlet pressure, in psia, use the value of $P_1/2$ in place of $(\Delta p P_2)^{0.5}$ in the denominator of the relation.

To size control valves for vapors other than steam, use the relation $C_v = W(v_2/\Delta p)^{0.5}/63.4$, where W is vapor flow rate, in lb/h, v_2 is specific volume of the vapor at the outlet pressure P_2, in ft^3/lb; other symbols as before. When P_2 is less than $0.5P_1$, use the value of $P_1/2$ in place of Δp and use the corresponding value of v_2 at $P_1/2$.

When the control valve handles a flashing mixture of water and steam, compute C_v using the relation for liquids given earlier after determining which pressure drop to use in the equation. Use the *actual* pressure drop or the *allowable* pressure drop, whichever is smaller. Find the allowable pressure drop by taking the product of the supply pressure, in psia, and the correction factor R, where R is obtained from Fig. 19.19. For a further discussion of control-valve sizing, see Considine— *Process Instruments and Controls Handbook*. McGraw-Hill, and G. F. Brockett and C. F. King—"Sizing Control Valves Handling Flashing Liquids," Texas A & M Symposium.

19.14 CONTROL-VALVE CHARACTERISTICS AND RANGEABILITY

A flow control valve will be installed in a process system in which the flow may vary from 100 to 20 percent while the pressure drop in the system rises from 5 to 80 percent. What is the required rangeability of the control valve? What type of control-valve characteristic should be used? Show how the effective characteristic is related to the pressure drop the valve should handle.

Calculation Procedure

1. Compute the required valve rangeability.
Use the relation $R = (Q_1/Q_2)(\Delta P_2/\Delta P_1)^{0.5}$, where R is valve rangeability, Q_1 is valve initial flow, in percent of total flow, Q_2 is valve final flow, in percent of total flow, P_1 is initial pressure drop across the valve, in percent of total pressure drop, and P_2 is percent final pressure drop across the valve. Substituting, $R = (100/20)(80/5)^{0.5} = 20$.

2. Select the type of valve characteristic to use.
Table 19.14 lists the typical characteristics of various control valves. Study of Table 19.14 shows that an equal-percentage valve must be used if a rangeability of 20 is required. Such a valve has equal stem movements for equal-percentage changes in flow at a constant pressure drop based on

TABLE 19.14 Control-Valve Characteristics

Valve type	Typical flow rangeability	Stem movement
Linear	12–1	Equal stem movement for equal flow change
Equal percentage	30–1 to 50–1	Equal stem movement for equal-percentage flow change[†]
On-off	Linear for first 25 percent of travel; on-off thereafter	Same as linear up to on-off range

[†]At constant pressure drop.

the flow occurring just before the change is made.[†] The equal-percentage valve finds use where large rangeability is desired and where equal-percentage characteristics are necessary to match the process characteristics.

3. *Show how the valve effective characteristic is related to pressure drop.*
Figure 19.20 shows the inherent and effective characteristics of typical linear, equal-percentage, and on-off control valves. The inherent characteristic is the theoretical performance of the valve.[†] If a valve is to operate at a constant load without changes in the flow rate, the characteristic of the valve is not important, since only one operating point of the valve is used.

Figure 19.20*a* and *c* give definite criteria for the amount of pressure drop the control valve should handle in the system. This pressure drop is not an arbitrary value, such as 5 lb/in^2, but rather a percent of the total dynamic drop. The control valve should take at least 33 percent of the total dynamic system pressure drop if an equal-percentage valve is used and is to retain its inherent characteristics. A linear valve should not take less than a 50 percent pressure drop if its linear properties are desired.

There is an economic compromise in the selection of every control valve. Where possible, the valve pressure drop should be as high as needed to give good control. If experience or an economic study dictates that the requirement of additional horsepower to provide the needed pressure is not worth the investment in additional pumping or compressor capacity, the valve should take less pressure drop with the resulting poorer control.

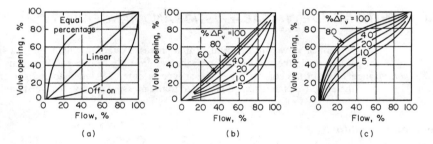

FIGURE 19.20   (*a*) Inherent flow characteristics of valves at constant pressure drop; (*b*) effective characteristics of a linear valve; (*c*) effective characteristics of a 50:1 equal-percentage valve.

19.15 CAVITATION, SUBCRITICAL, AND CRITICAL FLOW CONSIDERATIONS IN CONTROLLER SELECTION

Using the sizing formulas of the Fluid Controls Institute, Inc., size control valves for the cavitation, subcritical, and critical flow situations described below. Show how accurate the FCI formulas are.

Cavitation: Select a control valve for a situation where cavitation may occur. The fluid is steam condensate; inlet pressure P_1 is 167 psia (1151.5 kPa); ΔP is 105 lb/in^2 (724.0 kPa); inlet temperature T_1 is 180°F (82.2°C); vapor pressure P_v is 7.5 psia (51.7 kPa).

Subcritical gas flow: Determine the valve capacity required at these conditions; fluid is air; flow Q_g is 160,000 std ft^3/h (1.3 std m^3/s); inlet pressure P_1 is 275 psia (1896 kPa); ΔP is 90 lb/in^2 (620.4 kPa); gas temperature T_1 is 60°F (15.6°C).

Critical vapor flow: A heavy-duty angle valve is suggested for a steam pressure-reducing application. Determine the capacity required and compare an alternate valve type. The fluid is saturated steam; flow W is 78,000 lb/h (9.8 kg/s); inlet pressure P_1 is 1260 psia (8688 kPa); and outlet pressure P_2 is 300 psia (2068.5 kPa).

[†]E. Ross Forman—"Fundamentals of Process Control," *Chemical Engineering*, June 21, 1965.

Calculation Procedure

1. Choose the valve type and determine its critical-flow factor for the cavitation situation.

If otherwise suitable (i.e., with respect to size, materials, and space considerations), a butterfly control valve is acceptable on a steam-condensate application. Find, from Table 19.15, the value of the critical flow factor $C_f = 0.68$ for a butterfly valve with 60° operation.

2. Compute the maximum allowable pressure differential for the valve.

Use the relation $\Delta P_m = C_f^2(P_1 - P_v)$, where ΔP_m is maximum allowable pressure differential, in lb/in^2, P_1 is inlet pressure, in psia, and P_v is vapor pressure, in psia. Substituting, $\Delta P_m = (0.68)^2(167 - 7.5) = 74$ lb/in^2 (510.2 kPa). Since the actual pressure drop, 105 lb/in^2 (724.0 kPa), exceeds the allowable drop, 74 lb/in^2 (510.2 kPa), cavitation *will* occur.

3. Select another valve and repeat the cavitation calculation.

For a single-port top-guided valve with flow to open plug, find $C_f = 0.90$ from Table 19.15. Then $\Delta P_m = (0.90)^2(167 - 7.5) = 129$ lb/in^2 (889.5 kPa).

In the case of the single-port top-guided valve, the allowable pressure drop, 129 lb/in^2 (889.5 kPa), exceeds the actual pressure drop, 105 lb/in^2 (724.0 kPa), by a comfortable margin. This valve is a better selection because cavitation will be avoided. A doubleport valve might also be used, but the single-port valve offers lower seat leakage. However, the double-port valve offers the possibility of a more economical actuator, especially in larger valve sizes. This concludes the steps for choosing the valve where cavitation conditions apply.

4. Apply the FCI formula for subcritical flow.

The FCI formula for subcritical gas flow is $C_v = Q_g/[1360(\Delta P/GT)^{0.5}][(P_1 + P_2)/2]^{0.5}$, where C_v is valve flow coefficient, Q_g is gas flow, in std ft^3/h, ΔP is pressure differential, in lb/in^2, G is specific gravity of gas at 14.7 psia (101.4 kPa) and 60°F (15.6°C), and T is absolute temperature of the gas, in R; other symbols as given earlier. Substituting, $C_v = 160,000/[1360(90/520)^{0.5}][(275 + 185)/2]^{0.5} = 18.6$.

5. Compute C_v using the unified gas-sizing formula.

For greater accuracy, many engineers use the unified gas-sizing formula. Assuming a single-port top-guided valve installed open to flow, Table 19.15 shows $C_f = 0.90$. Then, $Y = (1.63/C_f)(\Delta P/P_1)^{0.5}$, where Y is defined by the equation and the other symbols are as given earlier. Substituting, $Y = (1.63/0.90)(90/275)^{0.5} = 1.04$. Figure 19.21 shows the flow correlation established from actual test data for many valve configurations at a maximum valve opening and relates Y and the fraction of the critical flow rate.

Find from Fig. 19.22 the value of $Y - 0.148Y^3 = 0.87$. Compute $C_v = Q_g(GT)^{0.5}/[834C_f \times (Y - 0.148Y^3)]$, where all the symbols are as given earlier. Or, $C_v = 160,000(520)^{0.5}/[834(0.90)(275)(0.87)] = 20.4$. This value represents an error of approximately 10 percent in the use of the FCI formula.

6. Determine C_f for critical vapor flow.

Assuming reduced valve trim for a heavy-duty angle valve, $C_f = 0.55$ from Table 19.15.

7. Compute the critical pressure drop in the valve.

Use $\Delta P_c = 0.5(C_f)^2 P_1$, where P_c is critical pressure drop, in in^2; other symbols are as given earlier. Substituting, $\Delta P_c = 0.5(0.55)^2(1260) = 191$ lb/in^2 (1316.9 kPa).

8. Determine the value of C_v.

Use the relation $C_v = W/[1.83C_fP_1]$, where the symbols are as given earlier. Substituting, $C_v = 78,000/[1.83(0.55)(1260)] = 61.5$. A lower C_v could be attained by using the valve flow to open, but a more economical choice is a single-port top-guided valve installed open to flow.

For a single-port top-guided valve flow to open, $C_f = 0.90$ from Table 19.15. Hence, $C_v = 78,000/[1.83(0.90)(1260)] = 37.6$.

A lower capacity is required at critical flow for a valve with less pressure recovery. Although this may not lead to a smaller body size because of velocity and stability considerations, the choice of a

TABLE 19.15 Critical Flow Factors for Control Valves at 100 percent Lift

Split body

A	Flow to close plug	0.80
	Flow to open plug	0.75
	Parabolic plug only	
B	Flow to close plug	0.50
	Flow to open plug	0.90
	Parabolic plug only	

Single-port, globe body

A	Flow to close plug	0.85
	Flow to open plug	0.90
	Parabolic plug only	
B	Flow to close plug	0.50
	Flow to open plug	0.90
	Parabolic plug only	

Angle body

A	Flow to close plug	0.40
	Flow to open plug	0.90
	Parabolic plug only	
B	Flow to close plug	0.55
	Flow to open plug	0.95
	Parabolic plug only	

Double-port, globe body

A	Parabolic plug	0.90
	V-port plug	1.00
B	Parabolic plug	0.62
	V-port plug	0.95

Butterfly

$D/d = 1$	$\alpha = 60°$	0.68
	$\alpha = 90°$	0.58
$D/d = 2$	$\alpha = 60°$	0.62
	$\alpha = 90°$	0.50

(A) Full-capacity trim, orifice diameter ~ 0.8 valve diameter

(B) Reduced capacity trim, 50% of (A) and less.

NOTE: The listed values apply for equal port-area valves only and do not include corrections for pipe friction.

Source: Henry W. Boger and *Chemical Engineering.*

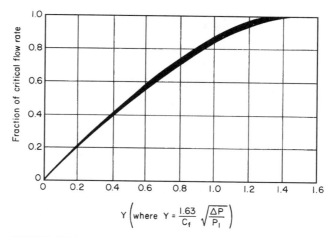

FIGURE 19.21 Flow correlation established from actual data for many valve configurations at maximum valve opening.

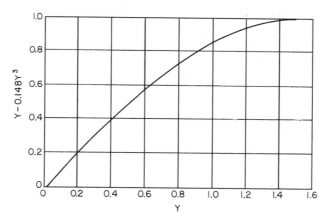

FIGURE 19.22 Correction-factor values.

more economical body type and a smaller actuator requirement is attractive. The heavy-duty angle valve finds its application generally on flashing-hydrocarbon liquid service with a coking tendency.

This procedure is the work of Henry W. Boger, Engineering Technical Group Manager, Worthington Controls Co.

19.16   USE OF A HUMIDITY CHART[†]

The temperature and dew point of the air entering a certain dryer are 130 and 60°F (328 and 289 K), respectively. Using a humidity chart (Fig. 19.23), find the following properties of the air: its humidity, its percentage humidity, its adiabatic-saturation temperature, its humidity at adiabatic saturation, its humid heat, and its humid volume.

[†]Adapted from McCabe and Smith—*Unit Operations of Chemical Engineering*, 3d ed., McGraw-Hill, Inc.

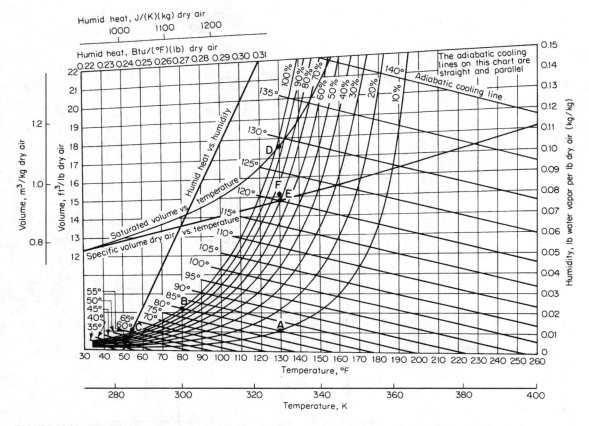

FIGURE 19.23   Humidity chart (air-water at 1 atm). (*From McCabe and Smith—Unit Operations of Chemical Engineering.*)

Calculation Procedure

1. Find the humidity.
In Fig. 19.23, the humidity is the ordinate (along the right side of the graph) of the point on the saturation line (the 100 percent humidity line) that corresponds to the dew point, the latter being read from the abscissa along the bottom. In the present case, the humidity is found to be 0.011 lb water per pound of dry air (0.011 kg water per kilogram of dry air).

2. Find the percentage humidity.
Find the dry-bulb temperature, that is, 130°F, along the abscissa, erect a perpendicular to intersect the 0.011-lb humidity line (at point A in Fig. 19.23), and find the percentage-humidity line (interpolating a line if necessary) that passes through that intersection. In this case, the 10 percent line passes through, so the percentage humidity is 10 percent.

3. Find the adiabatic-saturation temperature.
Find the adiabatic-cooling line (these are the straight lines having negative slope) that passes through point A, interpolating a line if necessary, and read the abscissa of the point (point B) where this line intersects the 100 percent humidity line. This abscissa is the adiabatic-saturation temperature. In the present case, it is 80°F (300 K).

4. Find the humidity at adiabatic saturation.
The humidity at adiabatic saturation is the ordinate, along the right side of the graph, of point B. Its value is 0.022 lb water per pound of dry air (0.022 kg water per kilogram of dry air).

5. Find the humid heat.
Find the intersection (point C) of the 0.011-lb-humidity line with the humid-heat-versus-humidity line, and read the humid heat as the abscissa of point C along the top of the graph. This abscissa is 0.245 Btu/(°F)(lb dry air) [or 1024 J/(K)(kg dry air)].

6. Find the humid volume.
Erect a perpendicular through the abscissa (along the bottom of the graph) that corresponds to 130°F, the dry bulb temperature. Label the intersection of this perpendicular with the saturated-volume-versus-temperature line as point D, and the intersection of the perpendicular with the specific-volume-dry-air-versus-temperature line as point E. Then, along line ED, find point F by moving upward from point E by a distance equal to

$$\overline{ED}\,[(\text{percentage humidity})/100]$$

or, in the present case $(\overline{ED})(10/100)$, where $\overline{E\,D}$ is the length of line segment $\overline{E\,D}$. The humid volume is the ordinate of point F as read along the left side of the graph. In this case, the humid volume is 15.1 ft^3/lb dry air (0.943 m^3/kg dry air).

Related Calculations.   Do not confuse percentage humidity with relative humidity. "Relative humidity" is the ratio of the partial pressure of the water vapor to the vapor pressure of water at the temperature of the air, this ratio usually being expressed as a percent. "Percentage humidity" is the ratio of the actual humidity to the saturation humidity that corresponds to the gas temperature, which is also usually expressed as a percent. At all humidities other than 0 or 100 percent, the percentage humidity is less than the relative humidity.

19.17   BLOWDOWN AND MAKEUP REQUIREMENTS FOR CHEMICAL-PLANT COOLING TOWERS[†]

A cooling tower handles 1000 gal/min (0.063 m^3/s) of circulating water that is cooled from 110 to 80°F (316 to 300 K). How much blowdown and makeup are required if the concentration of dissolved solids is allowed to reach three times the concentration in the makeup?

Calculation Procedure

1. Set out material-balance equations for the cooling tower.
When the system is at equilibrium, the makeup must equal the losses, so, by definition,

$$M = E + B + W \tag{19.9}$$

where M is makeup, E is evaporation loss, B is blowdown, and W is windage loss, all being expressed as percent of circulation.

Since the evaporation water will be essentially free of dissolved solids, all solids introduced with the makeup water must be removed by the blowdown plus windage loss, or

$$M p_m = (B + W)p_c$$

[†]Adapted from *Chemical Engineering*, June 21, 1976.

where p_m is concentration of solids in the makeup and p_c is concentration of solids in the circulating water, both in parts per million.

For cooling towers, the concentration in the recirculating water is arbitrarily defined as "cycles of concentration" C, namely, $C = $ (concentration in cooling water)/(concentration in makeup water). Thus,

$$M = \frac{(B + W)p_c}{p_m} = (B + W)C \qquad (19.10)$$

2. Make appropriate assumptions about windage and evaporation losses and set out and solve an equation for blowdown.

Windage losses will be about 1.0 to 5.0 percent for spray ponds, 0.3 to 1.0 percent for atmospheric cooling towers, and 0.1 to 0.3 percent for forced-draft cooling towers; for the forced-draft towers in this example, 0.1 percent can be assumed. As for evaporation losses, they are 0.85 to 1.25 percent of the circulation for each 10-degree drop in Fahrenheit temperature across the tower; it is usually safe to assume 1.0 percent, so $E = \Delta T/10$, where ΔT is the temperature drop across the tower. Therefore, in the present case,

$$M = \frac{\Delta T}{10} + B + 0.1 \qquad (19.11)$$

Combining Eqs. 19.10 and 19.11 gives

$$B = \frac{\Delta T}{10(C - 1)}$$

In the present case, then $B = (110 - 80)/[10(3 - 1)] = 1.5$ percent. Thus the blowdown requirement is 1.5 percent of 1000 gal/min, or 15 gal/min (9.45×10^{-4} m³/s).

3. Find the makeup requirement.
From Eq. 19.9, $M = (110 - 80)/10 + 1.5 + 0.1 = 4.6$ percent, or 46 gal/min (2.9×10^{-3} m³/s).

19.18   CHEMICAL-PLANT WATER-SOFTENER SELECTION AND ANALYSIS[†]

Select a water softener that will treat 100 gal/min (22.7 m³/h) of water at 60°F (289 K) that has the following dissolved components (with concentrations in parts per million as $CaCO_3$): calcium, 300; sodium, 100; magnesium, 100; and total cations, 500. The unit must produce water with no more than 2 ppm hardness and must operate for 8 h between regenerations. Manufacturer's data on the water-softening resin to be used include the following: (1) a regeneration level of 4 lb NaCl per cubic foot (64 kg/m³) will result in 2 ppm hardness leakage and a capacity of 16 kgr/ft³ (36 kg/m³), assuming standard cocurrent operation; (2) pressure drop per linear foot of bed (per 0.305 m of bed) for 60°F water at a linear velocity equivalent to 7.1 gal/(min)(ft²) [17.3 m³/(h)(m²)] is 0.6 lb/in² (4.14 kPa); (3) a flow rate of 6.4 gal/(min)(ft²) [15.6 m³/(h)(m²)] will bring about a bed expansion of 60 percent; and (4) rinse requirements are 25 to 50 gal/ft³ (3.34 to 6.68 m³/m³). Determine the resin volume needed, the pressure drop, the backwash requirement, the regenerant requirement, and the required volume of rinse water.

[†]Courtesy of Rohm & Haas Co.

Calculation Procedure

1. Determine the amount of water to be treated per cycle and the amount of hardness to be removed.

Softening of water requires use of a cation-exchange resin operated in sodium form to exchange divalent hardness cations for sodium regenerated with aqueous sodium chloride solution. Total amount of water to be treated is (100 gal/min)(60 min/h)(8 h/cycle) = 48,000 gal/cycle (182 m^3/cycle).

In determining the quantity of hardness to be removed, neglect the 2 ppm allowable hardness in the effluent (this is a conservative simplification) and assume complete hardness removal. Since the influence hardness is expressed as parts per million (equivalents as CaCO$_3$), it is necessary to convert to units consistent with resin manufacturers' capacity data, usually expressed as kilograins (as CaCO$_3$) per cubic foot of resin. A total of 400 ppm hardness (calcium plus magnesium) is to be removed. Convert this to kilograins as CaCO$_3$. Thus, (400 ppm as CaCO$_3$)(48,000 gal/cycle)/(1000 gr/kgr) (17.1 ppm per grain per gallon) = 1120 kgr (73 kg) as CaCO$_3$. per cycle.

2. Establish regeneration level and resin capacity.

An optimal level of regeneration exists for each softening application. This relates level of regeneration (pounds of regenerant per cubic foot of softening resin), leakage (ions not exchanged and thus appearing in the effluent), and operating capacity. In the present case, the desired information is given (based on information from the resin manufacturer) in the statement of the example: The optimal regeneration level is 4 lb NaCl per cubic foot (64 kg NaCl per cubic meter).

3. Determine volume of softening resin needed.

The hardness load per cycle is 1120 kgr, from step 1, and the resin capacity is given as 16 kgr/ft^3. So the amount of resin needed is 1120 kgr/(16 kgr/ft^3) = 70 ft^3 (1.98 m^3).

However, if water production must be continuous, two softening units must be obtained, so that one can be regenerated while the other is in service. The alternative is to supply storage facilities for several hours' production of water at 100 gal/min.

4. Determine column dimensions, pressure drop, and backwash requirements.

In conventional water softening, an acceptable space velocity is usually between 1 and 5 gal/(min)(ft^3) [8 and 40 m^3/(h)(m^3)]. In the present case, space velocity is (100 gal/min)/70 ft^3 = 1.43 gal/(min)(ft^3), which is within the normal range and thus is acceptable.

Normal *linear* velocity in a softening unit is equivalent to the range 4 to 10 gal/(min)(ft^2) [9.75 to 24.4 m^3/(h)(m^2)]. If the velocity is too high, the pressure drop is excessive; too low a velocity can cause poor distribution of flow through the unit. As for bed depth, it should normally be 3 to 6 ft (0.9 to 1.8 m).

Given these norms, determination of column dimensions is usually done by trial and error. Thus assume a bed depth of 5 ft (1.5 m). Then, cross-sectional area will be 70 ft^3/5 ft = 14 ft^2, and linear velocity will be equivalent to (100 gal/min)/14 ft^2 = 7.1 gal/(min)(ft^2), which is acceptable because it falls in the normal range.

The column diameter is (area × $4/\pi$)$^{1/2}$ = (14 × $4/\pi$)$^{1/2}$ = 4.2 ft (1.28 m). In establishing the column height, allow adequate head space, or freeboard, to permit backwashing. A good allowance is 100 percent of the bed height. Thus the column height is twice the bed height, or 10 ft (3.05 m).

The pressure drop per foot of bed depth is given in the statement of the example as 0.6 lb/in^2. Thus total pressure drop for the resin bed is [0.6 lb/(in^2)(ft)](5 ft) = 3.0 lb/in^2 (21 kPa). This excludes the pressure drop due to the liquid distributors and collectors in the column, as well as that due to auxiliary fittings and valves.

Backwashing is necessary to keep the bed in a hydraulically classified condition, to minimize pressure drop, and to remove resin fines and suspended solids that have been filtered out of the influent water. Normal practice is to backwash at the end of each run for about 15 min, so as to obtain about 50 to 75 percent bed expansion. The flow rate required to achieve this expansion is obtained from the manufacturers' data. As noted in the statement of the example, an appropriate flow rate in this case is 6.4 gal/(min)(ft^2). The total backwash rate is thus [6.4 gal/(min)(fl^2)](14 ft^2) = 90 gal/min. The total water requirement, then, is (90 gal/min)(15 min) = 1350 gal (5.11 m^3).

5. *Determine regenerant requirement and flow rate.*
The sodium chloride regeneration level necessary to hold leakage to 2 ppm is 4 lb NaCl per cubic foot, as noted in step 2. The salt (100% basis) requirement, then, is $(4 \text{ lb/ft}^3)(70 \text{ ft}^3) = 280$ lb NaCl per cycle. Now salt is typically administered as a 10% solution at a rate of 1 gal/(min)(ft³). The density of such a solution is 8.94 lb/gal. Thus the volumetric requirement is (280 lb NaCl per cycle)/(0.10 lb NaCl per pound solution)(8.94 lb solution per gallon), or about 310 gal/cycle. This should be fed at a flow rate of $[1 \text{ gal/(min)(ft}^3)] (70 \text{ ft}^3) = 70$ gal/min (15.9 m³/h).

6. *Determine required volume of rinse water.*
The resin bed must be rinsed with water following regeneration with salt. Rinse-water requirements are obtained from manufacturers' literature; in the present case, they are 25 to 50 gal/ft³. At 35 gal/ft³, the total rinse required is $(35 \text{ gal/ft}^3)(70 \text{ ft}^3) = 2450$ gal /cycle. Normal practice is to rinse at 1 gal/(min)(ft³) (in this case, 70 gal/min, or 15.9 m³/h) for the first 10 to 15 min and then at 2 gal/(min)(ft³) (here, 140 gal/min, or 31.8 m³/h) for the remainder.

Related Calculations.   This example assumes standard cocurrent operation of the column with downflow feed and downflow regeneration. Countercurrent operation is a special case that is best handled by a manufacturer of ion-exchange equipment.

19.19   COMPLETE DEIONIZATION OF WATER[†]

Select a deionization system to treat 250 gal/min (56.8 m³/h) of water at 60°F (289 K) that has the following dissolved components (concentrations in parts per million as $CaCO_3$ equivalent): calcium, 75; sodium, 50; magnesium, 25; chloride, 30; sulfate, 80; bicarbonate, 40; and silica, 10 (as SiO_2). Maximum tolerable sodium leakage is 2 ppm; silica leakage is to be under 0.05 ppm. Service-cycle length must be at least 12 h. The system is to use sodium hydroxide (available at 120°F) and sulfuric acid for regeneration.

Manufacturers' data on the resins to be used include the following: For the cation-exchange resin: (1) a regeneration level of 6 lb H_2SO_4 per cubic foot (96 kg H_2SO_4 per cubic meter) will result in a sodium leakage of 2.0 ppm and an operating capacity of 15.6 kgr/ft³ (35 kg/m³); (2) pressure drop per foot (per 0.305 m) of bed depth for 60°F water at a linear velocity equivalent to 8.6 gal/(min)(ft²) [21 m³/(h)(m²)] is 0.75 lb/in² (5.2 kPa); (3) a flow rate of 6.4 gal/(min)(ft²) [15.6 m³/(h)(m²)] will bring about a bed expansion of 60 percent; and (4) rinse requirements are 25 to 50 gal/ft³ (3.34 to 6.68 m³/m³) using deionized rinse water. For the anion-exchange resin: (1) a regeneration level of 4 lb NaOH per cubic foot (64 kg NaOH per cubic meter) will result in a silica leakage of 0.05 ppm and an operating capacity of 15.3 kgr/ft³ (35 kg/m³); (2) pressure drop per foot (per 0.305 m) of bed depth for 60°F water at a linear velocity equivalent to 8.5 gal/(min)(ft²) [20.8 m³/(h)(m²)] is 0.85 lb/in² (5.9 kPa); (3) a flow rate of 2.6 gal/(min)(ft²) [6.34 m³/(h)(m²)] will bring about a bed expansion of 60 percent: and (4) rinse requirements are 40 to 90 gal/ft³ (5.34 to 12.0 m³/m³) using deionized water. Determine the resin volumes, the pressure drops, the backwash, regenerant and rinse-water requirements, and overall operating conditions.

Calculation Procedure

1. *Decide on the ion-exchange system to he used.*
Deionization requires replacement of all cations by the hydrogen ion, accomplished by use of a cation-exchange resin in the hydrogen form, as well as replacement of all anions by the hydroxide ion, accomplished by use of an anion exchanger in the hydroxide form. Since complete removal of all anions, including carbon dioxide and silica, is required, it will be necessary to use a strongly basic

---

[†]Courtesy of Rohm & Haas Co.

anion exchanger, regenerated with sodium hydroxide. The simplest system, a strongly acidic cation exchanger followed by a strongly basic anion exchanger, will be employed here. More elaborate and, in some cases, more efficient systems involving use of degassing equipment, stratified beds of strong- and weak-electrolyte resins, or mixed-bed units are beyond the scope of this handbook.

2. Specify the cation-exchange column.

a. *Determine quantity of water to be treated per cycle and quantity of cations to be removed.* The amount of water is (250 gal/min)(60 min/h)(12 h/cycle) = 180,000 gal (681 m^3). To determine the cation load, neglect the 2 ppm of sodium leakage and assume complete removal of all cations. Since the influence cation load is expressed as parts per million (equivalents as CaCO$_3$), it is necessary to convert to units consistent with resin manufacturers' capacity data, usually expressed as kilograins (as CaCO$_3$) per cubic foot of resin. Total cation load in this cast is 75 + 50 + 25 ppm. Converting, (150 ppm)(180,000 gal/cycle)/(1000 gr/kgr)(17.1 ppm per grain per gallon) = 1580 kgr (102 kg) as CaCO$_3$ per cycle.

b. *Establish regeneration level and resin capacity.* Using manufacturers' data will determine the least amount of regenerant that will produce water of acceptable quality. In the present case, the desired information is given in the statement of the example: The optimal regeneration level is 6 lb H$_2$SO$_4$ per cubic foot (96 kg H$_2$SO$_4$ per cubic meter). *Note:* Care must be taken in regenerating cation-exchange resins with sulfuric acid to avoid precipitation of calcium sulfate.

c. *Determine the volume of cation-exchange resin needed.* The cation load per cycle is 1580 kgr, from step 2a, and the resin capacity is given as 15.6 kgr/ft^3. So the amount of resin needed is 1580 kgr/(15.6 kgr/ft^3) = 101 ft^3 (2.86 m^3). However, if water production must be continuous, two units are needed, so that one can be regenerated while the other is in service. The alternative is to provide storage for several hours of production of water at 250 gal/min.

d. *Determine column dimensions, pressure drop, and backwash requirement.* In conventional water treatment, an acceptable space velocity is usually between 1 and 5 gal/(min)(ft^3) [8 and 40 m^3/(h)(m^3)]. In the present case, space velocity is (250 gal/min)/101 ft^3 = 2.5 gal/(min)(ft^3), which is thus acceptable.

   Normal *linear* velocity is equivalent to the range 4 to 10 gal/(min)(ft^2) [9.75 to 24.4 m^3/(h)(m^2)]. Given this norm, determination of column dimensions is usually trial-and-error. Thus, assume a bed depth of 3.5 ft (1.07 m). Then the cross-sectional area will be 101 ft^3/3.5 ft = 28.9 ft^2, and linear velocity will be equivalent to (250 gal/min)/28.9 ft^2 = 8.6 gal/(min)(ft^2), which is acceptable.

   If either space velocity or linear velocity had been considerably greater than the normal ranges, it would have been necessary to assign more resin.

   The column diameter is (area × $4/\pi$)$^{1/2}$ = (28.9 × $4/\pi$)$^{1/2}$ = 6.1 ft (1.86 m). In establishing the column height, allow adequate head space, or freeboard, to permit backwashing. A good allowance is 100 percent of the bed height. Thus the column height is twice the bed height, or 7 ft (2.13 m).

   The pressure drop per foot of bed depth is given in the statement of the example as 0.75 lb/in^2. Thus total pressure drop for the cation-resin bed is [0.75 lb/(in^2)(ft)](3.5 ft) = 2.6 lb/in^2 (17.9 kPa). This excludes the pressure drop due to valves, fittings, or liquid distributors or collectors.

   Backwashing is necessary to keep the bed in a hydraulically classified condition, to minimize pressure drop and provide for proper flow distribution, as well as to remove resin fines and suspended solids that have filtered out of the water. Normal practice is to backwash at the end of each run to achieve 50 to 75 percent bed expansion. The flow rate required for this expansion is given in the statement of the example as 6.4 gal/(min)(ft^2). The total backwash rate is, thus, [6.4 gal/(min)(ft^2)](28.9 ft^2) = 185 gal/min (42 m^3/h).

e. *Determine regenerant requirement and flow rate.* The sulfuric acid regeneration level to hold sodium leakage to 2 ppm is 6 lb/ft^3, as noted earlier. The total acid requirement, then, is (6 lb/ft^3) (101 ft^3) = 606 lb (275 kg) per cycle.

A typical technique to avoid precipitation of calcium sulfate is to administer half the regenerant as a 2% solution and then the rest at 4%. Thus, in this case, the first step would require ($\frac{1}{2}$ × 606 lb H_2SO_4)/(8.43 lb solution per gallon)(0.02 lb H_2SO_4 per pound solution), or about 1800 gal (6.8 m^3) of 2% acid solution. The second stage requires ($\frac{1}{2}$ × 606)/[8.54(0.04)], or about 890 gal (3.37 m^3) of 4% acid solution. (The 8.43 and 8.54 lb/gal are densities of the acid solutions.) Each stage should be fed at a rate of 1 to 1.5 gal/(min)(ft^3), or in this case about 100 to 150 gal/min (23 to 34 m^3/h).

f. Determine the required volume of rinse water. The column must be rinsed with water after regeneration. Rinse-water requirements, as noted earlier, are 25 to 50 gal/ft^3. In the present case, for 101 ft^3, the requirement is about 2500 to 5000 gal (9.5 to 19 m^3). The first portion should be administered at 1 gal/(min)(ft^3) [8 m^3/(h)(m^3)], and the rest at 1.5 gal/(min)(ft^3) [12 m^3/(h)(m^3)].

3. Specify the anion-exchange column

a. Determine quantity of water to be treated and quantity of anions to be removed. The amount of water, from step 2a, is 180,000 gal. Total anion load is 30 + 80 + 40 ppm. Converting, (150 ppm) × (180,000 gal/cycle)/(1000 gr/kgr)(17.1 ppm per grain per gallon) = 1580 kgr (102 kg) as $CaCO_3$ per cycle.

b. Establish regeneration level and resin capacity. As given in the statement of the problem, the optimal regeneration level is 4 lb NaOH per cubic foot (4 kg NaOH per cubic meter), associated with an operating capacity of 15.3 kgr/ft^3.

c. Determine the volume of anion-exchange resin needed. The anion load per cycle is 1580 kgr, from step 3a, and the resin capacity is 15.3 kgr/ft^3. So the amount of resin needed is 1580/15.3 = 103 ft^3 (2.91 m^3). However, if water production must be continuous, it is necessary to either install a second anion-exchange column in parallel, so that one can be regenerated while the other is in service, or else provide for water storage.

d. Determine column dimensions, pressure drop, and backwash requirement. Space velocity is (250 gal/min)/103 ft^3 = 2.4 gal/(min)(ft^3), which falls within the acceptable range (see step 2d). As in step 2d, assume a bed depth of 3.5 ft (1.07 m). Then the cross-sectional area of the bed will be 103/3.5 = 29.4 ft^2, and linear velocity will be equivalent to (250 gal/min)/29.4 ft^2 = 8.5 gal/(min)(ft^2), which is also acceptable.

The column diameter is (area × 4/π)$^{1/2}$ = (29.4 × 4/π)$^{1/2}$ = 6.12 ft (1.86 m). Allowing 100 percent head space for backwashing, the column height is twice the bed height, or 7 ft (2.13 m).

The pressure drop per foot of bed depth is given in the statement of the example as 0.85 lb/in^2. Thus total pressure drop for the anion-exchange bed is (0.85)(3.5) = 3 lb/in^2 (20.7 kPa). This excludes the pressure drop due to valves, fittings, or liquid distributors or collectors.

As for the backwash requirement, as discussed in step 2d, the flow rate required is [2.6 gal/(min)(ft^2)](29.4 ft^2) = 76 gal/min (17.3 m^3/h).

e. Determine the regenerant requirement and flow rate. The sodium hydroxide regeneration level is 4 lb NaOH per cubic foot, as noted earlier. Total hydroxide requirement, then, is (4 lb/ft^3)(103 ft^3) = 412 lb (187 kg) NaOH per cycle. Regenerant concentration is typically 4% NaOH solution having a density of 8.68 lb/gal. Total regenerant-solution requirement, then, is (412 lb NaOH per cycle)/(0.04 lb NaOH per pound of solution)(8.68 lb solution per gallon), or 1190 gal per cycle (4.5 m^3 per cycle). This should be applied at about 0.5 gal/(min)(ft^3) [4 m^3/(h)(m^3)].

f. Determine the required volume of rinse water. Rinse-water requirements, as noted earlier, are 40 to 90 gal/ft^3. In the present case, for 103 ft^3, the requirement is about 4000 to 9000 gal (15.1 to 34.1 m^3). The first bed volume (i.e., first 103 ft^3 × 7.48 gal/ft^3, or 750 gal) should be applied at about 50 gal/min (11.4 m^3/h), and the remainder should be applied at about 150 gal/min (34.1 m^3/h).

19.20 CHEMICAL-PLANT COOLING-POND SIZE FOR A KNOWN HEAT LOAD

How many spray nozzles and what surface area is needed to cool 10,000 gal/min (630.8 L/s) of water from 120 to 90°F (48.9 to 32.2°C) in a spray-type cooling pond if the average wet-bulb temperature is 60°F (15.6°C)? What would the approximate dimensions of the cooling pond be? Determine the total pumping head if the static head is 10 ft (29.9 kPa), the pipe friction is 35 ft of water (104.6 kPa), and the nozzle pressure is 8 lb/in² (55.2 kPa).

Calculation Procedure

1. Compute the number of nozzles required.
Assume a water flow of 50 gal/min (3.2 L/s) per nozzle; this is a typical flow rate for usual cooling-pond nozzles. Then, the number of nozzles required equals 10,000 gal/(min)/(50 gal/min per nozzle) = 200 nozzles. If 6 nozzles are used in each spray group in a series of crossed arms, with each arm containing one or more nozzles, then 200 nozzles divided by 6 nozzles per spray group means that 33⅓ spray groups will be needed. Since a partial spray group is seldom used, 34 spray groups would be chosen.

2. Determine the surface area required.
Usual design practice is to provide 1 ft² (0.09 m²) of pond area per 250 lb (113.4 kg) of water cooled for water quantities exceeding 1000 gal/min (63.1 L/s). Thus, in this pond, the weight of water cooled equals (10,000 gal/min)(8.33 lb/gal)(60 min/h) = 4,998,000, say, 5,000,000 lb/h (630.0 kg/s). Then, the area required, using 1 ft² of pond area per 250 lb of water (0.82 m² per 1000 kg) cooled, is 5,000,000/250 = 20,000 ft² (1858.0 m²).

As a cross-check, use another commonly accepted area value: 125 Btu/(ft²)(°F) [2555.2 kJ/(m²)(°C)], based on the temperature difference between the air wet-bulb temperature and the warm entering-water temperature. This is the equivalent of (120 − 60)(125) = 7500 Btu/ft² (85,174 kJ/m²) in this spray pond, because the air wet-bulb temperature is 60°F (15.6°C) and the warm-water temperature is 120°F (48.9°C). The heat removed from the water is (pounds per hour of water)(temperature decrease, in °F)(specific heat of water) = (5,000,000)(120 − 90)(1.0) = 150,000,000 Btu/h (43,960.7 kW). Then, area required equals (heat removed, in Btu/h)/(heat removal, in Btu/ft²) = 150,000,000/7,500 = 20,000 ft² (1858.0 m²). This checks the previously obtained area value.

3. Determine the spray-pond dimensions.
Spray groups on the same header or pipe main are usually arranged on about 12-ft (3.7-m) centers with the headers or pipe mains spaced on about 25-ft (7.6-m) centers (Fig. 19.24). Assume that 34 spray groups are used, instead of the required 33⅓, to provide an equal number of groups in two headers and a small extra capacity.

Sketch the spray pond and headers (Fig. 19.24). This shows that the length of each header will be about 204 ft (62.2 m), because there are seventeen 12-ft (3.7-m) spaces between spray groups in each header. Allowing 3 ft (0.9 m) at each end of a header for fittings and cleanouts gives an overall header length of 210 ft (64.0 m). The distance between headers is 25 ft (7.6 m). Allow 25 ft (7.6 m) between the outer sprays and the edge of the pond. This gives an overall width of 85 ft (25.9 m) for the pond, assuming the width of each arm in a spray group is 10 ft (3.0 m). The overall length will then be 210 + 25 + 25 = 260 ft (79.2 m). A cold well for the pump suction and suitable valving for control of the incoming water must be provided, as shown in Fig. 19.24. The water depth in the pond should be 2 to 3 ft (0.6 to 0.9 m).

4. Compute the total pumping head.
The total head, expressed in feet of water, equals static head + friction head + required nozzle head = 10 + 35 + 8(0.434) = 48.5 ft of water (145.0 kPa). A pump having a total head of at least 50 ft of water (15.2 m) would be chosen for this spray pond. If future expansion of the pond is anticipated, compute the probable total head required at a future date and choose a pump to deliver that head.

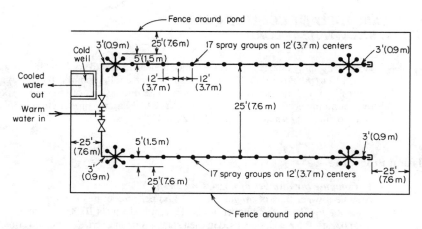

FIGURE 19.24   Spray-pond nozzle and piping layout (Example 19.14).

Until the pond is expanded, the pump would operate with a throttled discharge. Normal nozzle inlet pressures range from about 6 to 10 lb/in² (41.4 to 69.0 kPa). Higher pressures should not be used, because there will be excessive spray loss and rapid wear of the nozzles.

Related Calculations.   Unsprayed cooling ponds cool 4 to 6 lb (1.8 to 2.7 kg) of water from 100 to 70°F per square foot (598.0 to 418.6° C/m²) of water surface. An alternative design rule is to assume that the pond will dissipate 3.5 Btu/(h)(ft²) [19.9 J/(m²)(°C)(s)] of water surface per degree difference between the wet-bulb temperature of the air and the entering warm water.

19.21   *CHEMICAL-PLANT INDIRECT DRYING OF SOLIDS*

An indirect dryer consisting of a heating section, a constant-rate-drying section, and a falling-rate–drying section in series is to lower the water content of 1000 lb/h (454 kg/h) of feed from 20 to 5 percent. The feed temperature is 60°F (289 K), and the product leaves the dryer at 260°F (400 K). The specific heat of the solid is 0.4 Btu/(lb)(°F) [1.67 kJ/(kg)(K)]; that of water is 1.0 Btu/(lb)(°F) [4.19 kJ/(kg)(K)]. The heating medium is 338°F (443 K) steam. The heat-transfer rates in the heating, constant-rate-drying and falling-rate–drying sections are 25, 40, and 15 Btu/(h)(ft²)(°F), respectively. The surface loading in the three sections is 100, 80, and 60 percent, respectively. Thermal data for the moisture and solids are as follows:

Evaporation enthalpy:	970.3 Btu/lb (2257 kJ/kg)
Water enthalpy at 212°F:	180.2 Btu/lb (419 kJ/kg)
Water-vapor enthalpy, averaged over 212 to 260°F:	1159.0 Btu/lb (2696 kJ/kg)
Solids temperature during constant-rate drying:	212°F (373 K)
Product moisture at start of falling-rate drying:	10 percent

Determine the heat load for each section of the dryer, as well as the area required for each.

Calculation Procedure

1.   *Calculate the flow rate of dry solids W_S.*
As the feed moisture is 20 percent and the total flow rate is 1000 lb/h, $W_s = 1000(1 - 0.20) = 800$ lb/h (363 kg/h).

2. Determine the product flow rate W_P.
Since W_S is 800 lb/h and the product moisture is 5 percent, $W_P = 800/(1 - 0.05) = 842.1$ lb/h (382 kg/h).

3. Calculate the total amount of liquid to be removed.
Because the feed rate is 1000 lb/h and the product rate 842.1 lb/h, the liquid removed (i.e., evaporated) is $1000 - 842.1 = 157.9$ lb/h (71.7 kg/h).

4. Determine the amount of moisture in the in-process material as it passes from the constant-rate section into the falling-rate section.
As the moisture content at this point is 10 percent and the dry-solids rate is 800 lb/h, the flow rate for the in-process material at this point is $800/(1 - 0.10)$ lb/h. The moisture in the material is, then, $[800/(1 - 0.10)][0.10]$, or 88.9 lb/h (40.4 kg/h).

5. Calculate the moisture leaving in the final product.
The product rate is 842.1 lb/h and its moisture content 5 percent, so the moisture leaving in the final product is $(842.1)(0.05)$, or 42.1 lb/h (19.1 kg/h).

6. Determine the moisture removed in the falling-rate zone.
Because the moisture entering this zone is 88.9 lb/h and the amount leaving in the product is 42.1 lb/h, the amount removed from the in-process material in this zone is $88.9 - 42.1$, or 46.8 lb/h (21.2 kg/h).

7. Find the amount of moisture removed in the constant-rate zone.
The total amount removed in the dryer is 157.9 lb/h and the amount removed in the falling-rate zone is 46.8 lb/h, so the amount removed in the constant-rate zone is $157.9 - 46.8$, or 111.1 lb/h (50.4 kg/h).

8. Calculate Q_{HS}, the heat load for the solid in the heating zone.
Now, $Q_{HS} = W_S C_S \Delta T$, where W_S is the solid flow rate, C_S the specific heat of the solid, and ΔT the temperature rise for the solid in this zone. Thus, $Q_{HS} = (800)(0.4)(212 - 60) = 48,640$ Btu/h (14.25 kW).

9. Calculate Q_{HL}, the heat load for the liquid in this zone.
Use the same procedure as in step 8, but with the liquid flow rate and specific heat. Thus, $Q_{HL} = [(0.20)(1000)][1.0][212 - 60] = 30,400$ Btu/h (8.91 kW).

10. Find Q_H, the total heat load for this zone.
It is the sum of the solid and liquid heat loads: $Q_H = 48,640 + 30,400 = 79,040$ Btu/h (23.16 kW).

11. Determine Q_C, the heat load in the constant-rate–drying zone.
The heat in this zone serves solely for evaporation, with no sensible heating. The amount of water evaporated in this zone was found in step 7. Then, $Q_C = (970.3)(111.1) = 107,800$ Btu/h (31.58 kW).

12. Calculate Q_{FS}, the heat load for the solid in the falling-rate zone.
As in step 8, $Q_{FS} = (800)(0.4)(260 - 212) = 15,360$ Btu/h (4.5 kW).

13. Find Q_{FE}, the heat load for evaporation in the falling-rate zone.
The amount of water removed in this zone was found in step 6. For the heat load per pound of evaporated water, use the difference in enthalpy between that for water at 212°F and the averaged value for water vapor between 212 and 260°F, as stated at the beginning of the problem. Thus, $Q_{FE} = (46.8)(1159 - 180.2) = 45,808$ Btu/h (13.4 kW).

14. Calculate Q_{FL}, the heat load for unevaporated liquid in the falling-rate zone.
This is the liquid that remains as moisture in the final product. The amount was calculated in step 5. In the falling-rate zone, it becomes heated from 212 to 260°F. As in step 8, $Q_{FL} = (42.1)(1.0)(260 - 212) = 2021$ Btu/h (592 W).

15. Calculate Q_F, the total heat load in the falling-rate zone.
Sum the quantities calculated in the previous three steps. Thus, $Q_F = 15,360 + 45,808 + 2021 = 63,189$ Btu/h (18.51 kW).

16. Calculate ΔT_{mH}, the log-mean temperature difference in the heating zone.
Let T_i and T_o be the heating-medium (i.e., steam) temperature at the zone inlet and outlet, respectively, and t_i and t_o correspondingly be the temperature of the in-process material at the zone inlet and outlet. Then

$$\Delta T_{mH} = [(T_i - t_o) - (T_o - t_i)]/\ln[(T_i - t_o)/(T_o - t_i)]$$
$$= [(338 - 212) - (338 - 60)]/\ln[(338 - 212)/(338 - 60)] = 192.1 \text{ Fahrenheit degrees.}$$

17. Determine A_H, the surface area required in the heating zone.
Use the equation $A = Q/U\,\Delta T\,L$, where A is the required surface area, Q the heat load (from step 10), U the overall heat-transfer coefficient, ΔT the log-mean temperature difference, and L the surface loading. Thus, $A = 79,040/(25)(192.1)(1.0) = 16.46 \text{ ft}^2 (1.53 \text{ m}^2)$.

18. Calculate A_C, the surface area required in the constant-rate–drying zone.
In this case, the temperature difference between the 338°F steam and the 212°F in-process material is constant. Thus, via the equation from step 17, $A = 107,800/(40)(338 - 212)(0.8) = 26.74 \text{ ft}^2 (2.49 \text{ m}^2)$.

19. Calculate ΔT_{mF}, the log-mean temperature difference in the falling-rate zone.
As in step 16, $\Delta T_{mF} = [(338 - 260) - (338 - 212)]/\ln[(338 - 260)/(338 - 212)] = 100.1$ Fahrenheit degrees.

20. Calculate A_F, the surface area required for the falling-rate zone.
As in step 18, $A = 63,189/(15)(100.1)(0.6) = 70.14 \text{ ft}^2 (6.52 \text{ m}^2)$.

21. Determine the total surface area required for the dryer.
From steps 17, 18, and 20, the total is $16.46 + 26.74 + 70.14 = 113.3 \text{ ft}^2 (10.54 \text{ m}^2)$.

22. Find the total heat load.
From steps 10, 11, and 15, the total is $79,040 + 107,800 + 63,189 = 250,029$ Btu/h (73.3 kW).

Related Calculations. This example is adapted from *Process Drying Practice* by Cook and DuMont, published in 1991 by McGraw-Hill. More details are available in that source. Similar calculation for direct dryers is far more complex, involving the psychrometric relations of moist air. Trial-and-error loops are required, and manual calculation is not only time-consuming but also error-prone. A sequence of equations suitable for setting into a computer program can be found in the aforementioned Cook and DuMont.

19.22 CHEMICAL PLANTS VACUUM DRYING OF SOLIDS

A material having a wet bulk density of 40 lb/ft³ (640 kg/m³) and containing 30% water is to be fully dried in a rotary vacuum batch dryer. The dryer is 5 ft (1.5 m) in diameter and 20 ft (6.1 m) long, and has a working volume of 196 ft³ (5.5 m³) and a wetted surface of 206 ft² (19.1 m²). The maximum product temperature is 125°F (325 K). Cooling water at 85°F (302 K) is available to condense the water vapor removed, in a shell-and-tube surface condenser. The condenser is to maintain a vacuum of 85 torr, at which level the vapors will condense at 115°F (319 K). Pilot studies indicate that the dryer will dry the material at an effective rate of 1.0 lb/(h)(ft²) [4.9 kg/(h)(m²)] and a peak rate of 2.0 lb/(h)(ft²) [9.8 kg/(h)(m²)]. On average, how many pounds per hour of dry product can it produce? How much condenser surface is required?

Calculation Procedure

1. Determine the charge that the dryer can handle.
Multiply the working volume of the dryer by the wet bulk density of the feed. Thus, $(196 \text{ ft}^3)(40 \text{ lb/ft}^3) = 7840$ lb (3559 kg).

2. Determine the rate at which the dryer can remove vapor.
Multiply the wetted surface by the effective drying rate. Thus, $(206 \text{ ft}^2)[1.0 \text{ lb/(h)(ft}^2)] = 206$ lb/h (93.5 kg/h).

3. Determine the amount of water to be removed per batch.
Multiply the amount of material charged by its water content. Thus, $(0.30)(7840 \text{ lb}) = 2352$ lb (1068 kg) of water.

4. Calculate the drying time per batch.
Divide the water to be removed by the rate at which the dryer can remove it. Thus, 2352 lb/(206 lb/h) = 11.4 h.

5. Determine the amount of dry product produced per batch.
Subtract the amount of water removed (see step 3) from the amount of material charged (step 1). Thus, 7840 lb − 2352 lb = 5488 lb (2492 kg) dry product.

6. Determine the average production rate in pounds per hour.
Divide the amount of dry product per batch by the drying time per batch. Thus, 5488 lb/11.4 h = 481 lb/h (218 kg/h).

7. Determine the amount of water removed under peak drying conditions.
Multiply the wetted surface area by the peak drying rate. Thus, $(206 \text{ ft}^2)[2.0 \text{ lb/(h)(ft}^2)] = 412$ lb/h (187 kg/h).

8. Determine the required condenser surface.
Enter the graph in Fig. 19.25 with 412 lb/h peak water removal as abscissa and system operating pressure of 75 torr as the parameter. From the ordinate, read 190 ft² (17.7 m²) as the required condenser surface.

Related Calculations.   Figure 19.25 is valid for condensers of shell-and-tube design, employing cooling water at 85°F. This example is adapted from "Vacuum Dryers," *Chem. Eng.*, January 17, 1977.

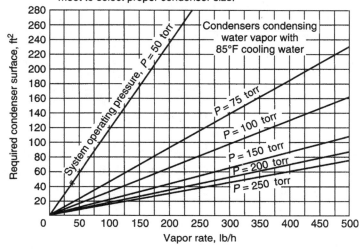

FIGURE 19.25   Shell-and-tube condenser surface as a function of vapor rate and system operating pressure (cooling water at 85°F). (*From Chemical Engineering, January 17, 1977, copyright 1977 by McGraw-Hill, Inc., New York. Reprinted by special permission.*)

19.23 ESTIMATING THERMODYNAMIC AND TRANSPORT PROPERTIES OF WATER

Estimate the following properties of liquid water at 80°F: (1) vapor pressure, (2) density, (3) latent heat of vaporization, (4) viscosity, (5) thermal conductivity. Also, estimate the following properties for saturated water vapor at 200°F: (6) density, (7) specific heat, (8) viscosity, (9) thermal conductivity. And calculate the boiling point of water at 30 psia.

Calculation Procedure   For each of the estimates, the procedure consists of using correlation equations that have been derived by regression analysis of the properties of saturated steam, as discussed in more detail under "Related Calculations." The temperatures are to be entered into the equations in degrees Fahrenheit and the pressure (for the boiling-point example) in pounds per square inch absolute. The results are likewise in English units, as indicated below. These correlations are valid only over the range 32 to 440°F.

1. Estimate the vapor pressure.
The equation is

$$P = \exp[10.9955 - 9.6866 \ln T + 1.9779(\ln T)^2 - 0.085738(\ln T)^3]$$

Thus, $P = \exp[10.9955 - 9.6866(\ln 80) + 1.9779(\ln 80)^2 - 0.085738(\ln 80)^3] = 0.504$ psia.

2. Estimate the liquid density.
The equation is

$$\rho_L = 62.7538 - 3.5347 \times 10^{-3}T - 4.8193 \times 10^{-5}T^2$$

Thus, $\rho_L = 62.7538 - 3.5347 \times 10^{-3}(80) - 4.8193 \times 10^{-5}(80)^2 = 62.16$ lb/ft^3.

3. Estimate the latent heat of vaporization.
The equation is

$$\Delta H_{vap} = 1087.54 - 0.43110T - 5.5440 \times 10^{-4}T^2$$

Thus, $\Delta H_{vap} = 1087.54 - 0.43110(80) - 5.5440 \times 10^{-4}(80)^2 = 1049.5$ Btu/lb.

4. Estimate the liquid viscosity.
The equation is

$$\mu_L = -0.23535 + 208.65/T - 2074.8/T^2$$

Thus, $\mu_L = -0.23535 + 208.65/80 - 2074.8/(80)^2 = 2.05$ lb/(h)(ft).

5. Estimate the liquid thermal conductivity.
The equation is

$$k_L = 0.31171 + 6.2278 \times 10^{-4}T - 1.1159 \times 10^{-6}T^2$$

Thus, $k_L = 0.31171 + 6.2278 \times 10^{-4}(80) - 1.1159 \times 10^{-6}(80)^2 = 0.369$ Btu/(h)(ft)(°F).

6. Estimate the vapor density.
The equation is

$$\rho_V = \exp(-9.3239 + 4.1055 \times 10^{-2}T - 7.1159 \times 10^{-5}T^2 + 5.7039 \times 10^{-8}T^3)$$

Thus, $\rho_V = \exp[-9.3239 + 4.1055 \times 10^{-2}(200) - 7.1159 \times 10^{-5}(200)^2$

$\qquad + 5.7039 \times 10^{-8}(200)^3]$

$\qquad = 0.0301$ lb/ft^3.

7. Estimate the specific heat of the vapor.
The equation is

$$C_p = 0.43827 + 1.3348 \times 10^{-4}T - 5.9590 \times 10^{-7}T^2 + 4.6614 \times 10^{-9}T^3$$

$$\text{Thus, } C_p = 0.43827 + 1.3348 \times 10^{-4}(200) - 5.9590 \times 10^{-7}(200)^2 + 4.6614 \times 10^{-9}(200)^3$$

$$= 0.478 \text{ Btu/(lb)(°F)}.$$

8. Estimate the vapor viscosity.
The equation is

$$\mu_V = 0.017493 + 5.7455 \times 10^{-5}T - 1.3717 \times 10^{-8}T^2$$

$$\text{Thus, } \mu_V = 0.017493 + 5.7455 \times 10^{-5}(200) - 1.3717 \times 10^{-8}(200)^2 = 0.028435 \text{ lb/(h)(ft)}.$$

9. Estimate the thermal conductivity of the vapor.
The equation is

$$k_V = 0.0097982 + 2.2503 \times 10^{-5}T - 3.3841 \times 10^{-8}T^2 + 1.3153 \times 10^{-10}T^3$$

$$\text{Thus, } k_V = 0.0097982 + 2.2503 \times 10^{-5}(200) - 3.3841 \times 10^{-8}(200)^2$$

$$+ 1.3153 \times 10^{-10}(200)^3 = 0.01400 \text{ Btu/(h)(ft)(°F)}.$$

10. Estimate the boiling point.
The equation is

$$T_B = \exp[4.6215 + 0.34977 \ln P - 0.03727(\ln P)^2 + 0.0034492(\ln P)^3]$$

$$\text{Thus, } T_B = \exp[4.6215 + 0.34977(\ln 30) - 0.03727(\ln 30)^2 + 0.0034492(\ln 30)^3] = 248.6°F.$$

Related Calculations. These equations are presented in Dickey, D.S.,[†] Practical Formulas Calculate Water Properties, Parts 1 and 2, *Chem. Eng.*, Sept. 1991, pp. 207, 208 and November 1991, pp. 235, 236. That author developed the equations via regression analysis of the properties of saturated steam as presented in *Perry's Chemical Engineers' Handbook*, 6th ed., McGraw-Hill. The maximum error (compared to tabulated values) and sample standard deviation of the correlations are as follows:

Property correlation	Maximum error, %	Sample standard deviation
Vapor pressure	3.64	1.84
Liquid density	0.26	0.10
Latent heat	0.63	0.19
Liquid viscosity	20.23	6.13
Liquid thermal conductivity	0.43	0.20
Vapor density	7.81	1.99
Vapor specific heat	0.49	0.13
Vapor viscosity	0.37	0.19
Vapor thermal conductivity	1.09	0.29
Boiling point	2.66	1.43

The original reference also includes correlations for liquid and vapor specific volume, liquid thermal expansion coefficient, liquid and vapor enthalpy, liquid specific heat, and liquid and vapor Prandtl numbers.

[†]Dickey is the author of Sec. 12 on liquid agitation in this present handbook.

*19.24 ATMOSPHERIC CONTROL SYSTEM INVESTMENT ANALYSIS FOR CHEMICAL PLANTS

An engineering atmospheric pollution-control system to protect the public against environmental pollution is under consideration by a chemical processing plant. The project will have an incremental operating cost of $100,000. If the pollution were uncontrolled, the damage to the public would have an estimated incremental cost of $125,000. Would this atmospheric control be a beneficial investment?

Calculation Procedure

1. Write the cost-benefit ratio for this investment.
The generalized dimensionless cost-benefit equation is $0 \leq C/B \leq 1$, where $C =$ incremental operating cost of the proposed atmospheric control, $, or other consistent monetary units; $B =$ benefit to the public of having the pollution controlled, $, or other consistent monetary units.

2. Compute the cost-benefit ratio for this situation.
Using the values given, $0 \leq \$100,000/\$125,000 \leq 1$. Or, $0 \leq 0.80 \leq 1$. This result means that 80¢ spent on environmental control will yield $1.00 in public benefits. Investing in the control would be a wise decision because a return greater than the cost of the control is obtained.

Related Calculations.   In the general cost-benefit equation. $0 \leq C/B \leq 1$, the upper limit of unity means that $1.00 spent on the incremental operating cost of the atmospheric control will deliver $1.00 in public benefits. A cost-benefit ratio of more than unity is uneconomic. Thus, $1.25 spent to obtain $1.00 in benefits would not, in general, be acceptable in a rational analysis. The decision would be to accept the environmental pollution until a satisfactory cost-benefit solution could be found.

A negative result in the generalized equation means that money invested to improve the environment actually degrades the condition. Hence, the environmental condition becomes worse. Therefore, the technology being applied cannot be justified on an economic basis.

In applying cost-benefit analyses, a number of assumptions of the benefits to the public may have to be made. Such assumptions, particularly when expressed in numeric form, can be open to change by others. Fortunately, by assigning a number of assumed values to one or more benefits, the cost-benefit ratios can easily be evaluated, especially when the analysis is done on a computer.

*19.25 ENVIRONMENTAL POLLUTION PROJECT SELECTION FOR CHEMICAL PLANTS

Five alternative projects for control of environmental pollution are under consideration. Each project is of equal time duration. The projects have the cost-benefit data shown in Table 19.16. Determine which project, if any, should be constructed.

Calculation Procedure

1. Evaluate the cost-benefit (C/B) ratios of the projects.
Setting up the C/B ratios for the five projects by the cost by the estimated benefit shows—in Table 19.16—that all C/B ratios are less than unity. Thus, each of the five projects passes the basic screening test of $0 \leq C/B \leq 1$. This being the case, the optimal project must be determined.

2. Analyze the projects in terms of incremental cost and benefit.
Alternative projects cannot be evaluated in relation to one another merely by comparing their C/B ratios, because these ratios apply to unequal bases. The proper approach to analyzing such a situation is:

TABLE 19.16 Project Costs and Benefits

Project	Equivalent uniform net annual benefits, $	Equivalent uniform net annual costs, $	C/B ratio
A	200,000	135,000	0.68
B	250,000	190,000	0.76
C	180,000	125,000	0.69
D	150,000	90,000	0.60
E	220,000	150,000	0.68

Each project corresponds to a specific *level* of cost. To be justified, *every* sum of money expended must generate at least an equal amount in benefits; the step from one level of benefits to the next should be undertaken only if the incremental benefits are at least equal to the incremental costs.

Rank the projects in ascending order of costs. Thus, Project D costs $90,000; Project C costs $125,000; and so on. Ranking the projects in ascending order of costs gives the sequence D-C-A-E-B.

Next, compute the incremental costs and benefits associated with each step from one level to the next. Thus, the incremental cost going from Project D to Project C is $125,000 − $90,000 = $35,000. And the benefit going from Project D to Project C is $180,000 − $150,000 = $30,000, using the data from Table 19.16. Summarize the incremental costs and benefits in a tabulation like that in Table 19.17. Then compute the C/B ratio for each situation and list it in Table 19.17. This computation shows that Project E is the best of these five projects because it has the lowest cost—75¢ per $1.00 of benefit. Hence, this project would be chosen for control of environmental pollution in this instance.

Related Calculations. There are some situations in which the minimum acceptable C/B ratio should be set at some value close to 1.00. For example, with reference to the above projects, assume that the government has a fixed sum of money that is to be divided between a project listed in Table 19.16 and some unrelated project. Assume that the latter has a C/B ratio of 0.91, irrespective of the sum expended. In this situation, the step from one level to a higher one is warranted only if the C/B ratio corresponding to this increment is at least 0.91.

Closely related to cost-benefit analysis and an outgrowth of it is *cost-effectiveness analysis*, which is used mainly in the evaluation of military and space programs. To apply this method of analysis, assume that some required task can be accomplished by alternative projects that differ in both cost and degree of performance. The effectiveness of each project is expressed in some standard unit, and the projects are then compared by a procedure analogous to that for cost-benefit analysis.

Note that cost-benefit analysis can be used in any comparison of environmental alternatives. Thus, cost-benefit analyses can be used for air-pollution controls, industrial thermal discharge studies, transportation alternatives, power-generation choices (windmills vs. fossil-fuel or nuclear plants), cogeneration, recycling waste for power generation, solar power, use of recycled sewer sludge as a fertilizer, and similar studies. The major objective in each comparison is to find the most desirable alternative based on the benefits derived from various options open to the designer.

TABLE 19.17 Cost-Benefit Comparison

Step	Incremental benefit, $	Incremental cost, $	C/B ratio	Conclusion
D to C	30,000	35,000	1.17	Unsatisfactory
D to A	50,000	45,000	0.90	Satisfactory
A to E	20,000	15,000	0.75	Satisfactory
E to B	30,000	40,000	1.33	Unsatisfactory

For example, electric utilities using steam generating stations burning coal or oil may release large amounts of carbon dioxide into the atmosphere. This carbon dioxide, produced when a fuel is burned, is thought to be causing a global greenhouse effect. To counteract this greenhouse effect, some electric utilities have purchased tropical rain forests to preserve the trees in the forest. These trees absorb carbon dioxide from the atmosphere, counteracting that released by the utility.

Other utilities pay lumber companies to fell trees more selectively. For example, in felling the 10 percent of marketable trees in a typical forest, as much as 40 to 50 percent of a forest may be destroyed. By felling trees more selectively, the destruction can be reduced to less than 20 percent of the forest. The remaining trees absorb atmospheric carbon dioxide, turning it into environmentally desirable wood. This conversion would not occur in these trees if they were felled in the usual foresting operation. The payment to the lumber company to do selective felling is considered a cost-benefit arrangement because the unfelled trees remove carbon dioxide from the air. The same is true of the tropical rain forests purchased by utilities and preserved to remove carbon dioxide which the owner-utility emits to the atmosphere.

Recently, a market has developed in the sale of "pollution rights" in which a utility that emits less carbon dioxide because it has installed pollution-control equipment can sell its "rights" to another utility that has less effective control equipment. The objective is to control, and reduce, the undesirable emissions by utilities.

With a potential "carbon tax" in the future, utilities and industrial plants that produce carbon dioxide as a by product of their operations are seeking cost-benefit solutions. The analyses given here will help in evaluating potential solutions.

*19.26 SAVING ENERGY LOSS FROM STORAGE TANKS AND VESSELS

Fuel oil at 12° API with a viscosity of 50 SSF (0.01068 mm^2/s) at 122°F (50°C) is stored at 300°F (148.9°C) in a 20-ft (6.1-m) diameter by 30-ft (9.1-m) high carbon-steel tank at atmospheric pressure. The oil level in the tank is 18 ft (5.4 m); the air temperature is 70°F (21.1°C). Determine the heat loss to the environment for two situations: (*a*) Total surface of the tank is uninsulated and black in color. Wind velocity is 0 mi/h (0 km/h). Surface emissivity of the tank is 0.9. Thermal conductivity of the ground under the tank is 0.8 Btu/(h·ft^2·°F·ft) [1.38 W/(m·K)]. (*b*) Roof of tank is uninsulated and is coated with aluminum paint. Sidewall is insulated with calcium silicate, or equivalent, and has a surface emissivity of 0.8. Wind velocity is 30 mi/h (48.0 km/h). The tank contents are not agitated. Thermal conductivity of the ground is 0.8 Btu/(h·ft^2·°F·ft) [1.38 W/(m·K)].

Calculation Procedure

1. Determine the heat loss from the wetted surface inside the tank.
For situation (*a*), the wetted area inside the tank is $A_L = \pi D H_L$, where A_L = wetted area, ft^2 (m^2); D = tank diameter, ft (m); H_L = liquid height, ft (m). For this tank, $A_L = \pi 20(18) = 1130.9$ ft^2 (105.1 m^2).

2. Find the temperature difference between the stored liquid and the atmospheric air.
The oil temperature T_i is 300°F (148.9°C), and the air temperature T_A is 70°F (21.1°C). Hence, $\Delta T_W = \Delta(T_i - T_A) = 300 - 70 = 230$°F (127.8°C).

3. Determine the heat loss from the tank.
Enter Fig. 19.26 at the bottom with the temperature difference of 230°F (127.8°C), project vertically upward to the unit heat loss curve q_T, and read $q_T = 664$ Btu/(h·ft^2) (2086.6 W/m^2). Then the total heat loss from the tank $Q_L = A_L q_T = 1130(664) = 750,320$ Btu/h (219,896.3 W).

4. Compute the heat loss from the dry inside surface and tank roof surface.
The area of the vessel $A_V = \pi D H_V + \pi D^2/4$, where the symbols are as above except that they apply to the dry surfaces of the tank. Then $A_V = \pi(20)(30 - 18) + \pi(20)^2/4 = 1068$ ft^2 (99.2 m^2), where A_V = area.

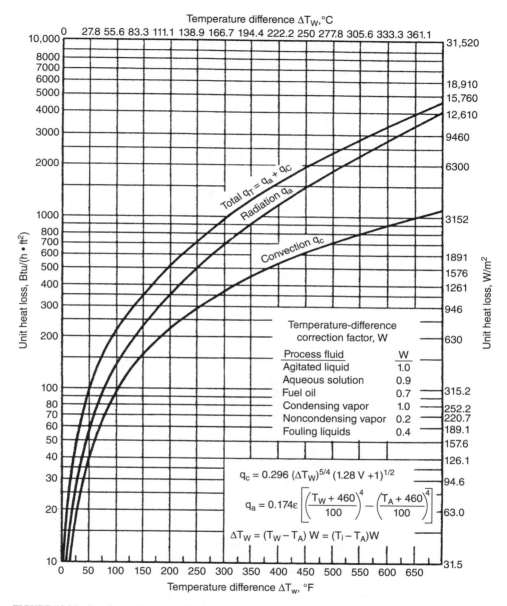

FIGURE 19.26 Heat losses from uninsulated tanks depend on nature of tank contents. Values in chart are for wind velocity of zero, surface emissivity of 0.9, and ambient air temperature of 70°F (21.1 °C). (*Chemical Engineering.*)

The temperature difference, by using the correction factor W from Fig. 19.26 for a noncondensing vapor and the temperature differences in step 2, is $\Delta T_W = (T_i - T_A) W$, or $\Delta T_W = (300 - 70) (0.2) = 46°F(7.7°C)$.

Next, the unit heat loss is found from Fig. 19.26 to be 84.3 Btu/(h·ft²) (265.9 W/m²) for a temperature difference of 46°F(7.7°C). Then the total heat loss from the dry inside surface in contact with the vapor is $Q_V = 1068(84.3) = 90,032.4$ Btu/h (26,385.8 W).

5. Compute the heat loss through the tank bottom to the ground.
The heat loss to the ground through the tank bottom, in Btu/h is $O_G = 2dk_G(T_L - T_G)$, where k_G = thermal conductivity of the ground, Btu/(h·ft²·°F·ft); T_L = liquid temperature, °F; T_G = ground temperature, °F; other symbols as before. Assuming that the ground temperature equals the air temperature, and with $k_G = 0.8$, we see $O_G = 2(20)(0.8)(300 - 70) = 7360$ Btu/h (2156.4 W).

6. Compute the total heat loss from the tank.
The total heat loss from the tank will be the sum of the losses from the liquid, vapor, and ground areas of the tank, or $Q_T = Q_L + Q_V + Q_G = 750,320 + 90,032 + 7360 = 847,712$ Btu/h (248,438.9 W).

7. Compare the results by using exact equations.
Figure 19.26 gives the exact algebraic equations for the total heat loss from uninsulated tanks. Substituting in these equations gives an exact total heat loss of 894,122 Btu/h (261,977 W). This is a difference of 4.7% from the approximate solution obtained by using Fig. 19.26. Most working engineers would be willing to accept such a difference in view of the savings in time and labor obtained by using the graphic solution.

8. Determine the insulation thickness needed for the tanks.
For situation (*b*), the wetted surface of the tank $A_L = 1130$ ft² (105.1 m²). This represents the interior circumferential area of the tank wetted by the fuel oil to a height of 18 ft (5.4 m).

Heat loss from the tank to the ambient air is a function of the temperature difference between the tank wall and the air, or $\Delta T_I = T_W - T_A = 300 - 70 = 230$°F (110°C).

From Fig. 19.27, find the recommended insulation thickness as 1.5 in (3.8 cm) and the wind correction factor for a 30-mi/h (48-km/h) wind and 1.5-in (3.8-cm) insulation as 1.10.

9. Correct the unit heat loss for wind velocity.
From step 4, $\Delta T_W = 46$; then $q_L = 46(1.10) = 50.6$ Btu/(h·ft²) (159.6 W/m²).

10. Compute the heat loss from liquid and vapor in the tank.
Heat loss from the liquid is $Q_L = A_L q_L = 1130(50.6) = 57,178$ Btu/h (16,757.2 W).

Now the heat loss from the insulated dry-side surface which contacts the vapor is computed from $A_V = \pi(20)(30 - 18) = 754$ ft² (70.0 m²). Then $Q_V = A_V q_V = 754(50.6) = 38,152$ Btu/h (11,181.2 W).

11. Determine the heat loss from the uninsulated roof.
The area of the roof is $A_R = \pi D^2/4 = \pi(20)^2/4 = 314$ ft² (28.3 m²). Correcting the temperature difference for a noncondensing vapor by using Fig. 19.26, we find $\Delta T_W = (T_i - T_A)W = 230(0.20) = 46$°F (7.77°C), where T_i = tank contents temperature, °F.

Using this corrected temperature difference and Fig. 19.26, we see the unit heat loss for radiation and convection can be found after the emissivity is determined. From the ASHRAE *Guide*, the surface emissivity is 0.40 for bright aluminum-painted surfaces at temperatures in the 50 to 100°F (10 to 37.7°C) range.

The unit heat loss from the roof for radiation and convection is found by using the corrected temperature difference and Fig. 19.26 and correcting the radiation loss for an emissivity of 0.4 and the convection loss for the wind velocity: $q_a = 54.2(0.4) = 21.7$ Btu/(h·ft²) (68.4 W/m²); $q_c = 35.5[(1.28)(30) + 1]^{0.5} = 222.8$ Btu/(h·ft²) (702.3 W/m²). Then $q_R = 21.7 + 222.8 = 244.5$ Btu/(h·ft²) (770.7 W/m²). Hence, for the roof, $Q_R = A_R q_R$, where A_R = roof area, ft². Or, $Q_R = 314(244.5) = 76,773$ Btu/h (22,499.9 W).

12. Determine the total heat loss from the tank.
The total heat loss from the tank is the sum of the component heat losses. Or, since the heat loss to the ground is the same as in situation (*a*), $Q_T = 57,543 + 38,529 + 76,773 + 7360 = 180,405$ Btu/h (52,871.3 W).

Using the algebraic equations gives $Q_T = 147,945$ Btu/h (43,358.2 W). This is a difference greater than in the first situation, but the time savings accrued from using the approximations are significant.

Related Calculations. This procedure can be used for a variety of insulated and uninsulated tanks and vessels used to store oil, chemicals, food, water, and similar liquids in almost any industry.

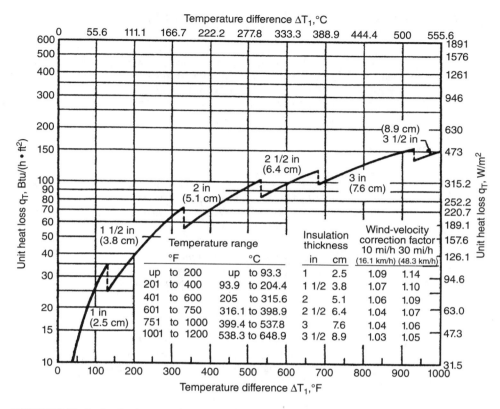

FIGURE 19.27 Insulated tanks, covered with calcium silicate, have heat losses based on negligible resistance to heat flow on process side. Values in chart are for wind velocity of zero, emissivity of 0.8, ambient air temperature of 70°F (21.1°C). (*Chemical Engineering.*)

TABLE 19.18 Air-Pollution Control Criteria

Substance	Maximum ground concentration, ppm[†]	Lower explosive limit, ppm	Odor threshold, ppm
Acetylene	—	2.5	—
Ammonia	100	15.5	53
Amylene	—	1.7	2.3
Benzene	50	1.4	1.5
Butane	—	1.9	5000
Carbon monoxide	100	12.5	Odorless
Ethylene	—	2.8	—
Hydrogen sulfide	30	4.3	0.1
Methanol	200	6.7	410
Propane	—	2.1	20,000
Sulfur dioxide	10	—	3.0

[†]8-h exposure.

Table 19.18 lists the typical conditions encountered with such tanks and vessels, the factors which can be neglected in the insulation calculations, and the exact procedure to follow for both graphical and algebraic methods. The methods given here are the work of Richard Hughes and Victor Deumango of the Badger Company and reported in *Chemical Engineering* magazine.

*19.27   ENERGY SAVINGS FROM VAPOR RECOMPRESSION

Determine the energy savings possible in a chemical plant where 15-lb/in^2 (gage) (103.4-kPa) steam is vented to the atmosphere while 5000 lb/h (2250 kg/h) of 40-lb/in^2 (gage) (275.8-kPa) steam is used from the boiler in another process, if the vented steam is recompressed in an electrically driven compressor to the 40-lb/in^2 (gage) (275.8-kPa) level. The boiler feedwater temperature is 80°F (26.7°C), the boiler efficiency is 80 percent, the boiler operates 8000 h/year, the cost of 150,000-Btu/gal no. 6 fuel oil is $1.00/gal, and electricity costs $0.03/kWh.

Calculation Procedure

1. Compute the annual heat input to the boiler.
The annual heat input to the boiler $H = W(\Delta h)T/e$, where H = heat input, Btu/year (W); W = weight of steam used, lb/h (kg/h); Δh = enthalpy change in the boiler = enthalpy of steam − enthalpy of the feedwater, both expressed in Btu/lb (J/kg); T = annual operating time of the boiler, h; e = boiler efficiency, expressed as a percentage. Substituting, we find $H = 5000(1176 − 48)(8000)/0.80 = 5.6 \times 10^{10}$ Btu/year (5.91×10^{10} J/year).

2. Find the annual fuel cost for generating the steam in the boiler.
The annual fuel cost $C = HP/h_v$, where C = annual fuel cost; P = price per gallon of fuel oil; h_v = fuel heating value, Btu/gal; other symbols as before. Substituting gives $C = (5.6 \times 10^{10})(\$ 1.00)/150,000 = \$373.333$ per year.

3. Determine the recompression energy input.
When 15-lb/in^2 (gage) (103.4-kPa) waste steam is compressed to 40 lb/in^2 (gage) (275.8 kPa) by an electrically driven compressor, the compressor ratio is $c_r = P_d$ = discharge pressure, lb/in^2 (abs) (kPa)/p_i = inlet pressure, lb/in^2 (abs) (kPa). Substituting gives $c_r = (40 + 14.7)/(15 + 14.7) = 1.84$. To find the recompression energy input, enter Fig. 19.28 at the computed compression ratio and project vertically to the inlet pressure curve. At the left read the energy input as 66 Btu/lb (153.5 kJ/kg) of steam recompressed.

4. Compute the energy cost of recompression.
The energy cost of recompression = (Btu/lb for recompression) (WT) (0.000293 kWh/Btu) ($0.03 per kWh) = $23,000 per year.

5. Find the annual energy cost saving from recompression.
The annual energy cost saving for this installation would be $373,333 − $23,200 = $350,133.

Related Calculations.   It is common practice in many industrial plants to vent any steam at pressures below 20 lb/in^2 (gage) (137.9 kPa) to the atmosphere. At the same time, there may be several users that require somewhat higher-pressure steam, 30 to 50 lb/in^2 (gage) (206.9 to 344.6 kPa). Rather than reducing high-pressure boiler steam to supply these needs, it is possible to compress the waster low-pressure vapor to a higher pressure so that it can be reused. Although energy must be supplied to the compressor to raise the steam pressure, this operation typically requires only 5 to 10 percent of the energy necessary to generate the same steam in a boiler. In practice, this principle is limited to situations where the compressor inlet pressure is above 14.7 lb/in^2 (abs) (101.4 kPa) and the compression ratio is less than about 2.0, owing to physical limitations of the compressors used.

Vapor recompression has been used for years as a means of lowering the steam requirements of evaporators. In this application, the overhead vapors are compressed and recycled to an evaporator steam chest where they evaporate more liquid. In this way, a single-effect evaporator can achieve a steam economy equivalent to an evaporator with up to 15 effects.

Figure 19.28 indicates the energy required to compress steam as a function of the compression ratio and inlet pressure. It is based on adiabatic compression with a compressor efficiency of 75 percent. Since the steam leaving the compressor is superheated, it is also assumed that water at 80°F (26.7°C) is sprayed into the steam to eliminate the superhead. The graph can be used in conjunction with standard steam tables to estimate the energy saving possible from employing vapor recompression.

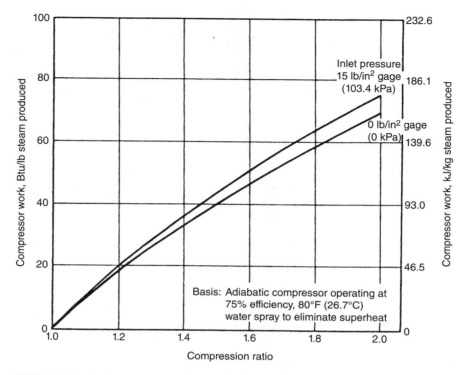

FIGURE 19.28   Recompression work input. (*Chemical Engineering.*)

The procedure given here can be used for any application—industrial, commercial, residential, marine, etc.—in which recompression of steam might prove economical. This procedure is the work of George Whittlesey and John D. Muzzy, School of Chemical Engineering, Georgia Institute of Technology, as reported in *Chemical Engineering* magazine.

*19.28 SAVINGS POSSIBLE FROM USING LOW-GRADE WASTE HEAT FOR REFRIGERATION

An industrial plant presently exhausts low-pressure steam to the atmosphere. What would the annual savings be if a mechanical chiller having a coefficient of performance (COP) of 4.0 producing an average of 150 ton/year (527.4 kW) of refrigeration for 4000 h were replaced by an absorption refrigeration unit using the exhaust steam? The cost of electricity is 3.0 cents per kilowatthour.

Calculation Procedure

1. Sketch the refrigeration system being considered.

Figure 19.29 shows the absorption refrigeration unit being considered. The heat input from the exhaust steam is indicated as Q_C.

In an absorption refrigeration unit, the mechanical vapor compressor is replaced by the generator which uses steam or hot water to revaporize the refrigerant. But absorption refrigeration units are characterized by a low COP, compared to the mechanical type. Hence, to be competitive, an

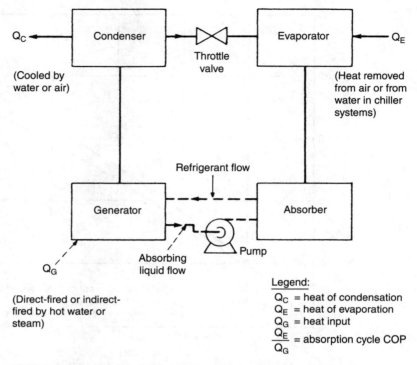

FIGURE 19.29   Absorption refrigeration system. (*Chemical Engineering.*)

absorption unit must use low-grade waste heat to power the generator. The low-pressure steam available in this plant would be ideal for this purpose.

2. Determine the hourly savings possible.
Use Fig. 19.30 to find the hourly savings. Enter at the refrigeration load of 150 ton (527.4 kW) on the left, and project vertically to the COP value of 4. From the intersection with this curve, project horizontally to the right to intersect the electricity cost curve of 3 cents per kilowatthour. At the bottom read the saving as $4 per hour.

3. Compute the annual savings.
Use this relation: annual savings, $ = (hourly savings)(annual number of operating hours) = ($4)(4000) = $16,000 per year.

Related Calculations.   This procedure can be used for an absorption refrigeration system by using waste heat in the generator. The heat can be in the form of exhaust steam, hot waste liquids, warm air, etc. The source of the heat is not important provided (1) the temperature of the heating medium is high enough for use in the generator, (2) the supply of heat is steady, and (3) the heat is not chargeable to the refrigeration process.

Given the above criteria as guidelines, the procedure given here can be used for absorption refrigeration machines in industrial plants, commercial building, ships, and domestic applications. Where the COP of an equivalent mechanical refrigeration system is not known, it can be approximated by applying data from known similar installations. Where the electricity cost exceeds the plotted values, use half the actual cost and multiply the result by 2.

This procedure is the work of Guillermo H. Hoyos, Universidad de los Andes, and John D. Muzzy, School of Chemical Engineering, Georgia Institute of Technology, as reported in *Chemical Engineering* magazine.

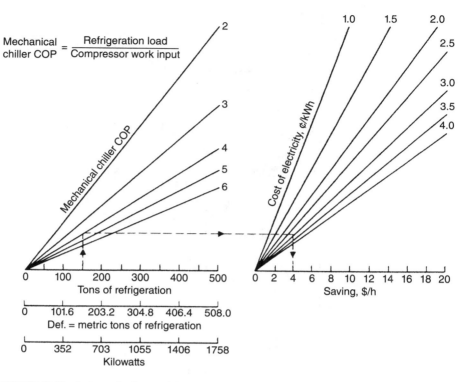

FIGURE 19.30   Savings using low-grade heat (*Chemical Engineering.*)

*19.29   ENERGY DESIGN ANALYSIS OF SHELL-AND-TUBE HEAT EXCHANGERS FOR CHEMICAL PLANTS

Determine the heat transferred, shellside outlet temperature, surface urea, maximum number of tubes, and tubeside pressure drop for a liquid-to-liquid shell-and-tube heat exchanger such as that in Fig. 19.31, when the conditions below prevail. This exchanger will be of the single tube-pass and single shell-pass design Table 19.19, with countercurrent flows of the tubeside and shellside fluids.

Conditions	Tubeside	Shellside			
Flow rate, lb/h	307,500	32,800	kg/h	139,605	14,891
Inlet temperature, °C	105	45	°F	221	45
Outlet temperature, °C	unknown	90			
Viscosity, cp	1.7	0.3			
Specific heat, Btu/h/°F	0.72	0.9	kJ/h°C	3.0	3.7
Molecular weight	118	62			
Specific gravity with reference to water at 20°C (68°F)	0.85	0.95			
Allowable pressure drop, psi	10	10	kPa	68.9	68.9
Maximum tube length, ft	12	12	m	5.45	5.45
Minimum, tube dia., in	5/8	5/8	mm	15.9	15.9
Material of construction	steel	(k = 26)			

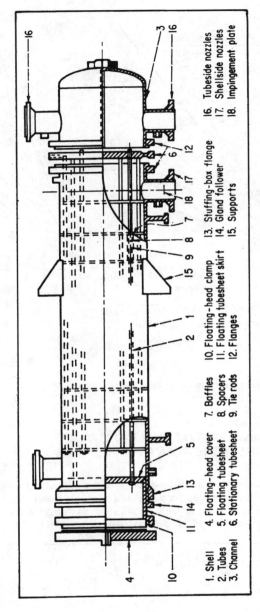

FIGURE 19.31 Components of shell-and-tube heat exchanger. This unit an outside-packed stuffing box. (*Chemical Engineering*.)

1. Shell
2. Tubes
3. Channel
4. Floating-head cover
5. Floating tubesheet
6. Stationary tubesheet
7. Baffles
8. Spacers
9. Tie rods
10. Floating-head clamp
11. Floating tubesheet skirt
12. Flanges
13. Stuffing-box flange
14. Gland follower
15. Supports
16. Tubeside nozzles
17. Shellside nozzles
18. Impingement plate

TABLE 19.19 Design Features of Shell-and-Tube Heat Exchangers

Design features	Fixed tubesheet	Return bend (U-tube)	Outside-packed stuffing box	Outside-packed lantern ring	Pull-through bundle	Inside split-backing rir
Is tube bundle removable?	No	Yes	Yes	Yes	Yes	Yes
Can spare bundles be used?	No	Yes	Yes	Yes	Yes	Yes
How is differential thermal expansion relieved?	Expansion joint in shell	Individual tubes free to expand	Floating head	Floating head	Floating head	Floating head
Can individual tubes be replaced?	Yes	Only those in outside rows without special designs	Yes	Yes	Yes	Yes
Can tubes be chemically cleaned, both inside and outside?	Yes	Yes	Yes	Yes	Yes	Yes
Can tubes be physically cleaned on inside?	Yes	With special tools	Yes	Yes	Yes	Yes
Can tubes be physically cleaned on outside?	No	With square or wide triangular pitch	With square or wide triangular pitch	With square or wide triangular pitch	With square or wide triangular pitch	With square or wide triangular pitch
Are internal gaskets and bolting required?	No	No	No	No	Yes	Yes
Are double tubesheets practical?	Yes	Yes	Yes	No	No	No
What number of tubeside passes are available?	Number limited by number of tubes	Number limited by number of U-tubes	Number limited by number of tubes	One or two	Number limited by number of tubes Odd number of passes require packed joint or expansion joint	Number limited by number of tubes Odd number of passes require packed joint or expansion joint
Relative cost in ascending order, least expensive = 1	2	1	4	3	5	6

Calculation Procedure

1. Determine the heat transferred in the heat exchanger.
Use the relation, heat transferred, Btu/h = (flow rate, lb/h)(outlet temperature − inlet temperature) (liquid specific heat)(1.8 to convert from °C to °F). Substituting for this heat exchanger, we have, heat transferred = (32,800)(90 − 45)(0.9)(1.8) = 2,391,120 Btu/h (2522.6 kJ/h). This is the rate of heat transfer from the hot fluid to the cool fluid.

2. Find the shellside outlet temperature.
The temperature decrease of the hot fluid = (rate of heat transfer)/(flow rate, lb/h)(specific heat) (1.8 conversion factor). Or, temperature decrease = (2,391,120)/(307,500)(0.72)(1.8) = 6°C (10.8°F). Then, the shellside outlet temperature = 105 − 6 = 99°C (210.2°F). Then, the LMTD = (54 − 15)/ln (54/15) = 30.4°C (86.7°F) = ΔT_m.

3. Make a first-trial calculation of the surface area of this exchanger.
For a first-trial calculation, the approximate surface can be calculated using an assumed overall heat-transfer-coefficient, U, of 250 Btu/(h) (ft^2) (°F) (44.1 W/m^2°C). The assumed value of U can be obtained from tabulations in texts and handbooks and is used only to estimate the approximate size for a first trial:

$$A = 2,391,000/(250 \times 30.4 \times 1.8) = 175 \text{ ft}^2 (16.3 \text{ m}^2)$$

Since the given conditions specify a maximum tube length of 12 ft and a minimum tube diameter of 5/8 in, the number of tubes required is:

$$n = 175(12 \times 0.1636) = 89 \text{ tubes}$$

and approximate shell diameter will be:

$$D_a = 1.75 \times 0.625 \times 89^{0.47} = 9 \text{ in (228.6 cm)}$$

With the exception of baffle spacing, all preliminary calculations have been made for the quantities to be substituted into the dimensional equations. For the first trial, we may start with a baffle spacing equal to about half the shell diameter. After calculating the shellside pressure drop, we may adjust the baffle spacing. Also, it is advisable to check the Reynolds number on the tubeside, Table 19.20, to confirm that the proper equations are being used.

4. Find the maximum number of tubes for this heat exchanger.
To find the maximum number of tubes (n_{max}) in parallel that still permits flow in the turbulent region ($N_{Re} = 12,600$), a convenient relationship is $n_{max} = W_i/(2d_i Z_i)$. In this example, $n_{max} = 307.5/(2 \times 0.495 \times 1.7) = 183$. For any number of tubes less than 183 tubes in parallel, we are in the turbulent range and can use Eq. (1) of Table 19.20.

From Table 19.20, the appropriate expressions for rating are: Eq. (1) for tubeside, Eq. (11) for shellside, Eq. (18) for tube wall, and Eq. (19) for fouling. Eqs. (21) and (25) from Table 19.21, respectively, are used for tubeside and shellside pressure drops.

5. Compute the tubeside and shellside heat transfer.
Using the equations from Table 19.20, Tubeside, Eq. (1):

$$\frac{\Delta T_i}{\Delta T_M} = 10.43 \left[\frac{1.7^{0.467} \times 118^{0.222}}{0.85^{0.89}} \right] \times \left[\frac{307.5^{0.2} \times 6}{30.4} \right] \left[\frac{0.495^{0.8}}{89^{0.2} \times 12} \right]$$

$$= 10.43 \times 4.27 \times 0.621 \times 0.0193 = 0.535$$

Shellside, Eq. (11):

$$\frac{\Delta T_o}{\Delta T_M} = 4.28 \left[\frac{0.3^{0.267} \times 62^{0.222}}{0.95^{0.89}} \right] \times \left[\frac{32.8^{0.4} \times 45}{30.4} \right] \left[\frac{1^{0.282} \times 5^{0.6}}{89^{0.718} \times 12} \right]$$

$$= 4.28 \times 1.89 \times 5.98 \times 0.00872 = 0.424$$

Tube Wall, Eq. (18):

$$\frac{\Delta T_w}{\Delta T_M} = 159 \left[\frac{0.72}{26} \right] \times \left[\frac{307.5 \times 6}{30.4} \right] \left[\frac{0.625 - 0.495}{89 \times 0.625 \times 12} \right]$$

$$= 159 \times 0.0277 \times 60.7 \times 0.000195 = 0.052$$

Fouling, Eq. (19):

$$\frac{\Delta T_s}{\Delta T_M} = 3820 \left[\frac{0.72}{1000} \right] \left[\frac{307.5 \times 6}{30.4} \right] \left[\frac{1}{89 \times 0.625 \times 12} \right]$$

$$= 3820 \times 0.00072 \times 60.7 \times 0.00150 = 0.250$$

$$(SOP)^{\dagger} = 0.535 + 0.424 + 0.052 + 0.250 = 1.261$$

Because SOP is greater than 1, the assumed exchanger is inadequate. The surface area must be increased by adding tubes or increasing the tube length, or the performance must be improved by decreasing the baffle spacing. Since the maximum tube length is fixed by the conditions given, the alternatives are increasing the number of tubes and/or adjusting the baffle spacing. To estimate assumptions for the next trial, pressure drops are calculated.

6. Make the pressure-drop calculation for the heat exchanger.
Tubeside, Eq. (21) of Table 19.21:

$$\Delta P = (17^{0.2}/0.85)(307.5/89)^{1.8}[(12/0.495) + 25]/(5.4 \times 0.495)^{3.8}$$

$$= 14.3 \text{ lb/in}^2 (98.5 \text{ kPa})$$

Shellside, Eq. (25):

$$\Delta P = (0.326/0.95)(32.8^2)[12/(5^3 \times 9)] = 3.9 \text{ lb/in}^2 (26.9 \text{ kPa})$$

To decrease the pressure drop on the tubeside to the acceptable limit of 10 lb/in^2 (68.9 kPa), the number of tubes must be increased. This will also decrease the SOP. In addition, shellside performance can be improved by decreasing the baffle spacing, since the pressure drop of 3.9 on the shellside is lower than the allowable 10 psi (68.9 kPa). Before proceeding with successive trials to balance the heat-transfer and pressure-drop restrictions, Table 19.22 is now set up for clarity.

7. Perform the second-trial computation for heat-transfer surface and pressure drop.
As a first step in adjusting the heat-transfer surface and pressure drop, calculate the number of tubes to give a pressure drop of 10 lb/in^2 (68.9 kPa) on the tubeside. The pressure drop varies inversely as $n^{1.8}$. Therefore, $14.3/10 = (n/89)^{1.8}$, and $n = 109$.

Each individual product of the factors is then adjusted in accordance with the applicable exponential function of the number of tubes. Since the tubeside product is inversely proportional to the 0.2 power of the number of tubes, the product from the preceding trial is multiplied by $(n_1/n_2)^{0.2}$, where n_1 is the number of tubes used in the preceding trial, and n_2 is the number to be used in the new one.

---

†Sum of the Product—see *Related Calculations* for data.

TABLE 19.20 Empirical Heat-Transfer Relationships for Rating Heat Exchanger

Eq. no	Mechanism of restriction	Empirical equation	Numerical factor	Physical-property factor	Work factor	Mechanical-design factor
	Inside the tubes					
(1)	No phase change (liquid), $N_{Re} > 10{,}000$	$\dfrac{h}{cG} = 0.023(N_{Re})^{-0.2}(N_{Pr})^{-2/3}$	$\Delta T_i/\Delta T_M = 10.43$	$\times \dfrac{(Z_i^{0.467} M_i^{0.22})}{s_i^{0.89}}$	$\times \dfrac{W_i^{0.2}(t_H - t_L)}{\Delta T_M}$	$\times \dfrac{d_i^{0.8}}{n^{0.2}L}$
(2)	No phase change (gas), $N_{Re} > 10{,}000$	$h = 0.0144\, G^{0.8}(D_i)^{-0.2} c_p$	$\Delta T_i/\Delta T_M = 9.87$		$\times \dfrac{W_i^{0.2}(t_H - t_L)}{\Delta T_M}$	$\times \dfrac{d_i^{0.8}}{n^{0.2}L}$
(3)	No phase change (gas), $2100 < N_{Re} < 10{,}000$	$h = 0.0059[(N_{Re})^{2/3} - 125]$ $[1 + (D/L)^{2/3}](c_p/D_i)(\mu_f/\mu_b)^{-0.14}$	$\Delta T_i/\Delta T_M = 44{,}700 \times (Z_f/Z_b)^{0.14}$		$\times \dfrac{W_i(t_H - t_L)}{\Delta T_M}$	$\times \dfrac{1}{[(N_{Re})^{2/3} - 125][1 + (d_i N_{PT}/12L)^{2/3}]nL}$
(4)	No phase change (liquid), $2100 < N_{Re} < 10{,}000$	$\dfrac{h}{cG} = 0.166\left[\dfrac{(N_{Re})^{2/3} - 125}{N_{Re}}\right]$ $[1 + (D/L)^{2/3}](N_{Pr})^{-2/3}(\mu_f/\mu_b)^{-0.14}$	$\Delta T_i/\Delta T_M = 2260$	$\times \left(\dfrac{M_i^{0.22}}{s_i^{0.89}Z_b^{1/3}}\right)\left(\dfrac{Z_f}{Z_b}\right)^{0.14}$	$\times \dfrac{W_i(t_H - t_L)}{\Delta T_M}$	$\times \dfrac{1}{[(N_{Re})^{2/3} - 125][1 + (d_i N_{PT}/12L)^{2/3}]nL}$
(5)	No phase change (liquid), $N_{Re} < 2100$	$\dfrac{h}{cG} = 1.86(N_{Re})^{-2/3}(N_{Pr})^{-2/3}$ $(L/D_i)^{-1/3}(\mu_f/\mu_b)^{-0.14}$	$\Delta T_i/\Delta T_M = 17.5$	$\times \left(\dfrac{M_i^{0.22}}{s_i^{0.89}}\right)\left(\dfrac{Z_f}{Z_b}\right)^{0.14}$	$\times \dfrac{W_i^{2/3}(t_H - t_L)}{\Delta T_M}$	$\times \dfrac{1}{n^{2/3}L^{2/3}(N_{PT})^{1/3}}$
(6)	Condensing vapor, vertical, $N_{Re} < 2100$	$h = 0.925\, k\, (g\rho_i^2/\mu\Gamma)^{1/3}$	$\Delta T_i/\Delta T_M = 4.75$	$\times \dfrac{(Z_i\lambda_i)^{0.333}}{s_i^2 c_i}$	$\times \dfrac{W_i^{4/3}\lambda_i}{\Delta T_M}$	$\times \dfrac{1}{n^{4/3}d_i^{4/3}L}$
(7)	Condensing vapor, horizontal, $N_{Re} < 2100$	$h = 0.76\, k\, (g\rho_i^2/\mu\Gamma)^{1/3}$	$\Delta T_i/\Delta T_M = 2.92$	$\times \dfrac{(Z_i M_i)^{0.333}}{s_i^2 c_i}$	$\times \dfrac{W_i^{4/3}\lambda_i}{\Delta T_M}$	$\times \dfrac{1}{n^{4/3}d_i L^{4/3}}$ (See Note 1)
(8)	Condensate subcooling, vertical	$h = 1.225(k/B)(cB\Gamma/kL_B)^{5/6}$	$\Delta T_i/\Delta T_M = 1.22$	$\times \left[\dfrac{(Z_i M_i)^{0.333}}{s_i^2}\right]^{-1/6}$	$\times \dfrac{W_i^{0.222}(t_H - t_L)}{\Delta T_M}$	$\times \dfrac{1}{(n^{4/3}d_i^{4/3}L)^{1/6}}$
(9)	Nucleate boiling, vertical	$\dfrac{h}{cG} = 4.02(N_{Re})^{-0.3}$ $(N_{Pr})^{-0.6}(\rho_L\sigma/P^2)^{-0.425}\Sigma$	$\Delta T_i/\Delta T_M = 0.352$	$\times \left(\dfrac{Z_i^{0.3} M_i^{0.2} \sigma_i^{0.425}}{s_i^{1.075} c_i}\right)\left(\dfrac{\rho_v^{0.7}}{P_i^{0.85}}\right)$	$\times \dfrac{W_i^{0.3}\gamma_i}{\Delta T_M}$	$\times \left(\dfrac{1}{n^{0.3}L^{0.3}}\right)\Sigma'$ (See notes 2 and 5)

Outside the tubes

	Item	h equation	ΔT equation
(10)	Nucleate boiling, horizontal	$\dfrac{h}{cG} = 4.02(N_{Re})^{-0.3}(N_{Pr})^{0.6}(p_i;\sigma/P^2)^{-0.425}\Sigma$	$\Delta T_o/\Delta T_M = 0.352 \times \left(\dfrac{Z_o^{0.3}M_o^{0.2}\sigma_o^{0.425}}{s_o^{1.075}c_o}\right)\left(\dfrac{\rho_v^{0.7}}{P_o^{0.85}}\right) \times \dfrac{W_o^{0.3}\lambda_o}{\Delta T_M} \times \left(\dfrac{1}{n^{0.3}L^{0.3}}\right)\Sigma$
(11)	No phase change (liquid), crossflow	$\dfrac{h}{cG} = 0.33(N_{Re})^{-0.4}(N_{Pr})^{-2/3}(0.6)$	$\Delta T_o/\Delta T_M = 4.28 \times \dfrac{Z_o^{0.267}M_o^{0.222}}{s_o^{0.89}} \times \dfrac{W_o^{0.4}(T_H - T_L)}{\Delta T_M} \times \dfrac{N_{PT}^{0.282}P_B^{0.6}}{n^{0.718}L}$ (See Note 3)
(12)	No phase change (gas), crossflow	$h = 0.11 G^{0.6}D^{-0.4}c_p^{(0.6)}$	$\Delta T_o/\Delta T_M = 7.53 \times \dfrac{W_o^{0.4}(T_H - T_L)}{\Delta T_M} \times \dfrac{N_{PT}^{0.282}P_B^{0.6}}{n^{0.718}L}$ (See Note 4)
(13)	No phase change (gas), parallel flow	$h = 0.0144 G^{0.8}D^{-0.2}c_p^{(1.3)}$	$\Delta T_o/\Delta T_M = 21.7 \times \dfrac{W_o^{0.2}(T_H - T_L)}{\Delta T_M} \times \dfrac{d_o^{0.8}N_{PT}^{0.685}}{n^{0.315}L}$
(14)	No phase change (liquid), parallel flow	$\dfrac{h}{cG} = 0.023(N_{Re})^{-0.2}(N_{Pr})^{-2/3}(1.3)$	$\Delta T_o/\Delta T_M = 22.9 \times \dfrac{Z_o^{0.467}M_o^{0.22}}{s_o^{0.89}} \times \dfrac{W_o^{0.2}(T_H - T_L)}{\Delta T_M} \times \dfrac{d_o^{0.8}N_{PT}^{0.685}}{n^{0.315}L}$
(15)	Condensing vapor, vertical, $N_{Re} < 2100$	$h = 0.925k(g\rho_L^2/\mu\Gamma)^{1/3}$	$\Delta T_o/\Delta T_M = 4.75 \times \dfrac{(Z_o M_o)^{0.333}}{s_o^2 c_o} \times \dfrac{W_o^{4/3}\lambda_o}{\Delta T_M} \times \dfrac{1}{n^{4/3}d_o^{4/3}N_{PT}^{1/3}L}$
(16)	Condensing vapor, horizontal, $N_{Re} < 2100$	$h = 0.76k(g\rho_L^2/\mu\Gamma)^{1/3}$	$\Delta T_o/\Delta T_M = 2.64 \times \dfrac{(Z_o M_o)^{0.333}}{s_o^2 c_o} \times \dfrac{W_o^{4/3}\lambda_o}{\Delta T_M} \times \dfrac{N_{PT}^{0.177}}{n^{1.156}L^{4/3}d_o}$

Tube wall

	Item	h equation	ΔT equation
(17)	Tube wall (sensible-heat transfer)	$h = (24 k_w)/(d_o - d_i)$	$\Delta T_w/\Delta T_M = 159 \times c/k_w \times \dfrac{W(t_H - t_l)}{\Delta T_M} \times \dfrac{d_o - d_i}{nd_oL}$
(18)	Tube wall (latent-heat transfer)	$h = (24 k_w)/(d_o - d_i)$	$\Delta T_M/\Delta T_M = 88 \times 1/k_w \times \dfrac{W\lambda}{\Delta T_M} \times \dfrac{d_o - d_i}{nd_oL}$

Fouling

	Item	h equation	ΔT equation
(19)	Fouling (sensible-heat transfer)	$h = assumed$	$\Delta T_s/\Delta T_M = 3820 \times c/h \times \dfrac{W(t_H - t_l)}{\Delta T_M} \times \dfrac{1}{nd_oL}$
(20)	Fouling (latent-heat transfer)	$h = assumed$	$\Delta T_s/\Delta T_M = 2120 \times 1/h \times \dfrac{W\lambda}{\Delta T_M} \times \dfrac{1}{nd_oL}$

Notes:
1. If $W_l/(ns_l d_i^{2.56}) > 0.3$, multiply $\Delta T_l/\Delta T_M$ by 1.3.
2. Surface-condition factor (Σ') for copper and steel = 1.0; for stainless steel = 1.7; for polished surfaces = 2.5.
3. For square pitch numerical factor = 5.42.
4. For square pitch, numerical factor = 9.53.
5. $G = W_o/(A\rho_v)$.

TABLE 19.21 Empirical Pressure-Drop Relationship for Rating Heat Exchangers

Eq. no.	Mechanism of restriction	Empirical equation
	Inside the tubes	
(21)	No phase change, $N_{Re} > 10{,}000$	$\Delta P = \dfrac{(Z_i)^{0.2}}{s_i}\left(\dfrac{W_i}{n}\right)^{1.8}\dfrac{N_{PT}[(L_o/d_i)+25]}{(5.4d_i)^{3.8}}$ (See note 1)
(22)	No phase change, $2100 < N_{Re} < 10{,}000$	$\Delta P = \left(\dfrac{Z_i}{s_i}\right)\left(\dfrac{W_i}{n}\right)\dfrac{N_{PT}[(L_o/d_i)+25][(N_{Re})^{2/3}-25]}{(50.2d_i)^3}$ (See note 1)
(23)	No phase change, $N_{Re} < 2100$	$\Delta P = \dfrac{(Z_b)^{0.326}(Z_f)^{0.14}}{s_i}\left(\dfrac{W_i}{n}\right)^{4/3}\dfrac{N_{PT}(L_a)^{2/3}}{(5.62d_i)^4}$
(24)	Condensing	$\Delta P = \dfrac{(Z_i)^{0.2}}{s_i}\left(\dfrac{W_i}{n}\right)^{1.8}\dfrac{N_{PT}[(L_o/d_i)+25]}{(5.4d_i)^{2.8}}\times 0.5$ (See note 1)
	Shellside	
(25)	No phase change, crossflow	$\Delta P = \dfrac{0.326}{s_o}(W_a)^2\,\dfrac{L_o}{P_B^3 D_o}$
(26)	No phase change, parallel flow	$\Delta P = \dfrac{(Z_o)^{0.2}}{s_o}\left(\dfrac{W_o}{n}\right)^{1.8}\left[\dfrac{n^{0.366}L_o}{(N_{PT})^{1.434}(4.912\,d_o)^{4.8}}+\dfrac{0.31n^{0.0414}(W_o)^{0.2}L_o}{d_o(N_{PT})^{1.70}(4.912\,d_s)^4 Z^{0.2}B_o^2}\right]$ (See notes 2 and 3)
(27)	Condensing	$\Delta P = \left(\dfrac{0.081}{S_o}\right)(W_o)^2\left(\dfrac{L_o}{P_B^3 D_o}\right)$

Notes:
1. For U-bends, use $[(L_o/d_i)+16]$ instead of $[(L_o/d_i)+25]$.
2. B_o is equal to fraction of flow area through baffle.
3. Number of baffles $(N_B) = 0.48\,(L_o/d_o)$.

The shellside product of the preceding trial is multiplied by $(n_1/n_2)^{0.718}$, and the tube-wall and fouling products by n_1/n_2. New adjusted products are then calculated as follows:

$$\text{Tubeside product} = (89/109)^{0.2} \times 0.535 = 0.514$$
$$\text{Shellside product} = (89/109)^{0.718} \times 0.424 = 0.367$$
$$\text{Tube-wall product} = (89/109) \times 0.052 = 0.042$$
$$\text{Fouling product} = (89/109) \times 0.250 = 0.204$$
$$\text{SOP} = 1.127$$

8. Make the last trial calculation.

For the third trial, baffle spacing is decreased to 3.5 in (88.9 cm) from 5 in (127.0 cm). Only the shellside product must be adjusted since only it is affected by the baffle spacing. Therefore, the shell-side factor of the previous trial is multiplied by the ratio of the baffle spacing to the 0.6 power:

$$(3.5/5.0)^{0.6} \times 0.367 = 0.296$$

The sum of the products (SOP) for this trial is 1.056. The shellside pressure drop $\Delta P_a = 3.9 \times (5.0/3.5)^3 = 11.4$ lb/in^2 (78.5 kPa).

Because we have now reached the point where the assumed design nearly satisfies our conditions, tube-layout tables can be used to find a standard shell size containing the next increment above 109 tubes. A 10-in-dia (254 cm) shell in a fixed-tubesheet design contains 110 tubes.

Again, correcting the products of the heat-transfer factors from the previous trial:

$$\text{Tubeside product} = (109/110)^{0.2} \times 0.514 = 0.513$$
$$\text{Shellside product} = (109/110)^{0.718} \times 0.296 = 0.293$$
$$\text{Tube-wall product} = (109/110) \times 0.042 = 0.042$$
$$\text{Fouling product} = (109/110) \times 0.204 = 0.202$$
$$\text{SOP} = 1.050$$

Tubeside pressure drop:

$$\Delta P_i = 10 \times (109/110)^{1.8} = 9.8 \text{ lb/in}^2 (67.5 \text{ kPa})$$

The shellside pressure drop is now corrected for the actual shell diameter of 10 in (25.4 cm) instead of 9 in (228.6 cm).

$$\Delta P_o = 11.4 \times (9/10) = 10.1 \text{ psi} (69.6 \text{ kPa})$$

TABLE 19.22 Results of Trial Calculations

	1st trial	2nd trial	3rd trial	4th trial
Number of tubes	89	109	109	110
Shell diameter, in	9	9	9	10
Baffle spacing, in	5	5	3½	3½
Product of factors:				
Tubeside	0.535	0.514	0.514	0.513
Shellside	0.424	0.367	0.296	0.293
Tube-wall	0.052	0.042	0.042	0.042
Fouling	0.250	0.204	0.204	0.202
Total sum of products	1.261	1.127	1.056	1.050
Tubeside ΔP, psi	14.3	10	10	9.8
Shellside ΔP, psi	3.9	3.9	11.4	10.1

Any value of SOP between 0.95 and 1.05 is satisfactory as this gives a result within the accuracy range of the basic equations; unknowns in selecting the fouling factor do not justify further refinement. Therefore, the above is a satisfactory design for heat transfer and is within the pressure-drop restrictions specified. The surface area of the heat exchanger is $A = 110 \times 12 \times 0.1636 = 216$ ft^2 (20.1 m^2). The design overall coefficient is $U = 2,391,000/(30.4 \times 1.8 \times 216) =$ Btu/(h) (ft^2) (°F) (35.6 W/m^2 °C).

The foregoing example shows that the essence of the design procedure is selecting tube configurations and baffle spacings that will satisfy heat-transfer requirements within the pressure-drop limitations of the system.

Related Calculations.   The preceding procedure was for rating a heat exchanger of single tube-pass and single shell-pass design, with countercurrent flows of tubeside and shellside fluids. Often, it will be necessary to use two or more passes for the tubeside fluid. In this case, the LMTD is corrected with the Bowman, Mueller, and Nagle charts given in heat-transfer texts and the TEMA guide. If the correction factor for LMTD is less than 0.8, multiple shells should be used.

Bear in mind that n in all equations is the number of tubes in parallel through which the tubeside fluid flows; N_{PT} is the number of tubeside passes per shell (total number of tubes per shell $= nN_{PT}$); and L, the total-series length of path, equals shell length (L_o) (N_{PT}) × (number of shells).

The above procedure can be used for any shell-and-tube heat exchanger with sensible-heat transfer— or with no phase change of fluids—on both sides of the tubes. Also, N_{Re} on the tubeside must be greater than 10,000, and the viscosity of the fluid on the shellside must be moderate (500 cp. maximum).

As pointed out, the designer should assume as part of his/her job the specification of tube arrangement that will prevent the flow in the shell from taking bypass paths either around the space between the outermost tubes and the shell, or in vacant lanes of the bundle formed by channel partitions in multipass exchangers. He/She should insist that exchangers be fabricated in accordance with TEMA tolerances.

By using the appropriate equations from Tables 19.20 and 19.21, the technique described for rating heat exchangers with sensible-heat transfer can be used also for rating exchangers that involve boiling or condensing. The method can also be used in the design of partial condensers, or condensers handling mixtures of condensable vapors and noncondensable gases; and in the design of condensers handling vapors that form two liquid phases. However, for partial condensers and for two-phase liquid-condensate systems, a special treatment is required.

In addition to designing exchangers for specified performances, the method is also useful for evaluating the performance of existing exchangers. Here, the mechanical-design parameters are fixed, and the flow rates and temperature conditions (work factor) are the variables that are adjusted.

The two process variables that have the greatest effect on the size (cost) of a shell-and-tube heat exchanger are the allowable pressure drops of streams, and the mean temperature difference between the two streams. Other important variables include the physical properties of the streams, the location of fluids in an exchanger, and the piping arrangement of the fluids as they enter and leave the exchanger. (See design features in Table 19.19.)

Selection of optimum pressure drops involves consideration of the overall process. While it is true that higher pressure drops result in smaller exchangers, investment savings are realized only at the expense of operating costs. Only by considering the relationship between operating costs and investment can the most economical pressure drop be determined.

Available pressure drops vary from a few millimeters of mercury in vacuum service to hundreds of pounds per square inch in high-pressure processes. In some cases, it is not practical to use all the available pressure drop because resultant high velocities may create erosion problems.

Reasonable pressure drops for various levels are listed below. Designs for smaller pressure drops are often uneconomical because of the large surface area (investment) required.

Pressure level	Reasonable ΔP
Subatmospheric	1/10 absolute pressure
1 (6.89 kPa) to 10 lb/in^2 (gage) (68.9 kPa)	1/2 operating gage-pressure
10 lb/n^2 (gage) (68.9 kPa) and higher	1/2 in^2 (34.5 kPa) or higher

In some instances, velocities of 10 to 15 ft/s (3 to 4.6 m/s) help to reduce fouling, but at such velocities the pressure drop may have to be from 10 (68.9 kPa) to 30 lb/in² (206.7 kPa).

Although there are no specific rules for determining the best temperature approach, the following recommendations are made regarding terminal temperature differences for various types of heat exchangers; any departure from these general limitations should be economically justified by a study of alternate system-designs:

- The greater temperature difference should be at least 20°C (36°F).
- The lesser temperature difference should be at least 5°C (9°F). When heat is being exchanged between two process streams, the lesser temperature difference should be at least 20°C (36°F).
- In cooling a process stream with water, the outlet-water temperature should not exceed the outlet process-stream temperature if a single body having one shell pass—but more than one tube pass—is used.
- When cooling or condensing a fluid, the inlet coolant temperature should not be less than 5°C (9°F) above the freezing point of the highest freezing component of the fluid.
- For cooling reactors, a 10°C (18°F) to 15°C (27°F) difference should be maintained between reaction and coolant temperatures to permit better control of the reaction.
- A 20°C (36°F) approach to the design air-temperature is the minimum for air-cooled exchangers. Economic justification of units with smaller approaches requires careful study. Trim coolers or evaporative coolers should also be considered.
- When condensing in the presence of inerts, the outlet coolant temperature should be at least 5°C (9°F) below the dewpoint of the process stream.

In an exchanger having one shell pass and one tube pass, where two fluids may transfer heat in either cocurrent or countercurrent flow, the relative direction of the fluids affects the value of the mean temperature difference. This is the log mean in either case, but there is a distinct thermal advantage to counterflow, except when one fluid is isothermal.

In cocurrent flow, the hot fluid cannot be cooled below the cold-fluid outlet temperature; thus, the ability of cocurrent flow to recover heat is limited. Nevertheless, there are instances when cocurrent flow works better, as when cooling viscous fluids, because a higher heat-transfer coefficient may be obtained. Cocurrent flow may also be preferred when there is a possibility that the temperature of the warmer fluid may reach its freezing point.

These factors are important in determining the performance of a shell-and-tube exchanger:

Tube Diameter, Length. Designs with small-diameter tubes [5/8 (15.8 cm) to 1 in (25.4 cm)] are more compact and more economical than those with larger-diameter tubes, although the latter may be necessary when the allowable tubeside pressure drop is small. The smallest tube size normally considered for a process heat exchanger is 5/8 in (15.8 cm) although there are applications where ½ (12.7 cm), 3/8 (9.5 cm), or even ¼-in (6.4 cm) tubes are the best selection. Tubes of 1 in (25.4 cm) diameter are normally used when fouling is expected because smaller ones are impractical to clean mechanically. Falling-film exchangers and vaporizers generally are supplied with 1½. (38.1 cm) and 2-in (50.8 cm) tubes.

Since the investment per unit area of heat-transfer service is less for long exchangers with relatively small shell diameters, minimum restrictions on length should be observed.

Arrangement. Tubes are arranged in triangular, square, or rotated-square pitch (Fig. 19.32). Triangular tube-layouts result in better shellside coefficients and provide more surface area in a given shell diameter, whereas square pitch or rotated-square pitch layouts are used when mechanical cleaning of the outside of the tubes is required. Sometimes, widely spaced triangular patterns facilitate cleaning. Both types of square pitches offer lower pressure drops—but lower coefficients—than triangular pitch.

Primarily, the method given in this calculation procedure combines into one relationship the classical empirical equations for film heat-transfer coefficients with heat-balance equations and with relationships describing tube geometry, baffles, and shell. The resulting overall equation is recast into three separate groups that contain factors relating to physical properties of the fluid, performance

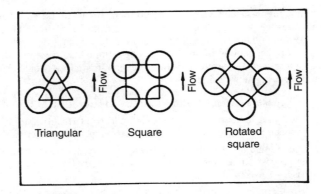

FIGURE 19.32 Tube arrangements used for shell-and-tube heat exchangers. (*Chemical Engineering*.)

or duty of the exchanger, and mechanical design or arrangement of the heat-transfer surface. These groups are then multiplied together with a numerical factor to obtain a product that is equal to the fraction of the total driving force—or log-mean temperature-difference (LMTD or ΔT_M)—that is dissipated across each element of resistance in the heat-flow path.

When the sum of the products for the individual resistance equals one, the trial design may be assumed to be satisfactory for heat transfer. The physical significance is that the sum of the temperature drops across each resistance is equal to the total available LMTD. The pressure drop on both tubeside and shellside must be checked to assure that both are within acceptable limits. As shown in the sample calculation above, usually several trials are necessary to obtain a satisfactory balance between heat transfer and pressure drop.

Tables 19.20 and 19.21, respectively, summarize the equations used with the method for heat transfer and for pressure drop. The column on the left lists the conditions to which each equation applies. The second column lists the standard form of the correlation for film coefficients that is found in texts. The remaining columns then tabulate the numerical, physical-property, work, and mechanical-design factors, all of which together form the recast dimensional equation. The product of these factors gives the fraction of total temperature drop or driving force ($\Delta T_f/\Delta T_M$) across the resistance.

As described above, the addition of $\Delta T_i/\Delta T_M$, tubeside factor, plus $\Delta T_o/\Delta T_M$, shellside factor, plus $\Delta T_s/T_M$, fouling factor, plus $\Delta T_w/\Delta T_M$, tube-wall factor, determine the heat-transfer adequacy. Any combination of $\Delta T_i/\Delta T_M$ and $\Delta T_o/\Delta T_M$ may be used, as long as a horizontal orientation on the tubeside is used with a horizontal orientation on the shellside, and a vertical tubeside orientation has a corresponding shellside orientation.

The units in the pressure-drop equations (Table 19.21) are consistent with those used for heat transfer. The pressure drop pin psi is calculated directly. Because the method is a shortcut approach to design, certain assumptions pertaining to thermal conductivity, tube pitch, and shell diameter are made.

For many organic liquids, thermal conductivity data are either not available or difficult to obtain. Since molecular weights (M) are known, for most design purposes the Weber equation, which follows, yields thermal conductivities with quite satisfactory accuracies:

$$k = 0.86\,(cs^{4/3}/M^{1/3})$$

An important compound for which the Weber equation does not work well is water (the calculated thermal conductivity is less than the actual value). Figure 19.33 gives the physical-property factor for water (as a function of fluid temperature) that is to be substituted in the equations for sensible-heat transfer with water, or for condensing with steam.

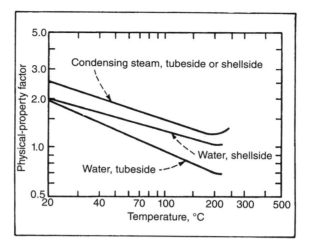

FIGURE 19.33 Physical-property factors for water and steam vs. temperature. (*Chemical Engineering.*)

If the thermal conductivity is known, it is best to obtain a pseudo-molecular weight by:

$$M = 0.636\,(c/k)^3 s^4$$

This value is substituted in the applicable equation to solve for the physical-property factor.

Tube pitch for both triangular and square-pitch arrangements is assumed to be 1.25 times the tube diameter. This is a standard pitch used in the majority of shell-and-tube heat exchangees. Slight deviations do not appreciably affect results.

Shell diameter is related to the number of tubes (nN_{PT}) by the empirical equation:

$$D_o = 1.75\,d_o(nN_{PT})^{0.47}$$

This gives the approximate shell diameter for a packed floating-head exchanger. The diameter will differ slightly for a fixed tubesheet, U-bend, or a multipass shell. For greater accuracy, tube-layout tables can be used to find shell diameters.

The following shows how the design equations are developed for a heat exchanger with sensible-heat transfer and Reynolds number 10,000 on the tubeside, and with sensible-heat transfer and cross-flow (flow perpendicular to the axis of tubes) on the shellside. Equations with other heat-transfer mechanisms are derived similarly.

For the film coefficient or conductance, h, and the heat balance, these equations apply:

Value of h	Heat-balance Eq.
$h_i = \dfrac{0.023c_iG_i}{(c_i\mu_i/k_i)^{2/3}(D_iG_i/\mu_i)^{0.2}}$	$W_ic_i(t_H - t_L) = h_iA\Delta T_i$
$h_w = 24k_w/(d_o - d_i)$	$W_ic_i(t_H - t_L) = h_wA\Delta T_w$
$h_o = \dfrac{0.33c_oG_o(0.6)}{(c_o\mu_o/k_o)^{2/3}(D_oG_o/\mu_o)^{0.4}}$	$W_oc_o(T_H - t_L) = h_oA\Delta T_o$
h_s = assumed value	$W_ic_i(t_H - t_L) = h_sA\Delta T_s$

Since the resistances involved in a tube-and-shell exchanger are the tubeside film, the tube wall, the scale caused by fouling, and the shellside film, then:

$$\Delta T_i + \Delta T_w + \Delta T_s + \Delta T_o = \Delta T_M$$

Therefore,

$$\Delta T_i/\Delta T_M + \Delta T_w/\Delta T_M + \Delta T_s/\Delta T_M + \Delta T_o/\Delta T_M = 1$$

or,

$$\underbrace{\frac{W_i c_i (T_H - t_L)}{h_i A \Delta T_M}}_{\substack{\text{Tubeside} \\ \text{product}}} + \underbrace{\frac{W_i c_i (T_H - t_L)}{h_w A \Delta T_M}}_{\substack{\text{Tube-wall} \\ \text{product}}} + \underbrace{\frac{W_i c_i (T_H - t_L)}{h_o A \Delta T_M}}_{\substack{\text{Fouling} \\ \text{product}}} + \underbrace{\frac{W_s c_o (T_H - t_L)}{h_o A \Delta T_M}}_{\substack{\text{Shellside} \\ \text{product}}} = 1$$

This last equation is obtained by dividing each heat-balance equation by ΔT_M and solving for $\Delta T_f/\Delta T_M$. The design equations are derived by substituting for h, the appropriate correlation for the coefficient; for k, the value obtained from the Weber equation; for A, the equivalent of the surface area in terms of the number of tubes, outside diameter and length, according to the relation $A = \pi n(d_o/12)L$; for mass velocity on the tubeside, $G_i = 183\ W_i/(d_i^2 n)$; and for mass velocity on the shellside, $G_o = 411.4\ W_o/(d_o n N_{PT}^{0.47} P)$.

The resulting equation is rearranged to separate the physical-property, work, and mechanical-design parameters into groups. To obtain consistent units, the numerical factor in the equation combines the constants and coefficients. The form of the equations shown in Table 19.20 as Eqs. (1), (11), (18) and (19) omits dimensionless groups such as Reynolds or Prandtl numbers, but includes single functions of the common design parameters such as number of tubes, tube diameter, tube length, baffle pitch, etc.

The individual products calculated from the four equations are added to give the sum of the products (SOP). A valid design for heat transfer should give SOP = 1. If SOP comes out to be less or more than one, products for each resistance are adjusted by the appropriate exponential function of the ratio of the new design parameter to that used previously.

More sophisticated rating methods are available that make use of complex computer programs; the described method is intended only as a general, shortcut approach to shell-and-tube heat-exchanger selection. Accuracy of the technique is limited by the accuracy with which fouling factors, fluid properties, and fabrication tolerances can be predicted. Nevertheless, test data obtained on hundreds of heat exchangers attest to the method's applicability.

This procedure is the work of Robert C. Lord, Project Engineer, Paul E. Minton, Project Engineer, and Robert P. Slusser, Project Engineer, Engineering Department, Union Carbide Corporation, as reported in *Chemical Engineering* magazine. SI values were added by the handbook editor.

Nomenclature

A	Out side surface area, ft^2		t	Temperature on tubeside, °C
B	Film thickness, $[0.00187 Z\Gamma/g_c s^2]^{1/3}$, ft		ΔT_M	Logarithmic mean temperature difference (LMTD), °C
c	Specific heat, Btu/(lb) (°F)		U	Overall coefficient of heat transfer, Btu/[(h)(ft^2) (°F)]
D_i	Inside tube diameter, ft		W	Flow rate, (lb/h)/1000
D_o	Inside shell diameter, in		Z	Viscosity, cp
d	Tube diameter, in		Γ	Tube loading, lb/(h)(ft)

f	Fanning friction factor, dimensionless		λ	Heat of vaporization, Btu/lb
G	Mass velocity, lb/(h)(ft^2 cross-sectional area)		θ	Time, h
g_c	Gravitational constant, (4.18×10^8) ft (h)^2		μ	Viscosity, lb/(h) (ft)
h	Film coefficient of heat transfer, Btu/(h) [(ft^2) (°F)]		P_v	Vapor density, lb/ft^3
k	Thermal conductivity, Btu/[(h) (ft^2)]		P_L	Liquid density, lb/ft^3
L	Total series length of tubes, $(L_o N_{PT} \times$ number of shells), ft		σ	Surface tension, dynes/cm
L_A	Length of condensing zone, ft *Subscripts*		Σ, Σ'	Surface condition factor, dimensionless
L_B	Length of subcooled zone, ft		o	Conditions on shellside or outside tubes
L_o	Length of shell, ft		i	Conditions on tubeside or inside tubes
M	Molecular weight, lb/(lb-mol)		b	Bulk fluid properties
N_{PT}	Number of tube passes per shell, dimensionless		f	Film fluid properties
n	Number of tubes per pass (or in parallel) dimensionless		H	High temperature
P	Pressure, lb/in^2 (abs)		L	Low temperature
P_B	Baffle spacing, in		s	Scale or fouling material
ΔP	Pressure drop, lb/in^2		w	Wall or tube material
Q	Heat transferred, Btu		*Dimensionless groups*	
s	Specific gravity (referred to water at 20°C), dimensionless		N_{Re}	Reynolds number, DG/μ
T	Temperature on shellside, deg C		N_{Pr}	Prandtl number, $c\mu/k$
			N_{st}	Stanton number, h/cG
			N_{Nu}	Nusselt number, hD/k

See Procedure for SI values for the variables above.

INDEX